Ecology of Butterflies in Europe

Due to the importance of butterflies as indicators of environmental quality and their usefulness as model systems to address ecological and evolutionary questions, butterfly biology has become a focus of research, especially within Europe. This book synthesizes all relevant and recent knowledge in the field, making this a definite work for those making use of this taxonomic group as a model system. It is divided into five major parts which deal with habitat use, population biology and genetics, evolutionary ecology, distribution and phylogeny, and global change and conservation. There are growing numbers of scientific projects and networks in Europe in which the use of butterflies as tools and targets for conservation is central, and application of knowledge is closely related to European cultural landscapes. However, the material can also be applied to a wide geographic scope. Written by an international team of experts, this timely book is suitable for students, researchers and enthusiasts.

JOSEF SETTELE currently works at the UFZ – Helmholtz Centre for Environmental Research. He is Head of the Animal Ecology section in the Department of Community Ecology, and Adjunct Professor of Ecology at the University of Halle. He was initiator and coordinator of the EU projects 'MacMan' (Maculinea butterflies as indicators and tools for conservation management) from 2002 to 2006, and 'ALARM' (Assessing LArge scale environmental Risks for biodiversity with tested Methods) from 2004 to 2009. His general research focus is on population and conservation biology of animals (in particular, butterflies). He has published more than 200 scientific papers. He is a Fellow of the Royal Entomological Society and Editor in Chief of the open access journal *BioRisk*. In 2009 he started a 4-year term as chairman of Butterfly Conservation Europe.

TIM SHREEVE is a Reader of Ecology at Oxford Brookes University. His current work in butterfly ecology encompasses three main areas: behaviour and activity in relation to microclimate and climatic constraints on population persistence; the role of wing morphology in relation to predation, mate attraction and thermoregulation; and the analysis of biogeographical patterns of Palaearctic species and the roles of ecological attributes of species and their associated strategies in determining occurrence and conservation status. He is a Fellow of the Royal Entomological Society and Editor in Chief of the *Journal of Insect Conservation*.

MARTIN KONVIČKA is Adjunct Professor at the Institute of Entomology of the Czech Academy of Sciences. He has major interests in butterfly ecology and conservation biology, and in using evolutionary and behavioural information for conserving endangered species. He uses butterflies as a model for the study of biodiversity. He has participated in many butterfly-related projects launched in the Czech Republic since the mid-1990s, including the butterfly monitoring scheme of the Czech Republic and the publication of the Czech butterfly distribution atlas.

HANS VAN DYCK is a Professor at the Biodiversity Research Centre of the Belgian Université catholique de Louvain (UCL). Since 2004, he has been the head of a new research team, the Behavioural Ecology and Conservation Group. His main interest is to combine both basic and applied research in ecology and evolution in order to better understand changing organisms in changing anthropogenic environments and what this means for conservation. He is also lecturer in behavioural ecology, landscape ecology and entomology.

Ecology of Butterflies in Europe

Edited by
Josef Settele
Helmholtz Centre for Environmental Research – UFZ, Germany

Tim Shreeve
Oxford Brookes University, UK

Martin Konvička
Czech Academy of Sciences, Czech Republic

Hans Van Dyck
Biodiversity Research Centre (UCL), Belgium

CAMBRIDGE
UNIVERSITY PRESS

CAMBRIDGE UNIVERSITY PRESS
Cambridge, New York, Melbourne, Madrid, Cape Town, Singapore, São Paulo, Delhi, Dubai, Tokyo

Cambridge University Press
The Edinburgh Building, Cambridge CB2 8RU, UK

Published in the United Kingdom by Cambridge University Press, Cambridge

www.cambridge.org
Information on this title: www.cambridge.org/9780521766975

First published 2009

Printed in the United Kingdom at the University Press, Cambridge

A catalogue record for this publication is available from the British Library

ISBN 978-0-521-76697-5 Hardback
ISBN 978-0-521-74759-2 Paperback

Additional resources for this publication at www.cambridge.org/9780521766975

Contents

The colour plates are situated between pages 340 and 341.

Contributors

MICHEL BAGUETTE
Muséum National d'Histoire Naturelle
Département Ecologie et Gestion de la Biodiversité
France

ROBERT BIEDERMANN
Landscape Ecology Group
Institute of Biology and Environmental Sciences
University of Oldenburg
Germany

BIRGIT BINZENHÖFER
Helmholtz Centre for Environmental Research – UFZ
Department of Conservation Biology
Germany

ROGER L. H. DENNIS
Institute for Environment, Sustainability and Regeneration
Staffordshire University
and
NERC Centre for Ecology and Hydrology
Cambridgeshire
UK

HENRI DESCIMON
Marseilles
France

MATTHIAS DOLEK
Büro Geyer und Dolek
Ecological Research and Planning
Germany

JOHN DOVER
Institute for Environment and Sustainability Research
Staffordshire University
UK

ANDREAS ERHARDT
Department of Environmental Sciences
University of Basel
Switzerland

THOMAS FARTMANN
Institute of Landscape Ecology
University of Münster
Germany

RICHARD FOX
Butterfly Conservation
UK

ENRIQUE GARCÍA-BARROS
Departamento de Biologia (Zoologia)
Universidad Autónoma de Madrid
Spain

DAVID GUTIÉRREZ
Área de Biodiversidad y Conservación
Escuela Superior de Ciencias Experimentales y Tecnología
Universidad Rey Juan Carlos
Spain

JANE K. HILL
Department of Biology
University of York
UK

THOMAS HOVESTADT
Field Station Fabrikschleichach
University of Würzburg
Germany

BENGT KARLSSON
Department of Zoology
Stockholm University
Sweden

MARTIN KONVIČKA
Faculty of Sciences
University South Bohemia
Institute of Entomology and
Czech Academy of Sciences
Czech Republic

DIRK MAES
Research Institute for Nature and Forest (INBO)
Belgium

JAMES MALLET
Galton Laboratory
Department of Biology
University College London
UK

JOSÉ MARTÍN CANO
Departamento de Biologia (Zoologia)
Universidad Autónoma de Madrid
Spain

JOVANNE MEVI-SCHÜTZ
Department of Integrative Biology
University of Basel
Switzerland

MIGUEL L. MUNGUIRA
Departamento de Biologia (Zoologia)
Universidad Autónoma de Madrid
Spain

GABRIEL NÈVE
Institut Méditerranéen d'écologie et de paléoécologie
Université de Provence
France

MARKO NIEMINEN
Faunatica Oy
Lansantie 3 D
Finland

SÖREN NYLIN
Department of Zoology
Stockholm University
Sweden

RALF OHLEMÜLLER
Institute of Hazard and Risk Research (IHRR) and
School of Biological and Biomedical Sciences
Durham University
UK

ADAM PORTER
Department of Plant, Soil and Insect Sciences and
Graduate Program in Organismic and Evolutionary Biology
University of Massachusetts
USA

DAVID B. ROY
NERC Centre for Ecology and Hydrology
Oxfordshire
UK

THOMAS SCHMITT
Biogeographie, FB VI
University of Trier
Germany

BORIS SCHRÖDER
Institute of Geoecology
University of Potsdam
Germany

NICOLAS SCHTICKZELLE
Quantitative Conservation Biology Group
Biodiversity Research Centre
Université catholique de Louvain
Belgium

JOSEF SETTELE
Helmholtz Centre for Environmental Research – UFZ
Department of Community Ecology
Germany

MARK SHAW
Honorary Research Associate
National Museums of Scotland
UK

TIM SHREEVE
School of Life Sciences
Oxford Brookes University
UK

CONSTANTI STEFANESCU
Butterfly Monitoring Scheme
Museu de Granollers–Ciències Naturals
Spain

BARBARA STRAUSS
Landscape Ecology Group
Institute of Biology and Environmental Sciences
University of Oldenburg
Germany

CHRIS D. THOMAS
Department of Biology
University of York
UK

HANS VAN DYCK
Behavioural Ecology and Conservation Group
Biodiversity Research Centre
Université catholique de Louvain
Belgium

SASKYA VAN NOUHUYS
Department Biological and Environmental Sciences
University of Helsinki
Finland
and
Department of Ecology and Evolutionary Biology
Cornell University
USA

CHRIS A. M. VAN SWAAY
Dutch Butterfly Conservation
The Netherlands

MARTIN S. WARREN
Butterfly Conservation
UK

PER-OLOF WICKMAN
Department of Education in Mathematics and Science
Stockholm University
Sweden

ROBERT J. WILSON
Centre for Ecology and Conservation
University of Exeter
UK

JACK J. WINDIG
Animal Breeding and Genomics Centre (ABGC)
Animal Sciences Group
The Netherlands

Preface

WHY?

The last 20 years has seen a dramatic increase in research on butterflies, driven from two directions. The first has been academic, with a realisation that butterflies can be model organisms to study evolutionary, behavioural and biogeographic processes. The second has been, and continues to be, an urgency to tackle extensive declines in butterfly species, a phenomenon not just restricted to Europe. As recently as 50 years ago there were no major issues concerning the conservation of butterflies and biodiversity was not regarded as an issue, though even at that time a few species of butterfly were in decline within Europe. About 30 years ago the study of butterflies was often regarded as an academic sideline, since when the contribution of butterfly biologists to mainstream biology has increased in importance. Over the last 10 years there has been a remarkable synthesis of ideas and an increasing realisation that academic studies can contribute much to conservation. The science of conservation is becoming increasingly evidence-based and a true dialogue between academics and conservationists is emerging, including a scientific approach to using butterflies as biodiversity indicators. It is thus timely to produce a European-scale book about butterfly ecology, with the potential to make a clear statement about the 'state of the art' and the emerging issues. Not least, such a text should clearly indicate where the current research is taking place and which scientists and research teams are focusing on what. In producing this book it has been our intention to do just this.

WHO?

The idea for this book was conceived by Roger Dennis almost 10 years ago in September 1999 at the time of 3rd International Symposium of Butterfly Conservation, organised by Butterfly Conservation in Oxford, UK. The original three editors were Roger Dennis, Tim Shreeve and Andrew Pullin. For health reasons Roger Dennis stepped down from his lead role and the team was then joined by Josef Settele. Andrew Pullin left because of other commitments and the editorial team 'evolved' into a combination of colleagues from four different European countries, namely Josef Settele, Tim Shreeve, Martin Konvicka and Hans Van Dyck. During the gestation period, chapter authors changed, with the majority of author combinations reflecting the international dimension of butterfly ecology in Europe. Surprisingly, the editorial team as a whole physically only met once, to finalise arrangements, at Wageningen (Netherlands) in April 2008 during the 'Future of Butterflies II' conference, organised by the Dutch Butterfly Conservation. All other communication between the editors has been electronic.

Although Roger Dennis left the editorial team, all the editors thought it appropriate that he be asked to write an introductory overview, firstly as a tribute for his originating the concept and secondly for his long-standing contribution to the development of butterfly ecology.

WHAT?

Within this book there is occasional evidence of differences of opinion and interpretation between the authors of some chapters. These differences are a healthy sign that there is room for debate, and as editors we have taken the view that authors are experts and that the differences of opinion reveal that the science of butterfly ecology is permanently developing and in a productive state. The subjects within each chapter cannot be viewed in isolation; there is thus an element of necessary overlap between some chapter parts. As editors, we have had to make decisions on some points; we have added cross-referencing between chapters to indicate where different areas merge; and for nomenclature, which is not consistent over Europe, we have been relatively conservative and adopted Karsholt & Razowski (1996) as the core reference.

We have attempted to provide a wide coverage of the many aspects of butterfly ecology that are being undertaken by European researchers. The text is broadly thematic and clearly states where there are knowledge gaps. Where appropriate authors also indicate where cross-disciplinary studies can contribute to the advancement of knowledge. It is, however, inevitable in a text of this size that some areas have not received as much attention as some readers would prefer. Reference to these areas, such as the major advances being made in the study of ant–lycaenid interactions, or the painstakingly detailed work on metapopulation models using *Melitaea* species, is made where relevant but we also think that these topics either warrant complete new exhaustive treatments or are covered elsewhere.

WHERE TO IN FUTURE?

It is evident from the contributions to this book that there are emerging new areas of butterfly research and we hope that the synthesis provided within the chapters will provide a stimulus to further work. Each chapter raises important questions. Despite extensive research there are many gaps in our existing knowledge; for example, we do not even know the host-plant ranges of many species, our grip on what determines a habitat for most species is rather limited, and our knowledge of population dynamic processes and interactions at the community level is in its infancy. We now have a range of techniques for studying processes from the molecular to the community and biogeographic levels. Integration of these techniques to provide novel approaches to studying academic questions and addressing immediate conservation issues is now possible and the levels of co-operation between different research groups to achieve these goals are also clear from the contributions to this book.

Research on butterflies within Europe is now transnational, and it is surprising how much has been achieved within the last two decades, often with limited financial resources. The emergence of Butterfly Conservation Europe, as yet poorly funded, provides a potential future mechanism for increasing European co-operation, in providing a forum for raising key issues and bringing subject experts and end-users together. Our plea to politicians and grant-awarding bodies is to guarantee the continuation of basic research and enhance funding for future collaborations. Within Europe there is a wealth of talent working on the ecology of butterflies, and their work is making major contributions to understanding the state of biodiversity, the mechanisms controlling it, and increasing the power of predictive models in a rapidly changing world. The work contained in this book is a testimonial to what has already been achieved. It is our hope that this book provides a stimulus for further work, in particular that it encourages the next generation of butterfly ecologists. Our one hope is that theory and practice can combine quickly enough to ensure a secure future for butterflies and the species they are associated with in a dynamic European landscape.

ACKNOWLEDGEMENTS

Throughout the years numerous colleagues have been involved in the making of this book. At the risk of having forgotten a few of them – for which we want to apologise – we want to mention at least those who have volunteered as referees of one or more chapters: Andreas Erhardt, Bengt Karlsson, Chris van Swaay, Christer Wiklund, Constanti Stefanescu, Darrell Kemp, Dirk Maes, Enrique García-Barros, Gabriel Nève, Henri Descimon, James Mallet, Jane Hill, Jens Roland, Klaus Fischer, Martin Warren, Michel Baguette, Michiel WallisDeVries, Miguel L. Munguira, Nicolas Schtickzelle, Niklas Wahlberg, Per-Olof Wickman, Robert B. Srygley, Roger L. H. Dennis, Sören Nylin, Thomas Merckx, Thomas Fartmann, Thomas Hovestadt, Thomas Schmitt and Tim New.

Furthermore we are indebted to Mandy Riemer and Alexander Harpke for their continuous technical support in the finalisation as well as the different book editors of Cambridge University Press for their never-ending patience and positive attitude throughout the genesis of the book, we are particularly indebted to Richard Marley and Dominic Lewis.

Josef Settele
Tim Shreeve
Martin Konvička
Hans Van Dyck

1 • Ecology of butterflies in Europe – where are we now and where to go?

ROGER L. H. DENNIS

To understand where we are now with butterfly research, and where to go next, we need first to clarify what we are 'about'. Inevitably opinions will be as varied as there are individual perspectives; we study butterflies for different reasons, but I doubt for one moment that any of us would belittle the contribution of those who study them simply for the joy of it – much of our knowledge base is founded entirely on enthusiasm for butterflies and our own research is fired by butterfly aesthetic beauty. A little more grist to the mill can be applied by trying to qualify objectives with purpose; butterfly research these days can no more escape an economist's cost–benefit analysis than anything else that is funded. Even so, it brings focus to what we do, to consider human benefits other than our own, and some like me would argue that we have an obligation, a duty, to future generations to do so. So, what should we be 'about'? Butterfly research has already provided the answer. Exponential human population growth conjures up visions of massive planetary changes, among them biome losses, ecosystem failures and climatic shifts, all with serious consequences for humanity itself: starvation, disease, disaster and warfare press home that despite technological advances, globally things are going seriously amiss. Butterflies, along with other biota, provide sufficiently sensitive barometers for monitoring the changes and predicting such outcomes in their initial stages (Thomas 2005); they are signposts to human well-being, indicators of environmental changes and ultimately with losses in other organisms herald a pending global extinction crisis of geological proportions and failure in sustainability. Aside from this vital role, butterflies are model organisms for numerous research areas in biology (Ehrlich & Hanski 2004), extending from physiology and development, gene–environment interactions, population and spatial dynamism, ecological order, community organisation to key themes in evolution, particularly speciation. They do so owing to their unique combination of attributes: their widespread occurrence throughout biomes, variation in abundances and range sizes, their short, discrete generations but variety of life histories, their array of patterns, diversity of taxa, multiplicity of species, their rapid response to conditions and assortment of adaptive modes for different aspects of phenotypes, variation in mobility, variety of habitat scaling and sheer aesthetic beauty. As such, not only do they present a variety of sensitive markers for environmental change at different scales and for different purposes, but are easily recognisable, attract attention, have wide popular appeal, are sufficiently numerous to be easily monitored using simple methods and can therefore be recorded at low cost.

It is hardly surprising then that the scientific literature on butterflies dating back to the nineteenth century is extensive and growing rapidly. The present book (EBIE) is another, important milestone along the path to understanding them. It runs in the vein of thematic texts dealing with topics and issues in butterfly biology and has direct predecessors in *The Ecology of Butterflies in Britain* (EBIB: Dennis 1992a) and *The Biology of Butterflies* (Vane-Wright & Ackery 1984), both founded on E. B. Ford's *Butterflies*. More recent specialist texts are *Butterflies: Ecology and Evolution Taking Flight* (Boggs *et al.* 2003) and *On the Wings of Checkerspots* (Ehrlich & Hanski 2004). These thematic books have counterparts in systematic texts which provide detailed accounts species by species. Thus, EBIB was married to *The Butterflies of Great Britain and Ireland* (Emmet & Heath 1990), *The Millennium Atlas of Butterflies in Britain and Ireland* (Asher *et al.* 2001) and *The State of Butterflies in Britain and Ireland* (Fox *et al.* 2006). For EBIE, there is as yet no direct partner covering the continent's butterfly fauna, but a most exacting model is Frits Bink's (1992) *Ecologische Atlas van de Dagvlinders van Noordwest-Europa* a painstaking audit of attributes of north-western European species. Nevertheless, continent-wide texts are growing in number and the details they convey (Tolman & Lewington 1997, Lafranchis 2004). In the UK, species' distributions are effectively illustrated in the systematic texts; in Europe, this is a singular, independent entity, *The Distribution Atlas of European Butterflies* (MEB: Kudrna 2002) with a second edition soon to be forthcoming. Both EBIB and EBIE are underpinned by numerous regional texts and atlases. In the UK, these take the form of county or multiple county

Ecology of Butterflies in Europe, eds. J. Settele, T. Shreeve, M. Konvička and H. Van Dyck. Published by Cambridge University Press. © Cambridge University Press 2009, pp. 1–5.

treatments and a single national text for Scotland (Thomson 1980). On mainland Europe an increasing number of books for individual countries is emerging, splendid in production, revealing in details, invaluable for data, many with thematic cores, engaging contexts and condition of the fauna (e.g. Henriksen & Kreutzer 1982, Geiger 1987, Stoltze 1996, Maes & Van Dyck 1999, Lafranchis 2000, Settele *et al.* 2000). Many texts are appearing for the countries once making up the Soviet Union, including Russia itself (e.g. Tuzov *et al.* 1997, 2000), and a growing number of colourful texts covering smaller regions and islands are appearing (e.g. Lake Garda region: Sala 1996; Schleswig-Holstein: Kolligs 2003; Cyprus: Makris 2003). On other continents, there are parallel, if less advanced, developments (USA: Scott, 1986; Australia: Braby 2001).

EBIE is built on the same pattern as EBIB. Throughout the text there is a shift from individual behaviour through populations to species and communities, and from ecological-scale issues to evolutionary-scale outcomes. The book too has as its crux, conservation. Indicators butterflies most certainly are, but such urgency as exists to conserve the indicators themselves impresses on us all that the trends are not being taken seriously enough or being adequately reversed by politicians and policy makers. Nevertheless, the differences that are evident between the two books are a measure of what has been achieved over the past one and a half decades. In several respects EBIE moves towards an order of magnitude larger than EBIB – in words, concepts, authors, references, ecological intimacy, model intricacy and evolutionary reach; all this reflects on the shift in focus from island to continent – in spatial scale, altitudinal range, biomes and in climate. There is also the expanding perspectives of researchers from different countries, cultures, contexts, facing different crises, each bringing their variety of talents and reminding us of work done by our predecessors closed off to us in the past by barriers of language and limits to intellectual exchange – the impressive work of Petersen (1947) on morphology gradients, Kostrowicki (1969) on faunistic patterns, Warren's (1926, 1936) monographs on the Hesperiidi and *Erebia*, and Lorkovic's extensive practical work on speciation, just four examples among many. In detail, a glance through the contents reflects three basic changes, advances in EBIE: many topics in EBIB now have their own chapters in EBIE (i.e. adult feeding, mate location behaviour, oviposition, thermoregulation, spatial gradients in attributes and species richness); others have been substantially updated with new models, findings and viewpoints (i.e. population structure and dynamics,

dispersal, population genetics, functional significance of wing morphology, conservation status). In EBIE, readers will find defence of and merit argued for some older concepts (e.g. subspecies in Chapter 16; faunal elements in Chapter 17), but a key feature is the advent of new research areas; such are predictive species modelling, population spatial structure and dynamics, parasitoids, the evolutionary ecology of fecundity, speciation, and responses to climatic warming. The findings on wing morphology–behaviour–substrate relationships promise an exciting future for research, the result of many years of data collection by Tim Shreeve and his team (Chapter 13). In EBIB we looked back to the Pleistocene glaciations to account for evolutionary adaptations and gross distributional patterns. Following the projections of climatic change on butterfly populations (Dennis 1993), in EBIE, the vantage is forwards to predict the outcome. Retrodictions, though fascinating in providing evolutionary perspective (Dennis 1977, 1993), are hamstrung by the ultimate lack of fossil proof; in direct contrast, predictions are directly testable and germane to the current biodiversity crisis (see Chapter 20). There is also a greater readiness in this new text to explain techniques and give direction (e.g. Chapter 7: Predictive species modelling; Chapter 19: Hybrid zone analysis; Chapter 22: Population viability analysis). With the increased sophistication in methods, this is essential. In each chapter the reader will find guidance to potential research areas: no Ph.D. student in butterfly biology can declare he/she lacks problems to solve; the contrary perhaps, they may well be overwhelmed with several lifetimes of work to be done.

Advances in EBIE and future direction can perhaps be best sought through consideration of three ploys: expansion of data and knowledge, development of models and experiment, and construction of explanatory systems. The reality of models and the completeness of systems depend on the quality of data on butterfly biology.

At its simplest 'data' is another word for the knowledge or information we can acquire on an object of interest. It is the data that have accumulated on butterflies that make them such valuable indicators of change and model organisms for evolutionary studies. Data for diverse taxa have spawned groups, distinct in life history, that act as markers for discrete processes (e.g. migrants that flag up an increase in aliens and climatic changes, specialists that warn of landscape fragmentation, generalists that monitor landscape toxicity). A prominent feature of EBIE is the signalling of an exponential increase in knowledge in butterfly biology; inevitably, therefore, some tasks highlighted in EBIE are already in progress (e.g. species diversity on European islands: Dennis *et al.* 2008a, Dapporto & Dennis 2008a, b) and making a difference for conservation – as a result of Leonardo Dapporto's biogeography work, butterflies have been given priority within the vicinity of Monte Capanne on Elba within the Legambiente, and the National Park of the Tuscan Archipelago and the Italian energy agency (ENEL) have funded the project. Much of the development in butterfly biology is in the demonstration of links and associations among traits or between traits and environmental agents (e.g. Chapter 2: nectar use; Chapter 6: modes of thermoregulation; Chapter 10: genetics; Chapter 12 and 13: morphology and environment; Chapter 14: fecundity; Chapter 15: gradients). A growing trend is the development of species databases (Chapter 5: larval host plants; Chapter 11: parasitoids). It cannot be emphasised enough how important database development is on butterflies, as for other organisms: their co-ordination, centralisation, documentation and universal access (Dennis *et al.* 2008b). Currently, the accumulation of data is uneven and some aspects of butterfly biology are poorly known (e.g. butterfly roosts, hibernation sites, flight threshold temperatures, use of scent and sight for interactions and resource use, wing patterning, impact of pathogens). Database development is also being unnecessarily duplicated by different research groups (host plants, nectar sources, phylogenies) and there is both lack of direction and universal access, especially to atlas databases, despite the fact that these data have been supplied by the goodwill of myriads of recorders. The most crucial aspect of databases for future progress is the quality of documentation. Although the key feature of butterflies is their space-time variability, most data on traits in butterfly biology lack explicit space-time co-ordinates. Much as it is unthinkable that museum specimens, atlas and monitoring scheme records exist without data for location and time, so too should it be equally inconceivable for data on resource use, behaviour, life history, phenotypic differences and all else to be bereft of such basic recording fields. Spatial and temporal variability are key aspects of butterfly biology; without a space-time context for observations (e.g. for nectar feeding, host use) comparative studies lack the most vital components of that variability that makes them effective indicators and model organisms. Turning this around, behaviour and substrate use of observed butterflies firms up the status of atlas and monitoring scheme records (Dennis 2004, Dennis *et al.* 2008b).

A crucial part of developing future databases is adequate sampling strategies, as urged by Mark Shaw and his colleagues for parasitoid records. Despite the failure of atlases to achieve universal cover even over relatively small regions such as the UK, there is still a reluctance to be weaned off the aspiration of continuous recording cover. Faced with the vast territory of Europe, the expanding conservation activity in its component states and the need for continually updating spatial records of species at finer scales, this is an issue that national conservation organisations and co-ordinators of atlases will inevitably have to confront, particularly if we move to establishing better the status of the records we are making (Dennis *et al.* 2008b). A key to developments is the adoption of stratified, fixed sampling designs and nested studies. Another part of this process of data acquisition is adequately defining terms. Some concepts are elusive and readers should give close attention to the call for pragmatic solutions as Descimon and Mallet (Chapter 16) have done for designation of species; as discussed below this is highly relevant to conservation which ultimately depends on identifying an appropriate unit, the habitat.

An essential part of the sampling strategy for data is regional development, that is, development of butterfly biology throughout more European states – a primary aim being that more regional work generates more data on spatial variation in species traits. EBIE demonstrates well that this is fast happening with many new centres for research. In 1972 when I was hoping to convert undergraduate observations on the Creuddyn Peninsula in North Wales to a Ph.D. studentship, I was informed that there was not a single institute in the UK that could provide 'adequate supervision'! Now, young researchers can look to Sweden, Norway, Britain, Holland, Belgium, France, Germany, Spain, the Czech Republic and many other countries, for centres, multiple ones in some states, to engage in butterfly research. A key to regional development, however, goes beyond research centres or organisations, to enthusing popular interest. This has formed the foundation of success for Butterfly Conservation UK and there is no reason why it should not spread to other parts of Europe. Many more students are now emerging from new universities, trained in the biological sciences with a deep interest in nature. They may not have careers in butterfly biology, and have busy lives in their own professions, but their training, coupled with their interest, especially when they are released at retirement, is an energy and knowledge we would do well to harness. Many others have no training in biology at all, but they bring a variety of other skills and their capacity for advancing butterfly conservation and biology should not be underestimated or overlooked. If we want to build up databases of butterfly sites, resources, behaviour, life history elements over the length and breadth of a continent, and firm up the platform for butterfly conservation, the starting point is to encourage the current and future generations to take part in the quest; the list of acknowledgements at the end of the parasitoid chapter is an indication of what can be achieved.

A significant advance for butterfly biology evident in EBIE is the development of models (aids for thinking and analysis): statistical, explanatory, experimental and predictive, many simulated in computer-driven algorithms. In the last two decades predictive modelling has become most prominent in spatial population dynamics owing to the superlative research by Ilkka Hanski and colleagues on Åland island butterflies such as *Melitaea cinxia*. These models have formed the basis for modern thinking in conservation, a move from single to multiple population units. In EBIE, we see the beginnings of the dialogue generating hybridisation between metapopulation models and landscape models; chapters on 'Predictive species modelling in butterflies' by Boris Schröder and colleagues (Chapter 7), metapopulation dynamics by Robert Wilson and David Roy (Chapter 8), and population viability analysis (PVA) by Nicolas Schtickzelle and Michel Baguette (Chapter 22) are particularly important. Link these up with landform influences on butterfly geography and conservation (Dover & Settele 2008). Another important area for modelling is projection of the impact of climate change and biotope fragmentation on species dynamics (persistence, range expansion, extinction). The historical branch of this does not escape scrutiny; retrodiction for evolutionary history cannot rely entirely on a molecular clock for determining vicariance, origins and dispersal pathways; to do so would be to fall into the same trap disparaged by this author in early attempts to determine origins of butterflies into Britain based on species' ('racial') physical distinctions (Dennis 1977). All these new approaches, sets of models, require a new interpretation of habitats involving a shift from simple notions of habitat equating with vegetation units to species resources (complementary resource use). Engaging finer resource units is the real ecological equivalent – not niche which lacks spatial co-ordinates – of subatomic particles to the atom in physics; much as physicists wouldn't think now of truncating their lower limit of attention to an atom, no longer can ecologists restrict their thinking to a vegetation unit as habitat. Referring to a biotope as a species' habitat is as bad as referring to that species by its higher taxon, or worse still, naming it as a mix of two or more higher taxa.

Fractionation to finer units will become a feature of butterfly biology; it is an inevitable part of moving issues to finer resolutions, of attaining greater accuracy and precision and of making better predictions. Such models are expected to become space-time explicit and depend on parameters generated from acquisition of space-time explicit data.

A parallel trend is that statistical modelling is becoming a great deal more sophisticated. The distinction between appropriate and inappropriate designs is becoming blurred by multiplicity in approaches and alternative methods. Adjustments for bias are now commonplace, dealing with problems of spatial autocorrelation, apparency, phylogeny and recording effort. We have reached a stage where we are forced to inquire whether the data we are collecting are apt and match the models for sophistication. Despite all this complexity, EBIE authors have provided a pain-free introduction to developments in thinking aids that will continue to grow; in this respect Chapter 7 on species predictive modelling is a splendid example.

The authors also disclose an increased preparedness to experiment and urge more to be done; for example, there is nothing in European butterflies to match the lifetime work on *Zygaena* genetics by Gerry Tremewan (2007) or on *Heliconius*. Experiment has long been integral to laboratory work (e.g. host choice in oviposition and herbivory; pheromone studies) and fieldwork (e.g. mark–release–recapture in population dynamics and dispersal); tests are essential to erase speculation. But, much as solutions cannot all be acquired through direct observation of nature, nor can they be achieved entirely through experiment – in several chapters it is revealed how experimental results are frustrated by interactions in real landscapes where real things happen. This should not deter experiments, just ensure that findings are transferred to real-world situations. In EBIE are accounts of simple but highly effective experiments to solve problems. Per-Olof Wickman describes the neat, groundbreaking example in the release of unmated females to determine mating cues (Chapter 3); as he demonstrates, direct observation cannot distinguish mated from unmated females which behave very differently. Another is the manipulation of wing morphology to test aspects of sexual and natural selection, a field greatly advanced by Paul Brakefield and his research team. John Dover and Garry Fry's (2001) use of artificial sightlines and obstacles to study corridors and barriers is delightfully inventive. EBIE introduces burgeoning ingenuity in large-scale experiments – the use of outdoor (semi-natural) cages has been highly productive, already the generator of important papers (see

Chapter 12 and 14); I look forward to the next stage, the manipulation of whole landscapes to better understand the impact of landscape and habitat components on population persistence and integration.

We come now to a most important question; how are we going to make the best use of all this new knowledge? Sören Nylin (Chapter 15) makes the essential point: 'We are still far from a situation where we really understand how the environment shapes the features of butterflies and other organisms.' In my view this stems from a preoccupation with single functions, paired associations and simplistic notions of cause and effect. Hitherto, this has been necessary and invaluable; understanding starts from isolating key variables, establishing links, distinguishing proximal from distal agents and determining levels – bottom–up versus top–down. However, full understanding has to be more than this. Butterflies and their components are part of natural systems. In EBIB, I tried to draw attention to this issue by producing a number of simple process–response models and warned against 'tunnel' reasoning. The time has come for butterfly biologists to become systems biologists. This requires a dedicated, co-operative effort to hardwire the components influencing key issues in butterfly biology, to locate feedbacks, key switches, trade-offs, rheostats, synergisms and equifinal routes within systems; from this should emerge new principles. Only with this approach will it become evident why conundrums, the many unexpected results reported in the following chapters, happen and keep happening (e.g. sawtooth clines in size; inverse egg size – host quality relationships). Of course, this is not easy; butterfly biologists are faced with bundles of reaction norms (e.g. plastic responses), genuinely complex patterns and processes, and even the arsenal of supposedly heuristic multivariate techniques are no panacea or substitute for originality and insight. Butterfly biologists, then, have to be prepared for a future where at one end variables are being fractionated, finer definitions are being made, and at the other, links are coming together into structured webs. The idea is that running the systems in a computer will inform us how butterflies will 'behave' in different circumstances; we then build landscape experiments to check them out. In this process, it cannot be overemphasised how important it is to get the basic units right; if we do not select real units we will generate unrealistic expectations. That is why habitat as a vegetation unit has outlived its usefulness.

The primary objective in all this research is difficult for readers to miss; it is compounded in the four chapters of the last part on 'Global change and conservation'. There is a simple way of appreciating the crisis: we have long had water-deficient red deserts and energy-deficient white deserts; now we have species-deficient green ones – marinated in herbicide and fertiliser. Green deserts pave the way to ultimate sterility as red deserts. The notion is foreign to much of the public and many politicians who pay lip service to biodiversity. The European landscape is indeed productive, but in diversity of plants and animals quite moribund. It is inevitable that if butterfly biologists are to make a difference for future generations, then they will have to become political animals; they will be taken seriously only if the science is watertight, predictions are accurate and actions responsible ones. Faced with human demands and continual changes to land and climate, we need to find solutions where the natural world can be subtly dovetailed into an economic structure that works for long-term sustainability. In this process tests are everything and evidence-based procedures the Holy Grail (Pullin & Knight 2001). There are powerful arguments for maintaining diversity and heterogeneity; loss and blandness equates to erosion of genes and gene complexes, of naturally occurring chemicals and structures, of ecosystem stability, planetary resources and ultimately economic well-being and the human spirit. The threats are well known and addressed in the final chapters. To combat Europe-wide problems, a Europe-wide organisation becomes the lynchpin: to map, monitor populations, co-ordinate regional studies, to encourage co-operation among researchers and make best use of available talent, to accumulate sufficient resources, to tackle nested research programmes, and to press for apt legislation. Butterfly Conservation Europe aims to conserve butterflies and their habitats across Europe – from the above account, the reader will gather that this author feels that a first step is to be absolutely clear about what we mean by habitat. It is the context for populations, for evolutionary changes, between which individuals transfer, and the base for management. Unless we isolate the elements in habitat – resources and conditions – and understand how they impact on individuals, and amalgamate to integrate populations, our grasp of butterfly biology will remain deficient, artificial, in which case the identification of prime sites (Chapter 21), reconstitution of traditional techniques (Chapter 23) or adoption of political solutions (Chapter 23) will fail to have the desired effect.

ACKNOWLEDGEMENTS

Many thanks to Peter Hardy, Tim Shreeve and Tim New for their helpful suggestions.

Part I
Habitat use: resources and constraints

2 • Adult food resources in butterflies

ANDREAS ERHARDT AND JOVANNE MEVI-SCHÜTZ

SUMMARY

Adult butterflies can substantially increase their longevity, and in most cases fecundity, by engaging in feeding. For some species adult feeding supplements the resources gained during the larval stage, but for others it is critical for egg-production and ensuring mating success, being the most important for long-lived species and for females that emerge as adults with few or no developed eggs. Butterflies feed from a variety of resources, and, within species, males and females may differ in their feeding patterns and substrates exploited, as they have different energy and nutrient requirements. The most commonly used substrate is floral nectar, which varies in composition and quantity between and within species, according to location, flower age, time of day, weather and number of pollinator visits. However, besides nectar sugars, nectar amino acids seem to play a more important role for butterfly fecundity than so far acknowledged. Butterflies primarily find nectar resources using vision, and secondarily using olfaction. Species may vary in their colour preferences, but there is ample evidence that foraging involves learning, and the range of plants exploited by individuals may be limited by memory constraints, with individuals displaying constancy for particular flower types. Visited flowers tend to be upright, with long tubes, or spurs, and are most frequently radially symmetrical. Adult butterflies may be efficient plant pollinators and their selectivity may play a role in the persistence of particular plant species and influence the evolution of flower morphology. When adult flight periods coincide with the flowering time of their larval host plants there are strong advantages for adults to exploit the nectar of their host plants. The spatial coincidence of larval and adult resources leads to the simplification of search images for adult food resources and egg-laying locations and the possibility that adult feeding ensures the reproduction of the larval host plants. Adult food resource distribution plays a key role in determining habitat quality and the suitability of landscapes for butterfly persistence; food resource quality and distribution play important roles in the dynamics of butterfly populations.

FOOD USE OF BUTTERFLIES AND CONSTITUENTS OF THESE FOOD SOURCES

Adult butterflies spend a great proportion of their lives foraging for and feeding on a broad variety of resources. The utilization of various nutrient sources by butterflies depends on the availability of resources at any given time or site (Brakefield 1982, Dennis 2004, Tudor *et al.* 2004). Weather, anthropogenic habitat alteration and competition for nutrients can result in variation of food quantity and quality. Basically, butterflies require water, carbohydrates, proteins, the essential amino acids (arginine, histidine, isoleucine, leucine, lysine, methionine, phenylalanine, threonine, tryptophan and valine), sterols, vitamins and minerals such as sodium, magnesium, calcium and phosphorus. Some of these requirements may be met, or partly met, in the larval stages, but adult feeding is known to contribute to adult energy requirements, longevity and reproductive success. Adult requirements may differ between species and also between sexes, but for the majority of species adult feeding is essential for success.

Floral nectar is by far the most common and widespread butterfly food source (Norris 1936, Gilbert & Singer 1975, Boggs 1987). The chemical composition of floral nectar is highly variable, with different plant species producing different quantities and various constituents (Baker & Baker 1983). These differences include the concentration and composition of sugars, amino acids, lipids, vitamins, alkaloids, phenolics and glycosides (Baker & Baker 1975, Baker & Baker 1977, Baker & Baker 1983, Rusterholz 1998).

Butterflies are able to utilize several components of nectar. Water is the primary constituent of nectar, with sugar concentrations ranging between 15% and 53% (Baker & Baker 1983, May 1985, Rusterholz 1998). Sucrose, fructose and glucose are the three main sugars in nectar and butterfly-pollinated flowers are characterized by high proportions of sucrose compared to fructose and glucose (Baker & Baker 1983). Amino acids, the building blocks of proteins, are found in higher concentrations in butterfly-visited

Ecology of Butterflies in Europe, eds. J. Settele, T. Shreeve, M. Konvička and H. Van Dyck. Published by Cambridge University Press. © Cambridge University Press 2009, pp. 9-16.

flowers compared to bee-visited flowers (Baker & Baker 1975). In addition, nectar amino acid composition shows a high degree of constancy within plant species, perhaps giving the nectar a distinct taste, thus making it recognizable for specific butterflies (Baker & Baker 1977, Gardener & Gillman 2001). All ten essential amino acids are found in the nectar of one plant species or another, as well as several quasi- and non-essential amino acids. The amino acids alanine, arginine and hydroxy-proline are present in almost all nectars, and asparagine, proline, serine, threonine and glycine are also common (Baker & Baker 1975, Gottsberger et al. 1989, Lanza et al. 1995, Rusterholz 1998).

The quality of nectar can change depending on flower age, time of day, weather and activities of previous nectar feeders (Corbet 1978, Gottsberger et al. 1990). Adult nectar stress within butterflies may be quite common due to plants' responses to weather conditions; high temperatures and low humidity can cause evaporation of water thus making the nectar too viscous to extract, whilst rain can dilute nectar (Boggs 2003). Furthermore, the expected rise in atmospheric CO_2 concentration has been shown to reduce the amount of nectar and alter the amino acid composition of some plants (Rusterholz & Erhardt 1998; but see Lake & Hughes 1999 and Erhardt et al. 2005). Pollen contamination of nectar caused by pollen being knocked into the nectar by floral visitors has profound effects on nectar. The resultant increase in amino acid concentration provides an even greater reward for butterflies (Erhardt & Baker 1990, Gottsberger et al. 1990).

Pollen provides certain species in the Neo-tropical *Heliconius* genus with amino acids and proteins (Gilbert 1972, Dunlap-Pianka et al. 1977, Boggs 1981a). Deliberate pollen feeding was also suggested for Neo-tropical *Parides* and *Battus* butterflies (DeVries 1979), but is not known to occur in other butterfly genera.

Adult butterflies are known to use a wide variety of food sources aside from nectar and pollen. Puddling is a common activity where primarily male butterflies frequent mud puddles, edges of ponds, damp sand, carrion and animal dung (Adler & Pearson 1982, Pivnick & McNeil 1987, Sculley & Boggs 1996, Beck et al. 1999, Hall & Willmont 2000). Males acquire nutrients such as sodium, potassium, trace elements and possibly also organic and inorganic nitrogenous compounds (Arms et al. 1974, Gilbert & Singer 1975, Boggs & Dau 2004). Rotten fruit, though more common in tropical climates, also attracts certain butterflies. Fermenting fruit offers a variety of nutrients including sugars, proteins and in some cases ethanol (Norris 1936, Brakefield 1994,

Braby & Jones 1995, Miller 1997). Honeydew, the sweet secretion from aphids, is an alternative source of nourishment for both male and female butterflies and particularly among woodland butterflies (Norris 1936, Porter 1992). However, the ingestion of honeydew from polluted leaves may adversely affect butterfly populations (Corke 1999). Butterflies occasionally alight on perspiration and saliva in search of water and salts, and in at least one case appear to prefer saliva over a sugar solution (Tumler 1885, Norris 1936, Arms et al. 1974). Additional, less commonly used resources include tree sap, wood ash, litter, dry earth, rotten plants, stones, ant-bird droppings, fungal exudations on grass flowers and in one documented case the blood of a wounded horse (Seitz 1894, Norris 1936, Ray & Andrews 1980, Brown 1984b, Shreeve 1992b). These resources provide the butterflies with various proteins, minerals and in some cases secondary metabolites used for defence against predators.

Furthermore, water, in the form of water droplets on plants, as well as being acquired through puddling, nectar intake and a variety of food sources, is essential for butterflies (Norris 1936, Watanabe 1992, Braby & Jones 1995, Miller 1997).

EFFECTS OF FOOD ON SURVIVAL AND REPRODUCTION

Life history strategies, including the number of generations per year, duration and timing of each developmental stage, number and size of offspring and longevity, are affected by both larval and adult feeding (Boggs 1987). Some butterflies emerge with all eggs yolked, whereas other species have only a few or no eggs ready upon emergence. Adult feeding becomes increasingly important for egg production in those long-lived species which emerge with few or no eggs (Leather 1995). Furthermore, there is a clear link between butterfly biological traits, resource use and host plant strategies (Dennis et al. 2004, Stefanescu et al. 2005b). Nutritional resources allocated for longevity and fecundity by butterflies are obtained in three ways: (1) larval uptake and storage in fat bodies, (2) adult feeding, and/or (3) nuptial gifts (male nutrient investments passed to females with sperm via the spermatophore at mating) (Boggs 1981a, Boggs 1990).

Nutrients acquired during larval development were believed to be the primary determinant of longevity and fecundity in butterflies, especially in those which do not feed as adults (Labine 1968, Dunlap-Pianka et al. 1977, Boggs 1987, Svärd & Wiklund 1988a, Baylis & Pierce 1991, Karlsson 1998, García-Barros 2000c, Hughes et al. 2000).

However, the role of adult nutrition in many butterfly species, in the form of direct uptake or as nuptial gifts, may be of substantial importance in determining longevity and fecundity in butterflies. Norris (1935), a pioneer in this field, showed that longevity and fecundity in *Pieris rapae* was greatly increased by adding sugar to water. The addition of sugars to adult food has been shown to increase longevity and fecundity in a number of other species (Stern & Smith 1960, David & Gardiner 1962, Murphy *et al.* 1983, Leather 1984, Moore & Singer 1987, Hill 1989, Hill & Pierce 1989). There is a significant increase in the number of mature eggs produced by *Papilio xuthus* L. when the females were fed sugar solutions (Watanabe 1992) and adult feeding in *Thymelicus lineola* can increase egg production by 27 times (Pivnik & McNeil 1985b). Rotting fruit and honeydew uptake have been shown to prolong female lifespan and oviposition rates (Miller 1989, Braby & Jones 1995). Adult fruit feeding is essential for the onset of oviposition in the tropical frugivorous butterfly *Bicyclus anynana* (Fischer *et al.* 2004b). Further studies have shown that adult derived sugars are incorporated into eggs and that in some cases may make up for 50–60% of egg carbon (Boggs 1997a, O'Brien *et al.* 2000).

The uptake and utilization of amino acids from pollen greatly increases the number of eggs and duration of egg laying in tropical pollen-gathering *Heliconius* species (Gilbert 1972, Dunlap-Pianka *et al.* 1977). Rotten fruit appears to supply females with an adequate supply of proteins to maintain constant egg production in *Mycalesis terminus* (Braby & Jones 1995) and radiotracer studies on *Speyeria mormonia* have clearly demonstrated that amino acids obtained in the adult stage are incorporated into eggs (Boggs 1997a). Honeydew, another nitrogen source, has also been shown to prolong female lifespan and enhance reproduction (Miller 1989). Despite this evidence, the role of nectar amino acids as a beneficial source of nitrogen for adult butterflies remains controversial. Although some butterfly species show a distinct preference for nectar containing amino acids, suggesting some form of dependency and utilization of this nectar constituent (Alm *et al.* 1990, Erhardt & Rusterholz 1998, Ruehle 1999, Rusterholz & Erhardt 2000, Mevi-Schütz & Erhardt 2003a, Mevi-Schütz & Erhardt 2004), the actual benefits of nectar amino acid have been difficult to demonstrate. Murphy *et al.* (1983) have shown that *Euphydryas editha* receiving amino acids lay larger eggs in later masses. These results were refuted by Moore & Singer (1987) who found no amino acid effect on the maintenance of egg weight over time for the same species.

The presence of amino acids in food stimulated the butterfly *Jalmenus evagoras* to feed, but did not lead to an increase in longevity and fecundity, and amino acids in the adult food actually reduced fecundity in *Euploea core corinna* (Hill 1989, Hill & Pierce 1989). Amino acids in a *Lantana camara* nectar mimic fed to *Lasiommata megera* butterflies raised under benign greenhouse conditions had no effect on any of the measured fecundity parameters. Egg number and egg weight decreased at the same rate regardless of diet treatment (Mevi-Schütz & Erhardt 2003b). However, more recent work has shown that *Araschnia levana* raised under natural larval food conditions laid more eggs when they were fed nectar containing amino acids, whereas nectar amino acids had no effect on the number of eggs laid by butterflies raised on larval food rich in nitrogen (Mevi-Schütz & Erhardt 2005a). Therefore, nectar amino acids should be recognized as essential resources affecting butterfly fecundity and butterfly populations under natural conditions.

Puddling and the uptake of various nutrients, especially sodium, increase the number of matings of male butterflies (Pivnick & McNeil 1987). Males contribute indirectly to egg production by passing essential sodium and potassium to females at mating (Shreeve 1992b). Amino acids acquired during puddling were incorporated into somatic tissue of *Papilio glaucus* L. (Arms *et al.* 1974) and males receiving electrolytes plus amino acids fathered seven times more offspring than those that did not (Lederhouse *et al.* 1990). The attraction of male butterflies to nitrogen-rich resources such as rotting carrion and albumin baits suggests that these nutrients may increase reproductive success (Beck *et al.* 1999, Hall & Willmont 2000). Males pass a substantial amount of nutrients in the spermatophore to the female during mating (Rutowski *et al.* 1983) and these nuptial gifts are known to increase female longevity and fecundity (Kaitala & Wiklund 1994, Boggs 1995). Although larval reserves predominate in the first spermatophore, male butterflies utilize adult resources for multiple matings and reproductive potential is severely limited if adult feeding is restricted (Lederhouse *et al.* 1990). Female butterflies receiving a large first spermatophore have a higher lifetime fecundity and multiply mated females live longer, thereby utilizing a larger part of their reserves for egg production (Ward & Landolt 1995, Oberhauser 1997, Karlsson 1998, Hughes *et al.* 2000). Males and females may use different food sources depending on the demands of egg and sperm production. Trade-offs in either sex between adult feeding habits and the effects of large or multiple spermatophores on adult food preference have yet to be examined.

BUTTERFLIES AS POLLINATORS

Flowers from the viewpoint of butterflies

The perceptual mechanisms of butterflies to locate flowers include vision in the first place and to a lesser degree, olfaction (Boggs 1987, Omura et al. 1999, Andersson 2003, Andersson & Dobson 2003, Omura & Honda 2005). The visual spectral sensitivity of butterflies varies across taxa, but extends from ultraviolet to red in some species, the widest known among animals, including humans (Silberglied 1984, Weiss 2001). Butterflies have different innate colour preferences, which can vary between genera within a family, between species in a genus, and even between the sexes in a species (Weiss 2001 and references therein). For instance, the European swallowtail *Papilio machaon* has a clear preference for purple (Ilse 1928), whereas the New World papilionid *Battus philenor* strongly prefers yellow, with a minor preference for blue/purple (Weiss 1997). These different colour preferences may be related to sexual behaviour, although evidence for this is not unambiguous (Silberglied 1984). Possible evolutionary explanations could also include niche differentiation of different butterfly species for different flower types and/or evolved colour preferences for the main nectar plants in the primordial biotopes of the different species. Different floral colours could also be correlated with different types of nectar, which in turn could have led to the evolution of the observed colour preferences of butterflies.

Butterflies are often considered to be opportunistic foragers which visit a variety of available flowers (e.g. Sharp et al. 1974, Courtney 1986, Grundel et al. 2000). If a species produces more than one generation per season, different generations are likely to encounter different available nectar plants and have therefore to be opportunistic. However, butterflies do not visit flowers randomly, but often show distinct flower preferences which can differ between species and even between conspecific males and females, according to their different energy needs and reproductive requirements (Watt et al. 1974, Wiklund & Ahrberg 1978, Pivnic & McNeil 1987a, Erhardt 1991, Erhardt & Thomas 1991, Porter et al. 1992, Rusterholz & Erhardt 2000). Research in subalpine meadows in the Swiss Central Alps (Erhardt 1995 and references therein) showed that butterflies and other diurnal Lepidoptera such as burnet moths visited only ca. 20 out of ca. 170 potential nectar plants in the study area. Most preferred were purple and/or yellow capitula of Compositae and Dipsacaceae, with white capitula

being preferentially visited by a few species (e.g. *Erebia melampus* or *Heodes virgaureae*). Furthermore, flower preferences of single species varied according to the flowers present in a particular biotope. Nectar of preferred flowers was characterized by either high volumes, moderate to high proportions of sucrose to fructose and glucose and by relatively low amino acid concentrations, or by low volumes with little sucrose but high concentrations of amino acids. The clear preference for a small, restricted number of nectar plants suggests that these nectar plants could play a particularly important role for the population dynamics of butterflies, and may thus function as keystone nectar plants. The findings of this study are also likely to apply to other areas and other biotopes and habitats (e.g. Douwes 1978, Brakefield 1982, Holbeck et al. 2000, Tudor et al. 2004).

Butterflies can also show high degrees of flower constancy (e.g. Watt et al. 1974, Kay 1978, Kay 1982, Erhardt & Thomas 1991). This can be important for the visited plants because it favours outcrossing. Lewis (1986) showed that memory constraints favour flower constancy in *Pieris rapae*. However, butterflies rapidly learn to associate a sugar reward with colour (Weiss 2001 and references therein). In summary, factors responsible for different flower visitation patterns and flower preferences of butterflies include innate colour preferences, learning, proboscis length, body mass and wing loading (i.e. a measure of body mass per wing area: for details see Corbet 2000), corolla tube length of flowers, flower colour and floral scent, nectar quality and quantity, and energy requirements and reproductive needs (Watt et al. 1974, Erhardt 1991, Porter et al. 1992). The ability of certain butterflies to dilute highly concentrated nectar with saliva (e.g. *Erebia* species: A. Erhardt pers. obs.) is another important factor for the different flower preferences of butterflies. Apart from these proximate causes, spatial factors clearly influence flower visitation patterns of butterflies (see below). The range and relative abundances of flower species available for exploitation vary over the geographical and ecological range of a particular butterfly species, and different flowers will be used according to geographical and ecological conditions, pointing to the possibility of geographical and ecological vicariants of keystone nectar plants for butterflies. For example, bramble (*Rubus* sp.) may play such a key role as nectar source in agricultural landscapes of Britain (Dover 1996, Dover & Sparks 2000, Dover et al. 2000), although *Rubus* flowers are certainly not adapted to and dependent on butterfly pollination (see below).

Since nectar is the resource butterflies get from the flowers they visit, its quality is likely to play a key role for observed flower preferences. Experiments testing the actual response of butterflies to different nectar types cover different aspects of preference. Responses of butterflies to sugar solutions of different concentrations have been carried out in order to assess rates of maximum energy gain by the tested butterflies. Based on a mechanistic model, Kingsolver & Daniel (1979) stated that net rate of energy gain in butterflies is optimal at sugar concentrations of 20–25% (weight to total weight). However, Pivnic & McNeil (1985) found a maximum rate of sucrose intake at a concentration of 40% in *Thymelicus lineola*. Similar maximum rates of energy intake were also found in non-European butterfly species (May 1985, Boggs 1988), and *Vanessa cardui* butterflies also consistently preferred a highly concentrated sugar solution (Hainsworth 1989). However, sugar concentrations in nectar of butterfly-pollinated flowers are usually lower (ca. 25% w/w: Baker & Baker 1982) than the optimum for maximum rate of energy intake. Explanations for this discrepancy could be water stress of butterflies (Pivnic & McNeil 1985, Boggs 1987), a trade-off between benefits of plants derived from increased visitation due to increasing sugar concentrations and costs required to produce more concentrated nectar, and/or physiological constraints in the nectary for secreting highly concentrated nectar.

Preference tests with butterflies for nectar sugars and amino acids are still limited. However, butterflies seem generally to prefer sucrose over fructose, and fructose over glucose, in plain solutions with only one sugar as well as in mixtures of these three sugars. This pattern was basically found (with some minor deviations) in all tested species (Rusterholz & Erhardt 1997 and references therein, Ruehle 1999). This corresponds well with the findings of Baker & Baker (1983) that butterfly-pollinated flowers produce sucrose-rich nectar. Responses of butterflies to nectar amino acids are less consistent and may be more plastic. Some species do not prefer nectar mimics (man-made nectar imitation) containing amino acids versus corresponding plain sugar solutions, whereas in other species (e.g. *Pieris rapae* and *Inachis io*) females show a distinct preference for nectar mimics containing amino acids (Alm *et al.* 1990, Erhardt & Rusterholz 1998, Ruehle 1999). However, it remains doubtful if butterflies can discriminate between different amino acids as Diptera can (Potter & Bertin 1988, Erhardt & Rusterholz 1998). Recent experiments have shown that larval food conditions affect *Arashnia levana* nectar preferences. Female butterflies raised on a nitrogen-poor larval diet showed a significant preference for nectar mimics containing amino acids compared to females which

had been raised on a nitrogen-rich larval diet (Mevi-Schütz & Erhardt 2003a, 2005b).

Butterflies from the viewpoint of flowers

Butterflies are certainly an important group of pollinators. Flowers pollinated mainly by butterflies share a number of common features which have been summarized in a 'syndrome' (Faegri & van der Pijl 1979). These flowers are open during the day, have different colours including ultraviolet and red, and are relatively weakly scented with a fresh and sweet odour. They are upright with a long tube or spur, and are usually radially symmetrical with a flat rim to provide a landing platform, although some butterfly-pollinated flowers, particularly orchids, are also monosymmetrical. Nectar is produced in moderate amounts, and is rich in sucrose and amino acids. Plant species with such features include *Dianthus carthusianorum*, *Centranthus ruber* and *C. angustifolius* and the orchids *Gymnadenia conopsea* and *Anacamptis pyramidalis* (Proctor *et al.* 1996). Jennersten (1988) showed that *Dianthus deltoides* is butterfly pollinated and that habitat fragmentation can significantly reduce visitation frequency by butterflies and consequently seed set in this plant.

The body parts of butterflies to which pollen grains adhere and which are therefore responsible for pollen transfer include the proboscis, labial palps, thorax and legs but also the wings (Cruden & Hermann-Parker 1979, Murphy 1984). The number of pollen grains carried by butterflies has only in a few cases been investigated (e.g. Kay *et al.* 1984, Murphy 1984), and knowledge on pollination efficiency is scanty (Scoble 1992). Pollination efficiency was low in one of the few studies in which it was thoroughly investigated (Levin & Berube 1972), but was higher than that of honeybees (!) when visiting wild radish, *Raphanus raphanistrum* (Conner *et al.* 1995). Butterfly-mediated pollen flow within plant populations appears to be restricted to short distances of mostly less than 50 metres due to the sedentary behaviour of many butterfly species (Thomas 1984b). However, Kay *et al.* (1984) suggested that pierid butterflies could effect some longer-range pollen flow, and migratory butterflies such as *Vanessa cardui* or *V. atalanta* can cover thousands of kilometres and have therefore a high potential for long-distance pollination, even though the extent of long-distance pollen transfer is still hardly known (e.g. Levin & Kerster 1974).

Some butterfly species identify isolated conspecific plants with suitable floral rewards in their habitats and visit these plants repeatedly along a defined route. This habit, known as 'trap-lining', has been reported mainly from trop-ical *Heliconius* butterflies (Gilbert 1980), but European but-terflies could potentially also trap-line, though there is no evidence to support, or universally refute this idea. Trap-lining of pollinators enhances cross-pollination over long distances, but also results in non-random, assortative polli-nation of the visited plants (Levin & Kerster 1974).

Butterflies can discriminate between the flowers of geno-typically different morphological and colour forms within the same plant species, and preferences for different flower colour morphs can differ even between individuals of the same species (Kay 1978, Kay 1982). Thus butterflies can have an important effect on the evolution of floral colours. Different colour preferences of butterflies can lead to segre-gation and subsequent genetic isolation of different colour morphs of a plant species and therefore affect plant speci-ation (Dronamraju 1960). Different butterfly species with different proboscis lengths could potentially also cause assortative mating in conspecific flower morphs differing in corolla tube or spur lengths, and could hence represent an example of pollinator-mediated sympatric speciation in plants (Bloch & Erhardt 2008).

During the last years, a generally observed decline in pollinators has raised concerns about the future sexual reproduction of crop plants as well as of wild plants, and was coined as the 'pollination crisis' (Kearns *et al.* 1998, Biesmejer *et al.* 2006). Butterfly-pollinated plants may be particularly at risk, as recently shown in a field study of *Dianthus carthusianorum* (Bloch *et al.* 2006).

COEVOLUTION OF BUTTERFLIES AND FLOWERS

The term coevolution, originally coined by Ehrlich & Raven (1964), has been used differently by different authors. Janzen (1980) provided a restrictive definition which has been adopted by most authors. After Janzen (1980), true coevolution means that a trait of one species has evolved in response to a trait of another species, the trait itself having evolved in response to the trait of the first. This definition requires specificity – the evolution of each trait is due to the other – and reciprocity – both traits must evolve (Futuyma & Slatkin 1983). Coevolution in this restrictive sense is difficult to show even in the most striking examples in pollination biology. As an example, the famous Madagascar star orchid *Angraecum sesquipedale*, with an exceedingly long spur, was predicted by Darwin to be pollinated by an (at his time) unknown hawkmoth with a correspondingly long

proboscis to reach the nectar. A candidate hawkmoth was not discovered until 40 years later: *Xanthopan morgani praedicta* (Nilsson 1988). Even so, it has never been documented that *X. morgani praedicta* actually pollinates *A. sesquipedale* (Nilsson 1998)!

By relaxing the criterion of specificity, the evolution of a particular trait in one or more species in response to a trait or suite of traits in several other species is included within coevolution, and this is called 'diffuse coevolution' (Futuyma & Slatkin 1983 and references therein). In this less restrictive sense, butterflies and flowers have certainly coevolved as suggested by the floral syndrome mentioned above, and the evolution of the proboscis of butterflies from originally chewing mouthparts in primordial Lepidoptera.

Butterflies: mutualists or floral parasites?

In spite of the strong evidence that butterflies can be impor-tant pollinators, their general role as pollinators has been doubted by some recent workers (see Courtney *et al.* 1982 and references therein). It has been emphasized that butter-flies often visit flowers and rob nectar without pollinating them (e.g. Lazri & Barrows 1984). This led Wiklund *et al.* (1979, 1982) to the hypothesis that the proboscis of Lepidoptera evolved as a means to rob nectar, and that butterflies are basically floral parasites. Of course, butterflies visit flowers only for their own needs (see above) and are clearly not always effective pollinators. The floral adapta-tions for pollination by butterflies could consequently be seen as a way to escape nectar predation and to counteract lepidopteran parasitism. Whether this was the main selective force or whether mutualistic coevolutionary processes led to the evolution of the proboscis of Lepidoptera and to the striking floral features of butterfly pollinated flowers must remain open and further research is certainly needed to clarify this question (Plates 1a and 1b).

It has long be questioned why butterflies do not visit the flowers of their larval hosts more often (Hering 1926). This question has never been systematically investigated. However, there are several examples of European and American butterfly species which not only visit the flowers of their larval hosts, but even seem to prefer them (Erhardt 1995). A precondition for visitation of the flowers of the larval host by adult butterflies is that the host plant is in bloom and that the flowers produce nectar during the flight period of the adults. Once this condition is fulfilled, there are strong advantages for butterflies to use the flowers of their larval host as nectar resources. These are:

(1) If a butterfly species also pollinates the flowers of its larval host, it enhances the reproductive success of its host plant and thus increases also its own potential for reproductive success.

(2) If the host plant flowers profusely enough, it also covers the needs of the adults. Consequently, a butterfly species may only depend on its larval host plant and not on additional plants as adult resources, which would enable a species to expand its geographical range. This may particularly apply to biotopes with few other available nectar plants during the flight period of the adults. (However, if butterflies are generalists for nectar sources, this may lower competition or pressure on plants that are also being used for larval hosts, particularly if flowers are fed on by larvae, as e.g. in *Anthocharis cardamines*, and may lead to less harassment of females by males, if females can use other flowers as nectar sources.)

(3) Only one search image for a plant species for oviposition and feeding has to be formed by females.

(4) Secondary compounds which larvae acquire from the host plant might also be accessible in the nectar for the adult insects.

Using the same plant as a larval host and as an adult nectar resource might have been the initial condition for the evolution of some of the most intriguing examples in pollination biology in which adult Lepidoptera pollinate the flowers whose seeds are later preyed upon by the larvae. The most well known among these examples are certainly the yucca plants and the yucca moths (e.g. Baker 1986, Gottsberger 1996, Althoff *et al.* 2005), but looser relationships showing potential transitional evolutionary steps have also been observed in Lycaenidae (e.g. *Maculinea telejus* and *M. nausithous*, *Polyommatus damon*, *Cupido minimus*, *Scolitantides baton* and *S. vicrama*) and in *Perizoma* and *Hadena* moths and Caryophyllaceae (Erhardt 1988, Westerbergh 2004, Kephart *et al.* 2006).

DISPERSION AND HABITAT USE OF BUTTERFLIES IN RESPONSE TO FOOD DISTRIBUTION

Nectar plants and other resources are as important habitat components as larval host plants (Dover 1997, Dennis 2004, Dennis *et al.* 2003, Dennis *et al.* 2004, Dennis *et al.* 2006). Thus, distribution of nectar sources in space and time plays a crucial role in the location and dispersal of butterfly populations. Where the distribution of nectar plants does not overlap with that of larval host plants or mating areas, the area occupied by a butterfly population is extended to include the nectar plants, and dispersal is increased (Boggs 1987). This may be the case for *Leptidea sinapis* in Sweden, where this species oviposits in open meadows, but forages on flowers growing in adjacent woodland (Wiklund 1977a), provided that this study was not confounded by the recent distinction between *Leptidea sinapis* and *L. reali*. Accordingly, Warren (1986) reports that habitat selection by *L. sinapis* is related to the abundance of nectar resources. Likewise, nectar plants were important for habitat selection in a population of *Gonepteryx rhamni* studied in Sweden (Jennersten 1980). Similarly, the distribution of the nymphalid *Stichophthalma louisa* was primarily determined by adult food sources, sap exuding from wounded trees, in a rain forest in Vietnam (Novotny *et al.* 1991). Nectar plant abundance was also highly correlated with total butterfly abundance and species richness in reclaimed coal surface mines (Holl 1995). In a study of habitat use by *Mellicta athalia* in the southern Swiss Alps, females used a range of different grassland types for nectar foraging according to where the flowers were in bloom, whereas males seemed to separate the study area into mating (hay meadows and recently abandoned grasslands) and feeding habitats (mature abandoned grasslands rich in flowers) (Schwarzwälder *et al.* 1997).

The presence of nectar plants can also affect oviposition sites. Thus, the location of hosts receiving eggs correlated significantly with locations of nectar plants in *Papilio glaucus* in New Jersey, USA (Grossmueller & Lederhouse 1988), and the distribution of nectar sources affected not only the distribution of adult *Euphydryas chalcedona*, but also their oviposition sites and hence their offspring (Murphy *et al.* 1984).

The microdistribution of butterflies is often influenced by the distribution of nectar plants. This was found in a mosaic-like habitat in the southern Swiss Alps (Loertscher *et al.* 1994). Intraseasonal changes and differences between years in the microdistribution of *Maniola jurtina* in Great Britain were also correlated with changes in the distribution of clumps of flowering plants (Brakefield 1982). A similar situation has been documented for *Heodes virgaureae* in Sweden where the presence of butterflies was correlated with nectar plants in warm, sheltered microhabitats (Douwes 1975). Microdistribution of female *Carterocephalus palaemon* butterflies was also correlated with their favoured nectar plants in western Scotland, and the restricted abundance of adult resources seemed to limit population distribution, which has important implications for management and conservation of this species (Ravenscroft 1994b). Population size of the endangered Oregon butterfly *Icaria icarioides fendrei* was not only limited by the density of suitable host plants, but also by the availability of nectar sources (Schultz & Dlugosch 1999). The strong decline of *Parnassius apollo* in the Swiss Jura Mountains and other lowland areas in Central Europe could at least partly also have been caused by the dramatic decline in nectar plants in these areas (A. Erhardt pers. obs.). Thus providing enough suitable nectar plants might be an essential, if so far not adequately practised, conservation measure. Indeed, Tudor *et al.* (2004) showed that management of woodland sites for butterfly conservation should give as much consideration to nectar sources as to host plant sources. Last, but not least, availability of nectar resources might be an essential factor determining spatial and temporal migration patterns of butterflies, including the success of individual migration events.

CONCLUSIONS

Butterflies (as well as most moths) are critically dependent on adequate food sources also in their adult life. Availability of suitable nectar plants for butterflies has only recently received adequate attention and may have led to some of the most intriguing examples known in pollination biology. These interactions between butterflies and flowers are, however, increasingly threatened by human-induced environmental changes, putting at risk both butterflies and plants depending on their service as pollinators. Efforts are needed to provide adult feeding resources within specific locations and across landscapes. Without such effort, the prognosis for maintaining already diminishing butterfly species is bleak. In addition, greater understanding of all aspects of adult feeding within the study of the ecology of butterflies is needed in order to fully appreciate the contribution of adult feeding to species persistence.

ACKNOWLEDGEMENTS

This work was supported by grants of the Swiss National Science Foundation (Grants 3100–063562 and 3100A0–108271/1 to Andreas Erhardt). We thank Roger Dennis, Per-Olof Wickman and Martin Konvička for valuable comments on an earlier draft of this chapter.

3 • Mating behaviour in butterflies

PER-OLOF WICKMAN

SUMMARY

The complex mixture of conflicts and cooperation that characterize sexual interactions make them among the most intriguing topics in ecology and evolution. Systematic research during the past decades has produced detailed information on the costs and benefits of mating frequency, mating rate and ejaculate size in butterflies. Within the theoretical framework of sexual selection, this chapter focuses on how mates function as a resource and influence resource use, as well as on the factors that influence mate-location, courtship and mate-choice behaviour. The ejaculate is transferred to the female in the form of a spermatophore containing both sperm and a number of additional substances, including nutrients. Relative spermatophore mass varies considerably among species and relates to the level of sperm competition. Spermatophore envelopes remain intact in the female and can be used to estimate female mating frequency. There are monandrous or polyandrous species. Females have evolved adaptations for using the nutrients of ejaculates to produce eggs. The reason why polyandry evolved only in some species may be because nutritional nuptial gifts make a greater difference in these; but the exact basis remains obscure. Males spend a considerable amount of their time searching for mates. Variation in male mate-acquisition behaviour has been described for numerous species. Instead of long-range pheromones, indirect cues like microclimate, landmarks or host plants are used in locating mates. Conceptually, male mate-location encompasses flight activity, home range and territoriality. Different mate-location strategies correspond to flight morphology. Male–male conflicts and ownership conventions have also attracted much attention in butterflies. Mate-location systems include resource-based systems, lekking, pupae finding and dispersed adult mating systems. Current knowledge about courtship and mate choice is reviewed, addressing in particular pre-copulatory mate choice, cryptic female choice and mating plugs. Butterfly mating behaviour will continue to be a productive field for many years to come.

INTRODUCTION

The basic function of mating is sperm transfer and subsequent fertilization. According to this fundamental requirement, the other sex is a resource carrying gametes needed to produce offspring. Typically, sperm cells can be produced at a higher rate than eggs and females are therefore a limiting resource for males. This means that male reproduction and fitness are constrained by the number of copulations, whereas female fitness and reproduction are constrained by the number and quality of the eggs they can lay. In a typical case the sperm contributes only genes to the nutrient-rich egg. In this scenario of sexual selection, males will evolve traits that give them competitive advantage in finding and fertilizing females. However, active mate-location behaviour by females will not be favoured by selection because males willing to mate will be super-abundant. With little searching cost females can freely choose the male that offers the best genes for her offspring (see Andersson & Simmons 2006 for a review of sexual selection theory).

In butterflies, this general scheme of sexual selection is modified in various degrees mainly for two reasons. Firstly, in addition to sperm, males of many species also transfer nutrients during copulation (Boggs & Gilbert 1979, Wiklund 2003). Hence, in some species males are not merely a gene resource to females, but a nutrient resource as well. Secondly, butterflies are often widely dispersed, relatively short-lived and dependent on stochastically variable weather conditions for flight. Because of this, females of some species may not encounter males when needed without actively searching for them. It also means that females of some species may have restricted opportunities to choose among males (Wickman & Rutowski 1999).

Mating behaviour and sexual selection in butterflies have been reviewed from several different perspectives (Rutowski 1984, 1997, Silberglied 1984, Dennis & Shreeve 1988, Shreeve 1992b, Wiklund 2003). In this chapter I use sexual selection with the aforementioned modifications as the theoretical framework. The focus is on how mates function as a resource and influence resource use, as well as on

Ecology of Butterflies in Europe, eds. J. Settele, T. Shreeve, M. Konvička and H. Van Dyck. Published by Cambridge University Press. © Cambridge University Press 2009, pp. 17-28.

the factors that influence mate-location, courtship and mate-choice behaviour.

MATES AS A RESOURCE

Some basic biology

In Lepidoptera, the ejaculate is transferred to the female in the form of a spermatophore, which she stores in her *bursa copulatrix*. The spermatophore is opened in the bursa and the sperm is transferred to the *receptaculum seminis*. Sperm is then used to fertilize eggs successively before they are laid. Beside sperm, a number of additional substances are transferred at mating, namely (1) egg production stimulants, (2) female receptivity inhibitors, (3) anti-sperm substances against other males' sperm, (4) pro-sperm substances that aid the male's own sperm and (5) nutrients (Arnqvist & Nilsson 2000). The nutrients transferred by males of some species may therefore constitute resources for egg production (Boggs & Gilbert 1979, Boggs 1981a, 1990, 1995, Boggs & Watt 1981, Vahed 1998).

Spermatophore envelopes remain more or less intact in the bursa of many species, and can be used to estimate the mating frequency of females (Burns 1968, but see e.g. Cordero 2000a), which has been established for a number of European butterflies (Drummond 1984, Ridley 1988, Svärd & Wiklund 1989, Wiklund & Forsberg 1991). In monandrous species females rarely mate more than once, while in polyandrous species females mate twice or three times and occasionally even five to eight times (Wiklund & Forsberg 1991). Although males can be very persistent in courting females, in most species males cannot mate with a female without her cooperation (see Scott 1973b, Rothschild 1978, Svärd & Wiklund 1989, Wiklund & Forsberg 1991, Bergström & Wiklund 2005). This suggests that in the majority of species where females mate multiply, females gain by doing so.

As opposed to the decision to initiate mating, the termination and hence the duration of matings appear to be determined by males (Wickman 1985b, Wiklund 2003). The length of copulations varies considerably from about 10 minutes in *Coenonympha pamphilus* (Wickman 1985c) to 102 h in *Gonepteryx rhamni* (Lindfors 1998). Usually the mating duration increases the more recently males have mated, and more recently mated males also produce smaller spermatophores (Rutowski 1979, Sims 1979, Boggs 1981b, Svärd 1985, Svärd & Wiklund 1986, 1989, Oberhauser 1988, 1992, Bissoondath & Wiklund 1996a, 1996b). For instance,

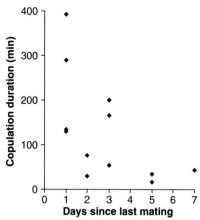

Figure 3.1. Copulation duration as a function of days since last mating in *Pararge aegeria* (data from Svärd 1985).

in *G. rhamni*, matings with males that had mated the day before on average take 35 times longer (34–102 h) than with virgin males (25 min to 4.5 h) (Lindfors 1998). In other species the increase in mating duration is smaller (Fig. 3.1). Also the time it takes for a male to recover and produce a virgin male sized spermatophore varies. In *Pararge aegeria* it takes 6 days (Svärd 1985) and in *Pieris rapae* and *P. napi* only 2 days (Bissoondath & Wiklund 1996a). *Papilio machaon* males never fully recover (Svärd & Wiklund 1986). Generally the fecundity of females mated to virgin males is higher than that of females mated to experienced males (Torres-Vila & Jennions 2005).

Spermatophore mass ranges from about 10–15% of total male body mass in for example *P. napi* (Bissoondath & Wiklund 1996a) to about 1.5% in *P. aegeria* (Svärd & Wiklund 1989). During the life of a polyandrous species like *P. rapae*, the spermatophores from subsequent matings may together add up to almost 30% of male total body mass (Bissoondath & Wiklund 1996b).

Mating frequency

Much effort has been directed to understanding the mating frequency of females. There is much evidence that female mating frequency can be viewed as a species-specific trait, and that the evolution of polyandry is the result of sexual conflict between males and females over the number of matings (Arnqvist & Nilsson 2000, Wiklund 2003). In polyandrous species females have evolved adaptations for using the nutrients of male ejaculates to produce eggs (Boggs & Gilbert 1979, Boggs 1981b, 1990, 1992, 1997a,

1997b, Boggs & Watt 1981, Oberhauser 1992, Wiklund *et al.* 1993, Karlsson 1998). By remating, females can replenish their nutrient supplies, and thereby increase the number of eggs laid. They may also live longer and lay larger eggs (Pivnick & McNeil 1987b, Rutowski *et al.* 1987, Watanabe 1988, Oberhauser 1989, Wiklund *et al.* 1993, 1998 – for exceptions see Jones *et al.* 1986, Svärd & Wiklund 1989, 1991, Kawagoe *et al.* 2001, Kemp & Rutowski 2004). Probably as a consequence, *Pieris napi* females receiving small ejaculates remate more quickly than those receiving large ones (Kaitala & Wiklund 1994). In the monandrous *Pararge aegeria* where males do not provide nutrients, female fecundity is independent of male ejaculate investment (Wedell & Karlsson 2003). Even so, copulations with virgin males result in higher numbers of larval offspring (Lauwers & Van Dyck 2006).

Boggs & Gilbert (1979) were the first to suggest that males may be a limited resource for females and there is now evidence that polyandrous females 'forage for matings' (Kaitala & Wiklund 1994). Females behaving in this way favour males that make large nutrient contributions that delay female remating, and hence there is selection for larger male ejaculates. This evolutionary scenario concurs with a whole suite of comparative observations on male traits, viz. that the degree of polyandry of a species is positively correlated with relatively larger testes and longer sperm (Gage 1994), larger ejaculates (Svärd & Wiklund 1989, Bissoondath & Wiklund 1995), higher protein content of ejaculates (Bissoondath & Wiklund 1995), larger male-biased sexual dimorphism (Wiklund & Forsberg 1991, Nylin & Wedell 1994) and relatively larger abdomen mass and amount of nitrogen in the abdomen, which suggests more resources allocated to reproduction (Karlsson 1995, 1996). Finally the degree of polyandry is correlated also with female reproductive output, measured as cumulative egg mass divided by female mass (Wiklund *et al.* 2001).

Male rematings are favoured when there is at least some degree of last male sperm precedence (the proportion of eggs fertilized by the last male to mate), which generally seems to be the case in Lepidoptera (Drummond 1984, Ridley 1989, Platt & Allen 2001). In *Pieris napi*, sperm precedence is influenced by male size; the larger the first male the fewer fertilizations the second male obtains (Bissoondath & Wiklund 1997). Moreover, females remate sooner when mated to a smaller male (that delivers smaller ejaculates). This is in agreement with the interspecific trend that males are relatively larger in more polyandrous species. Interestingly, the nuptial gift received by female *P. napi*

affects their feeding behaviour, decreasing their need for nectar containing amino acids, as demonstrated by their preference for amino acid nectar being negatively correlated with mating rates (Mevi-Schütz & Erhardt 2004).

The reason why polyandry evolved only in some species may be because nutritional nuptial gifts make a greater difference in these; but the exact basis for this remains obscure. Leimar *et al.* (1994) suggested that in species that experience large food variability as larvae, there will also be large variability in adult size. In such species there will be some large males that have much to offer in terms of nutrients, as well as small females and females that mate with small males that have a lot to gain by mating repeatedly. Hence, polyandry is expected to evolve in species where food or any other environmental factors vary substantially in ways that affect adult size. In agreement with this, relative ejaculate mass and degree of polyandry correlate interspecifically with intraspecific female wing length variation (Leimar *et al.* 1994, Karlsson *et al.* 1997). However, there are complications to this argument. Contrary to expectations, large females in the wild mate more often than small ones (Bergström *et al.* 2002) and small as opposed to large females were not capable of increasing their mating rate to compensate for a low weight at eclosion (Bergström & Wiklund 2002). There is thus little evidence that the hypotheses of Leimar *et al.* (1994) can be applied generally to explain mating frequency.

The reason for a lower number of matings in some species may also depend on costs associated with matings. Possible mating costs to females involve time loss and energy costs (Thornhill & Alcock 1983), increased risk of predation (Gwynne 1989), risk of physical injury (Orr 1999) and parasite transfer (Hurst *et al.* 1995). However, there is little evidence that such costs have influenced mating frequency in females. Nevertheless, a time cost of mating seems to have influenced female behaviour in *Pieris napi*. In this species, virgin females mate in the morning. However, females that mate again prefer to mate in the late afternoon, which should give females more time to lay eggs during the day (Forsberg & Wiklund 1989). Although there are costs associated with finding a male (Wickman 1992a, Wickman & Jansson 1997), there is nothing to suggest that these costs have influenced mating frequency. Hence, although research has demonstrated that polyandry is correlated to whole suites of traits as well as the numerous ways in which multiple mating affects fitness, little is still known about the selection pressures that in the first place determine mating frequency in butterflies.

Mating rate

The relative potential mating rate of the two sexes can be used for predicting the operational sex ratio (the ratio of receptive males to females) and hence which sex will be limited and which sex will compete for the other (Parker & Simmons 1996). Although spermatophores may represent a substantial proportion of male body mass, females generally have a lower mating rate compared to males that mate. Even in a polyandrous species like *Pieris napi*, where male nutrient expenditure in offspring on average equals that of females, males can mate every 6th hour, while females mate about once a week (Wiklund *et al.* 1998). Even so, males of many polyandrous species appear to recuperate faster and mate more times than males of many monandrous species (Bissoondath & Wiklund 1996b). This means that in most species of butterflies, males compete for females, and most females that males encounter and court are non-receptive. Reversed sex roles (females court choosy males and hilltop) occur in a few butterflies where sex ratios are female biased due to a male-killing bacterium (Jiggins *et al.* 2000, Jiggins 2002).

In most butterfly species, the probability that a male will encounter a receptive female is low. This is presumably the reason why recently mated males that produce small spermatophores readily mate with any receptive female, although it means spending a long time in *copula* (Bissoondath & Wiklund 1996a) and that the female may remate sooner (Kaitala & Wiklund 1994). The low probability of encountering another female has also favoured mate-guarding in some butterflies. In *Danaus plexippus* females do not start laying eggs until the day after copulation, and here copulations with males continue until after nightfall irrespective of when copulation commences (Svärd & Wiklund 1988b). Nightfall acts as a cue for sperm transfer and termination of *copula*.

FINDING MATES

Compared to females, males spend a considerable amount of their time searching for mates. Variation in male mate-acquisition behaviour – as opposed to female mate-acquisition behaviour – is therefore well known and has been described for numerous species (for reviews and classifications including European species see Magnus 1963, Dennis & Shreeve 1988, Rutowski 1991, Ebert & Rennwald 1991a, 1991b, Bink 1992, Shreeve 1992b, Wickman 1992b). In order to make systematic studies of receptive female behaviour, females need to be reared, released and followed in the field. This is rarely done, with the result that adaptive explanations for male-searching behaviour in relation to receptive female behaviour have been tested in only a few species. There is no reason to assume that receptive females behave in the same way as ovipositing females (Rutowski 1991, Wickman & Rutowski 1999), and it is not sufficient to examine female dispersion in the field and compare it with the dispersion of males, because ovipositing females will avoid males as opposed to receptive females. In many species receptive females have evolved specialized behaviours that help them to actively search out males (Rutowski 1980a, Rutowski *et al.* 1981, Wiklund 1982a, Stallings *et al.* 1985, Wickman 1986, 1988, 1992a). Although seemingly self-evident, the frequent omission of female mate-locating behaviour when studying mate-location suggests that it needs to be emphasized that male and female mate-locating behaviour have evolved through reciprocal coevolution (Wickman & Rutowski 1999, Ide & Kondoh 2000).

There is no evidence that long-range pheromones are used in mate attraction in butterflies. No chemical attractants are known that function over larger distances than a few centimetres (Rutowski 1991, 2003). Instead indirect cues like microclimate, landmarks or host plants are used in locating mates. Besides, scent is rarely used in close-range detection (but see Dennis & Williams 1987 for possible exceptions), and most butterflies use visual cues and are attracted to a wide range of objects with an appearance similar to their mates (see references below).

Male mate-location behaviour

Scott (1974) basically recognized two types of male mate-location behaviour, viz. perching and patrolling behaviour. Perching behaviour is characterized by males perching most of their time in a restricted area, and taking off to inspect passing butterflies. Extended aerial encounters between males involving circling, spiralling or chasing are common (Baker 1972, Bitzer & Shaw 1979, 1983, Thornhill & Alcock 1983, Wickman & Wiklund 1983, Cordero & Soberón 1990). Although the idea is opposed by Scott (1974), interactions are usually thought to be disputes over territory ownership (Baker 1972). Males showing patrolling behaviour fly widely in the habitat, and encounters between males are not as spectacular.

Conceptually patrolling and perching encompass at least three different variables, viz. flight activity, home range and territoriality (Rutowski 1991), which often covary in a particular male. For instance in *Coenonympha pamphilus* there are both patrolling and perching males. When perching, males spend about 10% of their time in flight and stay close to higher vegetation in restricted areas measuring a few square metres. When patrolling, males fly most of their time in the open field and may cover home ranges more than a hundred times as large as those of perching males. Individual males may switch between perching and patrolling. Aerial encounters involving a perching male last on average 12 s (maximum several minutes) whereas those between two patrolling males only last 3 s (maximum 11 s) (Wickman 1985a).

In many other perching species, patrolling also occurs, e.g. *Pararge aegeria* (Davies 1978), *Lasiommata megera* (Dennis 1982a), *Tarucus theophrastus* (Courtney & Parker 1985), *Ochlodes venata* (Dennis & Williams 1987) and *Kanetisa circe* (García-Barros 2000a). There are also species where only perching has been described, e.g. *Inachis io*, *Aglais urticae* (Baker 1972), *Lycaena phlaeas* (Suzuki 1976) and *Euphydryas maturna* (Wahlberg 1998). However, in these species also intruders are seen to come to and leave territories, which suggests that some males patrol between territories (Dennis 1982a). But this is not necessarily so. In *Oenis chryxus*, removal experiments suggest that there is no such floating population, but that perching males within a restricted area are competing for the most preferred spots (Knapton 1985). On the other hand, there are several species where only patrolling occurs, e.g. *Leptidea sinapis* (Wiklund 1977c), *Anthocharis cardamines* (Wiklund & Åhrberg 1978, Dennis 1982b), *Coenonympha tullia* (Wickman 1992a), *Thymelicus lineola* (Pivnick & McNeil 1987a), *Melitaea diamina* (Wahlberg 1997) and *Glaucopsyche nausithous* (Pfeifer *et al.* 2000).

It has been suggested that the different strategies used by males in perching and patrolling species would affect their morphological design for flight (Scott 1983). Wickman (1992b) predicted that in perching species sexual selection would favour traits associated with high acceleration and speed, while in patrolling species traits associated with flight endurance would be favoured. This was examined in a comparative study of 44 species. In agreement with this prediction it was found that males of perching species have larger thorax/body mass ratios, higher wing loadings and higher aspect ratios (more elongated and narrow wings) than males of patrolling species. This association still remained even when effects of common ancestry and covariation between the sexes were controlled for (Wickman 1992b).

Related results have also been found in a comparative study on riodinids in Ecuador (Hall & Willmott 2000) and in a study on *P. aegeria*, where perching males had relatively heavy thoraxes compared to patrollers (Van Dyck *et al.* 1997a, Van Dyck 2003). In experiments with *P. aegeria* it has been confirmed that males with relatively larger thoraxes and higher aspect ratios have enhanced acceleration capacity (Berwaerts *et al.* 2002). Such a capacity would seem crucial in competition between perching males to intercept females that enter territories.

Although flight activity, home range and territoriality often covary in a typical way, there are species that are difficult to categorize as either perchers or patrollers, because the three variables are associated in atypical ways (Rutowski 1991). For instance there are territorial species in which males fly most of their time (Rutowski *et al.* 1989, Orr 1999). In *Lycaena hippothoe* males perch in their territories at low temperatures, while inspecting them mainly in flight at high temperatures (Fischer & Fiedler 2001c). There are also species like *Euchloe simplonia* which incessantly fly in restricted areas on hilltops without any indication of territoriality (personal observation).

Fighting for territories

In aerial male–male encounters involving perching males, the outcome is usually that one male is expelled, suggesting that perching males defend territories. This is also strongly supported by the fact that aerial encounters involving perching males are longer than those involving only patrolling males (Wickman 1985a, 1992a, Dennis & Williams 1987, Alcock 1994). However, there has been little agreement between data coming from different species concerning how territorial disputes are settled (Hardy 1998, Field & Hardy 2000, Kemp 2000a), because butterfly fights do not involve weaponry (but see DeVries (1987: 71) on the Costa Rican canopy-inhabiting *Papilio* species that have serrations on the costal margin of the wing, which he suggested play a role in male–male interactions).

In many butterflies the resident often wins, but not always (Davies 1978, Wickman & Wiklund 1983, Kemp 2000b, Fischer & Fiedler 2001c), suggesting that butterfly contests often are determined conventionally by ownership. But winning and territory ownership are correlated with a number of other traits, even though the traits vary widely across species, e.g. large body size (Wickman 1985c, Rosenberg & Enquist 1991), small body size

(Hernandez & Benson 1998), old age (Kemp 2000b, 2002a, 2002b, Kemp *et al.* 2006a), young age (Fischer & Fiedler 2001c, Kemp 2003) and body temperature (Stutt & Willmer 1998). In an analysis and review of existing data on butterflies generally, Kemp & Wiklund (2001) concluded that the primary reason for males winning fights is ultimately because of fighting ability, although residency may also play a role.

The most cited study on ownership conventions is Davies (1978), who found that in *P. aegeria* residents always won fights over sunspots. Wickman & Wiklund (1983) & Shreeve (1984) in other field studies failed to repeat Davies's result, and found that males in territories were occasionally overthrown by intruders. In a suite of beautiful experiments with *P. aegeria*, Kemp & Wiklund (2004) recently demonstrated that fights over territories in this species are not settled by ownership but by fighting ability. The impression that ownership is used as a cue could be due to the accumulation of aggressive males on territories. The results of Kemp & Wiklund (2004) also refuted the hypotheses of Stutt & Willmer (1998) that the body temperature decides the outcome of disputes in *P. aegeria*. In another series of experiments it was established that age influences the outcome of contests in this species (Kemp *et al.* 2006a).

To explain the fact that traits associated with fighting ability vary to such a large extent across species, despite the fact that fights look so similar in different species, Kemp & Wiklund (2001) suggested that the cost of fights is ultimately measured in loss of residual reproductive value (Williams 1966, Parker 1974), which may be related to factors such as age, attractiveness and other options available and thereby influencing who wins (see also Kemp & Alcock 2003). Behind all the apparent variation in butterfly fighting behaviour there thus seems to be a common denominator explaining how fights are settled.

Mate-location systems in butterflies

Male mate-location behaviour is a convenient way of categorizing butterfly mate-location systems since receptive female mate-location behaviour has been described only for a few species. However, a more complete account of mate-location systems must encompass the behaviour of both sexes. One way to do this is to describe the distribution of matings with regard to other resources that influence butterfly dispersion (Wickman & Rutowski 1999). According to such a scheme mate-location systems can be categorized into aggregated and dispersed systems. Aggregated systems can in turn be either resource-based or non-resource-based. Non-resource-based mate-location systems are called 'lekking'. Aggregated systems usually involve territoriality as opposed to dispersed systems, although fighting for single females may occur also in the latter system. Dispersed systems are based either on female pupation sites or on encounters involving adult, receptive females.

RESOURCE-BASED SYSTEMS

In some species males appear to defend resources needed by females, or areas where females are likely to eclose, often involving host plants or nectar plants. One of the few examples of male defence of nectar plants is found in *Lycaena hippothoe*. In examining how a number of variables correlated with the location of territories, Fischer & Fiedler (2001c) discovered that male territories on average have twice as many nectar plants as randomly chosen patches, suggesting that nectar plants play a role in choice of territories. They also demonstrated that males on territories receive almost all matings.

In some nymphalids, e.g. *Asterocampa leilia* (Rutowski & Gilchrist 1988), *Anartia jatrophae* (Lederhouse *et al.* 1992) and *Hypolimnas bolina* (Kemp 2001), males occur near host plants. Nonetheless, there is no conclusive evidence that these instances constitute defence of oviposition or eclosion sites. Territories could just as well be conventional leks, as the distribution neither of matings nor of eggs or pupae were examined. In *Aglais urticae*, territories are often associated with host plants (Baker 1972), but how this affects mating success has not been examined. An obvious example of host plant defence occurs in the lycaenid *Tarucus theophrastus*, simply because males and females inhabit monocultures of bushes of the host plant. Although males seemed to distribute themselves in a way expected from game theory to maximize matings, this distribution was not checked against actual distributions of matings, as only six matings were observed (Courtney & Parker 1985). Another case of host-plant-based territoriality on sites where females emerge occurs in *Cressida cressida* (Orr 1999).

The studies show that more work on receptive female behaviour in relation to nectar resources and host plants is needed to better understand resource-based defence. Correlational evidence based on dispersions needs to be matched by behavioural data on preferences of males and females with experimentally offered different options to choose from (Wickman 1988, Singer & Thomas 1992, Wickman *et al.* 1995).

LEKKING

Even though clear examples of defence of nectar plants or host plants are rare in butterflies, territorial defence is not rare. However, in most species, male territories instead seem to be associated with visual features of the landscape, although access to nectar sources may sometimes play a role (Dennis & Williams 1987). In some species, however, feeding areas are completely decoupled from places were males defend territories (Rosenberg 1989). The mating system of many of these species can best be described as lekking, because territories contain no limiting resource to females (Lederhouse 1982).

A lekking species that has been studied extensively is *Coenonympha pamphilus* (Wickman 1985a, 1985c, 1986, 1992a, Wickman *et al.* 1995, Wickman & Jansson 1997). Censuses revealed that males of this species use trees and bushes as landmarks that determine where they will set up territories. The larger the landmark, the more males assemble around it. This was confirmed in experiments with males choosing between artificial trees of different size. Females disperse their eggs randomly in the habitat, and there is no association of female oviposition preferences, eclosion sites or nectar plants with the location of male leks (Fig. 3.2). Releases of reared virgin females showed that they fly toward landmarks (Fig. 3.3), and when they arrive they use a circling flight well above the field layer that attracts the males perching there (Wickman *et al.* 1995). In this way males at leks receive proportionally more matings than males patrolling in other parts of the habitat (Wickman 1985c). Hence, females of *C. pamphilus* actively search out males by using landmarks. On their way to territories they may be passed by several males, but they do not show the courtship soliciting behaviour until reaching a landmark. By not soliciting courtship from patrolling males in the open field (as occurs in the patrolling, non-lekking congener, *C. tullia*: Wickman 1992a), they incur a significant time cost in reduced fecundity of about 2.8% (Wickman & Jansson 1997). The benefits females receive to compensate these costs have not been examined. However, since this species is essentially monandrous and males have small spermatophores, female choice of males on leks by this species more probably is compensated by indirect genetical benefits rather than by nutritional benefits (Wickman & Jansson 1997).

In hilltopping species males congregate on the summits of high topographic landmarks (Plate 2). This involves

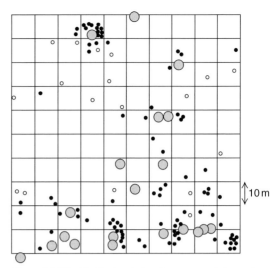

Figure 3.2. The distribution during six censuses of observations of males (solid circles) and females (open) of *Coenonympha pamphilus* on 1 hectare of grassland with trees and bushes (in grey). The area is surrounded by open grassland which explains the relative scarcity of females. Male distribution is aggregated as opposed to the random distribution of females. (From Wickman *et al.* 1995, courtesy Oxford University Press.)

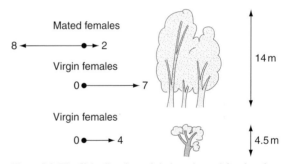

Figure 3.3. The flight directions of virgin and mated females of *Coenonympha pamphilus* released 20 m from two trees of different heights. All virgin females moved toward the trees, a preference which was significant. Mated females, on the other hand, seemed indifferent (Wickman *et al.* 1995). (From Wickman 1997, courtesy *Entomologisk Tidskrift*.)

receptive females and males actively searching out summits to mate (see discussion in Ehrlich & Wheye 1986, 1988, Baughman *et al.* 1990, Singer & Thomas 1992). Hilltops are widely used as landmarks not only in butterflies but in a wide range of insects (Shields 1967, Thornhill & Alcock 1983, Alcock 1984). In many hilltopping species there are no limited resources to be found on the summits used, and males compete for territories, which fits the general definition of leks (Alcock 1987). In *Papilio polyxenes* there is evidence that males on hilltops have a mating advantage (Lederhouse 1982). Receptive females of this species show a circling behaviour on hilltops very similar to that described for *C. pamphilus* virgin females at leks (Wickman 1986).

FINDING PUPAE

In for instance *Jalmenus evagoras* and *Heliconius* spp. males search for pupae and wait for females to eclose. These species are specialized host plant users and pupate on their host plants, making female eclosion sites predictable. In *Heliconius* spp. males are also aided by pheromones from the female pupa (Gilbert 1975, Brown 1981, Deinert *et al.* 1994). In *Heliconius hewitsoni* pupae are aggregated and males first compete for access to female pupae when they approach hatching (Deinert 2003). In the ant-tended *Jalmenus evagoras* the presence of ants aids males in finding females (Elgar & Pierce 1988). There are few species in which males search for pupae, as pupae with few exceptions are highly cryptic and widely dispersed. None is known from Europe.

DISPERSED ADULT MATING-SYSTEMS

In most patrolling species males search for already eclosed females and, as in lekking species, eclosion sites are widely dispersed (Fig. 3.4). In many patrolling species receptive females have evolved specific solicitation behaviours, which they use to enhance detection by males. In *C. tullia* newly eclosed females take off and approach males flying by. When the male discovers the female, he follows and courts her in a way typical for most butterflies (see below). Hence, virgin *C. tullia* females do not travel as far and mate closer to the eclosion site compared to the lekking *C. pamphilus* (Wickman 1992a). Similar approach behaviours of receptive females have been described for a number of patrolling species (Rutowski 1980a, Rutowski *et al.* 1981, Wiklund 1982a), and also for the territorial *Papilio glaucus*, where males patrol their territories (Krebs & West 1988, Rutowski *et al.* 1989). In most perching, territorial species, females instead solicit courtship by circling slowly above male territories (see above). However, there are cases where male and female mate-locating behaviours are associated in the reverse way, i.e. receptive females take off and approach passing males in territories (Stride 1958).

Even in strictly patrolling species, males do not use the habitat without discrimination. Male searching patterns are influenced by the distribution of flowering plants (Brakefield

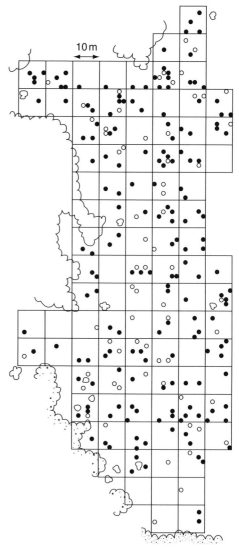

Figure 3.4. The distribution during four censuses of observations of males (solid circles) and females (open) of *Coenonympha tullia* on 1 hectare of wet grassland surrounded by trees and bushes (in grey). The distribution of both males and females is random (Wickman, 1992a). (From Wickman 1997, courtesy *Entomologisk Tidskrift*.)

1982) and host plants (Rutowski *et al.* 1997), microclimate/ edges/sunspots (Wiklund & Åhrberg 1978, Dennis 1982b) and eclosing females (Brakefield 1982, Rutowski *et al.* 1988). Temperature also influences male flight activity, and at cool temperatures males perch more often and their mobility decreases, although territorial defence is not seen (Sonntag 1981, Pivnick & McNeil 1985a, Hirota & Obara 2000).

Patterns of variation in mate-location behaviour

In some species mate-location occurs only during a certain part of the day, probably because females are only receptive during that time. Many Nymphalini species defend territories in the afternoon (Shields 1967, Baker 1972, Bitzer & Shaw 1979, 1983, 1995, Alcock & Gwynne 1988, Brown & Alcock 1991; see also Fric & Konvička 2000). These species typically wait to mate until they come to roost (Baker 1972, Mikkola 1976, personal observation). However, in many other species, males engage in mate-location during most of the day (e.g. Douwes 1975, Lederhouse 1982, Wickman 1985a, Ravenscroft 1994a), although they may change their mate-locating behaviour depending on a number of other variables (Kemp et al. 2006b, Takeuchi 2006).

Individual male mate-location behaviour may change depending on population density and weather. As population density increases, males shift from territory defence to patrolling (Alcock & O'Neill 1986). The reason may be that territorial defence is more costly at higher densities and that there is a higher probability that females are found and mated by patrolling males before they reach territories. The reason why weather influences male behaviour is that a higher thoracic temperature results in higher flight activity and permits males to search more efficiently for females and also to use parts of the habitat where microclimate or light levels otherwise would be suboptimal (Dennis 1982a, Wickman 1985a, McDonald & Nijhout 2000). For example, at high temperatures, P. aegeria males perching in sunspots in the wood may leave them to search also in shady locations (Davies 1978, Wickman & Wiklund 1983, Shreeve 1984). In P. aegeria territorial perching is more common in shady woodlands than in open landscapes (Merckx & Van Dyck 2005). However, relationships between male mate-location behaviour and external conditions are not often clear-cut (Dennis & Williams 1987).

The propensity of individuals to adopt a certain type of mate-locating behaviour may also be influenced by their phenotype. The already mentioned relationship in P. aegeria between morphology and thermoregulatory properties is also correlated with a greater tendency for paler and smaller males to perch in sunspots (Van Dyck et al. 1997a, 1997b, Van Dyck & Matthysen 1998). In a laboratory study it was found that paler males also were less asymmetrical, and it was suggested that success in territorial defence might be correlated with degree of wing symmetry (Windig & Nylin 1999).

In hilltopping species the microclimate of the hill complicates patterns. In the hilltopping Lasiommata megera (Wickman 1988) males perched at sheltered places below hills at low temperatures. As temperatures rose, males used increasingly higher elevations as perches. Concomitantly a higher proportion of males changed to patrolling. There was no effect of male density. At the highest temperatures males appeared to be dependent on convective cooling by the wind, with the top of the hill perhaps representing the coolest location due to highest wind speeds. When perching on bare rock, which is the substrate used by this species, males risk overheating in places sheltered from the wind. In this way weather appears to force males to adopt different mate-locating behaviours. A remarkable parallel exists in males of Lasiommata maera, which defend the shade of bushes on hilltops at hot weather (Wickman, unpublished data).

In Chlosyne californica, Alcock (1994) found that higher ambient temperatures resulted in more patrolling males, but population density also seemed to have an effect as patrollers were seen only at high densities. As opposed to L. megera, perching in C. californica is observed in males on summits only, while males downslope patrol. But just as in L. megera, matings may be seen at all levels, suggesting that patrolling is not necessarily the worse option in all conditions experienced by this species. Alcock found males that freely shifted behaviour and there were no size differences between males adopting the two types of behaviour. However, in L. megera it was shown that virgin females prefer to fly uphill (as opposed to mated females that are indifferent to slope), which should select against males searching for females at the lowest elevation to intercept ascending receptive females. The result may be a contest for an optimal position on the hill depending on receptive female behaviour (Courtney & Anderson 1986). The dispersion of males will be highly unstable though, because males can only have limited information about what the other males are doing. Also Pe'er et al. (2004) found a tendency for virgin females of the hilltopping species Melitaea trivia to fly uphill, although this occurred only in the absence of encounters with other individuals.

Explaining mate-location systems

It is obvious that both male and female butterflies search for mates. Thus, where the two sexes encounter each other will be a result of both female and male efficiency in mate-finding. Generally, males will be selected to maximize the number of matings, while females will be selected to find an acceptable male as soon as possible in order not to delay matings and lose time that could be spent laying eggs (Wickman 1992a, Wickman & Jansson 1997, Gotthard et al. 1999a).

Wickman & Rutowski (1999) used these ideas in a model to explain the dispersion of matings in species where eclosing females are widely dispersed, i.e. why aggregated mating systems like leks evolve in some species, while in others, matings are dispersed, as in many strictly patrolling species. By using the parameter 'receptive female home range' it could be demonstrated that the more efficiently the two sexes find each other, the closer to eclosion sites matings will occur, and the more dispersed matings will be. In species where mate-location is inefficient, females will travel far before finding a male. Females will be selected to search for males at locations where males are more likely to be encountered, for example due to microclimate, and a runaway process may be initiated that promotes aggregated matings in species with large receptive female home ranges.

Wickman & Rutowski (1999) showed that leks were increasingly favoured in a species when receptive female home range increases because (1) the mobility of receptive females as compared to males is relatively high (influenced by e.g. sexual dimorphism), (2) there is a greater unanimity in female flight directions and male habitat use (due to e.g. microclimate or thermoregulatory traits), (3) the shorter the distance of detection becomes (e.g. because of higher apparency: see Rutowski 2003) or (4) females are more choosy (because e.g. of a long life expectancy). The influence of population density and mating frequency will be slight, except under exceptionally high population densities and mating frequencies, which will select for dispersed matings (Scott 1970, Ide & Kondoh 2000).

One example of how receptive female home range may be related to the dispersion of matings is the genus *Coenonympha*. Eclosion sites of both *C. tullia* and *C. pamphilus* are highly dispersed. In the shorter-lived *C. tullia*, matings are dispersed because receptive females approach passing males and are quickly mated (often with the first male that passes) close to eclosion sites. In the lekking *C. pamphilus*, females are passed by numerous males before mating and females travel longer distances (Wickman 1992a, Wickman & Jansson 1997).

COURTSHIP AND MATE CHOICE

Before mating, butterflies display more or less elaborate courtship behaviour. However, before courtship can take place, the female and the male have to detect one another.

Table 3.1 *The number of times butterflies and other insect taxa (about the size of* Coenonympha tullia *or larger) passing within 1 m did and did not elicit approach by virgin females of* C. tullia *released in the field*

Taxon	Took off	Stayed
Argynnini	7	4
Pierinae	6	6
Polyommatini	0	1
Anisoptera	9	9

Source: Wickman unpublished data; see Wickman (1992a) for methods.

place, the female and the male have to detect one another. The first approach is usually the result of visual stimuli. Males and sometimes also females can be seen to approach objects resembling mates with little discrimination. Perching males approach tossed stones, flying leaves and so forth, although a general resemblance to females increases the likelihood of approach, and wing colour of females appears to play a role in identification (Tinbergen et al. 1942, Magnus 1958, Douwes 1975, Silberglied 1984). Also distance and perch location influence probability of detection (Rutowski et al. 2001). Perching virgin females of *C. tullia* also approach a wide range of stimuli, including Odonata, which may increase the risk of predation (Table 3.1).

When the two sexes meet, courtship ensues which involves visual, tactile and olfactory cues (Brower et al. 1965, Pliske 1975, Rutowski 1978a, 1983, 1984, Rutowski & Gilchrist 1987, Boppré & Vane-Wright 1989). Courtship has been described in detail for several European species (Poulton 1904a, Lederer 1939, 1960, Tinbergen et al. 1942, Magnus 1950, Lundgren & Bergström 1975, Wiklund 1977c, Pellmyr 1983, Pivnick & McNeil 1985a, Wiklund & Forsberg 1985, Forsberg & Wiklund 1989, Pivnick et al. 1992). Typically the male is the most active sex during courtship (Plates 3 and 4).

If the female is not receptive, she avoids mating in different ways. This may involve mate-refusal postures such as wing fluttering or elevating the tip of the abdomen (Tinbergen et al. 1942, Obara 1964, Rutowski 1978a, Wiklund & Forsberg 1985, Rutowski & Gilchrist 1987, Forsberg & Wiklund 1989). Females may also try to avoid the male by creeping away, by out-flying him or by ascending high in the air (Rutowski 1978b, Wickman 1986, 1992a). In some species olfactory cues are involved, which females disseminate by adopting specialized postures (Gilbert 1976, Wiklund & Forsberg 1985). In *Pieris napi* a volatile anti-aphrodisiac is transferred by males to the females during mating. The volatile substance is used subsequently by females to reduce male harassment during courtship (Plate 5a). The substance is so strong that most males refrain from mating with virgin females on which it has been applied experimentally (Andersson *et al.* 2000). Similar systems also occur in *P. brassicae* and *P. rapae* (Andersson *et al.* 2003). The effect diminishes with time from mating (Wiklund & Forsberg 1985, Forsberg & Wiklund 1989, Andersson *et al.* 2004). *Kanetisa circe* females are known to stridulate, which probably is part of their behaviour to reject males (García-Barros 1986). Female avoidance of males to reduce time spent in courtship or mating may also result in different habitat use of the two sexes (Wiklund & Åhrberg 1978).

Pre-copulatory mate choice

One important function of courtship is mate choice, making it possible for males and females to correctly identify the sex and the species, but also to assess the quality of the potential partner before copulation. However, there are few studies of how the details of courtship behaviour actually influence mate choice. Most studies instead have focused on the outcomes of courtship, that is, on the traits that distinguish preferred from non-preferred individuals by various standards like number of matings or courtship persistence. Such pre-copulatory mate choice has been demonstrated in both females and males.

In most species, females should benefit by choosing fresh virgin males over recently mated males (Torres-Vila & Jennions 2005). Rutowski (1979) demonstrated that virgin *Pieris protodice* males court females more persistently than recently mated males, and hence males act as 'honest salesmen'. In a study on *Colias eurytheme*, females preferentially mated with males of average size and with fresh males (less worn wings reflecting more ultraviolet light). However, male courtship persistence only paid off up to a point (Rutowski 1985), as prolonged courtships did not result in more matings. In *Pieris napi*, females did not discriminate between virgin and recently mated males (Kaitala & Wiklund 1995), while in *Dryas julia* old females seem to have this ability (Boggs 1995). In *Colias* spp. old females prefer to mate with PGI genotypes that are associated with advantages in flight capacity and survivorship (Watt *et al.* 1986; see also Watt *et al.* 1996). *Bicyclus anynana* females have been shown to prefer males either with larger or brighter eyespots on their

wings (Breuker & Brakefield 2002, Robertson & Monteiro 2005).

In some species where males incur a substantial cost at mating, males have been observed to discriminate between females. *Pieris protodice* males court young and large females for longer times than older and smaller females (Rutowski 1982). Although one possible reason may be that males prefer young and large females because they represent a larger return in terms of eggs, this explanation has not been examined further. Nevertheless, a preference for large females is not found in all species with large spermatophores; in *Colias eurytheme*, males preferred females of average size (Rutowski 1985). Males of *C. philodice eriphyle* prefer to mate with less melanized females, but this is probably merely a result of their higher detectability (Ellers & Boggs 2003).

The role of butterfly colours for mate choice and identification of sex and species have been discussed since Darwin (1871), who suggested that the often brilliant colours of males had evolved as a result of sexual selection through female choice. A support for this is found in species with sex-limited mimicry, which with few exceptions is limited to the female (Brower 1963, Turner 1978). This means that females regularly have the coloration of another, unpalatable species, whereas males retain the species specific wing pattern, allegedly preferred by the females.

Experimental removal of ultraviolet-reflective pigment from the wings of males has shown that these pigments are used by *C. eurytheme* and *P. protodice* females to discriminate males from females (Silberglied & Taylor 1978, Rutowski *et al.* 1981). Ultraviolet wing patterns of females derived from host plant substances are used by male *Polyommatus icarus* in mate selection (Knuttel & Fiedler 2001). Wiernasz (1989) manipulated the extent of melanization of *Pieris occidentalis* males, and showed that this influenced female choice. She also found a consistent female choice of certain wing patterns in the field across years, and suggested that this mate discrimination may function in species recognition and involves choosing co-specific males that look least like the sympatric *P. protodice* males. *Heliconius cydno* males use the polarized light reflected by the iridescent wings of females in mate recognition (Sweeney *et al.* 2003) and specific wing patterns have been shown to be used in species recognition also in two recently diverged lycaenid species (Fordyce *et al.* 2002). However, in species recognition, *Colias* butterflies also seem to use olfactory cues (Grula *et al.* 1980, Rutowski 1980b) as do *Hipparchia semele* (Tinbergen *et al.* 1942) and *Eurema hecabe* (Takanashi *et al.* 2001).

However, simply because mate choice can be demonstrated to be influenced by male wing coloration in some cases does not mean that it can be used to explain sexual dimorphism in wing coloration across all species. Silberglied (1984) suggested that intrasexual selection rather than female mate choice is the major cause. This hypothesis was tested on *Papilio polyxenes*, where males experimentally altered to resemble mimetic females were less successful in intrasexual competition for mating territories than unaltered males (Lederhouse & Scriber 1996). These altered males were involved in longer male–male encounters, and were less likely to take over territories from residents. On the other hand, released virgin females were equally likely to mate with altered and unaltered males. Rutowski (1992), however, found no effect in colour-altered males of the sexually dimorphic *Hypolimnas bolina* on either capacity to defend territories or on durations of fights. He argued by comparison with results on the significance of wing colour in *H. misippus* (Stride 1956, 1957, 1958) that wing colour in males is the result of intersexual selection. Similarly, colour alterations of males of the mimetic *Papilio glaucus* suggested that males must have a certain pattern of colour to attract and mate with females (Krebs & West 1988). Hence, male colour patterns appear to have different functions in different species (Ohsaki 2005).

Cryptic female choice

Female choice may also occur after copulation has begun, and females may use physiological or morphological means to favour one male's semen over another for paternity. Eberhard has termed this 'cryptic female choice', and has argued that it is widespread in insects (Eberhard & Cordero 1995, Eberhard 1996, 1997). Mechanisms might involve early remating or impeding plugging of the mating tract. Both are known to occur in butterflies, although their adaptive values have not been examined.

Mating plugs

After insemination is completed, the males of many butterfly species produce a so-called 'mating plug' that seals the copulatory opening (Ehrlich & Ehrlich 1978, Drummond 1984). Typically the plug is small and not easily observed. Mating plugs have been assumed to hinder additional matings by females, although many species that form mating plugs mate several times (Ehrlich & Ehrlich 1978). In *Euphydryas chalcedona* the mating plug impedes copulation, but reduces neither the females' attractiveness to males nor male courtship persistence (Dickinson & Rutowski 1989). However, in some butterfly taxa (Parnassiinae, Acreinae and some Satyrinae) the mating plug may form a large, visible structure called 'sphragis'. A sphragis also occurs in the Australian papilionid species *Cressida cressida*. Males encountering females with an intact sphragis follow them only briefly and rarely make contact. When the sphragis is removed, on the other hand, aerial courtship attempts become longer and more physical, suggesting that the sphragis acts as a visual signal (Orr 1999, 2002).

CONCLUSIONS

Systematic research during the past decades has produced detailed information on the costs and benefits of mating frequency, mating rate and ejaculate size in butterflies. Although females with few exceptions are a limiting resource for male reproductive success, there is also unequivocal evidence for Boggs & Gilbert's (1979) finding that males of many species contribute nutrients during copulation, which females need for realizing full reproductive output. At the same time, further research is needed on the ecological factors responsible for the variation among species in the extent to which males contribute nutritious copulatory gifts.

The range of male mate-location behaviours is well described and recent research has been able to answer the elusive question about how territorial disputes in males are settled. However, interspecific variation in female mate-location behaviour is little known though such information is required before the evolution of non-resource-based mate-location systems can be fully understood. Female mate-location behaviour is an important component of the reproductive cycle, and detailed knowledge about its requirements should help our understanding of population dynamics and aid species conservation.

Although numerous traits have been demonstrated to be involved in mate choice and mate refusal, critical questions remain. The relatively novel area of cryptic mate choice is largely an unexplored field. Also, the diversity and function of often elaborate courtship displays are still little examined, despite this being one of the oldest areas of study in butterfly mating behaviour. Obviously, butterfly mating behaviour will continue to be a productive field for many years to come.

4 • Butterfly oviposition: sites, behaviour and modes

ENRIQUE GARCÍA-BARROS AND THOMAS FARTMANN

SUMMARY

Where a female deposits her eggs is critical for the success of her offspring. Selecting the most appropriate plants in locations where not only survival, but offspring characteristics, are optimised is a complex process involving behavioural cues and assessment of habitat, patch and plant quality and context. Some butterflies are extremely precise in selecting where to lay eggs, and others are more generalist. There may be variation of strategies within populations and also between populations. Egg-laying females may not be able to perform optimally because of environmental conditions, interacting organisms, or shortage of time. Thus egg-laying represents the end product of a series of trade-offs between conflicting demands on females, operating in the context of the habitat(s) they find themselves in. Conflicts also exist between what is optimal for egg and larval development and what is optimal for female oviposition site selection. Many studies of oviposition have taken a reductionist view, breaking down oviposition site selection into a series of single-step processes. Whilst optimal oviposition and foraging theories have contributed much to understanding oviposition, a more integrative approach is necessary as optimal foraging by adults, optimal realised fecundity and optimal larval performance are not independent of each other. Egg production patterns (capital versus investment breeding) also have profound effects on realised fecundity and affect the intensity with which female non-oviposition behaviours influence their egg-laying patterns. Differences emerge between species that lay eggs singly and those that lay eggs in batches, which can be related to risk spreading and the predictability and stability of the environments in which oviposition resources occur. Surprisingly, there are gaps in our knowledge of larval host plants and egg-laying substrates, yet more comprehensive knowledge, alongside information about oviposition cues and behaviours, could contribute greatly to refining our understanding of habitat quality, and ultimately the conservation of species.

INTRODUCTION

In most phytophagous insects, dispersal by the pre-adult stages is negligible in comparison to that of the adults. The location of an egg will condition the survival and performance of the egg itself, and that of the larvae. Thus, oviposition behaviour and egg site selection by phytophagous insects have attracted considerable amounts of research during the last decades, for two main reasons. One is its applied nature: as far as an important number of insect species (including Lepidoptera) are of economic importance, understanding the cues needed to identify and accept potential larval hosts is valuable for designing strategies for agricultural pest control. This explains much of the work done on the chemical and visual aspects of host plant recognition by adult females, and on their acceptance by the immatures of Lepidoptera and other phytophagous insects (Prokopy & Owens 1983, Renwick 1989, Jones 1991, Bernays & Chapman 1994, Renwick & Chew 1994, Hilker & Meiners 2002). The second reason is related to the fact that, from ecological and evolutionary perspectives, the act of oviposition (egg-laying, or egg deposition) represents the spatial interface between three successive life stages (adult, egg and larvae) and the environment. An oviposition event implies the simultaneous expression of the complex features leading to egg site selection and density (hence, dispersal) in space and time. Perhaps not surprisingly, oviposition has been qualified as an outstanding milestone in the evolution of the Lepidoptera (Renwick & Chew 1994). Furthermore, the attributes of the habitat related to optimal oviposition sites (or at least those that are selected by a species or a demographic unit of it) may be crucial from a conservationist perspective. Thus as indicated by Dennis et al. (2003, 2006), in managed areas such as the strongly anthropogenic European landscapes, knowledge of the oviposition preferences by target species may be crucial to design adequate management practices (see also Thomas 1991b, Dennis & Eales 1997, Thomas et al. 2001b, Fleishman et al. 2002, Anthes et al. 2003b, Fred & Brommer 2003, WallisDeVries 2004).

Ecology of Butterflies in Europe, eds. J. Settele, T. Shreeve, M. Konvička and H. Van Dyck. Published by Cambridge University Press. © Cambridge University Press 2009, pp. 29–42.

This chapter focuses on the broad aspects and patterns of substrate selection for the oviposition of butterflies, with special reference to European species. Because of the complexity of the subject (which is intimately linked to the evolution of host plant selection preferences), we adopt a descriptive approach to summarise what is known on the main patterns and mechanisms. The way in which facts fit into current theory and the main theories themselves are addressed afterwards. For a broader perspective, former reviews on the subject within a similar or a wider geographic scope have been provided by Chew & Robbins (1984), Thompson & Pellmyr (1991) and Porter (1992). Preston-Mafham & Preston-Mafham (1993) is a good reference for a wide audience, and recent contributions include those of Singer (2003, 2004), Ehrlich & Hanski (2004) and Fartmann & Hermann (2006).

Ideally, oviposition sites should provide adequate host plant species in sufficient quantity and favourable quality, as well as adequate local microhabitat conditions (microclimate, low exposure to predators and parasitoids) (Fartmann & Hermann 2006); thus the conditions in which the insects have to thrive for most of their lives are primarily determined by the choice of the adult females. Evolutionarily, oviposition site selection should be linked to individual success through its relations with the fitness of three life-cycle stages (the 'adult–ovum–larva trade-off': Reavey & Lawton 1991). Ideally, females should benefit from investing in a careful choice that would ensure optimal properties of the host selected for larval development. The search for candidate oviposition substrates may impose costs (increased mortality due to adverse weather, exposure to predators, and the time needed for feeding) leading to a compromise between an optimal selection for the offspring and other factors.

AVAILABLE DATA

A substantial amount of information on the life histories of the European species (including oviposition sites) has been collected during decades although, with few exceptions (such as Bink 1992) the elementary aspects of butterfly life histories have seldom been tabulated explicitly. Thus, much detailed information is scattered across a high number of periodicals, and cannot be quoted here in detail (general references not quoted in other parts of the text include: SBN 1987, 1997, Thomas & Lewington 1991, Weidemann 1995, Tuzov et al. 1997, Settele et al. 2000, Lafranchis 2000, Fartmann 2004 and Eliasson et al. 2005). Larval food plants have been recorded, in broad terms, for a vast majority of the

European species. The same is not equally true, however, for all the European regions, and the data from South and East Europe probably are of comparatively lower quality at the host plant species level (while they may be satisfactory at the plant genus level). Even so, data quality is not yet fully satisfactory for Central Europe (Fartmann & Hermann 2006). This is of relevance for the study of geographical variation in host plant use by polyphagous and oligophagous species, and there are plenty of possibilities to increase the regional lists of food plants in relation to those normally quoted in the most popular field guides (such as Tolman & Lewington 1997). The quality of the data record is broadly similar with respect to the plant parts used for oviposition. A remarkable exception is that of the butterfly species with grass-feeding larvae, from both points of view (the plants used, and the oviposition sites selected). This may be a consequence of the difficulties in searching systematically for eggs and larvae of grass-feeders in contrast to other butterfly groups (Fartmann & Hermann 2006). Finally, however, the fine-grain level knowledge of the properties and conditions that drive oviposition site preferences beyond the taxonomic identity of the host are widely ignored except for a few species which have been intensively studied. These confirm the body of evidence on which much of the remaining text of this chapter relies.

BEHAVIOUR AND SPACE: GENERAL ASPECTS OF SITE LOCATION

Oviposition requires the female to follow a process of signal tracking through progressively narrower physical scales: habitat, microhabitat, plant and plant part (Dennis 1983a, Jones 1991, Porter 1992, Shreeve 1992b, Janz 2002). The typical sequence would consist of search, orientation and encounter, evaluation, and acceptance or rejection (Renwick & Chew 1994). The sequence depends on the spatial scale at which the butterfly operates, the species behaviour, and the physical features of the habitat and the plants used. The decisions are predominantly of an 'all or nothing' nature; each step of the process requires the evaluation of at least chemical or visual cues (Feeny et al. 1983, Jones 1991, Bernays & Chapman 1994, Rutowski 2003, Singer 2004).

Habitat location

Once mated, females will typically devote a substantial amount of time to host location and oviposition. When adult females emerge in an environment suitable for oviposition, the spatial distribution of the adults may broadly fit that of the larval hosts (Courtney 1982b, 1983), making large-scale habitat recognition unnecessary. However, different resources are often non-randomly distributed across habitats (Baker 1969, Wiklund & Åhrberg 1978, Courtney 1983, Ravenscroft 1994b, Dennis *et al.* 2006) and adults must locate spatially separated parts of their habitat for different activities (Wiklund 1977a). The necessity for intermediate-scale habitat location will grow as habitat heterogeneity increases (for natural reasons, or due to environmental disturbance). Directed flight (indicating willingness to reach a different location) will occur more often in unsuitable biotopes (Dennis 2004). In an extreme case, females of migrant species clearly have to operate at relatively large scales to locate suitable sites for reproduction. Such long- or medium-distance searching for distinct microhabitats requires reference landscape landmarks such as trees, clearings or edges that may either predict the presence of host plants or serve as references in a learning process (Renwick & Chew 1994, Conradt *et al.* 2000). Gross-scale features of the landscape are probably used in that way by the species capable of important inter-habitat movements (Gutiérrez & Thomas 2000).

From habitat to microhabitat

At low spatial scales, it may be difficult to differentiate between microhabitat search and detection of larval hosts in a strict sense. *Erebia ligea* and *Boloria euphrosyne* provide examples where microhabitat per se is selected as a first step. The former approaches oviposition sites by first flying towards wooded paths (Wiklund 1984), while the latter seeks open glades within woodland areas (Porter 1992). Apparently, the females of these butterflies detect oviposition sites (and presumably adequate microhabitats for their progeny) after macroscopic, probably physical cues at the within-habitat level. Intriguingly, the females of some *Parnassius* and *Argynnis* species fix the eggs to non-host substrates after the aerial parts of their herb hosts have been washed out (Wiklund 1984, Bink 1992, Porter 1992, García-Barros 2000b). This behaviour differs from that of oviposition on non-host material *after* effective host detection. It is to be investigated whether or not such butterflies actually detect traces of the hosts, or predict its presence based on indirect cues. Field work on a Nearctic *Argynnis* suggests an interesting mismatch between the location of eggs and that of the larval hosts (*Viola* spp.), potentially implying high early larval mortality (Kopper *et al.* 2000). Most clumped distributions of the immatures seen in the field may reflect adult female preferences, hence they might be assumed to be best suited for immature development (Thomas 1983b, Heath *et al.* 1984, Thomas *et al.* 1986, Webb & Pullin 2000, Martin & Pullin 2004a). However, similar patterns eventually result from non-adaptive canalisation of adult activity by microclimatic constraints such as temperature (Shreeve 1986) or edge effects (Courtney & Courtney 1982, Dennis 1983b, 1984, Jones 1991) (Fig. 4.1).

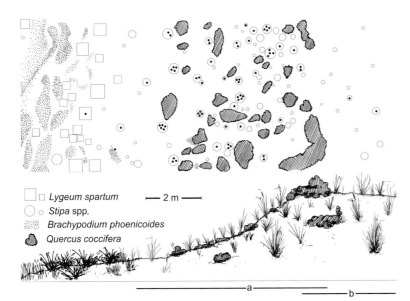

Lygeum spartum
Stipa spp.
Brachypodium phoenicoides
Quercus coccifera

— 2 m —

a   b

Figure 4.1. Distribution of eggs laid by *Melanargia occitanica* in a field path, illustrating a combination of circumstances that complicate the interpretation of female host selection. Eggs are represented by black dots. The females attach single eggs to the tips of leaves or inflorescences of thin-leaved *Stipa* grasses (see Fig. 4.2), in spite of other suitable species being present. Large plants are preferred to small ones, but slope (and perhaps edge effects) seem to determine plant choice: *S. offneri* and *S. lagascae* occur in sector a, while *S. parviflora* dominates sector b; few eggs were laid in the flat area dominated by the last species (unpublished observations made near Ciempozuelos, Madrid, May 1991).

Thus, in addition to the aspects related to plant identity, and plant part selection, it is interesting to address other equally relevant features of the oviposition sites such as plant quality or microclimatic conditions.

THE RIGHT HOST SPECIES

Upon the plant: identifying hosts

During the pre-alighting phase of host plant choice, butterflies visually assess the shape, size, colour and texture of the host leaves (Ilse 1937, Kolb & Scherer 1982, Stanton 1984, Figurny & Woyciechowski 1998, Kelber 1999a, b, Kelber *et al.* 2001, Stefanescu *et al.* 2006); thus colour vision needs to be sophisticated in these insects (Kelber 2001, Briscoe *et al.* 2003). This detailed assessment is generally done within intermediate to short distances

(10 to 100 cm: Rutowski 2003), but further host acceptance often requires a closer approach or physical contact (Wiklund & Åhrberg 1978, Wahlberg 1997, 1998). Females of the genera *Papilio* and *Anthocharis* often shift from a fast and directed flight pattern to a nearly stationary hovering-like flight when they approach a potential host (Petersen 1954, Wiklund 1974, Wiklund & Åhrberg 1978, Feeny *et al.* 1989), which suggests searching for volatile compounds before landing. Once upon the plant (post-alighting phase), acceptance or rejection depends on a positive or negative reaction towards surface allelochemicals (Feeny *et al.* 1983, Renwick *et al.* 1989, 1992, Schoonhoven *et al.* 1990, Dempster 1992, Bernays & Chapman 1994, Haribal & Feeny 2003, Nakayama *et al.* 2003, Nakayama & Honda 2004, Peñuelas *et al.* 2006). The fidelity to some secondary plant substances may be able to explain host selection better than plant taxonomy

Figure 4.2. Butterfly oviposition. (a) *Melanargia russiae* laying near the tip of a grass leaf (after Frohawk 1912). (b) *Maculinea arion* laying into the bud of *Origanum vulgare* (drawing by A. M. Schulte). (c) *Lopinga achine* dropping an egg among grasses (photograph by C. Wiklund). (d) Female of the lycaenid *Iolana iolas* ovipositing inside the calyx of a flower of *Colutea hispanica* (photograph by S. G. Rabasa). (e) Female of the papilionid *Parnassius mnemosyne* dropping an egg (from Wiklund 1984). (f) Three eggs of the skipper *Pyrgus sidae* laid on a flower of *Potentilla* sp. (SW Spain); the photograph is unusual since the females of this species generally lay one egg per flower.

(Wahlberg 2001, Bergström *et al.* 2004) or, in some instances, plant size (Bergström *et al.* 2006; see below). Such substances are perceived by means of chemoreceptor microsensillae located in the tarsi, antennae, proboscis or ovipositor (Feeny *et al.* 1983, Renwick & Chew 1994, Ono & Yoshikawa 2004). The time a butterfly spends in this stage of host discrimination varies from less than 1 second to 30 minutes (Singer 2004). In the final stage, a rather complex combination of stimuli may be required to lay the egg, e.g. tactile mechanoreceptors or even photoreceptors located in the cuticle surrounding the ovipositor (Arikawa & Takagi 2001). This mechanistic basis of oviposition explains the bizarre egg-laying postures of some species (Fig. 4.2).

Butterfly learning

Host plant preferences have an important genetic basis (Janz 2003a, Heinz & Feeny 2005, Nylin *et al.* 2005); a preference for hosts used during the larval life of the insect (Hopkins' host selection principle or HHSP) is not supported by the available evidence (Barron 2001). However, adult butterflies are capable of learning, which enhances oviposition success during the female's life (Wiklund 1974, Rausher 1978). Captive butterflies can be trained to associate colours to plant scents (Weiss & Papaj 2003), and associative learning permits the establishment of neural links between the basic plant-orientated search and peripheral information related to that plant (Snell-Rood & Papaj 2006). Such associations can be incorporated into search patterns, thus facilitating the location of further hosts (Stanton 1984, Traynier 1984, 1986, Weiss 1997, Kinoshita *et al.* 1999, Kroutov *et al.* 1999, Smallegange *et al.* 2006). The behavioural aspects of butterfly learning involved in oviposition are still poorly known, and certainly vary from one species to another. For example, the Nearctic *Euphydryas editha* appears to be able to learn about nectar sources, but not about oviposition substrates (Parmesan *et al.* 1995, McNeely & Singer 2001).

PLANT QUALITY AND VARIED CUES

Whenever there are limits set to adult life expectancy and host abundance, and when there is variation in host plant quality, theory predicts a trade-off between maximising either plant quality or realised fecundity: repeated failure to oviposit on available hosts would be expected to result in a reduction in the number of eggs laid per unit of effort invested (Agnew & Singer 2000, Nylin *et al.* 2000, Janz 2003b; see below). In spite of this, not every host of the

right species is invariably accepted, and some ovipositing females apparently behave overly choosily, using cues beyond species-level identity in assessing individual host plant suitability (Wiklund 1982b, 1984, Porter 1992, Haribal & Feeny 2003). Such features may include or relate to the plants' phenologic stage (Thomas & Elmes 2001), size, leaf colour (Stefanescu *et al.* 2006), presence of conspecific eggs or larvae, microclimatic conditions (discussed below) or chemicals avoided by parasitoids (Nieminen *et al.* 2003). Thus, although plant size and the presence of potential competitors may be evident to humans, the apparent causes for behaviour may be misleading. To illustrate this, recent work on the Nearctic *Pieris virginiensis* by Doak *et al.* (2006) demonstrates that although the females of that pierid discard a large proportion of the crucifer host ramets which the observers would judge to be adequate (e.g. on the basis of later plant senescence in the phenologically constrained habitat of that species), the plants selected were actually superior from the point of view of larval survival.

Plant size

Host size is often correlated with the amount of food available for the larvae, and large hosts provide high passive protection for immatures and can buffer against larval competition. Everything else being equal, large plants (or large, dense patches of plants when these are small) should be preferred over small ones, as in fact they are in most instances (Rausher *et al.* 1981, Courtney 1982a, 1986, Wiklund 1984, Forsberg 1987, Jordano *et al.* 1990, Thomas *et al.* 1991, Porter 1992, Dennis 1995, Dolek *et al.* 1998, Gutiérrez *et al.* 1999, Meyer-Hozak 2000, Anthes *et al.* 2003a, b, Küer & Fartmann 2005, Rabasa *et al.* 2005, Eichel & Fartmann 2008; see also Fig. 4.1). However, there are examples of selection of small hosts. This apparently striking alternative has often been explained in terms of plant quality, microclimatic optima or environmental constraints on female activity (Dennis 1984, 1985, Pullin 1986b, Forsberg 1987, Ebert & Rennwald 1991a, b, Porter 1992, Hermann & Steiner 1998, Dolek & Geyer 2001, Fartmann & Mattes 2003, Fartmann 2004). At least in relation to patch size, there are also mechanistic reasons related to the mode of detection (visual vs. contact) and the ratio of perimeter to area of the patches of host plants (Bukovinski *et al.* 2005). Females may also be flexible; those of some species can adjust the number of eggs laid to the size of the host (egg-load assessment), which has been recorded for some species that lay eggs in clusters (discussed below) such as the large white *Pieris brassicae* (Rothschild & Schoonhoven 1977, Le Masurier 1994) and *Hamearis lucina* (Porter 1992, Anthes *et al.* 2008); examples from other geographical regions are given in Pilson & Rausher (1988) and Vasconcellos–Neto & Monteiro (1993).

Ants and parasitoids

Myrmecophilous species would be expected to detect the presence of their associated ants (Jutzeler 1989a, b, Jordano *et al.* 1992, Pierce 1995, van Dyck *et al.* 2000, Wynhoff 2001), as some tropical lycaenids do (De Vries & Penz 2000). As yet, however, little is known about the assessment performed by the females of European myrmecophilous lycaenids. For example, the evidence available for *Plebejus argus* remains controversial (Seymour *et al.* 2003), and negative for *Maculinea* spp. where female choice seems to be guided by host plant apparency rather than by the presence of ants or ant-released chemicals (Thomas & Elmes 2001, Nowicki *et al.* 2005b, Arnyas *et al.* 2006, Musche *et al.* 2006).

Avoidance of parasitoids also plays a role in oviposition site selection (Porter 1992). Ohsaki & Sato (1994) found that the host plants selected by the females of three related *Pieris* species (*P. rapae*, *P. napi* and *P. melete*) did not represent an optimal choice, but resulted from a trade-off between parasitoid avoidance and host plant quality. Ovipositing females of *Melitaea cinxia* in Finland select plants with high concentrations of the iridoid glycoside catalpol, which is avoided by the parasitoids of the larvae (Nieminen *et al.* 2003). Virtually nothing is known about the ability of the female adults to detect risks of infection by viruses or bacteria which may induce locally high losses in the larval populations of some species (Cappuccino & Kareiva 1985).

Larval competence

Previous occupation of a plant by conspecific eggs or larvae does not necessarily have a negative impact. In some circumstances batch-laying females may benefit from laying in sites which have already been used because large numbers of feeding larvae may either benefit from overcoming plant defences in response to feeding damage, or provide a more powerful response to potential predators and parasites when in large numbers (as shown for the moth *Plutella*: Shiojiri *et al.* 2002). However, the potential advantages of high larval densities in mechanical facilitation of feeding and improved defensive performance should be tested empirically against non-adaptive alternatives. It is conceivable that plants

overdosed with larvae become more attractive for the adult females simply because leaf damage imposes a high rate of evaporation of volatile chemicals.

Overcrowded hosts may imply risks of competition or even cannibalism. This was reported for *Anthocharis* and *Callophrys* (Porter 1992, Watanabe & Yamaguchi 1993), and, interestingly, an oviposition-deterring pheromone was isolated from *A. cardamines* females (Dempster 1992). Olfaction of volatile chemicals which are released from the plant parts attacked by larvae is also known in *Pieris rapae* (Sato *et al.* 1999). Detection of eggs laid by conspecifics (or by taxonomically related species) can also be visual (Wiklund & Åhrberg 1978, Den Otter *et al.* 1980, Shapiro 1981, Courtney 1984b, Schoonhoven *et al.* 1990, Renwick & Chew 1994), and some tropical butterflies use conspecific aposematic larvae as negative cues for oviposition (Papaj & Newsom 2005). On these grounds, it has been hypothesised that the reddish spot-looking buds of some crucifers mimic pierid eggs (Shapiro 1981, Martín *et al.* 1990), which is a mechanism of plant defence similar to that found in tropical *Passiflora* vines on which heliconiine nymphalids oviposit (Benson *et al.* 1975, Williams & Gilbert, 1981).

Deterrent chemicals

One further interesting and overall poorly known item is the reaction to chemicals produced by the host plants as a response to herbivore attacks. Plants are known to respond to mechanical attack by secreting hormones like jasmonic acid. These may in turn induce the production of allelochemicals with deterrent effects on oviposition (glucosinates and perhaps others: Bruinsma *et al.* 2007). In some crucifers, the response is initiated by local cell death which can be induced by the deposition of single *Pieris* eggs (Little *et al.* 2007).

SELECTING MICROHABITAT TO FIT MICROCLIMATIC OPTIMA

Microclimate and microhabitat

Microclimatic conditions such as temperature and humidity may contribute to egg survival and development. Evidence for active selection of the putatively optimum conditions by the female adult has been noted in a number of species (Williams & Gilbert 1981, Dennis 1983c, Thomas 1983b, Heath *et al.* 1984, Grossmueller & Lederhouse 1985, Shreeve 1986b, Emmet & Heath 1989, Porter 1992, Clarke

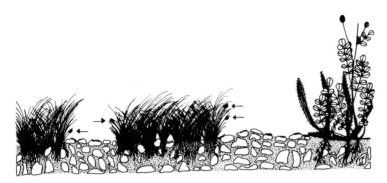

Figure 4.3. Illustration of a typical egg-laying site of *Hesperia comma* in calcareous grasslands in Germany. Eggs (see arrows) were mostly laid above bare ground or gravel on small and grazed tussocks of *Festuca ovina*. The right section of the drawing shows the typical higher vegetation with *Brachypodium pinnatum* and *Sanguisorba minor* (after Fartmann & Mattes 2003).

et al. 1997, Vogel 1997, Thomas *et al.* 1998b, Bourn & Thomas 2002, Roy & Thomas 2003, Fartmann 2006, Anthes *et al.* 2008). Most studies addressing microclimatic preferences in egg-laying have focused on species which oviposit on plants of the herbaceous layer. It is known that the females often choose sites of defined aspect, slope, sward height, vegetation cover or orientation, and oviposit in the proper height and exposure on the plant. Well-documented examples from Northern or Central Europe are *Hesperia comma* (Fig. 4.3) (Thomas *et al.* 1986, Hermann & Steiner 1997; Fartmann & Mattes 2003), *Maculinea arion* (Thomas 1980, 1991b, Pauler *et al.* 1995, Thomas *et al.* 1998b, Fartmann 2005) and *Polyommatus bellargus* (Thomas 1983b, Pfeuffer 2000, Thomas *et al.* 2001b, Roy & Thomas 2003). Higher swards, and vegetation with a cool and humid microclimate are preferred by females of *Euphydryas aurinia* (Anthes *et al.* 2003a, b), *Hamearis lucina* (Sparks *et al.* 1994, Fartmann, 2004, 2006, Anthes *et al.* 2008) and *Thymelicus acteon* (Thomas 1983b, Thomas *et al.* 2001b). The specific selection of microclimate is not always shown to be advantageous for the eggs or larvae, but there is some positive evidence including data from *E. aurinia* (Porter 1982, 1983, Anthes *et al.* 2003b).

Geographically and seasonally shifting microhabitats

Geographical and seasonal variation in the microhabitats selected is interesting because keeping constant microclimatic conditions across a macroclimatic gradient can be achieved only by adjusting site selection differently across the gradient. There are, in fact, known instances of geographically varying oviposition site preferences, for example in *Maculinea arion*. At its northern range in Northwest and Central Europe or at high altitudes, *M. arion* oviposits on *Thymus* spp. on warm south-facing slopes with short swards.

In southern France, by contrast, the eggs are laid on *Origanum* spp. growing in grasslands with swards taller than 20 cm on slopes that are not south-facing (Thomas 1980, 1989, 1995a, 1996, Pauler *et al.* 1995, Thomas *et al.* 1998b, Fartmann, 2004, 2006). Comparable data exist on *Hesperia comma* and *Polyommatus bellargus* (Thomas *et al.* 1986, Thomas 1993, Roy & Thomas 2003).

Seasonal (intra-generational) variation in oviposition niches (often also involving different hosts) may reflect the need for exact microclimatic conditions, but there are also other reasons including differential food availability (Dennis 1985, Porter, 1992). In *Polyommatus bellargus*, the second-generation autumn females select host plants in warmer microclimates than those of the first (summer) generation, with only marginal overlap between both microhabitats (Roy & Thomas 2003). Conversely, subtle microclimatic intra-habitat heterogeneity may affect low-scale differences in plant phenology to determine the actual distribution of eggs. In fact, an irregular topography can buffer phenological constraints on oviposition by increasing asynchrony in the bloom of the hosts, as shown for the lycaenids *Tomares ballus* and *Polyommatus semiargus* (Jordano *et al.* 1990, Rodríguez *et al.* 1993, 1994).

Shifts induced by human disturbance

Human-driven alterations of the landscape can affect species distribution ranges or abundances via spatial or temporal (e.g. phenological) effects on habitat features involved in oviposition behaviour. In at least some instances, changes in land use have expanded the 'essential' niche conditions required for oviposition. Such is probably the case of *Hesperia comma*, which in Northern and Central Europe used to be restricted to a narrow niche defined by south-facing grasslands with a loose grass cover (Thomas *et al.* 1986; Hermann & Steiner 1997, Fartmann & Mattes 2003,

Fartmann 2004). In England the species nowadays uses a wide range of slope aspects for oviposition and has expanded its geographical range northwards (Davies *et al.* 2006). Interestingly, some of the shifts in the geographical ranges of butterflies attributable to global warming (Warren *et al.* 2001, Parmesan 2003) are associated to changes in host plant selection and oviposition sites, as illustrated by the range expansion of *Aricia agestis* in England which matches a host plant shift from *Helianthemum* to *Geranium* (Thomas *et al.* 2001a). The relationship between habitat fragmentation and oviposition behaviour has received little attention (but see Krauss *et al.* 2004b, Rabasa *et al.* 2005, on *Cupido* and *Iolana* respectively). Unsurprisingly, connectivity (not patch size) was correlated to egg densities in the highly dispersive *Iolana*. However, it may prove to be difficult to differentiate between the effects of habitat fragmentation on oviposition behaviour itself from those on other aspects of adult dispersal.

MISTAKES AND MOTIVATION: INTER-GENERATIONAL CONFLICTS AND TIME CONSTRAINTS

Despite the sophisticated mechanisms involved in host location, oviposition on suboptimal substrates is not rare (Dethier 1959, Chew 1977, Courtney 1981, Larsson & Ekbom 1995, Stefanescu *et al.* 2005a). Such cases do not necessarily represent 'true' mistakes, for the range of potential hosts may be much wider than those used in the field. Oviposition on suboptimal hosts may well reflect differences between female choice and the properties of the plants for optimal larval performance, or the presence of other factors that constrain or canalise female choice. A perfect match between female preferences and the plants that ensure optimal larval performance is not universal (Rausher 1982, Singer 1984, 2003, Janz *et al.* 1994, Singer & Thomas 1996, Kuussaari *et al.* 2000, Hanski & Singer 2001, Forister, 2004). Wiklund (1974) argued that this asymmetry is a result of the different selective pressure experienced by adult females and their offspring: females should minimise oviposition on suboptimal hosts, while larvae should benefit from their ability to thrive on them unless they can correct the females' choice. In summary, female preference and larval performance must be balanced on the basis of their respective host plant specialisation (Thompson 1988). Future work should consider the female host plant range, ranking order and specificity (Janz 2003a, Singer 2004) combined with an assessment of differential larval success across hosts.

Limits set on the time available for locating hosts and laying eggs and stressful conditions during oviposition add potential sources for mismatch. According to the 'hierarchy theory model' (Courtney *et al.* 1989), the likelihood of a female accepting a potential host increases with the insect's motivation. Motivation (the relative willingness to lay an egg) should co-vary with the female's age, egg loading, host deprivation and the time elapsed since the last oviposition (Jones 1977a, Courtney & Courtney 1982, Courtney & Forsberg 1988, Singer *et al.* 1992b, Nomakuchi *et al.* 2001). Experiments with geometrid moths show that stressful conditions during the female adult life trigger an increased reproductive effort (Javois & Tammaru 2004) resulting in a less selective oviposition behaviour. In butterflies, substantially more 'bad choices' are made when the hosts are difficult to access (*Hamearis lucina*: Anthes *et al.* 2008), or when the females are 'pressed' by external circumstances such as a high density of courting males (Gibbs *et al.* 2005).

OFFSPRING VS. PARENT FITNESS: TOWARDS A UNIFIED THEORY

Much of the research done on oviposition site selection relates implicitly or explicitly to the 'optimal oviposition theory' (or 'oviposition preference–offspring performance hypothesis'). This theory states that a female adult will select hosts that optimise larval performance, as this will maximise her own fitness (Jaenike 1978; reviewed by Mayhew 1997). An alternative approach to oviposition site choice based in foraging theory (details in Stephens & Krebs 1986), is the 'theory of optimal foraging by adults' (or simply 'optimal foraging'). According to optimal foraging, the female adults should maximise fitness by optimising their own performance. In an extreme case the adult females may behave as 'bad mothers' and select substrates that are best for their own performance (namely those that represent optimal resources for adult nutrition), even at a cost for the performance of their offspring (Jaenike 1986, Mayhew 2001, Scheirs *et al.* 2004). More specific explanations such as the 'hierarchy theory model' (Courtney *et al.* 1989), or the proposal by Tammaru & Javois (2005) that a correlation between female age and choosiness might be selected for in low mortality environments, can be integrated in the broadest model frame in terms of time constraints (because 'motivation' relates to potential fecundity and the temporal and spatial constraints, e.g. Chew & Robbins 1984, Mangel 1987, Nylin *et al.* 1996b, Doak *et al.* 2006), and of the intensity of age-dependence of female choosiness.

Mangel & Clark (1986) argued that the spread of the theoretical basis of behavioural ecology in many directions (foraging, territoriality, predation, competition, life history tactics, clutch size and reproductive strategies) has led to a prevailing extreme reductionist approach, which is useful for work done under laboratory conditions but of limited application in the field. This is because the behaviour of animals in their natural environments is subject to multiple decisions related to the different aspects of their biology and, in practice, most of the resulting 'actions' are to some degree interdependent. However, reductionist thinking is just an expression of the 'constraints' on the activity of researchers. In the long term, competing scientific theories tend to be refined and their useful components are integrated and tend to coalesce. What is important is to keep in mind that most of the variables investigated can always be weighted from different points of view, and that the connections are taken into account. Thus an integration of the views of optimal oviposition and foraging theories is necessary as optimal foraging by adults, optimal realised fecundity and optimal larval performance are not independent of each other (Scheirs 2002).

As far as oviposition site selection by female butterflies is concerned, arguments for an integrated view come from two directions, i.e. the degree of overlap between adult and offspring optima, and the relationship between optimal choices across generations. The preferences of the adult females for nectar sources do not necessarily coincide with the optimal larval hosts. When there is an overlap, there are chances to evaluate the extent to which adult feeding determines oviposition site selection (Scheirs et al. 2004). Doing this with Polyommatus icarus, Janz et al. (2005) demonstrated that adult foraging largely determines host choice. A taxonomic overlap of that kind might in principle be regarded as an extreme example of spatial fit between adult and larval resources. However, it is more than just that, because the degree of taxonomic fit between the two optima might have a genetic basis and thus might be selected in one or another direction as a means to optimise total fitness. The extent to which the optima for the two life stages coincide should be considered as one further variable in this research program. When there is no overlap (Janz 2005), the integration between optimal foraging and optimal oviposition interpretations has to be done within a spatial as well as a temporal framework which takes into account that most activities of the female adult are actually needed for egg production and maturation. Such activities do not represent 'constraints'; it is the density and distribution of the resources which may impose the constraints. One further argument for integration concerns potential fecundity, which in butterflies is highly dependent upon performance of the host during the larval stage. In practice, larval success is frequently measured in terms of larval survival, development time and final size or weight (or pupal size). In these and other insects, final pre-adult size is often used as a surrogate for fecundity (e.g. Mayhew 1997). Thus some degree of feedback should be expected because larval performance on a host will set some parameters that determine, at least partly, the patterns of adult foraging within a habitat. For comparable arguments based on fitness, Mangel et al. (1994) proposed that fitness should be measured during a time-frame spanning from the ovipositing adult to the lifetime of her offspring. In conclusion, the cyclic nature of life histories has to be incorporated into models, since attributes such as fecundity and survival in one stage are importantly conditioned by, or condition, those in the earlier and later stages.

MODES OF OVIPOSITION

Fifty ways to lay an egg

Considered together, the west Palaearctic species attach their eggs to virtually any aerial part of the host plant, none has root-feeding larvae, and very few of them have truly endophytic larvae (miners or borers). In the latter the eggs are not properly 'injected' into the substrate but the neonate larva has to bore (Martín 1984, Nel 1985, Sarto & Masó 1991, Thomas et al. 1991). However, the eggs may be fixed in well-protected positions such as the inner edge of the flower calyx (Iolana spp.: Munguira & Martín 1993, Hesselbarth et al. 1995, Benyamini 1999; Fig. 4.2) or the depressions left by the fallen fruits (legumes) in the spikelets of Onobrychis plants (Polyommatus subgenus Agrodiaetus: Munguira 1988). Some species of Maculinea (Lycaenidae) attach the eggs tightly on the flowers in the flowerheads of Sanguisorba (Thomas et al. 1991), and the eggs of some hesperiine skippers like Thymelicus (Hesperiidae) are hidden within the leaf sheaths of grasses (Bink 1992).

Patterns of site selection and immediate use of the hosts

Wiklund (1984) claimed that eggs are most frequently laid directly on those plants which are subsequently consumed by the larvae. Additionally, hibernating eggs are more often laid on non-edible parts of the plant. Egg dropping is even

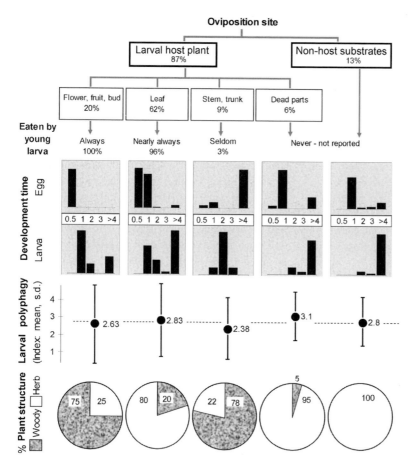

Figure 4.4. Patterns of plant use as oviposition sites by European butterflies (percentage of species whose larvae eat the substrate where the egg was laid), and its relationship with the duration of the egg and larval stages (in months), larval polyphagy and host plant structure. All the estimates given are percentages except the degree of polyphagy (mean and s.d.; the broken line indicates the average value). Polyphagy was measured as $P = (A \times B)^{1/2}$ where A = number of genera of host plants, and B = taxonomic diversity of the hosts (1 = one plant species, 2 = one plant genus, 3 = one plant family, 4 = one plant order, 5 = two or more plant orders). Data from 295 species from varied sources, except estimates of egg and larval development time (220 species).

less selective, and it is found in butterflies whose larvae feed on dense ground-covering plants. Here, the newly hatched larvae must locate their source of nourishment. Broadening the patterns proposed by Wiklund (1984), the oviposition sites used by the European butterflies fit into one of four categories:

(1) Flowers, fruits, young shoots or buds (polyommatine lycaenids, some pierids, and skippers like *Pyrgus* spp. (e.g. Nel 1984, 1985; Fig. 4.2)).
(2) Green leaves (most taxa).
(3) Inedible, sclerified, wilted or dead parts of the food plant (*Agrodiaetus* section of *Polyommatus* blues, several hairstreak lycaenids).
(4) Non-host plant materials (including egg-droppers such as *Parnassius mnemosyne* (Fig. 4.2), and several satyrines).

As shown in Fig. 4.4, the species that oviposit on highly nutritious substrates tend to have fast-developing eggs, and larvae that start feeding upon hatching. When eggs are laid on non-edible substrates, either the larval growth period is long or the substrate serves as a support for a prolonged egg development (diapausing eggs and pharate first instar larvae within the egg shell; e.g. some *Parnassius, Satyrus, Pyronia, Aricia*: Wiklund 1977b, 1984, Jutzeler & Leestmans 1994). No strong differences in average diet breadth are associated with the different modes of oviposition, although species laying eggs on wooden parts tend to be more selective and those ovipositing on dead plant parts are more polyphagous by comparison.

Oviposition behaviour is not invariant within each species, and a few instances of seasonal or geographic intraspecific variation are known despite the fact that this subject has received little attention. Known examples of butterflies of different broods using different oviposition substrates include *Polyommatus thersites* (Tolman & Lewington 1997) and *Spialia sertorius* (Hermann 1998, 1999). *Maniola jurtina* females can either attach the eggs to the substrate or jerk them away (Ebert & Rennwald 1991b), and the relative frequency of the two alternatives seems to vary geographically (cf. Wiklund 1984, García-Barros 1987, Bink 1992). A similar pattern is known for a Nearctic lycaenid (Fordyce & Nice 2003).

EGG CLUSTERS

Patterns

Egg clustering consists of laying several (or many) eggs together on the host, forming an egg cluster or batch. This is generally a fixed, species-specific trait (but see Pullin 1986a). An egg cluster may contain up to about five hundred eggs, depending on the species (Hinton 1981, Porter 1992), which are often arranged according to a species-specific pattern (Fig. 4.5). This behaviour is comparatively rare in butterflies: less than 7% of species worldwide, 6.9% in the Ethiopian Region (Stamp 1980), and 11% in Europe. The west Palaearctic egg-clustering species belong to the genera *Aporia* and *Pieris* (Pieridae) and *Inachis*, *Aglais*, *Araschnia*, *Nymphalis*, *Melitaea* and *Euphydryas* (Nymphalidae). A few species lay loose batches of fewer than ten eggs (*Heteropterus*, *Thymelicus*, *Archon*, *Boloria eunomia*), or both small loose batches as well as single eggs (*Zerynthia*, *Hamearis*, some *Lycaena* and *Satyrium*; examples in Fig. 4.5). Within the batch-laying species, the size of the egg clutch may partly

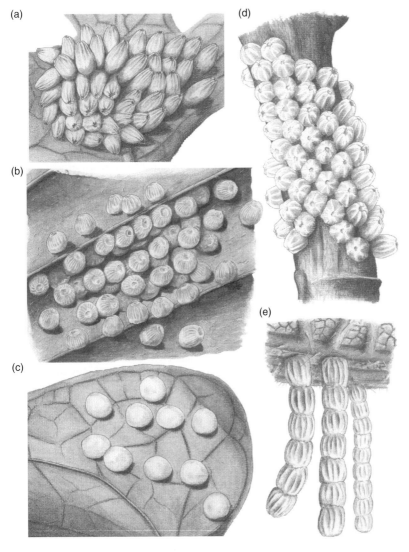

Figure 4.5. Examples of egg clusters. (a) *Aporia crataegi*, (b) *Melitaea cinxia* (partial view of an egg batch), (c) *Zerynthia polyxena*, (d) *Nymphalis polychloros*, (e) *Araschnia levana*. (a) Pieridae; (b), (d) and (e) Nymphalidae; (c) Papilionidae. Not drawn to the same scale.

depend on adult size and environmental conditions (Saastamoinen 2007 on *Melitaea cinxia*).

Advantages and disadvantages: adults

The relative rarity of the clustering habit suggests that it incurs negative effects, a reason why research on the evolution and ecology of egg clustering is attractive (Stamp 1980, Young 1983, Chew & Robbins 1984, Courtney 1984a, b, Parker & Courtney 1984, Skinner 1985, Clark & Faeth 1998). As far as the adult insect is concerned, egg clustering causes concentration of oviposition in space and time. Compared with other butterflies, the egg-clustering females might gain more time for selecting hosts (Singer 2004), or – conversely – minimise the time spent seeking them (thus lowering predation risk: Porter 1992). Clustering the eggs imposes that large numbers of oocytes mature synchronically, causing a high wing loading in young females which may restrict their dispersal (Boggs & Nieminen 2004).

Advantages and disadvantages: immatures

Egg clustering has benefits and costs for the immature stages. Possible advantages for the larvae concentrate on the consequences of larval aposematism and cooperative silk-web construction as defence against natural enemies and weather (e.g. Vulinec 1990, Fitzgerald 1993, Clark & Faeth 1997, Nieminen *et al.* 2001; further references in Saastamoinen 2007). Furthermore, it seems likely that this habit originated in connection with larval gregariousness, toxicity and aposematic apparency (Sillén-Tullberg 1988, Sillén-Tullberg & Leimar 1988). The advantages of egg clustering for the eggs themselves remain rather hypothetical. However, the number of eggs per batch is subject to individual variation and tends to decrease as the females get exhausted (Warren *et al.* 1994, Wahlberg 1995, Wahlberg *et al.* 2004). Findings on the Nearctic *Chlosyne lacinia* indicate that large egg batches endure desiccation better than small ones (Clark & Faeth 1998). Multilayered clusters might protect the innermost eggs from parasitoids (Stamp 1981b, Friedlander 1985). Finally, some egg-clustering species lay eggs of bright colours that suggest aposematism (Baker 1970, Aplin *et al.* 1975, Brower 1984), although it is not clear whether the eggs are actually protected in this way (Chew & Robbins 1984).

Egg clusters force larvae to share their food resources with their siblings (for most of the true egg-clustering species are gregarious during their early instars at least) and egg clustering would thus be predicted to require either large host plants, or dense patches of hosts (Courtney 1984, Davies & Gilbert 1985). This is, in fact, the case for *Nymphalis*, *Aporia* and the Mediterranean populations of *Euphydryas aurinia* that live on shrubs or trees, while herbs are used by the remaining species (however, *Inachis* and *Aglais* feed on plants such as the nettles *Urtica* spp. which often grow in dense formations). Small plants may soon become exhausted, forcing the larvae to disperse to new hosts and incur an increased risk of mortality (Dethier 1959, Porter 1992, Le Masurier 1994, Clark & Faeth 1998, Anthes *et al.* 2003b). Hence, not only plant size but also plant density, patchiness, and inter-plant connectivity are relevant for cluster-ovipositing species (Cain *et al.* 1985, Davies & Gilbert 1985). Sequential feeding on successive plants might benefit the larvae by circumventing the inhibition of growth that occurs in some plants as a response to larval feeding (Le Masurier 1994).

In summary, more comparative work is needed to test thoroughly the existing hypotheses on the origin of egg clustering including those relating to host size and density, cooperative silk-web construction, toxicity and aposematism, and female fecundity as related to dispersal ability. The available evidence suggests that egg clustering must have appeared under a combination of circumstances that probably involved large hosts, high fecundity and larval aposematism.

EGG DISPERSAL IN TERMS OF FECUNDITY, SURVIVAL AND FITNESS

Typologies in egg production

Egg dispersal largely predetermines some aspects of the species population structure, including the spatial distribution of adult emergence sites. With the aforementioned exceptions, egg dispersal depends on the adult female's movements, her longevity and daily egg production (hence fecundity) which are key items in life-history theory (Roff 2002). Since adult dispersal is dealt with elsewhere (Chapter 9), only a brief discussion of egg production patterns and female survival follows.

The elements of a specific average profile of egg production are: (1) onset, which determines the length of pre-oviposition, (2) daily egg production, (3) its decline over time, (4) concentration of ovipositing effort on a fine time-scale (e.g. daily) and (5) offset, i.e. the moment when oviposition would cease under average optimal conditions for the female (Boggs 1986). Ample variation occurs in all of these elements across species, and probably within them. The mean potential fecundity of European species varies between ca. 100 and 1500 eggs, although relatively few species have been thoroughly investigated (Richards 1940, Magnus 1958, Höeg-Guldberg & Jarvis 1970, Wiklund & Persson 1983, Karlsson & Wiklund 1984, Martín 1984, Wickman & Karlsson 1987, Wiklund *et al.* 1987, Svärd & Wiklund 1988a, García-Barros 1989a, 1992, 2000b, Karlsson & Wickman 1990, Bink 1992, Dennis 1992a, Wahlberg 1995). Bink (1992) made a valuable attempt to assess the potential fecundities of Northwest European species although, unfortunately, comparisons between independent estimates for the same species suggest that more refined counts are generally required. Oviposition may start within hours after emergence (e.g. *Maculinea nausithous*: Pfeifer *et al.* 2000, *Euphydryas aurinia*: Porter 1992), while a long pre-oviposition is found in species where the female adult migrates, hibernates (Pullin 1992) or undergoes summer reproductive dormancy (Scali 1971, García-Barros 2000a, b). The potential longevity of the females is believed to range between one and five weeks (excluding periods of dormancy or adult diapause), while the modal residence time from mark–release–recapture studies is below ten days (examples in Warren 1992a, Baguette & Nève 1994, Kuussaari *et al.* 1996, Munguira *et al.* 1997b, Fischer 1998, Baguette *et al.* 2000, Fric & Konvička 2000, Pfeifer *et al.* 2000). Assuming that residence time can be interpreted in terms of adult survival, the average realised fecundity must therefore account for only a small part of the potential egg complement. As a consequence subtle differences in adult survival represent strong selective advantages for the individual females (Wickman & Jansson 1997). Although in fact mean residence times from mark–release–recapture studies are not equivalent to adult survival under field conditions (except for 'closed' populations), there is some evidence that single-laying species of Pieridae have relatively low average realised fecundity in weather-constrained habitats (Courtney 1984a). A corollary is that more information on survival and dispersal rates from a vast proportion of the species is still missing. A more complete compilation of this kind of data will not ultimately negate the overall relevance of adult survival in terms of realised fecundity, but it would help to identify the differences between species' strategies.

Egg concentration vs. dispersal

The most extreme typologies in egg production (i.e. early egg production linked to strong spatial and temporal concentration, vs. delayed onset of oviposition associated with eggs laid singly and low daily egg production rates) can be explored within the scope of life history theory. For instance, larval age- or stage-dependent patterns of accumulation and use of reserves can be understood as responses to stress conditions experienced in other stages (e.g. Gatto & Ghezzi 1996). Early oviposition coupled to large egg masses might prevent negative effects of predictable 'catastrophes' during the female's life (e.g. bad weather: Douwes 1976b, Gossard & Jones 1977, Courtney & Duggan 1983, Shreeve & Smith 1992). The dichotomy of egg dispersal vs. egg concentration has also been discussed in terms of risk-spreading strategies: bet-hedging (that is, producing a varied range of clutch sizes) reduces variance in fitness (at the cost of a lower mean fitness) when the offspring are subject to unpredictable, density-independent mortality (Hopper 1999, Roff 2002; examples in butterflies in Root & Kareiva 1984, but see also Courtney 1986 or Roitberg *et al.* 1999). Tammaru & Haukioja (1996) proposed that lepidopteran life histories are ordered along a continuum whose extremes are capital breeding (egg production derived from larval-derived reserves) and income breeding (eggs mostly derived from adult-gathered nutrients). Capital breeding would be associated with high population eruptivity, low female dispersal, egg clustering and ephemeral non-feeding adults (Hebert 1983, Hunter 1995); some evidence for this pattern across Lepidoptera species is provided by Jervis *et al.* (2007). Adult female butterflies normally require some nectaring before starting to oviposit (Wickman & Karlsson 1989, Wheeler 1996) although, within these limits, the egg-clustering species would comparatively approach the pattern of capital breeding because of their fecundity and small eggs (Courtney 1984a, García-Barros 2000c), and the comparatively low dependence of early egg production on adult food income (Boggs *et al.* 2004, O'Brien *et al.* 2004, 2005).

Thus in summary, egg concentration in space and time (hence also egg clustering) seems associated with habitats that are overall more stable, predictable and perhaps safer for the immatures than for the adults, or where the adults assume a comparatively 'less active' role.

CONCLUSION: FACTS, METHODS, MECHANISMS

The complex nature of oviposition decisions makes some degree of step-by-step approximation necessary: reductionism should not be a major concern while good quality data are collected. From a factual point of view, a critical review and tabulation of basic life-history features such as host plants, parts of the host used, potential fecundity and temporal egg-laying schedules for the European taxa is needed as it would facilitate some preliminary conclusions on geographical variation and macroecological analyses. More precise descriptions of oviposition preferences and their environmental correlates would be required in most instances. Comparative work often suggests an important degree of integration between features of the habitat, the hosts plant, and behavioural, life-history or demographic features of the butterflies (Dennis 2004, Dennis *et al.* 2004, Karlsson & Wiklund 2005, Hamback *et al.* 2007, Jervis *et al.* 2007). However, there are few quantitative studies linking macroecological features of the species with their life history or ecological requirements (Dennis *et al.* 2004). Both broad-scale or phylogenetically based comparative studies (see Brooks & McLennan 1991, Harvey & Pagel 1991) may shed further light on some aspects of the evolution of oviposition behaviour, hosts and host parts selection and related items (further examples in Sillén-Tullberg 1988, Janz & Nylin 1998, Jervis *et al.* 2005).

The several types of adult activity which obviously interact with egg-laying should ideally be integrated in the analysis of adult oviposition behaviour. Some of these ('top–down forces' such as predation) may constrain the reproductive output, while others (e.g. adult feeding) actually contribute to, and/or are necessary to, the laying of eggs. The balance on the relative amount of energy to be distributed among them depends importantly on habitat structure at the several spatial levels; this may allow for reinterpretation of some frequently assumed trade-offs, such as that between dispersal and fecundity (Hanski *et al.* 2006). However, although the spatial structure of the data and spatial autocorrelation have risen as popular topics in other fields of ecology during the last years (e.g. Diniz-Filho *et al.* 2003, Kühn 2007), few studies have hitherto attempted to analyse oviposition behaviour within a hierarchical spatial context. Such an approach may help to integrate variables at different spatial levels and to identify the scales at which they are relevant (Kéry *et al.* 2001, Bukovinski *et al.* 2005, Rabasa *et al.* 2005). For comparable reasons, given the temporal structure of the oviposition process (and the links between female fecundity and the preceding and succeeding larval stages), a temporal component may also be justified.

Full integration of at least the most relevant variables in field studies requires explicit experimentation on a number of items which are frequently assumed based on indirect or intuitive evidence, such as the actual relevance of time constraints, larval performance, and the plants and situations preferred by the adult female. This may be important because females may actually be discriminating based on cues that cannot be perceived by the human eye (Doak *et al.* 2006). This, obviously, indicates that more precise knowledge of the cues used by each species and the physiological mechanisms to perceive them (vision, chemicals) might be of much help. More information on the genetic aspects of host recognition is a logical add-on to the list of queries, and may provide surprising insights in this field (e.g. Gibbs *et al.* 2006, Nygren *et al.* 2006, Quental *et al.* 2007). Putting all that together is not easy, but a long way has already been covered with some model organisms such as the comma butterfly *Polygonia c-album* (references in Janz 2003a, 2003b, Bergström *et al.* 2004, 2006, Nylin *et al.* 2005, Johansson *et al.* 2007) or some *Pieris* species.

Finally, for purposes of conservation and habitat management, knowledge (even at an elementary level) of the oviposition preferences may provide a short-cut to recording several potentially important features of the habitat, and the spatial relations between these and other resources should be considered (e.g. Fred *et al.* 2006).

ACKNOWLEDGEMENTS

We wish to thank Josef Settele and the editors for the invitation to write this chapter, as well as Per-Olof Wickman who set the initial ideas, Tim G. Shreeve, Nils Anthes and two anonymous reviewers for patient editing and constructive criticism, and Miguel Munguira, Tristan Lafranchis and Juan Hernández Roldán for access to their data on hosts and field experience. We are also indebted to the authors of some of the figures or photographs used for Fig. 4.2: Axel M. Schulte (drawing of *M. arion*), Sonia G. Rabasa (photograph of *I. iolas*), Christer Wiklund for the photos of *P. mnemonyne* and *L. achine*. The picture from Wiklund (1984) is used with kind permission of Springer Science and Business Media.

5 • Butterfly herbivory and larval ecology

MIGUEL L. MUNGUIRA, ENRIQUE GARCÍA-BARROS AND JOSÉ
MARTÍN CANO

SUMMARY

This chapter treats butterfly herbivory through a detailed
assessment of patterns of foodplant use, focusing on foodplant specificity, taxonomic patterns, growth form, selection
and geographical variation. This is followed by a treatment
of plant components and their influence on larval growth,
and by a section on butterfly caterpillars as herbivores,
where morphology, physiology and behavioural aspects are
dealt with. The last two parts focus on plants and phenology
and butterflies as pests.

European butterflies are specialised herbivores that select
species of only a third of the plant families as larval hosts.
They are predominantly herb feeders. Several strategies have
been developed by larvae to avoid noxious plant chemicals,
while specialisations of morphology, physiology and behaviour allow them efficiently to exploit plant resources. These
varied strategies have enhanced the appearance of a variety of
herbivore niches that are related to butterfly diversity. From
the phenological point of view larval overwintering is the
preferred strategy among European butterflies. Probably as
a result of changes in agricultural practices in Europe, very
few butterfly species can be regarded as pests.

PATTERNS OF FOODPLANT USE

There is a considerable amount of information on the foodplant use of European butterflies, and almost every guide or
manual dealing with the topic gives some information on the
plants used by the larvae. This is the case from the local to
the national or European scale. Databases with this kind of
information are being built by different butterfly specialists
(Fiedler 1998, Robinson 1999, Lafranchis 2000) and with a
few exceptions (e.g. Slansky 1976, Fiedler 1998), the information is limited to the list of plants used by a given butterfly
species with no further analysis. A good deal of information
is also available on specific aspects of the use of plants by
butterflies, but a treatment at a European scale is lacking,
while national efforts have provided a considerable amount
of information in some countries (Bink 1992, Dennis 1992a).

Larvae often play a secondary role in the selection of an
adequate plant species and even of a specific part of the
plant, because the main selection is achieved by the female
parent. However, larval activity is not devoid of relevance:
caterpillars must recognise the plant as suitable food, select
parts of it, and even move away in the search for new hosts.
Besides, the plant selection by the female may result in
choices that do not coincide with the preferences of the
larva (Chew & Robbins 1984). The lower mobility of larvae
limits the choice, not only to taxonomically close plants, but
also to those of similar ecological characteristics. This usually results in larvae accepting a broader range of foodplants
than the plants selected in the field, which is confirmed by
rearing experiences in the laboratory (Wiklund 1975, Singer
1984).

For the analysis of foodplants used by European butterflies we have built a database using 120 papers and books
with reliable and thorough information on the topic (some
examples are Henriksen & Kreutzer 1982, SBN 1987, Bink
1992, Dennis 1992a, Munguira *et al.* 1997a, Tolman &
Lewington 1997, Lafranchis 2000). The database was completed with references for rare species and with unpublished
observations from the authors. We obtained 2283 foodplant
records from a total of 348 butterfly species representing
76% of all European species (457 species according to
Karsholt & Razowski, 1996). The results are summarised
in Table 5.1.

Foodplant specificity

Butterfly larvae are usually specialised and selective in foodplant use. Several basic grades of specificity are traditionally
recognised (e.g. Cates 1980): monophagous (the larvae
accept a single species of plant), oligophagous (a few species
are accepted, usually close relatives but also often a few
species in different genera or families) and polyphagous (a
wide variety of plants from different families is used). The
oligophagous type can be subdivided into two categories
(Wiklund 1975, Courtney 1984b): the oligophagous–
polyphagous type (true oligophages), which will use several

Ecology of Butterflies in Europe, eds. J. Settele, T. Shreeve, M. Konvička and H. Van Dyck. Published by
Cambridge University Press. © Cambridge University Press 2009, pp. 43–54.

Table 5.1 *Foodplant use in European butterflies*

	Records	Total	Hesperidae (45)	Papilionidae (12)	Pieridae (48)	Lycaenidae (121)	Nymphalidae (231)
Poaceae (870)	621	27.21	44.7	–	–	–	51.3
Fabaceae (845)	395	17.31	2.7	–	22.5	52.9	–
Brassicaceae (651)	202	8.85	–	–	60.7	–	–
Rosaceae (477)	156	6.84	31.8	14.5	5.7	2.5	3.3
Violaceae (92)	82	3.59	–	–	–	–	8.1
Scrophulariaceae (516)	58	2.54	–	–	–	–	5.7
Lamiaceae (452)	52	2.28	8.8	–	–	5.2	0.01
Cyperaceae (263)	52	2.28	0.9	–	–	–	4.9
Polygonaceae (104)	50	2.19	–	–	–	7.8	0.4
Apiaceae (431)	49	2.15	–	41.9	–	–	–
Salicaceae (81)	46	2.02	–	–	–	–	4.5
Geraniaceae (74)	42	1.84	–	–	–	7.1	–
Asteraceae (1615)	41	1.80	–	–	–	–	4.0
Plantaginaceae (36)	37	1.62	–	–	–	–	3.6
Rhamnaceae (19)	36	1.58	–	–	5.7	2.9	–
Ericaceae (51)	36	1.58	–	–	0.9	4.2	0.8
Cistaceae (75)	36	1.58	4.4	–	–	4.4	–
Fagaceae (31)	25	1.10	–	–	–	3.9	0.01
Urticaceae (17)	24	1.05	–	–	–	–	2.4
Aristolochiaceae (14)	22	0.96	–	18.8	–	–	–
Caprifoliaceae (26)	21	0.92	–	–	–	0.17	2.0
Ulmaceae (10)	18	0.79	–	–	–	0.8	1.3
Crassulaceae (124)	18	0.79	–	10.3	–	1.01	–
Malvaceae (50)	17	0.74	5.8	–	–	–	0.4

Notes: The number of records in our database for each plant family and the percentage of the total is given for the plant families with more than 15 records. The percentage of the records in each butterfly family is also given for each family of the selection. Blank spaces are families with no records for the butterfly family with the blank. The number of species in each butterfly and plant family in Europe is given in parentheses. Butterfly nomenclature follows Karsholt & Razowski (1996) and plant nomenclature follows the Angiosperm Phylogeny Group (APG 1998).

plant species at the same locality and season, and the oligophagous–monophagous type, which may use only one plant species per population (or per season). Note that not all authors use these terms with the same meaning (Strong *et al.* 1984).

Feeding on plants from one genus is predominant in European Lycaenidae and Papilionidae, while feeding on one family predominates in the Hesperiidae, Pieridae and Nymphalidae and in butterflies overall, as can be seen in Table 5.2, where a comparative view of the butterfly families is given. The percentage of butterflies feeding on more than two plant families is very low (5% of the species). Our data are consistent with those from the West Palaearctic (Fiedler 1998) in which the average number of foodplant families used by butterflies is between one and two, being higher for the Nymphalidae and lower in the Lycaenidae (Table 5.2). Compared with North American data (Ehrlich & Murphy 1988), our data show a similarly low percentage of polyphagy and also a striking similarity with their statement: 'more than 80% of all North American butterflies for which host plant

Table 5.2 *Foodplant specificity in butterflies expressed as the percentage of species that feed on only one species of plant, on one genus, one family, two families and three or more plant families, respectively*

| | Butterfly families | | | | | | |
	Hesperiidae	Papilionidae	Pieridae	Lycaenidae	Nymphalidae	Total	*N*
One species	25.0	0.0	17.4	29.1	23.2	23.6	82
One genus	22.2	50.0	17.4	33.0	13.2	21.8	76
One family	33.3	25.0	47.8	25.2	29.8	31.0	108
Two families	16.7	25.0	13.0	7.8	27.2	18.4	64
Three or more families	2.8	0.0	4.4	4.9	6.4	5.2	18

Notes: Percentage is given for each family of Papilionoidea and Hesperioidea and for all European butterfly species considered in our study. *N* represents the sample size of species for each specificity category (total sample size is 348). The data have been taken from the database built for this study.

data exist are restricted to a single plant family' (77% in Europe). This latter figure may show that the predominance of butterflies feeding on plants from one family is commonplace throughout the Holarctic region.

Taxonomic patterns

A total of 56 plant families are used by the larvae of European butterflies. All of them are Angiospermae. No Gymnospermae or Pteridophyta have been recorded as hosts in Europe. These families account for just 32.4% of the European families of Angiospermae (Tutin *et al.* 1964), showing that butterflies are specialised on a fraction of the available plant taxa. In general, butterflies select the most diverse plant families, among which four families (Poaceae, Fabaceae, Brassicaceae and Rosaceae) account for 60% of the records (Table 5.1). The first 20 families in Table 5.1 appear in 90% of the records, leaving just 10% of the records amongst the other 36 families. The Asteraceae with only 41 records are a minor choice when compared with other speciose families. The following sections describe foodplant choice by the European members of the different butterfly families.

HESPERIIDAE
Records for this family total 226 for 80% of the European species. Eight plant families, 54 genera and 121 species are recorded. There are two groups within the family concerning foodplant use: Heteropterinae and Hesperiinae larvae feed on Poaceae and related monocotyledons (45% of the records), while Pyrginae use a range of dicotyledon plant

families (32% of the records are from Rosaceae). The grass-feeding species are less specific in their choice (e.g. *Ochlodes venata* has been recorded on 27 plant species). Most of the other Hesperiidae have a more restricted host plant range: 25% of the species are monophagous and 22% are restricted to one genus (see Table 5.2).

PAPILIONIDAE
Seven plant families, 44 genera and 99 species are used by the European Papilionidae. It is striking that 42% of the records belong to the Apiaceae. *Papilio machaon* with 47 records from the Apiaceae and Rutaceae is the species with the most records in Europe. *Papilio hospiton* and *Parnassius phoebus* are also oligophagous. Populations of *P. machaon* use plants of two different families in nearby habitats. This is probably a primitive trait because Dethier (1947) postulated that the selection of Apiaceae is secondary in species that previously selected Rutaceae. Aristolochiaceae, with 19% of the records, is used by five species of the subfamily Parnassiinae. Six Papilionidae feed on just one plant genus and three on one family.

PIERIDAE
The 334 records from this family give information for 96% of the species. A total of eight families, 74 genera and 191 species have been recorded. Relevant families are the Brassicaceae with 61% of the records, that represent the main food choice for the Pierinae, and the Fabaceae (Fiedler 1998) with 23% of the records within the Coliadinae (*Colias*) and Dismorphinae (*Leptidea*). The common species *Pieris brassicae* and *P. rapae* feed on four

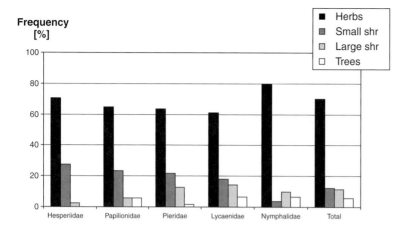

Figure 5.1. Growth form types of the plants used as foodplants by European butterfly species (Papilionoidea and Hesperioidea superfamilies). The percentage of foodplants classified as herbs, small shrubs (small shr), large shrubs (>1 m) (large shr) and trees is given for each family and for all butterfly species. The data have been taken from the database built for this study. Sample sizes as in the last column of Table 5.2.

different families (Capparaceae, Brassicaceae, Resedaceae and Tropaeolaceae) and can be considered polyphagous. Most of the other species only use plants from one family (48% of the Pieridae).

LYCAENIDAE

With 593 records, this family is the second in importance in our database. We have records from 84% of the species in the family, and the foodplants are from 29 families, 99 genera and 317 species. Most of the plant families have only a limited importance as foodplants, while the Fabaceae predominate with 53% of the records. Other important families do not exceed 8% of the records (Table 5.1). Lycaenids also feature the most polyphagous butterfly: *Celastrina argiolus* feeds on 12 different plant families. Other polyphagous species are *Callophrys rubi* and *Leptotes pirithous*, which feed on species from seven plant families each. On the other hand, 29% of the species feed on just one plant species and 33% on one genus.

NYMPHALIDAE

This is the largest family in Europe, for which we have 1014 foodplant records from 33 plant families, 134 genera and 376 species, available for 70% of the species. The number of recorded plant families is lower than in the West Palaearctic, where they feed on 55 plant families (Fiedler 1998). The major group of foodplants is the Poaceae (51% of the records), which are only eaten by the Satyrinae, which also feed on Cyperaceae and Juncaceae. The choice of these three families is comparable to other groups of grass-feeding Lepidoptera like some Hesperiidae (see above) and the Elachistidae (Powell 1980). Other important families are

the Violaceae (in the Heliconiinae, with 8% of the records) and the Scrophulariaceae (6%). Ten species can be considered polyphagous, including: *Euphydryas aurinia* which feeds on nine different plant families, *Polygonia c-album* (eight families), *E. maturna* (six families) and *Nymphalis polychloros* (six families). A high proportion (23%) of nymphalids feeds only on one plant species and 30% on just one family.

Growth form selection

We have categorised the growth form types of the plants from our database in four groups: herbs, small shrubs (up to 1 m), large shrubs and trees. The result of the analysis is shown in Fig. 5.1. Munguira *et al.* (1997a) state that all the Spanish Satyrinae and 49% of the Lycaenidae feed on herbs. Our data are very similar to those obtained for Spain although at the European scale the percentage of species using trees and shrubs decreases whilst the proportion of herb-feeding species increases. Our data run almost counter to the results for the world butterflies as obtained by Janz & Nylin (1998) for which trees and shrubs represent the host forms for 73% of the butterflies, vines for 21% and herbs a mere 31%.

That herb-feeders are significantly more abundant in Europe than elsewhere may be explained by four phenomena. First, diversity hotspots concentrate in mountains (Dennis & Williams 1995) where grassland habitats are clearly favoured. The Polyommatini and Satyrinae are an important part of the fauna of these species-rich areas and these two groups are specialised herb-feeders. Second, the predominance of herb-feeders certainly has a historical component. The human presence in Europe has had a strong impact and dates from very ancient times, favouring grassland habitats, fragmenting wooded areas and reducing forests to a small percentage of the landscape. This widespread and intensive management has probably increased the chance of survival of grassland species in comparison with woodland species (Thomas 1991b), and this might have had an effect on the percentage of species using herbs in the larval stage. Third, the general preference of European butterflies is for open biotopes (primordial grasslands, flood plains or steep slopes of gorges) which are richer in herbs than is the herb layer in closed biotopes (woodlands). A fourth reason is related to the diversity of plants of each group. While trees and shrubs make up a low proportion of the total number of European plant species, herbs represent the largest proportion. The latter clearly favours the predominance of herbs as foodplants of European butterflies.

Another feature of the selection of growth form types is the specificity of this trait in different species. Most butterflies are restricted to only one type of plant growth form. This may be a consequence of the fact that plant growth form probably plays an important role in foodplant choice, together with other features (Barbosa 1988). However, some species use two or more types, the most frequent case being to feed on herbs and small shrubs, which are the most similar growth form types. Nevertheless, *Lampides boeticus*, *Leptotes pirithous* and *Celastrina argiolus* use all four growth form types. It is remarkable that these species are also polyphagous, showing that their broad use of plants also implies a lack of specialisation in plant growth forms.

Geographical variation of foodplant choice

Even some of the species strictest in foodplant choice can vary in foodplant use in different locations (Martin & Pullin 2004a). Some species feeding on common or widespread plants show little variation, for example *Libythea celtis* always feeds on *Celtis australis* throughout its range. Strict monophagy is indeed more common in endemic butterflies: 49% of the butterflies restricted to one plant species are endemic, with a narrow distribution range. This is probably linked to the fact that monophagous species have narrower geographical ranges than polyphagous species. Other monophagous species are at the limit of their distribution ranges in Europe and the effect of distribution range is similar to that of endemic species. An example of this is *Tarucus theophrastus*. It is restricted in Europe to Southern Spain, but is a common species in Africa. In general, the size of the geographical range of the butterfly is correlated to that of its host (Quinn *et al.* 1998).

The shift from monophagy to oligophagy at a regional scale is another geographic aspect. In a detailed study for the Spanish Lycaenidae and Satyrinae, Munguira *et al.* (1997a) obtained a predominance of single plant species use for Lycaenidae and single family (Poaceae) for Satyrinae. For the whole of Europe, our data show that only 29% of lycaenids are strictly monophagous (Table 5.2) and many satyrids shift from one to two plant families. This illustrates a broader use of foodplants as geographical scale increases.

PLANT COMPONENTS AND LARVAL GROWTH

Quantitatively, nitrogen (proteins and soluble amino acids), carbohydrates and water are crucial for larval growth. Other substances and oligoelements are usually required as nutrients, including lipids, organic acids and ions. The distribution of such elements is not homogeneous across the plant anatomy, and thus specialisation in the exploitation of specific organs of plants may have drastic consequences for larval growth rate and other biological features asociated with larval nutrition (see Chapter 4). A broad-scope review on insect nutritional ecology is provided by Bernays & Chapman (1994), and much information concerning the nutrition of lepidopteran larvae is surveyed by Slansky (1993), so that these topics will only be addressed here briefly. For additional data regarding the nutritional properties of different plant kinds, or plant parts, see Slansky & Rodriguez (1987). Plant responses to environmental factors and the effect on larval feeding have also been considered in some studies (e.g. Bazin *et al.* 2002, Goverde *et al.* 2002). Also, mycorrhizal fungi have been shown to have a positive effect on larval growth (Goverde *et al.* 2000).

Together with supportive tissues, the leaves account for a large part of the plant material that is available as potential food. The leaves of most plants contain low amounts of nitrogen (at most 10% of the total dry weight: Mattson 1980, Slansky & Scriber 1985).

The highest nitrogen contents are found in herbaceous legumes, and in the reproductive organs and young tissues of other plant groups, while lignified tissues and grass leaves are generally poor in nitrogen. The conversion of ingested food is more efficient when the highest concentration of proteinaceous compounds is present, and variations of nitrogen content have been shown to have effects on larval growth rates, adult size, fecundity and even diapause induction (Bink & Siepel 1996 on *Lasiommata megera*). The efficiency of the caterpillars to accumulate protein resources is of key relevance, since adult diet will usually consist of nectar (in which carbohydrates predominate, although not the only relevant components, e.g. Mevi-Schütz & Erhardt 2005b: see Chapter 2). Thus, larval diet is responsible for a large part of the adult reproductive potential (Boggs 1986, Wheeler 1996). However, recent research on *Lycaena tityrus* shows that the overall performance (that is, growth and survival throughout the whole life cycle) may not be correlated to leaf nitrogen content (Fischer & Fiedler 2000a).

Carbohydrates are another essential constituent of the larval diet, and some sugars may act as phagostimulatory compounds. Their distribution within the plant varies and may also vary during the day/night cycle due to photosynthetic activity (Bernays & Chapman 1994). There may be a connection with the timing of larval activity, a subject that has not been investigated. Variation of carbohydrate contents may have stronger effects in female than in male caterpillars due to the use of carbohydrates for fat storage (e.g. Soontiens & Bink 1997).

Other compounds include ascorbic acid (which is more abundant in young tissues), and linoleic or linolenic acids. Lipids can represent an important resource accumulated by larvae that have to overwinter (Pullin 1987a). The low amount of some ions, like sodium, within plant tissue may explain the mud-puddling behaviour displayed by many male butterflies (Pivnick & McNeil 1987b, Shreeve 1987). Although seedling grasses may present relevant concentrations of alkaloid substances (Bernays & Chapman 1976), the secondary compounds found in mature grass leaves seem to occur at low concentrations and show little interspecific diversity (Bernays & Barbehenn 1987). This may be a reason why butterflies that feed on grasses as larvae are usually oligophagous (Bink 1985); this situation might be different for at least some tropical members of these groups (Singer 1984). However, grasses overall are characterised by their physical roughness and high content of silica (Vincent 1982, McNaughton & Tarrants 1983) as well as by a generally low nutritional value (Bernays & Barbehenn 1987 and references therein).

Water is essential for all physiological processes, and the water content of the plant tissue eaten may significantly affect the efficiency of the feeding process (Slansky 1993). Water accounts for 50% to 90% of plant weight. As a rule, water and nitrogen contents tend to covary, both topographically and seasonally. In general, young tissues have higher amounts of water and nitrogen than old ones. Most caterpillars can compensate for water stress when water droplets are available (e.g. from dew or rainfall), as can easily be observed in laboratory rearings.

BUTTERFLY CATERPILLARS AS HERBIVORES

Larval feeding is responsible for accumulating the materials used for adult body-building, including a large proportion of the reserves to be allocated to egg production. With few exceptions (aphytophagy, including entomophagous habits: e.g. Fiedler 1991, Pierce 1995) both materials and energy reserves are derived from plants. Most people would not regard phytophagy as a remarkable habit, due to the high species richness of the phytophagous animal taxa. Feeding on plants has potential advantages, such as the high availability of biomass in terrestrial ecosystems. However, these are counter-balanced by numerous problems: predictability to parasitoids, fixation to the host, poor protein content and high proportion of indigestible cellulose, toxic allellochemicals and mechanical plant defences (trichomes, latex, waxes and lignified tissues). Thus at a first sight at least, phytophagy involves problems that need to be overcome (Slansky & Scriber 1985, Reavey 1993, Slansky 1993). Some research has also focused on the limiting capacity of plants for butterfly populations, highlighting the importance of large plant populations to support viable butterfly populations (Kéry *et al.* 2001).

Morphology

Phytophagy is believed to be an ancestral trait in the Lepidoptera (Powell *et al.* 1998); butterfly larvae have the general structure of the macrolepidopteran Ditrysia (e.g. Bourgogne 1979, Scoble 1992). A brief description of the most relevant devices related to sensation, mechanical processing of food and fixation to substrate is given below.

Sight depends on six pairs of larval stemmata associated to a low number of photo-sensory cells. Visual stimuli alone are not likely to be crucial for plant detection (references in Scoble 1992). Mechanoreceptors (which often are hair-like sensilla) allow the caterpillars to detect sounds and air-driven vibrations (Minnich 1955, Fletcher 1978). Chemical stimuli are perceived through sensorial pits located in the maxillae and labium (Fiori 1956, Grimes & Neunzig 1986a, 1986b). Specialisation of the mouth parts has received little attention, but good examples are found in, for example,

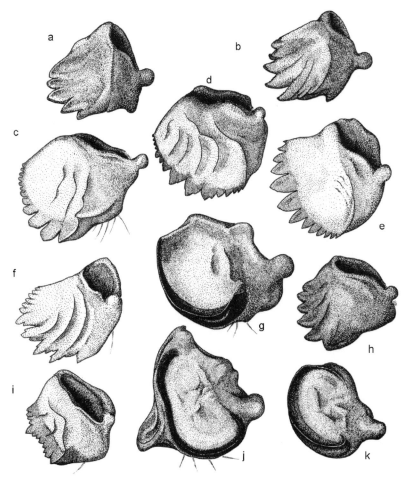

Figure 5.2. Right mandibles of butterfly larvae (oral view, final instar larvae except (f) and (i) = first instars; scale not constant). (a) *Iolana iolas*, (b) *Maculinea alcon* (Lycaenidae), (c) *Vanessa cardui* (Nymphalidae), (d) *Pieris brassicae* (Pieridae), (e) *Zerynthia rumina* (Papilionidae), (f) *Danaus chrysippus*, first instar, (g) *Charaxes jasius* (Nymphalidae), (h) *Muschampia proto* (Hesperiidae), (i) *Brintesia circe*, first instar and (j) last instar (Nymphalidae), (k) *Thymelicus sylvestris* (Hesperiidae). The host and host parts used are (a) seeds, (c), (d), (e), (f), (h) leaves of dicotyledon herbs (h also on inflorescences), (b) ant brood, (g) leaves of the small tree *Arbutus*, (i–k) grasses. Note similar morphologies in different families (e.g. (g)–(k), (a)–(h)), intrafamilial variation ((h)–(k)) and ontogenetic intraspecific changes ((i)–(j)). (Original unpublished drawings by Enrique García-Barros.)

mandible morphology (Fig. 5.2). Species adapted to feed on tough leaves such as those of Poaceae (De Vries *et al.* 1985, García-Barros 1989b) have specialised mandibles similar to those of moth larvae which feed on coarse leaves (Bernays & Janzen 1988, Godfrey *et al.* 1989). The combined action of silk (secreted by the silk glands and conducted by the spinneret), legs and prolegs is efficient for attachment to smooth substrates (these mechanisms are not universally used, e.g. the young larvae of some satyrines have poorly developed spinnerets and are rather inefficient upon smooth surfaces: García-Barros 1989b). Larval silk is essential to fix the pupae (Wojtusiak & Raczka 1983, Starnecker *et al.* 1998). Further modifications include the anal comb of some Pieridae (Stehr 1987), which propels the faeces away, thus avoiding deterioration of the plant and detection by predators.

Physiology

Plants contain secondary metabolites (allellochemicals) which are largely responsible for host dependence and identification. These include non-protein nitrogen compounds (glycosides, alkaloids, amines and non-protein amino acids), phenolic compounds (phenols, phenolic acids, flavonoids, tannins, quinones and coumarines), terpenoids (iridoids, cardiac glucosides, saponines, cardenolides and phytecdysteroids), other sulphur compounds (such as the crucifer glucosinates), organic acids and lipids. Some of these are toxic to the plant itself and are either confined to vacuoles or specialised reservoirs or achieve their active form only as a response to tissue damage (Singer 1984, Städler 1986, Bernays & Chapman 1994). Host acceptance often depends on a combination of chemicals, perhaps more heavily on the less volatile ones (Ma 1972, Städler 1984, Bernays & Chapman 1994). Host acceptance and subsequent feeding may depend on different chemical stimuli, and feeding behaviour can, to some extent, incorporate previous experience (Städler 1986, Karowe 1989).

Once in the larval gut, any noxious metabolites are deactivated by mixed function oxidase enzymes (MFO: Dowd *et al.* 1983, Brattsen 1988). The intensity of enzymatic activity is correlated to the diversity of secondary compounds ingested; thus polyphagous species are expected to make a comparatively higher investment in enzyme biosynthesis than are monophagous species. Hence polyphagy should have a physiological cost (Strong *et al.* 1984), while it might be rewarded by a more diverse income of oligoelements. The contrasting potential advantage of both extreme strategies probably explains why comparative research does not necessarily demonstrate different feeding efficiencies between caterpillars of the two types (cf. Scriber & Slanski 1981, Singer 1984).

Allelochemicals are eventually stored in some somatic tissues to fulfil special functions in the larval, pupal or adult stages, a process called 'sequestration' or 'pharmacophagy' (Boppré 1984, Bernays & Chapman 1994). Sequestered chemicals also include pigments (or their precursors) which will be incorporated into the adult wings (such as carotenoids: Wilson 1985, Wiesen *et al.* 1994, Geuder *et al.* 1997). Other sequestered chemicals participate in the chemical defence of the immatures and the adults, or are incorporated to pheromones (Boppré 1990, Bowers 1993, Dussourd 1993).

Further specialisation in relation to digestive processes may involve mid-gut pH (Berembaum 1980) and associated symbiotic microorganisms (Geuder *et al.* 1997). Finally, both food assimilation and moulting are markedly temperature-dependent (Heinrich 1993, Atkinson 1994), and in fact the optimum temperature range for assimilation is narrower than that required for motor activity. This explains the strong relationship between larval growth rates and ambient temperature (Stamp & Bowers 1990, Casey 1993, Gotthard 1998). Although caterpillars are basically ectothermic, they may control their body temperatures within limits, depending on microhabitat features, larval colour and basking behaviour (Casey 1993).

Behaviour: aggregation, time schedules and dispersal

Behaviour is a key element to circumvent host mechanical defences, and to interact with the physical and biotic environment (temperature, water balance, competitors, predators and parasitoids). Butterfly caterpillars can manipulate the hosts to counteract their defences, as reported for *Danaus* larvae which cut out the leaf veins of the host to reduce the flux of toxic latex before feeding (Dussourd 1993). Ravenscroft (1994c) has described a comparable behaviour for the skipper *Carterocephalus palaemon*: the larvae notch the blades of their grass foodplant *Molinia* at about halfway along their length, then feed on their tips. *Carterocephalus palaemon* larvae do not deactivate any flux of defensive latex (which does not occur in *Molinia*); instead, their behaviour seems to contribute to maintain nutrient concentrations in the part of the blade being eaten (Pullin 1987b). Locating, ingesting and processing food to grow involves a balance in terms of energy invested and gained per unit size, and of exposure to predators and parasitoids (Heinrich & Collins 1983, Reynolds 1990). The species-specific formula adopted to achieve pupation size within a limited time while minimising risks represents a foraging strategy (Hassel & Southwood 1978, Alcock 1989, Ide 2004). Larval strategies are marked by different behavioural traits, three of which deserve special attention, namely aggregation, timing of daily activity and dispersal ability.

Some solitary larvae may be indifferent towards the presence of conspecific larvae (e.g. SBN 1987), while others react aggressively against their presence (e.g. *Iphiclides*, *Papilio*: Weyh & Maschwitz 1982). The larvae of *Iphiclides podalirius* use different plant leaves for resting and feeding and depend on their silk threads to find their way on the plants. Weyh & Maschwitz (1982) have shown that silk trails of this species are embedded with volatile substances that are 'smelt' by the larvae; these distinguish an individual's own trails from those from conspecific larvae, which are avoided or attacked. In species with social or subsocial larvae, these maintain group size using tactile or olfactive stimuli and silk trails (Fitzgerald 1993). Communal feeding is advantageous in terms of thermoregulation, as gregarious basking helps to avoid convective heat loss and desiccation (Porter 1982, Casey 1993, Bryant *et al.* 2000). Communal feeding may also facilitate feeding for mechanical reasons (hard tissues are more easily harmed, and large silk coatings for easier support are produced with small individual effort: Stamp 1980, Young 1983) or because it enhances host plant suitability (Fordyce 2003).

Feeding activity follows species- and instar-specific rhythms of activity with a circadian basis. Wojtusiak (1979) demonstrated that feeding is often episodic, and is separated in time from basking and resting activities (when digestion occurs). Be it to avoid dangers or because of the low digestibility of the food, some caterpillars devote a small proportion of their time to actually ingesting food (ca. 15% in *Pieris rapae* larvae: Mauricio & Bowers 1990, Slansky 1993). This is surprising because, as discussed above, larval growth depends on the amount of food consumed, and the caterpillars must process an impressive total mass of plant material relative to their own body weight. The larvae of most European species are diurnal, but there are well-known exceptions such as the larvae of satyrine nymphalids, which are diurnal during their first instars but shift to nocturnal feeding in the final ones (Powell in Oberthür 1914, Jutzeler 1992). Further nocturnal larvae include those of *Argynnis*, *Thymelicus*, *Hamearis*, *Fixsenia* and *Lycaena* (e.g. SBN 1987).

Based on the spatial patterns of resource use, Fitzgerald (1993) proposed three main groups: nomadic foragers, patch-restricted foragers and central-place foragers. The ability to disperse is correlated to larval size within and between species (Weiss *et al.* 1987, Reavey 1993), which helps to better understand caterpillar movements in terms of the ratio between larval size and foodplant size. Having to seek out new plants should be more frequent when hosts are small and larval densities high (Jones 1977b, Le Masurier 1994), or when the hosts are large but the species' larvae are gregarious (e.g. *Euphydryas aurinia*: Templado 1975). In cases of dispersal forced by host depletion, the speed and directionality of the starved larvae can often be distinguished from those of satiated ones (Jones 1977b, Tsubaki & Kitching 1986).

PLANTS AND PHENOLOGY

Seasonality has marked effects on plant phenology. Butterfly larvae require not just one plant species, but a specific plant organ of (ideally) optimum quality. Leaf water and nitrogen contents, phenolic compounds, leaf toughness and other plant properties vary with the vegetative season (Slansky & Scriber 1985, Bernays & Chapman 1994, Brooks & Feeny 2004). Strong *et al.* (1984) also support the idea that by specialising on a type of tissue, the larvae specialise upon a period in the phenology of the plant. Seasonal variation in host plant quality has dramatic effects on larval survival and feeding performance (Stamp & Bowers 1990) and not surprisingly, this variation can severely restrict the length of the feeding season and even preclude the use of potential hosts (Williams & Bowers 1987, Jordano & Gomáriz 1994). The temporal variation

of foodplant quantity and quality may contribute to making a short larval development period advantageous, hence the widespread belief that butterflies should benefit from minimising the time spent as larvae (Dempster 1984; but see Wickman *et al.* 1990).

Insect phenology must ensure the temporal match of larvae and larval resource, which demands mechanisms that enable the insect to endure predictable adverse conditions (Fordyce & Shapiro 2003). The fact that related species often have similar phenologies suggests that historical factors must have strongly determined present patterns. Further, larval phenology is canalised by oviposition dates, hence by adult phenology. Since the thermal range for adult flight (e.g. Heinrich 1993) is narrower than the one required for larval activity, life cycle phenology is highly constrained by adult requirements. Insect seasonality is a complex field of research (Wolda 1988) and butterflies provide excellent experimental material (Lees & Archer 1981, Wickman *et al.* 1990, Nylin 1992, Stamp 1993). A detailed discussion of all the aspects of phenology that relate to butterfly herbivory is far beyond the scope of this chapter, and only broad patterns relative to butterfly–host plant synchronisation by the European taxa are briefly outlined here (see Slansky 1974, Tauber *et al.* 1986, Dennis 1993, Danks 1994).

About three-quarters of the European species overwinter as larvae (Fig. 5.3), and the same proportion is univoltine (their life cycle is completed in one year). The proportions of single-brooded species and of those with hibernating larvae are positively correlated to elevation and negatively to latitude (SBN 1987, Dennis 1993). Some sub-Arctic and Alpine butterflies take more than one year to complete their life cycle (biennial or semivoltine species, including several *Colias*, *Oeneis* and *Erebia* species). Most species have a constant number of broods per year across their range (in turn, most of these are single-brooded). This, together with the low number of multivoltine (roughly 'aseasonal') species, suggests a historical (biogeographical, taxonomic) faunal bias. Besides, host plant defences may have an effect on voltinism, as suggested by Cizek *et al.* (2006). Species that shift in voltinism along latitudinal or altitudinal clines (e.g. Nylin *et al.* 1993) are comparatively scarce. Winter is the adverse season throughout Europe and the length of the growing season increases southward. However, this trend is disrupted in the south by the dry summer of the Mediterranean region (Fig. 5.4), where summer water stress imposes an additional adverse season for plant growth. Thus butterfly

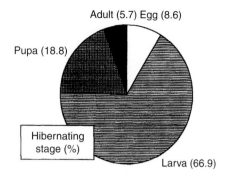

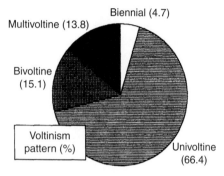

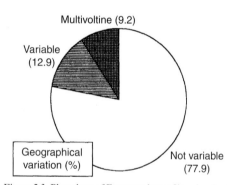

Figure 5.3. Phenology of European butterflies, showing the degree to which the European fauna is dominated by annual species that hibernate as larvae and have fixed voltinism patterns. The data are for 409 species, taken from several sources, expressed as percentages. Species with three annual broods were classified as multivoltines. The lower diagram shows the proportion of species in which the number of annual broods is subject to geographical (latitudinal or altitudinal) variation.

populations at intermediate latitudes and in south European mountains may have to face two adverse periods. There is an interesting contrast between the two adverse periods because, unlike winter, summer drought conditions are adverse for larval growth while they may remain suitable for adult life (see e.g. Masaki 1980). Mechanisms of summer arrested development or migration would be expected to feature in southern populations of the most widespread species. Whatever the adverse season is, it can be overcome by adult migration, or by dormancy in the adult, egg, larval or pupal stages.

Adult hibernation is restricted to several Nymphalinae, the brimstones (*Gonepteryx* spp.) and *Libythea*. Arrested ovarian development during summer has been reported for several satyrine nymphalids (Scali 1971, García-Barros 1994, 2000a) and for *Argynnis pandora* (García-Barros 2000b). Egg overwintering is found in lycaenids which oviposit on woody plants and in a few species of the other families. Some of these species hibernate as pharate first-instar larvae within the egg shell (e.g. *Brenthis hecate*, *Polyommatus thersites*, *Satyrus* spp.: Jutzeler 1994, Jutzeler & De Bro 1996). The only known example of summer diapause in the egg stage is the late spring satyrine *Pyronia bathseba*, where embryonic development is arrested until the end of the summer. Larval diapause is a flexible solution for hibernation: it enables the insect to gain some weight in autumn, to seek for safe overwintering sites by behavioural means and to use the early spring foliage as soon as it appears. In addition to larval overwintering, larval aestivation is known to occur in three main modes, namely:

(1) First-instar larvae that enter diapause prior to feeding (e.g. *Melanargia* spp.: Tilley 1983, Jutzeler *et al.* 1995, 1996).
(2) Intermediate larval instars that undergo summer diapause (such as *Plebejus hesperica*: Munguira 1988).
(3) Aestivo-hibernating larvae with quiescence at, or near, the final larval size (*Pararge aegeria*, *Aphantopus hyperantus*, *Hipparchia semele*, *Erynnis tages*: Wiklund *et al.* 1983, Tolman & Lewington 1997).

Finally, examples of pupal diapause are scattered across each butterfly family. Pupal-diapausing butterflies often feed as larvae on the flowering structures and young parts of herbs, or on leaves that are not available in winter (e.g. *Iphiclides*). In e.g. *Anthocharis*, *Zegris*, *Euchloe* and *Zerynthia*, pupal diapause is a means to survive both the dry summer and the subsequent cold winter. In these instances low temperatures are often required to resume development, and pupal diapause may be induced by long photophases (Caballero 1996; D. Jordano unpublished observations). Diapausing pupae have a thicker cuticle, and a different colour from directly developing ones (Templado 1981).

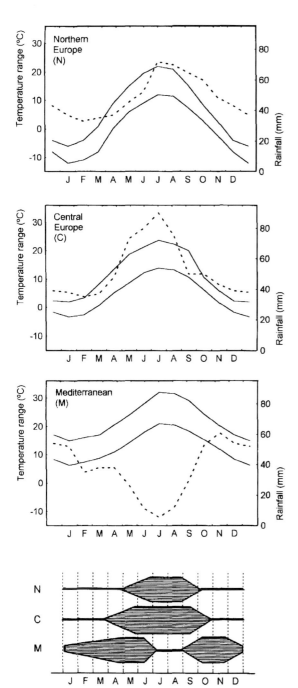

Figure 5.4. Ombrothermic diagrams representing the North, Central and South European (Mediterranean) climates, showing how the time window for larval growth is correlated to latitude in a probably non-linear manner. Solid lines, range of mean monthly temperatures; dotted line, average monthly rainfall. Lower graph: idealised time-frames for larval development, i.e. mean temperature above 10 °C and no water stress (drought period = monthly rainfall is lower than twice the mean temperature). Ombrothermic diagrams are based on a synthesis from several sources.

BUTTERFLIES AS PESTS

Overall, the European butterfly species cannot be considered as economic or health threats. There are few species that can in some cases damage agricultural crops, although generally to a limited extent. The species of interest as true or potential pests in Europe are the following: *Papilio machaon* and *Iphiclides podalirius* (Papilionidae); *Pieris brassicae*, *P. napi*, *P. rapae*, *Aporia crataegi* and *Colias crocea* (Pieridae); *Aglais urticae*, *Vanessa cardui*, *Inachis io*, *Polygonia c-album*, *Nymphalis antiopa* and *N. polychloros* (Nymphalidae); *Laeosopis roboris*, *Lampides boeticus* and *Cacyreus marshalii* (Lycaenidae); *Thymelicus lineola* and *Pelopidas thrax* (Hesperiidae). A number of the European species have tree-feeding larvae, but the level of damage they cause is so low that none of these forest butterflies can possibly be considered a pest. Among the insect species regarded as pests in European forests by Day & Leather (1997) there are no butterfly species. There are no butterfly species that feed during their life cycle on stored products nor are there butterfly species that directly or indirectly act as vectors or reservoirs of any disease affecting the human species or domestic animals.

It also seems clear that with the passing of time butterfly species have lost relevance as pest species in applied entomology handbooks, particularly during the second half of the twentieth century. The number of species considered and their inclusion has clearly diminished in this period. This is probably due to the fact that butterflies are conspicuous insects and their presence in crops can easily be overestimated. When the damage caused by butterflies has been precisely quantified, their importance was low in comparison with other Lepidoptera or other animals that are nocturnal, small or not colourful and therefore not so conspicuous. Changes in agricultural practices in Europe can also have important effects. During the twentieth century most European land has changed from extensive agriculture dominated by small plots and family-based yards to more intensive agriculture and farming. Crops are now intensively cultivated in large plots in which the use of chemical fertilisers and pesticides are now commonplace.

These changes in agricultural uses, which have been important for many butterfly species, have also caused a reduction of the potential of some species to act as pests. Species like *Piereis brassicae, P. napi* and *P. rapae* that feed on several crucifer species, or *Lampides boeticus*, which feeds on peas, have lost their damaging potential because the crops that supported them have largely disappeared or been confined to greenhouses.

Another factor responsible for pest status of several animal groups, such as the invasions caused by tourism or international commerce, has had a very limited impact on European butterflies. Only *Cacyreus marshalii* has evolved into an important pest of cultivated *Pelargonium* species. Other species that have established themselves (*Danaus plexippus*) or have been recorded occasionally (*Charaxes jasius*) in Europe during the second half of the twentieth century have not caused problems to crops or to the environment so far. However, the combination of factors like environmental change, climate change and the introduction of species can potentially be at the origin of pest status of previously non-damaging species. For example, the spread of the cultivation of *Vaccinium × corymbosum* in Spain combined with other changes might change the status of *Charaxes jasius* (which feeds on this plant: Molina 2000) into a pest of economic interest.

The species that can nowadays be considered pests in Europe are *Pieris brassicae* and the related *P. napi* and *P. rapae, Lampides boeticus* and *Cacyreus marshalii*. All other species could be considered potential pests at most, and they have, in some cases, been the cause of concern producing damage in localized areas.

Pieris brassicae

This is one of the most popular and well-known species in Europe (Feltwell 1981). It is one of the typical pest butterflies and has been referenced in most of the classical handbooks on applied entomology (Balachowsky 1962, Carter 1984, De Liñán 1998). Due to the fact that eggs are laid in clusters and that caterpillars are gregarious, the damage to cabbages and other cruciferous plants can easily be detected. However, its populations have traditionally been controlled by several parasitoid species. Nowadays the species is not so evident because it shares its habitat with two other Lepidoptera that also attack cabbages and other crucifers which are subject to strict control that keeps *P. brassicae* populations in low numbers. The co-generic *P. napi* and *P. rapae* are potential rather than real pests and have never caused serious problems, seemingly because they prefer wild crucifers.

Lampides boeticus

The females of this species lay their eggs singly on the flowers and developing seed pods of several Fabaceae such as trefoils, alfalfa and bladder senna (Martín 1984). It has been recorded several times as a pest of peas (*Lathyrus* spp.). Larvae bore into the developing legume and feed inside it on the seeds. Due to its endophagous character it is difficult to detect and control. In Europe, its incidence on peas seems to be limited because most of the production comes from greenhouses where the species has limited access. In alfalfa crops the larvae feed mainly on flowers, and because the crop is usually cut after flowering, the larvae cannot complete their development, resulting in a low incidence of the pest. It can only be regarded as a potential pest in alfalfa stands devoted to seed production.

Cacyreus marshalli

This can be considered the worst current butterfly pest in Europe. This butterfly originates from South Africa where it lives on several species of the Geraniaceae, some of which have been grown as ornamentals in Europe for a long time. Since its first record in the island of Majorca in 1990, it established itself in the Iberian Peninsula in 1993 and has been in a constant state of expansion ever since (Sarto 1998). It is now present in the Balearic Islands and most of Spain and Portugal, and has also reached France, Italy and other European countries, with its spread along the whole Mediterranean basin being probably over. It has caused serious damage in one of the most widespread ornamental plants belonging to the genus *Pelargonium*.

CONCLUSION

European butterflies are specialised herbivores that select just a third of the plant families in the area and are predominantly herb-feeders. Several strategies have been developed by the larvae to avoid noxious plant chemicals, while specialisations of morphology, physiology and behaviour allow them efficiently to exploit plant resources. These varied strategies have enhanced the appearance of a variety of herbivore niches that are related to butterfly diversity (Janz *et al.* 2006). From the phenological point of view larval overwintering is the preferent strategy among European butterflies. Probably as a result of changes in agricultural practices in Europe, very few butterfly species can be regarded as pests.

6 • Thermoregulation and habitat use in butterflies

PER-OLOF WICKMAN

SUMMARY

Butterflies strongly depend on weather and climate for virtually all aspects of their ecology. Rigorous flight is in many species restricted to body temperatures in the range 33–38 °C. Moreover, climate can also be viewed as a resource in its own right for a butterfly, since it is not evenly distributed in space. Most butterflies can be thought of as ectothermic, which means that the main heat source is external and not metabolic. Body temperature excess is mainly determined by air temperature, wind speed and solar irradiation. Butterflies can regulate their body temperature behaviourally by basking and also by microhabitat selection. There are four basic basking postures, viz. dorsal, lateral, body and reflectance basking. By changing body orientation relative to the Sun and by altering wing angle a butterfly can influence the area exposed to the Sun and thereby the amount of irradiation received. Body size, wing and body coloration and pubescence all influence thermal properties. But wing characteristics affecting thermoregulation depend on the basking mode. Under particular conditions, some species may also use endothermic heating of their thorax by wing-muscle shivering to reach body temperatures that allow active flight. Microclimate and local climate are known to affect habitat use by butterflies. Key factors – ambient temperature, wind speed and irradiation – vary in the habitat depending on substrate, insolation angle, orientation relative to higher objects, wind direction and height above the ground. In an open landscape, topography will affect local climate, and hence a butterfly's habitat use. As a result of more than 50 years of research on butterfly thermoregulation, we now have a clear picture of the various factors that influence butterfly activity and resource use. However, the details of how the distribution of microclimate in space interacts with a butterfly's possibility of optimal thermoregulation are still largely unresolved.

INTRODUCTION

Butterflies are highly dependent on weather and climatic conditions, with significant consequences for resource use and for their ecology. The term 'resource' is used by ecologists in a number of ways, but typically it designates something that can be reduced in quantity (Begon et al. 1990). Appropriate body temperature for flight is of course a basic condition for a butterfly to use resources (Willmer 1983, 1991, Boggs 1986). In addition, climate can be viewed as a resource also in its own right for a butterfly, since it is not evenly distributed in space. Local climate affects the body temperature of butterflies and thereby their metabolism and physiological functions such as flight activity and oviposition rate, both of which are fundamental components of an individual's fitness (Magnuson et al. 1979, Tracy & Christian 1986). It is thus not surprising that climate is closely connected to habitat use in butterflies. This chapter deals with the different ways in which butterflies thermoregulate, how climate varies in space, and how climate and weather determine butterfly activity. Previous reviews of this area are Heinrich (1981, 1993, 1996), Kingsolver (1985a), Shreeve (1992b) and Dennis (1993).

BUTTERFLY THERMOREGULATION

The body temperature (T_b) of a butterfly is the result of the balance between heat gain and heat loss (Parry 1951, Digby 1955, Porter & Gates 1969, May 1979) . Depending on this balance, the butterfly will attain a temperature excess (T_e) that is the difference between T_b and the ambient temperature (T_a). Most butterflies can be thought of as ectothermic, which means that the main heat source is external and not metabolic (endothermy). For a diurnal butterfly this implies that its body temperature increases with T_a, and that T_e is determined by irradiation, convection and conduction.

The amount of irradiation is a function of the height of the Sun above the horizon, cloud cover and the butterfly's location relative to shadowing objects. Convectional heat loss increases with wind speed and lowers T_e. At a given moment wind speed varies in space depending on the butterfly's position relative to wind obstacles and wind direction (Geiger 1965, Stoutjesdijk & Barkman 1992). Conductive heat gain or loss depends not only on air

Ecology of Butterflies in Europe, eds. J. Settele, T. Shreeve, M. Konvička and H. Van Dyck. Published by Cambridge University Press. © Cambridge University Press 2009, pp. 55-61.

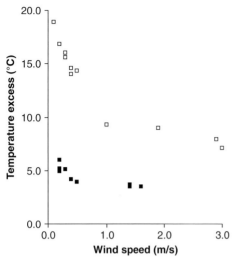

Figure 6.1. Values of T_e of mounted (see Wickman 1988) *Lasiommata megera* depending on wind speed and wing position. Open squares denote an individual basking to maximize heat gain with wings held flat and perpendicular to irradiation (14.32–14.45 h, 677 W/m², T_a = 21.0 °C). Filled squares denote an individual with wings folded in the heat avoidance posture (13.35–13.45 h, 680–712 W/m², T_a = 19.4 °C). Data collected on 12 August 1986 (data from Wickman 1988).

temperature but also on resting substrate temperature and its thermal properties (size, conductivity, heat capacity, etc.). However, except when the butterfly is pressed to the substrate, the most important factors determining T_e are air temperature, wind speed and solar irradiation (Kingsolver 1985a) (Fig. 6.1).

Basking behaviour

Although most butterflies are basically ectothermic, their T_b can be regulated behaviourally by basking. By changing body orientation relative to the Sun and by altering wing angle a butterfly can influence the area exposed to the Sun and thereby the amount of irradiation received. Clench (1966) suggested a classification of different stereotypic ways in which butterflies elevate T_b, which was later extended by Kingsolver (1985a), who identified four basic basking postures, viz. dorsal, lateral, body and reflectance basking. Although most butterfly species only adopt one mode, there appear to be species that use combinations, e.g. *Agriades glandon* in which most butterflies bask dorsally during midday, while 30–60% bask laterally in the morning and evening (Heinrich 1986a). As will be evident there is also

variation within modes. Irrespective of basking mode, all butterflies use the same heat-avoidance posture, with closed wings and with their longitudinal body axis parallel to sunlight. For a review of evolutionary constraints on basking posture see Dennis & Shreeve (1989) and Shreeve & Dennis (1992).

In dorsal basking, wings are opened 180°, and the dorsal side of the body and wings are exposed to the Sun. Typically the butterfly is facing away from the Sun (e.g. Wickman 1988) although this may depend on perch location (Rutowski *et al.* 1991). Regulation is achieved by opening or closing the wings. A variation on this theme occurs in hesperids, where the hind wings are fully opened, but the fore wings are only half open. Moreover, when wings are held open, air heated by the wings and the body can accumulate below the wings and reduce convective heat loss from the body (Vielmetter 1958, Rawlins 1980). Also, the body and wings may be pressed against the substrate, further reducing convective heat loss. Pressing wing margins to the substrate during low T_a has been described for *Hypolimnas bolina* (Kemp & Krockenberger 2002) but is also found in many European dorsal baskers (Plate 5b). In rocky habitats the substrate may contribute conductive heating, which is known for *Parnassius* species (Kingsolver 1985a). Dark substrates may reach temperatures well over 40 °C above ambient temperature (Stoutjesdijk & Barkman 1992), and high substrate temperatures may restrict butterfly activity (Wickman 1988).

In some dorsal baskers the position of the abdomen can be adjusted, so that it is exposed to the Sun above the open wings at low temperatures and concealed below them at high temperatures (Rawlins 1980). In *Argynnis paphia* the abdomen is exposed below the closed wings at high temperatures, apparently to increase convective heat loss. At the same time the butterfly extends its legs (stilting), raising the body, which lowers conductive heating (Vielmetter 1958). Thermoregulation in dorsally basking butterflies has been described for several other European species (Kevan & Shorthouse 1970, Wasserthal 1975, Shreeve 1984, Polcyn & Chappell 1986, Pivnick & McNeil 1987a).

Body basking may be seen as a special case of dorsal basking. The body position is similar, but the wings are only slightly opened (less than 10° between them). In this way the wings intercept very little irradiation and contribute almost nothing to T_e. Instead the body directly absorbs solar radiation (Kingsolver 1985a). According to Kingsolver (1985a) it occurs in lycaenids, satyrines and hesperids, but no examples of species are given.

In reflection baskers, the dorsal side of the body is oriented towards the Sun. This mode of basking is adopted by *Pieris* (Plate 6a) and *Pontia* species which have extensive white and highly reflective areas on their wings (Kingsolver 1985b, 1985c, 1987a, 1988). Wings are typically held at an angle between 10° and 60° reflecting solar radiation towards the body and the melanized basal parts of the wings. Shreeve (1992b) suggested that reflection basking also occurs in the European taxa *Anthocharis*, *Plebejus*, *Polyommatus*, *Meleageria* and *Melitaea cinxia*, although conclusive studies are needed. It may also contribute to T_e in *Lycaena virgaureae*, which rarely opens its wings more than 90° (Douwes 1976a).

In lateral basking, wings are not opened, and the lateral side of the body and ventral side of the wings are oriented perpendicularly to the Sun. Heat is absorbed by the exposed parts of the body but also by the basal part of the wings and conducted to the thorax (Watt 1968, Kingsolver 1985a). *Colias* butterflies rotate their longitudinal body axis relative to the Sun from perpendicular to random to parallel as their T_b increases (Kingsolver & Moffat 1983). In *Hipparchia semele* regulation in addition is achieved by tilting the wing plane relative to solar irradiation. In cool weather they lay their folded wings flat against the ground (Plate 6b). At high temperatures stilting occurs, apparently to avoid body contact with the hot ground (Findlay *et al.* 1983, Shreeve 1990a, Dreisig 1995). Lateral basking occurs in a few European pierid, eumaeine and satyrine genera (*Gonepteryx*, *Colias*, *Leptidea*, *Tomares*, *Callophrys*, *Satyrium*, *Neolycaena*, *Kirinia*, *Coenonympha*, *Hyponephele*, *Erebia embla*, *E. disa*, *Hipparchia*, *Brintesia*, *Chazara*, *Pseudochazara*, *Oeneis*) (Shreeve 1992b).

It might seem likely that butterflies do not only adjust their bodies relative to irradiation but also relative to wind direction. However, wind direction relative to body orientation has no effect on T_e in the laterally basking *Colias* (Kingsolver & Moffat 1983). However, in the dorsal basking *Vanessa cardui*, wind direction has an effect on orientation, especially when wings are held open; heat loss is lowest when the tip of the abdomen is oriented into the wind (Polcyn & Chappell 1986). The significance of these observations remains to be shown, as there is little evidence that any other butterfly species adjusts its body orientation in relation to the wind. This may be because turbulence close to the ground causes unstable wind directions, making it difficult for the butterfly to find an optimal orientation relative to the wind.

Influence of morphology and wing coloration

The T_e of butterflies is influenced by morphology. In particular body size, wing and body coloration and pubescence all influence T_e. Theoretically, a larger body size will result in slower heating and cooling and a higher equilibrium T_b, everything else being equal (Church 1960, Willmer & Unwin 1981, Willmer 1982, Stevenson 1985a). This has been confirmed in experiments with dried *Pararge aegeria* (Berwaerts *et al.* 2001) and dried *Hypolimnas bolina* (Kemp & Krockenberger 2004), although the effect of differences of body size on heating rate diminished with T_a in *P. aegeria* (Berwaerts *et al.* 2001). This suggests that small butterflies can adjust their body temperature more rapidly (Kemp & Krockenberger 2004), although the significance of these relationships for butterflies under field conditions needs to be examined further. For instance Heinrich (1986a) found, when comparing flight temperatures of 13 species of butterflies, that smaller size is generally associated with lower flight T_b. However, smaller butterflies also have lower wing loadings, which are known to be associated with lower T_b. In *Thymelicus lineola* males are smaller than females, and males heat and cool faster than females (Pivnick & McNeil 1986). Despite their smaller size males reach a higher equilibrium temperature, but this may be due to other physiological and morphological factors affecting T_b. In *Euphydryas phaeton* females are twice as heavy as males, but nevertheless the sexes show similar activity because they use different microhabitats (Gilchrist 1990).

Wasserthal (1975) showed experimentally that shadowing of the wings may reduce the thoracic temperature (T_{th}) of *Papilio machaon* by about 30%. But the whole wing does not contribute to elevate thoracic T_e. Heat from irradiation is mainly conducted from the innermost 15% of the wing area to the thorax, because the wings are poor conductors (Kingsolver & Koehl 1985), and because haemolymph flow between the wings and the body is slow (Kammer & Bracchi 1973, Wasserthal 1983).

Wing characteristics affecting thermoregulation depend on basking mode (Schmitz 1994). In lateral baskers like *Colias* the degree of melanization on the basal part of the ventral side of the hind wing affects thoracic temperature. Highly melanized individuals may increase T_e by as much as 15–80% compared to lighter individuals (Watt 1968) and darker individuals show higher flight activity (Roland 1982, 2005). In an experiment with *Colias*, Ellers & Boggs (2004a) showed that artificially darkening the basal part of the ventral side of hind wings resulted in higher flight activity in males and higher egg maturation rate in females. In the dorsal basker *Parnassius phoebus* individuals with more black on the basal parts of the wings reach higher T_{th} and can spend more time in flight (Guppy 1986). In reflectance baskers there is a correlation between the extent of white on the middle part of the dorsal side of the wing and the angle at which wings are held during basking in different sexes and species (Kingsolver 1985b). In the dorsal basking *Pararge aegeria* individuals with darker wings spend less time in sunshine and more time in flight and are more mobile than paler individuals (Van Dyck *et al.* 1997b; cf. Shreeve & Smith 1992). Darker individuals heat faster than paler ones, but there is no difference in take-off T_{th} (Van Dyck & Matthysen 1998). Similar relationships have been found in lycaenids (Biro *et al.* 2003) and a hesperid (Nakasuji & Nakano 1990). Also, iridescent wing colours can contribute significantly to heating depending on the thickness of the film layer responsible for this trait (Mialoulis & Heilman 1998). Pubescence has also been demonstrated to lower convective heat loss in wind tunnel experiments with laterally basking *Colias* (Kingsolver & Moffat 1983, Kingsolver 1985a, Jacobs & Watt 1994).

Endothermic heating

Although most butterflies are mainly ectothermic, some species are endothermic. In endothermy heat gain is produced from the metabolic activity of the flight muscles, which during pre-flight warm-up can be activated, which is known as wing-shivering (Vielmetter 1958, Kammer 1970, Rawlins 1980, Findlay *et al.* 1983). Shivering can produce a T_e of 18 °C in *Vanessa atalanta* (Krogh & Zeuthen 1941) and in *Inachis io* (Maier & Shreeve 1996). Shivering in the crepuscular Brassolinae butterflies *Opsiphanes* and *Caligo* can result in thoracic T_e of 15 °C and 10 °C, respectively. This behaviour helps males on mating territories to attain a thoracic temperature optimal for flight (Srygley 1994).

There is no evidence of endothermic heating of the abdomen. The capacity to regulate haemolymph circulation between the thorax and abdomen is known from *Papilio machaon* (Wasserthal 1980). By ligating the abdomen Kingsolver (1985b) showed that in *Pieris* the abdomen is used to intercept irradiation and the heat is transferred via the haemolymph to the thorax. In *Papilio polyxenes* abdominal pumping occurs at high temperatures apparently cooling the thorax (Rawlins 1980).

CLIMATE VARIATION IN SPACE

Because climate varies in space, the location of a butterfly affects its T_b. Microclimate and local climate (*sensu* Yoshino 1975) on a scale of about 10^{-2} to 10^2 m is known to affect the habitat use of butterflies. Butterflies' T_a, wind speed and irradiation vary in the habitat depending on substrate, insolation angle, orientation relative to higher objects, wind direction and height above the ground (Fig. 6.2).

Substrate choice has been studied in *Hipparchia semele*. During the day butterflies of this species shift from resting on dark lichens to sand, when sand becomes the warmest substrate (Shreeve 1990a). In hot weather sand is also avoided (Dreisig 1995). *Lasiommata megera*, a species that

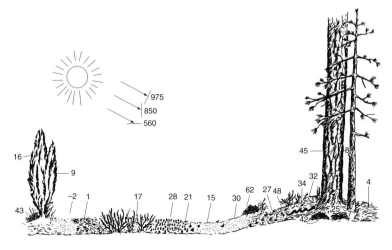

Figure 6.2. Surface temperatures (°C) on a heathland by a woodland edge on a sunny day in March at midday in the Netherlands. The intensity of solar radiation (W/m²) is indicated in three planes (adapted from Stoutjesdijk & Barkman 1992; courtesy Stoutjesdijk).

settles on rocks, uses shade in crevices as T_a increases (Wickman 1988) and in *Oeneis chryxus*, perching on rocks occurs more often on cool days, while perching on plants is more common on warm days (Daily *et al.* 1991). In spring *Inachis io* uses mini-landforms such as molehills in ways that raise T_b (Dennis & Sparks 2005).

Generally T_a is higher and wind speed is lower near ground. For example, Gilchrist (1990) showed that T_a was almost 5 °C lower at 35 cm as compared to 1 cm above ground, explaining why *Euphydryas phaeton* moves up in the vegetation as T_a increases. However, in dense vegetation, temperature may be lower closer to ground due to shading. *Hipparchia semele*, for example, seeks shade by creeping among grass stems (Findlay *et al.* 1983). The height at which *Thymelicus lineola* males fly in the field layer is negatively correlated to solar angle, possibly to maintain direct exposure to the Sun in flight (Pivnick & McNeil 1987a). Males of the dorsal basker *Asterocampa leila* can regulate T_b within a narrow interval as T_a increases, not only by closing their wings, but also by shifting from perching on the ground to tree foliage, and from the Sun to the shade (Rutowski *et al.* 1994, Rutowski 2000a).

On a somewhat larger scale, butterflies are affected by how higher vegetation like trees and bushes affects insolation. *Ochlodes venata* avoids shaded sides of trees and bushes (Dennis & Williams 1987) and it was experimentally shown that *Thymelicus lineola* avoids shade and prefers sunny places (Pivnick & McNeil 1987a). In *Lasiommata megera* activity is mostly confined to the warm, sunny, and wind-sheltered sites near vegetation edges (Dennis & Bramley 1985), while in *Pararge aegeria* an increasing number of butterflies use the shade of forest as opposed to sunny clearings with increasing temperature (Davies 1978, Wickman & Wiklund 1983, Shreeve 1984, Van Dyck & Matthysen 1998). In extreme high temperatures *Melanargia galathea* restricts activity to shade but at low temperatures it is active in sunlit areas (see Chapter 13).

In an open landscape, topography will affect local climate in a similar way as higher vegetation. Typically wind speed is higher on summits as compared to depressions. High arctic butterflies use gullies and creeks that are wind protected and have higher substrate temperatures compared to unsheltered areas (Kevan & Shorthouse 1970). *Lasiommata megera* preferentially uses sunny, lee sides of hills. Moreover, males use higher altitudes when wind speed decreases and T_a increases (Wickman 1988; see also Chapter 3).

Climate variation in space can be viewed as a resource that constrains butterfly habitat use. A butterfly regulates its body temperature not only by basking posture, but also by choosing a spot that permits optimal thermoregulation. Stevenson (1985b), who modelled the importance of the different mechanisms ectotherms use to regulate T_b, concluded that microhabitat selection is more important than postural adjustment for controlling T_b. Yet ectotherms like butterflies do adjust their postures, which suggests that butterflies often cannot choose microhabitat freely to optimize T_b. Further research is needed to understand how the use of microclimate as a resource in space interacts with the avoidance of predation and using other resources distributed in space such as oviposition sites, nectar plants and mates, and on how thermoregulatory behaviour in general affects performance, fitness and the population dynamics of a species (see also Chapter 13).

TEMPERATURE AND BUTTERFLY ACTIVITY

The range of T_b that permits survival is known as the tolerance zone. Within this zone metabolism increases with temperature, which generally means increased performance with temperature. However, there is an optimal T_b, which permits maximal performance in terms of fitness. This optimum is often located in the upper end of the tolerance zone (Huey & Kingsolver 1989). At high temperatures butterflies may overheat and this can result in decreased survivorship and fecundity (Rawlins 1980, Kingsolver & Watt 1983). Although the tolerance and performance characteristics of a species can be changed by acclimation (Hoffmann 1985, 1996), and although it takes time for irradiation to elevate T_b (Hirota & Obara 2000), T_b can change rapidly within a minute. This means that a butterfly may risk overheating when the wind stops or may soon be inactivated when a cloud covers the Sun (Kingsolver & Watt 1983). In terms of adult resource use, it is of special interest to see how T_b influences flight activity, food and water consumption, and oviposition rate.

Because flight muscles are in the thorax, T_{th} is relevant when studying the temperature needed for flight. Heat is also produced by the flight muscles during flight, but the contribution compared to sunlight is small in most butterflies. Medium-sized or large butterflies such as *Colias* spp., *Papilio polyxenes* and *Danaus plexippus* raise thoracic temperature only between 3 and 6 °C (Kammer 1970, Rawlins 1980, Tsuji *et al.* 1986). For several small species, that attain lower equilibrium temperatures and that cool more rapidly, the effect of metabolic heating seems negligible, and T_b falls and

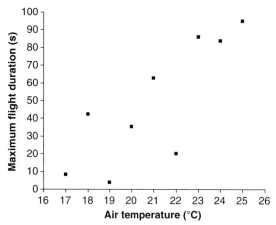

Figure 6.3. Maximum flight duration as a function of air temperature in male *Pararge aegeria* (data from Shreeve 1984).

approaches T_a during flight (Shreeve 1984, Shelly & Ludwig 1985, Heinrich 1986b).

In referring to studies of more than 40 different species of butterflies Kingsolver (1985a) concluded that a T_{th} between 28 and 42 °C is needed for flight, while rigorous flight is restricted to the range 33–38 °C (see also Vielmetter 1958, Douwes 1976a, Shreeve 1984, Dreisig 1995). Because flying often results in a lowered T_{th} (cooler microclimate and increased convective cooling when moving through the air) butterflies that cool in flight are forced to land and bask at regular intervals to regain the high T_{th} necessary for flight. The effect is that time spent in flight increases with temperature (Fig. 6.3). On the other hand, butterflies may have to cease activity during hot weather (Leigh & Smith 1959, Wickman 1988).

Kingsolver (1985a) concluded that the so-called flight space, i.e. the weather conditions when voluntary flight occurs, varies between species, but that species differences are mainly due to specific morphological (i.e. coloration and pubescence) and behavioural adaptations, rather than differences in T_{th} requirements for flight (Kingsolver 1983a, 1983b, Kingsolver & Watt 1983, 1984). There are thus weather conditions that may permit one species to fly, but not another. This was demonstrated by transplant experiments of *Colias* butterflies between habitats (Kingsolver & Watt 1984). There is also genotypic variation in the thermal properties of the glycolysis enzyme phosphoglucose isomerase in *Colias* butterflies along a geographical thermal gradient. These genotypes are correlated with different timings of flight in nature (Watt 1985). In addition there may be

differences between populations of T_{th} requirements for flight, as found in *Polyommatus icarus* (Howe 2004, Howe et al. 2007; see also Chapter 13).

Nevertheless there are butterflies that fly at lower T_{th} than those reported above. *Parnassius phoebus* flies voluntarily in the early morning at T_{th} of 17–20 °C (Guppy 1986). Guppy (1986) suggested that the low wing-beat frequencies of *P. phoebus* might permit this butterfly to fly at these low temperatures (Chaplin & Wells 1982). Low wing loading is known to permit flight at low temperature, because lower wing-beat frequency is needed for flight (Heinrich & Mommsen 1985). Lower wing loading means that lower T_{th} and metabolism is needed for muscle contraction. Wing loading and wing-beat frequency vary within a wide range in butterflies (Sotavalta 1952, Wickman 1992b, Grodnitsky 1993). Exceptionally low wing loading is found in *Leptidea* and *Aphantopus hyperanthus*, species that can be seen flying in cool, overcast weather. Individuals of *A. hyperanthus* regularly fly at a T_a of 16 °C and in drizzle (Wickman pers. observ.) In *Pararge aegeria* males as compared to females, and woodland individuals as compared to those of more open habitats, have longer flights under suboptimal temperatures. There was no clear-cut relationship to wing loading though (Merckx et al. 2006). The relationship between wing loading, wing-beat frequency and T_{th} needed for flight needs further examination.

Because T_{th} affects the efficiency of flight, the opportunity for a butterfly to attain optimal temperatures for flight may affect predation risk. It is known that in some species increased predation may be associated with weather that does not allow flight at all (Lederhouse 1983, Brower & Calvert 1985, Kingsolver 1987b). However, little is known about how predation has contributed to thermoregulatory behaviour and climate use of active butterflies. The major exception is a study of Neotropical butterflies by Srygley & Chai (1990a, 1990b). They demonstrated that palatable butterflies have higher T_e and T_{th} during flight compared to unpalatable species. Palatable species also are more constrained to sunny habitats and to the middle of day. They fly more quickly and erratically, and they evade predatory birds more easily (Chai & Srygley 1990). Besides, temperature does not only affect flight activity, but also acceleration (Berwaerts & Van Dyck 2004). Further studies of the causal pattern behind the suite of characters associated with predation is of great interest (Dudley 1991), especially since a number of selection pressures apart from predation – such as mate-locating behaviour and adult food preferences – may affect the traits involved (Wickman 1992b, Rutowski 1997,

Hill *et al.* 1999b, Van Dyck & Matthysen 1999). Interestingly, a study on a few Nearctic pierids suggested a similar association between palatability and capturability, although the T_b of the species was not examined (Srygley & Kingsolver 1998).

Female fecundity is a function of the time available for flight and the maturation rate of eggs. Both are temperature-sensitive and influenced by weather and climate. Longer periods of cold and cloudy weather may reduce fecundity (Boggs 1986, Pollard 1988). In *Pieris rapae* held in field cages the number of eggs laid per day increased linearly with ambient temperature. On overcast days females lay few eggs, but compensate by laying more eggs on the following sunny days (Gossard & Jones 1977). The oviposition potential of populations can adapt to local climatic conditions. *Colias* from high-altitude sites, which have less time for laying eggs (Kingsolver 1983b), are not capable of producing as many oocytes as those from low-altitude locations (Springer & Boggs 1986). In a similar manner the fact that woodland satyrines have lower potential oviposition rates compared to open-habitat satyrines was explained by restrictions on the time available for egg laying in the cooler woodlands (Wiklund & Karlsson 1988, Karlsson & Wiklund 2005). Besides, satyrine woodland species and populations have lower temperature optima for fecundity as compared to those of more open grasslands (Karlsson & Van Dyck 2005, Karlsson & Wiklund 2005). However, apart from these studies little is known about how optimal temperatures for oviposition vary between species and populations.

There is no general correlation between time spent feeding and T_b. In a few species examined feeding increases with temperature (Douwes 1976a, Pivnick & McNeil 1987a, Dreisig 1995), but in *Parnassius phoebus* time spent feeding decreases with increased temperature and irradiation (Guppy 1986). This is not surprising, because time spent feeding is not easily translated into amounts consumed, as nectar quantity and quality may change with temperature (cf. Willmer 1983, Boggs 1988, Muraoka & Watanabe 1994), and because nutrients can be stored for later use. In addition the relationship between time spent feeding and T_b is confounded by the ability of many species to bask whilst feeding.

CONCLUSIONS

As a result of more than 50 years of research on butterfly thermoregulation, we now have a clear picture of the various factors that influence butterfly activity and resource use. However, the details of how the distribution of microclimate in space interacts with a butterfly's possibility of optimal thermoregulation are still largely unresolved. For instance, vegetation structure may be decisive in this regard. Fewer clearings in a wood or fewer vegetation edges in an open field could have detrimental effects on butterfly thermoregulation and fitness. Hence, adopting well-founded habitat management regimes is critical in butterfly conservation, not only because they influence variables like host plant abundance or the availability of nectar plants, but also because microclimatic variation in space has a large impact on butterfly activity and on the rates of predation, mating and oviposition (see Chapter 4). Since weather and climate are known to be major influences on the population dynamics and the geographical distribution of butterflies (see Chapter 8), an optimal vegetation structure in terms of microclimate and thermoregulation resources could be critical to a population during years with bad weather. The study of such macro-scale interactions in space and time should be an important area for future research on butterfly thermoregulation. Such studies could make a major contribution to improving the conservation of species in the face of environmental change.

7 • Predictive species distribution modelling in butterflies

BORIS SCHRÖDER, BARBARA STRAUSS, ROBERT BIEDERMANN,
BIRGIT BINZENHÖFER AND JOSEF SETTELE

SUMMARY

The study of relationships between species occurrences and environmental characteristics depends on distribution modelling. Species distribution modelling has matured as a key field in ecology and conservation biology, with a growing range of techniques and statistical approaches. This chapter gives an overview of applications of predictive species distribution modelling using butterfly studies to illustrate general approaches, principles, limitations and perspectives. To understand the spatial distribution of species, a multi-scale, mechanistic and resource-based view of species–environment relationships is needed. This framework is equivalent to a multi-scale view of niche theory. Predictor variables describing relevant gradients of environmental resources can be consumable resources, direct gradients with physiological meaning, or indirect gradients without physiological meaning but that are easily measurable and may serve as proxies for combinations of resources and direct gradients. The basics of regression models and their applications to butterfly studies are discussed focusing on multiple linear regression, generalised linear models (GLM), logistic regression, GLM extensions and non-parametric regressions, and novel non-parametric machine-learning methods. State of the art species distribution modelling using logistic regression – being the most popular technique – is discussed in detail. Addressed issues include the importance of preliminary tests prior to model building, model selection (stepwise procedures and information theoretic criteria), model evaluation (calibration and model discrimination, model diagnostics and checking for residual spatial autocorrelation) and model validation (internal and external validation). Finally, the issue of species distribution models is illustrated with a detailed case study for the butterfly *Coenonympha arcania* and the burnet moth *Zygaena carniolica* in dry grasslands in southern Germany. Species distribution models yield a first and pivotal step for a more general mechanistic understanding of species–habitat relationships – with many fascinating butterfly examples.

INTRODUCTION

Habitat analysis and species distribution modelling, i.e. the analysis and modelling of functional relationships between species occurrences and environmental characteristics, provide the basis for different research questions including:

(1) Generating and testing hypotheses about essential habitat conditions for habitat selection and survival (e.g. Crozier 2003).
(2) Quantitative understanding of abiotic, biotic and disturbance-related factors driving habitat selection, species persistence and species richness in dynamic landscapes (Briers & Warren 2000, Thomas & Elmes 2001, Krauss *et al.* 2004a, Rudner *et al.* 2007) and in large-scale regions (Hawkins & Porter 2003, Stefanescu *et al.* 2004).
(3) Resource-based delineation of species habitats (Dennis *et al.* 2003, Vanreusel & Van Dyck 2007, Vanreusel *et al.* 2007).
(4) Quantification of habitat quality (Dennis & Eales 1997).
(5) Prediction of species distributions (Dennis & Eales 1999, Cowley *et al.* 2000; Fleishman *et al.* 2001) as well as species richness and biodiversity hotspots (Maes *et al.* 2003, Hortal *et al.* 2004) on different spatial and temporal scales.
(6) Generating habitat suitability maps, i.e. maps of predicted occurrence probabilities (Osborne *et al.* 2001, Austin 2002, Joy & Death 2004, Luoto *et al.* 2005, Strauss & Biedermann 2005).
(7) Forecasting species distributional patterns for changing environmental conditions (Beaumont & Hughes 2002, Dennis & Shreeve 2003) and assessing the effect of environmental change on extinction and persistence probabilities (Crozier 2003, Lavergne *et al.* 2005).

Species distribution modelling has matured as a key field in ecological research and conservation biology with a growing range of techniques and statistical approaches (see Guisan & Zimmermann 2000 for an excellent overview; Guisan & Thuiller 2005). It is applied to a wide range of specific

Ecology of Butterflies in Europe, eds. J. Settele, T. Shreeve, M. Konvička and H. Van Dyck. Published by Cambridge University Press. © Cambridge University Press 2009, pp. 62–78.

research questions in autecology, community ecology, spatial ecology, landscape ecology, macroecology, biogeography and conservation biology. This chapter gives an overview of applications of predictive species distribution modelling. We discuss its potential and limits and summarise the recent methodological state of the art. The focus is on hypothesis-driven, resource-based, scale-dependent, species-centred, quantitative statistical approaches. Hence, we concentrate on approaches that estimate functional relationships of species incidence and species richness as response variables (dependent variables) to scale-dependent environmental predictor variables (explanatory independent variables).

SPECIES DISTRIBUTION MODELLING: AIMS, BACKGROUND, ASSUMPTIONS, LIMITATIONS

Aims

Modelling of species–environment relationships relating species incidence (i.e. presence/absence) or species richness data to environmental predictors has two related and distinct aims: understanding and prediction (Mac Nally 2000). Explanatory modelling aims at understanding habitat selection or species ranges testing via hypotheses on the importance of different, scale-dependent predictors. Models that predict the potential spatial distribution of species (or species diversity) in current environments as well as accounting for the effect of changes in habitat, land use or climate are the most reliable if they are based on a thorough understanding of species–habitat relationships. Good explanatory models usually also yield good predictions. However, ideal models may require predictors that are frequently unavailable (e.g. fine-scaled data on vegetation structure or microclimate).

Habitat selection studies are a means to delineate habitat in a functional resource-based way (Dennis *et al.* 2003). Species distribution models that rely on direct predictor variables representing resource gradients can be applied to maps of these predictors. So, models can be extrapolated to entire landscapes, providing maps of habitat quality (habitat suitability maps) or maps of predicted spatial distributions (e.g. Luoto *et al.* 2005, Vanreusel *et al.* 2007).

Framework

To understand the spatial distribution of species, we suggest a multi-scale, mechanistic and resource-based view of species–environment relationships by combining the general frameworks proposed by Poff (1997) and Mackey & Lindenmayer (2001). Mackey & Lindenmayer (2001) identify a hierarchy of scales representing major changes in the processes that control the distribution and availability of primary environmental resources (i.e. light, heat, water, nutrients, shelter and nesting sites) as well as a hierarchy of individual and collective distributional behaviours. This framework is equivalent to a multi-scale view of niche theory (Hutchinson 1957, Guisan & Thuiller 2005). Similarly, Poff (1997) defines a hierarchical set of nested habitat selective forces, so-called 'landscape filters'. To survive in a landscape that exhibits a particular set of these constraining landscape features, individual species have to exhibit specific functional traits (Dennis *et al.* 2004). These two concepts are easily combined (Schröder & Reineking 2004), broadening Poff's filter approach to biogeographic scales by considering, for instance, bioclimatic envelopes. Following Mackey & Lindenmayer (2001), the species' response has to be empirically derived by using statistical relationships that correlate field observations of the species' distributions with key environmental attributes.

Scales

The research questions reviewed here focus on specific scales such as continental-scale (e.g. Europe: Hawkins & Porter 2003), regional-scale (e.g. Belgium: Maes *et al.* 2003), landscape-/meso-scale (e.g. calcareous grasslands of Dorset, UK: Thomas *et al.* 2001b) and small-scale (e.g. experimental plots in Swiss grasslands: Zschokke *et al.* 2000). Multi-scale studies simultaneously focus on more than one scale (e.g. on oviposition sites as well as on adult habitat selection on the plot and landscape scale; Rabasa *et al.* 2005). Studies frequently aim for comparing the importance of microhabitat vs. local (patch) vs. landscape (patch configuration) scale predictors for species distribution (e.g. Thomas *et al.* 2001b, WallisDeVries 2004, Heikkinen *et al.* 2005). Other studies aim for identifying the scale of perception or the scale of habitat use by considering habitat context at different spatial scales (e.g. Steffan-Dewenter *et al.* 2002, Binzenhöfer *et al.* 2005, Krauss *et al.* 2005). Regardless of the scale considered, from a methodological viewpoint, all these studies apply similar analyses.

Predictors

Predictor variables describe relevant gradients of environmental resources. Austin (1985) distinguishes (a) consumable resources, (b) direct gradients with physiological meaning, and (c) indirect gradients without physiological meaning that are easily measurable and may serve as proxies for combinations of resources and direct gradients. The stronger the relationship between predictor variables and relevant processes, the more robust and general are the resulting models (Poff 1997). Regarding applications in conservation biology as well as in large-scale studies, there is a trade-off between the costs of obtaining predictors and their ability to describe the underlying mechanisms in a direct way (Guisan & Zimmermann 2000, Schröder 2000, Hein *et al.* 2007b). Predictors should be closely related to underlying processes and selective forces to yield reliable models with which robust predictions can be made. The larger the scale, the more difficult this issue becomes. In large-scale models, mainly bioclimatic variables or variables derived from remote sensing are available, which are comparatively coarse and sometimes only indirectly related to the underlying processes (e.g. Luoto *et al.* 2002, Thuiller *et al.* 2005).

Assumptions and limitations

All methods used in species distribution modelling assume that the observed occupancy pattern results from habitat selective mechanisms maximising the individual fitness (Manly *et al.* 2002, Karlsson & Wiklund 2005) and that a quasi-equilibrium exists, i.e. environmental change is slow compared to the species lifetime (Guisan & Zimmermann 2000, Austin 2002). The resulting models are static models that yield potential species distribution patterns. They do not allow any inference on whether or how fast the potential habitat is occupied by a species. So, using these predictions to forecast distributions under changed environmental conditions is limited by the implicit assumption of an instantaneous realisation of a new equilibrium ignoring transient dynamics. Only spatially explicit, *dynamic*, process-based models can fulfil this task (e.g. Wahlberg *et al.* 1996, Schtickzelle & Baguette 2004, Schtickzelle *et al.* 2005a). However, species distribution models have been used to delineate the underlying geometry and quality of habitat patches for such dynamic models (Verboom *et al.* 1991, Akçakaya *et al.* 1995, Schröder 2000). This is even more important in dynamic landscapes, where habitat patches vary in quality, extent and connectivity (Stelter *et al.* 1997, Keymer *et al.* 2000, Wahlberg *et al.* 2002a, Biedermann 2004). However, if spatial and/or temporal autocorrelations are not explicitly modelled, the independence of samples is an important statistical assumption. Violating this assumption

results in an overestimation of the effective sample size, an underestimation of standard errors, and thus leads to an overoptimistic or even wrong interpretation of model results (Legendre 1993, Kühn 2007).

DATA ANALYSIS: STATISTICAL MODELS

There is a rapid development of statistical methods applicable for species distribution modelling (Guisan & Zimmermann 2000, Guisan *et al.* 2002, 2006, Latimer *et al.* 2006). All methods relate species incidence, species abundance, or species richness data to environmental predictors. However, they differ in implicit model assumptions (e.g. error distribution) and differ in their focus on the following contrasting aspects: model-driven vs. data-driven, parametric vs. semi-parametric vs. non-parametric, parsimony vs. flexibility (e.g. shape of response curves), simplicity (including the risk of being oversimplistic) vs. complexity (including the risk of overfitting), focusing mainly on explanation, interpretability and communicability vs. focusing mainly on prediction. The choice of an appropriate approach strongly depends on the research question and the type of model output desired.

Data acquisition

Since the approaches considered here are empirical statistical models, data acquisition is a crucial step. Several decisions regarding different trade-offs have to be taken. These decisions first concern the sample size (Stockwell & Peterson 2002, Reineking & Schröder 2003). As a rule of thumb, the number of events per predictor variable (EPV, i.e. the number of presences and absences (whichever is smaller) per predictor variable) should exceed ten. This is equivalent to having no more than $m/10$ predictors with m being the number of observation in the least represented category (Harrell *et al.* 1996). Reineking & Schröder (2006) show a strong effect of variable selection for low EPV producing larger optimism in case of small EPV due to overfitting, and better performance in case of larger EPV (see below). This relates to the definition of training data for model selection and model formulation and test data for validation (Hastie *et al.* 2001; see below).

Sampling design is another crucial point. We suggest random stratified sampling (Wessels *et al.* 1998, Bonn & Schröder 2001, Hirzel & Guisan 2002). The gradient length is another issue to be considered (Oksanen & Minchin 2002) which strongly depends on the scale. Another important decision concerns the scaling of predictor variables.

Generally, continuous predictors are preferred (Harrell 2001) as categorical predictors reduce the degrees of freedom of the model. Further, the selection of butterfly stages to be sampled (eggs, larvae or adults) is a question of detection probability and future model application.

REGRESSION MODELS

Multiple linear regression

We start with the simplest basic approach: multiple linear regressions. Although being limited by several implicit assumptions, this method is widely used to model species richness. This approach estimates the expected mean species number (or abundance) y depending on several environmental predictor variables ($x_1, x_2, ..., x_p$). The latter are combined to a linear predictor ($\beta_0 + \beta_1 x_1 + \beta_2 x_2, ..., + \beta_p x_p$). Regression coefficients β_i are estimated by means of ordinary-least square (OLS) estimation which minimises the squared differences between observations and predictions. Main assumptions (and limitations!) are homogeneity of variance (i.e. constant variance across all observations, also called homoscedasticity), normally, identically and independently distributed errors, as well as linearity in the predictors (Crawley 2002). Transforming the response variable and/or applying polynomial (e.g. x_i^2) and interaction terms (e.g. $x_i \times x_j$) of the predictors are used to deal with departures from these assumptions (Guisan et al. 2002).

There are many butterfly examples for multiple linear regression models: Kuussaari et al. (1996) applied multiple linear regressions to model the effect of patch characteristics on emigration rates, residence times, and the numbers of immigrants in Melitea cinxia on a Finnish island. Similarly, Välimäki & Itämies (2003) explained emigration and immigration rates by population size and patch area in Parnassius mnemosyne. To analyse the effects of habitat fragmentation on diversity, abundance and life history traits on calcareous grasslands, Steffan-Dewenter & Tscharntke (2000) also applied linear models with transformed variables. Similarly, Krauss et al. (2004b) and Pywell et al. (2004) analysed the effects of some environmental variables on butterfly abundances in intensively managed arable farmland. On a larger scale, Maes et al. (2003, 2005) applied multiple linear regression models to predict butterfly diversity from atlas data in Belgium, whereas Konvička et al. (2003) analysed Czech atlas data with respect to elevational shifts in butterfly distribution due to climate change. Stefanescu et al. (2004) assessed the effect of climate, topography, vegetation structure and human disturbance on butterfly species richness in Catalonia, and Kerr et al. (2001) as well as White & Kerr (2006) applied a similar analysis with variables derived from remote sensing to predict butterfly species richness in Canada. On a continental scale, Hawkins & Porter (2003) related butterfly diversity to the water–energy balance in Europe and northern Africa.

In the case of categorical predictor variables, analysis of variance (ANOVA) is the adequate analogue to the linear regression model. ANOVA estimates mean values for the respective categories (Quinn & Keough 2002, Crawley 2005). Huntzinger (2003), for instance, considered different soil-related or fire-related management plans as categorical predictors to explain butterfly diversity. Haughton et al. (2003) analysed the effect of different treatments of genetically modified herbicide-tolerant crops on butterfly abundance with an ANOVA approach. If predictors comprise both categorical factors and continuous covariates, analysis of covariance (ANCOVA: Crawley 2005, Karlsson & Wiklund 2005) can be applied. If both response and predictor variables are categorical, a contingency table approach is appropriate (Crawley 2002). Traditionally, such tables are analysed by applying the χ^2-test statistic (e.g. Ouin et al. 2004). As an alternative, Fred & Brommer (2003) used log-linear models to analyse the relationship between host-plant density, patch area, year and patch occupancy of Parnassius apollo in Finland.

Generalised linear models

Ecological data do not usually meet the implicit assumptions of simple linear models: for instance, count data are Poisson distributed and incidence data have a binomial distribution. To adequately model such data types, generalised linear models (GLMs) were developed (McCullagh & Nelder 1983). GLMs can model a wide range of distributions other than the normal distribution. This is made possible by linking the linear predictor to the response variable (or more precisely to its expected value) by a modified link function. (In case of the simple linear model, this is an identity link – hence linear models are a special case of GLMs.) These functions constrain the predicted values to be within a range of possible values – for incidence data, (using a logit link function), for instance, between 0 and 1. In the case of a Poisson distribution, the link function is the natural log (ln). In the case of the binomial distribution the logit link function is applied, which is given as $\ln[\mu/(1 - \mu)]$) where μ is the expected value. To calculate the predicted values, the inverse link function has to be applied, e.g. μ = exp (linear predictor) / (1 + exp (linear predictor)) for the logistic regression.

In GLMs, the regression coefficients are estimated by means of maximum likelihood estimation (Crawley 2002, 2005), i.e. by finding the parameters of a given model that provide the best fit to the observed data (Hilborn & Mangel 1997). The maximum likelihood of a model is an important measure of goodness-of-fit. It is equivalent to the residual sum of squares in OLS-regression and is often expressed as the residual deviance $D = -2$ log likelihood. The proportion of deviance explained in the model and the residual deviance can be assessed by an analysis of deviance analogue to ANOVA for non-normal error distributions (e.g. Pausas *et al.* 2003).

If the response variable is Poisson distributed – as in many data sets on abundance and species richness – Poisson regression should be applied, i.e. a GLM with a log-link and Poisson-distributed errors (Vincent & Haworth 1983). Mac Nally *et al.* (2003) as well as Hortal *et al.* (2004) applied Poisson regression to model butterfly species richness in the central Great Basin of western North America and in Portugal, respectively. Matter *et al.* (2005) modelled the number of immigrating butterflies depending on patch size and isolation, and Musche *et al.* (2006) modelled oviposition of *Maculinea nausithous* in relation to soil treatment and host plant traits.

GLMS FOR BINARY RESPONSE DATA: LOGISTIC REGRESSION

At present, the most popular procedure for predictive species distribution modelling is multiple logistic regression (also called logit regression). Many studies have applied logistic regression models at different spatial scales and with different binary response variables to model species incidence (see examples below), oviposition preference (Kuussaari *et al.* 2000), larval habitat use (Anthes *et al.* 2003b), response to habitat edges (Ries & Debinski 2001), local extinction and colonisation (Verboom *et al.* 1991), or occurrence inside or outside a butterfly species' range (Crozier 2003). If the predictor is used in its linear form, the corresponding response curve is sigmoid. To achieve a hump-shaped, unimodal response curve, the model has to consider a predictor both in its linear and its quadratic form (e.g. Thorne *et al.* (2006) estimating phenological models for butterflies). In this case, a negative coefficient has to be estimated for the quadratic term and a positive one for the linear term.

In most studies on species distribution at the landscape scale, the predictors depict one or several of the following items:

(1) Local habitat quality such as host plant abundance (Thomas *et al.* 2001b, Wahlberg *et al.* 2002a), host ant presence (WallisDeVries 2004), topographic variables (e.g. elevation: Wilson *et al.* 2005) or potential solar insolation (Fleishman *et al.* 2001), habitat type and land use (Franzén & Ranius 2004, Binzenhöfer *et al.* 2005), land use history (Pöyry *et al.* 2004), disturbance regimes (Rudner *et al.* 2007).

(2) Habitat patch metrics like patch area (Thomas *et al.* 2001b, Wahlberg *et al.* 2002a, Fred & Brommer 2003).

(3) Metrics about the functional spatial configuration of habitat patches depicting habitat connectivity (Hanski 1994a) or habitat isolation, expressed for example as the mean distance from the nearest patch (Thomas *et al.* 2001b, Krauss *et al.* 2005).

(4) The landscape context by calculating the amount of habitat within buffers of different radii characterising different spatial scales of habitat perception (Chardon *et al.* 2003, Bergman *et al.* 2004, Binzenhöfer *et al.* 2005) or other geographical variables (e.g. geographic coordinates: Dennis *et al.* 2002).

These studies frequently aim at discerning the importance of local-scale vs. landscape-scale predictors for species distribution (Thomas *et al.* 2001b, WallisDeVries 2004, Heikkinen *et al.* 2005). This is mostly done by means of partitioning the variance represented by the different predictors. In other conservation-related applications, predicted habitat patches are selected as reintroduction sites (Leon-Cortes *et al.* 2003b) or as parts of reserve networks (Cabeza *et al.* 2004). Other studies use model predictions to suggest management regimes for butterflies (Pywell *et al.* 2004).

GLM EXTENSIONS AND NON-PARAMETRIC REGRESSIONS

Simple GLMs do not account for not independently distributed errors, for instance due to spatial or temporal autocorrelation. However, some extensions of GLMs do account for autocorrelation. If logistic regression models include neighbourhood variables regarding observed or predicted occurrence in the neighbourhood, these models are called autologistic models (Augustin *et al.* 1996, Gumpertz *et al.* 1997, Osborne *et al.* 2001, Sanderson *et al.* 2005). Dennis *et al.* (2002) used such an autologistic approach to model French butterfly atlas data. For other examples of spatial linear models see Keitt *et al.* (2002), Lichstein *et al.* (2002), Segurado *et al.* (2006), Miller *et al.* (2007) and Dormann *et al.* (2007).

To account for repeated measures leading to statistically non-independent data, Van Dyck *et al.* (2000) applied a generalised linear mixed model approach (GLMM: Wolfinger & O'Connell 1993) to analyse the effect of ant nests on the egg load of a myrmecophilous *Maculinea* butterfly. A similarly sophisticated example is a hierarchical multi-scale study by Rabasa *et al.* (2005). Their mixed model accounts for the hierarchical structure in the data on egg presence surveyed on three levels: patch, plant and individual fruit (cf. multi-level modelling as applied by Bayliss *et al.* 2005).

Other extensions of GLMs as well as non-parametric methods enhance model flexibility regarding the shape of the response curve. Examples are semi-parametric generalised additive models (GAMs: Hastie & Tibshirani 1990, Yee & Mitchell 1991) as for instance implemented in the GRASP program (Lehmann *et al.* 2002b), non-parametric locally weighted regression (LOWESS: Trexler & Travis 1993), non-linear Huisman–Olff–Fresco models (HOF: Huisman *et al.* 1993, Oksanen & Minchin 2002), and multivariate adaptive regression splines (MARS: Leathwick *et al.* 2005) that use piece-wise linear segments to describe non-linear relationships. These upcoming methods are not frequently used yet, but there are some butterfly examples: to predict the spatial distribution of butterflies as a function of climate variables in Finland, Luoto *et al.* (2005) successfully applied GAMs. Hill *et al.* (2002) and Huntley *et al.* (2004) applied LOWESS to model butterfly distribution data relative to bioclimatic variables without any assumptions about the form of the relationship between predictors and species incidence. Wilson *et al.* (2005) modelled the altitudinal distribution of butterfly species in central Spain associated with climate change by logistic regression and non-linear HOF models.

Additional methods including machine-learning methods and methods for presence-only data

Recently, novel non-parametric machine-learning methods have been applied to species distribution modelling (for an excellent overview see Elith *et al.* 2006). The most widely used ones are classification and regression trees (CART: Efron & Tibshirani 1991, De'ath & Fabricius 2000) and artificial neural networks (ANN: Lek & Guégan 1999). ANNs are either multi-layer feed-forward neural networks trained by a backpropagation algorithm (i.e. backpropagation networks: Ripley 1996) or Kohonen self-organising maps (Céréghino *et al.* 2001, 2005). Berry *et al.* (2002) used an

ANN model for butterflies in Britain and Ireland. Koh & Sodhi (2004) successfully applied CART to compare the ecological traits (e.g. adult habitat specialisation, larval host plant specificity, adult body size and sexual dichromatism) of butterflies in different habitats in Singapore and found that trait sets varied between habitats. A well-known method for classifying locations due to climatic parameters is BIOCLIM (Houlder *et al.* 2001) which describes a species bioclimatic envelope. Beaumont *et al.* (2005) applied this approach to predict the spatial distribution of Australian butterfly species as a function of climate variables. The Genetic Algorithm for Rule-Set Prediction (GARP: Stockwell & Peters 1999) is a machine-learning method that induces 'if–then' rule sets to describe the species ecological niche in a multivariate way.

Comparatively new, but potentially promising in the context of species distribution modelling, is the concept of ensemble forecasting (Araújo & New 2007). Ensemble forecasting means that instead of selecting one single best model, several, or many, models (mostly regression trees) are used to produce an ensemble of predictions which can be combined. In the machine-learning literature, several algorithms are described for this kind of voting classifications (Bauer & Kohavi 1999). In bagging (i.e. bootstrap aggregation: Breiman 1996), multiple bootstrapped regression trees are fitted to the data and the mean prediction is finally used. The random forest algorithm is quite similar, but each single tree is grown with a randomised subset of predictors (Breiman 2001, Meinshausen 2006).

Boosted regression trees (BRT: Friedman 2002, Thuiller *et al.* 2006) are a combination of the boosting algorithm and regression trees. Here, regression trees are fitted sequentially on weighted versions of the data set, where the weights continuously adjust to take account of observations that are poorly fitted by the preceding models (Elith *et al.* 2006).

If only presence data are available, there is a set of methods applicable (Brotons *et al.* 2004, Pearce & Boyce 2006) including strategies to select pseudo-absence data (Lütolf *et al.* 2006), environmental niche factor analysis (ENFA with the BIOMAPPER software: Hirzel *et al.* 2001, Hirzel & Arlettaz 2003), GARP (Stockwell 2006, Pearson *et al.* 2007) as well as maximum entropy methods (MAXENT: Phillips *et al.* 2005). A number of studies comparing different modelling techniques could not identify any single method to be superior for all species and situations (Guisan *et al.* 1999, Manel *et al.* 1999, Moisen & Frescino 2002, Olden & Jackson 2002, Moisen *et al.* 2006, Potts & Elith 2006, Prasad *et al.* 2006). The most appropriate method does not only depend on the distribution being modelled, but also on the aim of the study (Segurado & Araújo 2004).

STATE OF THE ART IN SPECIES DISTRIBUTION MODELLING BY LOGISTIC REGRESSION

Logistic regression models currently are the most popular approach in species distribution modelling (Reineking & Schröder 2003). Logistic regression is easy to apply, available in all standard statistical packages, and there are well-developed methods for evaluating and interpreting the results. For this reason, in the following sections we focus on logistic regression and describe the state of the art of model selection, evaluation criteria and validation (Table 7.1). Most of these notions are also applicable to other methods.

Preliminary tests

We suggest some preliminary work before starting the model building. To detect inherent correlations between predictor variables (i.e. multicollinearity: Mac Nally 2002, Graham 2003), we suggest two steps: (1) make a scatterplot matrix (Fox 2002), and (2) calculate Spearman rank correlation coefficients for all pairs of predictors (Fielding & Haworth 1995).

Graham (2003) discussed different problems arising with multicollinearity (e.g. inaccurate model parameterisation, decreased statistical power, and exclusion of significant predictor variables during model formulation). The simplest way to deal with multicollinearity is to delete one of two correlated predictors, if the correlation coefficients exceed a certain threshold ($r_S >$ 0.7 according to Fielding & Haworth 1995). Other approaches to deal with multicollinearity are either residual and sequential regression (Graham 2003) or principal component regression as applied by Maes *et al.* (2003, 2005) and Stefanescu *et al.* (2004). By conducting a principal components analysis (PCA: Legendre & Legendre 1998) prior to model building, statistically independent linear combinations of predictors are estimated. These so-called principal components are used as predictor variables in the subsequent analysis.

For a preliminary analysis detecting the shape of the response, we suggest producing scatterplots of the response variable against each hypothesised predictor combined with univariate GAMs (Crawley 2002, Guisan *et al.* 2002, Luoto *et al.* 2005) or LOWESS curves (Huntley *et al.* 2004). This helps to decide whether or not to include quadratic terms of predictors in the model (Hosmer & Lemeshow 2000).

To get a first overview on the explanatory power of multiple predictor variables and possible interactions (Guisan *et al.* 2002), Crawley (2002) suggests applying classification and

Table 7.1 *Model building strategy for species distribution models*

The following list shows a sensible strategy for model building:
(1) Define problem and modelling target.
(2) Define dependent variable (target species, target stage(s) such as adults, larvae or eggs) and potential predictor variables.
(3) Formulate explicit hypotheses regarding the species–habitat relationships.
(4) Select sampling design – trade-offs between extent (size of the study area) and grain (resolution), costs for each predictor variable and sample size, generality and particularity.
(5) Data acquisition, i.e. field survey, GIS analyses, classification and interpretation of remote sensing data.
(6) Define training and test data for model selection, parameter estimation and model validation.
(7) Select the model structure – trade-offs regarding model complexity (number of predictors) and parsimony, model flexibility (shape of response curve) and parsimony, interpretability and predictive power.
(8) Model selection – seeking for a parsimonious model (cf. bias–variance trade-off), including checks regarding quadratic terms to model curvilinear relationships, and interaction terms.
(9) Parameter estimation and model evaluation including diagnostics.
(10) Model validation – either on test data or by resampling techniques – and model assessment including check for residual spatial autocorrelation.
(11) Repeat the last three steps (8–10) until result is adequate.
(12) Interpretation and application, e.g. extrapolating the predictions from survey plots to the landscape level.

regression trees. An alternative way to achieve this overview is hierarchical partitioning (Chevan & Sutherland 1991, Mac Nally 1996, 2000). In an instructive study, Heikkinen *et al.* (2005) applied variance and hierarchical partitioning to decompose the variation in *Parnassius mnemosyne* abundance and occupancy into independent and joint effects of larval and adult food resources, microclimate and habitat quantity. Brändle *et al.* (2002) applied hierarchical partitioning to assess the importance of body size, niche measures, resource availability, life-history traits and population parameters for geographical range sizes of butterflies on a regional and European scale. Mac Nally *et al.* (2003) found those predictors with the highest explanatory power with respect to butterfly species richness in the Great Basin of western North America. Hierarchical partitioning does not estimate a single best model, but gives an essential help in selecting predictors. However, hierarchical partitioning routines do not deal with nonlinear relationships such as hump–shaped response curves.

Model selection

Although some more or less automatic procedures and software tools exist (listed in Guisan & Thuiller 2005), model formulation, i.e. finding an appropriate set of predictor variables to derive a parsimonious species distribution model with high predictive power, is a difficult task requiring many iterations of model fitting (Lindenmayer *et al.* 1999). Ideally, the model structure would be known and for each coefficient to be estimated we would have many observations. However, usually we neither know the appropriate set of predictors and interactions, nor have large data sets.

STEPWISE PROCEDURES

If the number of potential predictor variables is high, there is a high risk of overfitting if many or all predictors are included in the model (Harrell 2001). Also, it is often not feasible to check all possible models in this case (all-possible-subset regression: Brennan *et al.* 1986, Crawley 2002). Therefore, we have to apply some kind of regularisation or model selection (Buckland *et al.* 1997, Reineking & Schröder 2006). The most prevalent – but not unproblematic (Whittingham *et al.* 2006) – approaches for model selection are stepwise procedures, with backward elimination being preferred over forward selection (Harrell 2001). A backward elimination procedure starts with a full model using all predictors and consecutively deletes the least informative predictor unless no more deletions are possible without a significant loss of model performance. In contrast, the

forward stepwise approach starts with the null model, which considers only the intercept and thus estimates the mean value (i.e. the prevalence in case of logistic regression). Consecutively, the most informative predictors are added to the model as long as they yield a significant model improvement. To decide which variable has to be considered in the next step, these procedures usually apply likelihood-ratio tests. This test calculates the difference between the deviance of the null model and the deviance of the actual model (which is the ratio of the log likelihoods of the respective models) to select either the next predictor to delete from (smallest difference) or to insert (largest value) into the model (Hosmer & Lemeshow 2000).

INFORMATION THEORETIC CRITERIA

Increasingly, information theoretic criteria are applied for variable selection (Burnham & Anderson 1998), i.e. mainly Akaike information criterion (AIC: Akaike 1974) as well as Bayes information criterion (BIC: Schwarz 1978). They account for model complexity by adding a penalty term to the deviance (D), that has to be minimised ($D = -2 \ln_e l$). In case of AIC this penalty term is simply $2p$, where p denominates the degrees of freedom in the model, i.e. the number of parameters to be estimated. This measure of model complexity is one for each metric and $k - 1$ for a k-class categorical predictor variable. In case of small data sets with $n/p < 40$, Hurvich & Tsai (1989) suggest a corrected AIC_C, where $2p$ is multiplied with the term $[1+(p+1)/(n-p-1)]$. For BIC the penalty term is $\ln_e(n)$ times the model's degrees of freedom p. Thus, choosing BIC means applying a heavier penalty that usually leads to a lower number of predictors in the model if n exceeds 7 (Harrell 2001, Reineking & Schröder 2003). In contrast, AIC is even less restrictive than the classical likelihood-ratio test with significance level $\alpha = 0.05$, because it is equivalent to applying $\alpha = 0.157$ (which is the quantile for $\chi^2_{df=1} = 2$) for predictors with one degree of freedom. The smaller AIC or BIC, the more parsimonious is the model. In butterfly studies, Binzenhöfer *et al.* (2005) applied AIC for model selection, whereas Fleishman *et al.* (2001, 2005) as well as Mac Nally *et al.* (2003) made use of BIC.

NEW PROMISING METHODS

Other promising but not yet extensively used regularisation methods use different measures to account for model complexity. The least absolute shrinkage and selection operator (lasso: Tibshirani 1996) uses the sum of the absolute values of the parameter estimates. Penalised maximum likelihood estimation (ridge regression according to le Cessie & van Houwelingen 1992) uses the sum of squares of the parameter estimates (Steyerberg *et al.* 2000). In a simulation study, Reineking & Schröder (2006) showed that lasso and ridge outperformed classical model selection at small sample sizes (EPV < 10). Other authors suggest applying resampling techniques like the bootstrap (see below) for model selection (Pan & Le 2001, Wisnowski *et al.* 2003). Lastly, ensemble methods such as Bayesian model averaging is a promising approach to deal with uncertainty in model selection (Chatfield 1995, Hoeting *et al.* 1999, Araújo & New 2007). Applying this method, one does not yield one specific 'best' model but calculates a weighted average over a set of plausible models to produce predictions. Wintle *et al.* (2003) present an application in the context of species distribution modelling.

MODEL EVALUATION

What is a 'good' model? There is a plethora of criteria for assessing model performance (Fielding & Bell 1997, Pearce & Ferrier 2000). The model formulation process seeks a parsimonious, generalisable model with high explanatory power and predictive performance (Mac Nally 2000). Therefore, model evaluation has to focus on different aspects of model performance, namely model calibration and refinement on the one hand, and model discrimination on the other (Murphy & Winkler 1992). Calibration refers to whether the predicted probabilities agree with empirical relative frequencies of occurrences. Refinement depicts the degree to which the predicted probabilities are close to 0 and 1. Discrimination refers to the model's ability to distinguish between presences and absences.

Calibration

Calibration of logistic regression models can best be evaluated by calculating and plotting the calibration curve showing the agreement between predicted probabilities of occurrence and observed proportions of sites occupied (Harrell 2001, Steyerberg *et al.* 2001a, Reineking & Schröder 2004). The slope of this curve is the regression coefficient of a logistic regression model with the logit of predicted probabilities as the only explanatory variable and the observed incidence as the dependent variable. In the case of a perfectly calibrated model, the calibration slope is 1 and the intercept 0. Whereas the intercept measures bias, i.e. systematic over- or underestimation of predicted probabilities, the slope is a measure of overfitting resulting from models that are too complex and have been fitted too closely to the training data (Harrell 2001).

If the slope is less than 1, the model is overconfident and systematically provides too extreme predictions: low predicted values are too low, high ones are too high. The potential for overfitting increases with higher model flexibility (e.g. higher for a GAM compared to its GLM-equivalent: Vaughan & Ormerod 2003), with a higher degree of model selection (e.g. in case of all-subsets regression: Harrell *et al.* 1996) and fewer training data available. According to Harrell (2001), overfitting is the major cause of unreliable models.

Another criterion for calibration is Nagelkerke's (1991) R^2_N. It simultaneously assesses model refinement. R^2_N is equivalent to the classical coefficient of determination R^2 in OLS-regression that characterises the amount of explained variation. It is calculated from the log likelihoods of the actual (l) and the null model (l_0):

$$R^2_N = \left(1 - \exp\left[-\frac{2}{n}(\ln l - \ln l_0) \right] \right) \Big/ \left(1 - \exp\left[\frac{2}{n} \ln l_0 \right] \right)$$

In the case of a perfectly calibrated and refined model, R^2_N reaches 1, but in realistic data sets its values are much lower than what is usually reached by OLS-R^2 (Hosmer & Lemeshow 2000).

Discrimination

The best way to assess model discrimination is to calculate the area under a receiver-operating characteristic curve (ROC curve: Swets 1988, Fielding & Bell 1997). A ROC curve (Fig. 7.1) is a scatterplot of sensitivity vs. the term 1 − specificity (see below) for each possible classification threshold P_{crit}. The area under this curve, AUC, which is equivalent to the concordance index c, does not require the dichotomisation of predictions into presences and absences. Thus, it is independent of the choice of a threshold probability P_{crit}. A value of 0.5 indicates random predictions. AUC = 1 indicates a perfect separation between occurrences and absences. Models that reach AUC values > 0.7 (Hosmer & Lemeshow 2000) or > 0.8 (Harrell 2001) are acceptable, > 0.9 outstanding (Hosmer & Lemeshow 2000). Note that AUC is not quite sensitive to poor calibration.

Threshold-dependent criteria are the correct classification rate, sensitivity (the amount of correctly classified presences), specificity (the amount of correctly classified absences), and Cohen's (1960) κ, a measure for the proportion of correctly classified cases after accounting for chance effects (Manel *et al.* 2001). According to Monserud & Leemans (1992), models yielding $\kappa > 0.4$ are acceptable and those yielding $\kappa > 0.55$ are good.

Table 7.2 *Species distribution model for* Zygaena carniolica*: after model selection applying AIC for model selection: coefficients with standard errors and p-value for a multiple logistic regression model. The corresponding response curve is given in Fig. 7.1*

	Coefficient	Standard error	*p*-value
Intercept	− 7.4971	1.75313	<0.0001
Moss cover	0.1758	0.08403	0.0365
Centaurea jacea	1.8549	0.63870	0.0037
Scabiosa columbaria	2.5779	0.98772	0.0091
Suitable habitat types within 25 m	0.0551	0.01702	0.0012

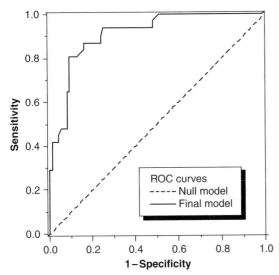

Figure 7.1. Receiver-operating characteristic (ROC) curve for the model given in Table 7.2. The area under curve (AUC = 0.910, after validation $AUC_{bootstrapped} = 0.892$) reveals an excellent discrimination between sites occupied and unoccupied by *Zygaena carniolica*.

It is obvious that these criteria strongly depend on the cut-off probability (Schröder & Richter 1999). The choice of this threshold is somewhat arbitrary (Fielding & Bell 1997, Pearce & Ferrier 2000), especially if the model is applied to other regions without any information on the prevalence (Strauss & Biedermann 2007). There are many approaches of determining a threshold; Liu *et al.* (2005) compared the performance of a variety of methods. Sometimes, the threshold is chosen by optimising specific criteria (e.g. Franklin

1998, Guisan *et al.* 1998, Bonn & Schröder 2001), e.g. by applying a program provided by Schröder (2006). In addition to its arbitrary character, dichotomising the gradient of predicted probabilities is tantamount to a loss of information (Pearce & Ferrier 2000). We strongly recommend the use of AUC for assessing discrimination, since it makes direct use of the occurrence probabilities (Vaughan & Ormerod 2005).

MODEL DIAGNOSTICS AND CHECKING FOR RESIDUAL SPATIAL AUTOCORRELATION

Besides assessing model performance, it is also wise to carefully check whether the model adequately represents the data (Nicholls 1989, Harrell 2001, Fox 2002). The distribution of residuals gives insight into any non-linear pattern and violations of distributional assumptions. Calculating Cook's distance helps to detect overly influential observations that have a great impact on estimated model parameters (Fox 2002; see Fig. 7.5 below).

To detect spatial autocorrelation in model residuals, which would violate the assumption of independence of sample sites and may lead to misleading interpretations (Kühn 2007), it is wise to calculate Moran's *I* (1950) as a global measure of spatial autocorrelation (as applied in Kerr *et al.* 2001, Hawkins & Porter 2003, Hein *et al.* 2007b). If this calculation is repeated for different distance classes between sample sites, then a correlogram is produced (see Fig. 7.6 below), depicting the isotropic or directional pattern of Moran's *I* (Keitt *et al.* 2002, Lichstein *et al.* 2002).

MODEL VALIDATION

Before applying a model, validation is a pivotal *sine qua non* (Vaughan & Ormerod 2005). Validation aims to objectively assess model performance (Reineking & Schröder 2003) and to test the model's applicability, credibility, generalisability and transportability (Rykiel 1996, Vaughan & Ormerod 2005). Usually, model performance on training data used for model estimation is higher than on test data that was not used to estimate model parameters (Verbyla & Litvaitis 1989). Due to the risk of overfitting, this effect is stronger the more complex the model is (Harrell 2001). Thus, model performance assessed on test data is overoptimistic. Since a model should also be generalisable and yield valid predictions when applied to different locations, model validation yields an unbiased assessment of model performance for independent test data. If the generalisability and transferability to other locations

Table 7.3 *Detailed internal validation results of the* Zygaena carniolica *model specified in Table 7.2 obtained by bootstrapping with 1000 replicates. 1st column: apparent model performance; 2nd column: mean values of apparent model performance on each bootstrap sample; 3rd column: mean of model performance of models built on bootstrap samples if applied to predict the original data set; optimism (4th column) is the difference between performance on training and test data which is subtracted from the original value to derive the corrected values (last column)*

	Original value	Performance on training data	Performance on test data	Optimism	Corrected value
AUC	0.910	0.9206	0.9024	0.0182	0.891
R^2_N	0.588	0.6248	0.5605	0.0643	0.524
Calibration intercept	0	0	−0.0545	0.0545	−0.0545
Calibration slope	1	1	0.8302	0.1698	0.8302

and/or years is not demonstrated, species distribution models allow only statements that are spatially and temporally restricted to the training data (Fielding & Haworth 1995, Schröder & Richter 1999). That is why testing for a model's generalisability and transportability is a crucial step before model application. Transportability describes the range of conditions over which conclusions are valid (Vaughan & Ormerod 2005). The other way round, it is a way to test for local adaptations (Kuussaari *et al.* 2000) or seasonal particularities (Kindvall 1995). In general, there are two approaches for validating predictive species distribution models: internal and external validation (Guisan & Zimmermann 2000).

Internal validation

For internal validation just one data set is used. In its simplest version, data splitting, the data set is randomly split into training and test data (e.g. 50 : 50, 67 : 33 or 80 : 20: Manel *et al.* 1999). Hastie *et al.* (2001) suggest partitioning the data into three sets (e.g. 50 : 25 : 25) for model estimation, model selection and model assessment. This approach is applicable only in data-rich situations and is not very effective in using the information available. More efficient alternatives are resampling methods like crossvalidation, jackknifing or bootstrapping (Verbyla & Litvaitis 1989, Chatfield 1995) that resample test data sets from the training data.

Crossvalidation is a generalisation of data splitting: in case of *m*-fold crossvalidation, the data is randomly split into *m* equally sized sets (at best maintaining prevalence). Then *m* − 1 parts are used as training data. Model performance is then evaluated on the remaining part as test data. This procedure is repeated *m* times. Overall model performance is then calculated from the *m* performance assessments on the *m* test data sets. Application in the field of species distribution

modelling are found in Schröder (2000), Bio *et al.* (2002), Boyce *et al.* (2002) and Lehmann *et al.* (2002a). An extreme case of crossvalidation is leave-one-out crossvalidation (also called jackknifing): in each test data set one single case is left out during model estimation (Manel *et al.* 1999).

These methods are usually outperformed by bootstrapping (Efron & Tibshirani 1993) which provides the most precise (least bias, least variance) estimates of model performance (Harrell 2001, Steyerberg *et al.* 2001b). When applying the bootstrap, the model is fitted to a large number of bootstrap samples (usually > 100). Each sample is randomly drawn with replacement from the original data set. Model parameters are estimated for each bootstrap sample, which represent the training data. Each time, this model is then applied to the original data (test data). Model performance is evaluated both for the training and the test data. The difference between these values is an estimate of optimism. Subtracting the average optimism over all replications from the final model's performance yields the internally validated performance (Efron & Tibshirani 1993, Harrell 2001, Steyerberg *et al.* 2001b; Table 7.3). If the model estimation on bootstrap samples is carried out with stepwise model selection, this method may also be used to check the reliability and robustness of selected predictors (Pan & Le 2001, Wisnowski *et al.* 2003, Hein *et al.* 2007b). Reineking & Schröder (2003), Oppel *et al.* (2004), Peppler-Lisbach & Schröder (2004) as well as Strauss & Biedermann (2005) apply bootstrapping for internal validation of species distribution models. Binzenhöfer *et al.* (2008) use it in the context of butterflies.

External validation

In contrast to internal validation, external validation uses independent test data. These may either be set aside from the original data before model building (equivalent to data splitting), or they may be collected at separate locations or separate periods of time. Validation is then done by applying the model to the test data which is equivalent to transferring the model in space and/or time (Thomas & Bovee 1993, Schröder & Richter 1999, Randin *et al.* 2006, Vanreusel *et al.* 2007). Therefore, the performance criteria as given above can be applied to test transferability. Schröder (2000) calculated AUC for the transferred model. If its confidence bounds contain a critical AUC (e.g. $AUC_{crit} = 0.7$ as in Binzenhöfer *et al.* 2005), the model is not transferable. This is equivalent to applying the critical ratio test following Beck & Shultz (1986): $z = (AUC - AUC_{crit})/SE_{AUC}$, where z is the quantile of the standard normal distribution and SE_{AUC} the standard error estimated for AUC. To calculate the confidence bounds for AUC, we recommend the bootstrap percentile method as performed in a software provided by Schröder (2006).

Examples for external model validation in studies on butterfly distribution are given by Dennis & Eales (1999), Fleishman *et al.* (2003), Luoto *et al.* (2002), Mac Nally *et al.* (2003), and Vanreusel *et al.* (2007). Binzenhöfer *et al.* (2008) applied both internal and external model validation regarding temporal and spatial model transferability. Bonn & Schröder (2001) applied a similar approach to test the representativeness of an umbrella species by transferring model predictions between species and species assemblages in carabid beetles. Analysing the transferability of distribution models between butterflies and bush crickets, Hein *et al.* (2007a) used the same approach.

CASE STUDY

In order to illustrate the above recommendations with real data, we briefly present predictive species distribution modelling for *Zygaena carniolica* and *Coenonympha arcania* in Northern Bavaria, Germany (Binzenhöfer *et al.* 2005).

Species distribution models for *Coenonympha arcania* and the burnet moth *Zygaena carniolica* in dry grasslands in southern Germany

Coenonympha arcania inhabits dry grasslands with bushes as well as grasslands near hedges and forest edges. The larvae feed on nutrient-poor grasses, the requirements of adults for nectar plants seem to be unspecific. *Zygaena carniolica* is a xerothermophilous species of fallow, moderately grazed or mown dry grasslands and within the research area (the

nature reserve Hohe Wann in Northern Bavaria, Germany), larvae feed on *Onobrychis viciifolia* and *Lotus corniculatus*. Adults prefer violet-flowering nectar plants such as *Knautia arvensis*, *Scabiosa columbaria* and *Centaurea* species. While agriculture still prevails in flat areas, slopes are mostly used less intensively and form a small-scale mosaic of semi-arid grasslands, warm fringes, scattered bushes and largely abandoned vineyards.

Within the study area, 118 sample plots of 30 m × 30 m were selected in a random stratified way (Hirzel & Guisan 2002) covering a range of six habitat types which represented the gradient of habitats within the study area. On each plot, a set of habitat parameters was quantified. Among others, it included parameters describing habitat type, management type, date of first annual management, inclination, exposition, level of shading, cover and height of vegetation, and landscape context (e.g. proportion of dry grassland in a radius of 100 m). All plots were surveyed for 15 minutes by two transect walks to record the presence or absence of both species.

MODEL DEVELOPMENT, EVALUATION AND VALIDATION FOR *ZYGAENA CARNIOLICA*

The final model for *Zygaena carniolica* after backward variable selection with AIC considers four parameters (Fig. 7.2 and Table 7.2; residual deviance = 71.696 on 101 degrees of freedom, null deviance = 128.119 on 105 degrees of freedom). The apparent model performance is quite good (AUC = 0.91, cf. Fig. 7.1, with 95% confidence interval 0.852 … 0.969, $R^2_N = 0.588$). If the cut-off probability is optimised for κ (yielding $P_{crit} = 0.4325$) then

- $\kappa = 0.71$ ('very good' according to Monserud & Leemans 1992),
- correct classification rate = 0.88 (altogether 93 out of 106 sites are correctly classified),
- sensitivity = 0.81 (25 out of 31 occurrences correctly classified), and
- specificity = 0.91 (68 out of 75 absences correctly classified).

Internal validation with 1000 bootstrap samples reveals only slight optimism: $R^2_N = 0.524$ and AUC = 0.892 (details see Table 7.3). A good calibration is also depicted in the calibration plot given in Fig. 7.3; the calibration line shows strong correspondence to the ideal straight line.

The related response curves are given in Fig. 7.2 showing increasing occurrence probabilities with increasing

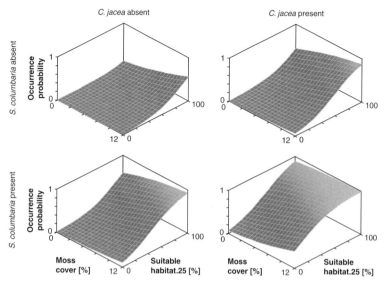

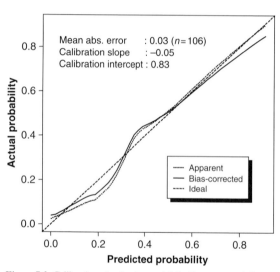

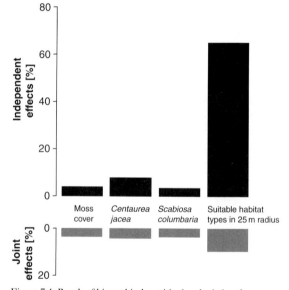

Figure 7.2. Response curves for *Zygaena carniolica* depending on cover of moss layer [%], coverage of suitable habitat types (i.e. semi-arid or extensively managed grassland, and thermophile fringes) within a radius of 25 m [%] and on the incidence of the host plant species *Scabiosa columbaria* and *Centaurea jacea* according to model estimates given in Table 7.2.

Figure 7.3. Calibration plot for the model for *Zygaena carniolica* given in Table 7.2.

Figure 7.4. Result of hierarchical partitioning depicting the independent and joint effects [% of total effect measured in terms of log likelihood] for each predictor variable on *Zygaena carniolica* incidence.

moss cover, increasing suitable habitat types within 25 m radius, and presence of both food plants, *Centaurea jacea* and *Scabiosa columbaria*.

A hierarchical partitioning reveals the independent and joint effects of these predictors on *Zygaena carniolica* incidence (Fig. 7.4) which cannot directly be derived from the coefficients in Table 7.2. The amount of suitable habitat types within 25 m radius has by far the strongest effect.

As an example for regression diagnostics, Fig. 7.5 enables the detection of overly influential observations that have a strong influence on model parameters and should therefore be checked for errors. This influence is measured as Cook's distance which integrates residuals and outliers in

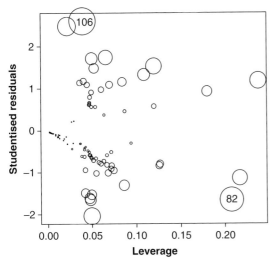

Figure 7.5. Regression diagnostic plot to detect overly influential observations (according to Fox 2002). The plot depicts for each study site the studentised residuals of the *Zygaena carniolica* model predictions and the leverages. Symbol size is related to Cook's distance which is calculated from residuals and leverage. In this case, sites no. 82 and 106 have the strongest effect on model estimation: site no. 82 due to a considerable residuum together with uncommon values for the predictors (high leverage), site no. 106 mainly due to its high residuum.

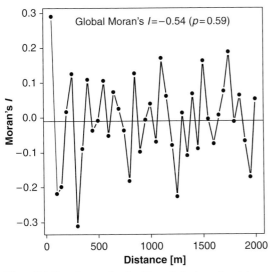

Figure 7.6. Correlogram showing Moran's *I* values of model residuals for different lag distances. The straight line indicates the expected value of Moran's *I* in case of no spatial autocorrelation. The global Moran's *I* does not significantly differ from this expected value. So, there is no residual spatial autocorrelation in case of the model for *Zygaena carniolica*.

Table 7.4 *Variance partitioning between geographical and environmental models of* Zygaena carniolica

Partitioning of R^2_N	
(1) Environmental factors alone	0.435 (trend and environment – trend alone)
(2) Spatial trend alone	0.060 (trend and environment – environment alone)
(3) Trend and environment	0.153 (trend alone + environment alone – trend and environment)
(4) Not represented	0.352 (1– trend and environment)

predictor values (measured as leverages). The two sites with highest values are marked. They exhibit either a large residual (site no. 106) or medium residual and high leverage. The authors did not find any spatial autocorrelation in the model residuals (predicted values – observed values): global Moran's *I* does not significantly differ from the expected value in absence of autocorrelation ($-1/(n-1) = -0.0095$). The correlogram does not depict an autocorrelation pattern deviant from the straight line (Fig. 7.6).

In a last step, geographical coordinates were additionally considered in the model to analyse the variance partitioning between the effect of geographic location and environmental predictors according to Borcard *et al.* (1992) and Heikkinen *et al.* (2005). Therefore two more models have been estimated: the spatial trend model considering only coordinates as well as the mixed model considering the environmental predictors shown in Table 7.2 and coordinates. The largest part of explained variance refers to environmental predictors (Table 7.4), the spatial trend alone is almost negligible.

COENONYMPHA ARCANIA: LANDSCAPE CONTEXT AND HABITAT SUITABILITY MAPS

The best model for *Coenonympha arcania* is shown in Fig. 7.7. It contains four parameters: proportion of hedges in suitable habitat types within 25 m radius, proportion of dry grassland within 100 m radius, presence of trees, and date of first annual management. Optimum habitats for *C. arcania* have 10% to 60% of dry grassland within 100 m. Furthermore, hedges contribute between 20% and 80% to suitable habitat types within 25 m. This span of optimum values for these two parameters is extended to a maximum in combination with late management and the presence of trees (lower right plot). The model shows high performance ($R^2_N = 0.64$, AUC = 0.92, $\kappa = 0.63$).

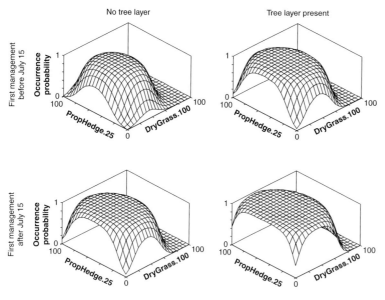

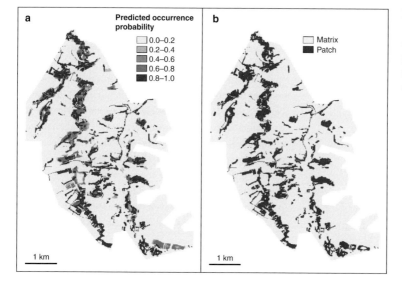

Figure 7.7. Response curves for *Coenonympha arcania* depending on tree layer (absent/present), date of first management (before/after July 15) and landscape context ('proportion of dry grassland in a radius of 100 m' and 'proportion of hedges within suitable habitat types in a radius of 25 m') (from Binzenhöfer *et al.* 2005, slightly modified).

Figure 7.8. Habitat suitability map of *Coenonympha arcania* in the Hassberge study area. (a) Predicted occurrence probabilities. (b) Predicted probabilities of map (a) dichotomised into a binary patch–matrix map (threshold P_{crit} = 0.57).

It has been widely recognised (Mazerolle & Villard 1999) that species' reactions cannot be fully understood without considering the influence of the surrounding landscape (i.e. landscape context). Most ecological processes depend not only on patch or plot characteristics, but on spatial scales much larger than a habitat patch (Holland *et al.* 2004). Effects of the landscape context have been demonstrated for a variety of taxa (Mazerolle & Villard 1999, Ricketts 2001).

The modelling of habitat suitability of *C. arcania* has shown that the inclusion of landscape context as an explanatory variable leads to a considerably better model. The performance of the habitat model for *C. arcania* based only on the plot parameters habitat type, date of first management, proportion of hedges and tree cover is fairly good (R^2_N = 0.47, AUC = 0.87, bootstrapped values). However, adding the landscape context parameters 'proportion of dry grassland in a radius of 100 m' and 'proportion of hedges in suitable habitat types in a radius of 25 m' into the modelling procedure resulted in a noticeable improvement of the habitat model.

The generation of habitat suitability maps requires that area-wide environmental data are available. Based on these data, a species distribution model can be used to produce a map of the predicted occurrence probabilities for a species. This is done by applying the regression equations to maps of the predictor variables. Habitat suitability maps may also depict predicted scenarios of future environmental conditions (Beaumont & Hughes 2002).

In *C. arcania* the 'best' model (see above) is not suitable for creating a habitat suitability map, as it contains environmental variables that are not available for the entire landscape. Consequently, a new model was specifically built for generating a habitat suitability map. It contains two parameters: 'proportion of dry grassland in a radius of 100 m' and 'proportion of hedges within suitable habitat types in a radius of 25 m'. This model performs slightly poorer than the 'best' model for *C. arcania* ($R^2_N = 0.59$, AUC = 0.90). The resulting habitat suitability map is shown in Fig. 7.8a. This map of continuous values of predicted occurrence probabilities can be dichotomised into a binary map (Fig. 7.8b) by applying the threshold probability $P_{crit.} = P_\kappa = 0.57$. This map, for instance, may serve as a map of habitat patches and surrounding matrix for the application of population viability analysis (PVA) or metapopulation models. However, since the matrix matters (Ricketts 2001), it is wise to avoid such simplification (Dennis *et al.* 2003) by considering the entire gradient of habitat qualities as predicted by the model. This results in a more realistic picture of both patch quality and matrix quality (Baguette & Mennechez 2004). Moreover, selecting an appropriate threshold when a model is applied to new data is problematic without information on the species' prevalence (Schröder 2000, Strauss & Biedermann 2007).

CONCLUDING REMARKS

The presented case study illustrates the recommendations given above with real data for *Zygaena carniolica* and *Coenonympha arcania* applying a standard logistic regression approach (Binzenhöfer *et al.* 2005). We emphasise the importance of model validation as well as model diagnostics to check for the violation of assumptions. For more general recommendations, we refer to the 'rules of thumb' presented by Austin (2007).

The presented methods yield valuable insights that help to understand species–habitat relationships, to test hypotheses regarding the effect of specific environmental properties on species distributions and to predict distributional patterns. They have several advantages such as their simplicity, their applicability on multiple scales, and the opportunity to predict the occurrence of comparatively large sets of species based on snapshot data. Nevertheless, species distribution models lack direct process implementation (Schröder & Seppelt 2006). But, since significant predictors are related to the underlying processes that are open for further testing by experiments or simulations (numerical experiments), species distribution models help to refine hypotheses of species–habitat relationships. The necessary next steps will be to determine the effects of different processes such as population dynamics, dispersal, non-equilibrium environmental conditions and biotic interactions on the performance of species distribution models as well as to couple species distribution models with dynamic process-based models (Guisan & Thuiller 2005, Zurell *et al.* (2009)). Generally, species distribution models yield a first and pivotal step for a more general mechanistic understanding of species–habitat relationships – with many fascinating examples regarding butterflies.

Part II
**Population biology: population structure,
dynamics and genetics**

8 • Butterfly population structure and dynamics

ROBERT J. WILSON AND DAVID B. ROY

SUMMARY

Butterfly population structure depends on interactions between the distribution of resources in the landscape, and the dispersal powers of the species in question. Resource limitation, natural enemies and climate affect the dynamics of local butterfly populations, and dispersal may cause separate populations of the same species to fluctuate in synchrony with each other. The survival or extinction of butterfly populations may depend on large-scale and long-term changes in the landscape that influence the quality, quantity and distribution of habitat. Relatively little is known about how climate could lead to changes in the phenological overlap between Lepidoptera and interacting species such as host plants or parasitoids and pathogens, and the ultimate effect on population occurrences and persistence. Testing the roles of interacting species, climate and habitat on butterfly life history and population dynamics can be done through field observations, field translocations and experiments in captivity. Detailed ecological data is essential to inform approaches involving metapopulation modelling and interpretation of molecular data, and these approaches in turn present valuable opportunities to predict large-scale or long-term effects of ecological change on species distributions and dynamics. The ability to use molecular techniques to identify genotypic variation among populations of species could increasingly be used to explore the evolutionary effects of environmental change, and the possibility that local adaptations could affect the ability of species to survive habitat fragmentation or climate change. Faced with widespread habitat fragmentation and the threat of climate change, conservation biologists must identify the factors that are pivotal to population persistence, and manage habitats and landscapes accordingly. Key areas for ongoing research into butterfly population structure and dynamics are suggested.

INTRODUCTION

The population structure of a species is the manner in which individuals are distributed in space, and the degree to which populations are linked by immigration and emigration. Population dynamics describes how numbers of the species change over time. Ecologists have always been interested in how populations respond to fluctuations in resources, natural enemies and environmental conditions, and can only succeed in safeguarding natural populations if they understand the processes that determine whether those populations increase or decline, persist, go extinct or colonise new areas. This chapter examines recent work on the structure and dynamics of butterfly populations, and considers its implications for butterfly conservation in the face of environmental change.

Species distributions consist of a number of populations, within which individuals interact through processes like courtship, mating and competition for food, mates and other resources. Some populations are defined by aggregations of individuals in discrete patches of suitable habitat, separated from other populations of the same species by unsuitable or unoccupied habitat. In sedentary species with localised habitat, migration events rarely link the separate populations. This type of population structure has been described as 'closed' because most individuals are born, live their entire lives, and die in the same 'patch' of habitat. At the other end of the spectrum, species are said to exhibit 'open' population structures if individuals range widely over the landscape, using widely distributed resources (Ehrlich 1984, Thomas 1984b, Warren 1992a). In reality, most butterfly populations are located somewhere between the two extremes: whether they are perceived as open or closed depends on the scale of observation, relative to the distribution of resources and the mobility of the species concerned, and population structure may change over time in relation to weather, climate and the distribution of resources.

In temperate Europe, most butterfly species have discrete, non-overlapping generations, and population dynamics represents fluctuations of abundance from generation to generation. Ecologists are interested in population dynamics as a quantitative measure of how species populations respond to changes in the environment. Short-term changes (e.g. variation in weather, host plant growth or the density of

Ecology of Butterflies in Europe, eds. J. Settele, T. Shreeve, M. Konvička and H. Van Dyck. Published by Cambridge University Press. © Cambridge University Press 2009, pp. 81-96.

natural enemies like parasitoids) lead to fluctuations in abundance from generation to generation, while longer-term changes (e.g. change in climate, habitat management or the distribution of natural enemies) can lead to trends in population size over numbers of generations. Since long-term changes affect overall population size, and short-term changes affect its variance, both can have important effects on the chances of population survival, and so are important in the context of conservation.

This chapter begins by showing how butterfly population structure depends on interactions between the distribution of resources in the landscape, and the dispersal powers of the species in question. We then examine the effects of resource limitation, natural enemies and climate on the dynamics of local butterfly populations, and how these factors and butterfly dispersal may cause separate populations of the same species to fluctuate in synchrony with each other. The penultimate section considers how the survival or extinction of butterfly populations may depend on large-scale and long-term changes in the landscape that influence the quality, quantity and distribution of habitat. Finally, we examine how population dynamics vary across species' geographical ranges, and the implications for butterfly responses to climate change, before suggesting key areas for ongoing research into butterfly population structure and dynamics.

THE DISTRIBUTION AND DENSITY OF BUTTERFLY POPULATIONS

The rarity or commonness of a species depends on both its distribution (range size and frequency of occurrence across its range) and the density or abundance of individuals within populations. Species can be rare because of localised distributions, low densities within populations or a combination of the two (Rabinowitz 1981). There is evidence that many taxa have positive density–distribution relationships: in other words that species with wider distributions tend also to have higher population densities (Gaston 1994, Gaston & Blackburn 2000). Such a pattern could be important because it implies that the survival of rare species is jeopardised both by small distribution size and low population density. Studies of the relationship in butterflies have often documented a positive association between density and distribution at one or more spatial scales, although results have been inconsistent (Hanski et al. 1993, Hodgson 1993, Gutiérrez & Menéndez 1995, Thomas et al. 1998c, Dennis et al. 2000b, Hughes 2000, Cowley et al. 2001a, b, Päivinen et al. 2005). Many of these studies have produced consistent associations

of either distribution or density with ecological variables like mobility, and host plant or habitat specificity, but have failed to show that distribution and density are themselves related.

Certain ecological variables can increase range size by enhancing migration, colonisation and persistence. Species with large wingspans, long adult lifespans or annual flight periods may be good colonists; and species with many or abundant host plants growing in a wide range of habitats may have widely available habitat (Dennis et al. 2000b). As a result, species with high mobility but low habitat specificity tend to have the broadest distributions, at a range of spatial scales (Hodgson 1993, Dennis & Shreeve 1996, 1997, Dennis et al. 2000b, Cowley et al. 2001a, Päivinen et al. 2005). Both high mobility and habitat generalism are important facets of the most widely distributed species, and the two patterns probably reinforce each other: selection may favour high mobility in generalist species, and generalism in mobile species. Dennis et al. (2004) showed how a wide range of life history characteristics in butterflies are associated with the strategy of their host plants (ruderal, competitive versus stress-tolerant species), with higher mobility and lower specialism in butterflies using ruderal and competitive species as host plants than those using stress-tolerant species.

On the other hand, butterfly abundance does not have such clear or consistent relationships with ecological variables. In general, butterfly species are found with the highest density in areas containing large amounts of their host plants (Quinn et al. 1998) and butterfly density at local or regional scales is closely related to host plant density (Gutiérrez & Menéndez 1995, Cowley et al. 2001b). However, no clear relationship is found between butterfly density and host plant density at a national scale in the UK (Hanski et al. 1993, Dennis et al. 2000b), and generalist species have lower population density than specialist butterflies in Finland (Päivinen et al. 2005). The relationship of abundance with larval host plant density could break down because the plants do not represent resources unless they grow in suitable microhabitats (e.g. Roy & Thomas 2003, Davies et al. 2006), and because other resources are also required for butterfly survival and reproduction (e.g. nectar for adult butterflies: Loertscher et al. 1994, Brommer & Fred 1999, Tudor et al. 2004). Therefore, a real understanding of butterfly population density may need to take account of all of the resources required through a species' life cycle (Dennis et al. 2003).

In conclusion, the type of relationship observed between distribution and density may depend on the main factors that influence the two variables. Cowley et al. (2001a, b) found that butterfly mobility had a positive effect on

Table 8.1 *Relationships (regressions) between distribution, density and mobility of British butterfly species at different spatial scales*

Scale	Local	Regional	National	European	Global
Location	Creuddyn Peninsula (Wales)	Creuddyn Peninsula (Wales)	England, Wales and Scotland	Europe	World
Total area	$0.25\,km^2(\times 5)$	$35\,km^2$	$229\,800\,km^2$	$10\,400\,000\,km^2$	
Grid cell area	$0.0025\,km^2$	$0.25\,km^2$	$100\,km^2$	$153\,000\,km^2$	
$N_{species}$	19	26	49	49	49
$N_{independent\ contrasts}$	17	24	45	45	45
Distribution versus density	+	(+)	+	+	−
r^2	0.368**	0.052^{NS}	0.278***	0.088*	0.170**
Distribution versus mobility	(+)	+	+	(+)	+
r^2	0.007^{NS}	0.203*	0.290***	0.033^{NS}	0.181**
Density versus mobility	(−)	−	(+)		
r^2	0.111^{NS}	0.271**	0.029^{NS}		

Notes: Distribution was the fraction of grid cells occupied at each scale, except global distribution, which was ranked according to the number of biogeographic zones occupied, from 1 = endemic to Europe, to 7 = occurring on five or more continents. Density was measured using transect walks at local, regional and national (British) scales. European and global distributions were related to national (British) density. Mobility was ranked on a 'consensus' classification, based on the proportion of 24 northern European respondents who classified each species as more, less or equally mobile than each other species. Analyses were controlled for the effects of phylogenetic relationships between species using comparative analysis by independent contrasts (CAIC) (Cowley *et al.* 2001a).
Significance: *** $P < 0.001$; ** $P < 0.01$; * $P < 0.05$; ^{NS} $P > 0.05$. Non-significant relationships shown in parentheses.

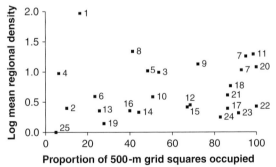

Fig. 8.1. Density and distribution of butterfly species on the Creuddyn Peninsula in north Wales (UK). Density is the mean number of butterflies counted during 1997 per occupied 300-m transect. Distribution is the proportion of 500-m grid cells ($n = 140$) where each species was recorded from 1996 to 1998. Points are labelled by relative mobility, from 1 (least mobile) to 25 (most mobile) (ties in rank mobility are given the same number). Species referred to in text: 1, *Plebejus argus*; 23, *Pieris brassicae*. (Reprinted from Cowley *et al.* 2001b, courtesy of Blackwell Science Ltd.)

distribution, but a negative effect on density at the regional scale (Table 8.1). Species that occurred at high densities

relative to their distributions were more sedentary and used more aggregated resources than species which occurred at low densities. For example, *Plebejus argus* was a sedentary species which occurred at high densities, but in few locations; whereas *Pieris brassicae* was a mobile species which was widespread but at low densities (Fig. 8.1). In fact, if high-quality habitat is localised for sedentary species, the isolated but high-density populations of these species can lead to a negative density–distribution relationship, as demonstrated for butterflies near their northern range margins in Finland (Päivinen *et al.* 2005; see 'Population dynamics at range margins' below).

LOCAL POPULATION DYNAMICS

Where butterfly populations have been monitored in the UK, population density of some species remains fairly constant over time, whereas other species show wide variations from one generation to the next (Table 8.2) (Thomas *et al.* 1998a). These changes in population size depend on birth and death rates, and rates of immigration and emigration. Birth, death and dispersal rates vary according to local

Table 8.2 *The variability of UK butterfly populations, compared with those of* Maculinea arion *and* M. rebeli

Species	Coefficient of variation (%)
Maculinea rebeli (c)	<55
Anthocharis cardamines (o)	55
Gonepteryx rhamni (o)	60
Polygonia c-album (o)	61
Maniola jurtina (c)	62
Pararge aegeria (c)	64
Aphantopus hyperantus (c)	66
Pieris napi (I) (o)	66
Argynnis paphia	69
Ochlodes venata (c)	69
Thymelicus sylvestris (c)	71
Coenonympha pamphilus (c)	71
Erynnis tages (c)	71
Inachis io (o)	72
Hipparchia semele (c)	74
Pieris rapae (I) (o)	75
Pieris brassicae (I) (o)	75
Pieris napi (II) (o)	75
Lycaena phlaeas (I) (c)	79
Melanargia galathea (c)	83
Polyommatus icarus (I) (c)	84
Lasiommata megera (I) (c)	86
Pieris rapae (II) (o)	86
Boloria selene (c)	88
Callophrys rubi (c)	88
Lysandra coridon (c)	90
Aglais urticae (o)	92
Argynnis aglaja (c)	92
Lycaena phlaeas (II) (c)	92
Maculinea arion (c)	**92**
Pieris brassicae (II) (o)	100
Lasiommata megera (II) (c)	113
Aricia agestis (II) (c)	115
Vanessa atalanta (o)	118
Polyommatus icarus (II) (c)	119

Notes: Relative amounts of variation in population dynamics are shown as coefficients of variation (CVs: the variance of a data set expressed as a percentage of the mean). To avoid statistical artefacts, CVs were calculated only for UK species with at least seven BMS sites which had annual population index values for at least eight consecutive years, and a mean index value exceeding 9 (Thomas *et al.* 1994). The value of CV was calculated for *M. arion* using data from four established UK populations (Thomas *et al.* 1998a); CV was estimated indirectly for *M. rebeli* using empirical models tested against field data (Elmes *et al.* 1996, Thomas *et al.* 1998a). (I) and (II) represent annual comparisons of the first and second generations of bivoltine species; (c) and (o) represent species with predominantly closed and open populations (Thomas *et al.* 1998a).
Source: After Thomas *et al.* (1998a).

population density (density dependence), and according to density-independent environmental factors like climate.

Density dependence

Population regulation occurs if population density increases when density is low and decreases when density is high. Such a pattern of negative feedback can result if death rate increases, if birth rate decreases or if net emigration increases at higher population density. Two agents of density dependence are resource limitation and natural enemies. If there are sufficient resources to support a population of a certain size but no larger ('carrying capacity'), then competition for resources limits population density when it reaches the carrying capacity. Alternatively, at high population densities, predators, pathogens or parasites may reach such high densities themselves, or may find potential prey or hosts with such ease, that they suppress further population growth. Density dependence is not easy to detect over short periods (Dempster 1983, Stiling 1988, Dempster & McLean 1998), but long-term data sets suggest that it is a widespread phenomenon in the Lepidoptera. Woiwod & Hanski (1992) tested for density dependence in British populations of moths, using data from the Rothamsted Insect Survey. They analysed year-to-year changes in the number of individuals of 263 species, at sites which had been monitored continuously for at least ten years (4073 species–site time series). Of these time series, 47% showed significant density dependence (tending to increase when low and decrease when high). Over short time periods density dependence can be obscured by environmental effects like annual changes in the weather. Thus, when the analysis was restricted to time series longer than 20 continuous years of monitoring, 79% of the 486 time series showed significant density dependence. The pattern was even more prominent (88%, $n = 446$) when time series showing significantly increasing or decreasing trends were excluded from the analysis. Long-term trends in population density reflect changes in carrying capacity that can be caused by changes to climate or habitat. When the local environment becomes progressively better, population density may increase to a new, higher level; if conditions become progressively worse the final outcome can be local population extinction.

Density dependence has also been demonstrated in several species of butterflies whose annual population growth rate declines as population density increases (Baguette & Schtickzelle 2006). Newly founded populations increasing towards a possible carrying capacity provide further

evidence. *Pararge aegeria* has been expanding at the edge of its range in Britain for the past century (Hill *et al.* 1999c, 2001, Asher *et al.* 2001), and has colonised a number of monitored sites. Pollard *et al.* (1996) analysed annual counts for ten sites in Eastern England that were colonised by *P. aegeria* between 1985 and 1993. Newly formed populations increased in size by three to four times one year after colonisation, but populations grew more slowly in later years, and ceased to increase six to eight years after colonisation. This pattern of growth could be due to an increase up to a density where available resources limit further growth, or where interference between butterflies increases population emigration rates. Alternatively, the decline in population growth could result from increased predation or parasitism, if species-specific natural enemies themselves colonised the new *P. aegeria* populations some time after their host.

The roles of different sources of mortality and density dependence in butterfly population dynamics have been investigated for some butterfly species using life tables and key factor analysis (see Gilbert & Singer 1975, Dempster 1983, Stiling 1988, Warren 1992a). The following sections consider some recent examples of the effects of resource limitation and natural enemies on lepidopteran populations.

Resource limitation

The dynamics of phytophagous Lepidoptera often track those of their host plants, suggesting resource limitation (Dempster & Pollard 1981, Dempster & McLean 1998). For many species, the critical factor is not the overall amount of host plant, but the availability of plants that are growing in suitable conditions for egg-laying and larval development. Species at their cool, upper-latitudinal range margins can be restricted to warm microclimates (Thomas 1983b, 1993, Thomas *et al.* 1986, 1999). This is exemplified by *Polyommatus bellargus* which has two generations per year at its northern range margin in the UK. Adults fly in May–June and August–September. Females from the second generation lay eggs on the host *Hippocrepis comosa* growing in the hottest, short-turfed (0–3 cm) microhabitats, so that larvae experience sufficiently warm temperatures for development during autumn and early spring. However, a wider range of microhabitats provides comparably warm temperatures for offspring to develop in summer, so that females of the early summer generation lay eggs in longer vegetation (2–6 cm) and even avoid the hottest short turf sites (Roy & Thomas 2003). As a result, there are approximately twice as

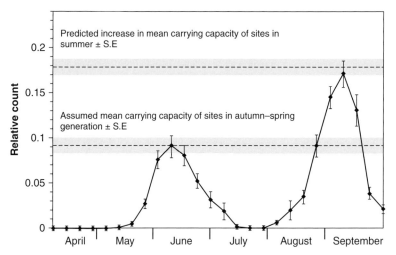

Fig. 8.2. Population size and carrying capacity of *Polyommatus bellargus* in Britain. Solid lines show mean weekly counts expressed as a proportion of annual site totals in 1976–2000. Dashed lines (± SE in grey) show seasonal differences in predicted carrying capacity, based on the availability of *Hippocrepis comosa* in suitable habitat for egg-laying. The autumn–spring carrying capacity is fitted to the peak May–June butterfly count, on the assumption that this represents an annual population bottleneck (after Roy & Thomas 2003).

many resources for larval development in summer than in autumn/spring for British populations of *P. bellargus*, and the adult population size of the species in the second generation is approximately twice the size of the first generation (Fig. 8.2).

At or near carrying capacity, resource limitation affects population growth rate through intraspecific competition. The outcome for the population depends on the mechanism of competition: in pure 'contest' competition, the number of surviving competitors matches the number that resources can support; in pure 'scramble' competition, the competitors deplete resources to such an extent that they all perish. In reality, contest and scramble competition are two ends of a continuum, but some species show characteristics that are more similar to one than the other.

Anthocharis cardamines shows contest competition for oviposition sites (Courtney & Duggan 1983, Thomas 1984a, Dempster 1992, 1997). Larvae feed on the developing seed pods of crucifers, usually *Alliaria petiolata* or *Cardamine pratensis* in the UK. Only one larva can usually survive per flower head, because larvae are cannibalistic. In addition, females avoid laying eggs on flower heads where an egg is already present (eggs are conspicuous, orange and coated in a pheromone which deters oviposition). At Monks Wood (UK), the number of eggs laid per year closely followed the availability of 'suitable' *C. pratensis* flower heads (unshaded, with more than seven flowers), where larval mortality was least. When the number of suitable flower heads was experimentally increased, the number of eggs increased correspondingly. Egg shortfall (i.e. female butterflies not laying their full complement of eggs because of

insufficient egg-laying sites, or insufficient good weather for flight to find suitable sites) is the major factor regulating populations of *A. cardamines* (Courtney & Duggan 1983). This system contrasts with the scramble-like dynamics of the moth *Tyria jacobaeae* whose populations regularly crash following complete defoliation of its host plant *Senecio jacobaea* (Dempster & Pollard 1981, Van der Meijden *et al.* 1998).

To maintain butterfly populations, it is important to identify the resources which impose critical limits on species population dynamics. *Maculinea* spp. are some of the most endangered butterflies in Europe (Thomas 1995a, Van Swaay & Warren 1999), and their population dynamics has been studied in detail (Thomas 1995a, Thomas *et al.* 1998a). In each species, the first three larval instars feed on species-specific host plants, and the fourth and final instar develops inside the nests of a species-specific ant host (*Myrmica* spp.). High and variable mortality within host ant nests is the key factor determining population density and dynamics, but density regulation within ant nests depends on the specific life history of the *Maculinea* species.

Maculinea arion caterpillars are obligate predators of the brood of *Myrmica sabuleti*. *Myrmica* species rear two ant broods per year, with large ant larvae present both in late summer and spring. *Maculinea arion* caterpillars feed preferentially on large ant larvae in autumn, and often exhaust the supply. Their winter diapause ends up to a month after their ant hosts, so that small ant larvae have grown larger before the caterpillars resume feeding. By monitoring the weights of *M. arion* caterpillars and pupae, Thomas & Wardlaw (1992) calculated that each caterpillar would have to consume 230 large ant larvae to complete development

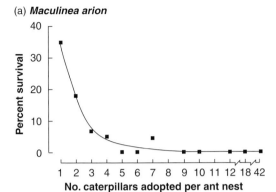

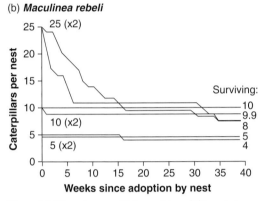

Figure 8.3. The outcomes of (a) scramble and (b) contest competition in two *Maculinea* butterflies. (a) Survival to pupation of *M. arion* caterpillars adopted by *Myrmica sabuleti* nests, against numbers of caterpillars adopted per nest. When the number of caterpillars exceeds carrying capacity, most or all caterpillars starve (from Thomas & Wardlaw (1992), courtesy of Springer-Verlag). (b) The numbers of *M. rebeli* caterpillars surviving in laboratory ant nests against time since adoption. Figures show numbers of caterpillars adopted per nest, and numbers surviving after 40 weeks. *Maculinea rebeli* is fed by *Myrmica* workers. When carrying capacity is exceeded, survival matches resource availability, and similar numbers of caterpillars survive (after Thomas *et al.* 1993).

(86 in autumn, and 144 in spring), and that ant broods of this size would only be supported by colonies with about 350 workers. When the number of caterpillars adopted into the ant nest exceeded the carrying capacity (based on ant larvae availability), scramble competition led to the starvation of all or most caterpillars (Fig. 8.3a).

In contrast, *Maculinea rebeli* adopts a 'cuckoo feeding' strategy in the nests of its host ant *Myrmica schencki*. The caterpillars' behaviour and surface chemistry mimic those of *M. schencki* larvae, and they are fed directly by worker ants

with regurgitations and nutritious trophic eggs (Elmes *et al.* 1991a, b, Akino *et al.* 1999). The cuckoo-feeding strategy is more energy-efficient than predation: the caterpillars feed at a lower trophic level because their food is not already digested and assimilated by the ant larvae. Consequently, fewer workers are required for *M. rebeli* to complete development in its host ant nests (Elmes *et al.* 1991a, b; Thomas & Elmes 1998). Furthermore, when too many *M. rebeli* caterpillars are adopted by a nest, workers preferentially feed larger larvae, mediating contest competition and increasing the chances that the nest will support as many caterpillars as its resources allow (Fig. 8.3b) (Thomas *et al.* 1993). Thus, cuckoo-feeding *Maculinea* species both feed and compete more efficiently than predacious species. Consequently, Thomas & Elmes (1998) found on average 5.3 *M. rebeli* pupae but only 1.2 *M. arion* pupae in the respective species' host ant nests. The drawback for cuckoo-feeding species is increased productivity at the cost of increased specificity: *M. rebeli* caterpillars adopted into non-host *Myrmica* nests experience about 30 times the mortality of those adopted by *M. schencki* (the main host for *M. rebeli*), whereas *M. arion* caterpillars adopted by non-hosts experience only three times the mortality of those adopted by *M. sabuleti* (the main host for *M. arion*) (Thomas & Elmes 1998).

The population dynamics of *M. rebeli* have been modelled using parameters based on detailed rearing experiments, life-table analysis and monitoring of habitat at a large population of the butterfly in the Spanish Pyrenees (Hochberg *et al.* 1992, 1994, 1996, Clarke *et al.* 1997). The models, and their field tests (Thomas 1995a, Elmes *et al.* 1996), demonstrate that larval survival in the nests of *M. schencki* is the key factor determining *M. rebeli* population density and dynamics. For example, mortality of the first three larval instars on the host plant *Gentiana cruciata* was 17.6% (caused by density-independent predation and density-dependent starvation and cannibalism). Model results showed that reducing host plant density by two orders of magnitude would reduce butterfly density by 85%, but the population would still persist, as long as there were sufficient *M. schencki* nests. In contrast, mortality in ants' nests was 80.4%, and reducing *M. schencki* nest density by 85% or more caused all modelled populations of *M. rebeli* to go extinct (Hochberg *et al.* 1992). The importance of host plant availability would increase if *M. schencki* was no longer the limiting resource, i.e. if *M. schencki* became more abundant, or if *G. cruciata* became much rarer.

Cuckoo-feeding *Maculinea* species are expected to have more stable population dynamics than obligate predators, because contest competition matches survival to resource availability in the former, whereas mortality caused by scramble competition overcompensates for resource shortages in the latter. Modelled population dynamics and their field tests appear to support this prediction (Elmes *et al.* 1996, Thomas *et al.* 1998a; Table 8.2), since the estimated coefficient of variation (CV) of *M. rebeli* populations (<55%) is much lower than that of *M. arion* (92%) (*M. arion*'s high CV may also reflect high variability in the climatically marginal sites in England where population size was measured). Interestingly, another species in which there is contest competition at the key stage of its life cycle (*Anthocharis cardamines*: see above) also has a low population CV (55%).

Natural enemies

Natural enemies like predators, parasites and pathogens are further important sources of mortality in butterfly populations (Gilbert & Singer 1975, Dempster 1983, Ehrlich 1984, Stiling 1988, Warren 1992a). Different natural enemies affect different stages of the life cycle: typically, vertebrates affect later stages than invertebrates. For example, *Lycaena dispar* is a univoltine butterfly which hibernates as a larva. Experiments using cages to exclude different types of natural enemy suggest that, before larval diapause, invertebrate predators or parasitoids are the dominant causes of mortality (~30% of larvae), whereas after diapause vertebrate predators may cause almost 50% larval mortality (Webb & Pullin 1996).

Rates of predation and parasitism depend on population structure and population dynamics. Migration may provide refuges from natural enemies, which could explain why sedentary and continuously brooded populations of *Danaus plexippus* have higher prevalence of the protozoan parasite *Ophryocystis elektroscirrha* than migratory populations (Altizer *et al.* 2000).

Parasitoids can cause high and variable rates of mortality in butterfly populations (Porter 1983, Lei & Hanski 1997). In two populations of *Maculinea rebeli*, 6–23% of mortality was caused by the specialist parasitoid *Ichneumon eumerus* (Thomas & Elmes 1993). Female *I. eumerus* locate *Myrmica schencki* nests, but may only enter the nests to parasitise *M. rebeli* caterpillars if they can be sure that the benefits (of finding a host) exceed the costs (of attack from worker ants). Some ant nests may act as refuges from parasitism because parasitoids may be unlikely either to detect or to enter nests containing one or few *M. rebeli* caterpillars. Modelling the dynamics of this system readily reproduced

stable dynamics with similar levels of parasitism to those observed in the field, as long as sufficient nests were parasitoid-free refuges to caterpillars (simulated by increasing the threshold number of caterpillars needed for a parasitoid to enter a host nest). As the fraction of parasitised nests and caterpillars increased, the system became less stable, eventually to the extent that both *M. rebeli* and *I. eumerus* went extinct (Hochberg *et al.* 1996). As predicted by general theory, host–parasitoid systems may be unstable unless there is sufficient heterogeneity in the risk of parasitism.

Populations in which there are very high rates of parasitism may be prone to local extinction. In the Åland islands in southern Finland, *Melitaea cinxia* occurs as a network ('metapopulation') of small local populations (Hanski *et al.* 1994, 1995a). In this system, risk of local extinction is directly related to rates of parasitism by the braconid wasp *Cotesia melitaearum* (Lei & Hanski 1997). Because *C. melitaearum* increases the extinction risk of *M. cinxia* populations (and hence the extinction risk of its own populations), the host–parasitoid interaction is unstable and can only persist at the metapopulation level. The host *M. cinxia* may survive in populations with no (or few) parasitoids. These populations may act as sources for the recolonisation of habitat patches where *C. melitaearum* had previously driven *M. cinxia* extinct, providing new colonies for *C. melitaearum* to colonise in turn. The parasitoid is found more commonly in large *M. cinxia* populations, where each species is less likely to go extinct, and in host populations close to other *C. melitaearum* populations, which are more likely to be colonised by parasitoids. *Cotesia melitaearum* has a large impact on the dynamics of *M. cinxia* because it is gregarious and produces three generations of adults per single host generation. *Hyposoter horticola* is another specialist parasitoid of *M. cinxia*, but produces one generation of solitary offspring per generation of its host. *Hyposoter horticola* is an inferior competitor in that it causes lower rates of parasitism than *C. melitaearum*, but it occupies a higher proportion of local populations, because it disperses more widely and does not destabilise the local dynamics of *M. cinxia* to such an extent (Lei & Hanski, 1998). The necessity of parasitoids to locate suitable hosts, but not to exploit so many that their hosts suffer local extinction, means that host–parasitoid systems exhibit often complex dynamics. These dynamics require further detailed research, not least because specialist parasitoids can be considered at a greater risk of extinction than their lepidopteran hosts (e.g. Thomas & Elmes 1993, Hochberg *et al.* 1996, van Nouhuys 2005).

The role of pathogen natural enemies in the regulation of butterfly populations is less well studied. The potential importance of fungal diseases as mortality agents of Lepidoptera, particularly in wet summers, has long been recognised (Bierne 1955) but rarely evaluated. Similarly, although most organisms are host to specialist and generalist microbial natural enemies, the role of these diseases in population regulation in natural systems is uncertain (Lafferty & Gerber 2002).

Climate conditions

Climate may affect butterfly population dynamics directly, for example, larval mortality caused by flooding in *Lycaena dispar* (Webb & Pullin 1996), catastrophic mortality caused by unseasonal frosts in *Euphydryas editha* (Singer & Thomas 1996, Thomas *et al.* 1996), or egg shortfall caused by poor weather during the flight period of *Anthocharis cardamines* (Courtney & Duggan 1983). Climate may also act indirectly, by affecting interactions between butterflies and other species. The second generation of *Aglais urticae* at sites in southern England is more abundant in years when rainfall is above average during May and June (Pollard *et al.* 1997), because high water and nitrogen contents in its host plant *Urtica dioica* increase larval growth rates (Pullin 1987b). Rates of larval mortality in two species of checkerspot butterflies have been shown to depend on the extent to which spring weather conditions synchronise larval growth rates with those of their specialist parasitoids. In cool but sunny springs, dark post-diapause larvae of *Euphydryas aurinia* and *Melitaea cinxia* are able to speed up their developmental rates by basking, whereas the white immobile cocoons of their parasitoids *Cotesia bignellii* and *C. melitaerum* develop more slowly, and most adult parasitoid wasps fail to emerge before the checkerspot larvae have pupated (Porter 1983, van Nouhuys & Lei 2004). Consequently, parasitism rates are much lower in cool sunny springs (e.g. 7.7% at an English population of *E. aurinia* in 1979) than warm but cloudy springs (74.5% at the same site in 1980: Porter 1983).

Climate can also affect resource availability from generation to generation (Roy & Thomas 2003). The same variation may produce contrasting effects among species with different phenologies and habitat requirements. In Britain, most bivoltine and some univoltine species are more abundant during warm, dry summers, possibly because conditions become more suitable for the development of young stages and the emergence of adults. On the other hand, species like *Aphantopus hyperantus*, *Polygonia c-album*, *Pieris napi* and *Pararge aegeria*, which breed in partially shaded habitats, tend to be more abundant in years following cool, moist summers, possibly because reduced host plant water content adversely affects larval growth (Pollard 1988, Pollard & Yates 1993, Roy *et al.* 2001). Following extremely unfavourable conditions, populations may retract to 'core' habitat types which persistently fulfil their niche requirements (Thomas *et al.* 1998c). At Monks Wood (UK), all local populations of *A. hyperantus* declined after a drought in 1976, but those in the most open and most shaded habitats went extinct, whereas populations survived in partially shaded habitat (Sutcliffe *et al.* 1997b).

When conditions favour populations in different habitats from year to year, persistence may require a variety of habitat types. The dynamics of *Euphydryas editha bayensis* at Jasper Ridge in California depend on complex interactions between rainfall, temperature and the timing of host plant growth and senescence (Singer 1972, Ehrlich *et al.* 1975, 1980, Dobkin *et al.* 1987, Weiss *et al.* 1988, 1993). The primary larval host plant, *Plantago erecta*, is an annual which germinates at the start of the rainy season (~December) and which completes flowering and sets seed by the beginning of the dry season (~May). Larvae undergo diapause during the dry season, and resume feeding after the germination of *P. erecta*. The starvation of pre-diapause larvae, if they fail to reach their fourth instar before host plant senescence, is the major cause of mortality, and the primary determinant of adult population size the following year (Singer 1972, Ehrlich *et al.* 1975, Hellmann *et al.* 2003). Eggs laid on north-facing slopes benefit from reduced mortality in dry years, because host plants do not senesce as quickly as on hotter slopes. However, during cooler years, post-diapause larvae may develop so slowly on north-facing slopes that the resulting adults emerge too late to lay their eggs on non-senescent plants (Weiss *et al.* 1988, 1993). Thus, climate and topography determine the degree of synchrony between the phenology of *E. editha* and *P. erecta*: extremes of rainfall (either very hot and dry conditions, or very cool and wet conditions) increase mortality by reducing the temporal overlap between larvae and their host plants. Since different slopes and aspects are favoured from year to year, topographic diversity buffers population dynamics against conditions which affect the suitability of different local sites antagonistically (Hellmann *et al.* 2003). Precipitation variability has increased so much at Jasper Ridge since the 1970s that two populations of *E. editha* have been driven extinct: a population in a relatively large (9.8 ha) but topographically homogeneous area went extinct in 1991, whereas a population in a smaller (2.5 ha) but more topographically variable patch survived another seven years before dying out in 1998

(McLaughlin *et al.* 2002a, b). This work on *E. editha bayensis* is an important case study of the importance of habitat heterogeneity in preventing butterfly population extinctions in a changing climate.

Spatial synchrony in population dynamics

Butterfly numbers depend on climate, natural enemies and the quality and distribution of resources. When these factors are correlated in space, different populations of the same species may fluctuate synchronously. British butterfly species show low levels of synchrony among populations at least 200 km apart, probably related to regionally correlated weather conditions (Pollard 1991, Pollard & Yates 1993, Pollard *et al.* 1993, Sutcliffe *et al.* 1996). Population synchrony can also be increased by dispersal among populations: large-scale levels of population synchrony in butterflies are lower than in moths and aphids, perhaps because longer-distance dispersal in moths and aphids synchronises populations over a wider area (Hanski & Woiwod 1993).

At shorter distances, correlated environmental conditions and dispersal can greatly enhance population synchrony (Sutcliffe *et al.* 1996) (Fig. 8.4). The relative importance of environmental effects versus dispersal depends on the spatial scale over which environmental factors are correlated relative to species mobility. In sedentary species, fine scale population dynamics may be highly synchronised because local population density corresponds closely to local habitat conditions, but dispersal declines steeply with distance, and only enhances population synchrony up to about 2 km (Fig. 8.4b). Very sedentary species like *Melitea athalia* and *Plebejus argus* may show asynchronous fluctuations as little as 1 km apart because of local variation in habitat quality and the near absence of dispersal between local populations (Warren 1987b, c, Thomas 1991a). In more mobile species, dispersal appears to enhance synchrony up to about 4 km on average (Fig. 8.4b).

EXTINCTION RISK

The conservation of butterfly species requires an understanding of the reasons why local populations sometimes suffer extinction. Extinction can result from long-term declines in population size caused by changes in habitat or other environmental conditions like climate. In addition, chance events or fluctuating unpredictability in natural systems can lead to local extinctions, particularly in small populations. We now consider how habitat quality and population size influence population dynamics and extinction risk, and how these factors can be important in landscapes where habitat is fragmented.

Habitat quality

In addition to year-to-year changes in population density, long-term trends resulting from deterministic changes in the quality or quantity of available habitat can be superimposed on population dynamics. If there are consistent increases or decreases in resource availability from year to year, there will be accompanying trends in population size. In species that use larval host plants growing in early successional stages (e.g. *Melitaea athalia*: Warren 1987c, *Plebejus argus*: Thomas 1991a), population density declines as succession proceeds, and eventually populations may go extinct if available resources no longer support a population.

Population growth rates and carrying capacities differ between areas of different quality habitat (Singer 1972, Thomas *et al.* 1998a). Birth rate exceeds death rate in suitable, 'source' habitats, whereas death rate exceeds birth rate in unsuitable 'sink' habitats. In 'source–sink dynamics', sink populations may persist purely because of net immigration from sources. However, it can be difficult to unequivocally identify sources and sinks in the field, with implications for conservation. Populations that consistently show negative growth in the presence of immigration may do so because population density has been elevated above their habitat's carrying capacity, but these populations ('pseudo-sinks') may be viable at lower densities in the absence of immigration, and may have an important role in regional species survival. For example, populations of *Euphydryas editha* on rocky outcrops in California had poor breeding success when their numbers were elevated by immigration from forest clearings. However, when extreme weather conditions caused the extinction of the clearing populations, outcrop populations survived but at reduced densities (Singer & Thomas 1996, Thomas *et al.* 1996). In this example, immigration from the clearings had formerly increased population density in the outcrops to levels beyond carrying capacity, and breeding success was poor as a result. When immigration ceased, the outcrop populations survived and acted as sources for the recolonisation of the clearings (Boughton 1999).

Population size

One common theoretical prediction is that small populations are more likely to go extinct than larger populations of the

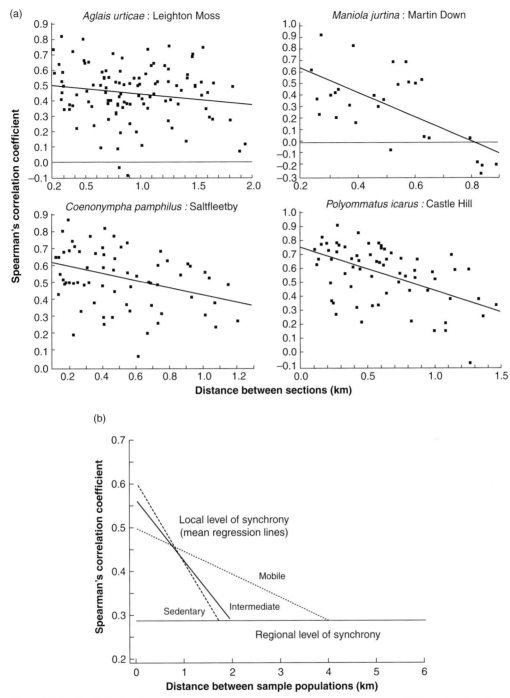

Figure 8.4. Declines in local-scale synchrony with increasing distance between local populations. (a) Examples of synchrony between populations of species monitored by the British Butterfly Monitoring Scheme. Spearman's correlation coefficients between population indices were plotted against distance between transect sections. Regression lines were calculated, and grouped according to species mobility (sedentary / intermediate / mobile: Pollard & Yates 1993). (b) The mean regression line for each mobility group, against distance between populations, plotted until it reached the approximate regional level of synchrony, which declined negligibly up to ca. 200 km (from Sutcliffe *et al.* 1996, courtesy of Blackwell Science Ltd).

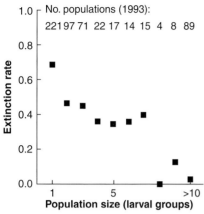

Figure 8.5. Rates of extinction of populations of *Melitaea cinxia* in the Åland islands, against population size (number of larval groups). Number of populations of each size class in 1993 is shown; extinction rate is the fraction of populations that went extinct by 1994 (after Hanski *et al.* 1995b).

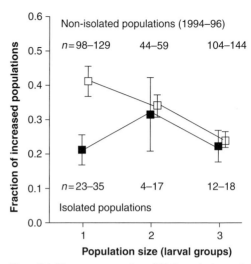

Figure 8.6. The mean proportion of *Melitaea cinxia* populations of different sizes that increased from year to year, grouped by extent of isolation. Isolated populations were more than 1 km from a pooled area of 0.5 ha of occupied habitat; *n* = range of sample sizes in each class, 1994–1996; bars show ± 1 SE (from Kuussaari *et al.* 1998, courtesy of Munksgaard International Publishers Ltd).

same species, because of unpredictable variation (stochasticity) in the environment, or in death and reproduction from year to year (Hanski 1999). Changes in mortality of a similar magnitude are more likely to reduce a small population to extinction, and small population size can also reduce fecundity and increase mortality because it increases rates of inbreeding (Saccheri *et al.* 1998). Small habitat patches can also suffer increased extinction risk through chance environmental variation, because an environmental perturbation or change in management is more likely to cause catastrophic mortality over a small area than a large area. In addition, per capita rates of emigration are higher from small habitat patches (Hill *et al.* 1996, Kuussaari *et al.* 1996, Sutcliffe *et al.* 1997a), so that the loss of individuals through emigration may increase extinction risk in small populations (Thomas & Hanski 1997, Thomas *et al.* 1998c). At low population density, butterflies may also have difficulty finding mates: in *Melitaea cinxia* on the Åland islands in Finland, emigration rate was higher and female mating success lower in very small, low-density populations, and population size decreased as a result (Kuussaari *et al.* 1998). Related to many of these problems, extinction risk in *M. cinxia* was inversely related to population size (Hanski *et al.* 1995b) (Fig. 8.5).

Habitat fragmentation

Habitat fragmentation occurs when habitat patches become smaller and further apart. Reducing habitat area can increase extinction risk through reductions in resource availability

and a greater vulnerability to chance environmental variation. Habitat quality may also decline when habitat area decreases, because of increased 'edge effects' from adjoining habitats. For example, lepidopteran population dynamics in field margins are affected by a wide range of agricultural practices, which may cause mortality directly (e.g. by the use of insecticides or the cutting of host plants) or by their effects on host plant cover or nutritional quality (e.g. because of fertiliser and herbicide use) (Feber & Smith 1995).

Any within-population process which increases the risk of local extinction adds to the potential importance of the 'rescue effect'. That is, very small populations, populations breeding in poor or variable quality habitat and populations with highly variable dynamics may only persist if they are bolstered by immigrants from nearby populations of the same species. For example, when isolated and non-isolated populations of *Melitaea cinxia* were analysed separately, the very smallest isolated populations tended to decrease from year to year, possibly related to high emigration (Fig. 8.6). In non-isolated populations, population growth rate decreased as population density increased, probably because the smallest populations were 'rescued' by emigrants from other populations nearby. Habitat fragmentation is a problem because it not only reduces the size and increases the extinction risk of individual populations, but, by increasing the

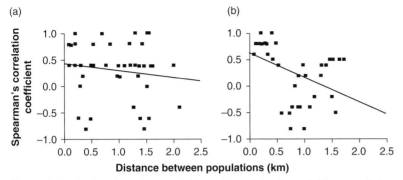

Figure 8.7. Levels of synchrony between *Aricia agestis* populations in two habitat networks in north Wales (UK). Spearman's correlation coefficients between four generation indices (2nd brood 1996 – 1st brood 1998) were plotted against distance between transects (km). (a) Ten site transects in an unfragmented habitat network, Mantel test for the relationship of synchrony with distance between populations, $n = 45$ correlations, $r = -0.16$, $p = 0.368$ (regression not significant – line shown for comparison with (b); (b) Nine transects in a fragmented habitat network, Mantel test, $n = 36$ correlations, $r = -0.43$, $p = 0.023$. Synchrony declined significantly with increasing distance (from Wilson 1999b).

distance between patches, it also reduces the likelihood that immigration will rescue declining populations.

'Metapopulations' are groups of local populations linked by migration, where colonisations of empty patches of suitable habitat can maintain a network of populations of a species in a landscape, despite occasional extinctions of some local populations (Thomas & Hanski 1997, 2004, Hanski 1999). However, persistence of metapopulations is favoured only if the constituent local populations fluctuate asynchronously (i.e. if one population decreases, others may increase, acting as sources of immigrants). Asynchrony in local population dynamics can result from interactions between weather and local habitat type (Weiss *et al.* 1988, Sutcliffe *et al.* 1996, 1997b). In *Plebejus argus*, asynchronous local dynamics dampen overall population variability among groups of variable linked sub-populations (Thomas 1991a).

Habitat fragmentation could affect levels of synchrony between local butterfly populations, because it affects rates of dispersal between populations (Hill *et al.* 1996, Kuussaari *et al.* 1996, Thomas & Hanski 1997). The population density of *Aricia agestis* was monitored for four generations on the Creuddyn Peninsula in north Wales, across two population networks: one in a large expanse of coastal grassland, and the other in fragmented habitat in an agricultural landscape. Levels of population synchrony declined more markedly in the fragmented network, possibly because populations were linked less by dispersal (Wilson 1999b) (Fig. 8.7). But other factors could also be important (e.g. effects of habitat fragmentation on dispersal by parasitoids). The patterns of synchrony in *A. agestis* were also related to changes in the density of suitable host plants. Females require lush

Helianthemum chamaecistus plants for oviposition (Bourn & Thomas 1993). Over summer, the quality of *H. chamaecistus* plants declines more on dry, south-facing slopes than on cooler slopes. Areas of habitat whose host plant quality fluctuated more synchronously also had more synchronous butterfly population dynamics (Wilson 1999b). Therefore habitat quality and habitat fragmentation interact to determine levels of synchrony between populations.

POPULATION DYNAMICS IN THE FACE OF CHANGE

The changes in butterfly population density and distribution that result from climate change can be partly understood by consideration of how population dynamics vary across species' geographical ranges, which we consider in the following section. We then look at evidence for the effects of recent climate change on butterfly population dynamics, before in the final section considering some of the priorities for butterfly population research in a rapidly changing environment.

Population dynamics at range margins

Moving from the core to the margins of a species range, fewer locations satisfy its ecological requirements. As a result, populations of sedentary species may be restricted to increasingly small, isolated areas of habitat at range margins (Thomas 1993, Thomas *et al.* 1998c, 1999), but they may maintain high population densities if persistent populations are only supported by locations with high resource densities and low risk of extinction (Päivinen *et al.* 2005). Dispersive species that sample the environment over wide areas, 'averaging out' the quality of resources may potentially suffer a reduction in population density towards range edges (Thomas *et al.* 1998c).

At species range margins, small temperature changes could push ectothermic insects beyond the limits of their physiological tolerance, and minor environmental changes could have major impacts on habitat availability. Species that are restricted to warm microhabitats at their cool range limits may be able to breed in a wider variety of locations during warm years, but may struggle to find suitable habitat in cool years (Thomas 1993, Thomas *et al.* 1998c, 1999). In contrast, hot or dry conditions at warm margins may be associated with increased mortality. As a result, habitat use may change over climatic gradients. For example, in central Spain *Aporia crataegi* lays egg batches on cool, north-facing branches of its host plant at hot, low elevations, but on warm, south-facing branches at cool, high elevations (Merrill *et al.* 2008). Nearer to the centre of the range, relative resource availability may not vary so much from year to year, potentially leading to more stable population dynamics. This could explain why the population dynamics of some butterflies show increased fluctuations near range margins (e.g. *Maniola jurtina*: Thomas *et al.* 1994a) (Fig. 8.8), potentially exposing them to increased local extinction risk.

Population dynamics and climate change

Considerations of butterfly population dynamics at range margins allow predictions to be made about the responses of butterfly populations to climate change. Under improving conditions (e.g. by climate warming at cool margins), populations may become less prone to extinction, because they are more stable (Thomas *et al.* 1994a), larger (Pollard *et al.* 1995) and less isolated from other suitable areas of habitat or populations of the same species (Thomas *et al.* 1998a, 2001a). On the other hand, under deteriorating conditions (e.g. by climate warming at warm range margins), extinction risk may increase as populations become more unstable, smaller and more isolated.

There is already evidence that climate change has affected the dynamics of butterfly populations in Britain, where the population sizes of most common, widespread species have increased since the 1970s (Pollard & Eversham 1995, Warren *et al.* 2001, Fox *et al.* 2006). Most of these species showed greater increases in the east than in the west, implying that a

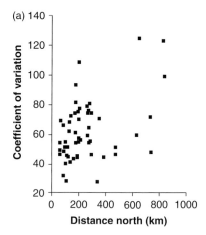

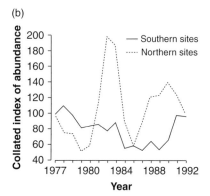

Figure 8.8. Levels of fluctuation of British populations of *Maniola jurtina*. (a) Coefficient of variation of population size against distance north in Britain (km, Ordnance Survey national grid, $r = 0.438$, $n = 54$, $p < 0.001$). (b) Collated fluctuations of the six most northerly and six most southerly populations, 1977–1992 (from Thomas *et al.* 1994a, courtesy of Munksgaard International Publishers Ltd).

regionally correlated factor like climate was related to the changes (Pollard *et al.* 1995). Many butterfly species reach their north-western range margins in Britain, probably because of climatic constraints (Turner *et al.* 1987, Dennis 1993, Thomas 1993). As climate warms, many 'marginal' British sites may become more climatically similar to sites nearer to the core of species' distributions. As a result, a larger area of microhabitat at each habitat patch may meet species' ecological requirements, thereby increasing population size (Thomas *et al.* 1994, 1998c, 2001a).

Recent research provides evidence of how climate change can affect survival and fecundity in butterfly populations. Warmer temperatures have increased overwintering survival at the cool range margin of *Atalopedes campestris* in

North America, allowing the species to expand its distribution northwards (Crozier 2003, 2004). On the other hand, increased temperatures at the low elevation limit of *Aporia crataegi* in central Spain may be linked to increased egg and young larval mortality, leading to a retraction of the species distribution to higher elevations (Merrill *et al.* 2008). The effects of increasing temperatures on fecundity depend on climatic conditions relative to the thermal adaptations of the species concerned. When females of two species of open grassland (*Hipparchia semele* and *Coenonympha pamphilus*) and two shade-dwelling species (*Aphantopus hyperantus* and *Pararge aegeria*) were kept at temperatures of 20–40 °C, lifetime egg production was greatest at intermediate temperatures. However, peak egg production occurred at higher temperatures for the open grassland species (30 °C) than the shade-dwelling species (25 °C) (Karlsson & Wiklund 2005). In British populations of *Hesperia comma*, egg-laying rates increase with ambient temperature from 18 to 28 °C, and egg-laying at warmer temperatures is less restricted to hotter, short-turfed microhabitats (Davies *et al.* 2006). Temperatures in south-east England during *H. comma*'s July–August flight period have increased by 2 °C since the 1980s, leading to increased fecundity, increased resource availability (i.e. more suitable habitat) and increased population sizes of the butterfly (Davies *et al.* 2006, Wilson *et al.* 2007). Reflecting increased extinction at low latitudes and elevations, and/or increased colonisation at high latitudes and elevations, many butterfly species distributions have recently shifted to higher latitudes and elevations (e.g. Parmesan 1996, Parmesan *et al.* 1999, Hill *et al.* 2002, Konvička *et al.* 2003, Wilson *et al.* 2005, Franco *et al.* 2006).

PRIORITIES FOR BUTTERFLY POPULATION RESEARCH

Rapid environmental change provides opportunities to identify the mechanisms that control species distributions and dynamics (Parmesan *et al.* 2005), but also increases the urgency to monitor and conserve populations that are suffering the effects of change. Understanding the contributions of interacting species (host plants and natural enemies), climate and the size and isolation of areas of suitable habitat to butterfly population dynamics will be critical to conserve species as the climate changes and habitats are increasingly modified.

Key areas for research include the relative importance for butterfly populations of direct climatic effects on mortality (Crozier 2003, Merrill *et al.* 2008), fecundity (Karlsson &

Wiklund 2005, Davies *et al.* 2006) or dispersal, versus indirect effects of climate on butterflies via interacting species. In particular, relatively little is known about how climate could lead to changes in the phenological overlap between Lepidoptera and interacting species such as host plants (Weiss *et al.* 1988) or parasitoids (van Nouhuys & Lei 2004) and pathogens. Climate effects on host plant condition or phenology could determine the importance of habitat heterogeneity for the conservation of butterflies (McLaughlin *et al.* 2002a, b), whereas climate effects on host–natural enemy dynamics could influence population variation and local extinction risk, and therefore the importance of maintaining networks of suitable habitat (Lei & Hanski 1997). At a wider scale, potential changes to the distributions of host plants (Araújo & Luoto 2007) or natural enemies will affect the ability of butterfly distributions to track the availability of suitable climates.

A range of approaches has already been applied to test the roles of interacting species, climate and habitat on butterfly life history and population dynamics, including field observations (Davies *et al.* 2006), field translocations (Crozier 2003, Merrill *et al.* 2008) and experiments in captivity (Karlsson & Wiklund 2005). In the context of conservation, it will be important to apply data such as these to life tables or population viability analysis to make predictions about the consequences of habitat changes for population survival (e.g. Thomas *et al.* 1998a, Schtickzelle *et al.* 2005a). Detailed ecological data are essential to inform approaches such as metapopulation models and molecular techniques, and these modern approaches in turn present valuable opportunities to predict large-scale or long-term effects of ecological change on species distributions and dynamics. Exemplary work on *Melitaea cinxia* in the Åland islands has combined detailed ecological information with modelling approaches to identify the processes that influence the landscape-scale dynamics of the species (e.g. Hanski & Meyke 2005). Molecular research on the *M. cinxia* system has now shown that the allelic composition of one enzyme (phosphoglucose isomerase, PGI) has a significant influence on population growth rate (Hanski & Saccheri 2006). This enzyme has been shown to be associated with flight metabolic performance and fecundity (Haag *et al.* 2005), and as a result different alleles seem to be favoured in large, well-connected habitat patches versus small, isolated patches. The ability to use molecular approaches to identify genotypic variation among populations of species could increasingly be used to explore the evolutionary effects of environmental change, and the possibility that local adaptations could affect the ability of species to survive habitat fragmentation or climate change (e.g. Schmitt & Hewitt 2004b, Zakharov & Hellmann 2008).

In summary, the structure and dynamics of butterfly populations reflect complex interactions between life histories, habitat requirements and environmental conditions. Faced with widespread habitat fragmentation and the threat of climate change, conservation biologists must identify the factors that are pivotal to population persistence, and manage habitats and landscapes accordingly. Understanding butterfly populations has never been more important as they are increasingly being used as indicators of the impact of environmental change and to represent trends in other insect taxa (Thomas 2005). The work discussed in this chapter represents a valuable beginning to this important and challenging process.

9 • Costs and benefits of dispersal in butterflies

THOMAS HOVESTADT AND MARKO NIEMINEN

SUMMARY

The roles of various factors promoting emigration in butter-flies are summarized, namely inbreeding avoidance, habitat succession, population density, habitat quality, patch size, predation and diseases, and temporal variability. Relevant theoretical arguments are discussed in relation to empirical evidence on emigration in butterflies.

Theory and evidence indicate that individual dispersal decisions are context-dependent and triggered by specific cues. Transition through the landscape is greatly affected by the structure of the landscape and observed dispersal pat-terns are the outcome of the interplay between specific environmental conditions, the landscape setting, and the individuals' set of behavioural rules. Thus dispersal patterns will be unique to individuals, landscapes and time. We need to learn more about the behavioural rules responsible for the emergence of population-level dispersal patterns. Difference in dispersal ability, or willingness, may be maintained in heterogeneous landscapes and lead to the emergence of population specific dispersal rules. Identifying simple 'rules of thumb', e.g. emigration decisions based on the time interval between consecutive encounters with conspe-cifics or resources, which allow individuals to make reason-able guesses about the status of their habitat, is a challenge in the future. The transition from phenomenological approaches describing dispersal patterns at the level of pop-ulations, to behavioural approaches attempting to under-stand the underlying mechanisms responsible for the emergence of these patterns has been fostered by both the advancement of theory on the evolution of dispersal strat-egy, and careful and cleverly designed field observations. This route of investigation is demanding, but likely to pro-vide real benefits for understanding the fate of butterflies in changing landscapes.

INTRODUCTION

The significance of butterfly mobility for population dyna-mics has been recognized for a long time (Ehrlich 1984,

Dempster 1991, Thomas 1994, Wilson *et al.* 2002). In fact, butterfly studies linking mobility to spatial population dynamics played a significant role in the development of metapopulation and landscape ecology (Thomas & Hanski 1997, Hanski 1998, 1999, Hanski & Gaggiotti 2004). More recently, increasing fragmentation of natural habitats has raised more practical concerns. How capable are butterflies in moving between suitable habitat patches in fragmented landscapes? And what are the consequences of habitat frag-mentation for population dynamics at the landscape level? The general conclusion from these studies has been that for most butterflies, distribution areas are currently decreasing because butterflies are not able to cope with increasing distances between suitable habitat patches (Thomas & Hanski 1997, Warren *et al.* 2001, Baguette *et al.* 2003).

Insects in general have vast potential for mobility and lepidopteran species are no exception (e.g. Lokki *et al.* 1978). Nonetheless, some species move very little and seem highly reluctant to leave their natal habitat (Thomas & Hanski 1997). These sedentary species include many lycaenid species. For example, in a study on *Polyommatus bellargus* in the UK, individuals covered distances up to 250 m within all parts of their natal site, but they were not observed to cross gaps of only 100 m between habitat patches (Thomas 1983b). Similar results were obtained for *Plebejus argus* in the UK: 89% of the same-day movements were less than 20 m, none exceeded 50 m, and all were within the focal patches (Thomas 1985). But why are some species highly dispersive while others are so sedentary?

Among all life-history attributes, 'dispersal' may argu-ably be the one with the widest spectrum of fitness conse-quences: untimely emigration may lead to certain death, but a successful colonizer may become a founder of a whole new population. Thus, evolution is likely to favour individuals that make the 'right decisions' and emigrate, when their expected fitness elsewhere is larger than that at 'home' (Bowler & Benton 2005). To be more precise, evolution should balance emigration rates such that the fitness of philopatric and dispersing individuals is similar (cf. Holt & Barfield 2001, Metz & Gyllenberg 2001, Poethke & Hovestadt 2002), an argument derived from the theory of

Ecology of Butterflies in Europe, eds. J. Settele, T. Shreeve, M. Konvička and H. Van Dyck. Published by Cambridge University Press. © Cambridge University Press 2009, pp. 97-106.

ideal-free distribution (Doncaster 2000). It is thus unlikely that an individual's dispersal decisions are random but rather that they are context-dependent and triggered by external and internal cues. This statement is supported by strong theoretical arguments and a multitude of empirical observations which cast doubts on the simplistic approaches of most current metapopulation models, which assume dispersal to be random (Baguette 2004). We will come back to this issue in the conclusions.

For the purpose of this chapter, we define 'dispersal' as a trans-generational, significant spatial shift of the site of reproduction away from the current/natal site; we exclude return-migration events from further consideration, where individuals return to the place of origin even if this migration period covers several generations. Dispersal becomes successful only if the individual survives dispersal *and* reproduces thereafter. A population geneticist will call displacement 'significant' only if an individual moves into a group of individuals with a different genetic background, a behavioural ecologist if an individual moves from an environment it knows to one it does not know, and a population biologist if it transfers an individual into a site which is either empty (colonization) or harbours a population following a dynamic independent from that it originated from (usually the criterion in metapopulation studies). In all cases, we can assume that dispersal is motivated by an individual's expectation to increase its inclusive fitness. It is this underlying motivation which – according to many ecologists – separates dispersal from everyday movement (Kennedy 1961, Dingle 1996, 2001, Van Dyck & Baguette 2005, Schtickzelle et al. 2007). Consequently, the dispersal frequency and distance of a species usually cannot be estimated from measures of local (daily) movements (Shreeve 1995) even though a net displacement may also come about by everyday movements.

Conceptually, it is helpful to separate dispersal into three different phases (Ims & Yoccoz 1997, Clobert et al. 2001b). The first is 'emigration', i.e. the decision to leave the current habitat patch or population – typically the natal population. It is followed by the period of 'transition' and is completed with 'immigration' or 'settlement' into a new habitat/population. In this chapter we take an evolutionary perspective and analyse what is known about the benefits and costs of dispersal in butterflies. Fundamentally, we expect more individuals to disperse if either the expected benefits of dispersal are large, or the costs of dispersal are low. The expected benefits and costs may differ between species – depending on their life history – and landscapes, but also among individuals within a population.

In the first section, we summarize the role of various factors promoting emigration in butterflies. For each of these potential benefits we briefly discuss the relevant theoretical arguments. Because there are recent books and reviews with a strong focus on the theory of dispersal and its measurement (Turchin 1998, Clobert et al. 2001a, Woiwod et al. 2001, Bullock et al. 2002, Hanski & Gaggiotti 2004, Bowler & Benton 2005) we focus on the empirical evidence on emigration in butterflies. In the second section we discuss the potential costs of dispersal associated with transition, again focusing on the empirical evidence collected from butterflies. In the final section we review the role of individual differences for variability in dispersal strategies within species or populations. We do not address issues concerning immigration/settlement as such.

THE BENEFITS OF DISPERSAL: WHY EMIGRATE?

In an ideal world, individuals would emigrate from their natal site when the expected benefit is larger than the expected cost of dispersal. To make such a truly rational decision, an individual would need to have complete information about the situation not only in its actual site but also about potential targets for immigration. The great benefit of information-gathering abilities for the evolution of adaptive behaviour has recently been discussed by Dall et al. (2005). However, as the world is usually less than ideal the critical question becomes: 'What, if anything, can individuals know about the benefits and costs of dispersal?' Before emigration, the only specific information individuals can collect is about the habitat in which they currently live. Thus, an evolving information-dependent strategy should weight the conditions in an individual's current habitat with the long-term average condition in the world at large. Information about the latter cannot be 'known' by an individual, but can be adjusted evolutionarily by, for example, the adjustment of trigger values required to set off dispersal behaviour (Hovestadt et al. 2001, Metz & Gyllenberg 2001, Poethke & Hovestadt 2002). In the following sections we discuss briefly the factors which have been suggested to promote emigration.

Kin competition and inbreeding avoidance

Theory predicts that kin competition will maintain a minimum amount of emigration in the absence of other factors promoting emigration (Hamilton & May 1977, Gandon & Michalakis 2001, Lambin et al. 2001). The underlying basis for this prediction is the assumption that population growth is density-regulated (cf. Hassell 1998). Under such conditions, an individual's own sacrifice in direct fitness, when leaving its natal population, may be compensated for by the added offspring its philopatric kin produce because the individual's emigration reduces the intensity of local competition (Ronce *et al.* 2000, Poethke & Hovestadt 2002, Poethke *et al.* 2007). A second genetic argument in favour of dispersal (Perrin & Goudet 2001, Roze & Rousset 2005) is the avoidance of inbreeding depression (e.g. Gaggiotti 2003). An effect of inbreeding depression has been found repeatedly in butterflies, e.g. in *Melitaea cinxia* (Saccheri *et al.* 1998, Nieminen *et al.* 2001, Haikola 2003). Saccheri & Brakefield (2002) even demonstrated that the genes of females introduced in inbred populations spread rapidly in their new population, i.e. that they had a disproportionate reproductive success compared to resident females. In practice, however, it may be very difficult to decide whether dispersal is motivated by kin competition or by inbreeding avoidance because they are both likely to be strong under the same conditions.

These arguments do not imply that dispersal is in general driven by the need to avoid kin competition or inbreeding. The lack of such evidence for butterflies may be due to the fact that it is not easy to document the kin structure of a population in the first place, and that dispersal is more strongly affected by some of the other factors discussed below (Perrin & Goudet 2001). Because substantial kin competition and inbreeding occurs only if a population develops a distinct genetic structure, it becomes a dominant factor in the decision to emigrate if successful dispersal is rare (Poethke & Hovestadt 2002, Poethke *et al.* 2007).

Emigration driven by kin competition does not require that individuals can actually determine their relatedness to other individuals (Bowler & Benton 2005): an individual can always expect to be more closely related to individuals in its natal site compared to individuals somewhere else. However, the magnitude of the difference in the relatedness will depend primarily on dispersal itself. Thus, selection should favour the evolution of certain emigration propensity up to the level where the expected fitness benefit of kin-driven emigration is compensated for by the expected costs of dispersal (Poethke *et al.* 2003).

Habitat succession

Naturally or anthropogenically induced habitat change is an important factor leading to local population extinction

(Thomas 1994). It is obvious that species inhabiting habitats of an early successional state must emigrate sooner or later in order to persist (Ronce & Olivieri 1997). Consequently, Southwood (1962) and Shreeve (1995) as well as Bergman & Landin (2002) proposed that butterfly species inhabiting successional habitats must have good dispersal and habitat-locating capabilities. This agrees with the observation that species feeding on early successional plants had the farthest, and species feeding on trees and shrubs the shortest flight distances (Shapiro & Carde 1970, Scott 1975). For *Melitaea athalia* females, Warren (1987b, 1992b) reported shorter residence times in patches where the habitat was deteriorating, implying that emigration increases as a patch becomes more unsuitable.

The statement that species living in short-lived habitats should be more dispersive than those occurring in stable habitats does, in itself, not implicate that a context-dependent emigration strategy should evolve. However, individuals would clearly benefit if they do not emigrate when the habitat still provides good reproductive opportunities. The tendency to emigrate should thus be low in the early successional phase but increase as succession proceeds (Olivieri & Gouyon 1997, Ronce & Olivieri 2004, Ronce *et al.* 2005). For a long-lived organism, its own age could provide a reasonable cue for the right time to emigrate (Ronce & Olivieri 2004), but this has not been tested for butterflies and is probably not a relevant option for such short-lived organisms. The observations reported by Warren (1987b) suggest that females of *Melitaea athalia* respond instead to some cues indicating the quality or the successional status of the habitat.

Population density

All populations must occasionally be regulated by density-dependent mechanisms (Hassell 1998). This implies that the fitness for an individual living in a high-density population is lower than that of individuals at low density. Thus, leaving a high-density population with the perspective to settle in another with lower density is presumably an adaptive strategy. Several theoretical studies indicate that density-dependent emigration should be characterized by a certain threshold density below which individuals should not emigrate at all and above which emigration probability should increase gradually (Janosi & Scheuring 1997, Travis *et al.* 1999, Metz & Gyllenberg 2001, Poethke & Hovestadt 2002).

Among butterflies, there are many examples of an increasing tendency to emigrate as population density increases. For example, high density (presumably associated with a shortage of food) facilitates emigration in *Ascia monuste* (Nielsen 1961), *Pieris protodice* (Shapiro 1970) and *Boloria eunomia* (Baguette *et al.* 1996). Thomas (1983b) observed that *Polyommatus bellargus* crossed habitat borders more frequently at high population density. The effect of high population density on emigration has been demonstrated experimentally for *Melitaea harrisii* (Dethier & MacArthur 1964), with nectar presumably being the limiting resource (Gilbert & Singer 1973). A special case is the tendency of the larvae of *Metisa plana* to disperse by ballooning as population density becomes high (Rhainds *et al.* 1997).

For female butterflies encounters with harassing males sometimes provoke emigration (Shapiro 1970, Ehrlich & Murphy 1981, Odendaal *et al.* 1985, 1989). Examples include *Boloria eunomia* (Baguette *et al.* 1996, 1998, Petit *et al.* 2001), and *Colias eurytheme* in which the amount of harassment and subsequent emigration is colour-morph-dependent (Gilchrist & Rutowski 1986). In addition, female *Aphantopus hyperantus* actively avoid males after copulation (Wiklund 1982a). Experimental evidence for emigration induced by harassment has also been reported for *Boloria aquilonaris* (Brunzel 2002). These cases indicate that individuals are able to assess density based on the number of encounters with conspecifics, or on reliable correlates of their presence.

It has also been reported that butterflies tend to emigrate from populations of low density. Examples include *Euphydryas chalcedona* (Brown & Ehrlich 1980) and *Parnassius smintheus* (Roland *et al.* 2000). Low female density increases emigration rate of males in *Boloria eunomia* (Baguette *et al.* 1998). For *Euphydryas editha* (Gilbert & Singer 1973) and *Melitaea cinxia* (Kuussaari *et al.* 1996, Enfjall & Leimar 2005) actually positive and negative density-dependent emigration have been reported. Several factors may explain such reversed density-dependence in emigration. First, the effect of population density may be confounded with that of habitat quality or patch size (Roland *et al.* 2000) (see below). Yet an Allee effect (Allee *et al.* 1949, Kuussaari *et al.* 1998, Menendez *et al.* 2002, Taylor & Hastings 2005) could in some conditions be the ultimate factor driving the evolution of such behaviour, if, for example, finding mates becomes too difficult (Kuussaari *et al.* 1998).

Habitat quality

Clearly, individuals should prefer 'good', i.e. habitat that provides the critical resources the animal needs, over 'poor' habitat. However, if individuals also compete among each other, habitat quality cannot be considered independently from population density. It may still pay to reproduce in a low-quality 'sink' population if population density is sufficiently lower there compared to that in high-quality patches (Sutherland *et al.* 2002). Thus the argument in favour of the evolution of density-dependent emigration can simply be expanded to account for the effect of habitat quality: with declining habitat quality the density threshold at which emigration is elicited should also decline. If we compare habitat patches of different quality we could thus observe a reversed density-dependent pattern of emigration across patches. This is the case for *Euphydryas editha* (Boughton 2000), even though, within each patch, emigration is positively density-dependent.

For several butterfly species, individuals leave patches if critical resources are rare or absent. For example, *Euphydryas editha* tends to leave patches that lack preferred hosts (Thomas & Singer 1987) and stay in preferred habitat (Boughton 2000). Also, a larger proportion of *Colias palaeno europome* left poorer than better-quality patches (Ruetschi & Scholl 1985). Emigration rate of male *Melitaea cinxia* may decrease with increasing amount of nectar plants in a patch (Kuussaari *et al.* 1996). The 'judgement' about what constitutes a good habitat may also vary within species or populations. For example, *Veronica*-preferring *Melitaea cinxia* females had higher emigration rate than *Plantago*-preferring females in non-*Veronica* patches although there was no such difference for males (Hanski & Singer 2001). In a system of meadows separated by small distances, *Parnassius smintheus* immigrated preferentially into meadows with a natural abundance of flowers compared to meadows where flowers had been removed artificially (Matter *et al.* 2004). These empirical data indicate that at least some butterflies are capable of estimating the availability of some critical resources (Schneider *et al.* 2003). In addition the presence of conspecifics may be used as an indicator of habitat quality, a trait which could lead to negative density-dependent emigration (cf. Boulinier & Danchin 1997, Grevstad & Herzig 1997, Odendaal *et al.* 1999, Välimäki & Itämies 2003, Reed 2004).

Patch size

Emigration rates often decline with patch size (e.g. Hill *et al.* 1996, Kuussaari *et al.* 1996), which can proximately result from the higher probability of edge encounters for individuals moving in small patches (Thomas & Hanski 1997,

Crone & Schultz 2003). However, there are also arguments for an underlying adaptive explanation behind this phenomenon. First, relatedness among individuals in a small patch may be higher and thus promote emigration of individuals (Poethke & Hovestadt 2002). Second, the proportional number of immigrants expected to arrive in a small population may be higher in small patches as has been demonstrated for *Aphantopus hyperantus* (Sutcliffe *et al.* 1997a), *Aporia crataegi*, *Melanargia galathea* (Baguette *et al.* 2000), *Maculinea nausithous* and *M. teleius* (Hovestadt *et al.* unpublished data). The great importance of a disproportionate 'rescue effect' for small populations has in fact been documented in *Melitaea cinxia* (Hanski 2003). However, from the perspective of an individual residing in a small patch, strong immigration implies disproportionately more competitors in the future! Theory thus predicts that selection should promote higher emigration rates from smaller habitat patches (McPeek & Holt 1992, Poethke & Hovestadt 2002).

Species observed to emigrate in higher proportions from small habitats are quite numerous, e.g. *Hesperia comma* (Hill *et al.* 1996), *Melitaea cinxia* (Kuussaari *et al.* 1996), *Aphantopus hyperantus* (Sutcliffe *et al.* 1997a), *Colias palaeno europome* (Ruetschi & Scholl 1985), *Aporia crataegi* and *Melanargia galathea* (Baguette *et al.* 2000). In *Iolana iolas* only males are more likely to emigrate from small patches (Rabasa *et al.* 2007). Similarly, *Boloria eunomia* males leave small patches and stay in large patches with higher probability than do females (Petit *et al.* 2001), a pattern which may be due to the patrolling behaviour of males.

Predation/disease

The avoidance of predators, e.g. birds, lizards, spiders or dragonflies (Moore 1987, Brakefield *et al.* 1992), may be a substantial component of butterfly movement. Clearly, direct escape flights (which will rarely lead to dispersal) are adaptive, but for aphids it has been suggested that they start to produce winged offspring following predator attacks (Weisser *et al.* 1999, Weisser 2001, Kunert *et al.* 2005). Such a delayed response to a predation signal could be adaptive only if a recent encounter with a predator is indicative of an elevated predation risk in the future (Poethke *et al.* 2007). For butterflies, this may not be the case with predation, but might be so with respect to encounters with parasitized or diseased conspecifics. Females, at least, may spread the risk for their offspring among several patches when there is a high risk of parasitism. The existence of such

a behaviour has been suggested for *Boloria eunomia* (Petit *et al.* 2001) and *Euphydryas chalcedona* (Brown & Ehrlich 1980), but in general there is little information for butterflies about the effect of predation, parasitism or diseases on emigration decisions.

Temporal variability

The relative importance of factors promoting dispersal may differ greatly among species. Such factors are part of the environment in which the species live, and their own life history. Apart from the effect of kin competition, dispersal is only worthwhile if the emigrating individual ends up in a patch where conditions are more favourable than in the patch in which it currently resides. Thus, the principal requirement for the evolution of dispersal is the presence of uncorrelated spatial variability in conditions (Comins *et al.* 1980, Hastings 1983); emigration from unfavourable conditions makes sense only if there is a chance of immigrating into a patch of better conditions. The observation that both variability in population density (Thomas *et al.* 1994a) and dispersal activity (Hill *et al.* 1999a) increase towards the edge of a species range is consistent with this prediction. In contrast, populations of the parasitic *Maculinea* species are renowned for their comparatively low population fluctuations, presumably due to the protection of larvae against environmental disturbances when developing inside the nests of their host ants (Hochberg *et al.* 1992, 1994, Thomas *et al.* 1998a). They are also considered to be species with a low tendency to emigrate (Nowicki *et al.* 2005a). High spatial variability is thus a general precondition for the evolution of dispersal. We need to keep in mind, however, that this variability is in part an emergent consequence of dispersal itself. In other words, there will be a complicated feedback between dispersal and the factors favouring its evolution (Poethke *et al.* 2003).

THE COSTS OF DISPERSAL: THE TRANSITION PHASE

Despite the many potential benefits of dispersal, usually only a small fraction of individuals emigrate from a population. To understand this, we need to consider the potential disadvantages of dispersal, i.e. the costs and risks involved in emigration, transfer and settlement. The costs of dispersal include two components: internal costs, i.e. the commitment of resources (energy or developmental time) to dispersal, and external costs exerted by the environment (predation risk, failed success, transition time). Some of the former costs are invested even if an individual never disperses (Rankin & Burchsted 1992, Zera & Denno 1997). Investment in flight capability and emigration propensity is therefore correlated in some insect species. In dimorphic populations with a low fraction of winged individuals, the willingness of the winged individuals to emigrate is lower than in populations with a high fraction of winged individuals (Roff & Fairbairn 1991).

In the previous section we based our arguments in favour of condition-dependent emigration on the assumption that individuals are capable of collecting information about the conditions in their natal patch, but that they cannot have information about the external costs of dispersal they are going to pay. Consequently, dispersal strategies cannot be conditional to the actual costs of dispersal (for the role of internal costs see below). However, over generations, selection can respond to the long-term average costs associated with dispersal in a certain landscape and thus lead to the evolution of landscape-specific dispersal strategies (cf. Hovestadt et al. 2001). Further, individuals may respond to triggers such as weather conditions, which indicate favourable conditions for dispersal. Such triggers are often the proximate factors leading to emigration.

Investment costs

Evidence from the study of dimorphic insects shows that investment in dispersal capability such as large wings or flight muscles, and the energy invested in actual movement itself, comes at a cost (Rankin & Burchsted 1992, Zera & Denno 1997). These costs involve a delay in reproduction, reduced lifetime reproductive success, increased developmental time and reduced longevity. Wing dimorphism is rare in butterflies but there may be more graded differences in the dispersal ability of individuals and it is plausible that such differences would also come at a cost. For *Pararge aegeria*, dispersal is associated with reduced investment into reproduction at the expanding range of the species (Hughes *et al.* 2003). Similar conclusions can be drawn from the comparative study of five butterfly species (Baguette & Schtickzelle 2006) while a study of *Melitaea cinxia* (Hanski *et al.* 2006) suggests a trade-off between dispersal and longevity.

Landscape-mediated costs of transfer

The greatest cost of dispersal is probably the risk of its failure, either by the individual dying in the matrix during transition (Schtickzelle *et al.* 2006) or by its failure to reproduce after immigration. This may be due to an increased predation risk (e.g. Schtickzelle & Baguette 2003), an inability to find nutrition while travelling or, most importantly, failure to find a new suitable habitat. In summary, these costs are determined by the landscape through which the transition takes place. Not surprisingly, theoretical models predict that emigration tendency should decline as the costs of dispersal increase (e.g. Travis & Dytham 1998, Metz & Gyllenberg 2001, Poethke & Hovestadt 2002). Evolution, however, could increase investment into dispersal abilities by increasing the flight ability of a species, thus allowing individuals to cope better with the hazards of dispersal itself (Hill *et al.* 1999a, b). The costs of dispersal are thus clearly landscape- and species-specific (Baguette *et al.* 2000).

Many butterflies do not willingly cross non-habitat (e.g. Wood & Samways 1991) and quickly return should they happen to do so (Schultz & Crone 2001). This is particularly true for species inhabiting open habitats, all of which are very reluctant to cross even narrow strips of trees or woodland, such as *Boloria selene* (Thomas 1991b), *Aphantopus hyperantus* (Sutcliffe & Thomas 1996), *Parnassius mnemosyne* (Meglécz *et al.* 1997), *Boloria eunomia* (Schtickzelle & Baguette 2003), *Plebejus argus* (Read 1985, Ravenscroft 1986) and *Parnassius smintheus* (Roland *et al.* 2000). Proof of a barrier effect of forest has also been found by genetic analyses in *Erebia medusa* (King 1998, Schmitt *et al.* 2000) and *P. smintheus* (Keyghobadi *et al.* 1999). In contrast, emigration may increase with increasing fraction of open patch boundary, as in *Melitaea cinxia* (Kuussaari *et al.* 1996) and *Boloria aquilonaris* (Mousson *et al.* 1999). For some woodland butterflies the opposite effect has been observed. For example, open fields are an effective movement barrier for *Pieris virginiensis* (Cappuccino & Kareiva 1985), and both woodland and farmland can be barriers for *Melitaea athalia* (Warren 1987b, 1992b).

Landscape features, such as corridors, can also channel movement, an effect recently studied in several butterfly species. Open clearings and rides in woodlands facilitate the movement of numerous species: *Melitaea athalia* (Warren 1992b), *Aphantopus hyperantus* (Sutcliffe & Thomas 1996) and *Euptoieta claudia*, *Eurema nicippe*, *Junonia coenia* and *Phoebis sennae* (Haddad 1999a, b, 2000, Haddad & Baum 1999). For *Boloria eunomia* a preference to follow river basins has been deduced from genetic studies (Neve *et al.* 1996b). Thus, features functioning as barriers can at the same time be corridors as butterflies tend to fly along the edges of these structures, e.g. hedges (Dover & Fry 2001).

It is not clear whether movement over different habitat types incurs different costs in terms of predation risk or energetic costs of flight but in *Parnassius smintheus* mean dispersal distance was lower through forest than through meadows (Matter *et al.* 2004). Butterflies may, however, also fly away from certain structures because these structures are, on average, less likely to 'channel' their flight towards suitable habitat. The different preferences by open-habitat specialists, woodland species or habitat generalists described above may indicate that this is at least in part responsible for the different behavioural responses to certain landscape elements.

Environmental conditions

Dispersal or migration often occurs only under certain environmental conditions. For example, individuals of *Pieris brassicae* and *Danaus plexippus* migrate only at higher temperatures than usually required for flight (Johnson 1969). In general, flight requires appropriate temperature; cloudy conditions often cause butterflies to cease flight and bask in sunlight (e.g. Watt *et al.* 1977, Roland 1982, Shreeve & Smith 1992, Petit *et al.* 2001, Berwaerts & Van Dyck 2004). The prevalence of suitable environmental conditions is thus either a physiological precondition for dispersal, or these conditions signal to any 'dispersal-motivated' individual that dispersal can take place at comparatively low costs.

Local adaptation/experience

In addition to the genetic benefits of dispersal, i.e. the avoidance of kin competition and inbreeding, one of the more subtle costs of dispersal may be the transfer from a local habitat which an individual is well adapted to, into a habitat it is less well adapted to (Whitlock 2001, Singer 2003). For example, Hanski & Singer (2001) showed that the colonization success in *Melitaea cinxia* depends on the 'match' between the relative host plant composition and the oviposition preference of dispersing females. It has been suggested that host specialization will in general be tighter in stable habitats (see Singer 2003). As already discussed, habitat stability favours the evolution of low emigration rates. Given low dispersal to begin with, the evolution of local adaptation is most likely. Under this scenario, the benefit of dispersal would be even lower and add further to selection against dispersal.

INDIVIDUAL VARIABILITY

Due to genetic, environmental and developmental differences individuals within populations are not identical. Fundamentally, differences between individuals can have implications for both the expected costs and the benefits of a dispersal decision. In this section we briefly examine what is known about individual differences in dispersal behaviour.

Gender

Differences in life history between the two sexes are often prominent, including dispersal behaviour. The ability and motivation to move and/or disperse may thus differ between males and females (Baguette *et al.* 1996, Kuussaari *et al.* 1996).

For males, the most important motivation to move is finding mates, while females also have to consider the consequences of egg placement for the survival and development of their offspring. Suitable locations can, for example, be places with low population density (see Baguette *et al.* 1996). Not surprisingly, in many species ovipositing flights largely determine the activity of female butterflies; *Maniola jurtina* females fly mainly to lay eggs or to feed (Brakefield 1982), and *Pararge aegeria* females distribute eggs over a large area with extended flights after laying each egg (Shreeve 1986b).

Many studies report that males fly more frequently than do females within habitats (e.g. Scott 1975, Brakefield 1982, Kingsolver 1983a, Väisänen & Somerma 1985, Baguette & Nève 1994, Pfeifer *et al.* 2000) but that females tend to emigrate more than males do (Baguette & Nève 1994, Baguette *et al.* 1996, Petit *et al.* 2001). The latter is also true in *Iolana iolas* but males of this species disperse farther than females (Rabasa *et al.* 2007).

Morphological dimorphism and general condition

Dispersal polymorphism is a widespread phenomenon among insects (Roff 1990, Zera & Denno 1997). However, in spite of some examples, e.g. the seasonal emergence of brachypterous morphs in the lymantrid moth *Orgyia thylleina* (Kimura & Masaki 1977), or the development of a 'dispersal syndrome' in density-stressed noctuid moth *Spodoptera exempta* (Gunn & Gatehouse 1987), outright morphological dimorphism is rare in butterflies. This rarity does not rule out that dispersing individuals differ from resident individuals in a more gradual way. Under experimental conditions at least, wing loading has a substantial effect on wing-beat frequency in hovering *Pontia occidentalis* (Kingsolver 1999), presumably affecting the cost of flight.

More generally, dispersing individuals are predicted to be larger than average (Roff 1991). This is true for a number of species: large *Heliconius erato* individuals move longer distances (Mallet 1986), and emigrating females of *Melitaea cinxia* are on average larger than philopatric individuals. For males of this species, such a difference could not be detected (Kuussaari *et al.* 1996). Colonizing individuals of *Pararge aegeria* are heavier and have larger thoraxes than do individuals from continuously occupied sites (Hill *et al.* 1999a). We must be careful, however, to separate between the effect of such large size on dispersal propensity itself (see Rankin & Burchsted 1992) and on dispersal success.

Age and mating status

Females emigrate usually after mating, which is a reasonable strategy apart from avoiding inbreeding. If avoidance of kin competition or the successional status of the habitat is driving emigration, however, females should leave soon after mating. If, on the other hand, risk spreading over several patches is important, or females 'intend' to reduce competition between own offspring, depositing a fraction of eggs in the natal patch is a reasonable strategy. Such patterns have not been studied in detail, but both scenarios seem to occur in butterflies. For example, older individuals of *Parnassius phoebus* move less than do younger ones (Scott 1973a), and freshly emerged *Heliconius erato* move the longest distances (Mallet 1986) – observations consistent with the first hypothesis. However, practically all *Melitaea cinxia* females emigrate if they live long enough, while ca. 30% of males stay in their natal patch (Kuussaari *et al.* 1996). Also, male *Melitaea athalia* tend to become less mobile as they age, whereas females become more mobile (Warren 1992b). These two observations are more consistent with the second hypothesis. In *Lopinga achine*, females place about two-thirds of their eggs in the natal patch but tend to move between patches as they age. This effect has been interpreted as a 'risk-spreading strategy' as the habitats of this species senesce (Bergman & Landin 2002).

Genotype

Little is known about the genetic basis of variability in dispersal strategies but as noted before, individuals originating from different populations often behave differently. For example, females of *Melitaea cinxia* from newly established populations are more dispersive than females from older populations, but no such difference exists between males (Hanski *et al.* 2004). Such correlative differences, however, do not unambiguously demonstrate that the difference between individuals originating from different populations has a genetic basis. The correlation may also be a phenotypic response to different growth conditions as in *Pararge aegeria* (Merckx & Van Dyck 2006). Yet Hanski *et al.* (2004) found that the mean flight tendency of *M. cinxia* populations correlated with the average ATP/ADP ratio in the flight muscle of individuals after performing standardized muscle activity. At least one locus involved in the coding of the enzyme phosphoglucose isomerase has been identified as a candidate gene contributing to this variability (Watt *et al.* 2003, Haag *et al.* 2005).

Another case in question may be the difference in behaviour at habitat borders expressed by *Pararge aegeria* individuals originating from populations located either in woodland or in an agricultural surrounding. Because the individuals tested were raised under controlled conditions, it is likely that the observed difference has in this case a genetic basis (Merckx *et al.* 2003). When released in the matrix, individuals from fragmented agricultural landscape also moved towards woodland habitat from a greater distance than individuals taken from woodland (Merckx & Van Dyck 2007).

CONCLUSIONS

In this chapter we have presented a brief review of the forces involved in the evolution of dispersal strategies. We have based our arguments on the assumption that dispersal is a distinct episode in a butterfly's life characterized by typical behavioural rules. Evidently, this may not always be the case, and to some degree dispersal can also be the consequence of random drift deriving from everyday activities. The latter is likely to occur only where costs of dispersal are low (Van Dyck & Baguette 2005). For example, *Danaus chrysippus* (Boppré 1983, Smith & Owen 1997) or *Erebia epipsodea* (Brussard & Ehrlich 1970) form effectively panmictic populations over hundreds of square kilometres.

Dispersal is often studied as a population level phenomenon, such as in mark–release–recapture studies or population genetic investigations. When it comes to conservation issues, the consequences of dispersal for metapopulation persistence is certainly important. However, there is compelling evidence and strong theoretical arguments that individual dispersal decisions are context-dependent and triggered by specific cues. Further, transition through the landscape is greatly affected by habitat type, borders between habitats, or corridors channelling movement direction: any specific dispersal pattern observed is thus the outcome of the interplay between specific environmental conditions, the landscape setting, and the individuals' set of behavioural rules. Documented dispersal patterns themselves will usually be unique, and we will quite certainly fail to predict dispersal patterns if we base them directly on those observed elsewhere (Baguette *et al.* 2000, Mennechez *et al.* 2004, Cizek & Konvička 2005). Instead, we need to learn more about the behavioural rules causally responsible for the emergence of population-level dispersal patterns. Such behavioural-based approaches have gained increasing interest over the last years, e.g. Conradt *et al.* (2000, 2001), Conradt & Roper (2006), Merckx *et al.* (2003), Schtickzelle & Baguette (2003) or Heinz (2007).

To study dispersal efficiently, it would be helpful to identify those individuals among the moving ones that are truly dispersing. Unfortunately, moving butterflies do not carry a flag saying 'I am dispersing.' Thus, in typical mark–release–recapture studies a mixture of individuals that are either performing everyday movement or that are truly dispersing will be collected. In general, dispersal events are probably rare during the lifetime of an organism. Consequently, if we pool movement distances of individuals performing everyday movement with those truly dispersing we may grossly underestimate mean dispersal distance. In a recent analysis we found statistical evidence for a 'mixed dispersal kernel' in a mark–release–recapture data set collected for *Maculinea nausithous* (Hovestadt *et al.* unpublished data). Of the observed displacement distances, 94% were attributed to negative exponential distribution with a mean distance of ca. 120 m, and 6% to a kernel with a mean distance of ca. 1300 m. It is at least tempting to assume that only the latter fraction includes the truly dispersing individuals.

Fortunately, the empirical data collected, and the theoretical arguments presented, can give us clues about the circumstances under which individuals are likely to disperse. In addition, specific morphological and/or physiological traits may characterize 'dispersal-ready' individuals. It may be worthwhile to focus on the identification of likely 'dispersal candidates' so that their behaviour can be studied in a more targeted way and in more detail.

However, even the behavioural rules relevant for dispersal decisions may be subject to strong selection and rapid change, e.g. at an expanding species range (Thomas *et al.* 2001a, Parmesan 2006, Baguette & Van Dyck 2007). For example, for *Hesperia comma* the highest relative investment in thorax size is in areas with rapid population expansion (Hill *et al.* 1999a); an analogous pattern has been found in *Pararge aegeria* (Hughes *et al.* 2007). After introduction, a rapid differentiation in wing morphometry in comparison to the mother population, and between newly founded local populations, was observed in *Boloria eunomia* (Barascud *et al.* 1999), presumably with effects on flight performance and the willingness to emigrate (cf. Berwaerts *et al.* 2002). Thus, difference in dispersal ability, or willingness, could be selectively maintained in heterogeneous landscapes (Hanski *et al.* 2004) and lead to the emergence of population-specific dispersal rules. Habitat fragmentation has also been found to correlate with investment in flight muscles (Berwaerts *et al.* 1998), but dispersal rates can also decline in more fragmented landscapes (Schtickzelle *et al.* 2006).

The ultimate factors driving the evolution of dispersal rules may not be identical with the environmental cues sparking dispersal. The latter could be a reliable correlate of the 'hidden' ultimate factors. For example, the encounter rate with conspecifics gives a fairly direct estimate of population density but a correlated indicator could be food shortage in the larval phase. In addition, the information that an individual collects about its environment will neither be complete nor accurate. Dispersal decisions will therefore rarely be optimal. Identifying simple 'rules of thumb', e.g. emigration decisions based on the time interval between consecutive encounters with conspecifics or resources, which allow individuals to make reasonable guesses about the status of their habitat, may be one of the big challenges in the future. Such cues may not only exist with respect to fitness expectation in the current patch, but also with respect to travel costs.

The evidence that dispersal is triggered by certain environmental conditions raises another important issue. Studies on dispersal must anticipate that dispersal may be an episodic event occurring only under specific conditions (Baguette *et al.* 1998). Consequently, we must expect substantial yearly variation in dispersal activities (e.g. Scott 1975, Arnold 1983a, Ehrlich & Murphy 1987, Baguette *et al.* 1998, Schtickzelle *et al.* 2006) and short-term studies may severely under- or overestimate the amount of emigration. Consequently, Shreeve (1995) argues that seemingly very sedentary species may be more mobile than expected on occasions where specific weather conditions favour or even trigger long distance dispersal. For example, the average colonization speed of *Boloria eunomia* introduced in central France was 0.4 km/year but 1–3 km in years with favourable weather conditions (Nève *et al.* 1996a).

Further, emigration of butterflies can significantly influence the dynamics of local populations. Substantially more losses may be due to emigration than to within-patch mortality, as in *Boloria eunomia* (Petit *et al.* 2001). If individuals leave patches because their fitness expectations there are comparatively low, however, the predicted effect of such losses for metapopulation persistence may be much weaker than originally thought. This is particularly obvious in the case of habitat succession; individuals emigrate from habitats and populations that will go extinct anyway. For density-dependent emigration, individuals emigrate preferentially from populations at a high density. Compared to random emigration, such a strategy clearly reduces the risk of population extinction due to demographic stochasticity (Hovestadt & Poethke 2006), often the final factor leading to the loss of small populations (Dempster & Hall 1980, Thomas 1980, Hanski 2003).

These are certainly exciting times to study dispersal in general, and specifically for butterflies. In the past years we have witnessed a transition from phenomenological approaches describing dispersal patterns at the level of populations, to behavioural approaches attempting to understand the underlying mechanisms responsible for the emergence of these patterns. As we have outlined in this review, this development has been fostered by both the advancement of theory on the evolution of dispersal strategy, and careful and cleverly designed field observations. This route of investigation is demanding, but also very promising, especially if (population) ecologists and evolutionary biologists as well as field workers and theoreticians learn to cooperate even more closely than they currently do.

ACKNOWLEDGEMENTS

We are greatly indebted to Hans Van Dyck, Michel Baguette, Nicolas Schtickzelle and Jens Roland for helpful discussions and valuable comments on this manuscript. Andreas Gros and Brenda Pfenning kindly read the manuscript to spot logical and typographic errors. TH was financially supported by the European Union within its RTD project EVK2-CT-2001-00126 (MacMan; Settele & Kühn 2009).

10 • Population genetics of butterflies

GABRIEL NÈVE

SUMMARY

Populations of butterflies generally show a relationship between genetic variation and geographical distance, but this relationship is not uniform because of migration, selection, drift and mutation. For many species, recent outright habitat loss has led to increasing isolation of extant population units and reductions of population sizes. Not only does this mean that remaining populations are vulnerable to local extinction via stochastic effects on small populations, but gene flow may also be reduced. The combination of small population sizes and restricted gene flow can lead to loss of alleles and an increasing importance of bottlenecks and drift on the genetic composition of populations. The consequences of this for already vulnerable species are only recently being recognised and genetic analyses of population structures are critical for understanding how species will react to further changes. For some species, there is evidence that small population sizes and scarcity of habitats have selected against traits associated with mobility, with consequences for colonising potential. In others, which naturally occur in small populations, continuous gene flow is essential to mitigate against reduced fitness caused by inbreeding. Molecular techniques have greatly enhanced our ability to determine population structures and enabled differentiation to be related to events that have occurred over long timescales. They are also providing key evidence to describe current changes, including selection of genotypes for specific environmental conditions and isolation and demographic effects. Whilst genetics is providing a powerful tool in understanding how population structures have changed, novel insights into the genetic structure of populations are being provided by understanding the effects of landscape structure and resource availability on mobility and hence gene flow. The lessons from genetic studies should provide powerful arguments against landscape simplification and provide conservationists with concrete evidence for the necessity for landscape wide approaches to conservation.

INTRODUCTION

In a world where natural habitats are fast disappearing and where the landscape is intensively managed, even in nature reserves, it is important to understand the consequences of management and large-scale processes on populations. In this respect, studies of butterflies have been seminal in understanding how landscape structure affects populations (Ehrlich & Hanski 2004; Chapters 12 and 13). From a genetic perspective, the population structure of a given species in a particular landscape depends on a series of parameters: population size in each patch of habitat, movements between these patches, and the level of immigration and emigration to and from the system, together with their respective points of origin or destination. The two key factors determining the population structure of butterfly populations are the spatial distribution of their habitats and the ability of each species to disperse through the different components of the habitat matrix in the landscape. Different species may respond differently to the same change in habitat structure, depending on their movement ability and their habitat choice mainly, but not exclusively, determined by their choice of host plants.

When examining populations, it is important to define the scale at which processes occur. For example, post-glacial dispersal does not occur at the same spatial and temporal scales as emigration between habitat patches within a meta-population. Depending on the scale, different population structures emerge. Population geneticists interested in spatial structure may ask different questions, which may be classified according to the scale at which they are tackled. These are:

(1) Local scale – what are the causes of within population variation? Not two butterflies are exactly alike and how is this variation maintained?
(2) Landscape scale – are populations of a given species distinct entities within a landscape, or do they form a single genetic unit? If populations are different within the landscape, what could be the causes of such differences?

Ecology of Butterflies in Europe, eds. J. Settle, T. Shreeve, M. Konvička and H. Van Dyck. Published by Cambridge University Press. © Cambridge University Press 2009, pp. 107-129.

(3) Regional scale – to what extent do populations from different landscapes (e.g. river systems or mountain ranges) interact within a given region?

(4) Continental scale – most species show differences of phenotypes between different regions. What are the causes of such differences?

At the local scale, population genetics theory recognises four parameters to explain the local make-up of any given population: mutation, genetic drift, migration and selection. Mutation is generally regarded as of minor importance in natural populations over short timescales. Genetic drift is the result of random changes of allelic frequencies due to the small size of an isolated population. Ultimately, drift may lead to loss or fixation of alleles. Migration involves movement of individuals originating in one population to another where reproduction occurs. This results in the immigration or emigrations of individuals to or from the population of interest. If different genotypes in a given population reproduce at different rates, then selection will be operating. This can be a major factor in shaping the population structure of organisms. The effect of selection or drift may be counterbalanced by migration processes but if the population is small and immigration is not occurring, genetic drift is likely to be an important factor. Thus the ability of butterflies to disperse is a key factor in shaping population structure. The individuals of some species rarely move more than a few hundred metres from their natal location (e.g. *Cupido minimus*: Baguette *et al.* 2000; *Plebejus argus*: Lewis *et al.* 1997), whereas the individuals of other species may move hundreds or thousands of kilometres (e.g. *Colias crocea*, *Cynthia cardui* and *Aglais urticae*: Roer 1968).

From behavioural and ecological studies, species have then been ranked according to their apparent dispersal ability (e.g. Thomas 1984b). Relationships between observed dispersal behaviour and gene flow are difficult to establish, as rare emigration events may have profound genetical consequences. However, the occasional foundation of a population far away from previously occupied patches may give some indication on the effective dispersal pattern. For example, intensive mark–release–recapture studies on *Proclossiana eunomia* gave the longest movement as 4 km, but colonisation movement on the same species was observed up to 6 km from established populations (Nève *et al.* 1996a). Due to the difficulty of directly assessing long-distance dispersal, other approaches are necessary and genetics may help in understanding current and past links between populations. For many species, even if they seem sedentary according to

mark–release–recapture studies, genetic approaches have demonstrated that populations may be linked by migration. In the American checkerspot *Euphydryas anicia*, movement studies gave mean movement of 75 m for males, with a maximum of 1 km whereas genetic studies, using nine allozyme loci, showed that the populations 2 to 58 km apart within the studied mountain peak system did not differ from each other (Cullenward *et al.* 1979). As a consequence care is needed in extrapolating observed movement patterns to genetic structuring, although both approaches provide complementary insights. The key review of Ehrlich & Raven (1969) demonstrated that when butterflies cannot move between populations, population differentiation occurs.

GENETICS WITHIN POPULATIONS

Population genetics aims to understand how populations change in genetic make-up through space and time. Ever since butterflies were scientifically described, it has been recognised that there is within-species variation. For example, wing patterns of *Parnassius apollo* vary greatly among mountain ranges, leading to the description of numerous subspecies (e.g. Capdeville 1978). Caution must be applied when looking at subspecies recognised only on the basis of morphological characters: in the Australian butterfly *Ogyris amaryllis*, the different subspecies were found to be less relevant than host plant choice in the partitioning of the among-populations genetic variability (Schmidt & Hughes 2006). Individual aberrations, more or less frequent, were often formally named (e.g. Courvoisier 1907; review in Russwurm 1978), and their genetic basis has sometimes been described (Robinson 1990, Harmer 2000). How are these forms first described by ardent collectors related to the population structure of these butterflies ? The first to give a biological interpretation of individual variation were Ford & Ford (1930). They showed that, in *Euphydryas aurinia*, the individual variation was dependent on local population trends: during phases of population increase, phenotypic variation increased and included a series of aberrant individuals, whereas in periods of population stasis, the individuals were much closer to a uniform phenotype. This process was interpreted as an increase of genetic variability due to a decrease of selection during population increase phases. Thus, selection seems to play a key role in the variability of individuals within populations. The other key role is played by genetic drift, especially in small populations. Drift has two major impacts on populations: it is one of the main factors differentiating populations among which there is no gene flow, and through an increase in homozygosity of individuals it may have a deleterious effect on individual fitness. This was shown to be the case in at least two European butterflies of which populations had undergone bottlenecks: *Melitaea cinxia* and *Coenonympha hero* (see below).

GENETIC DIFFERENTIATION AMONG POPULATIONS

The second level of variation is among populations. For mountain species, numerous subspecies have often been described, and their distribution corresponds roughly to the distribution of mountain ranges (e.g. *Parnassius apollo*: Glassl 2005). In Europe, many species have a wide range and show wing pattern variation between northern and southern populations, such as the Mediterranean and northern subspecies of *Pararge aegeria* (Sbordoni & Foresterio 1985, Brakefield & Shreeve 1992), or variations within France of *Melanargia galathea* (Descimon & Renon 1975, Mérit 2000). These species are classically described as sedentary, moving at the most a few kilometres out of their habitats. For migratory species, such as *Aglais urticae* or *Vanessa cardui*, hardly any within Europe variation can be phenotypically recognised, apart from some island forms (e.g. *A. urticae ishnusa* of Corsica and Sardinia). This contrast of migratory habits corresponds to the ecological classification of butterflies into erratic or migrant species, and sedentary species. The latter have local populations, which may persist year after year. By contrast, migratory species usually occur in a wide range of habitat, but their presence at any given locality is more difficult to predict. Thomas (1984b) stated that about 85% of British butterfly species have 'closed' populations, i.e. have viable colonies in distinct habitat patches, whereas the remaining 15% have open or migratory populations. Long-term studies on the distribution of these butterflies have shown that species which have 'closed' populations may disperse out of their habitat patches, as in the case of *Hesperia comma*, which recolonised many habitat patches from remnant populations (Davies *et al.* 2005). This questions the relationship between ecological data, either from population survival data or from mark–release–recapture studies and the genetic make-up of population in a spatial context. By essence, dispersal events are rare and difficult to record; this is a major drawback to comparative data among species (Bennetts *et al.* 2001). The tools of population genetics may be used in this context to assess the levels of gene flow among populations.

From a genetics point of view, a population is a group of individuals which share a common gene pool (Dobzhansky 1950) and populations will be different if they do not share a common gene pool. Such differences may be quantified using genetical and statistical techniques. From a statistical point of view, two populations are different from each other if their allele frequencies, as estimated from the samples, are statistically different.

Without the exchange of individuals, populations may become differentiated. In cases of complete isolation, each population has its own history of genetic drift and/or selection and over time the populations become more and more differentiated (Box 10.1). Movements among populations do not need to be abundant to counteract the effect of genetic drift; an exchange of only one individual per generation is sufficient to avoid population differentiation (Hartl & Clark 1989). Obviously, such a low movement rate is difficult to detect in the field by ecological studies. Furthermore, it is not possible to infer the probabilities of long-distance dispersal from the analysis of within-patch short-distance dispersal, as such movements follow different ecological clues. Usually an individual engaged in dispersal behaviour outside its preferred habitat flies higher and more quickly (Baguette *et al.* 1998). Such movements are hard to detect in the field by direct observation. Two kinds of data may be useful in this respect. Firstly, ecological data on colonisation gives evidence that a movement from an occupied to an unoccupied patch has occurred. Colonisation of empty patches is a key component of the metapopulation dynamics of many butterfly species (see Chapter 8). Secondly, genetic data may tell how different populations are from each other. Generally, the more the populations are differentiated, the fewer individuals they have exchanged, directly or indirectly.

Genetic indices of population differentiation may be used to infer the level of migration of individuals between populations (Box 10.1). As such, the relationships between the genetic differentiation of populations and their geographical distances may be compared between areas within a species, or between species. Using this approach, Britten *et al.* (1995) showed that the isolation by distance in *Euphydryas editha* populations was much stronger in the Rocky Mountains than in the Great Basin, resulting from stronger barriers to dispersal in mountain areas compared to the plains.

Isolation by distance

In a particular species, isolation by distance (IBD) may be observed or not at the same scale, depending on geographical area. In *Parnassius apollo*, Descimon *et al.* (2001) showed

---

Box 10.1   Genetics drift and population differentiation

The population genetics of population differentiation is based on the simple principle that two opposite forces act on this phenomenon: genetic drift tends to differentiate populations, whereas gene flow tends to homogenise populations. If two small populations are isolated from each other, they tend to diverge in allele frequencies more quickly than do large populations (Fig. 10.1). This principle is used backwards to evaluate how populations are different from each other, from their allele frequency differences using different methods of calculating a genetic distance between populations (Hartl & Clarck 1989, Hedrick, 2000). Within populations, individuals may mate at random, in which case the number of heterozygotes in the population will be dependent solely on allele frequency. For an allele frequency of p, the frequency of heterozygotes is $2p(1-p)$. If populations do not show heterozygote deficiency (in which cases the individual samples may actually result from a local deviance of random mating), the degree of difference among populations may be measured by means of Wright's F statistics, now easily computed using programs such as Hierfstat (Goudet 2005) or Genepop (Rousset 2008). When a large number of populations are studied, F statistics may be ranked to study hierarchical clustering of populations, to study differentiation within a cluster of habitat patches, or among them, which may be river systems, mountain tops or otherwise discrete habitats.

In particular, patterns of allozyme genotype frequencies allow the use of Sewall Wright's F_{ST} index, which gives a concise way of expressing the degree of population differentiation between populations (Wallis 1994). In a group of populations, the population differentiation is measured by the fixation index (symbolized F_{ST}), which is estimated as $F_{ST} = (H_T - \bar{H}_s)/H_T$, with $\bar{H}_s$ the mean expected heterozygosity of an individual in an equivalent population mating randomly, and H_T the expected heterozygosity of an individual in a total population mating randomly. Wright (1978) suggested some guidelines to interpret the resulting values. A value of F_{ST} smaller than 0.05 may be considered as indicating little genetic differentiation, the range 0.05 to 0.15 indicates moderate genetic differentiation, 0.15 to 0.25 great genetic differentiation and over 0.25 very extensive genetic differentiation. The higher the value of F_{ST} is, the lower the number of migrants between populations per generation would be.

In an island model, where individuals moving out of a population may move with an equal probability to any other population, not just to the ones nearby, the migration rate among populations may be expressed as

$$Nm \approx \frac{1}{4}\left(\frac{1}{F_{ST}} - 1\right)$$

(Slatkin 1993)

with m the migration rate and N the effective population size.

This equation allows an estimate of the number of migrating individuals from the F_{ST} differentiation statistics. The given value always has to be taken with caution. The island model assumptions are rarely upheld with field data; the model assumes that (1) there is no selection, (2) there is no mutation, (3) all populations host the same number of individuals and contribute equally to the migration pool, (4) migration is random (i.e. irrespective of the distance between the populations), (5) the system is at equilibrium (Whitlock & McCauley 1999). In practice most of these assumptions are violated with field studies, but comparative analyses using F_{ST} nevertheless yield valuable information on among-population migration (Bohonak 1999). Furthermore, if the number of populations from which F_{ST} is estimated is small, the variance of F_{ST} is large (Douwes & Stille 1988), rendering the estimate of the number of migrants impossible to infer. However, as F_{ST} increases or decreases monotonically with the number of migrants, F_{ST} values or the estimate number of variants may still be used to compare different population system, either within a species or among species (Wang & Whitlock 2003). Other indices of population differentiation have been implemented since the description of the F_{ST} index: G_{ST} is a generalisation of F_{ST} for the study of different loci simultaneously (Hartl & Clark 1989). These indices both quantify population differentiation irrespective of the relative spatial positions of the relevant populations.

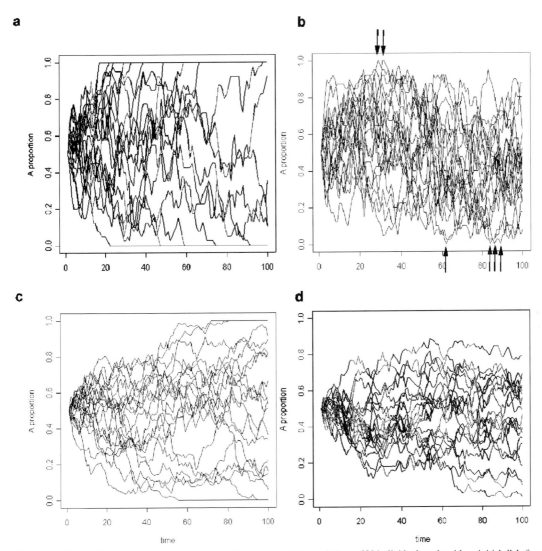

Figure 10.1. (a) and (b) Example of simulation of allele frequencies of 20 populations of 20 individuals each, with an initial allele frequency of 0.5. (a) With no migrant between populations, most populations lose one of the two alleles within 100 generations. In this run only 4 of the 20 populations retain both alleles, with a highly variable proportion. (b) With one migrant per population per generation, all populations retain their variability in the long run, as each event of allele fixation (arrows) is subsequently rescued through migration of an individual from another population. (c) and (d) Example of simulation of allele frequencies of 20 populations of 100 individuals each, with an initial allele frequency of 0.5. (c) With no migrant between populations, some populations lose one of the two alleles within 100 generations. In this run 16 of the 20 populations retain both alleles, with a highly variable proportion. (d) With one migrant per population per generation, all populations retain their variability in the long run, with allele frequencies closer to each other than if no migration occurred.

that isolation by distance was highly significant in the high Alps, but that populations from the southern Alps do not present such a pattern. This is because at some point in a recent past, the populations from the southern Alps were linked with each other in a single neighbourhood, and recent barriers between these populations have not yet led to isolation by distance. This is due to the large size of populations, and occasional migrations between them. By contrast, populations from the high Alps are more differentiated, due to the individual history of each of these populations, and post-glacial colonisation occurring in a stepping-stone fashion, leading to a greater isolation by distance (Box 10.2).

Box 10.2   Spatial aspect of population genetics

The two main parameters of spatial population genetics are the genetic neighbourhood and isolation by distance. The isolation by distance approach compares populations two at a time, and assesses whether there is a correlation between geographical distance and genetic differentiation of populations. The slope of the curve is then an indication of the level of gene flow between the populations, and this parameter varies both among species and among regions within species, according to the distribution of habitat patches (Nève *et al.* 2008).

If migration occurs more frequently between neighbouring populations than between distant ones, one would expect the allele frequencies to be more similar between neighbouring populations than between more distant ones. The genetic neighbourhood is the area within which the individuals mate at random (Wright 1978). Isolation by distance occurs if the genetic relatedness between populations decreases as the geographic distance between them increases. This may be estimated using Moran's *I* (Legendre & Legendre 1998), which computes the correlation between the allelic frequencies in pairs of populations according to the geographical distance between them. In cases of isolation by distance, the correlation decreases continuously with distance, going from a strong positive correlation for populations close to each other to a nil or negative correlation for populations the furthest away (Fig. 10.2a). The pattern showing a decrease of autocorrelation with an asymptotic approach to zero suggests that populations from a given area are similar to each other, but with no correlation at a wider scale (Fig. 10.2c).

The spatial pattern of differentiation may also reveal the different scales at which the different kinds of movements occur: in sedentary species most individuals reproduce within a small genetic neighbourhood, whereas the dispersal of a few individuals out of this area, and the distance to which they will eventually move, determines the level of isolation by distance between populations. The local effective population size, and hence the neighbourhood size, may be estimated by using the

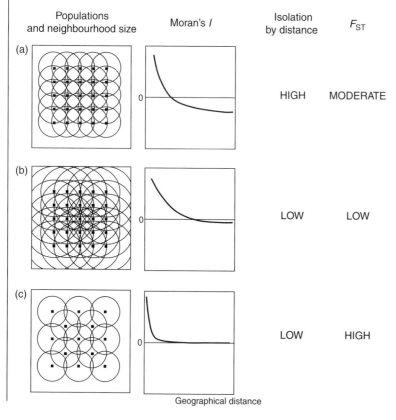

Figure 10.2. Spatial population structures. Black squares depict suitable habitat patches, and circles the neighbourhood sizes. (a) Each of the 25 populations exchanges individuals with its direct neighbours at each generation. (b) Each of the populations exchanges individuals with its direct and indirect neighbours, resulting in a much higher gene flow throughout the system. (c) The neighbourhood size is the same as in (a), but with a matrix of half of the habitat patches, each population ends up being isolated, and is subjected to genetic drift independently of its neighbours.

y-intercept of the relationship between Nm (see below) and geographical distance, using logarithmic scales on both axes (Slatkin 1993). For sedentary species, this would be equivalent to their natal habitat patch. The slope of the isolation by distance, on the other hand, reveals the extent to which individuals move away to other patches. In migratory or erratic species, both values are of high interest. As it would be expected that populations which are closer to each other would be more similar than more distant ones, there is a need to address the relationship between the degree of population differentiation and the distance between populations. The first approach on this topic is to assess the isolation by distance, by a study of the relationship between the geographical distance between populations and their index of differentiation. This is usually done by using the estimated number of migrants per generation (Nm) between any two populations, using the relationship $G_{ST} = 1/4(Nm+1)$ and the geographical distance between these two populations, usually both log-transformed (Slatkin 1993). The estimated number of migrants should not be taken literally; Whitlock & McCauley (1999) stressed that the relationship between Nm and G_{ST} or F_{ST} is based on a number of assumptions, most of which are unrealistic in natural populations. The most important concern here is that this relationship is based on an island model, where the number of migrants between populations is not correlated with the distance between populations. This is obviously not true in most butterfly populations. In consequence, the estimated number of migrants between two populations should be taken as an index of their differentiation, and not at face value (Whitlock & McCauley 1999, Neigel 2002).

This approach is usually applied on each studied allele separately (e.g. Porter & Geiger 1995, Nève et al. 2000, Descimon et al. 2001), but may also be applied using the information from the different loci simultaneously (Smouse & Peakall 1999), as performed by Harper et al. (2003) on English populations of Polyommatus bellargus.

The spatial structure of butterfly populations depends on where individuals of each sex have come from when they mate, and where females lay eggs. As the adult stage is usually the only one when long-distance dispersal is possible, population structure is strongly related to dispersal in the adult stage. In a habitat patch network where individuals all have the same movement potential, if there is a negative relationship between the distances from emergence to reproduction sites and their frequencies, a pattern of isolation by distance emerges (Wright 1943, Epperson 2003). When a large number of populations have been sampled, it is possible to infer the spatial structure of the population from the genetic make-up of the individual populations and their geographical location (Box 10.2). Spatial statistics give information on how alike populations are to each other depending on their location. Because most adult butterflies move more or less freely within their natal habitat patch, population structure is generally studied at the between-population level. The isolation by distance model is only applicable if movements between neighbouring patches are more frequent than between patches further apart, as would be expected in sedentary species. Indeed, many studies were started suspecting that movement out of the natal patch would be rare, as very few individuals were ever sighted outside the preferred habitat (e.g. Proclossiana eunomia, Boloria aquilonaris, Plebejus argus). At the broader scale, butterflies that are known to move a lot raise interesting questions. Aglais urticae has a migratory habit (e.g. Roer 1968), so to what extent do individuals actually move? This question was recently answered using a combination of techniques on a series of samples coming from the whole of Eurasia. In species with such wide distributions, the population structure would be expected to occur only at a very large scale. A study of nine populations from the Netherlands, Belgium and southern France showed that these populations were hardly differentiated from each other ($G_{ST} = 0.03$) and had a high heterozygosity (mean expected heterozygosity $= 0.248$: Vandewoestijne et al. 1999), without any isolation by distance effect. As local density of this species is usually low, a high heterozygosity can only be maintained if individuals disperse over large distances. A phylogeography study, based on the cytochrome oxidase (COI) gene and the control region of the mitochondrial DNA of this species, showed that from Europe to Japan, there is a high genetic diversity with wide distribution of both common and rare haplotypes (Vandewoestijne et al. 2004). This corroborates high gene flow, and hence the strong dispersal power of this species. In Maniola jurtina, local densities are generally high, and the global fixation indexes (F_{ST}) are in the range of 0.015 to 0.065, without any isolation by distance effect either at the local (Birmingham, Isles of Scilly) or at the continental scales (Europe) (Table 10.1); this implies frequent individual migration between populations of this species.

Porter & Geiger (1995) used the genetic approach to assess movements in the erratic species Pieris napi. In their study of 38 populations distributed throughout Europe, the isolation by distance model followed the relationship

Table 10.1 *Population genetics studies of butterflies conducted in Europe*

Species and study area	Methods	N populations	F_{ST}	Number of polymorphic loci	Approximate longest distance	Isolation by distance (IBD)	Reference
Hesperidae							
Carterocephalus palaemon							
Białowieża forest, Poland	Allozymes	2	0.191	16	(sympatry)	NA	Ratkiewicz & Jaroszewicz 2006
Thymelicus acteon							
GD Luxemburg + neighbouring areas	Allozymes	12	0.053	18	110 km	NA	Louy *et al.* 2007
Thymelicus sylvestris							
GD Luxemburg + neighbouring areas	Allozymes	12	0.023	18	110 km	*	Louy *et al.* 2007
Thymelicus lineola							
GD Luxemburg + neighbouring areas	Allozymes	12	0.008 NS	18	110 km	NA	Louy *et al.* 2007
Papilionidae							
Parnassius mnemosyne							
Hungary	Allozymes	8	0.075	3	80 km	*	Meglécz *et al.* 1999
Hungary	Microsatellites	8	0.051	3	80 km	NA	Meglécz *et al.* 1998
Alps + Pyrenees	Allozymes	24	0.135	9	500 km	NA	Napolitano & Descimon 1994
S Alps (France)	Allozymes	4	0.035 – 0.175	9	90 km	NA	Napolitano *et al.* 1988
Parnassius apollo							
French Alps + Jura	Allozymes	17	0.0548	14	350 km	**	Descimon *et al.* 2001
French southern Alps	Allozymes	10	0.041	14	120 km	NS	Descimon *et al.* 2001
Jura	Allozymes	2	0.124	12	100 km	NA	Descimon *et al.* 2001
Pyrenees	Allozymes	7	0.087	14	220 km	NA	Descimon *et al.* 2001
Massif central	Allozymes	3	0.266	10	200 km	NA	Descimon *et al.* 2001
Spain	Allozymes	2	0.152	12	NA	NA	Descimon *et al.* 2001
Parnassus phoebus							
French Alps	Allozymes	12	0.255	11	200 km	NA	Descimon 1995
Papilio hospiton							
Corsica	Allozymes	6	0.015 NS	8	100 km	NS	Aubert *et al.* 1997
Papilio machaon							
England	Allozymes	3	0.107	7	10 km	NS	Hoole *et al.* 1999
England	RAPD	3	NA	109	10 km	NS	Hoole *et al.* 1999

Pieridae

Pieris napi napi							
Europe	Allozymes	16	0.0226	9	1900 km	*	Porter & Geiger 1995
Pieris napi bryoniae							
Alps	Allozymes	8	0.0277	9	300 km	*	Porter & Geiger 1995
Pieris napi meridionalis							
South Europe	Allozymes	7	0.0052	9	650 km	NS	Porter & Geiger 1995
Pieris napi britannica							
Britain	Allozymes	6	0.1322	9	500 km	*	Porter & Geiger 1995
Birmingham	Allozymes	8	0.042 (NS)	5	30 km	NS	Angold et al. 2006
Pieris napi adalwinda							
Scandinavia	Allozymes	2	0.1010	9	1100 km	*	Porter & Geiger 1995
Pieris napi napi + bryoniae + meridionalis							
Europe	Allozymes	31	0.0258	9	1900 km	*	Porter & Geiger 1995

Lycaenidae

Maculinea teleius							
Poland	Allozymes	3	0.041	3	300 km	**	Figurny-Puchalska et al. 2000
Hungary, Romania, Slovenia	Allozymes	13	0.008	14	700 km	NA	Pecsenye et al. 2007b
Maculinea nausithous							
Poland	Allozymes	4	0.153	5	300 km	**	Figurny-Puchalska et al. 2000
Hungary	Allozymes	3	0.013	14	16 km	NA	Pecsenye et al. 2007b
SW Germany	Microsatellites	14	0.068	7	45 km	***	Anton et al. 2007
Maculinea alcon							
Denmark	Allozymes	13	0.09	5	330 km	NA	Gadeberg & Boomsma 1997
Hungary, Slovenia, Romania	Allozymes	13	0.138	14	1000 km	NA	Pecsenye et al. 2007b
Maculinea arion							
Hungary, Romania	Allozymes	3	0.040	5	650 km	NA	Pecsenye et al. 2007b
Plebejus argus							
Britain	Allozymes	9	0.07	12	30 km	NA	Brookes et al. 1997
South Spain	Allozymes	5	0.016	7	47 km	NS	Péténian & Nève 2003
Finland	Allozymes	9	0.015	10	16 km	NS	Péténian & Nève 2003
Polyommatus bellargus							
Britain	Microsatellites	26	0.127	4	200 km	***	Harper et al. 2003
Polyommatus icarus							
Portugal to Germany	Allozymes	29	0.0187	19	3200 km	*	Schmitt et al. 2003
W Germany	Allozymes	15	0.0041	19	220 km	NS	Schmitt et al. 2003

Table 10.1 (*cont.*)

Species and study area	Methods	N populations	F_{ST}	Number of polymorphic loci	Approximate longest distance	Isolation by distance (IBD)	Reference
Polyommatus coridon							
W Europe	Allozymes	18	0.021	17	1200 km	NA	Schmitt & Seitz 2001a
E Europe	Allozymes	18	0.028	17	800 km	NA	Schmitt & Seitz 2001a
Europe	Allozymes	36	0.060	17	1700 km	NA	Schmitt & Seitz 2001a
W Germany	Allozymes	22	0.014	20	150 km	NS	Schmitt & Seitz 2002a
Lower Saxony (Germany)	Allozymes	17	0.013	19	30 km	NS	Krauss et al. 2004a
France, Italy, Germany	Allozymes	39	0.021	20	1100 km	NA	Schmitt et al. 2002
Aricia artaxerxes issekutzi							
Bükk (Hungary)	Allozymes	2	0.022	13	5 km	NA	Pecsenye et al. 2007a
Aggtelek (Hungary)	Allozymes	6	0.024	13	25 km	NA	Pecsenye et al. 2007a
Aricia agestis							
Britain	Allozymes	6	0.1857	12	350 km	NA	Lai & Pullin 2005
Nymphalinae							
Proclossiana eunomia							
Belgium	Allozymes	26	0.123	3	120 km	*	Nève et al. 2000
Belgium	RAPD	4	0.0887	24	12 km	NS	Vandewoestijne & Baguette 2004
Pyrenees (France)	Allozymes	12	0.099	3	40 km	***	Descimon et al. 2001
Asturias (Spain)	Allozymes	4	0.123	10	73 km	*	Nève et al. 2008
Morvan (France)	Allozymes	11	0.105	3	24 km	NS	Barascud et al. 1999
Czech R.	Allozymes	11	0.035	3	70 km	***	Nève et al. 2009
Boloria aquilonaris							
Belgium	RAPD	8	0.179	18	90 km	**	Vandewoestijne & Baguette 2002
Belgium	Allozymes	8	0.105	4	90 km	NS	Vandewoestijne & Baguette 2002
Aglais urticae							
Europe	Allozymes	9	0.030	5	1000 km	NS	Vandewoestijne et al. 1999
Euphydryas aurinia							
Britain	Allozymes	17	0.1621	6	690 km	NA	Joyce & Pullin 2003
South Britain	Allozymes	8	0.0522	6	244 km	NA	Joyce & Pullin 2003
S France	Allozymes	35	0.113	10	600 km	***	Descimon et al. 2001

N France	Allozymes	26	0.065	10	730 km	**	Descimon et al. 2001
Melitaea cinxia							
Åland Is	Allozymes + microsatellites	369	0.1	6 + 2	20 km	'strong'	Saccheri et al. 2004
Meltiaea didyma							
C Germany	Allozymes	21	0.006 – 0.090	25	210 km	NS	Johannesen et al. 1996
Hammelburg (Germany)	Allozymes	14	0.015	25	15 km	NS	Johannesen et al. 1996
Mozel (Germany)	Allozymes	4	0.044	25	2 km	NA	Johannesen et al. 1996
Satyrinae							
Coenonympha pamphilus							
Birmingham	Allozymes	10	0.075	10	30 km	NS	Angold et al. 2006
Monte Baldo (Italy)	Allozymes	4	0.005	17	20 km	NS	Besold et al. 2008
Coenonympha hero							
Sweden, Estonia, Russia	Allozymes	13	0.141	5	2600 km	*	Cassel & Tammaru 2003
Maniola jurtina							
S England	Allozymes	15	0.049	2	135 km	NS	Handford 1973a§
Isles of Scilly	Allozymes	13	0.047	2	7 km	NS	Handford 1973b§
SE England	Allozymes	14	0.015 (NS)	12	165 km	NS	Goulson 1993
Europe	Allozymes	48	0.034	20	2700 km	NS	Schmitt et al. 2005b
Birmingham	Allozymes	10	0.048 (NS)	10	30 km	NS	Angold et al. 2006
Sardinia	Allozymes	5	0.057(NS)	15	140 km	NS	Grill et al. 2007
Europe	Allozymes	12	0.065	15	1930 km	NS	Grill et al. 2007
Maniola nurag							
Sardinia	Allozymes	6	0.04(NS)	15	120 km	NS	Grill et al. 2007
Pyronia tithonus							
Birmingham	Allozymes	11	0.068	6	30 km	NS	Angold et al. 2006
Erebia epiphron							
Europe	Allozymes	16	0.291	18	1600	NA	Schmitt et al. 2005c
Erebia embla							
Sweden	Allozymes	4	0.024	5	80 km	NA	Douwes & Stille 1988
Erebia medusa							
Germany, Hungary	Allozymes	53	0.149	19	1200 km	NA	Schmitt & Seitz 2001d
NE Hungary	Allozymes	6	0.005	20	7 km	NS	Schmitt et al. 2000
Balkan peninsula	Allozymes	28	0.137	17	730 km	NA	Schmitt et al. 2007
Erebia triaria							
Spain	Microsatellites	6	0.07	7	720 km	NS	Vila et al. 2006

Table 10.1 (*cont.*)

Species and study area	Methods	N populations	F_{ST}	Number of polymorphic loci	Approximate longest distance	Isolation by distance (IBD)	Reference
Melanargia galathea							
Belgium	Allozymes	7	0.016	6	110 km	*	Vandewoestijne *et al.* 2004
South France	Allozymes	7	0.038	4	100 km	NS	Nève (unpubl)
C Europe	Allozymes	11	0.061	11	1400 km	NA	Habel *et al.* 2005
South east Europe	Allozymes	16	0.070	11	1100 km	NA	Schmitt *et al.* 2006a
Morocco	Allozymes	4	0.088	11	400 km	NA	Habel *et al.* 2008
Chazara briseis							
Germany	Allozymes	9	0.022	15	10 km	**	Johannesen *et al.* 1997

Notes: NA, not available; NS, not significant; *, $P < 0.05$; **, $P < 0.01$; ***, $P < 0.001$; §, F_{ST} and isolation by distance computed by present author.

$F_{ST} = 0.03 - 0.45/(4x + 1)$, which for $F_{ST} = 0$ gives an estimated value of $x = 3.5$ km, which is thus the estimated radius of the neighbourhood area for this species, and the estimated F_{ST} for the whole continent was 0.0887 (SE = 0.0076), giving estimated numbers of migrants between populations at 2.6 (CI: 1.6–5.5). Such values suggest that there is a significant gene flow across the whole continent, which is not surprising, given the erratic behaviour of individuals and the wide distribution of their reproductive habitats. However, the populations of the Nordic montane subspecies *Pieris napi adalwinda* and the lowland subspecies (*P. napi napi*) are ecologically and genetically separated (Espeland *et al.* 2007).

The situation for many species of European butterflies is very different from this. Most occur in patchily distributed habitats, from which dispersal is a rare event. If a species is distributed in a series of discrete patches with metapopulation dynamics (see Chapter 8), the total effective size of its population will be much smaller than if each population were long-lived. Even a low extinction probability will have a dramatic effect on the total effective population size (Whitlock 2003).

Selection

Population differentiation may be caused by substantially different selection pressures occurring in different habitat patches. The identification of the cause of the selection pressure is always difficult (Manly 1985, Endler 1986). In butterflies the identified causes of selection concern primarily temperature and host plant availability, and there is evidence that habitat structure may also be a selective factor. *Maniola jurtina* in different habitat structures in the Isles of Scilly was studied by E. B. Ford. Populations from small islands (<16 ha) were either unimodal with 0 or 2 wing spots, or bimodal at 0 and 2 spots, whereas the populations from the bigger islands were more evenly distributed. Evidence of constancy of spot pattern distribution in individual populations, even after a bottleneck, strongly suggested that spot pattern was under selection pressure rather than the result of random genetic drift in small island populations. Spot pattern frequencies changed after habitat changes, such as the removing of cattle grazing, rather than with population bottlenecks (studies summarised in Ford 1975, Brakefield 1984, 1990). More recently, other evidence was found for selection in this species. Different phosphoglucose mutase (PGM) alleles were favoured in different areas of its English distribution (Goulson 1993). Indirectly, this explains why the relationship

between genetic similarity and geographical distance between populations is steeper in Britain than for the whole of Europe (Thomson 1987). As several of the loci studied by the latter author are probably under selection, similarity between regions under equivalent ecological conditions is expected to occur, thus counterbalancing the general isolation by distance effect, which has not been found for this species in more recent studies at the regional or continent scales (Goulson 1993, Schmitt *et al.* 2005b, Grill *et al.* 2007).

In the American species *Euphydryas editha*, the natural host plant at Schneider's Meadow (Nevada, USA) used to be the native plant *Collinsia parviflora*. Over a decade, the European plant *Plantago lanceolata* spread through the habitat. Singer *et al.* (1993) showed conclusively that the host plant choice switched from the native species to the introduced one. Associated with this host plant switch will be changes of selection regimes related to host plant quality and phenology.

For several American *Colias* species, temperature is a major factor affecting the polymorphism of the enzyme phosphoglucoisomerase (PGI). Watt *et al.* (1983) showed that the alleles present in different individuals were related to the temperature at which they fly, according to the optimal temperature of the PGI enzyme, as checked in vitro. At a broad scale, populations of the Alpine *Colias meadii* have different PGI polymorphisms depending on the habitat they occupy (above tree line tundra vs. below tree line steppe), irrespective of the distance between these populations (Watt *et al.* 2003). The populations of these different habitats may be either isolated or exchange many individuals each year, including between the two habitat types. Nevertheless there is consistent 10–20% difference in PGI frequencies between the two habitat types; such a pattern may only be explained through continuous strong selection at the PGI locus. This key enzyme of glucose metabolism affects flight capacity. In *Melitaea cinxia*, the dispersal ability of individuals bearing the different PGI alleles was significantly different, and this in turn affected population growth and dispersal pattern (Haag *et al.* 2005, Hanski & Saccheri 2006). Using single nucleotide polymorphisms, Saastamoinen & Hanski (2008) showed that the two most common alleles for PGI in the Åland Islands (Finland) populations of *M. cinxia* were linked with different temperature preferences; the individuals with the PGI-*f* genotypes flew at lower temperature and laid 32% larger clutch sizes than PGI-non-*f* females because they tend to initiate oviposition during the warmest part of the day when clutches tend to be larger. As this leads to a strong selection against PGI-non-*f* alleles, the question then remains as to what favours the PGI-non-f alleles in the population system.

Behaviour may also be subject to selection: individuals from isolated patches of habitat from which emigration would be extremely unlikely to be successful may be selected against. In the UK, severe isolation of the last remnant populations of both *Maculinea arion* and *Papilio machaon* resulted in reductions of thoracic sizes in recent museum specimens compared to older ones (Dempster 1991). A more thorough study on the effect of isolation on flight ability was conducted in UK populations of the silver-spotted skipper (*Hesperia comma*), which was once widespread in southern and eastern England. It declined to its smallest range in the 1970s and 1980s, because a decrease of grazing rabbit populations, due to myxomatosis, led to a loss of habitat areas. With the recovery of rabbit populations from the beginning of the 1980s, the species has recolonised some areas (Thomas & Jones 1993). Hill *et al.* (1999b) related morphology to colonisation and demonstrated that thorax size was bigger in the area where recolonisation had been the quickest (East Sussex) than where it was slower (Surrey). They suggested that selection had operated more strongly against large thorax size and mobility in Surrey where the species had persisted in small (<1 ha) isolated refuges, compared to East Sussex where the population had persisted in a large (18 ha) refuge. The stronger flight ability of East Sussex populations resulted in a higher colonisation rate and gene flow, whereas the lower flight ability of Surrey populations resulted in a higher isolation by distance effect.

The identification of selection pressure implicitly asks the question of what populations are. The East Sussex and Surrey populations of *H. comma* had suffered different selection pressures according to the characteristics of the two regions. However, such clear-cut situations are infrequent, as most butterflies exhibit gene flow between habitat patches. The question of the identification of what constitutes a population is central to many problems in ecology, and – as seen above – the genetic make-up in a population is under selection from its environmental conditions. The scale at which selection will affect butterfly populations will depend on gene flow among these populations. *Hesperia comma* displays discrete populations which suffered a bottleneck in the 1970s and 1980s (Hill *et al.* 1999b), and a differential effect of selection could be detected between Surrey and East Sussex. In widely distributed species with strong flight abilities, such as *Pieris napi*, such a phenomenon does not occur. The case of *Colias meadii* where there is a significant difference of PGI allele frequencies according to altitude (Watt *et al.* 2003) may be due to individuals actively seeking a habitat according to their individual temperature requirements. The question remains open as to how often this may occur in other species. The number of generations per year for *Aglais urticae* may also be under a similar selection pressure, although at a larger scale, as *A. urticae* is trivoltine in central France, mostly bivoltine in England and univoltine in northern Scotland, with local variation according to altitude (Brakefield & Shreeve 1992b).

TECHNIQUES IN POPULATION GENETICS

Wing pattern

The first techniques used in population genetics concerned phenotypic variations, using these as surrogates for genetic information. Several species of Satyrinae often display variable numbers of spots on their wings. The number and size of spots in *Coenonympha tullia* vary with sex (females have more spots than males) and with locality (Turner 1963, Dennis *et al.* 1984). Spot pattern also varies among localities in *Maniola jurtina* and has been used to study population differentiation (Dowdeswell & Ford 1953, Dowdeswell 1981). It was assumed that the wing spotting was heritable, on the basis that the pattern was consistent among years. However, heritability of this character was only formally demonstrated later, and was found to be sex linked: heritability was first tentatively estimated at 0.14 in males and 0.63 in females (McWhirter 1969), and later at 0.66 in males and 0.89 in females (Brakefield & van Noordwijk 1985).

Protein electrophoresis

Upon the general availability of protein electrophoresis from the 1960s (Johnson 1971), this technique has been widely used for population genetic studies of butterflies, from the pioneering studies of Handford (1973a, b) to date. Nowadays this technique still remains the most widely used in population genetics studies of butterflies. The main reason for this choice is a combination of relative ease of scoring, and a fairly low cost (Wynne *et al.* 1992). The scoring of the resulting zymograms is usually straightforward (Richardson *et al.* 1986), and the Mendelian basis of the observed polymorphism may be checked using the known quaternary structure of the given protein, by experimental crosses or by Hardy–Weinberg equilibrium expectations. This technique has proved powerful as many species exhibit a high degree of polymorphism. Studies usually focus on three to 25 polymorphic loci. Most of the studies on protein electrophoresis assume that allele variation is neutral, or at least that no selection could be detected (Besold *et al.* 2008). As the proteins of interest have all definite functions, this is unlikely to be true (van Oosterhout *et al.* 2004) but population differentiation based on protein electrophoresis has been, and still is, widely studied.

Some species (e.g. *Lycaena helle*) or life stages (e.g. caterpillars) have sometimes proved difficult to study by protein electrophoresis, due to toxic compounds (e.g. oxalic acid, phenols) which interfere with enzyme activity. In this case the homogenisation procedure should extract or neutralise these compounds. Polyvinylpolypyrrolidone (PVPP) and a few grains of instant coffee have been mentioned as compounds which may improve enzyme stability by removing phenolic compounds during the grinding and homogenisation procedures (Hebert & Beaton 1993). For adult butterflies, I have used the following homogenising solution: 50 mM Tris-HCl, 0.5% (v/v) triton X-100 (optional), 15% (w/v) sucrose, adjusted to pH 7.1 with HCl (Wynne & Brookes 1992). For fourth and fifth instar *Melitaea cinxia* larvae, the following homogenising solution has been used: 100 ml distilled water, 10 mg NADP, 100 µl β-mercaptoethanol (I. J. Saccheri pers. comm.; solution from Richardson *et al.* 1986).

Molecular techniques

Protein electrophoresis has the major drawback that a common protein migration rate may occur in different alleles, hiding heterogeneity (Johnson, 1977). Recent genetic studies on butterflies often rely on DNA-based molecular techniques. These are rapidly improving tools, and the choice of a method depends primarily on the questions asked, the scale of the study and on the chosen organism. Several good reviews of methods are currently available (e.g. Parker *et al.* 1998, Avise 2004, Behura 2006).

Mitochondrial DNA sequencing

Variation of mitochondrial DNA (mtDNA) is studied by sequencing one or several genes of the short strand of mtDNA. The most studied parts are the control region (CR), and cytochrome oxidase I or II (COI and COII). The region of interest is amplified by polymerase chain reaction (PCR) and then sequenced for each individual (Avise 2004). Sequences of mitochondrial DNA are usually scored for phylogeography studies aiming at understanding the pattern of colonisation at the continental scale (e.g. Vandewoestijne *et al.* 2004). Variation in mtDNA sequence has also been used to assess levels of genetic variation in cases where the conservation of frozen specimens would have been difficult. Diversity of populations of *Mycalesis orseis* within forest fragments in Malayan Borneo, assessed by the number of mt haplotypes, was negatively affected by isolation of their habitat patch, but not by population size or patch size (Benedick *et al.* 2007). The low level on mutation within the mitochondrial DNA allows the study of long-term processes: in the North American *Parnassius smintheus*, the local variation of mtDNA haplotypes could be linked with the range expansion and retraction during glacial–interglacial cycles. During warm periods, populations persisted at mountain tops, whereas they expanded during cold spells. As a result, populations from an area within a mountain range have a series of possible refugia during warm periods, and end up being more diverse than those from an area with fewer refugia (DeChaine & Martin 2004).

Randomly amplified polymorphic DNA (RAPD)

This technique uses short PCR primers (ca. 10 bp) to amplify DNA fragments. This primer length is short enough to find several annealing sites in the genome by chance alone, but long enough as not to amplify too many fragments. Usually several possible primers are tested, and the ones yielding recognisable and repeatable banding patterns are then selected for the study. The major drawback of this method is that it is not possible to identify from which genome region each band is amplified. Furthermore, the banding pattern is very sensitive to laboratory conditions. Due to these drawbacks, RAPD results are difficult to replicate. Zakharov *et al.* (2000) managed to amplify DNA from museum specimens of *Atrophaneura alcinus* and four *Parnassius* species, but these authors did not publish further studies based on RAPD. Vandewoestijne & Baguette (2002) showed that RAPD on 18 polymorphic loci in *Boloria aquilonaris* yielded significant isolation by distance, while isozymes (four polymorphic loci) on the same populations did not. This draws attention to the fact that the lack of genetic differentiation found with one marker does not necessarily mean that the populations are not differentiated. From a statistical point of view, it is simply that the hypothesis that the populations are not differentiated may not be rejected (Bossart & Prowell 1998).

Amplified fragment length polymorphism (AFLP)

To avoid the drawbacks of RAPD, primer pairs are used to amplify known regions of the nuclear genome and then digested by restriction enzymes. These generate a series of bands, according to the length of the amplified regions. Three primer pairs used on 190 specimens from across the North American range of *Lycaedes melissa* yielded a total of 143 bands ranging in size from 71 to 481 bp (Gompert *et al.* 2006).

Microsatellites

Given the highly functional nature of enzymes studied by allozyme electrophoresis, and the proven selection which may occur on these loci, a neutral marker is desirable for population studies. Microsatellites are tandem repeats of 1- to 6-bp motifs such as ACACACAC. This marker seemed to be the 'Holy Grail' for population geneticists, as microsatellites are non-coding lengths of repetitive DNA. It was thought that these would be neutral and, by providing high levels of polymorphism, could end up as excellent tools for population genetics. Although it has been shown that they may be linked by hitch-hiking to a gene under selection pressure or that they may even be under selection themselves (Estoup & Cornuet 1999), they are still regarded as the first choice of neutral markers (Goldstein & Schlötterer 1999). Compared with allozymes, which require fresh or frozen material, microsatellite analysis can be conducted with dry material. This facilitates the study of museum material (Meglécz *et al.* 1998, Harper *et al.* 2006) or the use of non-lethal sampling (Lushai *et al.* 2000, Keyghobadi *et al.* 2005). From a practical point of view, each microsatellite locus is specifically amplified by PCR using locus specific primer pairs which recognise the flanking region of each side of the studied loci. The identification of these microsatellite loci with the design of primers is the most time-consuming task for the set-up for microsatellite based studies. This has to be done for each new species studied, but workable pairs of loci in one species are typically tested in other congeneric species, with various success. From a series of 17 loci identified for *Papilio zelicaon*, between five and 14 loci could be amplified in other *Papilio* species, but their polymorphism in these other species still remains to be tested (Zakharov & Hellmann 2007). In the 1990s, many butterfly biologists tried to develop microsatellite methods for butterfly population biology studies. By the end of the decade, it became apparent that the recurrent problems faced by the butterfly geneticists might be linked to the structure of the Lepidoptera genome rather than to the expertise of the laboratories involved (Meglécz & Solignac 1998, Nève & Meglécz 2000, Sunnucks 2000a, b). Many researchers tried to apply microsatellite techniques to butterflies, but gave up because of the low number of usable microsatellite loci. An analysis of the flanking regions of the microsatellite of *Euphydryas aurinia* and *Parnassius apollo* showed that many microsatellite loci could be grouped by similar flanking regions. Thus the numbers of microsatellites with unique flanking regions were drastically reduced (Meglécz *et al.* 2004). Subsequently this was found to be the case for many other Lepidoptera species, and also other insects (Meglécz *et al.* 2007).

Single-nucleotide polymorphism

Due to problems in the use of microsatellite loci in Lepidoptera, other markers useful for population genetics markers were desirable. Orsini *et al.* (2007, 2008) identified a series of single-nucleotide polymorphisms (SNPs) in *Melitaea cinxia*. Among these, two could be identified with known variants of the PGI locus, a key enzyme in the glycolysis cycle, which was already known to affect dispersal rate in this species (Haag *et al.* 2005). As the use of SNPs could be done without killing the individuals (a 2-mm diameter part of the hind wing was enough), the PGI genotypes could be studied on individuals which were later followed for their behaviour, with data collected on their flight body temperature, oviposition time and clutch size.

The new partial sequencing of the coding region of the *M. cinxia* genome gave sequence information on over half of the genes of this species (Ellegren 2008, Vera *et al.* 2008). As the method used a pool of ca. 80 individuals (caterpillars, pupae and adults from eight families), this approach provided unprecedented access to *M. cinxia* genome polymorphism, leading to the identification of numerous SNPs and to future detailed quantitative trait loci studies. No doubt the future of butterfly population genetics will increasingly use SNPs in its approach.

DIFFERENTIATION AMONG BUTTERFLY POPULATIONS

In Europe, a total of 87 studies of spatial aspects of population genetics have been located (Table 10.1). Of these, the great majority involved allozyme electrophoreses (80 cases), three studies involved RAPD and five involved microsatellites.

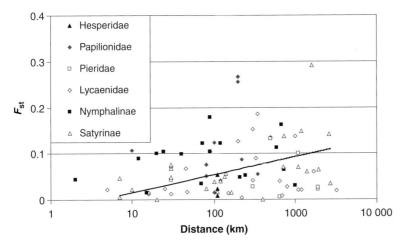

Figure 10.3. Fixation index (F_{ST}) as a function of the longest geographical distance involved in studies of European butterflies. There is a significant correlation only within the subfamily Satyrinae.

Generally the authors give a value of population differentiation, either F_{ST}, or one of its derived estimates (θ, G_{ST}). Each study species has its own ecological needs, and history of postglacial colonisation, from one or several refugia. These species may have widespread populations with frequent movements, as is the case of migratory species such as *Aglais urticae* (Vandewoestijne *et al.* 1999), whereas others are very sedentary, such as *Plebejus argus* (Brookes *et al.* 1997).

Most of the values of the global fixation index (F_{ST}) among populations of European butterflies show that generally populations show little genetic differentiation (*sensu* Wright 1978): the median F_{ST} value is 0.053. The lowest F_{ST} value (0.004) is found in *Polyommatus icarus*, a common and widely distributed butterfly which thus shows numerous movements among its populations. The highest value (0.291) is found among isolated populations of the mountain species *Erebia epiphron*, showing the ancient separation of its populations and the lack of movements between mountain massifs. The 86 studies of spatial population genetics on European butterflies show a general relationship between F_{ST} and the size (log scale) of the study area for all species combined, which is close to significance ($t = 1.94$, 84 df, $P = 0.055$) (Fig. 10.3). However, global F_{ST} tends to vary among the five butterfly (sub)families (excluding the four species of Hesperidae) ($F = 2.88$, 4 and 79 df, $P = 0.03$). Probably due to the narrow habitat choice of many species, butterflies of the families Papilionidae and Nymphalinae tend to have higher F_{ST} than those of the families Lycaenidae, Pieridae and Satyrinae (Fig. 10.4). This is also due to a bias in the studied species, as e.g. many Lycaenidae species have narrow ecological requirements, and probably a low colonisation power. Within four of these families (Lycaenidae, Papilionidae, Pieridae, Nymphalinae), there is no trend between the size of a study area and the observed F_{ST} ($P > 0.3$). For Satyrinae, however, the size of the study area, in logarithmic scale, is correlated with the observed F_{ST} ($t = 3.07$, 22 df, $P = 0.006$) (Fig. 10.3). Such a trend is probably the consequence of similar open structures of most studied Satyrinae populations, and therefore may be linked with a global isolation by distance process affecting Satyrinae species in a similar way, despite the absence of isolation by distance in *Maniola jurtina*. The choice of study species may also have biased these results. It is noteworthy that only two genetic studies, involving four species, could be found on Hesperidae, despite numerous studies on their ecology and distribution. Within Papilionidae, only isolated populations or mountain species have been studied. The genetic structures of e.g. the widespread *Iphiclides podalirius* and *Papilio machaon* have not been worked out. The only study of *P. machaon* has been carried out in Britain where the species is localised and threatened. In Pieridae the bias is the other way: only the widespread species *Pieris napi* has been thoroughly investigated. The diverse Lycaenidae family has been studied both in localised threatened species, such as the *Maculinea* species, and in widespread species such as *Polyommatus icarus* and *Aricia agestis*. With the exception of the widespread and migratory *Aglais urticae*, studies on the Nymphalinae have focussed on species with localised populations, often with vulnerable and decreasing distributions. The various studies on Satyrinae, like those of Lycaenidae, have involved both common species such as *Maniola jurtina* and *Coenonympha pamphilus*, and species with very restricted ranges such as *Erebia triaria* and *Coenonympha hero*.

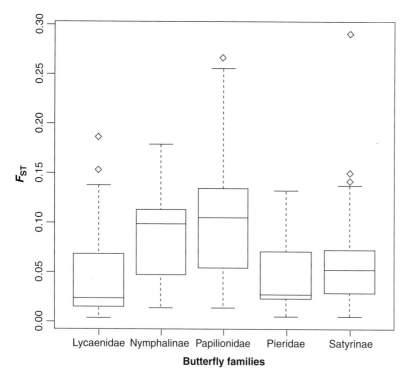

Figure 10.4. Box plots of the global fixation index (F_{ST}) for all studied species (Table 10.1). The horizontal bars indicate the median values, the boxes include 50% of the values and the vertical bars indicate the range of the values. Outliers are indicated by diamonds. Studies on Nymphalinae and Papilionidae clearly indicate higher values than those of Lycaenidae, Papilionidae and Satyrinae.

DISPERSAL ABILITY AND POPULATION DIFFERENTIATION

Species with low dispersal abilities show larger F_{ST} values. *Parnassius apollo* is known to be very vagile and it is therefore not surprising that populations from the same mountain massif are hardly different from one another, with a non-significant isolation by distance effect in the southern Alps. However, when all French populations are included, the slope of F_{ST} against distance is −0.54 indicating that areas between mountain ranges act as effective barriers to dispersal (Descimon *et al.* 2001), even if vagrants sometimes occur there (Lafranchis 2000). In the case of large-scale disturbances (such as fires or drought) the local genetic diversity of a population will depend on the scale at which migration and colonisation events take place. In the tropical species *Drupadia theda*, populations in areas near to undisturbed habitats tend to be more diverse than more isolated populations (Fauvelot *et al.* 2006).

When a range of species within a single habitat network are studied, their dispersal abilities may effectively be compared. With a mark–release–recapture survey, Baguette *et al.* (2000) showed that *Cupido minimus* had much less dispersal ability than the sympatric *Melanargia galathaea*

and *Aporia craetegi*. Both mark–release–recapture and genetic approaches showed that *Euphydryas aurinia* is less prone to movement between patches than *Melitaea phoebe* (Wang *et al.* 2003, 2004). Genetic studies of three *Thymelicus* species in Luxembourg and Germany showed that the three species have very different genetic structures as a result of their different dispersal ranges and habitat requirements. *Thymelicus lineola* displays a high dispersal ability and has broad habitat requirements, resulting in a panmictic genetic structure at the regional scale; *T. sylvestris* displays a lower dispersal ability in the same habitat matrix, which results in an isolation by distance effect. The third species, *T. acteon*, has narrow habitat requirements in combination with a low dispersal ability, resulting in populations being more isolated from each other, as reflected by this species having the highest F_{ST} value of the three species in the same habitat patch network, without any isolation by distance effect (Louy *et al.* 2007). *Thymelicus acteon* has declined in many European countries, and is of conservation concern, while the other two are stable (van Swaay and Warren 1999). Populations in southern mountain areas tend to be more variable than low-elevation ones. *Pieris bryonae* populations within the Swiss Alps are more isolated from each other than

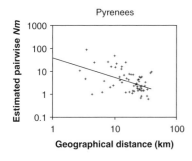

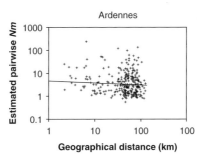

Figure 10.5. The estimated pairwise number of migrants *Nm*, as a function of the geographical distances in *Proclossiana eunomia* populations in Ardennes and Pyrenees. The slopes indicate the levels of isolation by distance (from Nève *et al.* 2008).

Pieris napi populations between south France, Germany and Hungary (Porter & Geiger 1995). Similarly, populations of *Proclossiana eunomia* from the Pyrenees display more differentiation than those from the Ardennes, as the slope of isolation by distance is −0.91 in the Pyrenees and −0.53 in the Ardennes (Fig. 10.5).

Population differentiation generally occurs when population are isolated from each other in space. In some cases, however, there seems to be isolation through ecological preferences. In *Carterocephalus palaemon*, two morphotypes (*Carterocephalus palaemon palaemon* and *C. p. tolli*), probably originating from distinct glacial refugia, are found in Białowieża primeval forest (northeast Poland). These are maintained because of their ecological differentiation both in habitat and phenology, resulting in assortative mating (Ratkiewicz & Jaroszewicz 2006). The two subspecies of *Proclossiana eunomia* at Białowieża (Krzywicki 1967) may show the same phenomenon.

POPULATION ISOLATION

For conservation biology, the consequences of population isolation set challenges. An isolated population may evolve locally according to local conditions, but it may also undergo genetic drift, leading to loss of genetic variability, or to stochastic demographic extinction. If populations are long-lived, the total effective size of the population is higher if it is the sum of a series of isolated demes (Whitlock & Barton 1997). However, in a metapopulation system (see Chapter 22), as local populations result from an equilibrium between extinction and colonisation, there is high gene flow between the populations, and little room for local adaptation to take place; selection will then operate more at the metapopulation scale. Genetic drift and selection are difficult to distinguish in population differentiation. In *Erebia triaria*, the isolated population of Xistral (northwest Spain), named *Erebia triaria pargapondalense*, is as different from Cordillera

Cantabrica populations (ca. 120 km apart) as from Pyrenean populations (720 km away), according to a study using four microsatellite loci and mitochondrial DNA. In this case, the genetic approach confirmed ecological and morphological data (Vila *et al.* 2005, 2006).

POPULATION SIZE

Small and isolated populations undergo genetic drift, due to the low numbers of reproducing individuals, and the homozygosity of such populations tends to increase. Isolated populations also suffer from several ecological effects, which affect their reproduction and hence their long-term survival: demographic stochasticity, a low buffering effect due to the small habitat patch size and microhabitat diversity, behaviour alterations due to low population density or close proximity to habitat boundaries, and often an overall lower habitat quality. These phenomena are often exarcerbated by the position on the edge of the relevant species' range (see Chapter 17). Such situations lead to increased homozygosity in individuals, as observed on westernmost populations of *Coenonympha hero* (Cassel & Tammaru 2003) or *Polyommatus bellargus* (Harper *et al.* 2008). By contrast phenotypic variation may increase, due to lower canalisation in populations with low genetic variation (Debat & David 2001), as seen in peripheral populations of *Polyommatus icarus* (Artemyeva 2005).

In the Åland Islands, the main factors affecting *Melitaea cinxia* population survival are population size, density of neighbouring populations, patch size and cattle grazing (Hanski *et al.* 1995b). A genetic study conducted on individuals caught in 1996 showed that heterozygosity had a significant extra effect on the extinction risk of the 42 genetically studied *M. cinxia* populations: the seven populations that went extinct between 1995 and 1996 had both ecological factors affecting their survival and a lower than average heterozygosity (Saccheri *et al.* 1998). A low heterozygosity was thus shown to be a significant extra factor affecting

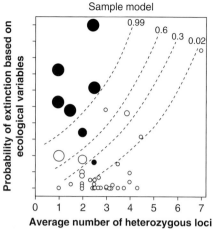

Figure 10.6. The probability of extinction in *Melitaea cinxia* populations in the Åland Islands (Finland) depends on both ecological and genetic factors. The vertical axis gives the probability of extinction predicted by a model including several ecological factors. The horizontal axis gives the average number of heterozygous loci per individual (from eight enzyme and microsatellite loci). The size of the symbol indicates the extinction probability. Black dots indicate populations that went extinct in one year (from Saccheri *et al.* 1998).

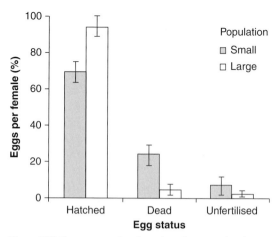

Figure 10.7. Percentages of total number of eggs per female *Coenonympha hero* females (± 1 SE) that hatched, died as zygotes or remained unfertilised, according to whether they came from large or small Swedish populations (from Cassel *et al.* 2001).

population survival (Fig. 10.6). Further evidence that population heterozygosity affects survival was given by an experiment in which the founder individuals of each populations were either full sibs, generating individuals with an inbreeding coefficient of 0.25, or outbred individuals from parents from two different populations, thus having a zero inbreeding coefficient. Three larval groups from either outbred or inbred individuals were introduced into one of 12 unoccupied habitat patches. Only two of the inbred populations attained adulthood and reproduction, and went extinct by the next year, while four of the six outbred populations survived until the next year (Nieminen *et al.* 2001). As inbreeding is deleterious, mate choice could avoid inbreeding, but experimental results showed that individual butterflies are unable to recognise sibs from non-sibs for mating (Haikola *et al.* 2004). As a result, inbreeding is more likely in small than in large populations, and this affects egg hatching rate, larval surviving rate (Haag & de Araú 1994, Haikola *et al.* 2001) and adult survival rate (Saccheri *et al.* 1998); furthermore, a second generation of inbreeding, by pairing full sibs, further decreased clutch size (Haikola 2003). A comparison of inbreeding effect conducted on French and Finnish individuals showed that a reduction of hatching rate from full sibs occurred in all cases, but was

more pronounced in French individuals than in Finnish individuals. The genetic load thus seems to be less severe in Finnish populations than in French ones. This is probably due to the repeated bottlenecks that Finnish populations have been through, which have purged them of a number of deleterious recessives (Haikola *et al.* 2001). Data on low reproductive output in isolated populations of *Parnassius apollo* indicates that inbreeding was the main factor affecting small populations of this species as well (Witkowski *et al.* 1997); in some cases the reproductive power of an isolated population was enhanced by the introduction of individuals from nearby populations with which exchanges are now unlikely to occur (Nakonieczny *et al.* 2007).

Coenonympha hero is a species which has already vanished from most of its former European range (van Swaay & Warren 1999). In Sweden, some local populations are extremely small, with an estimated size of seven to 15 individuals during the flight period, while others are larger, with estimated population sizes from 51 to 128 individuals. Furthermore, large populations are connected to each other by suitable habitat corridor structures, such as grassy roadbanks. Cassel *et al.* (2001) collected eggs from the different populations, and placed them in semi-natural conditions on tussocks of *Festuca ovina*. The hatching rate of eggs from small populations was lower, and their death rate and proportion of unfertilised eggs larger, compared to eggs from larger populations (Fig. 10.7). This phenomenon is most likely to be related to the increased homozygosity of the small populations, due to local inbreeding. Furthermore, females from small populations had a higher probability of not being mated, and thus to be effectively infertile. It is unlikely that small populations of this species will remain viable, both for the genetic reasons of increased homozygosity, and for ecological reasons such as a reduced microhabitat variability which will not effectively buffer against environmental variation. Furthermore, Swedish populations of *C. hero* already have a heterozygosity which is much lower ($H_{obs} = 0.017$) than that in the more central populations of Estonia or the Urals ($H_{obs} = 0.052$) (Cassel & Tammaru 2003).

The studies of inbreeding effects on *Melitaea cinxia* and *Coenonympha hero* indicate that inbreeding depression can have a significant effect on small and isolated populations. As reintroduction schemes are being considered for a number of species in parts of their range from which they have disappeared, there is therefore a requirement for genetic diversity to be considered to avoid inbreeding effects.

HOST PLANT RANGE

Populations which use a number of host plants tend to be more varied than ones using a single host plant. In this respect the case of *Euphydryas aurinia* in France is spectacular. In the Atlantic and continental part of the country it feeds on one or two Dipsacaceae species: mainly *Succisa pratensis* and sometimes on some *Knautia* species. Sampled populations, up to 700 km apart, show a low F_{ST} value of 0.0648, whereas populations from South France (up to ca. 650 km apart) show an F_{ST} value of 0.112. Descimon *et al.* (2001) concluded that the high F_{ST} obtained from southern populations is due to the large number of food plants used in this part of the range: Dipsacaceae (*Cephalaria, Knautia, Scabiosa, Succisa*), Caprifoliaceae (*Lonicera*), Valerianaceae (*Centrentus*) and Gentianaceae (*Gentiana*). Each local population seems to use only one food plant in a given locality (Mazel 1986). In the South of France, a neighbour-joining dendrogram (Fig. 10.8) shows that populations mainly cluster according to their geographical origin, i.e. according to distance. However, some populations show strong difference from this tendency. The population from Sommail is more similar to ones in southwest France, than to others in Languedoc, and this population is the only sampled Languedoc population which feeds on *Succisa pratensis*, like the ones in the southwest, and unlike the others in Languedoc, which feed on *Cephalaria*. Similarly, two populations from the Pyrenees, found feeding on *Succisa*, do not group with the other ones from the same area feeding on *Lonicera*, but with the ones from Languedoc, also feeding on

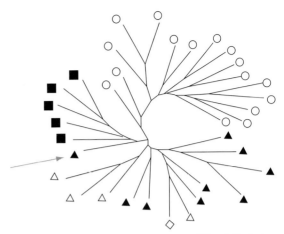

Figure 10.8. Neighbour-joining dendrogram of French *Euphydryas aurinia* populations, based on genetic distances between populations calculated for 10 enzyme loci. Circles; Provence; black triangles, Languedoc; open triangles, Pyrenees; squares, southwest France. The population from Sommail (Languedoc) is indicated by an arrow (from Descimon *et al.* 2001).

Succisa. In a study of 11 populations of *E. aurinia* from south France and north Spain, scored by AFLP markers, larvae of *E. aurinia* found on *Succisa* or on *Lonicera* at the same site were shown to be as different as two allopatric populations feeding on different hosts are (Singer & Wee 2005). Differences between individuals at the same site, whether on the same or on different hosts, are generally smaller within than between populations. A high level of differentiation between larvae found on different host plants will be more likely to occur in allopatry than in sympatry, but the latter may be relevant in some cases (Wee 2004, M. C. Singer unpubl.) This tends to confirm Descimon's interpretation that *E. aurinia* may be undergoing speciation in South France. However, such a differentiation pattern is not the rule. *Euphydryas editha* caterpillars at Sonora Junction (California, USA), were not genetically different whether they came from eggs laid on *Penstemon* or *Castilleja*, thereby discarding any speciation event between the individuals on the two hosts. Furthermore, the selection pressure put forward by the different hosts may affect the variability of the population. Singer & Wee (2005) showed that larvae of *E. editha* developing on *Castilleja* (from eggs naturally laid on this host) had a higher heterozygosity (0.137, SE = 0.007) than larvae from *Pedicularis* (0.119, SE = 0.007) ($P < 0.001$, Mann–Whitney U test) at the same site (T-junction, California). As female *E. editha* from this site all accept both hosts to lay their eggs, a process of sympatric speciation may be ruled out. The only cause seems to be that the two hosts induce different mortality rates between egg hatch and time of sampling, thereby indicating different selection pressure on the populations.

BARRIERS TO DISPERSAL

In Germany, populations of *Chazara briseis* have declined. Present populations on igneous and calcareous hills are separated from each other by farmland and urban development. Most individuals of this species stay in their natal patch, with a mean distance between capture events of 80 m ($n = 191$), while about 2% ($n = 5$) of individuals moved more than 1000 m. The question then is: are these occasional long-distance movements efficient in maintaining gene flow within the landscape? An analysis of 165 individuals from nine populations up to ca. 10 km apart gave a mean F_{ST} of 0.022 for the 15 polymorphic loci. Johannesen *et al.* (1997) suggested that these populations show limited substructure and that the agricultural landscape does not constitute a barrier for this species. Dispersive individuals moving out of their preferred habitat tended to move to a neighbouring habitat patch, leading to an isolation by distance effect, which was indeed detected.

POST-GLACIAL DISPERSAL PATTERN

Since the last glacial maximum (18 000 yr BP), the climate of Europe has changed dramatically, and this has led to the colonisation of northern areas from southern refugia (Huntley & Webb 1989). This change in distribution of organisms can have lasting consequences for the genetic make-up of populations. For species with a low dispersal ability, post-glacial colonisation events took place slowly, and usually by stepping-stone patterns (see Chapter 17). The main consequence for the genetic diversity is that the centres of origin of this post-glacial northern migration are in southern Europe or Asia. Consequently there is generally a decrease of genetic diversity from the centre of origin to the edge of a species range (Hewitt 1996). In *Polyommatus bellargus*, for example, a study of the mitochondrial control region and of a section of the 12S rRNA gene (totalling 722 bp) showed a much lower variation within the UK (mean pairwise difference between two individuals: 0.000 in non-Dorset populations and 0.295 in Dorset) than in France (mean pairwise difference: 7.42). Such a small variation is unlikely to have remained over a long period. It is concluded that the British populations of *P. bellargus*

probably originated from western European stock during historical times, and that this colonisation was subjected to a bottleneck (Harper *et al.* 2008).

OPEN QUESTIONS AND CONCLUSION

Generally, populations of butterflies show differentiation according to the geographical distance separating them, but the pattern may differ widely between species within a given landscape. As expected, widely distributed and highly mobile species display less differentiation than more sedentary species. The high fragmentation of habitats decreases gene flow between populations. As most species of European butterflies occur in discrete populations, this effect is of major consequence, and may lead to the quick decline of populations once a critical threshold of habitat connectivity has been reached (With & Crist 1995). However, the type of habitat between patches also has crucial importance in the dispersal ability of individuals. For example, forest habitats are effective barriers to dispersal for open-habitat species such as *Erebia medusa* (Schmitt *et al.* 2000) and *Parnassius smintheus* (Keyghobadi *et al.* 2005), while the reverse is true for several forest species, such as the moth *Operophtera brumata* (Van Dongen *et al.* 1994) or *Pararge aegeria*, as it seems that individuals from forest areas need to perceive the presence of a forest to move towards it (Merckx *et al.* 2003). How other habitat structures affect butterfly dispersal remains an open, but critical, question. The response of individuals will also depend on regional habitat structure, as dispersal behaviour is expected to be selected against in very fragmented habitats, while such behaviour will be more common in more continuous habitats (Baguette *et al.* 2003). As dispersal ability is heritable – it was recently shown to be heritable from mother to female offspring in *Melitaea cinxia* (Saastamoinen 2008) – the selection pressure of landscape structure on individual dispersal abilities may occur, and result in differential dispersal abilities according to regional landscape structure (Van Dyck & Baguette 2005, Schtickzelle *et al.* 2006), as was first suggested by difference in thorax width among *Hesperia comma* populations (Hill *et al.* 1999b).

In detecting significant barriers to dispersal, Bayesian methods may prove to be very useful. The emerging field of landscape genetics uses techniques of Bayesian clustering to identify clusters of individuals, and hence the barriers between clusters (software GENELAND: Guillot *et al.* 2005, or EASYPOP: Balloux 2001). Bayesian methods may also help to delineate populations in very mobile species such as *Aglais urticae* or various *Colias* species. Graphical methods using multivariate analyses offer some grouping, but do not allow confidence to be made about the barriers. In contrast, Bayesian methods may prove useful in this respect, as these may combine the information from the genotypes of the sampled individuals and test which of several barriers may be the most likely (Corander *et al.* 2004). Unfortunately, such important studies involving population genetics and landscape ecology need high research investment, and may be carried out only in a few, carefully chosen study systems.

The inbreeding consequences shown in Finnish *Melitaea cinxia* populations most probably occur generally. As more and more populations become isolated, they end up occurring in non-equilibrium metapopulations, where extinctions are more frequent than colonisation, eventually leading to total extinction, due to both local inbreeding and demographic disequilibrium. In extreme cases where active management aims at rescuing a declining population, on top of habitat restoration, 'genetic restoration' should also take place by pairing individuals which come from populations which used to be part of the same metapopulation, even if populations in between have now become extinct.

The distribution of many European butterflies is likely to change dramatically over the coming decades as a result of global warming. Several species have already started a northward shift (Parmesan *et al.* 1999). The future of this pattern is likely to have far reaching consequences, as populations of the mountains of southern Europe will move upwards, which will result in their increased isolation, and in many cases eventual extinction (Wilson *et al.* 2005). Furthermore, species should not be viewed generally as adapted to one particular environment; they generally include populations genetically adapted to a range of environmental conditions (see Chapters 6 and 12). Populations from the middle of the range will also be affected, as these will have to adapt to changing conditions, in a typical 'Red Queen' fashion: they

will have to adapt to new environmental conditions to stay in the same place (Lythgoe & Read 1998), through selection of genotypes adapted to warmer temperatures, as has been shown for numerous other species (Parmesan 2006). In erratic species, alleles adapted to warm conditions presently present in the south of the range will migrate north, through a change in the selective gradient due to climatic conditions getting warmer. For more sedentary species, this process may be more difficult, as most species occur in patchy habitats, which are now isolated from each other (Bridle & Vines 2007). In such species, the genetic variation on the edges of the distribution is generally lower than at its centre, as shown in *Erynnis propertius* and *Papilio zelicaon* (Zakharov & Hellmann 2008). This in turn may affect the future shift in distribution, as the species with less gene flow (*E. propertius*) will be more affected than the more vagile species (*P. zelicaon*).

For species relying on photoperiod for the timing of specific life stages, this adaptation is under a strong selection pressure. With climate change the environmental conditions associated with specific photoperiods will alter, imposing new selection pressures and requirements for changes of responses to specific photoperiods. Populations of butterflies will then be affected in numerous ways, involving a large set of ecological and physiological characters, ultimately depending on their genetic make-up governing their response to photoperiod, dispersal behaviour and temperature range for adult and larval activities. How these strong selection pressures will globally affect European butterflies is largely unknown, as such diverse and large-scale environmental change has not been observed before.

ACKNOWLEDGEMENTS

This chapter benefited from the constructive comments of Emese Meglécz, Sören Nylin, Thomas Schmitt and the editors.

11 • Parasitoids of European butterflies

MARK R. SHAW, CONSTANTÍ STEFANESCU
AND SASKYA VAN NOUHUYS

SUMMARY

Parasitoids are insects (mainly Hymenoptera and Diptera) whose larvae develop by feeding on the bodies of other arthropods (mainly immature insects) and the adults are free living. They are of immense importance in practically all terrestrial ecosystems because of their impact on the population levels of their hosts. In contrast to the considerable attention paid to European butterflies their parasitoids have received very little scrutiny, and the recognition that some reared parasitoids are in fact hyperparasitoids (parasitoids of parasitoids) has often been lacking in published records. A review of the biology and taxonomy of European butterfly parasitoids is presented, including a simple key to the various parasitoid families that include relevant species, followed by a brief discussion of the authors' current knowledge of their use of butterfly hosts. Quantitative case studies on the butterfly taxa *Iphiclides podalirius*, *Pieris* spp., *Maculinea rebeli*, *Aglais urticae*, *Melitaea* and *Euphydryas* spp. and *Thymelicus lineola* illustrate the range of parasitoid assemblages and the extent to which parasitoids account for butterfly mortality. The needs and difficulties inherent in moving from a simple understanding of what parasitoids are using a host population to a more sophisticated assessment of the impact they may have on the host's population dynamics are outlined. Parasitoids can be expected to have many important effects on the ecology of butterfly species and, on the evolutionary timescale, defence against parasitoids has shaped aspects of the physiology of the immature stages and the behaviour of both larval and adult butterflies. Butterfly researchers are encouraged to help build better knowledge about the host relations and effects of parasitoids. To better understand the place of parasitoids in the lives of butterflies there is still much to be gained from close attention to natural history, careful experimentation and thorough taxonomic investigation.

INTRODUCTION

Parasitism affects the egg, larval and pupal stages of butterflies and its importance as a mortality factor in butterfly populations is both general and indisputable. Some of the parasitoids involved are extreme specialists but others are more generalist; however, mortality is often high even in the absence of specialists (e.g. Stefanescu *et al.*, 2003b), and almost certainly no European butterfly species altogether escapes.

In view of the considerable attention paid to European butterflies over the years, first by collectors and more recently by conservationists, it is anomalous that we have such poor knowledge of their parasitoids (Shaw, 1990). Apart from a very few, such as pest *Pieris* and *Thymelicus* species, and flagship conservation groups like the *Maculinea* and Melitaeini species which have also attracted population ecologists, almost none has been subjected to extensive sampling to investigate parasitism. Such deliberate effort would be needed firstly to establish which parasitoids regularly attack a particular butterfly species, and in which parts of its range, and secondly to investigate the effect these parasitoids have on their host populations.

However, there is a complementary need for parasitoid-based knowledge. Host–parasitoid complexes are not closed systems, and there is no *a priori* way of assessing whether a parasitoid reared from a particular species also uses related hosts, or perhaps unrelated ones occurring in similar environments. Each parasitoid species must be fitted into a continuum between absolute monophagy at one extreme or using the butterfly only marginally as part of a diffuse or differently focused host range at the other. Therefore we need to establish comprehensive knowledge not only of which species attack butterflies, but also of the host associations of each one. This is gradually accruing, at least for common parasitoid species, especially through small-scale rearings involving a large number of host species.

We begin this chapter by explaining why so little is known about parasitoids of butterflies, highlighting the sources of error that undermine the knowledge base (see also Shaw, 1990). Next we summarise the general biology of parasitoids of European butterflies. A simple key, including biological characteristics, links this with the following taxonomically organised section on their life history and host

Ecology of Butterflies in Europe, eds. J. Settele, T. Shreeve, M. Konvička and H. Van Dyck. Published by Cambridge University Press. © Cambridge University Press 2009, pp. 130-156.

ranges. The chapter then examines the few existing case studies of parasitoid communities associated with particular butterfly taxa, the contribution of parasitoids to mortality and, to some extent, their potential roles in butterfly population dynamics.

UNDERSTANDING AND IMPROVING OUR KNOWLEDGE

The historical passion for butterfly-collecting was unconducive to a systematic study of their parasitoids, which were generally seen as a troublesome scourge and too obscure to be interesting. With the more recent decline of butterflies in Western Europe, collecting has decreased, and science-based conservationists have become the more active group. Most effort focuses on getting the 'bottom–up' aspects of the habitat (vegetation composition and quality, thermal properties of sites, etc.) right for the dwindling butterfly population, and (with some notable exceptions) parasitoids and other 'top–down' influences have generally been ignored. However, this is to neglect the strong possibility that the butterfly population will harbour specialist parasitoids even more at risk than their host, that might also play important roles in the host's population dynamics. Thus, it is more enlightened to see the butterfly's conservation in broader terms, including the trophic level above it (Shaw & Hochberg, 2001). Gradually our currently low knowledge might then be enriched by conservation biologists as well as by the scientifically minded enthusiasts who contribute important data on such a wide front.

The difficulty of finding eggs, larvae and especially pupae of most butterfly species in comparison with adults has also kept knowledge on parasitism low. This contrasts strongly with some groups of microlepidoptera whose early stages are the most easily found, resulting in abundant data on the host associations of their parasitoids (e.g. Askew & Shaw, 1974, 1986; Askew, 1994; Shaw & Horstmann, 1997).

One of the most important properties of a parasitoid is how host-specific it is. In practice rather few parasitoids are absolutely host-specific (though locally they may be, if only one of the possible hosts occurs), but rather they usually have a host range that either comprises a group of phylogenetically related hosts or a group of ecologically similar ones, or some balance between the two. It is useful to define *host range* conceptually, such that the host range includes only those species of potential hosts that the parasitoid is usually able to attack successfully, following a pattern of searching behaviour enabling it to encounter them regularly (Shaw,

1994). Thus an abnormal success in developing on a host that is usually rejected (or unsuitable if accepted), or the occasional discovery of a suitable host that is usually absent from the parasitoid's searching environment, will not alter the concept of host range for the parasitoid concerned. Importantly, if host range is expressed in the quantitative terms implied by the definition, then abnormal events and misidentifications will gradually become marginalised.

Many of the parasitoids of European butterflies are univoltine and use univoltine hosts, or are plurivoltine in synchronisation with plurivoltine hosts, or sometimes the parasitoid may be plurivoltine and use a univoltine host generation for successive broods. In all these cases the parasitoid has a potential to be absolutely host-specific, but there are also plurivoltine parasitoids that depend on different host species at different times of year. This complicates the parasitoid's population dynamics enormously, adds another dimension to the concept of host range and places an additional habitat demand because representatives of more than one set of hosts must be present (see also Shaw, 2006). In some cases the alternate hosts of parasitoids of European butterflies are not themselves butterfly species, but other Lepidoptera. Gaining insights into host ranges in these terms is particularly challenging, but of fundamental importance.

Many case studies (e.g. Shaw, 1982, 1990, 1993, 1994, 2002a; Askew & Shaw, 1986; Noyes, 1994) demonstrate that compilations of host–parasitoid records abstracted from the literature are so full of misinformation and ambiguity that they are useless for understanding either the nature or the breadth of host ranges. In particular, misidentifications have been rife, of not only the parasitoid but also the host (often an extraneous insect overlooked during rearing) and, additionally, either the compilations are non-quantitative (giving equal weight to rare and common occurrences) and/or they are corrupted because authors have reiterated host records from already published sources without making it clear that they are not giving new records (thereby leading to multiple scoring: see Shaw, 1993). The only way round these problems is for the focus to shift from literature records to an assessment of extant specimens by competent taxonomists, and for host data for parasitoids to be given in key works, reviews, etc. in quantitative form (as 'host mortalities', scoring gregarious broods as one) based on reared specimens actually seen. Even then, host misidentification may still be a problem, especially if host remains have not been preserved with the adult parasitoid. Ideally, each parasitoid (host mortality) that is reared will have come through rigorous rearing protocols (see Shaw, 1997), to be preserved with its carefully recovered and assessed host remains and deposited in an active research collection where it will come to the attention of taxonomists. Much can then be understood about realised host ranges (e.g. Shaw, 1994, 2002a; Shaw & Horstmann, 1997) and the extent to which parasitoids of a particular host are engaged in other parasitoid complexes (e.g. Shaw & Aeschlimann, 1994).

Advice on appropriate rearing protocols and preservation techniques can be found in Shaw (1990, 1997) (the latter can also be applied to Tachinidae, with the rider that adults are best killed once their wings have hardened and then direct-pinned) and, for the special techniques needed for preserving Chalcidoidea and other 'microhymenoptera', in Noyes (1982, 1990).

A frequent aim might be to assess the percentage parasitism in a host population, but it is important to understand that this is extremely difficult to do for several reasons (Shaw, 1990). Firstly, parasitised and unparasitised caterpillars often do not behave in the same way, so one category becomes more amenable to whatever sampling method is used; and secondly, parasitised caterpillars often develop at a different rate from unparasitised ones (parasitised hosts are often retarded, though in some cases parasitism speeds the host to its next stage), so that a disproportion of hosts may have left the sampling arena (e.g. as healthy hosts pupate, leaving parasitised ones behind). Indeed, whenever exceptionally high levels of parasitism are recorded in collections of final-instar larvae, these two possibilities should be considered. Further, at whatever time sampling is done, some hosts may have already been killed by parasitoids (and been lost to the sampling process), and some parasitoid attack may not yet have happened. Thus it is usually impractical to sample all stages effectively enough to cover generational parasitism quantitatively.

GENERAL PARASITOID BIOLOGY

The following account covers only insect taxa that are fairly regular as parasitoids of European butterflies per se: for fuller accounts see Gauld & Bolton (1988) or a summary in Shaw (1997) for the biology of parasitic Hymenoptera, Shaw & Huddleston (1991) for Braconidae, Herting (1960), Belshaw (1993, 1994) and Stireman *et al.* (2006) for Tachinidae, Shaw & Askew (1976) for parasitism of (British) Lepidoptera as a whole, and classic texts such as Clausen (1940) and Askew (1971) for the biology of wider groups of insect parasitoids. Godfray (1994) and Quicke (1997) give more evolutionary treatments of different aspects of behaviour and biology.

Parasitoids of butterflies fall into two insect orders, Diptera and Hymenoptera. In both cases the adults are free-living and the larvae develop (whether solitarily or gregariously) by feeding on a single immature host which is killed as a result (cases of survival have occasionally been reported, especially involving Tachinidae). Some other organisms such as Mermithidae (Phylum Nematoda) have life-styles comparable to insect parasitoids (see Eggleton & Gaston, 1990) and may occasionally be associated with European butterflies, but obligate specific relationships are not known and this chapter will not deal with them further.

Several families of Diptera behave as parasitoids but only one, the large family Tachinidae, includes regular and important parasitoids of butterflies (in addition, some species of Bombyliidae are very occasionally reared). All Tachinidae attacking Lepidoptera parasitise the larval stage, though some do not kill the host until it has pupated. They do not have piercing tubular ovipositors as such, and employ a wide variety of strategies to get their larvae into the host.

Many families of Hymenoptera are parasitoids, but relatively few include parasitoids of European butterflies. Only Braconidae and Ichneumonidae (together comprising the superfamily Ichneumonoidea) regularly, and a very few Eulophidae (superfamily Chalcidoidea) rarely, are parasitoids of the larval stage. While Braconidae attacking European butterflies all kill the larval stage, a few Ichneumonidae attack the larval stage but do not complete their development until the host has pupated. Some other Ichneumonidae attack the pupal stage, as do a few Pteromalidae and Chalcididae (both superfamily Chalcidoidea). The eggs are attacked by minute chalcidoids (in several families, especially Trichogrammatidae) and Scelionidae (superfamily Platygastroidea), which complete their development to adulthood within the host egg. Although other access strategies are known, all the Hymenoptera that parasitise European butterflies have tubular piercing ovipositors by means of which the egg is placed either inside or (in rare cases) onto the host directly.

The trophic relationship that a reared parasitoid has to its apparent host may not be as straightforward as it seems. *Primary parasitoids* attack and eventually kill the host itself, but surprisingly often they are themselves subject to attack from *secondary parasitoids*, also known as *hyperparasitoids*, which are parasitoids of the primary parasitoid. Although Tachinidae and Braconidae are all essentially primary parasitoids, several groups of both Ichneumonidae and Chalcidoidea contain species that function as hyperparasitoids. There are two main categories of hyperparasitoids. *True hyperparasitoids* attack the primary parasitoid while it is still growing inside (or occasionally on the outside of) the host, which is usually still alive at this time (Plate 7a). Thus the searching behaviour of true hyperparasitoids is, at least initially, focused on finding the same hosts as the primary parasitoids it will attack. Virtually all true hyperparasitoids are completely specialised and cannot function as primary parasitoids. Usually the primary parasitoid is not killed until after it has killed the host and made its own apparently normal preparations for pupation, such as spinning a cocoon – from which the adult hyperparasitoid will eventually emerge. *Pseudohyperparasitoids*, on the other hand, attack the primary parasitoid only after it has completed its feeding, by which time the host is dead or moribund. The primary parasitoid will typically be attacked in its cocooned or pupal stage (Plate 7b). Although not sampled by collecting living caterpillars, pseudohyperparasitoids affect host population dynamics by reducing the population of primary parasitoids just as true hyperparasitoids do. Many pseudohyperparasitoids opportunistically use a range of hosts in small cases and cocoons, some of which just happen to be those of primary parasitoids, but others are more specialised. Some parasitoids are capable of functioning as either a primary parasitoid or as a pseudohyperparasitoid of a given host, in which case they are said to be *facultative hyperparasitoids*.

Whether primary or secondary, parasitoids whose larvae feed from the outside of the host are called external parasitoids or *ectoparasitoids*, while those that feed from within the host's body are internal parasitoids or *endoparasitoids*. All tachinids are endoparasitic. Though many parasitic Hymenoptera are ectoparasitic, their frail larvae are in general unsuited to ectoparasitic existence unless they can develop in concealment (exceptions relevant to butterflies occur in Eulophidae). Thus ectoparasitism of butterflies is rare, but might be expected when the host larva is quite deeply endophytic (e.g. some Polyommatini). However, many groups of braconids start their life as endoparasitoids but have a final ectophagous phase, and there are just a very few butterfly species (e.g. some Hesperiidae and *Vanessa atalanta*) whose larvae rest concealed in retreats that are robust enough to support this.

There is another way to categorise parasitoids by their developmental characteristics than simply as ectoparasitoid or endoparasitoid, which correlates better with certain parameters such as the potential for breadth of host range (see Haeselbarth, 1979; Askew & Shaw, 1986). In this case the emphasis is on the immediate effect on the host's development. If it is permanently arrested or killed at the time of parasitism, the immature parasitoid need not accommodate to a living host, and is said to be an *idiobiont*. If, on the other hand, the host continues to develop, or move around and look after itself, for at least some time following parasitisation (so that the immature parasitoid has to withstand the various challenges mounted by a living host), the parasitoid is said to be a *koinobiont*. The latter are generally constrained to relatively narrow host ranges, while idiobionts – at least potentially – can use a broader range of hosts found within their searching environment. For example, true hyperparasitoids are all koinobionts, while pseudohyperparasitoids (many of which are facultative hyperparasitoids, thereby having host ranges spanning at least two insect orders) are idiobionts. There is a weak correlation between endoparasitism and koinobiosis on the one hand and ectoparasitism and idiobiosis on the other: importantly, however, when ectoparasitic koinobiosis (rare for parasitoids of European butterflies) or endoparasitic idiobiosis (frequent in parasitoids of butterflies) occur, the host range parameters follow the koinobiosis and idiobiosis dichotomy rather than endoparasitism and ectoparasitism. Tachinidae, however, fit rather uncomfortably into this: they are all endoparasitic koinobionts, yet some have immensely wide host ranges. There are important differences between Tachinidae and parasitic Hymenoptera, regarding both the means of accessing the host and the capacity of Tachinidae to escape the host's encapsulation defences, that seem likely to account for this (see later).

Parasitoids may be *solitary*, when a single individual develops in or on each host, or *gregarious* when a brood of two or more develops from one host. Some Tachinidae parasitising butterflies are strictly solitary, but others typically develop in brood sizes ranging from about one to four. In Hymenoptera, usually a given species is consistently either solitary or gregarious (though for egg parasitoids host size may be a factor) but a few essentially gregarious species of *Cotesia* (Braconidae) parasitise differently sized hosts according to season, and occasionally develop solitarily. In gregarious species there are often several tens of individuals in a brood. The braconid subfamily Microgastrinae furnishes the only gregarious koinobiont endoparasitoids of butterfly larvae, but gregarious idiobiont endoparasitoids of pupae are found in both the ichneumonid subfamily Cryptinae and the chalcidoid family Pteromalidae.

Multiparasitism refers to two species parasitising the same host, and *superparasitism* to more than one oviposition attack from the same parasitoid species. In both situations supernumeraries are usually eliminated, and the first-instar larvae of many species of solitary koinobiont endoparasitoids, in particular, are adapted for fighting competitors.

Most primary parasitoids attack their hosts at a fairly precise life-history stage, and can be categorised to reflect this. Thus all *egg parasitoids* oviposit into and kill insect eggs, and (in this strict sense of the term) always emerge as adults from them. *Larval parasitoids* attack and also kill the host in its larval stage; if it is an endoparasitoid the parasitoid larva may leave the host to pupate elsewhere (Ichneumonidae and Braconidae make cocoons; Tachinidae pupate inside their tanned last larval skin, known as a puparium), or in some groups (Rogadinae (Braconidae) and a few Campopleginae (Ichneumonidae)) the parasitoid pupates inside the host's larval skin (hardened, or sometimes strengthened by a clear additional cocoon). As butterflies on the whole pupate either in the open, or at least without the benefit of a strongly enclosing cocoon, the rather large number of groups of ichneumonoid parasitoids that depend strongly on Lepidoptera cocoons (including many koinobionts that finally kill prepupal hosts) do not enter the parasitoid complexes of European butterflies. *Pupal parasitoids* (for European butterflies involving only Ichneumonidae and Chalcidoidea) oviposit in or on host pupae and, if they are endoparasitoids, emerge as adults from them. Idiobionts, by definition, always kill the stage attacked, but some koinobionts invariably kill the host at a stage later than the one attacked. In these cases they are known as *egg–larval parasitoids* or *larva–pupal parasitoids*. Egg–larval parasitism of European butterflies is seldom recorded, but it does occur in some species of the braconid subfamily Microgastrinae, in which it is probably only facultative (see Johansson, 1951), and in at least one species of *Hyposoter* (Ichneumonidae: Campopleginae) (Plate 8a) in which it appears to be obligatory (van Nouhuys & Ehrnsten, 2004). In both cases, however, the host larval embryo is so well advanced by the time it can successfully be attacked that oviposition is essentially into a first-instar larva that has not yet hatched. Larva–pupal parasitism of European butterflies is practised by several Tachinidae and a few groups of Ichneumonidae. In the case of the Ichneumonidae pupation is always inside the host pupa, but while some tachinids form their puparium within the host pupa the majority leave the host beforehand.

In Tachinidae the egg must be fertilised to develop, and the female has no control over the sex of her progeny. In Hymenoptera, however, an unusual form of sex determination, called haplodiploidy, prevails. In this, unfertilised (haploid) eggs develop and become males (a process known as *arrhenotokous parthenogenesis*) while fertilised (diploid) eggs become females. It is usual for mated female parasitic Hymenoptera to regulate access of the stored sperm to the egg as it passes down the oviduct, and so control its sex. Both the overall sex ratio and, for gregarious species, the sexual composition of broods thus reflect reproductive strategies that have presumably been optimised by natural selection. In the case of solitary idiobionts in particular, female progeny are often invested in the larger-sized hosts attacked and males tend to result from the smaller hosts (see Luck *et al.*, 1992; Godfray, 1994). However, in a good many Hymenoptera species males are practically unknown, and diploid females develop from unfertilised eggs by a process known as *thelytokous parthenogenesis* or thelytoky (thelytoky is sometimes, but not always, mediated by microorganisms: Stouthamer *et al.*, 1992).

Adult parasitoids generally feed, at least on sugars which they can obtain from flowers and honeydew, but also in some cases on protein needed to mature their eggs. Some Hymenoptera use their piercing ovipositors or biting mouthparts to wound and often kill hosts (or sometimes other insects) in order to imbibe haemolymph (Plate 8b). This *host-feeding* is practised by parasitoids attacking all host stages, but it is more often done by idiobionts than koinobionts (see Jervis & Kidd, 1986), and it can cause substantial mortality. Another source of protein is pollen (e.g. Zhang *et al.*, 2004).

TAXONOMIC REVIEW

In this section we will firstly provide a simple key to the families of Hymenoptera and Diptera that include known parasitoids of European butterflies, secondly overview these families in so far as they parasitise butterflies, and thirdly within each parasitoid group attempt to give a list or annotated table of butterfly–parasitoid associations (but only for primary parasitoids) that we believe to be reliable. While all of this is complicated by insufficient knowledge, we hope the first two parts can be done fairly confidently: additions and exceptions will undoubtedly arise, but we believe they will be relatively minor.

Regarding host associations, however, we can give only a skeletal account, best seen as a stimulus for further study. The tables we provide for groups of Hymenoptera are based on specimens we have seen (material in the National

Museums of Scotland (NMS) unless otherwise indicated), or in a few cases those seen by careful taxonomists who have passed reliable information on to us (indicated pers. comm. – but in these cases often with no statement of depository). Records from the literature are included only in the very few cases where we believe both host and parasitoid determinations to be particularly reliable. Thus we deliberately exclude a very large number of literature records which (though they may be correct) we cannot directly confirm or corroborate. This involves several taxa that were described from butterfly hosts; if the circumstances are beyond our assessment we felt it inappropriate to enter them in tables, though we have sometimes mentioned the supposed association in the text.

A parasitoid's name is bracketed if we believe it is only casually and infrequently associated with butterflies. In a few cases we have included at the generic level parasitoids for which we cannot give a specific identity, but only when the association seems regular or is noteworthy for some other reason. Butterfly names (other taxa in the parasitoid's host range are not included) follow Karsholt & Razowski (1996) except as subsequently revised, and are entered in bold if the association is known from at least two separate occasions (rearings from numerous host individuals collected in the same place and year are reckoned as only one occasion). If putting the entry in bold depended on NMS specimens reflecting a single occasion combined with those from another source, a '+' is given in the superscript referencing the latter (as all unattributed records depend on NMS specimens). Most host associations in bold are likely to be at least fairly regular – many of those not in bold will be too, but we cannot say so with such confidence. It is, however, extremely important that the host–parasitoid associations mentioned here are NOT used as a means to identify new material, which must always be done more rigorously.

In addition to columns giving host and parasitoid names, a central column is used to give an informal estimation of host range. This depends partly on data given in the table, but also on wide sources of both positive and negative information which, it rapidly became apparent, could not be accounted or referenced in any consistent or purposeful way. The 'less than' symbol < should be read as 'some, but not all, of the species in' [the taxon following].

This whole section looks at parasitism only qualitatively, and from the viewpoint of the parasitoids. A later section will review the published quantitative studies that exist for particular host species.

Simple key to families of insect parasitoids of European butterflies

This key is restricted to taxa likely to be reared from European butterflies. It does not cover parasitoids of all European Lepidoptera, let alone parasitoids in general. Attempting to use it in a wider context may therefore often lead to mistakes. Further, it does not include all of the groups of Chalcidoidea that might be reared as hyperparasitoids unless they are also likely to arise as primary parasitoids.

1. One pair of membranous wings; antenna inconspicuous, shorter than length of head. *(Host killed as larva or pupa; never making a silken cocoon for pupation)* . . . . .Diptera 2

    – Two pairs of membranous wings (or wingless); antenna conspicuous, longer than length of head. *(Host killed as egg, larva or pupa; egg parasitoids can be minute (<0.5 mm) but otherwise from ca. 3–30 mm; primary parasitoids and hyperparasitoids)* . . . . .Hymenoptera 3

2. Body bristly; legs relatively short, robust and with strong bristles; antenna 3-segmented; colour of wing membrane usually uniform; pupation (usually outside the host remains) in an ovoid immobile puparium formed from the hardened and darkened last larval skin (in which the paired posterior spiracles remain discernible). *(Host killed as larva or pupa; 6–12 mm; solitary or in small broods; always primary parasitoids)* . . . . .Tachinidae

    – Body at least partly furry; legs long, slender and with fine bristles; antenna with more than 3 segments; wing membrane usually partly darkened (at least near costal margin); pupa exarate, conspicuously spiny, mobile. *(Host usually killed as pupa, inside which the parasitoid pupates; 6–15 mm; solitary; primary parasitoids or sometimes hyperparasitoids)* . . . . .Bombyliidae

    [Other families of Diptera may be saprophagous on the moribund or dead bodies of butterfly larvae or pupae; especially Phoridae, pupating in brown boat-shaped puparia, often numerous, adults (ca. 3 mm) run rapidly with jerky movements and often make short flights.]

3. Antenna with more than 13 segments; fore wing venation including closed cells (or wingless); pupation inside or outside host remains, if outside then a cocoon is made. *(Host killed as larva or pupa)* . . . . .Ichneumonoidea 4

    – Antenna with at most 13 segments; fore wing venation only near costal margin or absent, lacking clear enclosed cells (if minute and from host egg may be wingless); pupation usually inside host remains (or in primary

Figure 11.1. Typical fore wing of Ichneumonidae with second recurrent vein (2m-cu) present (except for apterous or brachypterous species).

Figure 11.2. Typical fore wing of Braconidae with second recurrent vein (2m-cu) absent (species always fully winged).

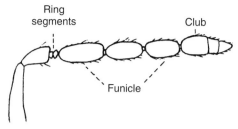

Figure 11.3. Stylised antenna of Chalcidoidea (ring segments are present in all except Encyrtidae). Scelionidae (Platygastroidea) are broadly similar but lack ring segments.

parasitoid cocoon if a hyperparasitoid) and usually no cocoon made [some Chalcidoidea pupate naked or under a slight net outside the host as small obtect pupae]..5

4. Occasionally apterous or brachypterous, otherwise fore wing with second recurrent vein (2m-cu) present (Fig. 11.1) [disposition of other veins, especially costad of this, variable]. *(Host killed as larva (then pupates in cocoon which can be inside or away from host's skin) or pupa (then pupates inside it). Almost always solitary (one gregarious species attacks pupae). As well as primary parasitism, both true hyperparasitism and pseudohyperparasitism occur)*................................Ichneumonidae

 – Always fully winged and fore wing with second recurrent vein (2m-cu) absent (Fig. 11.2) [otherwise venation, especially costad of this area, variable]. *(Always primary parasitoids, killing host in larval stage. Either a mummy is made from the host body, inside which the parasitoid pupates (Rogadinae:Aleiodes), or one or more cocoons are formed externally to the host. Gregarious development frequent (in Microgastrinae))*......................Braconidae

5. Parasitoids of the egg stage. *(Small to minute, emerging as adults from the host egg)*....................................6

 – Parasitoids of other stages..............Chalcidoidea 9

6. Metasoma (= the most posterior of the 3 clear body divisions) with some parallel longitudinal grooves towards base; female ovipositor apical (but concealed); antenna without minute ring segments (see Fig. 11.3) following

the second segment (pedicel); pronotum extends back to meet tegulae. *(Usually black)*...........................
Scelionidae (Platygastroidea)

 – Metasoma without basal longitudinal grooves; female ovipositor issuing before apex of metasoma (discernable ventrally if concealed); antenna usually with ring segments (Fig. 11.3); pronotum separated from tegulae. *(Often pale coloured or metallic)*...
Chalcidoidea 7

 [There are more groups that might possibly be reared from butterfly eggs than are covered by the following couplets to the most regularly found taxa.]

7. Tarsi with 3 segments; if fully winged then fore wing with lines of hairs radiating from the wing base; antenna with not more than 2 funicle segments. *(Body relatively pale, non-metallic, weakly sclerotised; particularly small (0.3–1.2 mm) and often gregarious)*.....................Trichogrammatidae

 – Tarsi with 5 segments; if fully winged then fore wing lacking radiating lines of hairs; antenna with 6 or 7 funicle segments. *(Body more or less dark and at least partly metallic)*...8

8. Antenna with 1 ring segment and 7 funicle segments (or 5 very short ones, totalling less than half the length of the strongly elongate club, in male of *A. bifasciatus*); female fore wing strongly banded (even if brachypterous), in male usually hyaline...............Eupelmidae (*Anastatus*)

 – Antenna with no ring segments and 6 funicle segments; wings hyaline and antennal club much shorter than funicle in both sexes...............Encyrtidae (*Ooencyrtus*)

9. Hind femur swollen, extensively toothed on lower (posterior) margin; hind tibia markedly curved to same profile. *(Moderately large (ca. 4–8 mm) solitary parasitoids of pupae; also (usually smaller) pseudohyperparasitoids ex Ichneumonoidea cocoons; black with yellow (occasionally red) leg markings; very heavily sculptured (except metasoma))*............................Chalcididae (*Brachymeria*)

– Hind femur and tibia unmodified. *(Usually smaller, ca. 2–4 mm; usually at least partly metallic greenish or bronze)* 10

10. Antenna with 5 or 6 funicle segments; tarsi with 5 segments; front tibial spur well developed and distinctly curved. *(Gregarious parasitoids of pupae; also pseudohyperparasitoids ex Ichneumonoidea cocoons)*. [N B Several of the chalcidoid families not known as primary parasitoids of butterflies that do behave as pseudohyperparasitoids will run here (e.g. Eurytomidae, Eupelmidae)] ....Pteromalidae

– Antenna with 2–4 funicle segments; tarsi with 4 segments; front tibial spur weak, practically straight. *(Primary ectoparasitoids attacking the larval stage (especially as gregarious koinobionts of exposed hosts, but possibly also as idiobionts of concealed ones) do occur, but not commonly. More often seen as solitary or gregarious true hyperparasitoids, and also pseudohyperparasitoids, ex Ichneumonoidea cocoons)* ....Eulophidae

Hymenoptera

ICHNEUMONOIDEA

This superfamily comprises just two extant families, Ichneumonidae and Braconidae, both of which are very large and diverse.

Ichneumonidae

The overwhelming majority of ichneumonids parasitise the larvae or pupae of holometabolous insects, though a few attack spiders and their egg sacs. About 35 subfamilies of Ichneumonidae occur in Europe but, although 14 contain at least some parasitoids of Lepidoptera, only five include parasitoids of butterflies. Species of two other subfamilies are obligatory true hyperparasitoids, and one of these (Mesochorinae, containing the very large genus *Mesochorus*; Plate 7a) is regularly associated with butterflies. These are solitary with respect to each primary parasitoid attacked, though often several in a gregarious brood can be affected. They are most often reared through endoparasitic koinobiont Ichneumonoidea attacking the host in its larval stage. Ichneumonid pseudohyperparasitoids are found in the subfamilies Cryptinae and Pimplinae, and indeed most of the involvement of Cryptinae with butterflies is through a few common genera that regularly behave as pseudohyperparasitoids. Literature on species of Ichneumonidae is traceable via Yu & Horstmann (1997) and current versions of Taxapad (see www.taxapad.com).

Subfamily Anomaloninae

All species of this relatively small subfamily are solitary koinobiont larva–pupal endoparasitoids, mostly of Lepidoptera. No genus specialises on butterflies, but most of the few species concerned are specialists with narrow host ranges.

Parasitoid	Suggested host range	Supporting host records
Agrypon anomelas (Gravenhorst)	<Lycaenidae	*Neozephyrus quercus*[1]; *Polyommatus coridon*[2]
Agrypon delarvatum (Gravenhorst)	<Satyrinae + <Hesperiinae	*Thymelicus lineola*[3]; *Lasiommata maera*; *Coenonympha* sp.
Agrypon flexorium (Thunberg)	Broad	*Callophrys rubi*
Agrypon polyxenae (Szépligeti)	<Parnassiinae	*Zerynthia rumina*; *Z. polyxena*; *Archon apollinus*
Clypeocampulum sp. nov.[4]	*Anthocharis euphenoides*	*Anthocharis euphenoides*
Erigorgus apollinis Kriechbaumer	*Parnassius*	*Parnassius apollo*[5]
Erigorgus foersteri (Mocsáry)	<Satyrinae	*Pyronia tithonus*; *Aphantopus hyperantus*
Erigorgus melanops (Foerster)	<Satyrinae + <low-feeding Noctuidae	*Maniola jurtina*; *Melanargia galathea*
Heteropelma grossator Shestakov (= *parargis* Heinrich)	*Lasiommata*	*Lasiommata maera*[5]

Species in NMS have been either determined or checked by H. Schnee.

Notes: 1. BMNH.   2. Horstmann *et al.* (1997).   3. Carl (1968).   4. To be described by H. Schnee.   5. H. Schnee, pers. comm.

Subfamily Campopleginae

All species in this large subfamily are koinobiont larval (very rarely larva–pupal or egg–larval) endoparasitoids. Most are solitary parasitoids of Lepidoptera, but a few genera attack other holometabolous orders. In Europe parasitoid species specialised to butterflies are known in several genera but *Benjaminia* is the only genus that apparently wholly specialises on butterflies. All *Benjaminia* probably parasitise Melitaeini but, while several Nearctic species have been reared, reliable host data exist for only one European and one other of the eight Palaearctic species (Wahl, 1989). Species of *Benjaminia* and most of the *Hyposoter* (Plate 8a) species that attack butterflies pupate inside the host's skin (a blotched cocoon shows through in some), killing it about an instar before full growth or in some cases as an advanced final-instar larva. *Casinaria* and *Phobocampe* species also generally kill the host before full growth, but in these cases the cocoon is spun outside the host remains. Some genera (*Campoplex*, *Diadegma*, *Enytus* and *Sinophorus*) specialise on concealed hosts that are usually killed as prepupae, and there are apparent specialists (*D. aculeatum* and *Sinophorus* sp. indet.) on concealed butterflies. Otherwise, rearings of these genera from butterflies probably represent abnormal (if sometimes repeated) events; for example, several species that usually attack concealed microlepidoptera larvae occasionally develop in small larvae of *Vanessa atalanta*, presumably attacked because of its similar resting position. Mostly the Campopleginae that attack butterflies overwinter as early-instar larvae inside host larvae and they spin their cocoons, which are often bird-dropping mimics, firmly attached to vegetation: the cocoons, which are very prone to pseudohyperparasitism, then produce adults quickly. *Phobocampe*, however, make unattached hard ovoid cocoons, which the parasitoid larva within can cause to 'jump' in response to heat and light until the cocoon finds a secluded place. Although often plurivoltine so that emergence from the cocoon can be rapid, these *Phobocampe* species spend the winter as a cocooned stage.

Parasitoid	Suggested host range	Supporting host records seen
Alcima orbitale (Gravenhorst)	>*Zygaena*[1]	*Hipparchia semele*; *H. statilinus*; *Arethusana arethusa*
Benjaminia fumigator Aubert	*Melitaea didyma*	**Melitaea didyma**
(*Campoletis annulata* (Gravenhorst))[2]	Includes *Autographa gamma*[3]	*Maniola jurtina*
(*Campoplex lyratus* (Thomson))[2]	<Microlepidoptera	*Vanessa atalanta*[4]
(*Campoplex tumidulus* Gravenhorst (= *rufinator* Aubert))[2]	<Microlepidoptera	*Vanessa atalanta*[4,5]
Casinaria petiolaris (Gravenhorst)	Coenonymphini	**Coenonympha tullia**; **C. pamphilus**
Diadegma aculeatum (Bridgman)	*Cupido*	**Cupido minimus**
(*Diadegma* sp.)[6]	?	**Celastrina argiolus**[5]
(*Enytus apostatus* (Gravenhorst))[2]	<Microlepidoptera	**Celastrina argiolus**[4,5]; *Vanessa atalanta*[4]
Hyposoter ebeninus (Gravenhorst)[2]	<Pyrginae; <Pierinae[7]	**Carcharodus alceae**; *Anthocharis cardamines*; *Euchloe belemia*; *E. crameri*; **E. simplonia**; *E. ausonia*; *E. insularis*; **Pieris brassicae**; **P. rapae**[3,8]; *P. napi*; *Pontia daplidice*
Hyposoter ebenitor Aubert[2]	<Pierinae	**Euchloe ausonia**; *Pontia daplidice*; **P. chloridice**; **P. edusa**[8]

Hyposoter horticola (Gravenhorst)[2]	<*Melitaea*	***Melitaea cinxia***; *M. aurelia*
Hyposoter notatus (Gravenhorst)	<Polyommatini	*Cupido osiris*; *C. lorquinii*; *Scolitantides orion*[9]; *Plebejus hesperica*; *Aricia eumedon*[9]; *A. agestis*; *A. artaxerxes*; *Polyommatus icarus*; *P. abdon*; *P. hispana*[9]; *P. albicans*; *P. fabressei*; *P. ?ripartii*; *P. ?aroaniensis*
Hyposoter placidus (Desvignes)	Lycaenini	***Lycaena phlaeas***; *L. helle*; *L. dispar*; *L. hippothoe*
Hyposoter rhodocerae (Rondani)[2]	*Gonepteryx*	***Gonepteryx rhamni***; *G. cleopatra*
Hyposoter caudator Horstmann	<Polyommatini	***Plebejus pyrenaica***; *Polyommatus dorylas*[3]
Hyposoter sp.	<Maniolini	***Pyronia tithonus***; *Maniola jurtina*
Phobocampe confusa (Thomson)	<Nymphalini	***Inachis io***; *Aglais urticae*; ***Polygonia c-album***[4]; *Araschnia levana*; *Nymphalis polychloros*[4]
(*Phobocampe crassiuscula* (Gravenhorst))[2]	Broad	*Limenitis camilla*[4]
Phobocampe tempestiva (Holmgren)[2]	Broad	***Limenitis camilla***
Phobocampe quercus Horstmann	Theclini	*Thecla betulae*[8]; *Neozephyrus quercus*
Sinophorus sp.	<Pyrginae	***Carcharodus alceae***; *C. baeticus*; ***Muschampia proto***
(*Sinophorus turionus* (Ratzeburg))[2]	Includes *Ostrinia nubilalis*[3]	*Vanessa atalanta*[4]

Single rearings of unidentified parasitoids have been excluded from the table.

Notes: 1. *Alcima orbitale* is a regular genus-specific parasitoid of *Zygaena* in most parts of Europe; only in Spain have we seen it from a range of other hosts. 2. Determined by K. Horstmann. 3. K. Horstmann, pers. comm. 4. Infrequent host. 5. Only two rearings (separate occasions) from this extremely commonly reared host. 6. The condition of the specimens prevents a certain determination (K. Horstmann, pers. comm.). 7. Such a disparate host range suggests the possibility of two parasitoid species. 8. BMNH. 9. Horstmann *et al.* (1997).

Subfamily Cryptinae

This large subfamily has a wide host range, focused on cocoons or similar structures including Diptera puparia. Most species are solitary idiobionts, though both gregarious development and (separately) larva–pupal koinobiosis occur. Relatively few genera and species are associated with Lepidoptera and, as butterflies do not make cocoons, there are few Cryptinae that parasitise them. *Blapsidotes vicinus* (Gravenhorst) is a moderately common gregarious idiobiont endoparasitoid of exposed butterfly pupae. Brood sizes of up to a few tens have been seen from *Pieris rapae*, *Vanessa atalanta*, *Polygonia c-album* and *Euphydryas desfontainii*. We have seen a single *Agrothereutes parvulus* (Habermehl) reared from a pupa of *Euphydryas aurinia*, and Horstmann *et al.* (1997) record a single *Polytribax rufipes* (Gravenhorst) parasitising the pupa of *Polyommatus coridon*, but it is doubtful that these species regularly attack butterflies. The same applies to

several species of *Gelis* that occasionally parasitise Lepidoptera pupae including those of butterflies (see Schwarz & Shaw, 1999). The genus *Ischnus* includes species that specialise on Lepidoptera pupae, but again no regular association with butterflies is evident. However, many genera of small Cryptinae (e.g. *Acrolyta*, *Bathythrix*, *Gelis* and *Lysibia*) include pseudohyperparasitoids that regularly attack the cocoons of primary parasitoids of butterflies (Plate 7b). Some species are highly polyphagous but others are more specialised, for example on cocoons of Microgastrinae, though none is exclusively associated with butterflies (see Schwarz & Shaw, 1999, 2000, in prep.).

Subfamily Ichneumoninae

The entire subfamily, which is large, attacks Lepidoptera as solitary endoparasitoids. With very few exceptions (not concerning butterflies), all species emerge as adults from the host pupa. Rather few genera include species that parasitise butterflies, but some that do are specialised to them at the generic level (e.g. *Hoplismenus*, *Psilomastax*, *Trogus*), as is the entire tribe Listrodromini (genera *Anisobas*, *Listrodromus* and *Neotypus* in the table) which are larva–pupal parasitoids of Lycaenidae. *Listrodromus nycthemerus*, at least in Britain, appears to be responsible for cyclical population crashes in its host, *Celastrina argiolus* (Revels, 1994, 2006). Larva–pupal parasitism is frequent in the subfamily, but a majority of species attack hosts either as prepupae or as freshly turned pupae, apparently tracking semiochemicals associated with this moult (Hinz, 1983). Many ichneumonines (particularly in the genus *Ichneumon*) pass the winter as adult females, and most others do so within host pupae. However *Psilomastax pyramidalis* attacks young hosts and overwinters inside the diapausing host larva which is still small (Dell &

Burckhardt, 2004). The related genus *Trogus* (Plate 9a) can attack old or young larval hosts facultatively, and similarly delays larval development beyond its first instar until the host has pupated (Prota, 1963).

Several species of *Hoplismenus* and *Ichneumon* other than those tabulated have been recorded in the literature as parasitoids of butterflies (especially Nymphalidae). The determinations of tabulated *Anisobas*, *Ichneumon*, *Neotypus* and *Thyrateles* species have been either made or checked by K. Horstmann. The taxon given as *I. eumerus* (Plate 9b) might be an aggregate of two species (one from *M. rebeli* and the other from both *M. alcon* and *M. teleius*), but if so it is unclear at present which is the true *I. eumerus* (K. Horstmann, pers. comm.). The taxon listed as *I. gracilicornis* might also be an aggregate (Hilpert, 1992). Data on experimental hosts of *Ichneumon* species are given by Hinz & Horstmann (2007).

Parasitoid	Suggested host range	Supporting host records
Anisobas brombacheri Heinrich (= *martinae* Riedel)	*Glaucopsyche alexis*	***Glaucopsyche alexis***[1]
Anisobas cephalotes Kriechbaumer	*Iolana iolas*	***Iolana iolas***
Anisobas cingulatellus Horstmann (= *cingulatorius* (Gravenhorst), preocc.)	<Polyommatini	***Plebejus argus**; **Aricia agestis**; Polyommatus coridon; P. hispana*[1]
Anisobas hostilis (Gravenhorst)	*Neozephyrus quercus*	***Neozephyrus quercus***[1]
Anisobas platystylus (Thomson)	*Callophrys rubi*	***Callophrys rubi***[1]
Anisobas rebellis Wesmael (= *jugorum* Heinrich)	*Lycaena*	***Lycaena phlaeas***[1]; ***L. dispar***[1]; ***L. virgaureae***[1]; *Lycaena* sp.
Anisobas seyrigi Heinrich	>*Glaucosyche melanops*[2]	*Glaucopsyche melanops*
(*Cratichneumon fabricator* (Fabricius))	?Broad	*Neozephyrus quercus*
Hoplismenus axillatorius (Thunberg) (= *albifrons* Gravenhorst)	Coenonymphini	*Coenonympha tullia*; ***C. pamphilus***
Hoplismenus terrificus Wesmael	?	*Polygonia c-album*[3]
Ichneumon albiornatus Tischbein	*Limenitis*	*Limenitis populi*[4]; *L. camilla*[4]
Ichneumon berninae (Habermehl)	Includes high altitude Arctiidae[4]	*Boloria pales*[4]
Ichneumon caloscelis Wesmael	<Satyrinae	*Pyronia tithonus*; ***Maniola jurtina***; *Hipparchia semele*
Ichneumon cessator Müller	Nymphalini	*Aglais urticae*; *Nymphalis antiopa*[4]
Ichneumon cinxiae Kriechbaumer	<Melitaeini	***Melitaea cinxia**; M. britomartis; **M. athalia***
Ichneumon cynthiae Kriechbaumer	*Euphydryas cynthia*	***Euphydryas cynthia***[4]
Ichneumon eumerus Wesmael agg.	<*Maculinea*	***Maculinea teleius***[4]; ***M. alcon***[4]; ***Maculinea rebeli***[4]
Ichneumon exilicornis Wesmael	<Polyommatini	*Plebejus argus*[4]; *Polyommatus ?amandus*[5]; ***P. coridon***[+4]; *P. admetus*; ***P. ripartii***[5]; ***P. nephohiptamenos***[5]

Ichneumon fulvicornis Gravenhorst	<*Maculinea*	**Maculinea teleius**[4,6]
Ichneumon gracilicornis Gravenhorst	<Nymphalidae	?*Brenthis ino*; *Euphydryas aurinia*[4]; **Melitaea cinxia**; *M. ?didyma*; **M. diamina**[4]; **M. athalia**; *Maniola jurtina*
Ichneumon macilentus (Tischbein)	High-altitude Nymphalinae	**Boloria pales**[4]; *Euphydryas cynthia*[4]
Ichneumon novemalbatus Kriechbaumer	<Satyrinae	*Melanargia lachesis*
Ichneumon obliteratus Wesmael	*Euphydryas cynthia*	*Euphydryas cynthia*[4]
Ichneumon occidentis Hilpert	Includes high-altitude Geometridae and Arctiidae[4]	*Boloria pales*[4]; *Euphydryas cynthia*[4]
Ichneumon ?quadrialbatus Gravenhorst[7]	?	*Colias crocea*
Ichneumon quinquealbatus Kriechbaumer	?	*Boloria eunomia*
Ichneumon sculpturatus Holmgren	?	*Lycaena tityrus*[4]
Ichneumon silaceus Gravenhorst	<*Boloria*	*Boloria selene*; *B. pales*[4]
Ichneumon stenocerus Thomson	?	*Euphydryas aurinia*
Ichneumon vorax Geoffroy	*Apatura*	*Apatura iris*[4]
Ichneumon sp.[7]	?	*Euphydryas iduna*
Ichneumon sp.[7]	<Melitaeini	**Euphydryas maturna**
Listrodromus nycthemerus (Gravenhorst)	*Celastrina argiolus*	**Celastrina argiolus**
Neotypus coreensis Uchida	<*Maculinea*	**Maculinea arion**[8]
Neotypus intermedius Mocsáry	<Polyommatini	*Lampides boeticus*[9]
Neotypus melanocephalus (Gmelin) (= *pusillus* Gregor)	<*Maculinea*	*Maculinea teleius*[10]; **M. nausithous**[10]
Neotypus nobilitator (Gravenhorst)	*Cupido*	*Cupido minimus*
Psilomastax pyramidalis Tischbein	*Apatura*	**Apatura ilia**; *A. iris*
Syspasis scutellator (Gravenhorst)	*Thymelicus*	**Thymelicus lineola**[8,11]
Thyrateles camelinus (Wesmael)	Nymphalini	**Vanessa cardui**; **Nymphalis antiopa**
Thyrateles haereticus (Wesmael)	Nymphalini	*Inachis io*
Trogus lapidator (Fabricius)	*Papilio*	**Papilio machaon**
Trogus violaceus (Mocsáry)	*Papilio*[12]	*Papilio machaon*; **P. hospiton**
Virgichneumon callicerus (Gravenhorst)	<Lycaenidae	*Callophrys rubi*[8]; *Polyommatus bellargus*; *Polyommatus* sp.
Virgichneumon tergenus (Gravenhorst)	<Lycaenidae	**Satyrium w-album**; *S. pruni*[8]; *S. esculi*; *Plebejus argus*; **Polyommatus icarus**[+13]; *P. hispana*[13]

Notes: 1. Horstmann (2007). 2. The parasitoid is more widely distributed than this host (Horstmann, 2007). 3. HNSalzburg (K. Horstmann, pers comm.). 4. Hinz & Horstmann (2007). 5. Possibly a separate species from Greece; females are identical to *I. exilicornis* from central Europe, but in the males the tyloids differ slightly and the 2nd and 3rd gastral tergites are darker (K. Horstmann, pers. comm.). 6. The only certainly identified host is *M. teleius*, but it has been reared many times from co-occurring pupae of *M. teleius* or *M. nausithous*, which are morphologically indistinguishable. 7. The specimen(s) are male and determination is therefore uncertain. 8. BMNH. 9. Selfa *et al.* (1994). 10. K. Horstmann, pers. comm. 11. Carl (1968). 12. *T. violaceus*, a regular parasitoid of *P. hospiton*, is confined to Corsica and Sardinia where it also parasitises the much less frequently collected *P. machaon*. Wahl & Sime (2006) do not regard it as a species distinct from *T. lapidator*. 13. Horstmann *et al.* (1997).

Subfamily Pimplinae

Host associations and modes of development in this medium-sized subfamily are exceptionally wide. Only one rather limited tribe (Pimplini) of solitary idiobiont endoparasitoids of (mostly) Lepidoptera pupae attacks butterflies regularly. Most records involve the genera *Pimpla* (Plate 10a) and *Apechthis*, the latter being particularly adapted to oviposit in obtect pupae with overlapping sclerites through having a hooked ovipositor tip (Cole, 1959). One species in each of these genera is very regularly (but not exclusively) reared from butterfly pupae. Host range in Pimplini, most of which are plurivoltine, depends mostly on searching behaviour, host recognition (some, for example, prefer cocooned hosts, or those concealed in stems) and size suitability – in general, females will be reared from the large end of the host-size spectrum used. Another genus, *Itoplectis* (Plate 8b), is more often a primary parasitoid of semi-concealed or cocooned smallish moth pupae, but some could be expected to use small species of butterflies occasionally. *Itoplectis* species are also regularly pseudohyperparasitoids, and often reared from Campopleginae cocoons in particular, including those deriving from butterflies. A further species, *Theronia atalantae* (Poda), is regularly reared from largish Lepidoptera pupae including those of butterflies, but it seems to be an obligatory secondary parasitoid. Among idiobiont ectoparasitic Pimplinae, *Scambus signatus* (Pfeffer) (tribe Ephialtini) regularly develops as a solitary parasitoid of Tortricidae feeding in Fabaceae pods, and would be expected also to attack lycaenid larvae in the same situations; however, we have seen no reared specimens.

Braconidae

As a family, Braconidae has a wider insect host range than Ichneumonidae (Gauld, 1988), being associated also with some groups of hemimetabolous insects and in some cases attacking the adult stage of long-lived insects, though not Lepidoptera (Shaw & Huddleston, 1991). About 33 subfamilies of Braconidae occur in Europe, of which 22 include parasitoids of Lepidoptera. Species attacking macrolepidoptera are found in 11 of these, but there are substantial numbers of species doing it in only two, Microgastrinae and Rogadinae. Only these two and a few species of *Meteorus* (Euphorinae), and under exceptional circumstances possibly Braconinae, parasitise butterflies in Europe. However, many species of Microgastrinae (especially some *Cotesia*; Plates 10b and 11) specialise on butterflies and are among their most conspicuous natural enemies. No group of Braconidae functions as hyperparasitoids. The failure of many subfamilies of Braconidae to have radiated into parasitising butterflies, with their (typically) exophytic feeding biology and usually weakly concealed pupation habits, is biological rather than accidental. Many braconid taxa are either ectoparasitic idiobionts or (very often) have a

Parasitoid	Suggested host range	Supporting host records
Apechthis compunctor (Linnaeus)	Broad (often medium/large butterflies)	***Pieris brassicae***; *Colias crocea*; *Lycaena dispar*; *Satyrium pruni*; *Boloria titania*; ***Vanessa atalanta***; *Inachis io*; ***Aglais urticae***; *Nymphalis antiopa*; *Euphydryas maturna*; *E. desfontainii*; ***E. aurinia***
Apechthis quadridentata (Thomson)	Broad (especially Tortricidae, but butterflies possibly important over the winter)	*Pieris napi*; *Pararge aegeria*[1]
(*Itoplectis maculator* (Fabricius))	Broad (seldom small/medium butterflies)	*Satyrium w-album*
Pimpla aethiops Curtis	?Large fenland Lepidoptera[2]	*Papilio machaon*[2]
Pimpla rufipes (Miller) (= *instigator* (Fabricius), preocc.)	Broad (often medium/large butterflies)	***Papilio machaon***; ***Pieris brassicae***; ***P. rapae***; *P. napi*; *Colias crocea*; *Lycaena dispar*; ***Vanessa atalanta***; *Euphydryas aurinia*; *Charaxes jasius*
(*Pimpla turionellae* (Linnaeus))	Broad (seldom small/medium butterflies)	*Pieris rapae*; *Satyrium w-album*

Notes: 1. Cole (1967), as *A. resinator*, who considered it to be an important winter host. 2. Based on old British material in BMNH, but *P. aethiops* appears now to be extinct in Britain (Fitton *et al.*, 1988).

larval existence that starts as a koinobiont endoparasitoid but ends with an ectoparasitic final instar (Shaw & Huddleston, 1991). In both cases, concealed hosts – and/or physically well-protected host pupation sites, such as tough or subterranean cocoons – are a vital requirement for the parasitoid. This reality is clearly demonstrated by the kind of butterfly hosts suitable for the genus *Microgaster* (see later), and it also explains why genera such as *Cotesia* (which is fully endoparasitic), rather than some others in the subfamily Microgastrinae, have been such successful colonists of butterflies as a host group. The current version of Taxapad (see www.taxapad.com) includes a treatment of the literature on species of Braconidae.

Subfamily Braconinae

Except for the small non-European subtribe Aspidobraconina (which includes endoparasitoids in butterfly pupae: van Achterberg, 1984), all species of the large subfamily Braconinae are believed to be idiobiont ectoparasitoids and they attack a wide range of concealed holometabolous insects, usually as well-grown larvae. Although we have not seen reared specimens, it seems probable that some species of the large genus *Bracon* (most of which are gregarious) could facultatively attack butterflies that live endophytically as late-instar larvae, for example in the seed pods of Fabaceae, or as miners. While these habits are essentially restricted to a few Lycaenidae, other possible hosts could be Hesperiidae resting in strongly constructed retreats.

Subfamily Euphorinae

This diverse medium-sized subfamily contains koinobiont endoparasitoids of the adult stages of various insects (though not Lepidoptera), but also includes a group (dominated by the genus *Meteorus* and previously classified as a separate subfamily, Meteorinae) which parasitises larval Coleoptera and Lepidoptera. One, *Meteorus colon* (Haliday), is a regular parasitoid of *Limenitis camilla* in Britain (Shaw, 1981) though it also uses other hosts. Two others are occasionally reared from butterflies: we have seen *M. pulchricornis* (Wesmael) from *Iphiclides podalirius*, *Thecla betulae* and *Charaxes jasius*; and *M. versicolor* (Wesmael) from *Callophrys rubi* and *Maniola jurtina*, although both are much commoner from other macrolepidoptera. All the foregoing are solitary species that suspend their cocoons from the foodplant on a thread.

Subfamily Microgastrinae

This is one of the largest subfamilies of Braconidae and is practically restricted to Lepidoptera (though one species has been reared from terrestrial Trichoptera: van Achterberg, 2002). All are koinobiont endoparasitoids, killing the larval stage of the host, and many are gregarious. Several species of the very large genus *Cotesia* (which used to come under the old generic concept of '*Apanteles*') parasitise butterflies, usually having narrow host ranges. Some of the smaller genera also contain specialist parasitoids of butterflies, though the occasional use of butterflies as part of more diffuse host ranges is more often seen. Egg–larval parasitism sometimes occurs, and various modifications of host behaviour have been noted, including the selection of particular resting sites before the parasitoid larva egresses.

Many microgastrines are essentially haemolymph-feeders and leave the host little consumed when they vacate it, allowing it to remain alive for a few days afterwards. In the case of *Pieris brassicae* parasitised by *Cotesia glomerata*, the moribund host sometimes spins a layer of silk across the yellow cocoon mass of its gregarious parasitoid (Brodeur, 1992) – probably an induced behaviour affording protection from generalist predation (as it does not deter pseudohyperparasitoids significantly) rather than an attempt to entrap the *Cotesia*. Some microgastrines, including the genus *Microgaster*, are haemolymph-feeders at first but then erupt from the host to continue their feeding externally, consuming virtually all but the skin of the host (illustrated in Shaw, 2004). This needs the protection of a secluded environment: the fact that the only butterflies regularly parasitised by *Microgaster* species are those, such as Hesperiidae and *Vanessa atalanta*, that rest in retreats like the mainstream hosts of the genus (concealed microlepidoptera larvae) particularly clearly shows the role of life history and behaviour in determining a host's parasitoids.

Several *Cotesia* species that are plurivoltine, notably those associated with Melitaeini (Plate 10b), have successive generations on a single host generation, having progressively larger broods on older hosts. Other plurivoltine *Cotesia* species may use different hosts at different times of year: for example, *C. vestalis* is a solitary parasitoid that overwinters in Satyrinae larvae but attacks *Aglais urticae* and sometimes other Nymphalini in summer, as well as its economically important host, the plutellid *Plutella xylostella* (see Wilkinson, 1939). Similar alternations are presumably needed by several others, for example *C. saltator* and *C. risilis*, which are parasitoids of respectively Anthocharini and *Gonepteryx* species that feed as larvae in early summer. These parasitoids emerge from their cocoons quickly, but the hosts needed subsequently to complete the annual life cycle are unknown. *Cotesia gonepterygis*, another solitary parasitoid of *Gonepteryx*, spins a tough golden cocoon (invariably on a twig; Plate 11) which persists to emerge the following spring, and consequently this univoltine species needs no other host. Some *Cotesia* species can pass the winter in more than one way (for example *C. glomerata* does so in its cocoons if its late-summer host was a *Pieris* species, or as larvae inside the diapausing larva of *Aporia crataegi*), but no Microgastrinae are believed to hibernate as adults.

Adult Microgastrinae always emerge from the cocoon by cutting a neat circular cap from the anterior end. If a hyperparasitoid develops in the cocoon, however, the adult hyperparasitoid invariably chews its way out, making a much less regular hole, usually subapically. This difference can be used to assess the level of hyperparasitism suffered by Microgastrinae when their cocoons are collected post-emergence.

Parasitoid	Suggested host range	Supporting host records
Cotesia acuminata (Reinhard) agg.*	<Melitaeini	Molecular[1] and morphological data suggest there are 4 separate species on (a) **Euphydryas maturna**, (b) **Melitaea phoebe**, **M. telona** (c) **M. didyma** and (d) **M. athalia**
Cotesia ?amesis (Nixon)*[2]	<Polyommatini	*Plebejus glandon*
Cotesia ancilla (Nixon)*	*Colias*	*Euchloe charlonia*[3]; *Colias palaeno*[4]; *C. crocea*; *C. chrysotheme*[4]; *C. hyale*[4]; *C. alfacariensis*
Cotesia astrarches (Marshall)*	<Polyommatini	*Cupido minimus*; **Aricia agestis**; **A. artaxerxes**; *Polyommatus thersites*
Cotesia bignellii (Marshall) agg.*[5]	*Euphydryas aurinia*	**Euphydryas aurinia**
Cotesia cuprea (Lyle)*	*Lycaena*[6]	**Lycaena phlaeas**; *L. helle*[4]; **L. dispar**; **L. thersamon**
Cotesia cynthiae (Nixon)*	*Euphydryas cynthia*	**Euphydryas cynthia**[+4]
Cotesia glabrata (Telenga)*	<Pyrginae	**Carcharodus alceae**; *C. tripolinus*; *Pyrgus cirsii*
Cotesia glomerata (Linnaeus)*	<Pierini	**Aporia crataegi**; **Pieris brassicae**; *P. mannii*; **P. rapae**; *P. napi*; *Pontia daplidice*[4]
Cotesia gonepterygis (Marshall)	*Gonepteryx*	**Gonepteryx rhamni**; **G. cleopatra**
Cotesia inducta (Papp)	<Lycaenidae	*Tomares ballus*; *Callophrys avis*; *Satyrium w-album*; **Celastrina argiolus**; *Glaucopsyche melanops*
Cotesia lycophron (Nixon)*	<*Melitaea*	**Melitaea trivia**; **M. didyma**
Cotesia melitaearum (Wilkinson) agg.*	<Melitaeini	Molecular[7] and morphological data suggest that at least 5 species are included; provisionally assessed as on (a) **Euphydryas aurinia**, (b) *E. desfontainii* + *E. aurinia* (Spain), (c) **Melitaea cinxia** and probably *M. diamina*, (d) *M. trivia*, (e) *M. athalia* and probably both *M. deione* and *M. parthenoides*
Cotesia pieridis (Bouché)*	*Aporia crataegi*	**Aporia crataegi**
Cotesia risilis (Nixon)	*Gonepteryx*[8]	**Gonepteryx rhamni**; *G. cleopatra*
Cotesia rubecula (Marshall)	<*Pieris* spp. (not *P. brassicae*)	*Pieris rapae*
Cotesia saltator (Thunberg)	<Anthocharini[8]	**Anthocharis cardamines**; *A. euphenoides*; *Euchloe crameri*
Cotesia saltatoria (Balevski)	<Polyommatini	*Aricia agestis*; *A. artaxerxes*; **Polyommatus amandus**; *P. icarus*; *P. coridon*[9]
Cotesia sibyllarum (Wilkinson)*	<*Limenitis*	**Limenitis camilla**; *L. reducta*
Cotesia specularis (Szépligeti)*	<Polyommatini (especially in Fabaceae pods)	**Lampides boeticus**; *Glaucopsyche alexis*; **Iolana iolas**

Cotesia tenebrosa (Wesmael)*	<Polyommatini	*Cupido alcetas*; **Plebejus argus**; *Aricia morronensis*; **Polyommatus icarus**; *P. daphnis*; **P. bellargus**; *P. coridon*; **P. albicans**[10]; *P. admetus*
Cotesia tetrica (Reinhard)*	<Satyrinae	*Lasiommata megera*[4]; **Maniola jurtina**; *Erebia aethiops*
Cotesia tibialis (Curtis)*	<Noctuidae; <Satyrinae	*Pyronia tithonus*[4]; *Maniola jurtina*[4]
Cotesia vanessae (Reinhard)*[11]	<Nymphalini; <Noctuidae over the winter	**Vanessa atalanta**; *V. cardui*; **Aglais urticae**
Cotesia vestalis (Haliday) (= *plutellae* (Kurdjumov))	<Nymphalini; <Satyrinae; various others include *Plutella xylostella*[12]	*Vanessa cardui*; **Aglais urticae**[13]; *Nymphalis polychloros*; *Maniola jurtina*[4]; **Hipparchia semele**
Cotesia sp.*	<Polyommatini	*Scolitantides orion*; *Polyommatus coridon*; *P. caelestissima*
Cotesia sp. nov.*[14]	*Boloria eunomia*	**Boloria eunomia**
Cotesia sp. nov.*[14]	<Heliconiinae	*Argynnis aglaja*; *A. adippe*
Cotesia sp. nov.*[14]	<Heliconiinae	**Boloria selene**
Cotesia sp.*	<Pyrginae	**Carcharodus boeticus**
Diolcogaster abdominalis (Nees)	*Coenonympha*	*Coenonympha tullia*; *C. oedippus*
Distatrix sancus (Nixon)*	Theclini, Eumaeini and possibly Lycaenini	?*Lycaena* sp.; *Thecla betulae*; *Neozephyrus quercus*; *Callophrys rubi*[15]
Dolichogenidea sicarius (Marshall)	Broad (seldom butterflies)	*Carcharodus alceae*; *Vanessa cardui*
Glyptapanteles vitripennis (Curtis)	Broad (?seldom butterflies)	*Leptidea sinapis*[16,17]; *Limenitis populi*
Microgaster australis Thomson	<Pyrginae	*Carcharodus alceae*; *Muschampia proto*; *M. tessellum*; *Pyrgus serratulae*; *P. onopordi*; *P. armoricanus*; **Pyrgus sp.**
Microgaster nixalebion Shaw	*Anthophila, Prochoreutis, Pleuroptya* and *Vanessa atalanta*	**Vanessa atalanta**; *Aglais urticae*[16]
Microgaster nobilis Reinhard	<Pyrginae	**Carcharodus alceae**; *C. baeticus*; **Muschampia proto**
Microgaster subcompletus Nees*	<Pyralidae, including *Pleuroptya ruralis*, and *Vanessa atalanta*	**Vanessa atalanta**; *V. cardui*[16]; *Polygonia c-album*[16]
Microplitis retenta Papp	Anthocharini	*Anthocharis euphenoides*; **Euchloe sp.**
Protapanteles anchisiades (Nixon)	Broad (?seldom butterflies)	*Leptidea sinapis*[17,+18]
Protapanteles incertus (Ruthe) (= *caberae* (Marshall))	Broad (?seldom butterflies)	**Pararge aegeria**
Protapanteles sp.	?	**Limenitis populi**

Notes: *Gregarious species. Generic classification follows Mason (1981). Single rearings of unidentified species have been excluded from the table. 1. Kankare & Shaw (2004); Kankare *et al.* (2005a). 2. Reared specimens are from Spain and differ a little from the non-reared type series from Switzerland. 3. The parasitoid specimens differ a little and may belong to another species. 4. Nixon (1974). 5. There is evidence for the existence of two species, both using only *E. aurinia*. 6. Various authors give species of Polyommatini as hosts but we have not been able to verify any. 7. As 1; also Kankare *et al.* (2005b). 8. There must also be a late-summer host, as yet unknown. 9. Baumgarten & Fiedler (1998). 10. Including ssp. *arragonensis*. 11. Some populations in South Europe are thelytokous. 12. Lloyd (see Wilkinson, 1939) demonstrated, through rearing experiments, that the parasitoid of *A. urticae* and *P. xylostella* are the same species. 13. Also ssp. *ichnusa* from Corsica. 14. To be described by M. R. Shaw (in press). 15. Fiedler *et al.* (1995). 16. Infrequent host. 17. From England, where the only *Leptidea* is *L. sinapis*. 18. BMNH.

Subfamily Rogadinae

As currently restricted, Rogadinae are all koinobiont endo-parasitoids of Lepidoptera larvae and make characteristic shrunken and hardened 'mummies' from their hosts, inside which they pupate. Most species are solitary and the large genus *Aleiodes* dominates the subfamily, usually attacking the host as a small larva and killing it in its penultimate instar. Two species parasitise butterflies in Europe: *A. coxalis* (Spinola) (= *tristis* Wesmael) which has been reared from a range of Satyrinae (*Coenonympha tullia*, *C. pamphilus*, *Maniola jurtina*, *Erebia* sp., *Melanargia lachesis*, and several unidentified species) and is also a frequent parasitoid of the hesperiid *Thymelicus lineola*; and *A. bicolor* (Spinola) which parasitises polyommatine Lycaenidae (*Cupido alcetas*, *Plebejus idas*, *Aricia agestis*, *A. artaxerxes*, *Polyommatus icarus*, *P. eros* (in BMNH), *P. coridon*, *P. albicans*, *P. damon* and ?*Cupido minimus*). (A closely related species, *A. assimilis* (Nees), para-sitises *Zygaena* species (Zygaenidae) and has often been mis-identified in the literature as *A. bicolor*.)

CHALCIDOIDEA

All but the most esoteric of the approximately 20 families in this large and biologically diverse superfamily occur in Europe, but only a few species in the families Chalcididae, Pteromalidae, Eulophidae, and several groups of egg parasitoids (Trichogrammatidae, and small elements of Eupelmidae, Encyrtidae and perhaps others) are likely to be reared as primary parasitoids of butterflies. More families might arise as hyperparasitoids, with certain Eulophidae (particularly gregarious *Baryscapus* species) and Perilampidae acting as true hyperparasitoids, and a number of families (mostly not dealt with further) furnish-ing potential pseudohyperparasitoids.[1]

Pteromalidae

This is a very big and biologically diverse family. A few gregarious species in the large genus *Pteromalus* attack butter-fly pupae, preferentially ovipositing through the fresh cuticle while it is still soft. To be on hand for this, the female

Parasitoid	Suggested host range	Supporting host records
(*Coelopisthia caledonica* Askew)	Usually <Noctuidae	*Melitaea cinxia*[1]
Coelopisthia pachycera Masi	?	**Maniola jurtina**[+2]
(*Dibrachys* spp.)	Broad, mostly pseudohyperparasitoids	No direct records of primary parasitism, but see[3]
(*Psychophagus omnivorus* (Walker))	Large Lepidoptera, especially those pupating below ground	*Satyrium w-album*; *Polygonia c-album*
Pteromalus apum (Retzius)	Megachiline bees and <butterflies, especially Melitaeini	**Gonepteryx rhamni**; *Aglais urticae*[4]; *Euphydryas maturna*; *E. desfontainii*; **E. aurinia**; **Melitaea cinxia**; *M. didyma*; **M. athalia**
Pteromalus puparum (Linnaeus)	<Butterflies, especially Papilionidae, Pieridae and <Nymphalidae	**Papilio machaon**; **Aporia crataegi**; **Pieris brassicae**; **P. rapae**; *Pontia daplidice*; *Libythea celtis*; *Argynnis pandora*; *A. adippe*; **Vanessa atalanta**; *V. cardui*; *Aglais urticae*; **Polygonia c-album**; *Araschnia levana*; **Nymphalis antiopa**; *N. polychloros*; *Euphydryas desfontainii*; *E. aurinia*; *Melitaea cinxia*[5]; *Limenitis camilla*

Unreferenced records include those from R. R. Askew's collection as well as NMS.
Notes: 1. Probably abnormal host. 2. Honey (1998). 3. Askew & Shaw (1997). 4. Pyörnilä (1977) as *P. venustus*; see also Shaw (2002b). 5. Lei *et al.* (1997).

[1]   A database containing information on world Chalcidoidea is main-tained at http://www.nhm.ac.uk/research-curation/projects/chalcidoids by J. S. Noyes.

parasitoid often adopts a caterpillar before it chooses its pupation site, then sits inconspicuously on it until pupation occurs. While *Pteromalus puparum* seems only to attack Lepidoptera (and essentially butterfly) pupae and has a wide host range (Plates 12a and 12b), *P. apum* has a remarkable host range as it attacks both the cells of megachiline bees (Hymenoptera: Apidae) and the pupae of butterflies, especially Melitaeini (Askew & Shaw, 1997; Shaw, 2002b). Related genera contain similarly gregarious and endoparasitic pupal parasitoids of Lepidoptera, but although several have been reared from butterfly pupae mostly the relationship is more incidental (Askew & Shaw, 1997). The entire brood usually egress as adults through one or very few emergence holes chewed through the host pupa (Plate 12a). Different *Pteromalus* species, and also those of various other genera, are common pseudohyperparasitoids ex Ichneumonoidea cocoons or Tachinidae puparia.

Chalcididae
This is a rather small family in Europe, albeit of mainly relatively large species. A few species of *Brachymeria* are solitary primary parasitoids of Lepidoptera pupae, and two attack butterflies in grassland habitats regularly though not exclusively: *B. femorata* (Panzer) which we have seen from *Pieris brassicae*, *Melitaea didyma*, *M. deione* and *Maniola jurtina*; and *B. tibialis* (Walker) from *Euphydryas aurinia* and *E. desfontainii*. The latter (= *B. intermedia*) was recorded by Carl (1968) from *Thymelicus lineola*, and our records from Melitaeini probably reflect considerable sampling effort rather than any particular affinity. The pupa seems to be attacked soon after its formation. A few other Chalcididae (including further species of *Brachymeria*) arise as pseudohyperparasitoids ex Ichneumonoidea cocoons.

Eulophidae
Few records of Eulophidae attacking European butterflies as primary parasitoids exist, but we have seen one brood of the gregarious koinobiont ectoparasitoid *Euplectrus ?flavipes* (Fonscolombe) reared from *Charaxes jasius* (Plate 13a), and species of *Eulophus* with broadly similar biology might also occasionally use exposed butterfly caterpillars. Butterflies having endophytic feeding stages, or those resting in deep concealment, might be prone to parasitism from various genera (e.g. *Elasmus*, *Elachertus* and *Sympiesis*). Some gregarious *Baryscapus* species are true hyperparasitoids, and some regularly attack Microgastrinae (Braconidae) parasitising butterflies (Askew & Shaw, 2005). Pseudohyperparasitism could arise from a number of genera (in a few of which there seem to be specialist parasitoids of Ichneumonoidea cocoons).

Egg parasitoids (Chalcidoidea and Platygastroidea)
Parasitism of insect eggs, and full development to the adult stage therein, is practised by the entire family Scelionidae (Platygastroidea), two entire families of Chalcidoidea (Trichogrammatidae and Mymaridae), and one or more genera in several additional families of Chalcidoidea and also a few Platygastridae (Platygastroidea). Representatives of at least nine families altogether have been recorded (whether correctly or not) as egg parasitoids of Lepidoptera in Europe (B. Pintureau, pers. comm.), though most are much more strongly associated with the eggs of other insects. The most important parasitoids of butterfly eggs are in the genera *Trichogramma* (Trichogrammatidae) (Plates 13b and 14), *Anastatus* (Eupelmidae) and *Ooencyrtus* (Encyrtidae) in the Chalcidoidea, and also *Telenomus* (Scelionidae) and possibly related genera in the Platygastroidea.

With notable exceptions, Lepidoptera eggs are neither easy to find nor to identify. This, and the difficult taxonomy of major genera such as *Trichogramma* and *Telenomus*, has so hampered knowledge of host range in egg parasitoids that we do not attempt to give specific host data here. It does seem, however, that for many butterfly species egg parasitism is an extremely important cause of mortality, that in some cases it can outweigh all other parasitism (e.g. Stefanescu *et al.*, 2003b), and that parasitism by *Trichogramma* species is particularly prevalent.

Considerable biological and behavioural differences between groups of egg parasitoids exist, even though the host's egg stage is in general poorly defended. Most *Telenomus* species need to attack rather young eggs and, like *Anastatus* which also fails to develop in older eggs, usually develop solitarily with respect to each egg attacked, though a high proportion of batched eggs might be parasitised at a single visit. *Trichogramma* species are especially small, often developing in surprisingly large broods per egg, are capable of using eggs of almost any age (though young eggs may be the most suitable: Ruberson & Kring, 1993), and can even behave as hyperparasitoids (Strand & Vinson, 1984). These attributes suggest that *Trichogramma* species will have broader host ranges, perhaps determined more by searching environment than by the hosts themselves, than scelionid species, which does indeed seem to be the case (see Babendreier *et al.*, 2003a, b; Roemis *et al.*, 2005), although a *Telenomus* species has also been recorded as a hyperparasitoid (Viktorov, 1966). While *Ooencyrtus* species are often gregarious and quite small, *Anastatus* are relatively large and for them host size might also be a constraint.

Eggs parasitised by Chalcidoidea often darken abnormally, but early signs of parasitism due to Scelionidae are typically less evident. All emerging adults of gregarious species of *Trichogramma* usually leave the host egg through a single hole, in contrast with *Ooencyrtus* (Stefanescu *et al.*, 2003b). In both Trichogrammatidae and Scelionidae females of some species are known to locate and attach themselves to the parent host before she oviposits (Plate 14), through this phoresy being able to parasitise eggs that might otherwise be difficult to locate – also gaining access while eggs are young and before competitors. Male *Pieris brassicae* are almost as attractive to *Trichogramma brassicae* Bezdenko hitching a ride as are mated females, while virgin females are shunned, because the mounting allomone used is a male-derived host pheromone (Fatouros *et al.*, 2005). Some Trichogrammatidae and Scelionidae mark eggs into which they have oviposited, either physically or with pheromones that deter conspecifics.

Diptera

Tachinidae

This is one of the largest of approximately 125 families of Diptera in Europe. All Tachinidae are koinobiont endoparasitoids, and a wide range of medium to large terrestrial insects (and a few other arthropods) comprise the host spectrum overall, though Lepidoptera predominate. Some tachinids parasitise long-lived adult insects but all of those reared from Lepidoptera attack only the larval stage (though either obligatorily or facultatively the host is sometimes not killed until it has pupated). Out of four subfamilies in Europe, three (Dexiinae, Exoristinae and Tachininae) contain parasitoids of Lepidoptera – though none of them exclusively – but only the last two (especially Exoristinae) include species that regularly attack butterflies (Plate 15). Most species associated with Lepidoptera use only that order, but a few of the most polyphagous sometimes also develop in sawfly larvae.[2]

In general, Tachinidae parasitising Lepidoptera are substantially less narrowly host-specialised than koinobiont parasitic wasps, and the physiological interaction with the host is less sophisticated (Askew & Shaw, 1986; Eggleton & Gaston, 1992; Belshaw, 1994). Two attributes underlie their capacity for diffuse host ranges. Firstly, rather than placing

[2] Herting (1960), Belshaw (1993) and Stireman *et al.* (2006) provide biological overviews and rearing records, and there is a website covering especially British species (http://tachinidae.org.uk). Reared Tachinidae can be sent to H.-P. Tschorsnig (Naturkundemuseum, Rosenstein 1, 70191 Stuttgart, Germany; tschorsnig.smns@naturkundemuseum-bw.de) for identification.

an inactive egg into the host's haemocoel, incurring a major risk of encapsulation by the host's haemocytes, either (i) in the tribe Goniini (Exoristinae), specialised minute 'microtype' eggs (Salkeld, 1980) are laid on vegetation to be ingested by the host and hatch in its gut, or (ii) the active first-instar larva first penetrates the host, boring into it either straight from eggs laid externally on the host (Plate 16) or following the laying of well-incubated eggs (or even larviposition) onto vegetation nearby. Exceptionally (e.g. in *Blondelia* and *Compsilura*) the female tachinid gouges a wound on the host with a specialised abdominal piercing structure and places first-instar larvae or eggs close to hatching directly into the host's body. Secondly, the larva, once in the haemocoel, avoids suffocation by diverting the host's encapsulation response so that only a sheath is formed, which the tachinid larva keeps open in the region of its mouth by its feeding activity, and at the other end by connecting its large posterior spiracles directly to an air supply, either piercing through the host's integument or via its tracheal system. In contrast, koinobiont endoparasitic Hymenoptera typically not only place an immobile egg in the host, but also live as larvae free in the haemocoel and depend for respiration on diffused oxygen – in both cases making them highly vulnerable to suffocation in any host for which they do not have a specific way of disrupting the encapsulation response.

Because of the plasticity in many tachinid host ranges, it is pointless to list any but the rather few species that use

Tachinidae more or less specialised to butterfly species	Main hosts
Subfamily Exoristinae	
Aplomya confinis (Fallén)	Lycaenidae
Buquetia musca Robineau-Desvoidy	*Papilio*
Cadurciella tritaeniata (Rondani)	*Callophrys*
Epicampocera succincta (Meigen)	*Pieris*
Erycia fasciata Villeneuve	Melitaeini (*Melitaea*)
Erycia fatua (Meigen)	Melitaeini (*Euphydryas*; *Melitaea*)
Erycia festinans (Meigen)	Melitaeini (*Melitaea*)
Erycia furibunda (Zetterstedt)	Melitaeini (*Euphydryas*)
Sturmia bella (Meigen)	Nymphalini
Thecocarcelia acutangulata (Macquart)	Hesperiinae
Subfamily Tachininae	
Pelatachina tibialis (Fallén)	Nymphalini

butterflies significantly more than other hosts. As the records largely come from an unpublished private database (H.-P. Tschorsnig, pers. comm.), we tabulate just an indication of the hosts. Additionally, several polyphagous species are regularly reared from various butterflies, but as part of much broader host ranges. These include (all in the subfamily Exoristinae): *Compsilura concinnata* (Meigen), *Phryxe vulgaris* (Fallén), *P. nemea* (Meigen) and to a lesser extent *Blondelia nigripes* (Fallén), *Phryxe magnicornis* (Zetterstedt), *Exorista larvarum* (Linneaus), *E. segregata* (Rondani), *Bactromyia aurulenta* (Meigen), *Pales pavida* (Meigen) and *Masicera sphingivora* (Robineau-Desvoidy). Many of these are extremely abundant and more often reared from butterflies than the more specialised species. More than 20 further species have been recorded (perhaps correctly) from one or more butterfly species on at least one occasion.

Bombyliidae

This fairly large family of flies is widespread in Europe but most numerous in the Mediterranean area. They are all essentially parasitoids, though the overall host relations of the family are extremely broad and only partly understood (Yeates & Greathead, 1997). Relatively few species parasitise Lepidoptera. Eggs are laid, often individually though always in very large numbers, in sites likely to support hosts, which are actively sought by the first-instar larva. This is a specialised form (shared by several groups of parasitoids) called a planidium, being minute, highly mobile and adapted to withstand desiccation during a long period of host-seeking. The few species that have been (infrequently) recorded as parasitoids of Lepidoptera in Europe oviposit onto the soil, and consequently butterfly species with larvae that spend time at the soil surface, e.g. Satyrinae, are the most likely to be attacked, especially by polyphagous species of the genus *Villa*. While most Bombyliidae are ectoparasitoids, *Villa* is endoparasitic, and the biology of one species has been described by Du Merle (1964, 1979a, b, c). The planidial larva attaches itself externally to the caterpillar and is carried to the host's pupation site, where it will penetrate the fresh pupal cuticle to develop internally. Although pupation is inside the host, the tough and heavily spiny bombyliid pupa is highly mobile, and it breaks out of the host's pupal shell and makes its way to the soil surface where the adult fly emerges. Rearing records from European butterflies are scarce, but García-Barros (1989a) reported an unidentified species of *Villa* from both *Hipparchia statilinus* and *H. semele*. Bombyliidae can also behave as hyperparasitoids.

CASE STUDIES

Here we present case studies on parasitism of the six butterfly taxa which have been most quantitatively studied in Europe. Our aim is to illustrate the range of parasitoid assemblages and the extent to which parasitoids account for butterfly mortality.

Papilionidae: *Iphiclides podalirius*

With the exception of *Iphiclides podalirius*, information on parasitoids attacking the few species of European Papilionidae is mainly based on casual records. Surprisingly, in a 4-year study carried out in Norfolk, England, to determine pre-adult mortality of *Papilio machaon*, not a single parasitoid was recorded from monitoring over 300 eggs, 293 larvae and 40 pupae (Dempster *et al.*, 1976). However, the specialist larva–pupal parasitoid *Trogus lapidator* (Ichneumonidae: Ichneumoninae) (Plate 9a) was later found to be established in the general area (Shaw, 1978) and further sampling would presumably have revealed the generalist and regular pupal parasitoids *Pteromalus puparum* (Plates 12a and 12b) and *Pimpla rufipes* (Plate 10a).

The parasitoid complex of *I. podalirius* was studied at a site in Catalonia, Spain, from 1996 to 1999. From over 1000 eggs, 124 larvae and 32 pupae, eight species of parasitoids were reared, some of them regularly attacking the host and having an important impact on population size (Stefanescu *et al.*, 2003b). Egg parasitoids were by far the most abundant and diverse, perhaps a common pattern for Papilionidae (e.g. Watanabe *et al.*, 1984; Garraway & Bailey, 1992). Larval parasitoids were scarce and no parasitoids were recorded from the rather small sample of pupae.

The egg parasitoids comprised *Trichogramma cordubense* Vargas & Cabello and *T. gicai* Pintureau & Stefanescu (Trichogrammatidae), *Ooencyrtus telenomicida* (Vassiliev) and *O. vinulae* (Masi) (Encyrtidae) and *Anastatus bifasciatus* (Geoffroy) (Eupelmidae). With the possible exceptions of *O. vinulae* (recorded only once) and the poorly known *T. gicai*, all these are plurivoltine and broad generalists, and parasitise eggs of several insect orders.

In the study area *I. podalirius* has two, and a partial third, annual generations. Parasitism had a strong seasonal pattern. In summer, egg parasitoids killed 25–45% of the eggs, and vertebrate predation of the overwintering pupae caused most of the remaining mortality (Stefanescu, 2004). In the spring parasitism was negligible, and most mortality was due to invertebrate predation of young larvae and bird predation of old larvae and pupae (Stefanescu, 2000b; and unpublished data). This seasonal pattern of parasitism, whereby increasing levels occur as summer progresses, has been noted by Dempster (1984) and Askew & Shaw (1986), and may be a common feature in plurivoltine temperate Lepidoptera regularly attacked by generalist parasitoids.

The extremely polyphagous tachinid *Compsilura concinnata* was the only regular larval parasitoid, but it was restricted to summer generations when it caused widely varying mortality up to 20%. Two other polyphagous larval parasitoids, *Blepharipa pratensis* (Meigen) (Tachinidae) and *Meteorus pulchricornis* (Braconidae: Euphorinae), were recorded twice and once, respectively. *Blepharipa pratensis* is particularly a parasitoid of *Lymantria dispar* (but also other moths), and *M. pulchricornis* parasitises various macrolepidoptera.

Pieridae: *Pieris* spp.

Some *Pieris* butterflies are synanthropic as well as being conspicuous in the European landscape, so it is not surprising that some of the earliest documentation of parasitoid life history involves *Cotesia glomerata* parasitising *Pieris brassicae* (for example, Goedart, 1662; Ray, 1710; as reported in Feltwell, 1982). Much attention has been paid to the world-wide agricultural pests *Pieris rapae* and *P. brassicae*, and their commonest parasitoids are well known (also through laboratory experiments). However, there are relatively few parasitoid species regularly associated with European *Pieris* (summarised for *P. brassicae* by Shaw, 1982), and many of them are known to use a wide range of hosts.

Three large-scale studies on mortality of *P. rapae* and *P. brassicae* in Europe have been published. One by Moss (1933) details his collections of *Cotesia glomerata* and *Pteromalus puparum* (Plates 12a and 12b) for importation to Australia to control *P. rapae*. Incidentally, both proved to be bad choices: *C. glomerata* is better adapted to *P. brassicae*; and *P. puparum* has a wide host range. The second is a 4-year study in the UK by Richards (1940) which is outlined below. The third is a 3-year study of the effects of the insecticide DDT on *P. rapae* and its parasitoids (Dempster, 1967, 1968). Several smaller studies include pupal parasitism of *P. rapae* and *P. brassicae* (Bisset, 1938; Littlewood, 1983), and egg and larval mortality of *P. brassicae* and *P. rapae* in the UK (Baker, 1970) and *P. brassicae* in Denmark (Kristensen, 1994). In Europe there has been very little work on parasitism of *P. napi*, which is not a pest (but see Lee & Archer, 1974). However, it has been investigated in Japan, where *P. napi* co-occurs with *P. rapae* and *P. melete* Ménétriès (Ohsaki & Sato, 1999), as has parasitism of the closely related *P. virginiensis* Edwards in North America, which has

declined owing to non-target effects of the biological control of *P. rapae* (Benson *et al.*, 2003).

Richards' (1940) study was conducted near Slough in England between 1932 and 1936. *Pieris brassicae*, *P. rapae* and *P. napi* were present, but only the first two were abundant and sampled. Richards reared 3581 *P. rapae* and 949 *P. brassicae* larvae, collected in each instar and during all of the three annual generations, primarily from cabbage. He followed many fewer pupae and eggs. His text and tabulation illustrate clearly the complexities of reporting rates of parasitism as a continuing process.

The most important parasitoids were the braconids *Cotesia rubecula* and *C. glomerata*. *Cotesia rubecula* is solitary and parasitises *P. rapae*, while *C. glomerata* is gregarious and uses primarily *P. brassicae*. A total of 24% of the *P. rapae* reared were parasitised by *C. rubecula*. This differed between years (ranging from 3% to 61%) and, as is generally the case for plurivoltine parasitoids in temperate climates, it increased in successive generations within a year. About 17% of the *C. rubecula* were themselves parasitised by the gregarious secondary parasitoid *Baryscapus galactopus* (Ratzeburg) (misidentified as *Tetrastichus rapo*) (Eulophidae) and 4% by the solitary *Mesochorus olerum* Curtis (misidentified as *tachypus*) (Ichneumonidae: Mesochorinae). *Cotesia glomerata* parasitised *P. rapae* only sparingly (none most generations, up to 15% occasionally), but consistently parasitised a large fraction (on average 53%) of early-instar gregarious *P. brassicae* larvae. The same two true hyperparasitoids were found as from *C. rubecula*, but at a lower rate.

Of several tachinid species reared, only the polyphagous *Phryxe vulgaris* and to a lesser extent the specialist *Epicampocera succincta* were in any numbers, and they were at a competitive disadvantage to the *Cotesia* species. The polyphagous pupal parasitoid *Pteromalus puparum* (Pteromalidae) emerged from six out of the 155 *P. rapae* pupae collected, but another generalist, *Pimpla rufipes* (Plate 10a), which is a very regular pupal parasitoid of *Pieris* species, was not found. Polyphagous *Trichogramma* (Trichogrammatidae) egg parasitoids regularly use *Pieris* hosts in Europe (see Feltwell, 1982, for a review) but were apparently absent at Slough. Similarly, Kristensen (1994) found no parasitism of 48 *P. brassicae* egg batches observed in Denmark.

In Richards' study parasitism caused significant mortality of *P. rapae* and *P. brassicae*, on par with the effects of disease, climate and predators. On average, about 25% of the total mortality of *P. rapae* was due to parasitism, primarily (80%) by *C. rubecula*. For *P. brassicae* it was somewhat higher because the rate of parasitism of *P. brassicae* by *C. glomerata* was on average twice that of *C. rubecula* on *P. rapae*.

Moss (1933) and Dempster (1967) found broadly similar patterns to Richards', though recording respectively somewhat higher and lower rates of parasitism. Dempster (1968) showed that application of DDT increased the *P. rapae* population by reducing the impact of ground-dwelling predators and eliminating *C. rubecula*. In a small study, Kristensen (1994) found a high rate of larval predation on *P. brassicae*. Only 30 of 960 larvae escaped arthropod and avian predation, suggesting that at least under some conditions (see also Dempster, 1967, 1968; Baker, 1970) predation outweighs parasitism.

Differences in the learning and memory characteristics of the two *Cotesia* species have been elucidated by Smid *et al.* (2007), and other work on their behavioural ecology can be traced through this reference.

Lycaenidae: *Maculinea rebeli*

Although only the *Maculinea rebeli* parasitoid associations have been studied in detail, egg, larval and pupal parasitoids are known from several other European Lycaenidae (e.g. Bink, 1970; Fiedler *et al.*, 1995; Shaw, 1996; Horstmann *et al.*, 1997; Gil-T, 2001, 2004) and the few available data suggest that parasitism can be heavy. Bink (1970) gives quantitative data showing high parasitism in several populations of *Lycaena* species. Shaw (1996) found that 67% of 214 larvae of *Aricia artexerxes* from 13 sites in south Cumbria and southeastern Scotland were parasitised, mostly by *Hyposoter notatus* (Ichneumonidae: Campopleginae), and Gil-T (2001) recorded 55% parasitism in 58 larvae of *Iolana iolas* collected at a locality in southern Spain, mostly due to *Anisobas cephalotes* (Ichneumonidae: Ichneumoninae). In an unquantified report Martín Cano (1981) stated that between 70% and 100% of an unspecified number of larvae of *Glaucopsyche melanops* and more than 50% of those of *G. alexis* collected in several Spanish sites were parasitised by Ichneumonidae, Braconidae and Tachinidae. Such figures might seem surprising, as the larvae of these butterflies establish associations with ants (Fiedler, 1991; see also Gil-T, 2004), which have been presumed to provide protection against parasitoids (see Pierce & Easteal, 1986; Pierce *et al.*, 1987). These casual findings, along with those of Seufert & Fiedler (1999), call into question the generality of such conclusions, as specialised parasitoids can clearly flourish. Fiedler and collaborators (Fiedler *et al.*, 1995; Baumgarten & Fiedler, 1998)

have explored multi-species interactions between myrmecophilous lycaenid caterpillars, their tending ants and specialised parasitoids.

Even more surprisingly, carnivorous *Maculinea* butterflies, which spend most of their larval stage and then pupate inside ant nests, are regularly attacked by host-specific ichneumonid parasitoids of the subfamily Ichneumoninae (Thomas & Elmes, 1993; Munguira & Martín, 1999; Tartally, 2005). A compilation recording in detail recent findings on parasitism of *Maculinea* species (Anton *et al.*, in prep) underlies some of the information given below (and also in the Ichneumoninae rearing table). Unfortunately the taxonomy of some of the parasitoids concerned has been confused, and in one genus (*Ichneumon*) some uncertainty remains. Thus apparently three *Neotypus* species have been reported from the three European species of the predatory clade, *Maculinea arion*, *M. nausithous* and *M. teleius*, though partly as a result of confusion over the validity of the name *N. melanocephalus* and the occasional use of its junior synonym *N. pusillus* (see Horstmann, 1999). There is no doubt that *Neotypus coreensis*, parasitising *M. arion*, is distinct (though it is less certain that its name is correct), but the perception that there are two further species, differing in leg colour, has not been upheld following a recent review of all available material from a range of sources (K. Horstmann, pers. comm.). Specimens associated with *M. nausithous* could not be separated from those much less often reared from *M. teleius*, with colour differences in the plentiful *M. nausithous*-associated material not even proving to be consistent geographically (K. Horstmann, pers. comm.), and we here regard them as belonging to the same species, *N. melanocephalus* (= *pusillus*). In the genus *Ichneumon* uncertainty remains unresolved. One well-defined species, *I. fulvicornis*, parasitises certainly *M. teleius* and possibly *M. nausithous* (usually the host pupae from which it has been reared could not be separated, and the two butterflies very often co-occur). However, another, *I. eumerus* (Plate 9b), is reported to have populations parasitising *M. rebeli* in southwest Europe that overwinter in the host (Thomas & Elmes, 1993), unlike the central European populations (reared at several sites, from *M. alcon*, *M. teleius* and probably also *M. rebeli*) which overwinter as adults and thus conform to the general biology of the genus *Ichneumon*. The two supposed races are indistinguishable morphologically (Hinz & Horstmann, 2007; K. Horstmann, pers. comm.), but if the biological differences are consistent then they should be regarded as distinct species. If so, it is not clear which would take the name *I. eumerus*, but it is used in the sense of Thomas & Elmes (1993) in the following summary.

The complex relationships among the parasitoid, the host and the ants were studied in detail by Thomas & Elmes (1993) for the 'cuckoo' species *M. rebeli*, which is regularly parasitised by *Ichneumon eumerus* in France and Spain. Failure to detect parasitism of *M. rebeli* caterpillars during their free-living phase and when being adopted by *Myrmica schencki* or *M. sabuleti* workers, and the observation of several females of *I. eumerus* trying (unsuccessfully) to enter *M. sabuleti* nests known to contain *Maculinea* caterpillars, suggested that parasitism takes place inside the ant nests. This was confirmed in the laboratory. In contrast, the *Neotypus* species parasitising predatory *Maculinea* species attack the young caterpillars while they are still feeding on flower heads. Thomas & Elmes (1993) suggest that the specialised behaviour of *I. eumerus* precludes the use of other hosts. Females identify nests of the right ant species by odour, and can ascertain at the nest entrance whether a *Maculinea* caterpillar is inside. On entering the selected host-supporting nest (in which hosts tend to be clumped), the female is fiercely attacked, but is well armoured and soon spreads an allomone which induces the worker ants to fight each other (Thomas *et al.*, 2002). This sophisticated behaviour goes with an exceptionally low fecundity, most females entering just one or two suitable ant nests, and laying only five to ten eggs. The impact of the parasitoid on host populations seems to be slight (Hochberg *et al.*, 1998), with parasitism rates ranging from 6% to 23%. Considerable theoretical work on this extraordinary system (Hochberg *et al.*, 1992, 1994), aimed at the conservation of both *I. eumerus* and *M. rebeli* (Hochberg *et al.*, 1996, 1998), suggests that *I. eumerus* is slightly more vulnerable to extinction than *M. rebeli*, which is a species considered globally endangered (Munguira & Martín, 1999). Therefore, providing it is distinct from the race that has a broader host range, *I. eumerus* provides an excellent example of the need to consider parasitoid conservation specifically when dealing with biodiversity conservation (Shaw & Hochberg, 2001; Thomas *et al.*, 2002). This discussion of *Maculinea* parasitoids also shows the need to gain a clear understanding of the taxonomy of parasitoids in relation to their ecology, and the difficulty in doing so.[3]

Nymphalidae: *Aglais urticae*

Because caterpillars in the attractive tribe Nymphalini are conspicuous and easily reared, frequent small-scale rearings of parasitoids have led to many literature records. However,

although most of the Nymphalini are very common throughout Europe, only one detailed study of a parasitoid complex has been made.

Between 1971 and 1973, Pyörnilä (1976a, b, 1977) collected 387 egg clusters, 3908 larvae and 132 pupae from an *A. urticae* population in eastern Finland. Eight parasitoid species were reared. The most numerous larval parasitoids were *Pelatachina tibialis* and *Phryxe vulgaris* (both Tachinidae) and *Phobocampe confusa* (misidentified as *Hyposoter horticola*) (Ichneumonidae: Campopleginae). *Pteromalus puparum* (Pteromalidae) (Plates 12a and 12b) was a very important pupal parasitoid in the area, unlike single specimens of *Pteromalus apum* and three species of Ichneumonidae (*Ichneumon gracilicornis*, *Pimpla flavicoxis* Thomson (misidentified as *aquilonia*) and *Apechthis rufata* (Gmelin)). No egg parasitoids were detected.

The commonest larval parasitoid in three consecutive summers was *P. tibialis*, a near specialist of Nymphalini butterflies (sometimes also recorded from Noctuidae). It attacks larvae throughout their development, parasitisation rates increasing steadily until the last larval instar. *Phryxe vulgaris*, an extremely polyphagous species, also attacks larvae throughout their development, especially those occurring in late summer. Both kill the host in its last larval instar or, occasionally, soon after pupation. From 23.5% to 65.0% of larvae collected in their fourth instar were parasitised by these tachinids.

The ichneumonid *P. confusa* was reared each year from about 10% of the collected larvae. Its cocoons were subject to unquantified pseudohyperparasitism by *Gelis agilis* (Fabricius) (= *instabilis*) (Ichneumonidae: Cryptinae), which is highly polyphagous (Schwarz & Shaw, 1999). Finally, 35–58% of the collected pupae were killed by *P. puparum*, comparable to Julliard's (1948) findings in Switzerland.

Although there are many reports of the same and other species parasitising *A. urticae*, Pyörnilä's study is the only one that provides a fairly comprehensive picture of the parasitoid complex at a given locality. Interestingly, this common butterfly is believed to be decreasing in some parts of its range (e.g. Kulfan *et al.*, 1997), and it has been suggested that parasitoids such as *Sturmia bella* (Plate 15) increasing their geographical range could be partly responsible (Greatorex-Davies & Roy, 2005).

Nymphalidae: Genera *Melitaea* and *Euphydryas*

Melitaeini butterflies share many life-history characteristics that are relevant to their parasitoids: eggs are laid in clusters,

---

[3] Continuing research on this system can be traced via www.macman-project.de.

larvae live gregariously under a web for at least the first couple of instars, and there is larval diapause in the summer (where it is dry) or winter (where it is cold) (Kuussaari *et al.*, 2004). Most Melitaeini feed on herbaceous plants chemically defended by iridoid glycosides or secoiridoids (Wahlberg, 2001), and several species commonly co-occur in meadows or forest clearings, sometimes even sharing foodplants (Kankare *et al.*, 2005a). Thus we might expect the butterflies to have the same or closely related parasitoids, or ones with life-history characteristics in common.

Parasitoids of the European Melitaeini are among the best known of any butterflies (Porter, 1981; Komonen, 1997, 1998; Wahlberg *et al.*, 2001; Eliasson & Shaw, 2003; Kankare & Shaw, 2004; van Nouhuys & Hanski, 2004; Kankare *et al.*, 2005a; Stefanescu *et al.*, 2009). These parasitoid complexes are rather atypical in that they are so strongly dominated by specialists, notably gregarious *Cotesia* species (Plate 10b). In the Åland Islands in southwest Finland the parasitoid complex of *Melitaea cinxia* has been studied for more than a decade. There are two specialist primary larval endoparasitoids, each with their own secondary parasitoids, and several generalist pupal parasitoids (Lei *et al.*, 1997; van Nouhuys & Hanski, 2004).

The gregarious primary parasitoid *Cotesia melitaearum* (agg.; see table of Braconidae: Microgastrinae) has two or three generations per host generation and is locally monophagous. Despite its high reproductive potential and close relationship with *M. cinxia*, *C. melitaearum* (agg.) generally forms small populations in Åland, for three reasons. Firstly, the wasp has weak dispersal ability relative to the scale of the fragmented landscape. Secondly, the host has relatively fast dynamics for unrelated reasons (Lei & Hanski, 1997; van Nouhuys & Hanski, 2002). Finally, the parasitoid is constrained on a local scale by density-dependent secondary parasitism (van Nouhuys & Hanski, 2000; van Nouhuys & Tay, 2001). Nevertheless, where host density is high, *C. melitaearum* (agg.) can impact host population size, perhaps causing local host extinctions (Lei & Hanski, 1997). Similarly strong influence in *Euphydryas aurinia* population dynamics by *Cotesia bignellii* was observed by Porter (1981) in England. Though studied in less detail, in eastern North America the related *Euphydryas phaeton* (Drury) also sustains variable, sometimes high rates of parasitism by *Cotesia euphydridis* (Muesebeck) (Stamp, 1981a), as does western North American *Euphydryas editha* (Boisduval) by *Cotesia koebelei* (Riley) (Moore, 1987).

The second parasitoid of *M. cinxia* in Finland, *Hyposoter horticola* (Ichneumonidae: Campopleginae) (Plate 8a), is also restricted to Melitaeini hosts and uses only *M. cinxia* locally. It consistently parasitises about a third of host larvae at all spatial scales and natural densities. As a consequence, its impact is simply to reduce overall population size, leaving local populations prone to extinction by other means. This unexpected relationship results from the behaviour of the adult parasitoid (van Nouhuys & Ehrnsten, 2004; see caption to Plate 8a).

Each of the two primary parasitoids has abundant secondary parasitoids. The wingless and thelytokous *Gelis agilis* (Ichneumonidae: Cryptinae) is a generalist pseudohyperparasitoid of *C. melitaearum* (agg.). It aggregates in response to high *Cotesia* density, and apparently can drive local populations of *C. melitaearum* (agg.) to extinction (Lei & Hanski, 1998; van Nouhuys & Hanski, 2000). Two other species, *G. acarorum* (Linnaeus) and *G. spurius* (Foerster) (misidentified as *ruficornis*), have occasionally been reared from *C. melitaearum* (agg.) in this system, and *G. agilis* has been reared from *H. horticola*, but their importance is unknown. The true hyperparasitoid *Mesochorus* sp. (Ichneumonidae: Mesochorinae) predominantly uses *H. horticola* as a host, with only infrequent development through *Cotesia*. It parasitises about a quarter of the *H. horticola* in all but the most isolated local populations in the Åland Islands (van Nouhuys & Hanski, 2005). Because, like its host, it has a relatively uniform rate of parasitism at most spatial scales, it does not directly affect the dynamics of the primary parasitoid or the butterfly.

Melitaea cinxia in Finland also hosts pupal parasitoids. Lei *et al.* (1997) reared the solitary *Ichneumon gracilicornis* (Ichneumonidae: Ichneumoninae), and the gregarious pteromalids *Coelopisthia caledonica*, *Pteromalus apum* and *P. puparum* (Plates 12a and 12b). Experimental field exposure of fresh pupae has also revealed *Ichneumon cinxiae*. While the *Ichneumon* species are probably uncommon as well as being rather specialised, the pteromalids have broader host ranges and can be abundant, but the impact of pupal parasitoids on host dynamics is unknown. No egg parasitoids have yet been associated with *M. cinxia* in Finland, nor are there tachinid flies.

Other Melitaeini host some of the same or closely related parasitoids, especially gregarious species of *Cotesia* (Eliasson & Shaw, 2003; Kankare & Shaw, 2004) (Plate 10b). Recent studies using molecular, morphological and behavioural data have found surprising host specificity among these parasitoids, with several supposed *Cotesia* 'species' actually being groups of cryptic species (see table of Braconidae: Microgastrinae). However, a variety of patterns of host specificity are seen, suggesting ecologically and phylogenetically dynamic host–parasitoid relationships (Kankare & Shaw, 2004; Kankare *et al.*, 2005a, b).

In addition to *Hyposoter horticola* and *Cotesia* species, specialist parasitoids of European Melitaeini include large solitary ichneumonids in the genera *Benjaminia* and *Ichneumon*, and tachinid flies in the genus *Erycia* (for example studies see Wahlberg *et al.*, 2001; Eliasson & Shaw, 2003; Kankare *et al.*, 2005a; Stefanescu *et al.*, 2009). Being generally much less abundant than the *Cotesia* species, however, their host ranges remain poorly understood. Generalist parasitoids attacking pupae are sometimes quite abundant, and one, *Pteromalus apum*, has a strong (though not exclusive) relationship with the group. Because they typically rely on other hosts as well, generalists may strongly affect local population sizes, but are not themselves greatly affected by the densities of specific hosts.

Hesperiidae: *Thymelicus lineola*

Although a few firm parasitoid–host associations are known for butterflies belonging to the Hesperiidae (e.g. *Cotesia glabrata* attacking *Carcharodus alceae*), a moderately good knowledge of a whole parasitoid complex has been gathered only for *Thymelicus lineola*. This butterfly was accidentally introduced to Canada in the early twentieth century, and has since spread to the USA. In contrast to European populations, in North America *T. lineola* is a pest, causing extensive damage to hay-fields and pastures (Pengelly, 1961). This prompted a comparative study of its parasitoid complexes in both regions, in search of possible biological control of American populations (Arthur, 1962, 1966; Carl, 1968).

The parasitoid complex at sites in Switzerland, France and Austria was assessed by Carl (1968), in a 4-year study based on a sample of over 3400 larvae and an unspecified (but much lower) number of pupae. Egg parasitism was not included in the study.

Six primary and two secondary parasitoids were found. The most important were two larval parasitoids, the generalist *Phryxe vulgaris* (Tachinidae) and *Aleiodes coxalis* (= *tristis*) (Braconidae: Rogadinae), which is an oligophagous species attacking also Satyrinae (M. R. Shaw, unpublished data) but not the wider hosts ascribed to it by Carl (1968), and the specialised pupal parasitoid *Syspasis scutellator* (Ichneumonidae: Ichneumoninae), all of which had a high incidence in all sampled populations. More rarely, the larval parasitoid *Thecocarcelia acutangulata* (= *incidens*) (Tachinidae), the larval–pupal parasitoid *Agrypon delarvatum* (Ichneumonidae: Anomaloninae) and the pupal parasitoid *Brachymeria tibialis* (= *intermedia*) (Chalcididae) were

reared. In a few samples, an unidentified Mermithidae (Phylum Nematoda) was also detected.

Phryxe vulgaris was present in 3–44% of larvae collected, and was especially prevalent at high host densities. Being plurivoltine as well as polyphagous, *P. vulgaris* was inevitably also depending on other hosts. *Aleiodes coxalis* is also a plurivoltine koinobiont species, and parasitised 3–57% of the collected larvae. In 1–10% of cases it was hyperparasitised by *Mesochorus tetricus* Holmgren (= *macrurus*) (Ichneumonidae: Mesochorinae). *Syspasis scutellator* is univoltine and probably has a very narrow host range. It caused apparently high pupal mortality ranging from 30% to 50% (but sample sizes were low). The generalist pseudohyperparasitoid *Gelis cursitans* (Fabricius) (Ichneumonidae: Cryptinae) developed from 9% to 17% of host pupae harbouring *S. scutellator*.

In Canada up to 22 generalist species were recorded (Pengelly, 1961; Arthur, 1962), but parasitism was light and appeared to be largely incidental, with many sparse records despite heavy sampling. Carl (1968) suggested that the ineffective parasitoid complex in Canada might progress to become less species-rich but more specialised, and it would be interesting to resurvey Canadian *T. lineola* populations to see if there are signs of this 40 years on.

Overview of case studies

In spite of the work that has gone into each case study, our knowledge of these parasitoid assemblages is incomplete for several reasons. As well as low sampling of the pupal or egg stages and inadequate taxonomic and autecological knowledge of the parasitoids, the sampling has generally been conducted in only one or a few areas, where not all of the parasitoids that will regularly attack that host necessarily occurred. For example, although both *Cotesia vestalis* and *C. vanessae* are frequently abundant as parasitoids of *Aglais urticae* over most of Europe, they were not found in Finland by Pyörnilä (1976a, b, 1977), and the regular parasitoid of *Pieris* species, *Hyposoter ebeninus*, is absent from all British populations.

Finally, while it is relatively straightforward to tabulate the mortality factors for a host, it is more difficult to assess the effects of a parasitoid on the dynamics of host populations; that is, the extent to which the abundance of a butterfly over time is influenced by its parasitoids. It is important to appreciate that, however severe parasitoid-induced mortality may become, local host population dynamics will only be directly regulated by parasitoids if the mortality they cause depends, at some spatial scale, on the density of the host (see Hassell, 2000). Few studies have run for long enough to examine this, though it was attempted on a small scale by measuring the density dependence of parasitism of *Pieris* (Richards, 1940). At a regional (or landscape) scale the role of parasitods in determining host population dynamics will depend to a large extent on the relative rates of movement of the host and the parasitoids in the landscape, and local density dependence may not necessarily be pivotal for explaining population dynamics at larger spatial scales (Holt, 1997). The interaction of butterflies and parasitoids at both local and landscape scales has only been addressed for *Melitaea cinxia* and its parasitoids (see above, but see also Menéndez *et al.*, 2008).

Nevertheless, the case studies illustrate that parasitoid assemblage sizes differ, and contain species using different host stages and exhibiting differing degrees of host specificity. Clearly, some patterns of parasitoid assemblages result from the biology of the butterfly. For example the lycaenids *Maculinea* spp. host few parasitoid species, each with narrow host ranges, undoubtedly connected with the specialised life history of the hosts. In contrast *Iphiclides podalirius* (Papilionidae) has a generalised life history and is attacked by a set of parasitoids with wide host ranges. In each case study a few key parasitoids explain an important fraction of host mortality. However, their prominence can vary seasonally, as illustrated by plurivoltine *Pieris* species and *Iphiclides podalirius*; among years, for instance *Cotesia* parasitoids of *Melitaea cinxia* and *Pieris*; and among locations, as for *Thymelicus lineola*.

CONCLUDING REMARKS

In this chapter we have tried to do two things. The first is simply to open the way for a more informed interest to be taken by butterfly ecologists and enthusiasts in the parasitoids they encounter. We have exposed the paucity of reliable information partly in the hope that this will engage others to help build better knowledge about parasitoids. We encourage researchers and collectors to seek and preserve parasitoids reared from butterflies, and to pass them on to taxonomists and/or to deposit them in active research collections.

The second aim of this chapter is to assert that parasitoids can be expected to have many important effects in the ecology of butterfly species. In ways that differ substantially between host species, they can influence population size and can be a cause of strong fluctuations at different spatial scales. But the effects of parasitoids can be subtler. Host sex ratios can sometimes be distorted through disproportionate parasitism of one sex for a variety of reasons (see Shaw, 1975; Porter, 1984). Apparent competition (see Holt & Lawton, 1993) mediated by shared parasitoids might altogether exclude a butterfly from a particular region, and secondary parasitoids can have profound effects on primary parasitoid population sizes (e.g. van Nouhuys & Hanski, 2000; Stefanescu *et al.*, 2009) and thus on butterfly population dynamics. On an evolutionary timescale, defence against parasitoids has shaped aspects of the physiology of immature stages, and the behaviour of both larvae and adult butterflies. For example, Ohsaki & Sato (1999) suggest that *Pieris napi*, *P. rapae* and *P. melete* in Japan have each evolved to use habitats and host plant species that minimise parasitism by the braconid wasp *Cotesia glomerata* and the tachinid fly *Epicampocera succincta*.

For us to understand the place of parasitoids in the lives of butterflies there is still much to be gained from close attention to natural history, careful experimentation and thorough taxonomic investigation. The importance of parasitoids can also be dramatically revealed by changes caused by humans; for instance, the introduction of *Cotesia rubecula* to North America from Europe that was followed by the precipitate decline of native *Pieris virginiensis* Edwards (Benson *et al.*, 2003), and the increase in crop damage after the application of insecticide as shown experimentally with DDT in the *Pieris* and *Cotesia* system by Dempster (1967).

The decline of many butterflies and their shifting distributions with changing land-use practices and climate has surely led to corresponding changes among their parasitoids. It seems safe to say that parasitoids are likely to be more vulnerable to habitat degradation than their hosts (Komonen *et al.*, 2000; Shaw & Hochberg, 2001; van Nouhuys & Hanski, 2005). However, although there are indications of serious declines (Thirion, 1976, 1981), for the most part too little is known about parasitoids for us to see this clearly enough to address it.

ACKNOWLEDGEMENTS

We are grateful to Dick Askew, Klaus Horstmann, John Noyes, Bernard Pintureau, Juli Pujade-Villar, Heinz Schnee and Hans-Peter Tschorsnig for their helpful comments and provision of information about parasitoids in the groups of their special expertise, and to Sean Edwards, Nina Fatouros, Konrad Fiedler, Paul Mabbott, Geoff Nobes, Josep Planas, Josep Ramon Salas, Chris Raper and Jeremy

Thomas for generously supplying and allowing us to use the photos credited to them in the plates. Among the very great number of people whose kind donations of reared parasitoids to MRS for the NMS collection have enabled perceptions of host range to form, the following (in addition to the authors) have donated parasitic wasps reared from European butterflies: H. Abbasipur, R. Ainley, B. Aldwell, A. A. Allen, B. S. Angell, J. Asher, R. R. Askew, G. R. Ayres, K. E. J. Bailey, C. R. B. Baker, E. Balletto, H. Bantock, D. A. Barbour, B. Barr, R. Barrington, D. Beaumont, P. E. Betzholtz, F. Bink, D. S. Blakeley, K. P. Bland, A. Blázquez, K. M. G. Bond, M. R. Britton, J. P. Brock, M. Brooks, A. Buckham, C. Bulman, T. Burnhard, G. N. Burton, J. F. Burton, C. Bystrowski, V. Cameron-Curry, J. M. Chalmers-Hunt, D. Chanselme, J. Chavoutier, J. Choutt, I. C. Christie, S. G. Compton, J. Connell, M. W. Cooper, D. Corke, M. F. V. Corley, P. W. Cribb, A. R. Cronin, I. Cross, J. Dantart, A. W. Darby, D. Dell, R. L. H. Dennis, J. Dickenson, K. Dierkes, A. Duggan, R. Early, C. U. Eliasson, R. Elliott, H. A. Ellis, J. Feltwell, R. L. E. Ford, T. H. Ford, A. P. Fowles, B. Fox, E. García-Barros, M. Gascoigne-Pees, J. Gifford, W. Gilchrist, F. Gil-T, M. Ginés Muñoz, S. Glover, J. L. Gregory, M. A. Grimes, P. Gross, G. M. Haggett, C. Hallett, I. Hanski, A. Harmer, C. Hart, G. Hart, J. Hernández-Roldán, A. M. V. Hoare, R. N. Hobbs, M. A. Hope, A. Hoskins, M. Howe, V. Hyyryläinen, J. Jubany, M. Kankare, P. Kantonen, G. E. King, F. Kinnear, P. Kirkland, A. Komonen, M. Kuussaari, S. Labonne, T. Lafranchis, J. R. Langmaid, G. Lei, O. Lewis, A. Liebert, S. C. R. Littlewood, H. V. McKay, P. Marren, L. Martin, I. Mascall, T. M. Melling, H. Mendel, R. Menéndez Martínez, J. R. Miller, M. Miralles, M. J. Morgan, M. G. Morris, J. Muggleton, K. Murray, C. N. Nicholls, M. Nieminen, G. Nobes, U. Norberg, D. G. Notton, M. Oates, S. Oehmig, D. F. Owen, R. Parker, D. Parkinson, B. T. Parsons, J. E. Pateman, J. Paukkunen, J. H. Payne, E. C. Pelham-Clinton, H. G. Phelps, E. J. Philpott, A. Piper, J. Planas, R. Plumbly, K. Porter, A. Pullin, J. Quinn, R. Revels, I. Richardson, T. Rickman, J. Rowell, J. Roy, P. J. C. Russell, P. K. Schmidt, E. Seale, M. C. Shaw, W. Shaw, T. Shepherd, H. G. Short, M. J. Simmons, A. N. B. Simpson, M. Singer, L. Sivell, L. Slaughter, D. J. Smith, R. L. Soulsby, D. Stokes, P. Summers, R. Sutton, S. Swift, G. M. Tatchell, D. Teague, P. Tebbutt, W. J. Tennent, J. A. Thomas, S. Thoss, W. G. Tremewan, I. P. Tuffs, C. Turlure, S. Viader, J. C. Vicente, R. Vila, N. Wahlberg, A. Walker, I. W. Wallace, P. Waring, E. J. M. Warren, M. S. Warren, M. R. Webb, M. C. White, C. Wiklund, P. R. Wiles, K. J. Willmott, M. Wilson, R. Wilson, L. D. Young.

Part III
Evolutionary biology

12 • Adaptation and plasticity in butterflies: the interplay of genes and environment

HANS VAN DYCK AND JACK J. WINDIG

SUMMARY

Evolutionary ecology has developed from a broadly descriptive and partially quantified explanation of adaptation to a mature science in which the fitness components of phenotypic characters need to be quantified and adaptive hypotheses rigorously tested. The importance of genes for trait determination is now seen in the context of development and plasticity, with the recognition that a genotype may give rise to different phenotypes under different developmental conditions. This plasticity may confer advantages in variable environments, and the challenge for evolutionary ecologists is to understand selection and adaptation in the context of trade-offs between conflicting requirements and constraints on alternative developmental pathways. Butterflies vary in a number of characteristics and in order to understand adaptation in these organisms (and others for which they are often used as model systems) it is important to consider multiple rather than single traits, developmental processes and the effects of genes and regulatory mechanisms on all life stages. In effect, to understand adaptation and constraints on adaptation we need to fully explore gene × environment interactions at a range of levels, including molecular, cellular, physiological, behavioural and ecological processes. Butterflies have been used to illustrate micro-evolutionary changes within species, and the responses of species to resource changes and interlinked population processes. Comparisons between species may also illustrate how different taxa, even those that are closely related, may vary in their capacity to make short-term adaptive responses to environmental changes. By understanding the constraints on change within organisms, which operate as multifunctional integrative systems, butterfly evolutionary ecology will make an increasingly important contribution to understanding adaptation, and make vital contributions to understanding the response of these declining organisms to widespread and increasing environmental change, including changes of habitat availability and quality and climate change.

HOW DOES AN ADAPTATION EVOLVE IN THEORY?

Butterfly evolutionary ecology and genetics

Butterflies have attracted the attention of biologists from the onset of evolutionary thinking (e.g. Darwin 1859, Wallace 1865). Like any other organism, butterflies cope with the problems posed and exploit the opportunities offered by the environment through their phenotype. The phenotype encompasses any recognizable structural or functional trait (i.e. morphology, physiology, behaviour and life history), and not just wing morphology which was the characterstic of most concern to early evolutionary biologists. Life-history traits (Stearns 1992) describing the major features of an organism's life cycle (e.g. age and size at reproduction, fecundity, growth rate) do not fit well into the categories of morphology, physiology and behaviour and have aspects of all three (Nylin 2001). A phenotype is the result of a particular genotype interacting with a particular environment through developmental processes and variation in a butterfly's phenotype can be caused either by the genes, by the environment, or often by gene × environment interactions. When phenotypic variation for a trait is at least partly based on genetic variation, it has an evolutionary potential; it can be changed or maintained by the process of natural selection (but see also Chapter 10). Selection acts on the whole organism, and therefore mainly on the performance of an individual relative to that of its conspecifics in a population.

Butterflies – but also some moths – have played and still play a significant role in evolutionary biology and evolutionary genetics in particular (Beldade & Brakefield 2002, Watt & Boggs 2003). A pioneer in this field was E. B. Ford who studied the performance of genotypes with different wing patterns under natural conditions (Ford 1953, 1964). Butterfly wings have been attractive model systems to study how genotypes produce different phenotypes under different environmental conditions, i.e. phenotypic plasticity (Shapiro 1976, Brakefield et al. 1998, Nijhout 2003) and extensive knowledge about the development of butterfly

Ecology of Butterflies in Europe, eds. J. Settele, T. Shreeve, M. Konvička and H. Van Dyck. Published by Cambridge University Press. © Cambridge University Press 2009, pp. 159–170.

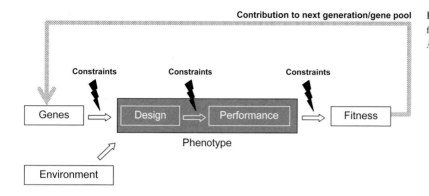

Figure 12.1. Scheme showing the major factors of the process of adaptation (after Arnold 1983b and Leimar *et al.* 2006).

wing patterns has contributed to their use in evolutionary biology (Nijhout 1991, Brakefield & French 1999). Image analysis systems have proved to be valuable and accurate tools to quantify series of butterfly wing traits including lengths, areas, shapes and different colour measures (Windig 1991, Chesmore & Monkman 1994). Butterflies have also featured as models for evo-devo work (e.g. Beldade & Brakefield 2002, Brakefield 2006, Joron *et al.* 2006) for studies on adaptation and plasticity in life-history traits (e.g. Nylin 1994, Nylin & Gotthard 1998) and behaviour (e.g. Merckx *et al.* 2003, Kemp & Wiklund 2004). There is a growing interest for integrative studies dealing with the evolutionary ecological understanding of morphology, physiology, behaviour and life-history traits in the same study system. A classic example in butterflies is the long-term work on *Colias* spp. in which gene variants have been identifed to probe adaptations from molecular to ecological levels (Watt 2003).

Natural selection generates adaptation, which means that a trait (or a particular value of a trait) enables an organism to survive and/or reproduce better in its environment (i.e. attain higher fitness) than without the trait, or when the trait has a different value. To evaluate whether a particular morphological, physiological, behavioural or life-history trait represent an adaptation, it is necessary to study different steps of the adaptation process, i.e. the selection gradient (Fig. 12.1): (1) How is genetic information within an environment translated to variation in phenotypic design? Design is a shorthand for any functional relevant measurable or countable aspect of the trait; (2) How is variation in design related to differences in performance for a particular function (i.e. performance gradient); and finally (3) How does variation in performance relate to variation in realized fitness (i.e. fitness gradient) (Arnold 1983b). At each step of the adaptation process, different factors may operate as constraints to prevent trait optimization (Arnold 1992b). Traits are called adaptive only when they are shown to have arisen due to natural selection for a particular function. From a design-oriented methodology it is often assumed, but rarely tested, that a trait has been at least partly shaped or modified by selection if it is beneficial for a function (Gotthard & Nylin 1995).

Besides natural selection, there is a second set of factors enhancing the probability of leaving offspring, i.e. sexual selection. This was clearly noticed by Darwin (1871) who described examples of sexual dimorphism in butterflies in which females preferred the more brilliant males. Sexual differences in morphology, beyond the gonads and genitalia, are indeed common in butterflies (Rutowski 1997). The importance and relevance of sexual selection is not limited to morphology, but also includes mate location behaviour, mating systems and aspects of life history (reviewed by Wiklund 2003). Sexual selection typically has an intrasexual and intersexual component (i.e. male–male competition and mate choice, respectively). Butterflies are in this context a fascinating group as the males of many species compete via pairwise intrasexual contests, yet lack any obvious morphological traits conventionally associated with animal aggression (Kemp & Wiklund 2001).

This chapter is inevitably a highly selective review on butterfly evolutionary ecology. Our main aim is stimulating students and researchers who are currently outside the field of evolutionary ecology to learn more about the study of the complex but intriguing interplay of genes and environment. This interplay is at the heart of integrative biology providing theory and empirical work to better understand (butterfly) life on Earth.

Variation: the fuel of evolutionary change

Individual variation can be seen as the fuel of evolutionary change by natural selection (Stearns 1989). Within a population, butterflies vary for several traits, although there are also fixed traits (e.g. number of legs). Traits may vary continuously (within a certain range they can have any possible value like wing length) or in a discrete way with a limited number of categories (e.g. colour morphs).

GENETIC VARIATION

Genetic variation encompasses variation between individuals caused by different genes. To study the effects of gene variants we need molecular techniques, or we need to compare individuals of known relatedness in order to know whether differences between individuals are genetic in origin. Butterflies reproduce sexually, so offspring inherit one half of their genes from each of their parents. Likewise, only half the genes of each parent are inherited by a particular offspring. All genetic analyses are based on this basic relationship, but things become more complex when many genes influence one or more traits.

Major genes: some genes with large effects
A major gene is one that has such a large effect on the phenotype that individuals with the gene can be distinguished from individuals without it. An example occurs in *Lycaena phlaeas* with a normal form (i.e. dominant wild-type with an orange border on the dorsal hindwings) and a recessive 'obsoleta' form (i.e. type with completely dark dorsal hindwings) (Brakefield & Shreeve 1992b). Another example of a one-gene trait is the female-limited wing colour polymorphism in *Colias* butterflies (Remington 1954): females with the dominant *A* allele have white wings ('alba' form), rather than male-like yellow or orange wings of homozygous-recessive *aa* females.

Discrete variation is not always related to major genes. It can be induced by an underlying, continuous factor (like hormone concentration) which is influenced by many genes with a small effect of each (Brakefield & French 1999). Discrete phenotypic traits may also result from environmental variation such as the phenomenon of seasonal polyphenism in which the expressed phenotype is determined by environmental conditions experienced during development. To unravel the origins of such variation, breeding experiments are needed. If a major gene is involved, frequencies of the offspring have to follow those as expected by Mendelian genetics.

Polygenic inheritance: many genes with small effects
Several traits, including life-history traits, do not occur discretely. There are two reasons for this. Firstly, environmental variation may blur the differences between discrete classes. Secondly, if many genes with small effects influence a trait, then the result may be partially overlapping phenotypic classes in which the effects of individual genes cannot be distinguished. The genetic analysis of such traits – i.e. quantitative genetics – uses the infinitesimal model, which implies that an infinitesimal number of genes influence traits (see Falconer & Mackay 1996, Lynch & Walsh 1998, and http://nitro.biosci.arizona.edu/zbook/book.html).

Heritability
In quantitative genetics the phenotypic variation (V_P) for a trait consists of three components which can be summarized as: $V_P = V_A + V_D + V_E$. V_E is the variation due to the environment. The genetic variation is split into a part that is directly inherited from the parents (i.e. additive component, V_A) and a part that is caused by the particular combination of genes in individuals (i.e. non-additive or dominance component, V_D). The only component of variation that is passed directly from parents to offspring is the additive variation and natural selection can only act on this component. The ratio of V_A to V_P determines the rate at which selection may change the average trait value in a population and is termed heritability (h^2). Given the heritability, the response to selection can be predicted. If S is the difference between the mean phenotypic value of selected parents and non-selected parents, and R is the difference between the offspring of the selected parents and the population mean before selection, then $R = h^2S$. Heritability provides a simple measure of how quickly a trait may evolve under selection.

There are several ways to determine heritability. In selection experiments the mean of the parental generation can be compared with the mean of offspring of selected parents. Such experiments need to be done under controlled conditions to ensure differences between means are not caused by changes of environmental conditions. Another frequently used method is to regress offspring values on parental values. For each parent (or the mean of the two) the average of its (or their) offspring is determined. The slope of offspring means on parental values is equal to half h^2 if only fathers or only mothers are used. If mid-parental values are used the slope equals the h^2. Other methods also exist, either with fixed groups of relatives (e.g. half sib/full sib analysis) or with mixtures of differently related individuals (e.g. REML analysis: Windig & Nylin 2002).

General patterns in butterfly studies on heritabilities agree well with those of other taxonomic groups. Morphological traits generally have higher heritabilities than life-history traits (including size at emergence). For eyespot size, high heritabilities have been found (e.g. 80% for *Bicyclus anynana* at high temperatures: Windig 1994; 90% for *Inachis io*: Windig 1998). Kingsolver & Wiernasz (1991), Van Dyck *et al.* (1998) and Ellers & Boggs (2002) report significant heritabilities for wing melanization, a trait associated with thermoregulation (see Chapter 6). Heritabilities for life-history traits are typically much lower. For example, heritabilities of adult size (e.g. wing size) are found of around 30% for *B. anynana* (Windig 1994), *Arashnia levana* and *I. io* (Windig 1998) and for *Pararge aegeria* (Van Dyck *et al.* 1998).

Molecular variation

Molecular techniques now offer the opportunity to examine variation in the genes themselves, rather than their effects. However, translating variation at the molecular level to variation at the phenotype requires considerable research effort. Firstly, a set of variable molecular markers has to be found. Once available, the relationship between the different markers has to be established, i.e. the different linkage groups or chromosomes have to be determined and recombination rates between the marker within the linkage groups have to be established. Once a linkage map is available it can be used to locate regions on the genome that are associated with variation in the phenotype. Often this is done by crossing lines diverging for a particular trait: if a certain allele of a certain marker is more often associated with a particular phenotypic value than expected by chance, a gene influencing that trait has to be at or close to the marker. When enough markers are available the gene can be located in a particular region; a quantitative trait locus (QTL). In butterflies the molecular genetic analysis of phenotypic variation is still in its infancy. One problem is that microsatellites, most frequently used in molecular studies in other organisms, tend to be scarce in butterflies (Nève & Meglécz 2000; Chapter 10).

Sexual variation

In butterflies, the heterogametic sex (XY) is the female. Many traits are known to be influenced by sex, including wing colour (e.g. sex-specific differences in the evolutionary genetics of dorsal melanization in *Colias*: Ellers & Boggs 2004b), eye dimorphism (e.g. sex-specific expression of visual pigments in the eyes of *Lycaena rubidus*: Sison-Mangus *et al.* 2006), flight morphology and flight performance (e.g. take-off performance in *Pararge aegeria*: Berwaerts *et al.* 2002, Berwaerts & Van Dyck 2004), life-history traits (e.g. sex-specific mortality rates in *P. aegeria*, *Inachis io* and *Polygonia c-album*: Gotthard 2000, Wiklund *et al.* 2003; sex-specific developmental time in *Lycaena tityrus*: Fischer & Fiedler 2000) and behaviour (e.g. sex-specific behaviour and polymorphic mimicry in *Heliconius numata*: Joron 2005), amongst others. X-linkage is common for quantitative traits in Lepidoptera and differences between closely related species tend to be concentrated on the sex chromosome (Sperling 1994, Janz 1998). Differences between reciprocal crosses of lines of *Bicyclus anynana* with large or small ventral eyespots are consistent with X-linkage for one or more loci (Wijngaarden & Brakefield 2000). To evaluate the evolutionary potential of sexual dimorphism in a trait, genetic correlations between the sexes need to be understood.

ENVIRONMENTAL INFLUENCES AND INTERACTIONS WITH GENES

Significance of environmental variation

Many ecological studies have demonstrated the functional significance of environmentally induced differences. Variation in environmental conditions during development can affect variation of the adult phenotype, even if genotypes do not differ among environmental conditions. For instance, variation in larval food quantity and quality strongly affects adult size (e.g. Fischer & Fiedler 2000a) whilst climatic factors are other sources of environmental effects (e.g. Descimon & Renon 1975). Some environmental influences have probably no evolutionary significance, but others have as they may influence the expression of the phenotype. Cold shocks (i.e. brief periods during which the pupae are exposed to low temperatures) cause aberrant wing-colour patterns in many butterfly species (Shapiro 1976). The higher degree of melanization of the wings in colder climates at higher altitudes and latitudes has a well-underpinned functional significance in thermoregulation in several pierids (Kingsolver & Wiernasz 1991, Ellers & Boggs 2003).

It is important that small changes in the environment (micro-environmental variation) do not disrupt the development and distort the phenotype completely. However, larger differences in the environment (macro-environmental variation) may require different phenotypes for functionality. Both aspects have received considerable attention in butterfly evolutionary biology within the fields of fluctuating asymmetry and phenotypic plasticity respectively.

Micro-environmental variation: fluctuating asymmetry

Minor changes in the environment may cause changes in a butterfly's phenotype. Asymmetry in otherwise symmetrical traits such as wing size has been hypothesized to reflect micro-environmental variation (i.e. fluctuating asymmetry: FA), and the ability of an organism to buffer its development against environmental distortions (i.e. developmental stability). This hypothesis has been heavily debated and we are generally lacking the understanding of the underlying mechanisms that would drive such relationship (see Van Dongen 2006 for a full review).

Studies of FA give heterogeneous and conflicting results. FA has been proposed as a monitoring tool for species under significant stress loads, as in threatened species in small declining populations (Cassel *et al.* 2001). As an example, Poulsen (1996) compared FA of wing length in pairs of *Coenonympha* species in Denmark. Each pair consisted of two closely related species, one declining and one stable. The FA turned out to be larger in the strongly declining *C. hero* than in the more stable *C. pamphilus*. Windig *et al.* (2001) compared FA of wing length and eyespot size in the same two species in Sweden, plus a third species, *C. arcania*. Surprisingly no difference in FA between *C. hero* and *C. pamphilus* was found, despite *C. hero* declining in Sweden while *C. pamphilus* is not. Even more surprising FA in *C. arcania* was clearly larger than in both other species, despite that it is not declining in Sweden. Such conflicting results seem typical for FA research at this stage.

Macro-environmental variation: phenotypic plasticity

Phenotypic plasticity is the ability of a genotype to produce alternative phenotypes according to the environment it experiences. Phenotypically, it may result in continuous or discrete variation (for a recent review on polyphenism in Lepidoptera, see Brakefield & Frankino 2009). Phenotypic plasticity may be an adaptation to spatially heterogeneous or temporally varying environments. In butterflies, seasonal polyphenism is best known (Shapiro 1976), but it also occurs in other contexts, for example in differences in flight morphology between populations in *Pararge aegeria* (Sibly *et al.* 1997) and between different types of landscape in the same species as revealed by reciprocal transplant experiments (Merckx & Van Dyck 2006). Among European butterflies, the most spectacular example of seasonal polyphenism is *Araschnia levana* with an orange spring and a black-and-white summer form. The environmental conditions causing genotypes to develop one of the forms have been known for a long time (Müller 1955) and the developmental mechanism

through hormonal control by the timing of ecdysteroid release has also been analysed (Koch 1992). However, the adaptive value of the two *Araschnia* forms still remains unresolved (see Chapter 13). Other examples of seasonal polyphenism include dry and wet season forms of tropical species like *Bicyclus* butterflies (Roskam & Brakefield 1999) and spring forms with higher degrees of wing melanization compared to summer forms in several Pieridae (Kingsolver & Wiernasz 1991).

Phenotypic plasticity does not exclude a trait from being under genetic control. Within a particular environment, the heritability of such a trait can be considerable and the response of a trait to the environment itself may be under genetic control. Two methods are central to the analysis of the genetics of phenotypic plasticity: reaction norms and genetic correlations within traits across environments. A reaction norm is the representation of the phenotype of a single genotype in different environments. Usually it is shown as a graph with an environmental variable on the *x*-axis and average phenotype on the *y*-axis (Fig. 12.2). Since clones are not available, butterfly researchers usually split families over different environments and use the means in different environments as approximations for genotypic values within each environment (e.g. Windig & Lammar 1999).

A population can be represented by a bundle of reaction norms. Such a bundle can be highly informative about different aspects of phenotypic plasticity. The form of the response and its variation can easily be observed from the bundle of reaction norms. The more the genotypes differ within an environment (i.e. the width of the bundle), the higher the genetic variance. Finally, if all reaction norms run parallel, there is no variation in the response of the different genotypes, and their ranking is the same in each environment. This means that selection in one environment will have a similar effect over the whole range of environments. If on the other hand reaction norms cross, there is variation in the response and the ranking of genotypes changes across environments. A genotype producing a relatively large phenotypic response in one environment may produce a relatively small one in another. Consequently selection in one environment may have an opposite effect in another environment. Genotype × environment interactions are essential to understand phenotypic variation.

Maternal and paternal effects

Butterflies do not only provide genes to their offspring. A fertilized egg also contains cytoplasm with mitochondria, nutrients and hormones. So, a butterfly's phenotype may

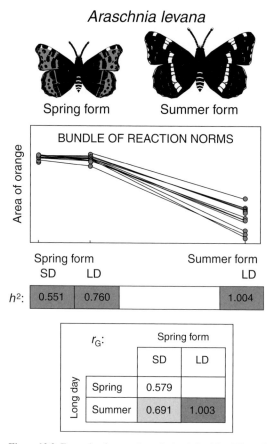

Araschnia levana

Spring form   Summer form

BUNDLE OF REACTION NORMS

Area of orange

Spring form   Summer form
SD   LD   LD

h^2:

| 0.551 | 0.760 | | 1.004 |

r_G:		Spring form	
		SD	LD
Long day	Spring	0.579	
	Summer	0.691	1.003

Figure 12.2. Example of a genetic analysis of plasticity. The trait measured is the relative amount of orange in wings of the males of the seasonally polyphenic map butterfly *Araschnia levana*. Full sib families were split between long day (LD, 16 h light) and short day (SD, 12 h light). Spring forms were produced both in LD and SD, summer forms only in LD. The bundle of reaction norms consists of lines connecting family means. Heritabilities are calculated with a REML analysis from a full-sib design (dark grey indicates significant difference from $0 < p < 0.01$). Note the correspondence of the width of the bundle and the value of the h^2. Values of r_G are genetic correlations across environments (dark grey: significant difference from $0 < p < 0.01$, light grey $p < 0.05$). Note that the low amount of crossing of reaction norms results in relatively high correlations (all correlations were not significantly different from 1 at the 5% level).

not only be influenced by its own genes and environment, but also by the genetic and environmental influences of its parents. These are maternal and paternal effects. In insects much of the maternal influence on offspring quality is thought to be mediated via egg size (Mousseau & Dingle

1991). In a number of butterflies egg size decreases with female age; hence later eggs have to manage with fewer nutrients. Experiments with *Pararge aegeria* (Wiklund & Persson 1983) and *Lasiommata megera* (Wiklund & Karlsson 1984) failed, however, to find any correlation between egg size and several fitness components like egg mortality, larval development time, pupal weight and starvation resilience.

Nutrients in eggs can be derived from three different sources: the reserves of the female that resulted from larval feeding, adult feeding or nuptial gifts from the male by a nutrient-rich spermatophore (see Chapter 14). Besides the effect of egg size, the cytoplasm of eggs may also vary qualitatively dependent on hormone level. Hormones can modify gene expression. However, in their study on the genetic basis of eyespot size in *Bicyclus anynana* with different selection lines and reciprocal crosses, Wijngaarden & Brakefield (2000) found no evidence for cytoplasmic effects.

Maternal effects through behavioural differences among mothers have often been neglected. However, the oviposition site choice largely determines the environment the early stages of the offspring will experience. According to Resetarits (1996) oviposition site choice must be under equally strong selection as egg size, egg number and the other 'hard' components of life histories to generate and optimize locally adapted or ecologically specialized life-history phenotypes. Variation in environmental conditions and resources, combined with variation in female responses to these may cause maternal effects on the phenotype of the offspring. Janz *et al.* (1994) show an example about the effect of host plant choice on offspring performance in *Polygonia c-album*.

Paternal effects may also play a role. During a copulation the male butterfly transfers a spermatophore to the female. Spermatophores contain both sperm and accessory secretions. Radioactive label experiments with *Polygonia c-album* showed that females use male nutrients from spermatophores for egg production, and for somatic maintenance (Wedell 1996). A positive correlation between male size and spermatophore size has been shown in butterflies with polyandry and hence sperm competition (e.g. Forsberg & Wiklund 1989). A male's mating status may also induce parternal effects. In the monandrous species *Pararge aegeria*, number of eggs laid and female lifespan were not affected by the mating status, but copulations with virgin males which provided large spermatophores resulted in higher number of offspring than copulations with non-virgin males providing small spermatophores (Lauwers & Van Dyck 2006). Such effects are typically more pronounced in polyandrous species (Torres-Vila *et al.* 2004). In the polyandrous species *Pieris napi*, Wedell (2006) recently showed that paternal nutrient investment in spermatophores is heritable and that females may derive both direct (increased lifetime fecundity and longevity) and indirect (sons with greater reproductive success) fitness benefits from paternal nutrient donations. At least in some butterfly species, male sodium gifts from puddling behaviour could be significant contributions during mating (Molleman *et al.* 2005).

Studies on adaptations between different environments typically need to control for environment-related maternal effects when comparing traits using laboratory-based breeding experiments. Therefore, comparisons are often based on the second generation of breeding to avoid such confounding effects. This was, for example, done by Karlsson & Van Dyck (2005) when comparing temperature-related life-history traits between populations of *Pararge aegeria* originating from different landscapes.

Natural selection: the motor of evolutionary change

Genetic variation provides the potential for evolutionary change, but natural selection is needed to realize the change. In other words, natural selection is the motor that sets evolution into motion (Stearns 1989). Determining the form and intensity of natural selection on phenotypic traits in natural populations requires that the values for fitness and phenotypic traits of a series of individuals are known (Endler 1986). Under field conditions, variation in survival is mainly used as a fitness measure as individual variation in realized fecundity is difficult to measure in butterflies. Capture–recapture methods have been used frequently to estimate survival for individuals in discrete phenotypic classes (Lebreton *et al.* 1992). Working with butterflies, Kingsolver & Smith (1995) developed a maximum-likelihood method for estimating selection coefficients for continuously varying (quantitative) phenotypic traits from capture–recapture data. Seasonal polyphenisms provide excellent material for the experimental study of how selection on particular traits varies within and between environments (Kingsolver 1995a, b, Brakefield 1996). To understand adaptation, understanding the ecological causes of natural selection is essential and experimental manipulations of phenotypes provide a powerful means to detect natural selection and causal relationships between traits and fitness in the wild.

Both polymorphism and polyphenism can enable an organism to perform optimally in two different environments.

But when can we expect polymorphism to evolve and when polyphenism? The most likely answer is that it depends whether there is a reliable cue present at the time of development to indicate the future environment. Daylength or photoperiod is a reliable indicator for seasonal polyphenism. In two satyrine butterflies, *Lasiommata maera* and *Lopinga achine*, larval development time is shortened with increasing daylength (signal of coming summer) or lengthened with decreasing daylength (coming winter) (Gotthard *et al.* 1999b). The extensive work on the alba polymorphism of *Colias* butterflies provides an interesting case on how a polymorphism may be maintained (Watt 2003). But recent theoretical evolutionary research advocates the conceptual integration of polymorphism and phenotypic plasticity (based on the idealization of development as a switching device with environmental and genetic cues as input) rather than the widely applied dichotomy (Leimar *et al.* 2006).

Constraints: grit in the motor of evolutionary change

The process of adaptation can be constrained because phenotypic traits may not be independent of each other, either because traits are coupled, or because there are trade-offs between them.

GENETIC CORRELATIONS

Lack of genetic variation prevents evolutionary change. Genetic correlations among traits are another type of genetic constraint as they prevent the independent change of different traits. It has also been demonstrated that selection on a single trait may cause another trait to change. Genes may influence more than one trait of a phenotype (pleiotropy) and genes influencing different traits may also be linked, although recombination tends to break down such relationships. There has been much discussion about evolutionary constraints but empirical data testing them directly are sparse (Beldade *et al.* 2002). Genetic correlations, either positive or negative, have important evolutionary consequences. Negative correlations are frequently proposed to explain why genetic variation persists for fitness-related traits despite directional selection. A negative correlation implies that if one trait is selected upward, the other trait automatically changes downward. Consequently the two traits cannot be maximized by selection; there is a trade-off. Such a negative correlation was, for example, found between larval development time and pupal coloration in

Inachis io: fast-growing larvae tended to have less melanized pupae than slow-growing larvae (Windig 1999). Many researchers have searched for negative genetic correlations between life-history traits, but often have not found them. The expectation is that if resources are used for one trait (e.g. larval growth) they cannot be allocated to another trait (e.g. fecundity). One reason for positive correlations between life-history traits may be that variation in acquisition can be more important than variation in allocation. If some individuals are better in obtaining resources than others they may invest more resources in all their traits compared to individuals that do not obtain many resources.

Genetic correlations may also play a role when dealing with sex-specific traits and selection. At the interspecific level, male butterflies differ in their flight morphology according to their mate-locating system (Wickman 1992b). Characters that have been selected for an extreme expression in males may be expressed to a lesser extent in females as well (Lande 1987). This was confirmed in the study of Wickman (1992b), but since male design still differed after removing the covariance with females, it suggested that selection has acted directly on males.

Although genetic correlations are of key significance to our understanding of phenotypic variation, traits can also be uncoupled by breaking genetic correlations. In a breeding experiment with *Colias*, Ellers & Boggs (2004b) found a positive phenotypic correlation between the degree of ventral and dorsal melanization in both males and females, but a positive genetic correlation was only found in males, but not in females. Hence, there is an uncoupling of both traits in females.

DEVELOPMENTAL CONSTRAINTS

The development of different traits (like investment in different body parts) may be in competition for energy or resources (e.g. Nijhout & Emlen 1998). The development of traits (or trait values) across successive life stages may also not be evolutionarily independent. More melanized final-instar caterpillars of *Araschnia levana* were found to grow more slowly in the early instars and *Inachis io* adults from melanized pupae were smaller and had less black, demonstrating trade-offs between melanization, development time and adult size (Windig 1999). Hence, the development of an adult phenotype may be constrained by developmental choices in response to selection earlier in its life. Several studies have exclusively focused on the adult stage, but it has became clear that evolutionary insect ecology has to take into account larval and pupal contributions to better understand adaptations and constraints (Reavey & Lawton 1991).

Although developmental constraints may play significant roles, they are not always the right explanation. Scaling relationships between appendages and body size show high interspecific variation but low intraspecific variation, which could result from natural selection for specific allometries or from developmental constraints on patterns of differential growth. Frankino *et al.* (2005) performed artificial selection on the allometry between forewing area and body size in *Bicyclus anynana* to test for developmental constraints. They showed that the short-term evolution of allometries is not limited by developmental constraints. Instead, scaling relationships are shaped by strong natural selection.

FUNCTIONAL CONSTRAINTS

A butterfly's phenotype can be considered as a compromise and not necessarily a multiple optimal design for several functions. Under particular environmental conditions, different functions like flying, foraging, roosting, mating, etc. may impose conflicting demands on the phenotype. Moreover, within a particular ecological relevant function like flight, there can be conflicts over optimization for different functions, such as flight speed versus flight endurance. Such a constraint can result from physiological or mechanical trade-offs.

A classic example of functional trade-offs is the oogenesis–flight syndrome, i.e. investment in reproduction versus flight ability (see Chapter 14). Gotthard (2000) showed evidence for the often assumed, but seldom tested trade-off between growth rate and predation risk. Fast-growing larvae of *Pararge aegeria* were more likely to be killed by a generalist predator than their slowly growing siblings; a four times higher relative growth rate was associated with a 30% higher daily predation risk. The proximate relationship behind this trade-off probably relates to the fact that larvae that can lower their feeding activities substantially decrease the risk of predation in a dangerous environment. Hence, growth rates of organisms are not necessarily maximized according to physiological capacities, but may be constrained by the costs associated with high growth rate. Thus, larval growth patterns are the result of 'strategic decisions' (Nylin & Gotthard 1998, Gotthard 2000).

We often lack detailed information on trade-offs among life-history traits and their consequences for evolution. Kemp & Rutowski (2004) showed, for example, a surprising effect of mating on female survivorship in the polyandrous pierid *Colias eurytheme*. Mated females had a reduced longevity compared to virgin females, possibly as a result of a toxic side effect of the male ejaculate.

ENVIRONMENTAL STRESS AND CONDITION-DEPENDENT CONSTRAINTS

Several constraints can be condition-dependent, such as those experienced under environmental stress. Talloen *et al.* (2004) showed that *Pararge aegeria* had less wing melanization when raised as larvae on drought-stressed host plants compared to siblings raised on high-quality host plants. In *Colias eurytheme*, developmental food quality stress affects nanostructural architecture of dorsal wing scales and, hence, UV reflectance which is used in mate choice (Kemp *et al.* 2006a).

Besides abiotic and food stress, there are also several sources of biotic stress relative to predation or social stress. Gibbs *et al.* (2005) demonstrated the influence of social factors on reproductive behaviour and oviposition site choice in *Pararge aegeria*. When males were present during egg-laying, females made a trade-off between egg size and number. Their presence also prevented females from selective oviposition on good-quality host plants. Males probably induce time-constrained conditions with females adopting a high-fecundity oviposition strategy to maximize host plant encounter rate. In the absence of males, females adopted a low-fecundity strategy by depositing each large egg at an optimal oviposition site, thus maximizing offspring survival (Gibbs *et al.* 2005).

Condition dependence of trait development and trade-offs between traits are important issues to evolutionary ecology and they may also confound the interpretation of laboratory experiments or the transferability of results among studies if not carefully controlled for.

CONSTRAINTS BY SEXUAL SELECTION

Sexual selection and natural selection may also be opposing selection forces. In *Papilio polyxenes* male–male intrasexual selection restricts the potential evolution of a more mimetic dorsal wing pattern in males. Lederhouse & Scriber (1996) released sibling males that were matched for size and age as pairs, one with an altered female-like mimetic wing pattern and one a marked but unaltered control. Experimental elimination of the male wing pattern had no effect on female choice, but increased the likelihood of holding a high-ranking territory. Hence, successful mate-location and predator escape by mimicry have conflicting interests in this species, with females being better Batesian mimics than males.

Experiments in which wing colour patterns are manipulated are interesting tools to evaluate sexual and natural selection (Brakefield 2003a). However, researchers need to be careful as artificially changed wing colour patterns may have differential effects outside the spectrum visible to the human observer (Berwaerts *et al.* 2001). Butterflies have proved to be excellent models to test for sexual selection as trait expression can be manipulated by changing developmental conditions or even by direct interference with trait development like eyespots on the wings (Breuker & Brakefield 2002, Brakefield 2006). Kemp *et al.* (2006a) took advantage of seasonal plasticity in flight morphology in *Pararge aegeria* (Van Dyck & Wiklund 2002) to test biophysical correlates of male contest success. Through different breeding conditions they reared individuals of markedly varying flight morphologies that were pitted against each other under semi-controlled conditions to test this morphology-associated contest-winning hypothesis.

THE STUDY OF ADAPTATION IN PRACTICE

Breeding butterflies

Studies on adaptation and plasticity require experimental work like controlled breeding. Amateur lepidopterologists have contributed to our knowledge (Friedrich 1986), but practical information often remains unpublished. Butterfly species vary considerably in their willingness to mate and oviposit under artificial conditions. Breeding designs of evolutionary studies are usually different from amateur breeding since the interest is not in short breeding bouts of several species, but rather in multiple generations of several families of a single species in reasonable numbers under controlled environmental conditions such as photoperiod, temperature, food quality or larval density (Bowden 1990) in contrast to daily fluctuating conditions (Brakefield & Mazzotta 1995). When families or individuals need to be reared separately, breeding programmes can become labour intensive. Experimental breeding programmes also face the risk of being destroyed by epidemics (e.g. *Bacillus thuringensis*).

Keeping butterflies for multiple generations may induce adaptations to this environment. Individuals of *Pieris brassicae* from a captive culture (kept for 100 generations) were heavier with smaller wings and lower wing aspect ratios than individuals in a recently established stock (Lewis & Thomas 2001). Females also laid many more eggs and had higher ovary mass at the time of peak egg production. Host plant preference in *Bicyclus anynana* is another example. Females preferred to lay eggs on the natural foodplant *Oplismenus* to maize which was used in the culture, but after several years in the laboratory their preference was reversed (Kooi 1996).

Reliable quantitative genetic estimates require large data sets. As a rule of thumb increasing the number of families is a better option than increasing the sample size per family. Butterflies typically lay sufficient numbers of eggs in order to split families over more than one environment (e.g. Windig & Lammar 1999). The minimum number of families needed in a quantitative genetic analysis is hard to tell as it depends on the heritability of the traits under study. Below 20 families, a quantitative genetic analysis becomes weak since only traits with a high heritability (>50%) will reach significance. Ten families may be sufficient at least with very large family sizes (e.g. >100 individuals: Windig 1998), but reliable estimation of genetic correlations may become problematic.

Measuring selection and fitness components

FITNESS COMPONENTS

Fitness is the essential, technical evolutionary biological term for the relative contribution that an individual makes to the gene pool of the next generation. Most studies measure particular fitness components like survival and reproductive success parameters (e.g. mating frequency, number of eggs). The use of these components is, however, often based on assumptions. Laying more eggs is, for example, assumed to result in a larger contribution to the next generation. This can be the case, but under particular conditions laying fewer eggs of higher quality or at higher-quality oviposition sites may have an equal or even larger pay-off than the opposite strategy.

In insects (including butterflies) female size is often used as a correlate of fitness (Honek 1993). However, laboratory experiments on egg output among individuals may not be relevant to variation of realized fitness under variable field conditions. Because butterflies are rather short-lived helio-therms, the time a female has available to select host plants and to oviposit is recognized to be of major significance for variation in fitness (Doak et al. 2006)

Field data on male butterfly fitness are very scanty. In several species, individuals are rarely observed copulating, with courtship behaviour and mating refusal behaviour being more commonly observed. Since females receive a spermatophore, the number of matings can be counted from the number of spermatophores in the bursa copulatrix after dissecting the female's abdomen (Burns 1968). Cordero (2000b) observed 21 out of 159 Callophrys xami males

copulating. It corresponded to a rate of 0.0028 copulations per male per h. But since three males were observed copulating more than once (one even four times), it confirms high variance in reproductive success among males.

Research on variation in fitness would benefit from applying genetic data to assign maternity and paternity within experimental and natural populations. Moreover, new techniques allow such analyses on the basis of minor non-lethal wing tissue samples (e.g. Lushai et al. 2000). Studies can also be done under semi-natural conditions with populations in large outdoor cages (Norberg et al. 2002; Hanski et al. 2006). Joron & Brakefield (2003) used a technique involving the transfer of marker dusts during copulation to track mating success in cage experiments. They actually showed that a small decrease in mating success of captive inbred male butterflies was greatly accentuated in conditions with unconstrained flight. So it suggests that patterns of mating can be profoundly influenced by a history of inbreeding or by any restraining experimental conditions.

MEASURING SELECTION

It is an intriguing but complex challenge to identify the targets of selection in the wild and measure it accurately. Strong directional or stabilizing selection will reduce the variance in a trait, so estimates of selection in progress on traits that have been shaped by selection tend to have lower statistical power since reduced phenotypic variance decreases the likelihood of detecting a relationship where one exists (Anholt 1991). Experimental manipulation of phenotypes under natural conditions provides the most direct manner to do so (Kingsolver & Schemske 1991). Since natural selection acts on phenotypic variation and is blind to the underlying genotype, the results of selection on experimentally produced phenotypic variation can be used to understand selection on potential variation (Anholt 1991).

Kingsolver (1995a, b) evaluated differences in fitness consequences and selection among seasonal forms of Pontia occidentalis. Kingsolver (1996) used manipulative experiments to measure selective differences. He altered the degree of melanization along the veins of the ventral hindwings by either blackening or yellowing particular parts, with the yellow treatment as a control for the addition of the pigment with permanent markers. As predicted from thermal associations, it was found that blackened individuals had consistently lower survival probabilities than yellowed individuals under warm, midsummer field conditions.

However, this turned out to be the case in males only. A-posteriori explanation of this sexual difference may be that the manipulation may have caused a greater change in melanization in males than in females since females are more melanized than males in the natural summer form (Kingsolver 1996). Other manipulative experiments to evaluate fitness consequences under field conditions using variation in survival probabilities include experimental reductions in wing area (Kingsolver 1999), and experimentally gluing weights on the thorax to alter wing loadings (Kingsolver & Srygley 2000, Srygley & Kingsolver 2000). Selection in each of these studies was estimated based on data gathered by multiple-release–capture methods to evaluate how survival or recapture probabilities vary with treatment. This was done by an extension of the Cormack–Jolly–Seber model, which allows estimation of selection coefficients on quantitative traits (Kingsolver & Smith 1995). The key to the model is defining a fitness (survival) function, which was next incorporated into a standard mark–release–recapture model.

Comparative studies: taking phylogeny into account

Studies comparing adaptations among species need to take into account phylogenetic relationships. Interspecific correlations between phenotypic traits and particular conditions can either be the result of independent condition-related selection events or of common ancestry. There are different approaches to control for common ancestry. One approach is the use of phylogenetic independent contrasts. Associations of traits are compared between pairs of related species rather than comparing groups directly. Wickman (1992b) used this approach to analyse relationships between mate location behaviour and flight morphology. From the 44 butterfly species, 25 were assigned to eight contrasts. After controlling for phylogeny, males of perching species had larger relative thorax mass, higher wing loading and higher aspect ratios than patrolling species (Wickman 1992b). Using largely the same selection of species and contrasts, Rutowski (2000b) could not find a relationship between mate-location behaviour and compound eye size, after controlling for phylogeny. Another, more recently developed approach to take into account phylogeny, is the use of phylogenetic analysis of covariance by empirically scaled computer simulation models (Garland et al. 1993). This technique requires a phylogenetic tree for the species of interest.

CONCLUSIONS AND PERSPECTIVES

Evolutionary ecological research on adaptation and plasticity in butterflies is a fascinating field where there is much to learn on how organisms change through the process of evolution. Narrow studies with a trait-by-trait, function-by-function and life-stage-by-life-stage approach can be interesting, but multidisciplinary work dealing with multiple traits, multifunctionality and different life stages is the real challenge to better understanding of the process of adaptation. Butterfly evolutionary ecological research would greatly benefit from increasing efforts to further integrate or combine different disciplines including quantitative genetics and molecular genetics, functional morphology, ecophysiology, life-history research, behavioural ecology and phylogeny.

Several evolutionary ecological studies have been criticized for their adaptive storytelling. Although the heritable basis of traits is sometimes tested, the relationships between design and performance are indeed frequently assumed, but seldom tested. Experimental tests of such relationships will provide better insight in the associated trade-offs and their mechanisms as they are key aspects of our understanding of the adaptation process. Variation in butterfly morphology is, for example, often directly interpreted in terms of adaptive differences relative to flight such as for dispersal (Van Dyck & Matthysen 1999). Moreover, physiological variation is only rarely taken into account in butterfly evolutionary studies (but see Kohane & Watt 1999, Hall & Willmott 2000). Evolutionary research would more generally benefit from a renaissance and integration of physiological research (Feder et al. 2000).

There are also fascinating perspectives being derived from the first results of finding linkages between traits at the genomic location, like, for example, linkage of butterfly mate preference and wing colour preference cues at the location of gene *wingless* (Kronforst et al. 2006).

There is still little comparative work on butterfly adaptations within an explicit phylogenetic context focusing on several species and genera. This requires detailed knowledge of phylogenetic relationships. It can be expected with increased applications of molecular techniques (e.g. using mitochondrial DNA) that more, reliable phylogenetic trees will become available. However, molecular data may not be the only requirement since collaborative work by several butterfly phylogeneticists showed synergistic effects of combining morphological and molecular data in resolving phylogenetic relationships in butterflies and skippers (Wahlberg et al. 2005).

Recent studies have demonstrated sufficiently fast rates of evolution of ecologically important phenotypes that have the potential to impact the outcome of ecological interactions while they are under way (Hairston *et al.* 2005). These analyses support the view that in order to understand temporal dynamics in ecological processes it is critical to consider rapid evolutionary changes. So, natural selection may affect population dynamics for which evidence is recently accumulating in butterflies (Haag *et al.* 2005, Hanski & Saccheri 2006, Saccheri & Hanski 2006). This issue has wide consequences for ecology and for conservation.

There is a growing and exciting interface between evolutionary ecology and conservation biology. Butterfly resource distributions change rapidly in human-dominated landscapes across Europe. Habitat fragmentation, modern agriculture and urbanization do not only alter the ecological settings for butterflies, but also the evolutionary settings by altered selection regimes (Van Dyck & Matthysen 1999,

Merckx *et al.* 2006). Similar reasoning applies for the ecological and evolutionary impact of climate change. Kohane & Watt (1999) pointed to implications for evolutionary physiological responses to environmental warming: if anthropogenic causes change thermal environments more quickly than butterflies can evolve in response, massive population decreases or even extinctions may occur (see Chapter 10). So, besides the significance of improving our insight of the evolutionary process for the sake of our fundamental understanding of life on Earth, evolutionary ecology will also contribute to illuminate conservation problems and hopefully answers as well.

ACKNOWLEDGEMENTS

This research was supported by a grant from UCL (FSR06 'Behavioural Ecology of Dispersal') and from FNRS (FRFC 2.4595.07 'Genetic Analysis of Dispersal') to HVD.

13 • Functional significance of butterfly wing morphology variation

TIM SHREEVE, MARTIN KONVIČKA AND HANS VAN DYCK

SUMMARY

The relatively large and colourful wings are by far the most apparent feature of adult butterflies. Butterfly wing morphology varies between species, but often also within species. This variation extends to colour, pattern, size, shape, thickness, distribution of mass and venation pattern, and has attracted much attention from evolutionary biologists. Butterfly wings have a range of interacting functions: flight, thermoregulation, intraspecific signalling and communication with predators. Wing size and shape, relative to body mass and thoracic muscle mass, are important determinants of flight dynamics. Changes in these traits may indicate changes in flight power, and probably the ability to sustain flight. This has particularly been studied in a context of habitat fragmentation. Manoeuvrability and the efficiency of flight is, however, also dependent on the way butterflies move their wings. The analysis of in-flight biomechanics warrants more detailed attention. Detailed studies of butterfly body temperatures indicate that for many species adult activity becomes optimal within the range 28–38 °C. However, some butterflies fly with lower body temperatures and differences may occur within species. Warming by the adoption of species-specific basking postures to absorb solar radiation is affected by pigment and scale properties of the wings. There can be a complicated interplay between morphological variation, geographical variation of temperature and weather-dependent microhabitat use, affecting activity. Butterfly wings have communicative functions to predators, but only a few studies have attempted to quantify background matching of butterfly wings. Once detected, the presence of startling devices (e.g. wing tails) alter the chance of consumption. The effectiveness of marginal eyespots on the wings warrants, however, further experimental testing. The quantitative nature of UV coloration extends to many species and to date, predation studies have ignored this component. There is also a UV component to butterfly vision. Hence, it is essential to consider this if we want to understand intraspecific communication and mating behaviour. The functioning of all the elements of wing morphology depends on habitat structure. Many species show geographical or altitudinal variation in wing morphology, and seasonal variation within species with more than one generation a year. With increasing information on wing morphology and flight on individual species it is becoming evident that generalities are becoming harder to make. There is an urgency to understand fully this aspect of butterfly ecology to help understand how increasingly vulnerable species function and persist.

INTRODUCTION

Butterfly wing morphology (colour, pattern, size, shape, thickness, distribution of mass and venation pattern) varies between species and most aspects also vary within species. This variation has a strong phylogenetic component and is a major tool for species identification. Of relevance to the ecology of butterflies are the interactions of species with their physical and biological environments mediated by wing morphology. Understanding these relationships may help to uncover evolutionary events that have given rise to differentiation and speciation, and also identify any current constraints that wing morphology imposes on species' biogeographical patterns, their spatial and temporal dynamics and community interactions.

Several aspects of wing morphology have been the subjects of intensive study in the past and continue to receive attention, often with a focus on particular geographical areas or taxonomic groups. Such studies have made important contributions to evolutionary theory, to the integration of physiology into studies of activity and ultimately population dynamic process, and to understanding the genetics of morphological expression. For example, concepts concerning the evolution and functional aspects of mimicry have largely been developed using Palaetropical butterflies (Turner 1987, Turner & Mallet 1996, Joron & Mallet 1998). Studies of pierid butterflies (*Colias* and *Pieris* species) in the Nearctic (e.g. Watt 1968, Roland 1982, Kingsolver 1985b) have established relationships between butterfly thermoregulation,

Ecology of Butterflies in Europe, eds. J. Settele, T. Shreeve, M. Konvička and H. Van Dyck. Published by Cambridge University Press. © Cambridge University Press 2009, pp. 171–188.

wing melanisation and adult activity, and potential links to predation (Kingsolver 1987b). Developmental studies (e.g. Nijhout 1991) have revealed the broad patterns of wing development, and genetic studies have demonstrated the underlying control of specific wing pattern elements (Koch *et al.* 2000, Beldade & Brakefield 2002, Brakefield 2003b, Marcus 2005). Other works have revealed how wing pattern and coloration function as defence mechanisms against visual hunting predators (Hill & Vaca 2004, Stevens 2005, Wiklund 2005, Vallin *et al.* 2006) and how wing shape, wing loading, flexion and wing stroke influence flight dynamics (Betts & Wootton 1988, Chai & Srygley 1990, Wootton 1993, Srygley 2004, Berwaerts *et al.* 2006).

Butterfly wings have a range of interacting functions including flight, thermoregulation, intraspecific signalling and communication with predators (Shreeve 1992b, Dennis & Shreeve 1989), the last involving both primary (avoidance of detection) and secondary defence (escape/evasion once detected) mechanisms. Each of these functions may be subject to different selection pressures. For any population, these individual functions may not be optimised and the expression of any particular wing morphology is best viewed as a possible compromise between different functions. Detailed examination of these different functions in the context of where species occur may facilitate an understanding of the importance of wing morphology to the ecology of butterflies, in particular to the success of particular morphologies in the contexts of environmental variation and population density.

WING MORPHOLOGY AND BUTTERFLY FLIGHT

Butterfly wings have an obvious function in flight, and wing size and shape, relative to body mass and thoracic muscle mass, are important determinants of butterfly in-flight dynamics, including manoeuvrability and flight persistence. However, wing area alone does not determine overall flight capacity. Some species with large wing areas may not engage in long-distance dispersal (e.g. *Apatura iris*), whilst some with intermediate-sized wings (e.g. *Colias croceus, Vanessa atalanta*) are regular migrants, and those with small wings may sometimes engage in long-distance dispersal (e.g. *Everes argiades*). Functionally, species potential mobility is related to the relationships of wing area to body mass, flight muscle mass and wing architecture. However, the capacity for flight is most likely to be a selected trait related to past resource distributions and the persistence of resources in

the landscape. Species that have experienced predictably widespread but ephemeral resources are likely to be more mobile than those whose resource sets are predictably spatially scarce, but persistent (Southwood 1962). There is thus an evolutionary component to species mobility patterns. However, changes of resource distributions throughout the Palaearctic during the Holocene, particularly within the last century, may mean that species mobility patterns are not optimised to current resource distribution patterns.

Changes of wing area to thoracic mass and total body mass ratios (wing loading) within species indicate that changes in flight power, and probably the ability to sustain flight, can occur. These may be in response to changes of resource distribution and abundance or when climate change shifts the trade-off between the benefits of dispersal flight (ex-resource patch) and its costs in survivorship and realised fecundity. For example, Thomas *et al.* (1988) identified differences of traits associated with potential flight performance (wing area, thorax and abdomen mass) in *Plebejus argus* in response to heathland and limestone fragmentation, suggesting that evolutionary changes in morphology occur in response to altered costs and benefits of migration. However, Hill *et al.* (1999b) found no differences in investment in flight (measured as wing area to thorax mass) by *Hesperia comma* in newly established versus refuge sites within metapopulations. Hanski *et al.* (2004) found physiological and biochemical differences associated with flight performance in association with patch distribution, connectivity and colony age in *Melitaea cinxia* and Breuker *et al.* (2007) found sex-specific differences in wing morphology and between-female differences of wing shape in this butterfly that could be related to dispersal characteristics. Thus, different species, and perhaps populations, may respond uniquely in characteristics that influence flight performance in relation to changing trade-offs associated with dispersal.

Wing loading has a role in flight performance. Berwaerts *et al.* (2002) found differences of wing loading in *Pararge aegeria* in relation to mate-location mechanisms. Territorial individuals were characterised by having a relatively low wing area to mass ratio, whilst individuals that were less able to defend territories and relied more on patrolling as a mate-locating mechanism had a higher wing area to thoracic mass ratio. These differences were explained (Berwaerts *et al.* 2002) by a need for rapid acceleration from a perch to intercept intruders or potential mates in territorial individuals and a higher demand for sustained flight in patrolling individuals. Whilst area to mass ratios have not been extensively investigated it is possible to identify a potential phylogenetic component to mate-location mechanisms on the basis of such ratios (Wickman 1992b). For example the Palaearctic Hesperiidae, characterised by relatively small wings and large thoracic mass employ perching (associated with territoriality) as their primary mate-locating mechanism (Dennis & Shreeve 1988). In comparison, most Pieridae (especially Coliadinae and Dismorphinae) have a large wing area to thoracic mass ratio and are patrollers (Dennis & Shreeve 1988), whilst the Lyceanidae, with intermediate area to mass ratios, tend to include some element of perching in their mate-locating repertoire. The most variable groups of butterflies are amongst the Nymphalidae and Satyrinae. As a group they have both variable area to mass ratios and mate-locating tactics, the latter even within individual species. Manoeuvrability and the efficiency of flight is dependent on the way butterflies move their wings, including whether the fore- and hindwings are coupled, wing torsion, and the generation of lift from vortices and flapping (Betts & Wootton 1988, Wootton 1993).

Flight is also affected by wing shape: in a study of Neotropical butterflies Chai & Srygley (1990) and Srygley & Chai (1990a) found a relationship between flight behaviour, palatability and wing shape. Unpalatable species had slower flights and rounder wing shapes than palatable species. This was related to in-flight manoeuvrability: speed and angular wings may help to evade predators, whilst a slow gliding flight may aid predator recognition of distastefulness. Studies of the palatability of Palaearctic butterflies have received little attention, making it difficult to equate wing shape and wing beat characteristics to protection from predators. However there are several studies of variation within species that demonstrate that variation of shape between populations may be of functional significance.

Polyommatus icarus is a widespread butterfly and within the British Isles there is a tendency for northern (univoltine) populations to occur at high density and in the south of the British Isles (bivoltine) for the first generation to occur at higher density than the second. There are also differences in wing shape between locations, between sexes within locations, and between generations in bivoltine populations (Howe 2004). Both sexes have more angular wings in the south than the north and females have more angular wings than males in the south, but not in the north. These differences can be equated with the demands of flight performance in relation to weather. Individuals with rounder wings are associated with slower, more prolonged flight while those with angular wings are associated with rapid acceleration. Cool conditions in the north are the least favourable for

flight due to increased exposure and cooler temperatures and may preclude rapid acceleration and manoeuvrability, features associated with wing angularity (Chai & Srygley 1990), and requiring relatively warm body temperatures which are not easily achieved in the north (Howe 2004). Differences between the sexes may also relate to the attractiveness of females. In this butterfly, angularity is also correlated with wing area (Howe 2004) and large females are more attractive to males than small females (Burghardt *et al.* 1997). In the north, population density is predictably higher than in the south. It can be argued that for females in the north the prime requirement for flight under poor weather conditions (small round wings) is more important than attractiveness to males, but in the south, with fewer thermal constraints on flight activity, the prime requirement is for female attractiveness at lower population density.

Differences in wing shape between sexes and with geographical location have also been identified for *Pieris napi* in a comparison of wing morphology in northern and southern parts of the British Isles and southern France (Wilcockson 2002, Wilcockson & Shreeve 2003). In both sexes, individuals from southern France had the most angular wings, whilst individuals from the north of the British Isles had, on average, the roundest wings. However, the variability of wing shape was the greatest in the north. This pattern may reflect overall differences in thermal regimes between the locations and constraints on flight imposed by body temperature, but the large variance for the north of England, it is argued, reflects the greater variability in weather patterns experienced by northern populations, precluding less directional selection on wing morphology.

That wing shape varies within species is also demonstrated by studies of *Pararge aegeria* in Belgium (Merckx & Van Dyck 2006). Populations of *P. aegeria* in heavily wooded landscapes comprised a higher frequency of individuals with an angular wing shape compared to individuals from more open landscapes. In wooded landscapes woodland resource patches are more abundant than in open landscapes and population density predictably higher. The wing morphology corresponds with that best fitted to the demands of short accelerating flights in wooded areas, whilst the morphology in open landscapes is that best suited for sustained flight between patchily distributed resources and low population density.

Whilst it is possible to associate both variation of wing shape and wing size to thoracic mass ratios to the requirements for mobility and flight performance, key elements in analysis are missing. These are the sustainability of flight, manoeuvrability and the appearance of flying individuals to conspecifics and potential predators. Appearance is partly determined by wing beat frequency and wing beat amplitude. Much remains to be learned with the development of appropriate field studies. For example, analysis of the wing beat frequencies and wing amplitude of flying male *Satyrus actaea* from video footage taken in the field (T. G. Shreeve, unpublished) reveals a potential link between behaviour and flight mode. Males engaged in patrolling mate-location flights hold the hindwings virtually horizontal whilst the forewing is used to power a relatively slow flight over open vegetation where females usually occur. In these flights each forewing goes through a 90° stroke, but the wing beat frequency is relatively slow. When males engage in foraging flight, moving between nectar sources, both the fore- and hindwings are coupled, with an amplitude of between 90° and 140° with a higher wing beat frequency than when mate-searching. Similar analysis of video footage of flight by males of *Melanargia galathea* and *M. russiae* also reveals differences of wing beat frequency and amplitude in relation to activity, and differences between two closely related species. Males of both species patrol in search of females, but *M. galathea* does this with a higher wing beat frequency than *M. russiae*. When engaged in patrolling flights, the stroke amplitude of *M. russiae* (up to 90°) is also less than that of *M. galathea* (up to 145°). When either species engages in movements between nectar patches they do this with gliding flights interspersed with periods of high amplitude (c. 170°), rapid wing beat flight.

Whilst the analysis of in-flight wing beat and posture is still in its infancy and there is relatively little data, this aspect of flight warrants more detailed attention. In the case of *Satyrus actaea*, *Melanargia galathea* and *M. russiae*, differences of wing stroke pattern may have implications for visual apparency. The early work of Chai & Srygley (1990) and Srygley & Chai (1990a, 1990b) indicated links between palatability, appearance and flight characteristics, mediated by wing beat amplitude and frequency (gliding versus rapid wing beats). In these three Palaearctic species it is evident that flight characteristics change in relation to behaviour, with possible variation of apparency to conspecifics and potential predators.

WING MORPHOLOGY AND THERMOREGULATION

As well as the obvious function of flight, wings are important in thermoregulation. Enzyme systems operate most efficiently within narrow temperature ranges (Watt 2003) and most detailed studies of butterfly body temperatures indicate that activity becomes optimal within the range 28–38 °C (Shreeve 1992b).

Within the Palaearctic, most butterflies rely on microhabitat selection and solar gain to elevate body temperatures to values that are usually greater than ambient air temperature. In most species warming by the absorption of solar radiation is achieved by adopting species-specific basking postures, usually accompanied by orientation to maximise the rate of warming from solar radiation (e.g. Shreeve 1992b). Three basking methods are usually described, and the thermoregulatory function of specific wing areas differs according to basking method. Some species bask with closed wings (*lateral baskers*). For these species, the most important wing areas for solar absorption are the hind- and forewing underside areas that overlie the abdomen and thorax when the wings are closed. Adjustments to the rate of solar gain are made by changing the angle of the wings in relation to solar radiation, effectively changing the angle of tilt. In *dorsal absorbance baskers*, the wings are held open and absorbed solar radiation is conducted to the body from the wings as well as being received directly by the body itself. Widely open wings maximise radiation gain, more closed wings reduce the warming rate. The areas of the wings most important for thermoregulation are the upper surface of fore- and hindwings closest to the body; other wing areas can be considered as thermally neutral (Wasserthal 1975). In some pierids which adopt open wing basking the mechanism of gaining thermal energy is not by absorption, but by reflection onto the body, which is a function of pigment and scale properties (Kingsolver 1985b). In these *reflectance baskers* the rate of warming is maximal when the wings are held at 45°, because the maximal reflective wing area is exposed and the body and basal wing area also gain solar radiation by absorption. In reflectance baskers the bulk of the upper wing surfaces have a thermoregulatory function.

Detailed studies of thermoregulatory behaviour of butterflies have revealed that variation in pigmentation in those wing areas of significance to thermoregulation influences warming rates (e.g. Clench 1966, Watt 1968, Kingsolver 1985b, Van Dyck & Matthysen 1998, Berwaerts *et al.* 2001). As a consequence, in some species there is geographical and altitudinal variation in pigmentation that can be correlated with climate and weather variables of significance to thermoregulation (Dennis 1993). However, not all species are dependent on solar radiation gain when at rest to elevate body temperature above ambient to near optimal. Alpine

Parnassius phoebus are active with relatively low body temperatures (17–18 °C) (Guppy 1986) and can fly when air temperatures are as low as 8–9 °C, providing there is a high radiation load (>c. 800 W m^{-2}). Flight at low air temperatures may be facilitated by gliding flight maintaining a thicker boundary layer around the body than would be maintained during slow flapping flight. Other species may employ alternative mechanisms to elevate body temperatures in marginal conditions. Observations of shivering have been made in a number of species and Maier & Shreeve (1996) found that *Vanessa atalanta*, *Inachis io* and *Vanessa cardui* could achieve thoracic temperature excesses of 13 °C, 18 °C and 16.5 °C, with rates of between 3 °C min^{-1} and 4.3 °C min^{-1} in conditions of very low radiation loads (<5 W m^{-2}). This is a potentially energy-expensive mechanism, which may be employed in marginal conditions experienced during migration or when moving from dark overwintering sites. Alternatively, species may engage in activity with body temperatures below optimal. Merckx *et al.* (2006) found that *P. aegeria* could fly with body temperatures in the range 10–21 °C, substantially below the optimal of 32–34.5 °C identified by Shreeve (1984).

Maintaining a body temperature suitable for activity involves more than temperature elevation. Body mass and insulating hairs affect thermal stability and warming and cooling rates (Wasserthal 1975, Kingsolver & Moffat 1983, Heinrich 1986b, Shelly & Ludwig 1985). Individuals with a large body may take longer to warm than small individuals, but may be more thermally stable. Konvička *et al.* (2002a) compared activity at the interspecific level between two related *Erebia* satyrids in a subalpine environment. Smaller-sized *E. epiphron* began activity in lower temperatures (consistant with rapid warm-up), but ceased their flight activity earlier in the evening and more quickly after weather worsened, compared to larger-sized *E. euryale* (consistent with a longer heat retention in the latter species). Well-insulated individuals are also more thermally stable than less-insulated individuals. If wing beats are not vigorous individuals may cool in flight (Shreeve 1984, Merckx 2005) and the requirements to maintain a body temperature that facilitates activity may have profound effects on flight duration and microhabitat use. There may also be interactions between morphology and behaviour (Van Dyck *et al.* 1997b, Van Dyck & Matthysen 1998).

Most studies of thermoregulation in butterflies have been made on species of moderate to large size, with an assumption, made on the basis of early work, that the norm is for individual species of butterfly to operate within narrow body temperature ranges and that direct solar radiation is the main source for thermal gain. It is worth asking whether this is the general case. Because size affects thermal stability, there has, in theory, to be a threshold body mass at which any thermal gain from basking is at equilibrium to loss via convection and re-radiation. This threshold will vary with the air temperature around the body, the degree of insulation and the effectiveness of wings as solar radiation collectors.

Comparative studies of the thermoregulation and inputs into body temperature elevation of *Polyommatus icarus* in a predictable cool location (Berneray, Outer Hebrides) and a warm location (Chiltern Hills, central southern England), indicate that there can be geographical differences in body temperature at which individuals are active. In this small butterfly, direct solar radiation inputs to the body need not be the primary source of thermal energy (Howe 2004, Howe *et al.* 2007). Direct measurements of thoracic temperatures at voluntary flight initiation indicate that the mean and minimum thoracic temperatures of males and females at take-off from Berneray are lower than in the Chilterns (Table 13.1). Flight durations are correspondingly shorter. There are no differences in the warming rates of individuals from the two locations, but individuals from the Chilterns population do not fly under the radiation loads and lower temperatures that are experienced by the Berneray population. These findings indicate either that there are differences in optimal body temperatures between individuals from the different populations, or that the body temperature at which individuals fly is a learned response. If the latter is the case then individuals from the warmer location may not fly at what could be suboptimal body temperatures because they frequently experience conditions under which they could achieve high body temperatures and will therefore 'sit and wait' until they achieve a high body temperature. On Berneray, with a much cooler and cloudier climate, flights have to be undertaken when body temperatures are potentially suboptimal; the infrequency of warm conditions with bright sunshine means that individuals rarely achieve optimal body temperatures and respond accordingly. Whether a learned response or selection is involved warrants further work.

Individuals of the two populations settle in similar microhabitats between flights (Howe 2004). These are small areas of bare ground sheltered from wind by vegetation up to 15 cm high. In a study of the response of thoracic temperature to settling site substrate temperature, air temperature adjacent to the body of a settled individual and solar radiation, thoracic temperature responded more to accumulated

Table 13.1 *Mean (±s.e), minimum and maximum thoracic temperatures and mean temperature excess (±s.e.) at voluntary flight initiation of* Polyommatus icarus *in the field at Berneray (Outer Hebrides) and Grangelands (Chiltern Hills) (from Howe 2004)*

Location and generation		N	Thoracic temperature (°C)			Mean temperature excess[a]
			Mean	Min.	Max.	
Berneray (univoltine)	♂	44	25.6 ± 0.6	18.3	33.2	11.7 ± 0.6
	♀	20	24.9 ± 0.8	18.1	30.2	10.8 ± 0.8
Grangelands	♂	26	30.1 ± 0.5	25.2	36.1	12.9 ± 0.9
Generation 1	♀	18	29.4 ± 0.8	24.1	35.4	11.9 ± 0.9
Generation 2	♂	37	32.3 ± 0.5	22.9	36.4	12.3 ± 0.6
	♀	20	30.3 ± 0.8	24.1	35.1	10.2 ± 0.7

[a] Temperature excess is above basking site temperature. Flight initiation temperatures differed between regions and generations in males ($F_{(2,104)} = 44.9$, $P < 0.001$) and females ($F_{(2,55)} = 13.3$, $P < 0.001$). Thoracic temperature excess did not differ between regions or generations for males ($F_{(2,104)} = 0.03$, $P = 0.97$) or females ($F_{(2,55)} = 1.2$, $P = 0.32$).

radiation load over 10-minute periods than to direct solar radiation over periods of 5–10 seconds (Howe 2004, Howe *et al.* 2007). Early work by Kingsolver & Watt (1983) on *Colias* butterflies established that when butterflies respond to direct solar radiation inputs, thoracic temperature responds within 5–10 seconds of changes in radiation load. In *Polyommatus icarus* there is no such short response time, indicating that in this small-bodied butterfly, body temperature is raised to a flight threshold by equilibrating with microhabitat temperature rather than by direct radiation input. Further evidence that direct radiation is not important for this small butterfly is provided by similar relationships between air temperature, ground temperature and solar radiation and flight and settling durations and warming rates for males and females with different wing coloration in both the Berneray and Chiltern Hills populations (Howe 2004).

The response of larger species to regional differences in climate may differ to that of *Polyommatus icarus*. For example, studies of *Pieris napi* in the same geographical regions as *P. icarus* (Lewis, Outer Hebrides, Oxfordshire, central-southern England) (Wilcockson 2002) and of *Maniola jurtina* in Argyll (northwest Scotland) and the Chiltern Hills and Oxfordshire (Maier 1998) have demonstrated that there are no geographical differences in the thoracic temperatures of individuals of each species when they engage in voluntary flight. Both species also respond quickly to variation in solar radiation intensity, indicating that they are dependent on direct radiation load for elevating body temperature. However, they differ in the magnitude of melanisation in relation to geographical climate variation. Detailed analysis

of wing variation in *Pieris napi* (Wilcockson & Shreeve 2003) reveals complex variation in a number of characteristics (Table 13.2 and Fig. 13.1) that can be related to variation in regional differences of weather and its predictability. Northwestern and central-southern populations of *P. napi* have a number of characteristics of significance to thermoregulation which differ in their means in predictable directions associated with average temperature, and sunshine hours. However a striking feature of northern populations is the large within-population variation of wing melanization and wing size. In addition, there is considerable overlap of phenotypes from different locations (Fig. 13.1a and b). In the region occupied by the northwestern population weather experienced by spring and summer flying broods is very variable, both within and between years but less so in central-southern England. Wilcockson & Shreeve (2003) argue that the variability in weather in northwest Scotland may constrain directional selection, with phenotypic fitness varying within and between years. Reduced weather variability in central-southern England imposes more directional selection and hence the population is less variable. The response of *Maniola jurtina* to climate gradients is rather different. There is little morphological variation in this butterfly in thermally important characteristics (e.g. basal wing melanisation) (Maier 1998) and warming rates do not differ between populations. Because of the absence of morphological variation of significance to thermoregulation and a geographically invariant body temperature for activity, *Maniola jurtina* is far more restricted in the vegetation structures it occupies in the north than are either *Pieris*

Table 13.2 Selected mean wing morphology measures ($\pm$ s.e.) of Pieris napi from Scotland, southern England and southern France and statistical comparisons between generations within regions and between regions and generations (from Wilcockson 2002)

Variable		Scotland		Southern England		Southern France	F and P (between region and generation)
		G1	G2	G1	G2	G2	
Forewing area (mm^2)	♂	209.2 ± 20.1	215.0 ± 22.0	182.6 ± 21.5	218.5 ± 23.0	252.7 ± 24.1	$F_{(4,95)} = 35.5\ P < 0.001$
	♀	203.6 ± 20.6	195.3 ± 9.4	176.0 ± 21.2	193.5 ± 17.3	242.9 ± 14.0	$F_{(4,88)} = 19.2\ P < 0.001$
Hindwing area (mm^2)	♂	226.3 ± 20.5	230.8 ± 30.9	205.3 ± 19.8	233.4 ± 15.4	250.8 ± 24.5	$F_{(4,94)} = 23.7\ P < 0.001$
	♀	213.7 ± 22.4	212.6 ± 17.5	190.9 ± 26.0	208.4 ± 21.1	262.5 ± 33.6	$F_{(4,88)} = 22.9\ P < 0.001$
Forewing dorsal proportion of wing covered by black scales	♂	0.08 ± 0.01	0.07 ± 0.01	0.08 ± 0.01	0.07 ± 0.01	0.11 ± 0.01	$F_{(4,95)} = 1.1\ P > 0.05$
	♀	0.12 ± 0.01	0.11 ± 0.01	0.11 ± 0.01	0.10 ± 0.01	0.09 ± 0.01	$F_{(4,88)} = 14.7\ P < 0.001$
Hindwing dorsal proportion of wing covered by black scales	♂	0.19 ± 0.08	0.17 ± 0.03	0.15 ± 0.02	0.11 ± 0.03	0.08 ± 0.03	$F_{(4,95)} = 18.75\ P < 0.001$
	♀	0.39 ± 0.18	0.32 ± 0.08	0.31 ± 0.12	0.15 ± 0.04	0.07 ± 0.03	$F_{(4,88)} = 32.5\ P < 0.001$
Forewing dorsal basal area brightness	♂	125.2 ± 5.0	136.3 ± 3.4	135.7 ± 4.4	138.0 ± 3.3	141.7 ± 3.6	$F_{(4,95)} = 48.0\ P < 0.001$
	♀	113.2 ± 19.0	133.2 ± 4.5	131.9 ± 4.8	135.0 ± 5.9	139.0 ± 4.5	$F_{(4,88)} = 20.6\ P < 0.001$
Hindwing dorsal basal area brightness	♂	125.2 ± 5.7	137.3 ± 3.4	135.1 ± 3.9	145.1 ± 17.3	140.2 ± 3.3	$F_{(4,94)} = 14.6\ P < 0.001$
	♀	130.4 ± 4.8	137.9 ± 3.1	136.7 ± 3.46	139.7 ± 3.5	137.1 ± 3.8	$F_{(4,88)} = 17.5\ P < 0.001$
Hindwing ventral surface vein brightness	♂	125.3 ± 3.8	127.3 ± 2.6	133.2 ± 2.7	130.6 ± 3.5	127.3 ± 4.3	$F_{(4,95)} = 16.6\ P < 0.001$
	♀	122.4 ± 3.6	122.5 ± 3.1	129.1 ± 2.9	126.8 ± 3.2	127.6 ± 4.3	$F_{(4,87)} = 6.6\ P < 0.001$
Hindwing ventral proportion of wing covered by black scales	♂	0.59 ± 0.05	0.52 ± 0.07	0.51 ± 0.04	0.36 ± 0.13	0.13 ± 0.08	$F_{(4,95)} = 106.9\ P < 0.001$
	♀	0.53 ± 0.09	0.46 ± 0.10	0.49 ± 0.06	0.23 ± 0.81	0.07 ± 0.03	$F_{(4,87)} = 112.3\ P < 0.001$
Hindwing ventral surface proportion yellow	♂	0.22 ± 0.02	0.26 ± 0.01	0.27 ± 0.02	0.33 ± 0.02	0.41 ± 0.04	$F_{(4,96)} = 10.7\ P < 0.001$
	♀	0.16 ± 0.03	0.13 ± 0.03	0.33 ± 0.03	0.54 ± 0.03	0.63 ± 0.02	$F_{(4,87)} = 68.4\ P < 0.001$

Sampling locations: Scotland: Breaclet, Isle of Lewis, bog edges; southern England: Chiltern Hills and Oxford floodplain, river margins; southern France: La Drôme, river margins.

All measurements were made under constant lighting conditions using image analysis. Brightness measurements range from 0 (black) to 255 (complete saturation)

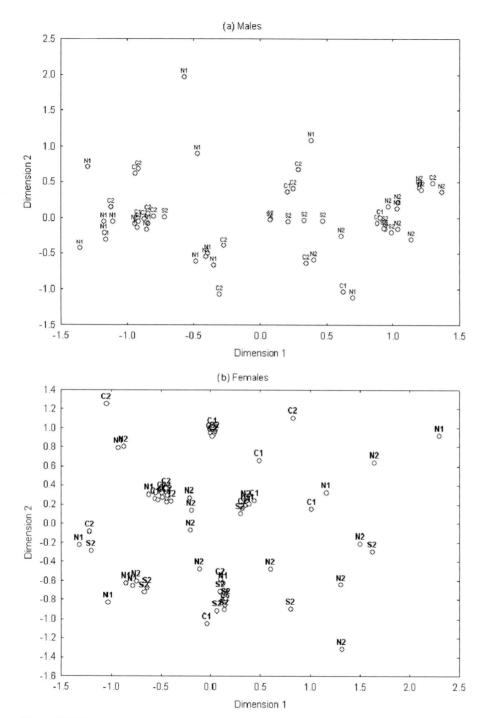

Figure 13.1. Non-metric multidimensional scaling plot of (a) male and (b) female *Pieris napi* for northern, central and southern locations. Sampling locations are given in Table 13.2. N1, generation 1 north; N2, generation 2 north; C1, generation 1 central; C2, generation 2 central; S2, generation 2 south (from Wilcockson 2002).

napi or *Polyommatus icarus*. In the north females spend less time in flight and tend to produce only a fraction of the egg output (c. 10%) of individuals from the south. As the butterfly is only active in bright sunshine in the north its whole activity there is limited, due to a combination of thermal constraints and little variation in morphological characters associated with thermoregulation.

In thermoregulatory behaviour, microhabitat use is important. If a butterfly can select a basking location (microhabitat) where the difference between the air temperature round the body and optimal body temperature is minimised, activity can be maximised. Microhabitat temperatures within any particular environment are highly variable, dependent on substrate, solar radiation and protection from wind. From detailed physiological studies of thermal receptors in antennae and wing bases (Schmitz & Wasserthal 1993) it is probable that butterflies are able to sense their thermal environment and probably assess the potential rate at which they can change temperature in a given location. Thermal receptors on the antennae protrude from any boundary layer around the body and probably function to monitor air temperature, whilst those on the wing bases will monitor body temperature and/or solar radiation heating effects. Differences between the two, if integrated over time, may allow individuals to monitor heating (or cooling) rates.

Substrate use by territorial perching nymphalids (Dennis 2004, Dennis & Sparks 2005) has been shown to vary with temperature. A study of eight widespread species (*Pieris rapae*, *P. napi*, *Vanessa atalanta*, *V. cardui*, *Aglais urticae*, *Polygonia c-album*, *Pararge aegeria*, *Maniola jurtina*) demonstrated that when these species bask to maximise their warming rate, they tend to do this on bare or short-herb substrates, but when they bask with postures that do not maximise heating rates, they do so on substrates such as tall herbs, shrubs and trees. These broad findings confirm more detailed studies of changes of microhabitat use in relation to thermal requirements (e.g. *Hipparchia semele*: Shreeve 1990b, Maes *et al.* 2006, *Pararge aegeria*: Shreeve 1984, Wickman & Wiklund 1983, *Maniola jurtina*: Maier 1998). As a rule, the warmest microhabitats are used when there is the greatest need to elevate body temperature as rapidly as possible. As a consequence, many species use a range of basking locations during different parts of the day. In addition, for species that can occur in a range of vegetation structures, there is a tendency to occupy a restricted range of (warmer) structures in cool seasons, at higher altitudes and in northern latitudes.

Most thermoregulatory studies have been conducted in conditions in which air temperature is lower than optimal body temperature. However, in some locations overheat is a possibility. At high air temperatures (>30 °C), morphological features and behaviours that maximise the gain of solar radiation are potentially disadvantageous. In this instance, elevation of body temperature above ambient may raise it above a lethal maximum. To avoid this, butterflies may restrict activity to shaded locations and be active only in cool vegetation structures. As an example, *Melanargia galathea* is usually described as being associated with open grasslands, but observations of heat avoidance under conditions of moderate solar radiation loads and high air temperature (Table 13.3) indicate that in warm locations, the butterfly requires the presence of shade. In this study flying individuals were also tracked whilst observations of overall microhabitat use were made. In the warmest part of the day, individuals only flew in shaded areas, turning back into shade immediately they experienced full solar radiation loads. Thus the thermal constraint of heat avoidance in this butterfly dictates that in potentially warm areas their resource requirements include shade, which can only be supplied by trees and shrubs. In cooler locations this requirement is absent. Therefore, thermal constraints influence the vegetation structures that are used, dependent on weather and climate. Requirements for maintaining body temperatures between lower and upper limits for activity in this and other species have profound effects on where individual species occur. This may influence their ability to encounter conspecifics and their interactions with predators, if the locations of these are associated with specific vegetation structures.

WING MORPHOLOGY AND PREDATION

Apart from flight and thermoregulation, butterfly wings have obvious communicative functions, extending from intraspecific communication to communication with predators. In the last case the most successful individuals will be those that predators ignore, and amongst those that are detected, those that can avoid being eaten. The major predators of adult butterflies are birds, lizards and to a lesser extent spiders, dragonflies and large crickets, whilst larvae are attacked by birds and invertebrates. Whilst there may be nocturnal predation by small mammals and by invertebrates, most predation of adults occurs during the day and the majority of predation events rely on visual recognition. Colour, pattern and the background against which any individual is perceived therefore play a critical role in defence.

Table 13.3 *Diurnal changes of location and activity of* Melanargia galathea *in relation to air temperature, solar radiation load and microhabitat temperature recorded on 30 June 2006 in a grassland, scrub and woodland matrix at St Andeol-en-Quint, near Die, La Drôme, France*

		Mean					Proportion				
Start time	Air temp	Ground temp.	Veg. temp	Solar radiation	Roosting	Open-winged basking	Closed-wing basking	Flying	% in sun	% in shade
0530	12.8	12.4	12.4	78	1.0	0	0	0	100	0
0600	13.1	14.7	16.3	103	0.44	0.44	0	0.12	100	0
0630	14.2	16.4	16.0	162	0.56	0.44	0	0	100	0
0700	14.4	17.1	16.3	173	0.56	0.22	0	0.22	100	0
0730	14.4	17.2	16.5	177	0.29	0.58	0	0.19	100	0
0800	14.6	26.4	20.7	285	0	0.20	0	0.80	100	0
0830	14.7	34.5	26.0	387	0	0.07	0	0.93	100	0
0900	25.5	40.4	27.0	506	0	0.05	0.18	0.77	90	10
0930	25.8	43.9	27.8	579	0	0.22	0.17	0.61	100	0
1000	26.5	47.1	28.6	664	0	0.14	0.23	0.63	75	25
1030	26.8	49.9	29.0	720	0	0.30	0.09	0.61	91	9
1100	26.9	51.7	29.1	780	0	0.09	0.43	0.58	69	31
1130	27.6	53.9	30.6	823	0	0.05	0.32	0.63	88	12
1200	28.0	53.5	30.3	851	0	0	0.56	0.44	33	67
1230	28.3	52.7	30.8	807	0	0	0.72	0.38	26	74
1300	28.5	52.9	31.3	881	0	0	0.67	0.33	24	76
1330	29.0	53.4	31.8	878	0	0	0.72	0.38	31	69
1400	29.3	52.4	31.5	862	0	0	0.71	0.39	33	67
1430	29.5	53.2	32.0	845	0	0	0.74	0.36	24	76
1500	29.4	49.6	31.9	814	0	0.05	0.64	0.31	8	92
1530	29.1	46.1	31.3	764	0	0.04	0.44	0.52	59	41
1600	29.4	43.0	31.8	709	0	0.24	0.55	0.21	48	54
1630	28.9	34.7	30.8	640	0	0.13	0.33	0.55	69	31
1700	29.1	37.3	31.3	555	0	0.23	0.31	0.46	54	46
1730	29.0	35.2	30.6	484	0.06	0.33	0.17	0.44	67	33
1800	28.6	32.7	30.3	401	0.03	0.20	0.28	0.49	65	35
1830	27.8	28.2	28.2	224	0.22	0.05	0.66	0.07	93	7
1900	27.1	25.7	26.7	78	0.55	0.20	0.20	0.05	85	15
1930	26.4	24.5	26.0	40	0.86	0	0.07	0.07	78	22
2000	26.2	23.9	24.9	38	1.0	0	0	0	75	15

Activity and location of *Melanargia galathea* was recorded in 10 10 × 10 m plots, from 0530 to 2000 GMT (approx. dawn to dusk). Each plot was visited at 30-minute intervals and the activity and location of uniquely identified *M. galathea* was recorded for a total of 2 minutes in each plot. Activity and location of individuals was recorded as the behaviour and location when first seen. Plots were selected to be representative of the available vegetation structures within the flight area of *M. galathea*. Plots ranged from short, cattle-grazed, unimproved grassland, to tall ungrazed unimproved grassland, and scrub and woodland edge. Weather and microclimate variables were recorded in one open plot, using a Skye Datahog, recording solar radiation intensity, air, ground and vegetation temperature at 0.3 m. Recordings were taken at 30-second intervals and are averaged over 30-minute periods.

However, where visual signals are ineffective, such as with overwintering adults in dark locations, predation by rodents may be significant and site specific. Wiklund *et al.* (2008) found no difference between predation rates on overwintering *Inachis io* and *Aglais urticae* (which differ in wing colour and pattern and secondary defence mechanisms) by climbing rodents, with predation rates of 53% and 58% respectively during the first 2 weeks of overwintering, declining thereafter to 98% survival in the last 16 weeks of overwintering.

In theory, if the wing colour and pattern of a butterfly is cryptic it should be maximally so against the background on which it occurs when it is the most vulnerable to predation. Likewise, an aposematic colour and pattern should be most obvious when the individual is the most vulnerable. Endler (1984) quantified background matching in terms of colour and patch size, with background matching (= crypsis) being maximal when both the colour and the patch size distributions of the organism and its background are identical. Unfortunately, only a few studies have attempted to quantify the background matching of butterfly wings (see Shreeve 1990b). However, there are direct studies of predation that indirectly link the probability of detection to the degree of background matching. The overwintering nymphalids *Polygonia c-album*, *Inachis io* and *Aglais urticae* are characterised by dark undersides, with *P. c-album* having a wing shape that mimics a dead leaf. Vallin *et al.* (2006) presented all three species to a bird, *Parus caeruleus*. In 40-minute trials, it took the predator much longer to find the leaf-mimicking *P. c-album* than either of the other two species. However once detected, the presence of secondary defence mechanisms (startling devices) significantly altered the chance of consumption. *Polygonia c-album* remained motionless once detected, but the other two species flicked their wings. The prominent eyespots of *Inachis io* afforded the most protection, with 100% survival of attacked individuals; the wing flicking by *Aglais urticae*, probably associated with weak secondary defence cues, only enabled 8% to survive, whereas the inert behaviour of *P. c-album* when attacked afforded more protection: 22% survived. Other secondary defence mechanisms, such as wing tails or marginal ocelli operate in a different manner, since they do not have a startling, but deflective properties. Wing tails, seen in a number of Theclinae and *Papilio* species are effective as deflecting devices (Brakefield *et al.* 1992). Whilst there are models which relate marginal wing ocelli of satyrine butterflies to predation (Dennis *et al.* 1986, Brakefield *et al.* 1992) their effectiveness warrants further experimental testing. Field evidence is scarce and Lyytinen *et al.* (2003) found

no evidence that marginal eyespots in the satyrine *Bicyclus anynana* had any influence on the chance of escape from two predators, the lizard *Anolis carolinensis* and the pied fly-catcher *Ficedula hypoleuca* or on the direction from which strikes were launched. Experimental work using artificial models of eyespots has demonstrated that overall conspicuousness and eyespot number are more effective as anti-predator devices than any specific resemblance to eyes per se (Stevens *et al.* 2008).

Predation pressure (the probability of being detected) varies with predator number and the extent of alternative food sources for specific predators. Wiklund & Tullberg (2004) presented darkish winter morphs and lighter summer morphs of *Polygonia c-album* to *Parus major* on tree trunks and nettles. The winter morph was less attacked on both backgrounds than was the summer morph. When summer and winter morphs were released to go to roosting sites, the majority of winter morphs chose tree trunks, branches and twigs, whist the summer morphs exclusively selected leaves. Wiklund & Tullberg (2004) concluded that the winter morph is protected from predation, but the summer morph is not constrained by predation, because predation pressure when it is active is low. However, the winter morph selected backgrounds against which it was probably the least visually obvious.

There is ample evidence that some colour and pattern forms are strongly aposematic and that predators can learn the visual signal of specific phenotypes. Studies of larval predation and recognition have also revealed that a simple division of visual signals into cryptic and aposematic is insufficient. The likelihood of the visual perception of an object is dependent on distance; Tullberg *et al.* (2005) effectively demonstrated that the natural coloration of *Papilio machaon* larvae is not maximally aposematic at short distances or maximally cryptic at long distances, providing empirical support for the idea that some colour patterns may combine warning coloration at a close range with crypsis at a longer range (see Stevens 2007 for further discussion). In addition, studies of resting moths and a few butterfly species (Brakefield 1984, Endler 1984, Shreeve 1990b, Brakefield *et al.* 1992) indicate that the visual apparency of any colour and pattern form is dependent on the background against which it is perceived and the distance from which it is perceived.

A complication to the human understanding of visual signalling comes from our inability to see in the same spectra as butterflies and their predators. Human vision is essentially trichromic, with peak sensitivities in the blue, green and red parts of the spectrum. Most birds, lizards and insects have tetrachromic vision, with peak sensitivities in the near ultraviolet (UV), blue, green and red parts of the spectra (Bennett *et al.* 1994, Burkhardt 1996, Honkavaara *et al.* 2002, Eaton & Lanyer 2003). Without an understanding of the UV component of vision, our understanding of the role of colour and pattern may be incomplete. For example, *Melanargia galathea* appears to the human observer to be highly apparent, but in the UV it can be cryptic (Plate 17) and the perception of this butterfly by organisms with peak spectral sensitivity in the UV (as some birds and lizards) is different to that of the human. In the case of larvae, Church *et al.* (1998) found that the supposedly cryptic larvae of the moth *Lithophane ornitopus*, are highly apparent and possibly aposematic to birds, because of UV brightness of the cuticle which contrasts against UV absorption by their normal resting background. However, in a series of experimental trials with *Parus major* Lyytinen *et al.* (2001) could not demonstrate that UV alone could act as an aposematic signal.

A UV component to wing colour and pattern is present in approximately 40% of examined Palaearctic butterflies, and is itself a variable characteristic (Plate 18), ranging from an extreme brightness over the whole of both wing surfaces (e.g. *Celastrina argiolus*), to very bright patches as part of the general pattern over single surface (e.g. *Argynnis paphia*), or to variable 'general' brightness in parts of the wing pattern (e.g. *Euchloe ausonia*). Even patterns apparently similar to the human eye may differ in the UV (e.g. *Anthocharis cardamines* versus *Euchloe ausonia*). In other species the UV reflective pattern may be non-existent (e.g. *Coenonympha pamphilus*), or be a low-brightness representation of the general wing pattern (e.g. *Brintesia circe*). The quantitative nature of UV coloration extends to many species and to date, predation studies on adult butterflies have ignored this component.

As well as variation in the UV components of wing colour and pattern in a significant number of butterflies, the backgrounds on which butterflies occur may also have a variable UV component. For example, most vegetation is UV-absorbing and a few flowers are UV-reflecting whilst some rock and mineral substrates are UV-reflecting and others not. Given the lack of comprehensive studies of predation that include the UV component of vision it is difficult to draw solid conclusions about how butterfly wing colour and pattern function as defence mechanisms. Where UV has been taken into account in studies of insect–bird interactions, it has questioned accepted explanations (e.g. Majerus *et al.* 2000). Further, appropriate studies are required.

WING MORPHOLOGY AND INTRASPECIFIC COMMUNICATION

Visual signalling by butterfly wings is not just confined to defence mechanisms, it also plays a role in male–male and male–female communication and recognition, including mate choice. Apart from those species that detect females entirely by scent, one of the primary cues for potential mate recognition is vision. As with predation, there is a UV component to butterfly vision. Early work by Rutowski (1977) and Silberglied & Taylor (1973, 1978) indicated that UV reflection can be used in mate attraction and recognition. In intraspecific communication the signal function of a butterfly wing has to be placed in the context of how and where the individual is perceived. For example, if the primary mate-locating mechanism is patrolling then flying males are attempting to locate potential females that may be either settled or flying. If the primary mate-locating mechanism is perching, males will respond to the visual signal of flying females. If the species engages in leks, then females may well be responding to the cue presented by a number of settled males. Key questions are the nature of the visual signals that individuals respond to and which signals maximise the response. For some individuals that perch, the primary response may be to movement against a background. For example, perched males of *Pararge aegeria* and *Ochlodes venata* fly up to and inspect almost any flying object (Shreeve 1984, Dennis & Williams 1987). For species that patrol the cue may be the contrast of the conspecific to the background.

Studies of the spectral sensitivity of butterflies have revealed a sensitivity from below 300 nm in the UV through to 700 nm in the red (Bernard 1986, Arikawa 2003, Briscoe *et al.* 2003, Qui & Arikawa 2003). Earlier studies have revealed that neurones in the medulla intermedia (the central part of the brain) respond maximally to visual signals that correspond to the individual's wing colour (Swihardt 1967). Thus intraspecific visual communication is likely to have an element that focuses on wavelengths that butterfly wings reflect. Despite pioneering work by, for example, Rutowski (1977) and Silberglied & Taylor (1973, 1978) this aspect of communication has not received significant attention. Studies by Lundgren (1977) have demonstrated that male *Polyommatus icarus* are attracted to male-like objects, and a more recent study by Howe (2004) has demonstrated that female *P. icarus* with blue suffusion which adopt an open-winged posture are more attractive to males than are brown females or females with closed wings.

Table 13.4 *The relationship between the presence and absence of UV reflection in females of 110 Palaearctic species in which males engage in predominantly perching or patrolling mate location, or in males which engage in lek assemblies, in relation to the presence of UV in the backgrounds against which conspecifics are usually perceived by mate-locating individuals*

Mate-locating site	Major mate-locating mechanism	Background against which female is perceived	% of species with females that reflect UV
Perching or patrolling species			
Tree canopy	Perching	Sky (UV bright)	30
	Patrolling	Leaves (UV dark)	80
Forest open patch (light gap and edge)	Perching	Sky (UV bright)	0
	Patrolling	Leaves (UV dark)	60
Grassland	Perching	Sky (UV bright)	0
	Patrolling	Leaves (UV dark)	75
		Mineral (UV bright)	0
		Mineral (UV dark)	100
Lek species			% of species with males that reflect UV
Shrubs and trees	Male aggregation	Leaves (UV dark)	80

The presence of UV in butterflies or backgrounds was determined using image analysis of images captured in bright sunshine with a UV pass filter. UV brightness of images was determined on a scale of 0 to 255 (none to saturation) and objects were classified as possessing a UV signal if any part of the object had a brightness greater than 96.

Additionally, blue females were as attractive as males. The precise role of coloration and pattern in intraspecific signalling in other Palaearctic species is largely unknown, but should be important as mate detection by most butterflies is primarily a visual-based system.

Because a significant number of species have a UV element to their wing colour and pattern, this UV signal may also play a role in communication. Under laboratory conditions Knuttel & Fiedler (2001) found that males of *Polyommatus icarus* were less attracted to the underside of flavinoid-poor and UV-bright females than to flavinoid-rich and UV-dull female undersides. However, it is difficult to extrapolate this to the field as their experimental cages were small and males would be approaching females from a closer range than in the wild. In addition, when males of this species are flying in search of females, they most frequently approach open-winged females from above; the role of the underside in mate detection in the wild remains to be disclosed.

Despite a lack of comprehensive data, it is possible to suggest that the possession of UV reflectance is of significance to how individuals locate mates in the context of the backgrounds against which individuals of the opposite sex are perceived (Table 13.4). In a study of the behaviour and microhabitat use by Palaearctic species in a range of vegetation structures, from woodland to scree slopes and grassland, in the French Alps and foothills (T. G. Shreeve, unpublished), the mate-locating mechanisms of 110 species (each with >50 observations) were divided into exclusive patrolling, the possession of perching, and the formation of male leks. From behavioural observations, the substrates against which perching males perceived females were categorised as UV-bright sky (for ground/ low vegetation or tree canopy male perchers) or UV-absorbing vegetation (some forest gap male perchers). For patrolling males conspecific female settling sites were categorised as UV-absorbing vegetation and mineral substrates or in a few cases UV-reflecting mineral substrates. For lek species the lek substrate, on which males settle, was always UV-absorbing vegetation.

There are clear patterns in the possession of UV reflectance on the basis of the background against which a conspecific is likely to be perceived. In most cases, the UV characteristic of the female is most likely to enhance contrast with the background against which they are most likely to be detected. Females of species in which perched males perceive them against the UV-bright sky tend to be UV-absorbing on all four wing surfaces. Those that may be perceived flying against UV-dark vegetation tend to have a UV-bright component to their wing colour and pattern. Settled (but not roosting) females that are most likely to be perceived by flying males tend to have a UV-bright element to their wing colour and pattern if they usually settle on UV-absorbing vegetation or mineral substrates whilst those that settle on UV-bright mineral substrates tend to have UV-dark wings. For species in which males engage in leks with individuals settled against UV-dark vegetation then males have a UV-reflective component. Whilst this analysis would benefit from further work on different species in different regions, there is an indication that UV may serve to enhance the visual contrast of an individual against its background. However, where a UV signal is present, it tends to be absent from the underside. This probably relates to signal function when the butterfly is roosting. For example, species that roost in prominent locations (e.g. grass heads) tend to be cryptic and lack a UV signal on the underside. Those few species that do have a prominent UV signal on the underside (e.g. *Argynnis paphia*, *Celastrina argiolus*) tend to roost in hidden locations, such as leaf undersides or within ivy (*Hedera helix*) bushes where any UV signal is not obvious, due to a lack of direct sunlight.

WING MORPHOLOGY AND HABITAT STRUCTURE

The functioning of all the elements of wing morphology depends on habitat structure and microstructure. Many trade-offs are likely to be involved. For example, backgrounds suitable for thermoregulation may expose individuals to both predators and conspecifics. In addition there may be trade-offs between mate attraction and conspicuousness to predators. Given the dependency of individual wing functions on the microhabitat, there is an expectation that species with similar morphologies may occur in similar microhabitats. Whilst there is a potential phylogenetic component to this because closely related species tend to have similar wing morphologies (and behaviours) it is possible that wing morphology (and perhaps phylogeny) acts as a constraint on where species occur.

In the study of mate location (see above), the microhabitat use by 110 species in a range of vegetation structures in the French Alps (T. G. Shreeve, unpublished) was also quantified: with the vegetation height, composition, colour and precise substrate use of settled individuals being recorded together with behaviour. Microhabitats were recorded in small plots of $0.25\,\mathrm{m}^2$, centred on the observed butterfly. In addition the wing morphology of 77 of these species was quantified using a range of characters (Table 13.5). Using these characters, principal components analysis (PCA) was used to identify morphological groups. For roosting butterflies, only the characteristics of the underside hindwing were used. For resting/basking butterflies the underside hindwing characteristics of lateral basking species and fore- and hindwing upperside characteristics of open-winged baskers (dorsal absorbance and reflective baskers) were used. To identify whether there were morphological groups of flying butterflies the characteristics of all four wing surfaces were used. The analysis identified four species groups for roosting, five for resting/basking and six for flying (Table 13.6). Membership of groups was not consistent over the three activities, a logical finding as different wing surfaces (potentially with different colours and patterns) may be exposed when individuals undertake different activities. With the exception of flight activity, the different groups mostly used different microhabitats (Table 13.6), although any one functional group used more than one microhabitat type. Whilst this work would benefit from more extensive data sets, it is apparent that there is a possible link between wing morphology and microhabitat use, indicating that the resources particular species can use may well be constrained by their visual appearance. If this is a real constraint, then it is likely that the primary visual functions of wings, acting via relationships with visual hunting predators and intraspecific communication, together with thermodynamic functions, limit microhabitat use. Because the presence and relative abundance of microhabitat types is related to overall vegetation structure and composition, it is possible that wing morphology may constrain the use of vegetation structures at larger spatial scales.

PATTERNS OF GEOGRAPHICAL AND SEASONAL VARIATION

Many species of butterfly have some degree of geographical or altitudinal variation in wing morphology, and, for species with more than one generation a year, there is frequently seasonal variation. Whilst this variation has been long recognised, there are relatively few studies that have quantified it. Where quantification has been undertaken, it is often of single characteristics. Some of this variation has been correlated with geographical and seasonal patterns of weather, for example wing melanisation in *Parnassius*, *Colias*, *Pieris* and

Table 13.5 *Quantified morphological characters of fore- and hindwing dorsal and ventral surfaces of Palaearctic butterflies used in a PCA to determine morphological groups on the basis of wing shape, size, colour and pattern*

Character		Scale	Dorsal forewing	Dorsal hindwing	Ventral forewing	Dorsal forewing
Area		mm^2	+	+		
Length		mm	+	+		
Perimeter length		mm	+	+		
Overall luminance[a]		0–255	+	+		
RGB colour 1–64[b]	Patch area	mm^2	+	+	+	+
	Number of patches		+	+	+	+
	Patch size variance		+	+	+	+
	Maximum patch size		+	+	+	+
Fractal dimension of colour			+	+	+	+
UV brightness 1–4[c]	Patch area	mm^2	+	+	+	+
	Number of patches		+	+	+	+
	Patch size variance		+	+	+	+
	Maximum patch size	mm^2	+	+	+	+

[a] Overall luminanace was measured as the brightness of a greyscale image (on a scale of 0 to 255) of a colour image of the relevant wing surface.
[b] RGB colour: each axis (0–255) of the red, green, blue colour cube was divided into 4 to give 64 minor colour cubes. Wing areas within each colour cube were detected using Optimas software and each detected area was measured.
[c] UV brightness was measured using a greyscale image (0–255) taken in UV. The single axis was divided into 4 units and wing areas in each brightness class were detected and measured.

Pararge species (Watt 1968, Kingsolver 1985b, Guppy 1986, Van Dyck & Wiklund 2002), underside lunulation in lycaenids (e.g. Høegh-Guldberg 1974, Väisänen *et al.* 1994a) and marginal spotting in *Coenonympha tullia* (Dennis *et al.* 1986). The emphasis of these studies has been on thermoregulation and activity, or predation in relation to weather and its thermal constraints. More recent studies have considered that wing morphology affects a number of fitness components (see Chapter 10). Wing morphology may be related to a number of potentially conflicting demands and also to landscape structure, resource availability and population history and origin, indicating that comprehensive explanations of wing morphology variation may be complex and that gradients of variation may not be simple (Dennis & Shreeve 1989; see Chapter 3). In addition analysis of mean traits without consideration of their variance may hide important insights into the function and adaptive significance of wing morphology. For example, within-population variation of both male and female wing morphology traits in *Polyommatus icarus* and *Pieris napi* is larger in northern than southern populations (Wilcockson 2002, Howe 2004), which might be associated with lower predictability of weather in the north. In addition morphology can be influenced by development history (e.g. Van Dyck & Wiklund 2002) and the extent of variation observed may be as much to do with development as selection. Given the magnitude of predicted climate changes and regional variation in these changes (e.g. IPCC 2002) studies of developmental plasticity would help our understanding of how butterflies will respond to future rapid change.

Plasticity is a selectable trait (Hazel *et al.* 1990, Gavrilets & Scheiner 1993) and seasonal polyphenism can be viewed as an extreme form of a plastic response to different, but predictable, conditions experienced during development. Within the Palaearctic, *Araschnia levana* has a pronounced seasonal polyphenism, having a reddish spring form and a disruptively coloured black-and-white summer form. The proximate cues for the production of spring and summer forms of this butterfly are well known (Koch & Buckmann 1987, Windig & Lammar 1999) but the ultimate reasons are less well understood. Early explanations of the ultimate cause of this seasonal polyphenism (Brakefield *et al.* 1992) centre on thermoregulation and predator avoidance. The

Table 13.6 *Membership of roosting, basking (including feeding) and flying morphological groups of 77 Palaearctic butterflies from La Drôme, France, for which there is adequate microhabitat use data* (n > 50 observations per category)

Activity	Group[a]	Species		Major microhabitats used[b]
Roosting[c]	1	*Spialia sertorius*	*Limenitis camilla*	Low-growing (<1 m) grass heads and drying
		Plebejus argus	*Limenitis reducta*	inflorescences of medium herbs
		Satyrium spini	*Apatura iris*	Tree canopy leaves, bramble (>5 m), liana (>5 m)
		Satyrium esculi	*Pararge aegeria*	Underhangs and rock crevices
		Polyommatus dorylas	*Lasiommata megera*	
		Polyommatus ripartii	*Coenonympha arcania*	
		Argynnis paphia	*Coenonympha dorus*	
		Argynnis aglaja	*Coenonympha pamphilus*	
		Argynnis adippe	*Pyronia tithonus*	
		Brenthis ino	*Erebia aethiops*	
		Brenthis daphne	*Erebia albergenus*	
		Boloria dia	*Erebia epistgyne*	
		Inachis io	*Melanargia russiae*	
		Euphydryas aurinia	*Melanargia galathea*	
		Melitaea phoebe	*Satyrus ferula*	
		Melitaea didyma	*Hipparchia semele*	
		Melitaea parthenoides	*Brintesia circe*	
	2	*Iphiclides podalirius*	*Polyommatus escheri*	Tall grass heads (c. 1 m) and drying
		Papilio machaon	*Polyommatus amanda*	inflorescences of medium herbs
		Papilio alexenor	*Polyommatus thersites*	Wood edge shrubs (2–5 m)
		Hamearis lucina	*Polyommatus icarus*	
		Lycaeana alciphron	*Lysandra bellargus*	
		Satyrium ilicis	*Issoria lathonia*	
		Cupido minimus	*Boloria euphrosyne*	
		Celastrina argiolus	*Aglais urticae*	
		Pseudophilotes baton	*Melitaea athalia*	
	3	*Pyrgus malvae*	*Melitaea cinxia*	Short perennial bushes (<2 m), undersurfaces
		Thymelicus lineola	*Lasiommata petroplitana*	of tall (>2 m) shrub leaves
		Thymelicus acteon	*Pyronia bathsheba*	Low-growing grass heads (0.2–05 m)
		Lycaena virgaureae	*Aphantopus hyperantus*	
		Everes argiadus	*Maniola jurtina*	
		Polyommatus daphnis	*Erebia triaria*	
		Vanessa atalanta	*Satyrus acteaea*	
		Polygonia c-album	*Hipparchia alcyone*	
		Nymphalis antiopa		
	4	*Leptidea sinapis*	*Gonepteryx rhamni*	Purple/blue flowers (<2 m high)
		Anthocharis cardamines	*Gonepteryx cleopatra*	Tree canopy leaves
		Aporia crataegi	*Lycaeana tityrus*	Tall herb leaves (c.1 m)
		Pieris brassicae	*Cupido osiris*	
		Pieris rapae	*Glaucopsyche alexis*	
		Pieris napi	*Glaucopsyche melanops*	
		Pontia daplidice	*Cyaniris semiargus*	
		Colias croceus	*Polyommatus icarus*	
		Colias hyale	*Lysandra coridon*	

Table 13.6 (*cont.*)

Activity	Group[a]	Species		Major microhabitats used[b]
Basking	1	*Iphiclides podalirius*	*Pararge aegerea*	Bare ground, rock, short herbs (< 0.1 m) <25% cover bramble banks (>1 m < 4 m) Grass flowers, knapweed seed heads (> 0.3 m < 1 m) Wood edge vegetation (>5 m)
		Gonepteryx rhamni	*Lasiommata megera*	
		Hamearis lucina	*Lasiommata maera*	
		Satyrium esculi	*Coenonympha arcania*	
		Satyrium acacea	*Pyronia bathsheba*	Grass leaves, herb leaves, unopened flower buds (>0.1 m < 0.3 m)
		Argynnis adippe	*Melanargia russiae*	
		Brenthis ino	*Melanargia galathea*	
		Brenthis daphne	*Satyrus acteaea*	
		Boloria euphrosyne	*Hipparchia fagi*	
		Euphydryas aurinia	*Hipparchia alcyone*	
		Melitaea phoebe	*Brintesia circe*	
	2	*Erynnis tages*	*Issoria lathonia*	Medium herb leaves and grasses (>0.3 m <1 m)
		Spialia sertorius	*Boloria dia*	Wood edge vegetation (<5 m)
		Thymelicus lineola	*Melitaea cinxia*	
		Thymelicus acteon	*Melitaea didyma*	
		Carterocephalus palaemon	*Melitaea parthenoides*	
		Ochlodes venata	*Melitaea athalia*	
		Lycaeana alciphron	*Coenonympha dorus*	
		Callophrys rubi	*Coenonympha pamphilus*	
		Satyrium spini	*Pyronia tithonus*	
		Satyrium ilicis	*Aphantopus hyperantus*	
		Cupido minimus	*Maniola jurtina*	
		Plebejus argus	*Erebia aethiops*	
		Plebejus argyrogonum	*Erebia triaria*	
		Aricia agestis	*Erebia alberganus*	
		Poyommatus semiargus	*Erebia epistgyne*	
		Argynnis aglaja	*Minois dryas*	
	3	*Carcherodus alcaea*	*Pieris rapae*	Medium herbs (<0.3 m)
		Carcherodus lavatherae	*Pieris napi*	Tall flowers and seed heads (>1.5 m <2 m)
		Pyrgus carthamni	*Pontia daplidice*	Bramble banks (>1 m < 4m) Short perennials (>1 m < 2 m) and medium herbs
		Pyrgus malvae	*Colias croceus*	
		Leptidea sinapis	*Colias hyale*	
		Anthocharis cardamines	*Everes argiadus*	
		Aporia crataegi	*Everes alcetus*	
		Pieris brassicae		
	4	*Lycaeana phlaeas*	*Polyommatus escheri*	Short herbs, low-growing flowers (<0.2 m)
		Lycaena virgureae	*Polyommatus dorylas*	Tall shrubs (>2 m)
		Lycaeana tityrus	*Polyommatus amandus*	
		Cupido osiris	*Polyommatus thersites*	
		Celastrina argiolus	*Polyommatus icarus*	
		Pseudophilites baton	*Polyommatus daphnis*	
		Glaucopsyche melanops	*Polyommatus bellargus*	
		Maculinea arion	*Polyommatus damon*	
		Plebejus idas		

Table 13.6 (*cont.*)

Activity	Group[a]	Species		Major microhabitats used[b]
	5	*Gonepteryx rhamni*	*Polygonia c-album*	Canopy leaves, tall shrubs and liana (>3 m)
		Gonepteryx cleopatra	*Nymphalis antiopa*	Tall herbs (>0.3 m)
		Argynnis paphia	*Limenitis camilla*	
		Vanessa atalanta	*Limenitis reducta*	
		Aglais urticae	*Apatura iris*	
		Inachis io		

[a] Group membership is determined by loadings in a PCA using morphological attributes listed in Table 13.5.
[b] Microhabitat use data were collected from eight major vegetation structures/locations: calcareous woodland/scree slope 1200 m, neutral flood plain woodland and mixed agriculture 800 m, acidic xeric scrub and woodland 1000 m, acidic and neutral grassland 1500 m, calcareous grassland scrub mosaic 1200 m, calcareous grazed grassland 1100 m, abandoned calcareous grassland 1500 m, grazed calcareous grassland 2100 m.
[c] Roosting morphologies comprise underside hindwing only and resting groups by exposed wing surfaces normally exposed when basking.

riddle is that the darker summer form flies in *warmer* summer conditions. The spring generation is usually less numerous and also less mobile; on the other hand, both forms employ an identical mate-locating tactic, whereby males perch in aggregations on prominent shoots of vegetation at or near patches of their host plant, the nettle (*Urtica dioica*) (Fric & Konvička 2000, 2002). Besides wing colour, spring butterflies have heavier thoraxes and more angular wings, as if selected for better manoeuvrability, whereas summer butterflies have more rounded and relatively larger wings, presumably advantageous for longer-distance flights (Fric & Konvička 2002). Fric *et al.* (2004) raise a key point about the evolutionary origins of the trait. By examining the phylogeny of *Araschnia* butterflies, they conclude that the reddish spring colour is the ancestral trait and the summer form the derived trait, and indicate that the origin of the seasonal polyphenism pre-dates the movement of *Araschnia* butterflies into the Palaearctic. Thus, it is the derived summer coloration that requires an adaptive explanation. Fric *et al.* (2004) speculate that it may confer advantage via disruptive concealment, particularly in dry-season environments. Such an advantage may be particularly useful for individuals dispersing beyond familiar habitat patches, which is consistent with the higher mobility of the summer form. Although detailed field studies of the activity and predation pressure on different forms remain to be conducted, the ultimate reasons for this seasonal polyphenism pre-date current conditions and should be sought in environments where the trait has evolved.

CONCLUSIONS

Clearly wing morphology has a number of functions, related to flight, thermoregulation, intraspecific communication (including mate choice) and predator avoidance and evasion. How effective a particular morphology is in any location is dependent on weather, resource availability and microhabitat. Recent studies have started to reveal the complexity of the relationship between different wing functions and the context in which they operate. Potentially, wing morphology may act as a constraint on where species may occur and all aspects of wing morphology warrant further studies. With increasing information on individual species it is becoming evident that generalities are becoming harder to make and there is an urgency to understand fully this aspect of butterfly ecology to help understand how increasingly vulnerable species function and persist. Key questions that need further study include the relationship of colour and pattern to microhabitat use, the extent to which morphological variation within species can buffer them against environmental variation, including changes of weather and landscape. In particular developments in monitoring long-distance flight and movements, including the use of harmonic radar (Cant *et al.* 2005) may facilitate much greater understanding of the link between wing morphology and its variation and the ability of butterflies to move within increasingly fragmented and modified landscapes.

14 • Evolutionary ecology of butterfly fecundity

BENGT KARLSSON AND HANS VAN DYCK

SUMMARY

Butterfly reproductive traits vary considerably between and within species. Examples of those traits include the number of mature eggs at eclosion, the lifetime number of eggs laid, egg production rate, size of eggs, the difference between potential and realized fecundity, oviposition rate and the degree of egg clustering. Egg-laying is not independent of other life-history traits, behaviours and habitat use, and it is subject to evolutionary change. Most butterflies are capital breeders since they mainly use nutrients from reserves that were accumulated during the larval stages for their reproductive output. Nitrogen is the limiting factor. However, the female's nutrient budget can be complemented by adult feeding and by male spermatophore donations. Size of spermatophores relative to male body size depends on the mating system, and consequently varies among species. Potential fecundity is determined by larval resources, whereas realized fecundity also depends on biological and physical factors. Hence, realized fecundity is clearly condition-dependent with a significant contribution from female body size and condition. Trade-offs between the number and the size of eggs have attracted much attention. Because butterflies are ectotherms and more particularly heliotherms in the adult stage, temperature can have a strong effect on their fecundity. Temperature affects growth rate and development time, and hence the nutritional reserves built up during the larval stages, but it also affects flight activity and oviposition time-budgets. Fecundity also depends on the allocation and reallocation of material to reproductive versus somatic tissue. There is experimental evidence for the increase of fecundity from flight-muscle-derived nutrients. Trade-offs between fecundity and mobility need to take into account such physiological dynamics. Although knowledge on butterfly fecundity has improved significantly over the last decades, it still remains a challenge to predict what will happen to fecundity patterns under anthropogenic environmental changes.

INTRODUCTION

The lifetime number of eggs a butterfly lays is a key factor for its reproductive success and, ultimately, for its fitness.

Resources gathered during the larval and adult stages need to be stored and dispatched into eggs and the way reproductive resources, as eggs, are distributed over time and space is subject to evolutionary change. Strategies of egg production and egg-laying are adaptations, but they can be constrained by environmental conditions experienced during both larval and adult stages (Chapter 12). Relationships between the number and the size of eggs are a classic life-history theme (Stearns 1992). Egg-laying is, however, not independent of other life-history traits, behaviours and habitat use. Hence, fecundity can be traded off against other traits like mobility or survival. We review the evolutionary ecology of butterfly fecundity, particularly highlighting female fecundity. Earlier reviews on the subject include Chew & Robbins (1984), Porter (1992) and Boggs (2003). Oviposition itself is reviewed elsewhere in this volume (Chapter 3).

THE BUTTERFLY REPRODUCTIVE SYSTEM

Female butterflies have eight egg-producing ovarioles. Each ovariole consists of lined-up, progressively larger oocytes in follicles. Each follicle consists of an oocyte and seven trophocytes or nurse cells. Scoble (1992) provides a wider description of the internal reproductive morphology. Vitellogenesis – i.e. the incorporation of yolk into the growing egg – is achieved by the uptake of vitellogenetic proteins and lipoproteins from the haemolymph and by released cytoplasm from the nurse cells (Telfer 1965). When the oocyte is full grown, the egg chorion is formed by the follicular epithelium and the egg is ready to be laid. This is the general scheme of the oogenesis process. There is, however, significant variation between and within species for reproductive traits like the number of mature eggs at eclosion, lifetime number of eggs, egg production rate, size of eggs, difference between potential and realized fecundity, oviposition rate and degree of egg clustering (Labine 1968, Dunlap-Pianka 1979). Variation in such traits can be correlated into integrated reproductive strategies, but species

Ecology of Butterflies in Europe, eds. J. Settele, T. Shreeve, M. Konvička and H. Van Dyck. Published by Cambridge University Press. © Cambridge University Press 2009, pp. 189–197.

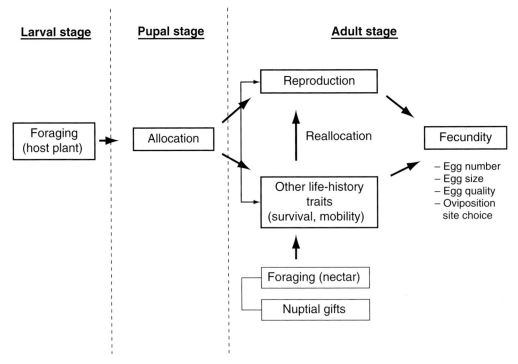

Figure 14.1. Diagram of the links between foraging, allocation, reallocation of resources and fecundity in butterflies.

with a higher oviposition rate do not necessarily have a higher oogenesis rate.

The number of mature eggs at eclosion varies considerably among species. This is expressed by the ovigeny index, i.e. the number of eggs females have ready to lay divided by the lifetime potential fecundity (Jervis *et al.* 2005). Some species emerge with a substantial number of mature eggs, but most butterflies start with no or only a few mature eggs. The total amount of eggs laid also varies considerably among species, ranging from <50 (e.g. *Phulia nymphula*) up to >1000 eggs (e.g. *Pieris brassicae*) (García-Barros 2000c). The most fecund species appear to cluster their eggs in batches (Courtney 1984a), although most species are single egg-layers. Butterflies also vary in the size and shape of their eggs (Porter 1992). Whether shape differences other than size represent adaptations has to our best knowledge not yet been addressed.

LARVAL CAPITAL AND ADULT SUPPLEMENTS TO REPRODUCTIVE OUTPUT

Most butterflies are capital breeders since they mainly use nutrients from reserves that were accumulated during the larval stages for their reproductive output (Boggs 1992). Nutrients are stored in the fat bodies, the haemolymph and to some extent in the reproductive organs as well. However, in several species there is also a resource income component from adult life. The significance of these supplements again varies amongst species (Fig. 14.1).

Nitrogen is the limiting nutrient for butterfly reproduction (Boggs 1981a, 1997b, Karlsson & Wickman 1989, O'Brien *et al.* 2004). García-Barros (2006) provides data on nitrogen contents of eggs of several European satyrines (6.8–9.3% of egg dry weight). The major part of the nitrogen-rich components (i.e. amino acids and proteins) for reproduction originates from larval feeding. Adult food sources typically consist of carbohydrates, like nectar, but also tree fluids and honeydew (Norris 1936, Boggs 1987) (see Chapter 2). Pollen-feeding butterflies are an exception as they have an alternative source for their adult nitrogen budget (e.g. *Heliconius* butterflies: Gilbert 1972, Dunlap-Pianka *et al.* 1977), but pollen intake is limited to absent in most European species (Jennersten 1984). The nectar of some plant species can be richer in amino acids than of others (Baker & Baker 1973, Erhardt & Rusterholz 1998) and uptake and utilization of nectar amino acids may function as compensatory mechanism to reduce or override the impact of nitrogen-poor larval food, as was experimentally shown for *Araschnia levana* (Mevi-Schütz & Erhardt 2005a).

Besides adult food, females may have other nitrogen sources to supplement their budget. During copulation a male delivers a spermatophore to the female's bursa copulatrix. The spermatophore obviously contains sperm, but also accessory gland products (Marshall 1982, Bissoondath & Wiklund 1995). Hormones that stimulate oviposition can be present in the spermatophore, but also other substances that hinder female remating (Gilbert 1976, Andersson *et al.* 2000). Butterfly spermatophores are viewed as nitrogen-rich 'nuptial gifts' which may be used by females for reproduction or somatic maintenance (e.g. Boggs & Gilbert 1979, Wedell & Karlsson 2003). Radiotracer studies by Boggs & Gilbert (1979), Boggs & Watt (1981) and Wiklund *et al.* (1993) showed that material from the spermatophore was transferred to the female and also to the eggs. It is also probable that spermatophores function not only as parental investment but also as mating effort.

There is considerable variation in the relative size of spermatophores among species; it varies from on average 1% to 15% of the male's body mass (Svärd & Wiklund 1989, Karlsson 1998). Males of monandrous species (i.e. those in which females usually mate only once) have a small investment in spermatophores that only constitute a few percent of body weight while males in polyandrous species (i.e. females mate several times) produce large and nutrient-rich spermatophores (Rutowski *et al.* 1983, Svärd & Wiklund 1986, Wiklund & Forsberg 1991, Bissoondath & Wiklund 1995, Hughes *et al.* 2000). The largest spermatophores documented among European species are found in the polyandrous species *Pieris napi* (up to 25% of the body mass with an average of 15%: Wiklund *et al.* 1993). Svärd & Wiklund (1989) and Karlsson (1995) reported a positive correlation between the degree of polyandry and spermatophore size at the interspecific level. Thus, the relative value of a spermatophore in terms of nitrogen supplement to the female differs with mating system. For a review on sperm competition in butterflies, we refer to Wedell (2005).

Colias eurytheme females that mated with recently mated males have lower fecundity and a shorter lifespan than females mated by virgin males (Rutowski *et al.* 1987). This is because recently mated males deliver spermatophores of reduced size, indicative of a cost of spermatophore production. However, Jones *et al.* (1986) found no effect of multiple

mating in two *Euphydryas* species and Svärd (1985) found no relationship between spermatophore size and longevity in the monandrous butterfly *Pararge aegeria*. Likewise, Svärd & Wiklund (1988a) could not detect any effect of the number of matings on reproductive output and longevity in *Danaus plexippus*. By contrast, Oberhauser (1997) found a correlation between the size of the spermatophore and egg production in the same species, and also between the number of matings and fecundity (Oberhauser 1989). Watanabe & Ando (1993) reported a positive effect of number of matings and fecundity in the polyandrous *Pieris rapae* and Wiklund *et al.* (1993) and Wedell & Karlsson (2003) showed a strong relationship between the number of matings and fecundity in the highly polyandrous *P. napi*. Karlsson (1998) confirmed this relationship in *P. napi*, but he also showed that nitrogen from the spermatophore must be included in the female budget in order to realize nitrogen levels for the observed egg production. The latter study ruled out the possibility that spermatophores would only increase fecundity by hormonal substances in the ejaculate.

Two comparative studies showed that male reproductive reserves measured as mass of the abdomen relative to total body mass and as the relative nitrogen content of the abdomen increase with degree of polyandry, whereas abdomen carbon, thorax nitrogen and thorax carbon did not (Karlsson 1995, 1996). Moreover, Bissoondath & Wiklund (1995) showed that polyandrous pierids had spermatophores with relative high protein content. Thus, in polyandrous species, males seem to have a body constitution favourable for producing large and nutritious ejaculates. Wiklund *et al.* (2001) also concluded that reproductive output increased with the degree of polyandry in eight pierids. There is therefore evidence that butterflies with a polyandrous mating system use spermatophores as nutrient gifts that help to increase female fecundity.

POTENTIAL AND REALIZED FECUNDITY

At the start of adult life a female has a certain reproductive potential. This is the number of eggs she could lay under suitable ecological conditions. Lifespan and the amount of time invested in egg-laying are important factors that may vary considerably within and between species (García-Barros 1998, 2000c). Several factors may prevent a butterfly from laying all eggs successfully. Potential fecundity is determined by larval resource quality and quantity, whereas realized fecundity also depends on biological and physical factors during adult life. Only under optimal environmental conditions will realized fecundity approach potential fecundity, and typically, butterflies only realize a proportion of their potential fecundity (Warren 1992a).

Size, body mass and fecundity

Body mass is most frequently used as a measure of resources available for egg production. Within a species, large individuals lay more eggs than do small individuals (Dunlap-Pianka 1979, Gilbert 1984, Kimura & Tsubaki 1986, Banno 1990, Karlsson & Wickman 1990, Karlsson 1998, Hughes et al. 2000) and interspecific studies have found similar patterns (Wiklund et al. 1987, 2001, García-Barros 2000c). From an evolutionary perspective, butterflies can increase their fecundity by growing larger. Butterflies have a compartmentalized body with the thorax containing flight muscles and the abdomen containing the reproductive and digestive system. Abdomen mass should therefore be a more precise correlate of potential fecundity than total mass. Wickman & Karlsson (1989) found an allometric increase of abdomen mass with total body mass in seven butterfly species; the proportion of abdomen mass to total mass increased with total mass. Karlsson & Wickman (1990) tested whether fecundity scaled in the same manner in *P. aegeria*, and larger females were found to lay proportionally more eggs.

Size is predicted to be correlated most strongly with realized fecundity in species with low (quality) adult food intake and less so in species with significant amounts of adult-derived nutrients, including nuptial gifts. Several studies indicate low correlations between wing length (measure of adult size) and fecundity (Boggs 1986), and although there are examples of strong correlations observed in the laboratory between size and fecundity (see above), few studies have measured size–fecundity patterns in the field (but see Saastamoinen 2007). Under field conditions females may, however, be constrained in their egg-laying and size effects can be masked by other factors, including longevity (Leather 1988) or the time available for egg-laying (Wiklund & Ahrberg 1978, Wiklund & Persson 1983). In species with nuptial gifts, females may only realize their full reproductive potential when they receive a sufficient amount of male material via the spermatophores (Karlsson 1998). What ultimately determines adult size under natural conditions remains a timely subject worth investigating in more detail. Little is known on what actually sets the limits to body size or, in other words, what are the costs of being large (Gotthard 2004)?

Condition and fecundity

Host plant quality is a significant factor for reproductive output in insects (Awmack & Leather 2002), including butterflies. Individuals feeding on poor-quality plants typically grow smaller with lower reproductive reserves (Leimar et al. 1994, Karlsson et al. 1997, Mevi-Schütz & Erhardt 2003a). Under such conditional constraint, we end up with larval-capital-related size effects on fecundity as discussed above.

Realized fecundity is clearly condition-dependent and this dependence includes the nutritional status of the female. Carbohydrates from adult feeding are incorporated in the eggs to a varying extent (Boggs 1997a, O'Brien et al. 2004) and a lack of nectar carbohydrates can severely reduce fecundity (Stern & Smith 1960, Murphy et al. 1983, Hill & Pierce 1989, Karlsson & Wickman 1990, Romeis & Wäckers 2002). However, there is a consensus that nitrogen is the nutrient that usually limits reproductive output (Boggs 1997a, 1981, Tabashnik 1982, Karlsson & Wickman 1990, Boggs & Ross 1993, Fisher et al. 2004a, O'Brien et al. 2005) since eggs have high nitrogen content (Karlsson 1998, García-Barros 2006). Butterflies from host plants with low nitrogen content show reduced size, as experimentally shown for *P. aegeria* (Talloen et al. 2004). Under natural conditions, growing larvae may actively search for better host plants to avoid or reduce shortage of nitrogen and to try to maintain an optimal balance between nitrogen and carbohydrates (Goverde et al. 2002).

Butterfly egg production usually starts at a high level; it peaks after a few days, and then decreases until the end of life (Stern & Smith 1960, David & Gardiner 1962, Labine 1968, Gossard & Jones 1977, Suzuki 1978, Zalucki 1981, Jones et al. 1982, Boggs 1986, Wiklund & Karlsson 1984, Karlsson 1987, Wickman & Karlsson 1987). The rate of decrease is a function of the female's feeding ability, food quantity and quality and number of matings which altogether determines her condition. Some *Heliconius* species show almost no decrease in egg production when feeding on pollen, but do show a decrease in their egg-laying rate when deprived of pollen (Dunlap-Pianka et al. 1977). When butterflies consume nectar, they ingest sugars and to a varying extent, nitrogen-rich amino acids. Thus, most adult butterflies complement their capital of larval resources with income from adult nectar feeding. Nectar is a nutritional complex food source containing primary monosaccharides (glucose + fructose) and disaccharides (sucrose) but also several amino acids (Baker & Baker 1973). Carbohydrates in nectar are an important supplement for egg manufacturing and adults deprived of carbohydrates will significantly reduce the number of eggs laid. Amino acids in nectar are also predicted to be of importance for fecundity (Baker & Baker 1973, Murphy et al. 1983, Hill & Pierce 1989, Alm et al. 1990), although several studies have failed to detect any significant effect of amino acids in nectar on fecundity (Hill & Pierce 1989, Mevi-Schütz & Erhardt 2003b). However, Mevi-Schütz & Erhardt (2005a) found an effect of amino acids in nectar on fecundity for females reared under poor conditions.

Trade-off between number and size of eggs

Egg size is a significant factor for insect fecundity as there is a trade-off between offspring size and number (Smith & Fretwell 1974, Roff 1992, Fox & Czesak 2000). Several intraspecific butterfly studies have found a negative correlation between egg size and number (e.g. Blau 1981, Fischer & Fiedler 2001b, Braby 2002). Seko et al. (2006) showed both a phenotypic and a genetic negative correlation between egg size and fecundity in the skipper Parnara guttata. At the interspecific level, García-Barros (2000c) concluded from his comparative study that evolutionary changes favouring increased egg size act as a constraint on fecundity (and vice versa).

Temperature may influence the trade-off between egg size and number. Females of the tropical butterfly Bicyclus anynana laid smaller, but more numerous, eggs at higher temperatures (Fisher et al. 2003) and also laid a considerable higher total egg mass compared to females under lower temperature treatment. Similar to many other insects, butterfly egg size decreases under laboratory conditions with maternal age (Jones et al. 1982, Murphy et al. 1983, Wiklund & Persson 1983, Karlsson & Wiklund 1984, 1985, Boggs 1986, Kimura & Tsubaki 1986, Hill & Pierce 1989, Braby & Jones 1995, Fischer et al. 2003). Decreasing egg size can help females to preserve high fecundity later in life. It could, however, be argued that it is an artifact of captivity in which females become depleted of certain nutrients that are easy to acquire in the field (Moore & Singer 1987). However, field studies have shown that reproductive reserves and egg weight decreased in a similar way to laboratory conditions (Karlsson 1987, Wickman & Karlsson 1987). Since egg size and caterpillar size are highly correlated (Karlsson & Wiklund 1984, Reavey 1992, Braby 1994, Fischer et al. 2002, 2003), higher investment in egg mass during early life suggests that 'females play it safe'. However, this is on the assumption of a positive correlation between egg size and fitness. This assumption is often made in theoretical studies

on optimal reproductive strategies (Smith & Fretwell 1974, Begon & Parker 1986, Roff 1992), but it does not always seem to hold empirically. In five Swedish satyrines this correlation was low, at least over a critical minimum value of egg size (Wiklund & Persson 1983, Karlsson & Wiklund 1984, 1985). These studies are of temperate-zone butterflies. Interestingly, Braby (1994) found a strong correlation between the size of young caterpillars (which relates to egg size) and survival in three tropical satyrine species and in another tropical satyrine, Bicyclus anynana; Fisher et al. (2003, 2006) found a correlation between egg size and offspring fitness. Wickman et al. (1990) did, however, find a positive relationship between egg size and offspring survival in the temperate-zone Coenonympha pamphilus, but there was an interaction with phenology. Larvae from relatively heavy eggs laid during the autumn survived the winter better than larvae from smaller eggs. So, for butterflies living in seasonal environments, predictions of egg size may change with season (Braby & Jones 1995, Seko & Nakasuji 2004). Egg size can also vary with thermal differences among vegetation and landscape types at the intra- and interspecific level (Karlsson & Van Dyck 2005, Karlsson & Wiklund 2005). Altogether, a complicated picture emerges about the adaptive significance of egg size variation and its relationship to fecundity when we address variation with foodplant quality and habitat use (Braby 1994, Fischer & Fiedler 2001b, Karlsson & Van Dyck 2005, Karlsson & Wiklund 2005), time constraints, seasons and voltinism (Wickman et al. 1990, Braby & Jones 1995), latitudinal and altitudinal patterns (García-Barros 1994, Fischer & Fiedler 2001b), egg to body size scaling and phylogeny (Wiklund et al. 1987, García-Barros 1994) and maternal lineage (Mevi-Schütz & Erhardt 2003b). Of course there are also several methodological issues that may generate differences among studies.

Since nectar-feeding butterflies cannot compensate to maintain reserves obtained in the larval stage, nitrogen and other body reserves reduce concomitantly with the number of eggs laid (Karlsson 1994, Stjernholm & Karlsson 2000, Stjernholm et al. 2005). Wiklund & Karlsson (1984) proposed the resource depletion hypothesis based on the idea that a physiological constraint operates on egg mass. This hypothesis was inspired by early work of Telfer & Rutberg (1960), Campbell (1962), Wellington (1965) and Richards & Myers (1980). Daily egg mass will decrease under continuous egg-laying as a consequence of the depletion of reproductive reserves. Thus, the proportion of the larval reserves expended for each egg is an unchanging function of the remaining reproductive reserves. Breakdown of this constraint will be

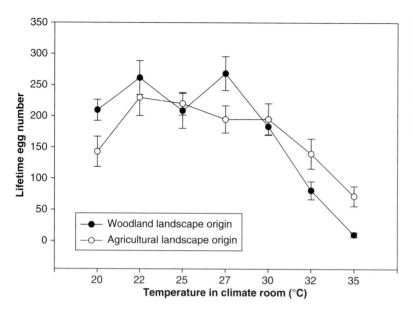

Figure 14.2. Total lifetime egg production of two populations of *Pararge aegeria*, one originating from an open agricultural landscape and one originating from a closed woodland landscape, under seven different temperature treatments (from Karlsson & Van Dyck 2005, courtesy by authors).

under different selective pressure depending on, first, how strongly females are limited by some factor for their egg-laying and, second, how strongly egg size is coupled to off-spring fitness (Wiklund *et al.* 1987). Wiklund & Karlsson (1984) found evidence for the resource depletion hypothesis in *Lasiommata megera* and results by Boggs (1997a) further supported this hypothesis.

Thermal constraints on fecundity

Temperature affects growth rate and development time of caterpillars and hence also the total mass and nutritional reserves (or 'capital') available for reproduction in the adult stage. But weather conditions, and in particular temperature, also interfere with realized fecundity. Body temperature is a crucial factor that affects activity and reproductive performance in flying heliotherms (Chapter 6). As in other ectotherms, the relationship between body temperature and performance typically shows an optimum bounded by critical limits (Chown & Nicolson 2004). As body temperature departs from the optimum, egg maturation and egg-laying performance decline gradually as well. Hence, poor weather can severely reduce fecundity by limiting flight activity and oviposition time-budgets (Dennis 1993). This is particularly significant in temperate-zone areas, and surely in northern Europe.

Studies by Watt and colleagues showed that genetic variation in phosphoglucose isomerase (*pgi*) in *Colias* butterflies

is associated with different adult activity levels (see Wheat *et al.* 2006 for references). Females with genotypes enabling flight under relatively low temperature may have a higher realized fecundity in relatively cool environments whereas other genotypes are favoured under high ambient temperatures (Watt 1992). Changes of temperature-related performance curves can trigger selective responses in life-history traits including fecundity (e.g. Partridge *et al.* 1995) and it can be expected that insects adapted to different temperature regimes exhibit optimal performance corresponding to the temperature prevailing in their habitat. In other words, females of species living in warm habitats should have a higher temperature optimum for egg-laying than females of cold habitats. Karlsson & Wiklund (2005) tested this prediction in two Swedish satyrine butterfly species adapted to dry open habitat (*Hipparchia semele* and *Coenonympha pamphilus*) and two adapted to forest habitat (*Pararge aegeria* and *Aphantopus hyperantus*). Woodlands are typically cooler and better buffered environments than more open land (Noss & Csuti 1997). The results showed that lifetime fecundity exhibited bell-shaped curves relative to constant ambient temperature with the open habitat species peaking at a higher temperature than the forest species (Fig. 14.2). Longevity decreased with increasing temperatures among all species, but the open landscape living species survived better at higher temperatures. These results support the idea that changes in habitat temperature profile alter temperature-related fecundity.

Different thermoregulation related behavioural and morphological adaptations may affect fecundity, but these relationships are not straightforward. Wing melanization may, for example, help females to raise their body temperature and increase flight activity and fecundity (Watt 1968, Gilchrist & Rutowski 1986, Kingsolver 1987a, Ellers & Boggs 2004a). *Colias* females with darker colour can heat up faster during cold weather and thereby raise their fecundity (Nielsen & Watt 1998, 2000). However, production of melanin is costly in terms of nitrogen, and females with pale wings can allocate more nitrogen to abdominal reserves (Graham *et al.* 1980) and thereby increase their potential fecundity. Under warm conditions paler females have a higher realized fecundity compared to darker females. Ellers & Boggs (2004a) found in *Colias eurytheme* significantly raised levels of egg maturation when the hind wings were experimentally darkened.

Time constraints and fecundity

Time for flight is a factor that strongly influences butterfly fecundity (Kingsolver 1983a, b, Springer & Boggs 1986) and the number of unlaid eggs or egg shortfall is recognized to be an important mortality factor in butterfly life-table analyses (e.g. *Leptidea sinapis*: Warren *et al.* 1986). Unsuitable weather conditions limit activity time-budgets of egg-laying females and even cloudy conditions with moderate temperatures will reduce or prevent adult activity in most species, although egg maturation continues. As a result, females will have higher egg loads ready to be laid for the next bout of suitable weather. Such conditions typically result in less discriminative oviposition behaviour. Oviposition choice is known to be an important life-history trait in butterflies (see Chapter 4). As a consequence, time constraints due to unsuitable weather may have a considerable impact on realized fitness.

ALLOCATION AND REALLOCATION TO REPRODUCTION

Resource allocation

As discussed above, there are different ways for adult females to increase their reproductive nutrient reserves, but these contributions are relatively small compared to collected reserves in the larval stage. However, larval reserves are not exclusively used for reproduction, but they are also needed for somatic maintenance. Similar to other holometabolous insects, butterflies have to decide during metamorphoses on how to allocate the larval collected resources, some of which cannot be replenished by the adult. An individual could maximize its fitness by finding the best balance in partitioning the resources between building a robust enough body to serve the animal throughout its life and putting aside as much as possible for reproduction (Boggs 1981a). A large proportion allocated to reproduction will increase potential fecundity but not necessarily realized fecundity. Boggs (1981a) proposed that butterflies will alter their allocation depending on expected differences during adult life. Thus, an expected poor adult diet or low income from male nuptial gifts would select for a high proportion of the larval reserves to be allocated to reproduction (Karlsson 1995). Short-lived animals would be expected to allocate relatively more to reproduction than those expecting a long life. *Polygonia c-album* is an interesting case here. Its winter morph hibernates and its summer morph develops directly leading to huge within-population variation in adult lifespan. This is expected to influence the relative pay-offs of allocation to somatic maintenance versus reproduction. As predicted, Karlsson & Wickman (1989) found that females of the short-lived summer morph allocate a higher share of their nitrogen reserves to reproduction than do overwintering females, resulting in increased fecundity (Karlsson *et al.* 2008).

Trade-off between fecundity and mobility: oogenesis–flight syndrome

Although some allocation decisions are irreversible, there is still an option of recycling somatic tissue for reproduction. The idea of the secondary use of flight muscles and flight fuel for reproduction is well established. The oogenesis–flight syndrome was proposed by Johnson (1969) who suggested that migratory insects delay reproduction until after their migratory flight and thereafter they utilize thorax resources for reproduction. A common denominator to species that break down their flight muscles is a life cycle with a pre-reproductive dispersal phase (Johnson 1969). Hence, most studies about this syndrome addressed the trade-off between flight versus reproduction in insects with dispersal dimorphism and in species that can selectively histolyse their flight muscles. In dimorphic species, the dispersive morph with clear flight ability has a reduced fecundity (and/or later sexual maturation) than flightless morphs (Zera & Denno 1997). Insects that histolyse their flight muscles achieve higher fecundity or an earlier peak egg production than those that retain their muscles. Most studies on the cost of flight ability have focused on females, but flight ability also reduces male reproductive success in some species at least (Zera & Denno 1997).

Histolysis of flight muscles is known from several insect orders, including Homoptera, Diptera, Coleoptera, Heteroptera, Orthoptera and Hymenoptera (see Stjernholm *et al.* 2005 for references). Recent findings support the view that it also happens in butterflies. Karlsson (1994), Stjernholm & Karlsson (2000) and Norberg & Leimar (2002) showed that thorax mass and thorax nitrogen content decreased with age in several butterfly species. Karlsson (1998) showed for *Pieris napi* that resources from the thorax were incorporated into the eggs. Although Vilora *et al.* (2003) found a suspected case of brachypterous females in an alpine satyrine, this seems to be an extremely rare option in butterflies, whereas it appears to be more frequent in moths, particularly in high alpine geometrid moths. In butterflies, both sexes rely on flight to locate nectar and suitable host plants, amongst other essential activities (Chapter 13). Since reproductive potential depends on the amount of resources available, a reallocation of resources from flight muscles to reproductive parts could increase reproductive output. Hence, despite obvious requirements for flight, butterflies may benefit from selective reallocation of resources from flight muscles to reproduction (Fig. 14.1).

Evidence for the increase of fecundity from flight-muscle-derived nutrients comes from experiments (e.g. Stjernholm & Karlsson 2006) and comparative studies (e.g. Stjernholm *et al.* 2005). Karlsson (1998) showed in a nitrogen budget study that thorax nitrogen was used for egg-laying. Stjernhom (unpubl. data) demonstrated a massive breakdown of muscle mass in *Pieris napi* females and eliminated the possibility that this nitrogen came from some sort of storage proteins. A field study by Stjernholm & Karlsson (2000) on the highly polyandrous *P. napi* suggests that males can also use flight muscles for the production of spermatophores. This was experimentally confirmed; mated males had significantly lower thorax nitrogen content than unmated males (Stjernholm & Karlsson 2006). Thus, males may use thorax nitrogen to manufacture nitrogen-rich spermatophores which in turn can boost female egg production.

Srygley & Chai (1990a) found that unpalatable butterflies had relatively smaller flight muscle mass compared to palatable butterflies. They concluded that the decreased flight muscle mass could incur loss of flight performance and manoeuvrability but that more of their total resources could be directed to reproduction. Hence, a reduction of flight muscle mass could affect flight ability and the ability to locate host plants. Interestingly, Stjernholm *et al.* (2005) found in a comparative study with 11 temperate-zone butterflies that although thorax mass decreased with age, all species had a thorax ratio (i.e. thorax mass divided by total body mass) that increased with age due to a faster use of abdominal resources. Although thorax flight muscle mass does decrease with age, flight performance – which relates to relative thorax investment rather than absolute investment – may be unchanged over the lifespan. Further studies are needed to confirm this prediction.

British work shows that *Pararge aegeria* females from permanent populations invested more in abdomen mass than females from recently colonized sites at expanding range margins (Hill *et al.* 1999a). They assumed that larger abdomens equate to higher fecundity. This was confirmed by direct fecundity measures from common garden experiments; females from resident populations laid significantly more eggs than females from colonized, more dispersive populations (Hughes *et al.* 2003). However, they repeated the measurements in two regions and fecundity differences were clear, while this was not always the case for abdomen mass. These data indicate reduced investment in reproduction in more dispersive populations, but a lack of any difference between populations in abdomen mass may not necessarily mean any differences in fecundity (Hill *et al.* 2005). However, Hanski *et al.* (2006) could not find a trade-off between dispersal and fecundity in *Melitaea cinxia* since these patterns can be confounded by differences in metabolic rates among dispersive and resident females. Such trade-off patterns have also been found in other Lepidoptera. Female *Cydia pomonella* moths of a sedentary selection strain laid, for example, more eggs during their lifespan and lived longer compared with a mobile selection strain (Gu *et al.* 2006).

BUTTERFLY FECUNDITY IN CHANGING LANDSCAPES

Habitat fragmentation changes resource distribution patterns. Several evolutionary ecological studies have addressed this issue relative to consequent selective changes in dispersal to track resources (e.g. Travis & Dytham 1999, Thomas 2000). Altered environmental conditions in fragments compared with continuous habitat may also change life-history traits other than dispersal. Karlsson & Van Dyck (2005) investigated the effect of temperature on fecundity and longevity in *Pararge aegeria* from continuous woodland landscape and from fragmented agricultural landscape. Parallel measurements show that temperature profiles of sites used by *P. aegeria* differ significantly between the landscapes: both average and maximal daily temperatures are higher in fragmented agricultural landscapes than in woodland landscapes (Merckx *et al.* 2008). In a common garden experiment with F_2 generation individuals, females of woodland landscape origin achieved a higher maximum daily fecundity and lifetime number of eggs than did agricultural landscape females at low ambient temperatures, but this reversed at high ambient temperature (Fig. 14.2). Additional to the temperature effect on egg size (i.e. larger eggs under cooler conditions), Karlsson & Van Dyck (2005) showed that eggs of females from agricultural landscape were heavier. Different, though not mutually exclusive, adaptive hypotheses were proposed to explain this pattern relative to differential desiccation risks and to differential time allocation for oviposition between contrasting landscapes.

CONCLUSIONS AND PERSPECTIVES

Factors influencing reproductive patterns and fecundity are essential ingredients to explain key ecological parameters like population size and distribution range. Therefore, studies examining resource acquisition, allocation patterns and potential and realized fecundity do not only make significant contributions to life-history research and evolutionary ecology, but also to population dynamics and other conservation-related issues. However, several demographic and conservation studies still ignore variance components in fecundity. As we have demonstrated in this chapter, butterfly fecundity is variable both at an ecological and also at an evolutionary scale.

Although our knowledge on butterfly fecundity and reproduction has improved a great deal over the last decades, it still remains a challenge to predict what will happen to fecundity patterns and ultimately to population dynamics under anthropogenic environmental changes as we face them now. Rising ambient temperature could increase fecundity per se, but other often indirect climatic effects and feedbacks may also affect fecundity negatively at the very same time. For instance, relative nitrogen content of host plants may decrease with elevated CO_2 levels which accompany higher temperatures (Mevi-Schütz *et al.* 2003). Future butterfly fecundity studies need to take up the complex issues of climatic factors on potential and realized fecundity. There are reasons to believe and some evidence to demonstrate that changes in the spatial configuration of ecological resources across landscapes interact with the responses to climatic changes. Butterfly populations may show contemporary evolution in fecundity patterns and respond to climate change and other anthropogenic disturbances by adaptive physiological or life-history responses (Karlsson & Van Dyck 2005). This is a field we need to explore more as it will help us to predict better how populations respond to fast-changing environmental conditions.

The broader scientific challenge is to improve the integrative nature of studies on butterfly reproduction and fecundity relative to other fields in biology. We need to improve the integration of results and concepts on fecundity with models trying to understand insect–plant relationships, population ecology, dispersal and life-history strategies.

15 • Gradients in butterfly biology

SÖREN NYLIN

SUMMARY

The extent of variation in a number of key characteristics differs between species and may relate to mobility patterns and gene flow. Variable attributes can include morphology, size, voltinism, resource use, life-history patterns, physiology, biochemistry and behaviour. Despite a long history of documentation of variation within species, it is only recently that fundamental questions about the mechanisms controlling and maintaining variation and the significance of this variation (or its lack) to population and species persistence have been asked. Environmental variation is the norm. Within Europe there are recognisable and distinct climatic gradients, with unique patterns of seasonal weather conditions, for which it is possible to predict how some species characteristics should respond. However, simple predictions often fall down and understanding why reveals much about the relative roles of selection, gene flow and butterfly developmental constraints. Our understanding of the nature and importance of variation has been advanced by the recognition that some traits change rapidly in relation to current selection pressures, and that observable changes of traits represent alterations of genotype × environment interactions during development. However, there are many unanswered questions, which fact precludes generalisations about the adaptive versus non-adaptive nature of variation. Gene flow may prevent local adaptation, and in some mobile species some traits may be 'generalistic' rather than adapted to specific environmental conditions. As human activity continues to change the spatial occurrence, abundance and quality of resources in the landscape, and is likely to increase in severity and rate in the future, understanding how variation is produced and what roles it has in enabling species persistence are becoming more pressing. Variation is an important contributor to evolution. Understanding the former is becoming critical if we are to understand the fate of butterflies in a rapidly changing world in which we impose new selection pressures.

INTRODUCTION

Ecological gradients

Environmental factors are ubiquitous and vary quantitatively over a range of scales. Gradients exist in the *chemical* composition of the environment (including nutrients as well as detrimental substances) and in *physical* parameters such as temperature, humidity or solar radiation. These gradients impact on organisms and hence cause *biological* gradients affecting other species. For example, the local population size of *Cupido minimus* in locations around Göttingen, Germany is dependent on the availability of the host plant *Anthyllis vulneraria*, which in turn occurs on calcareous grasslands (Krauss *et al.* 2004b). Other biological gradients could involve the risk of being attacked by predators and parasites, the strength of competition, or the availability of partners in mutualism. Thus the distribution of *Plebejus argus* in Doñana National Park, Spain is determined by the distribution of its mutualist ant species *Lasius niger*, which in turn is determined by vegetation and ground characteristics (Gutiérrez *et al.* 2005).

Ecological gradients can be more or less stable over time, depending on what causes them and can also occur at very different scales – from molecular distances upwards. For butterflies, the relevant scale of gradients goes from millimetres when it involves leaf chemistry affecting larvae, to the thousands of kilometres that individuals of some migrating butterflies cover in their adult life. In more sedentary species this geographical scale is still highly relevant; it is the scale of the distribution range of the species, subspecies and local populations. This chapter focuses primarily on the geographical scale of gradients and within-species effects rather than between-species effects.

Why study gradients?

A central research question in ecology is why a certain species occurs (or not) at a certain location, and at what density? It

Ecology of Butterflies in Europe, eds. J. Settele, T. Shreeve, M. Konvička and H. Van Dyck. Published by Cambridge University Press. © Cambridge University Press 2009, pp. 198–216.

follows from this that population dynamics, and the evolution of local adaptations that influence these dynamics, are central subjects in ecology. Hopefully, from 'pure research' on butterflies will come the knowledge and theoretical understanding that can eventually be put to use in more applied areas of ecology (Nylin 2001); see for instance Dennis (1993).

Population dynamics is the summary result of the specific characteristics of the studied organism plus the environmental factors that act upon these characteristics. This summing exercise is not a simple one (Nylin 2001), because of a complex hierarchy of interactions. Firstly, fluctuations at the population level are strongly influenced by processes at both lower (individuals) and higher (communities and ecosystems) hierarchical levels. Secondly, species characteristics are not identical in all individuals nor fixed for all time, and neither are those of the other species in the community – they vary and evolve. However, evolutionary theory provides a framework on which to base predictions of species characteristics expected in a given environment, and thus a basis for generalisation and theoretical understanding. Thirdly, all of the processes already mentioned interact in complex ways with environmental factors. For a complete understanding of population dynamics we therefore need to know not only about adaptations to the particular environment, but also exactly how organisms react to *variation* in environmental factors. Few traits are directly determined by genotype, usually the resulting phenotype (the traits that are eventually expressed) is determined also by the environment experienced during the development period, and to a lesser extent the adult environment. In other words, there is usually *phenotypic plasticity* (Gotthard & Nylin 1995, Nylin & Gotthard 1998; see also Chapter 12). The set of phenotypes observed for a single genotype over a range of environments is the *reaction norm* of that genotype. Adding even further to complexity, other species in the community that interact with the study species will also show phenotypic plasticity.

In addition, there is variation in the *behaviour* of the study species and other members of the community. Behaviour is generally conditional, i.e. contingent on the precise situation, and can be considered as an extreme form of phenotypic plasticity. Typical questions that arise include: How is host plant choice by females affected by variation in plant abundance, or plant quality (itself an example of plant plasticity)? Do enemies such as parasitic wasps shift to preferentially attacking the study species when it becomes very abundant, in which case there is density-dependent mortality that will dampen fluctuations in population number?

Because of complexity, our understanding of population dynamics in the wild, and hence of the core question in ecology, is still fragmentary and our ability to predict dynamics is weak. Even understanding dynamics with the benefit of hindsight is difficult (but see Pollard & Yates 1993, Pollard & Moss 1995, for promising butterfly studies). This is where the study of ecological gradients can make an important contribution; where there is an observable gradient in the chemical and physical characteristics of the environment, there is also a basis for understanding and perhaps even predicting variation in the traits of living organisms along that gradient. This is explored here, with a particular focus on adaptive and non-adaptive gradients along different latitudes and altitudes.

GENETICS

Theory

For local adaptation to evolve, genetic variation must be present in the population (or arise de novo from mutation, but this is a rare event) and there must be selection in favour of the variant which must be strong enough to counteract gene flow from other populations (see Chapter 10), with differing selection pressures. Gene flow and the possibility of local adaptation are linked to population structure and mobility, which varies between species. This variation partly follows phylogenetic patterns, but the degree of mobility can vary strongly, even among closely related species. For example, *Pieris brassicae* and *P. rapae* are mobile species (Petersen 1947, Tolman & Lewington 1997) with poor abilities to hibernate in northern Europe. They must be mobile, as they evidently recolonise northern areas yearly from the south. By contrast, the related *P. napi* is thought to be a much more sedentary species.

If gene flow is not too great, there can be genetic *isolation by distance* even in the absence of barriers (see Chapter 10), but when geographical barriers to dispersal exist, there is even greater isolation. In both circumstances, populations can diverge genetically. When they diverge without strong barriers, there may be a gradual change in the genetic composition of the species along a geographical gradient: a genetic *cline*. Even if the environment changes rapidly with distance, gene flow from one type of environment to the other will tend to lead to a cline being formed. With stronger barriers to gene flow, or in a very sedentary species, there can be more threshold-like changes in the genetic composition over distance. Part of the genetic differences among local populations may be adaptive, the result of selection fitting the population to local conditions and another part will be at least potentially non-adaptive, the result of random processes such as genetic drift. Whatever the reasons for the differences, they will sometimes be large enough to prevent reproduction between members of separate populations, in which case speciation has occurred.

Gene flow and phenotypic patterns

The observed phenotypic variation within species, among geographical locations, broadly follows the expectations from genetic theory. This was demonstrated in two classic works by Petersen (1947) and Danilevskii (1965). Petersen found evidence of geographical variation, within Fennoscandia, in wing coloration and size of *Pieris napi* but not in the more mobile *P. brassicae* and *P. rapae* (Table 15.1). The observed variation in *P. napi* was presumed to be genetically based and probably representative of local adaptation. That this is the case is even clearer from Danilevskii's studies of photoperiodic thresholds for diapause induction. Diapause is a hormonally controlled state of arrested development (Tauber *et al.* 1986), which permits

survival in extreme temperatures and when food is absent. Danilevskii reported that in the mobile *P. brassicae*, large areas are populated by a 'monotypic race' with the same photoperiodic diapause threshold at different latitudes, whereas clines adapting the population to local daylengths are evident not only in *P. napi* but also in *P. rapae*, a weaker flier than *P. brassicae*.

Petersen (1947) also found little evidence of geographical variation in the nymphalids *Nymphalis antiopa* and *Aglais urticae*, both of which are medium- to long-distance migrants in the tribe Nymphalini. In contrast, nine other less mobile nymphalid species in the Melitaeini and Argynnini showed geographical variation within the study area (Table 15.1).

Gene flow and population structure

The phenotypic patterns reported above constitute only indirect evidence of butterfly population structures, with the extremes of very *open* structures in migratory species preventing local differentiation, to *closed* structures in more sedentary species, promoting differentiation and speciation. More direct evidence of genetic variation (or lack of it)

Table 15.1 *Butterfly species studied in Scandinavia by Petersen (1947)*

Species	Migratory	Flight capacity	Pattern clines (to north)	Size clines (to north)	Voltinism
Pieris brassicae	Yes	Strong	No	No	1–3, opportunistic
Pieris rapae	Yes	Medium	No	No	1–3, opportunistic
Pieris napi	Rarely	Medium	Darker, more yellow	Sawtooth (see text)	Shifting in region, 1–2
Colias palaeno	No?	Strong	Less yellow	Smaller	Univoltine
Nymphalis antiopa	Yes	Very strong	No	Smaller	Univoltine
Aglais urticae	Sometimes	Strong	Darker?	No	Univoltine in region
Mellicta athalia	No	Medium	Darker	Smaller	Univoltine in region
Mesoacidalia aglaja	No	Strong	No	Smaller	Univoltine
Brenthis ino	No	Medium	No	Smaller	Univoltine
Argynnis aphirape = Proclossiana eunomia	No	Medium	Complex clines	Smaller	Univoltine
Boloria selene	No	Medium	Darker	Smaller	Shifting in region, 1–2
Boloria euphrosyne	No	Medium	Darker	Smaller	Univoltine in region
Boloria freija	No	Medium	No	No	Univoltine
Boloria frigga	No	Medium	Darker	No	Univoltine
Argynnis pales = Boloria napaea	No	Medium	Smaller dark spot	Smaller	Univoltine
Argynnis arsilache = Boloria aquilonarois	No	Medium	Darker	Smaller	Univoltine

among local populations of European butterflies, from e.g. allozymes or variable sections of DNA, also exists.

In a highly mobile species, *A. urticae*, Vandewoestijne *et al.* (1999) found considerable genetic variation; most occurred at the within-population level rather than between populations. This is consistent with a pattern expected with high mobility among adjoining patches and occasional long-distance migration (hence a lack of isolation by distance) in combination with rare extinctions and recolonisations. Only rarely have mobility and genetic differentiation actually been measured together, but Lewis *et al.* (1997) assessed the population structure in the lycaenid *Plebejus argus* using three independent methods: mark–release–recapture studies, observation of colonisation of newly created habitat over 7 years and allozyme investigations. The species is relatively sedentary with the maximum movement detected being 395 m, and 1.4% of individuals moved between patches separated by 12–200 m; however this low mobility was sufficient to prevent differentiation at the metapopulation scale.

Several studies of allozyme genetic diversity and patterns of gene flow in *Parnassius mnemosyne* have also been performed. Napolitano & Descimon (1994) found that apparent gene flow was not always related to geographical barriers and/or isolation by distance. Other factors, e.g. historical colonisation processes, are important for a full understanding of population structure in this species. Barascud *et al.* (1999) report the results of a case study of such a colonisation event, the introduction of *Proclossiana eunomia* in the 1970s to Morvan, central France, where it was previously absent. Both enzyme electrophoresis and wing patterns demonstrated a significant differentiation of the populations both from their source population (Ardennes) and within themselves, over fewer than 30 generations.

Several of the studies mentioned clearly demonstrate the power of forces opposing local adaptation, even when genetic variation is present. They also indicate that the isolation by distance (genetic differentiation as a more or less linear function of the distance between sites) may not be correlated with mobility in a straightforward manner (see also Peterson & Denno 1998).

Gene flow and selection

In some cases genetic variation among sites and geographical variation in selection pressures have been studied together in the same butterfly species, but selection has typically been studied only indirectly. Goulson (1993) studied allozyme variation in the satyrine *Maniola jurtina* across populations in southeast England. He found that geographically distant populations differed very little in gene frequencies, an unexpected result for a sedentary species if alleles are neutral to selection. Goulson concluded that similar selection pressures may be maintaining allele frequencies across the area sampled, a conclusion based on observed associations between patterns of variation at two loci and behavioural, morphological and environmental variables. Of particular interest with regards to the focus of this chapter is that the genotype at the locus for phosphoglucose mutase (PGM) affected the length of time for which individual butterflies could fly continuously, and PGM allozyme frequencies varied according to altitude. In the case of *M. jurtina*, allozyme variation gave little information on population structure but instead on patterns of selection.

The most thorough investigation of direct relationships between allozyme variation and selection in butterflies is the series of studies of American pierids by Ward Watt, Joel Kingsolver and colleagues, prominently involving adaptation to circumstances at different altitudes. At higher elevations, the lack of time suitable for flight may be a significant factor in reducing the reproductive success of butterflies. In *Colias* butterflies, counter-adaptations involve basking behaviour, wing pigmentation (e.g. Watt 1968; see below) and fur thickness (Kingsolver & Moffat 1983; and see also Kingsolver & Watt 1983). However, there is also adaptive variation in alleles of the enzyme phosphoglucose isomerase (PGI), which affects glycolytic metabolism and hence flight capacity at different times of the day and several consequent fitness components in the wild (Watt *et al.* 1996, 2003). Some evidence of similar patterns exists from *Colias alfacariensis* in Europe (Cleary *et al.* 2002).

WING PATTERNS

General

Wing patterns are the most often studied trait in investigations of geographical variation in butterflies, probably because of the comparative ease with which they can be studied nondestructively in collections of preserved insects. Known functions of wing patterns include thermoregulation, predator avoidance and other forms of communication, such as mate recognition (see Chapter 13 and Dennis 1993).

Wing colours are determined either by the structural characteristics of wing scales, such as the iridescent blues of lycaenids or *Apatura* butterflies, or by pigments synthesised in the metabolic pathways of the insect, such as the dark pigment melanin. Pigment synthesis and deposition is the more important source of actual wing patterns. Nijhout (1991) has demonstrated that despite the diversity of butterfly wing patterns there is also considerable phylogenetic conservatism. The constraints on wing patterns suggest that convergent and parallel evolution of wing patterns in similar environments – as well as multiple functions for a specific pattern – is likely to be common. The latter possibility remains largely unexplored, as researchers tend to focus on a single function (but see Kingsolver & Wiernasz 1987, Dennis 1993, and Chapter 13).

Thermoregulation

The best understood variation in wing patterns is undoubtedly the darker (melanic) pigmentation observed in pierid butterflies at higher latitudes and altitudes. This was noted by Petersen (1947), for instance in Fennoscandian *Pieris napi* (Table 15.1). The northern subspecies *adalwinda* is heavily pigmented dorsally and ventrally, especially along the veins. Similar patterns of pigmentation occur in several different pierid taxa in the Alps as well as in North and South America (Watt 1968, Shapiro 1976, 1984a, b, Kingsolver & Wiernasz 1987).

The adaptive significance of this wing pattern variation is believed to be in relation to thermoregulation. Darker insects should warm up more easily and hence take crucial advantage of sunlight at sites where ambient temperatures are low, whereas light insects run less risk of overheating at high temperatures. This interpretation is supported not only by the comparative evidence, but also by phenotypic plasticity in pierid wing pigmentation, with darker wings occurring in spring generations (Shapiro 1976, 1984b). Darker (and more hairy) butterflies warm up faster and maintain higher body temperatures than paler individuals, although probably only pigmentation of the body and the wing surface closest to it is adaptive in this respect (Kingsolver 1987a). The degree of pigmentation often differs between the sexes (Shapiro 1984a, Kingsolver & Wiernasz 1987), probably because of trade-offs with other selection pressures (Kingsolver & Wiernasz 1987).

Gradients in pigmentation related to thermoregulation seem to be present in most butterfly taxa. Although published studies detailing the evidence for such climatic adaptations are scarce, the general patterns are clear. In Scandinavia (Table 15.1) and the British Isles most variable species show clines in wing morphology in a north–south direction, most often with darker wings in the north.

Exceptions are found in the Lycaenidae (Dennis & Shreeve 1989, Dennis 1993). This could relate to the structural iridescent colours that are prominent in the Lycaenidae. Minute variation in the thickness and morphology of the structural films can have large effects on the absorption of solar radiation in non-iridescent wavelengths, and hence can be important for thermoregulation (Miaoulis & Heilman 1998, Biró *et al.* 2003). For a lycaenid butterfly with structural colours, it is not obvious whether a bluer or blacker insect absorbs most light in total. Also, some lycaenid species may be dorsal reflectance baskers, so that increased iridescence on the midpart of the wings may actually aid the insect in heating up the body (Dennis 1993). Lycaenids at high altitudes sometimes totally lack the blue structural colours and the associated wing morphology, perhaps because absorption of blue and UV light is important for thermoregulation in these cool environments, where solar radiation in this spectral range is strong (Biró *et al.* 2003). An alternative explanation not involving thermoregulation is that the latitudinal patterns are confounded by those shaped by other selection pressures, e.g. the need for rapid mate-finding when time is limited (Dennis 1993; and see below).

Among skippers, species that are darker towards the north include *Ochlodes venata* (Dennis & Shreeve 1989, Dennis 1993). In the Papilionidae, high-altitude populations of the holarctic *Parnassius phoebus* are darker (Guppy 1986). In the Nymphalidae, north–south clines of this kind are especially common among the satyrines, e.g. *Pararge aegeria* and *Maniola jurtina* (Dennis & Shreeve 1989, Dennis 1993).

Predation avoidance and communication

Besides thermoregulation, wing patterns can serve many functions that relate to predation avoidance and intraspecific communication (for reviews see Brakefield 1984, Nijhout 1991, Dennis 1993). Adaptive gradients should be present in such patterns as well, but evidence for clines caused by predation is not clear. However, for these kinds of presumed functions of wing patterns, the evidence from seasonal plasticity is suggestive. Brakefield & Larsen (1984) noted that many tropical butterflies differ in wing patterns between the dry and wet season forms, the former seemingly more cryptic and the latter with larger ventral eyespots that may serve to deflect predator attacks. These authors suggested that such seasonal variation in patterns could correlate to variation in the activity of the butterflies. Different generations of temperate butterflies often differ in wing patterns as well. In some cases, especially among pierids, this most probably relates to different requirements for thermoregulation (see above). In other cases, such as satyrine eyespot patterns, predation patterns differing between generations may play a role (Brakefield 1984). Latitudinal and other geographical clines in eyespots are also common in satyrines. In *M. jurtina* there is a latitudinal decrease in mean ventral hindwing ocellation, but an increase in size of forewing eyespots with latitude both dorsally and ventrally in Britain (Dennis 1993). Spots decrease in size and/or number towards the north in *Coenonympha tullia*, *C. pamphilus* and *Aphantopus hyperantus* (Dennis & Shreeve 1989, Dennis 1993). On the island of Corsica eyespot patterns vary with altitude in *C. corinna* (Brunton *et al.* 1991).

Geographical variation in predation could contribute to these clines. Undoubtedly predation from lizards and some invertebrate taxa will decrease with latitude and altitude, and bird communities will also change, perhaps with the net result of lower predation pressures. At a smaller scale, gradients may be caused more by the presence or absence of protective cover, and hence the need for deflective patterns rather than crypsis (Brakefield 1984). Nevertheless, it would seem likely that observed gradients are often the outcomes of complex trade-offs also involving selection pressures other than predation.

Aposematic coloration or mimicry in adults has not been clearly demonstrated for any native European butterflies, but it is likely that the *Aristolochia*-feeding *Zerynthia* of southern Europe signals unpalatability. In tropical areas both functions are ubiquitous and often form geographical gradients. This happens when one aposematic model for Batesian or Mullerian mimicry is replaced by another over distance, as in *Heliconius* (Mallet & Gilbert 1995).

Selection for patterns aiding mate-finding could be involved in shaping the gradients seen in some lycaenid species, with brighter blue dorsal coloration in northern localities. In *Polyommatus icarus* females show more blue colours in populations from northern Britain, and this could be due to the shorter seasons and lower population densities in the north (Dennis 1993, Howe 2004; and see below). Blue females are more apparent and preferentially approached by males, and so this phenotype may have a selective advantage in cool climates despite (presumably) increased risk of predation.

Wing patterns as adaptations

The adaptive interpretation of phenotypic gradients is not always straightforward. Non-adaptive historical events may also be involved and adaptation and history are not mutually exclusive. However, Dennis & Shreeve (1989) suggest that in the case of latitudinal clines in Britain the available evidence supports explanations involving recent adaptations to climatic variables. A case for such adaptation can also be made for *M. jurtina* across the western Palaearctic; genetic affinities split the species into Western and Eastern genotypes (reflecting recolonisation from different glacial refuges) but some wing patterns show latitudinal clines which seem to have evolved in parallel (Schmitt *et al.* 2005b).

Another alternative interpretation of observed gradients in the amount of dark pigmentation involves development. When insects are reared at lower temperatures this often results in darker individuals (e.g. Goulson 1994). Hence the variation could be environmentally determined, with temperature affecting enzymatic activity in metabolic pathways. Consequently, the observed variation may not always represent a 'true' adaptation for thermoregulation in the sense that its historical origin is related to this function, or even in the sense that it has been modified by selection.

Non-adaptive explanations for seasonal plasticity in butterfly spot patterns cannot be entirely ruled out either, especially when development time is the prime determinant of seasonal form (Gotthard & Nylin 1995) as in African *Bicyclus* (Windig 1994). A delayed ontogeny may clearly affect the induction of wing patterns (Nijhout 1991). In the temperate *M. jurtina*, where wing patterns vary seasonally, genes that affect development rate also pleiotropically affect spot number and placement (Brakefield 1984), indicating that selection on development rate may also cause variation in wing pattern expression.

LIFE CYCLES

Phenology

One of the most universally found ecological patterns is the shift in phenology seen along climatic gradients. With increasing latitude and altitude butterflies start to fly later in the season. For instance, Gutiérrez & Menéndez (1998) showed such a shift in the phenology of butterflies with elevation in a northern Spanish mountain area, Picos de Europa, associating this change with the direct effect of lower average temperatures on juvenile growth rates and adult flight thresholds (and hence in a sense possibly 'non-adaptive'). Phenology is also affected by temperature variation within locations, as observed with climate change (Sparks & Yates 1997, Stefanescu *et al.* 2003a). As a consequence, temperature determines the northern limits of bivoltinism (two generations per year) and to some extent the northern limits of species distributions (Bryant et al. 2002).

Shifting phenology to a later date may be adaptive if this means that butterflies fly when temperatures are highest (Gutiérrez & Menéndez 1998). In addition, host plant phenology will also shift with changes of climatic gradients, so the butterflies 'should' shift concurrently with the plants to synchronise their life cycles with them (Rodríguez et al. 1994, Sparks & Yates 1997). It is clear that temperature effects on growth and development are not universally the same in all species and populations, or for all developmental stages, but can evolve (Nylin & Gotthard 1998, Gotthard et al. 2000a, Fischer & Fiedler 2002; and see below). The studies on American pierids mentioned above show that adult flight thresholds also evolve. To the extent that emergence dates are determined by conditions that will break diapause in the spring, this is yet another trait that is very probably genetically variable among and within populations (Pratt & Ballmer 1993). Hence, some part of gradients in butterfly phenology is likely to represent adaptation to local conditions rather than non-adaptive 'direct' effects of temperature, but it is not a simple matter to distinguish them.

Peterson (1995) found that populations of the lycaenid Euphilotes enoptes in Washington State, USA, differed markedly in phenology with altitude, but flight times were not linearly correlated with altitude. Peterson suggested that this is consistent with gene flow that is biased from low to high elevations (a result of source size and oviposition behaviour), so that selection on optimal phenology is swamped at high altitudes. One way to investigate local adaptation more directly is by rearing samples from different altitudes in a common environment. Pratt & Ballmer (1993) did this for E. enoptes and another lycaenid, Philotiella speciosa. They demonstrated that the intensity of pupal diapause correlated positively with the flowering periods of the hosts of each species and negatively with the elevation of their sites of origin. This is evidence of adaptation to life at different altitudes, serving to synchronise the butterflies with their hosts.

Voltinism

Many butterfly species show gradients in voltinism within Europe, with an increase in the number of generations in response to favourable season length in terms of days or day-degrees. The exact number of generations possible for a species at a given site is determined by its life history, host plant preferences (see below) and by adaptations for thermoregulation (Bryant et al. 1998, 2002). However, if adding a generation becomes possible by evolutionary modification of one of these characteristics, and if this increases long-term fitness of the genotype, then such modifications may evolve – unless hindered by evolutionary constraints.

Some vagrant European species, e.g. Pieris brassicae, Colias crocea, Lampides boeticus, Issoria lathonia, Cynthia cardui or Vanessa atalanta, lack true diapause and hence the ability to hibernate at high latitudes and altitudes, and probably lack local adaptation to season length. However, they make opportunistic use of the season as long as it is favourable, fitting in as many generations as possible. This may result in clines in voltinism, but perhaps entirely due to phenotypic plasticity. In more sedentary species, there is probably always also a genetic component to shifts in voltinism via genetic differences in photoperiodic thresholds and in the propensity for other environmental factors to induce diapause.

Latitudinal gradients of voltinism differ between species. For example some species (Pieris napi, Lycaena phlaeas, Polyommatus icarus, Pararge aegeria and Coenonympha pamphilus) have two full generations per year as far north as well within Sweden and Fennoscandia (Dal 1978, 1981, Tolman & Lewington 1997), whilst others (e.g. Erynnis tages, Papilio machaon, Leptidea sinapis, Celastrina argiolus, Maniola jurtina and Polygonia c-album) have occasional or partial second generations at the same latitudes (if at all), and complete bivoltinism only further south. In a similar response to season length, some species have two or more generations in the central European lowlands but only one at higher altitudes in the Alps and other mountain ranges. Species in this category include Erynnis tages, Iphiclides podalirius, Cupido minimus, Pararge aegeria and Melitaea cinxia.

Many species depart from these gradients of increasing number of generations per year with increasing local season length. They may occur only at latitudes and altitudes where two or more generations are possible, and never seem to form univoltine populations. Some of these species are strictly Mediterranean, but among those with ranges stretching further to the north are Pontia daplidice, Lycaena tityrus, Aricia agestis, Everes argiades and Lasiommata megera. Whether the lack of univoltine populations in such species is a result of their relatively restricted distributions or whether the inability to form univoltine populations is actually what restricts their ranges is unclear. In the case of Aricia agestis, this species may have formed relatively recently as the sister species to the univoltine A. artaxerxes (Aagaard et al. 2002) and may not yet have evolved univoltinism. Similarly,

P. daplidice is a relatively vagrant species, which forms semi-permanent populations at more northern locations such as the Swedish island of Gotland. Possibly it is never resident in the north long enough (in terms of numbers of generations) to evolve univoltinism as an alternative strategy, which could enable it to become a permanent inhabitant of even more northern sites.

Many species have only a single generation per year throughout at least their European ranges. Some of them are restricted to northern locations, which by itself might explain their univoltinism (e.g. *Pyrgus centaureae*, *Colias nastes*, *Boloria polaris*, *Oenis bore*). Similarly, others occur only at high altitudes in mountainous areas of Central Europe (e.g. *Parnassius phoebus*, *Boloria pales*, *Oenis glacialis*, many species of *Erebia*, *Polyommatus eros*). Yet others have disjunct distributions in only short-season areas, occurring in the north and then also at high altitudes further to the south (e.g. *Pyrgus andromedae*, *Boloria napaea*, *Erebia pandrose*, *Pseudaricia nicias*). A few species in these short-season categories even have 2-year development periods.

Season length, especially in Mediterranean climates, may be shortened by high temperature and summer drought that makes host plants inedible. In American *Danaus plexippus* this results in a reverse cline in voltinism, with more generations at a cooler northern site (Malcolm *et al.* 1987). Such reversed patterns have not yet been documented from Europe, but they could well be present in e.g. Spanish *Melitaea cinxia* (E. García-Barros, pers. comm.). Several satyrines (*Hyponephele lupina*, *Maniola jurtina*, *Kirinia roxelana*) may also spend the hottest time of the year as adults in a summer diapause (aestivation). Aestivation in other developmental stages also occurs, as in the satyrine *Pararge aegeria* (Wiklund *et al.* 1983), the lycaenid *Cyaniris semiargus* (Rodríguez *et al.* 1994) or the pierid *Pieris brassicae* (Spieth 2002). However, this is most probably a secondary adaptation to such climates, rather than the explanation for univoltinism.

The relationship between voltinism and geographical range is often unclear, even in the case of butterflies occurring only at high latitudes and/or altitudes. There is a phylogenetic component. Species in genera such as *Thymelicus*, *Zerynthia*, *Anthocaris*, *Gonepteryx*, *Satyrium*, *Agrodiaetus* and *Nymphalis* seem to be more or less constrained to univoltinism even in southern areas. In addition, in the large nymphalid taxa Argynnini and Satyrinae univoltinism is a rule with few exceptions. Within these groups, species in genera like *Boloria*, *Erebia* and *Oeneis* tend to have ranges restricted to areas with short favourable seasons. Such patterns suggest that the voltinism of a butterfly may

be part of a co-adapted trait complex, involving for instance life-cycle regulation, life-history traits, hibernation strategy and host plant choice. The developmental stage used for hibernation is often also phylogenetically conservative, in turn constraining host plant use (Hayes 1982, Carey 1994a).

Many univoltine species feed on host plants of low nutritional quality (e.g. grasses or woody plants), so that larval growth is necessarily rather slow and a multivoltine life cycle hard to achieve (Cizek *et al.* 2006). Other species feed on plants that are of high quality only in spring (deciduous trees and bushes) or on plant parts that occur only part of the season (flowers, fruits or seeds) so that they are constrained to oviposit at a certain time of the year. Hence, winter diapause can become obligatory and strictly tied to other aspects of the life cycle. Oviposition could for instance happen before hibernation in the egg stage, on permanent structures close to the host plant rather than directly on it, as in some Nymphalidae (Wiklund 1984). Species in taxa with such a univoltine lifestyle may then be locked in to inhabiting short-season areas, so that the univoltinism causes the distribution rather than vice versa.

Undoubtedly both directions of causality can be involved in specific cases – and also the biogeographical history of the butterfly in question. A population of a widespread species with shifting voltinism can become geographically isolated in a marginal area, and over time adopt a lifestyle suitable to such an area and univoltinism. Such a history has been suggested for the North American *Papilio canadensis*, formerly *Papilio glaucus canadensis* (Scriber 1988). This speciation event is evidently rather recent, but in other cases such a newly formed univoltine species can, over evolutionary time, give rise to new species sharing the same suite of characteristics, thus forming a higher taxon constrained to univoltinism.

It is clearly possible that shifts in voltinism can play a causal role even in the initial speciation events, when species differing in voltinism are first formed. This is precisely because voltinism affects, and is affected by, so many other traits. In transitional areas, where two generations are just barely possible and some individuals follow a univoltine life cycle, there may be disruptive selection in favour of one or the other lifestyle, rather than a poor compromise. Comparative studies like the one on the *P. glaucus/canadensis* complex could illuminate the relationships between voltinism, other aspects of the life cycle, and speciation. A European example is the lycaenids *Aricia agestis* and *A. artaxerxes*, formerly believed to be a single species. The northern species *A. artaxerxes* is strictly univoltine whereas *A. agestis* has two to three generations per year, and the difference in voltinism exists also in areas where the species overlap (Burke *et al.* 2005). It may well have played a part in the speciation process.

Parallel cases can undoubtedly be found in the case of altitude, and should be sought among univoltine species constrained to mountains, but with bi- or multivoltine close relatives at lower elevations. There are no absolute clear-cut cases, because of a lack of phylogenetic resolution to determine sister species, but genera with varying voltinism and with univoltine species in mountains include *Colias*, *Polyommatus* and related genera, *Boloria*, *Mellicta / Melitaea*, *Coenonympha* and *Erebia*. An equally common pattern, however, is that univoltine mountain species have lowland relatives that are also univoltine despite a longer favourable season, in support of the idea that causality can work both ways.

Other types of comparisons are also of interest, if intriguing. Why, for instance, is *Neptis rivularis* univoltine and *N. sappho* bivoltine, despite very similar ranges in Europe? Is this related to their different host plants? They differ in their Asian distributions, with *N. rivularis* being the more northern species – should the answer be sought there? The ultimate answers to such questions will perhaps often remain hidden in history, but there is a good chance of finding at least the proximate reasons for obligatory univoltinism. Subtle differences in the life cycle may be enough to prevent another generation to be easily added, and without a comparative approach they may easily be overlooked. To date, not many studies have exploited such possibilities.

Life-cycle regulation

The 'choice' of developmental pathway, either diapause or direct development to sexual maturation, typically shows a reaction norm with a threshold at a certain photoperiod and/or temperature (Danilevskii 1965, Tauber *et al.* 1986). This is because in many insect systems it is not adaptive for the insect to engage in 'half a diapause'; reproduction should occur this season or the next, nothing in between. Consequently, theoretical models suggest a sharp increase in the frequency of individuals entering diapause at the date in the season when such a life cycle becomes the optimal one to follow (Cohen 1970). In real situations, however, reaction norms may be more gradual, as in *Pararge aegeria* (Nylin *et al.* 1989). There can also be mixed uni- and bivoltinism over an extended part of the season, as in *Pieris napi* (Wiklund *et al.* 1991). The explanation is probably that genetic variation for photoperiodic thresholds is maintained

by complex trade-offs between the respective advantages of following one or the other developmental pathway, coupled with climatic variation among years and gene flow among areas.

Nevertheless, relatively sharp thresholds for diapause induction are common, especially when closely related individuals (families of offspring from the same parents) are exposed to different environments. One of the most clear-cut predictions that can be made regarding how insects should adapt to geographical gradients is that in relatively sedentary species photoperiodic thresholds for inducing diapause should be longer at more northern sites. The reasons for this expectation are twofold. Firstly, between spring and autumn, equinox daylengths are longer at higher latitudes, so that a given daylength corresponds to a later date in the north than in the south. Secondly, the favourable season is shorter in the north, so that at a given date (and daylength) less of it remains. The second principle applies also to altitude; at higher elevations insects should enter diapause at longer daylengths to ensure that early frost does not kill them off.

In the classic investigation of the pitcher-plant mosquito, *Wyeomyia smithii*, a model including latitude and altitude explained 96% of the variation in critical photoperiod among local populations (Bradshaw 1976). Comparable investigations of butterflies involving many different populations are lacking, but the more variable juvenile environments as well as the stronger flight of adults, leading to more gene flow, suggest that gradients could be more 'noisy'. However, Danilevskii (1965) did find threshold gradients in *Pieris napi* and *P. rapae*, and the papilionid *Atrophaneura alcinous* shows a clear latitudinal gradient in Japan (Kato 2005). Patterns consistent with such gradients in Europe have been found in *Pararge aegeria* (Nylin *et al.* 1995b), *Lasiommata petropolitana* (Nylin *et al.* 1996a, Gotthard 1998), *Inachis io* (Pullin 1986c) and *Polygonia c-album* (Nylin 1989).

Photoperiod is not the only seasonal cue for life cycle regulation, although for European butterflies it is the most 'noise-free' signal telling the date of the season, and in temperate areas the date is reasonably well correlated to seasonal fluctuations in conditions. However, there is often also a strong influence of temperature, interacting with and shifting the photoperiodic threshold, as in *P. napi* (Wiklund *et al.* 1991). Another important role of temperature in insect life cycle regulation is through its effects on juvenile developmental rates. Even in a hypothetical situation of pure photoperiodic control of diapause induction, the percentage of individuals entering diapause in the population will depend on the fraction that are in each developmental stage at a given time of the year. A detailed study of the influence of weather on the life cycle (and fraction of the population entering diapause) of *Pararge aegeria* in Britain is given by Shreeve (1986a).

The combinations of reliable seasonal cues and direct effects of temperature may be the only way to adapt reasonably well to life in strongly seasonal environments over large geographical areas, especially when constrained from optimal local adaptation by gene flow. Temperature alone is not a good cue for diapause induction (the relative role of temperature is probably much higher for diapause maintenance and termination, but these aspects of life-cycle regulation have been much less well studied). High temperatures in late summer might mean that there is time for an additional generation, but continued growth will not preclude death from an unexpected frost – or from starvation when the host plants are killed by frost (Inouye 2000). Locally adapted photoperiodic thresholds probably always take this risk into account; selection will favour conservative thresholds that induce diapause well in advance of winter.

Danilevskii (1965) suggested that the influence of other environmental cues should be most important near critical photoperiods, where they may provide additional information to the insect. An extension of this reasoning is that they may be more important in transitional areas, with mixed voltinism. There are few butterfly studies where the interactions between photoperiod and factors other than temperature have been studied in the laboratory. However, Bink & Siepel (1996) found a tendency towards a higher propensity of diapause in *Lasiommata megera* when the amount of nitrogen in the grass offered to larvae was experimentally altered. Adult diapause (and seasonal form) in *Polygonia c-album* is determined mainly by daylength together with temperature, but larval host plant quality also has significant effects (Wedell *et al.* 1997). Wedell and colleagues proposed that any factor that reduces growth or developmental rates, be it low temperatures or poor host plants, could be selected to serve as a cue that future conditions will also be unlikely to support rapid development and hence that diapause may be a better option.

LIFE-HISTORY PATTERNS

Generation length

Life-history traits are those aspects of organisms that quantitatively describe frequencies and temporal patterns of transitions in the life cycle, i.e. traits associated with growth, development and reproduction (Nylin 2001). The most direct prediction that can be made about insect life histories in seasonal climates is that the generation length (here defined as the time needed to complete the entire life cycle from adult to adult, but excluding any time spent in diapause) should evolve to fit gradients in season length, within an area where all individuals adopt a single pattern of voltinism (for many species this would apply throughout their entire range). In other words, generation length should be progressively shorter at high latitudes and altitudes. However, there are several caveats even to this simple prediction, which means that patterns will not be the same in all species because of constraints apart from season length on life history.

In opportunistic insects such as *Drosophila*, scramble competition for transient resources directly selects for as short generation length as possible – given that a reasonable adult size must still be reached. Although this is a less important consideration in most butterflies (competition can be fierce in batch-layers, but the genetic relatedness within the clutch means weaker selection for competitive abilities), similar situations could arise in very dense populations. If this is the norm for the species, it might minimise development time and no clear patterns among populations might be seen with respect to season length (Chown & Gaston 1999). Another caveat is that once the season is long enough to easily permit one complete generation, there may be little fitness gain to be had in prolonging the generation length, so in some species perhaps only populations in extreme areas will show local adaptation to season length. A third caveat is that there are many ways to achieve a short generation time, so there may be unique solutions within species or a group of species.

One mechanism is by thermoregulation, through morphological, physiological or behavioural adaptations. Morphological adaptations that facilitate flight at colder ambient temperatures may shorten the life cycle because mating and oviposition can take place at a higher rate in marginal conditions. Lowering the physiological temperature thresholds for adult activity is an alternative or complementary solution. The relevant adaptations may also occur in the juvenile stages. Females can place eggs at warmer sites when needed, to speed up egg development and early larval growth and developmental rates. This has been observed in American *Papilio glaucus* (Grossmueller & Lederhouse 1985). Increased larval melanisation, especially together with basking behaviour (Porter 1984, Ayres &

Scriber 1994, Bryant *et al.* 1998), is another way. Bryant *et al.* (1998) calculated that a second generation of *Aglais urticae* would not be possible in Britain without larval basking.

Simple models of insect growth and development (see Tauber *et al.* 1986) usually include a threshold temperature, specific for each species, below which development stops. Above this threshold a more or less linear response is imagined, with growth and developmental rate increasing at higher temperatures, being dependent on the average temperature and days above the threshold (day-degrees). Such models have been relatively successful when trying to predict the seasonal occurrence of insect pests. Undoubtedly such calculations can be useful also when predicting butterfly phenology, voltinism and life-history patterns over climatic gradients (Scriber 1994). However, such models are simplifications; if larvae bask they effectively experience more day-degrees than conventially measured (Bryant *et al.* 1998). Additionally, too high temperatures can be detrimental and day-degree overestimated (Bryant *et al.* 1997). Moreover, as part of local adaptation to climate, mean population threshold temperatures and reaction norms to temperatures above the minimum threshold may be locally adapted.

Temperature reaction norms can also differ even within the same individuals, a plastic response according to the state of that individual (Gotthard *et al.* 2000a). Different populations of the same species can evolve genetically different reaction norms. In *Papilio canadensis* the threshold temperature is about 10 °C, and the thermal sums for the season are 583 degree-days in interior Alaska and 985 in northern Michigan (Ayres & Scriber 1994). Basing life-cycle predictions only on these facts does not, however, give a correct result because at a low temperature of 12 °C, larvae from Alaska grow much faster than those from Michigan.

Such population differences are likely to be the result of 'counter-gradient selection' (Conover & Schultz 1995): selection favouring reaction norms that counter the adversary effects of the environment. For this reason phenotypic difference among populations may not be observed when each is observed under their natural conditions, even though both the genotypes and the environment differ! Another consequence is that care needs to be exercised in performing and interpreting laboratory rearing experiments. For example, high-altitude populations may be well adapted to slow growth and have long development times in the field, but when they are compared with low-altitude populations at a higher common temperature they may grow and develop rapidly, probably also showing correlated life-history responses which are not representative of natural situations.

Differences in temperature reaction norms are expected to occur among species. Bryant *et al.* (1997) studied such variation in four European nettle-feeding nymphalid species (*Aglais urticae*, *Inachis io*, *Polygonia c-album* and *Vanessa atalanta*), and found that their differing temperature reaction norms and corresponding day-degree requirements could broadly explain their relative distributions and patterns of voltinism.

Another set of reaction norms may also evolve to fit butterflies to the season length at their sites, namely those that describe responses to daylength *within* developmental pathways (Nylin & Gotthard 1998). These reaction norms can serve to fine-tune development to the progress of the season, so that the developmental stage adapted for hibernation is reached at the correct time of the year. Such gradual responses have been observed in *Pararge aegeria* (Nylin *et al.* 1989) and the related *Lasiommata petropolitana* (Gotthard 1998), in *Polygonia c-album* (Nylin 1992) and in the lycaenid *Polyommatus icarus* (Leimar 1996). The likely adaptive nature of such reaction norms is demonstrated by the fact that they vary among species and individual states according to expectations from optimality (Abrams *et al.* 1996). In Swedish populations of *Pararge aegeria* development times are progressively shorter in shorter daylengths, probably because they signal that less time is left until winter (Nylin *et al.* 1989). These reaction norms are steeper in the univoltine central Swedish population than in the bivoltine south Swedish one, and absent in populations from England and further south where they would be maladaptive because the same larval stages occur before and after summer solstice in different generations (Nylin *et al.* 1995b).

Two additional ways of shortening the development time are also known. The first is by laying larger eggs, thus providing the offspring with more of the necessary resources for quickly completing growth and development. The second mechanism is by simply cutting development short, i.e. pupating at a lower mass at the cost of reaching a smaller adult size. Whatever the exact mechanisms, we can predict that there will be shorter generation lengths in areas of shorter season length, at least at extreme latitudes and altitudes. But what happens when there is a shift in voltinism?

In areas where there is a partial second generation, there are two possible life cycles, associated with very different optimal life histories. Insects which develop directly to sexual maturation and produce a generation of offspring in the same season, at the same site where other individuals, which

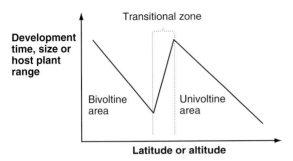

Figure 15.1. Hypothetical sawtooth patterns in development time and correlated traits (adult size, host plant range) across a shift in voltinism (number of insect generations per year). See Roff (1980, 1983), Nylin (1988), Nylin & Svärd (1991) and Scriber & Lederhouse (1992).

might be siblings, only complete a single generation, have a season length that is effectively cut in half (Roff 1980, 1983). This should lead to a sawtooth pattern for generation length (and traits expected to correlate with it, especially juvenile development time) over climatic gradients (Fig. 15.1). In the transitional area with a shift to bivoltinism there will be very variable development times, according to life cycles, and in the area of complete bivoltinism average development times well below those in adjacent areas of complete univoltinism. If development time and adult size are correlated, there should also be corresponding patterns in size (Roff 1980, 1983; and see below). The sawtooth patterns could be due entirely to genetic differences in life-history traits or entirely to phenotypic plasticity (e.g. extra instars added or slower growth as long as seasonal cues signal that there is still time before winter) but a combination is much more likely (Nylin & Gotthard 1998).

Sawtooth patterns have been most thoroughly studied in the pitcher-plant mosquito *Wyeomyia smithii* (Bradshaw & Holzapfel 1983), crickets (Bradford & Roff 1995, Mousseau & Roff 1989) and in the water strider *Aquarius remigis* (Blanckenhorn & Fairbairn 1995). These studies show clear evidence of local adaptation to season length, involving development time, photoperiodic thresholds and/or adult size, and in some cases there is also evidence of the larger variation expected in transitional areas. However, the studies also serve to display the complexity of observed patterns, because development time or size may vary among populations, but not necessarily both. There is also the problem created by counter-gradient selection on reaction norms. Some unexpected results are undoubtedly due to this category of genotype × environment interactions.

Egg size

Egg size (or weight) is often allometrically related to female size, and also involved in trade-offs with total fecundity; more eggs may mean smaller eggs (Fox & Czesak 2000, García-Barros 2000c). Beyond this, the adaptive and non-adaptive causes of variation in insect egg size are not well understood (Fox & Czesak 2000). Latitudinal and altitudinal clines in egg size, often with larger eggs (relative to female size) to the north and at higher altitudes, are common in insects (Azevedo *et al.* 1996, Fox & Czesak 2000, Armbruster *et al.* 2001) and have been found in Lepidoptera (Harvey 1983). In some cases this could be a mechanism for shortening development time in colder areas where season length is short; a large egg should result in a hatchling with a head start in development (Sibly & Monk 1987).

Geographical gradients in egg size along climatic gradients could sometimes be due to direct effects of the environment rather than to genetic clines, because low temperatures experienced by females often result in larger egg size (Fox & Czesak 2000, Fischer *et al.* 2003). The mechanism is not understood, and such reaction norms may or may not be adaptive. One possible adaptive explanation could be that low temperatures constrain the number of eggs that can be laid, because of reduced female activity, hence larger eggs may evolve in such environments if they somehow give higher offspring fitness (Wiklund *et al.* 1987, Karlsson & Wiklund 2005).

In butterflies, there is little or no evidence of a positive correlation between egg (or hatchling) size and development rate among individuals, as would be necessary for the hypothesis that geographical clines in egg size are caused by gradients in season length. Among species the correlation is in fact the opposite to expected (García-Barros 2000c), but this could more reflect species differences in general life history. For instance, high-latitude satyrine species lay larger eggs (García-Barros 1994) and can be expected to also more often be adapted to a slow life cycle with univoltinism or even 2-year development.

Some insect studies have found the opposite gradient with larger eggs in the south (e.g. in the water strider *Aquarius remigis*: Blanckenhorn & Fairbairn 1995). One source of conflicting results (in addition to gene × environment interactions) may be that shifts in voltinism occur within the area inhabited by the studied populations. García-Barros (1992) took this into account when choosing the univoltine butterfly *Hipparchia semele* for study. In this species, eggs were smaller in a northern population (Sweden) than in the south

(Spain), both in absolute terms and relative to adult body size. García-Barros suggested that large egg size may have been selected for in Spain; the dry late summer in Mediterranean areas will result in low-quality plants and a need for large hatchlings. Feeding ecology may be an important correlate to egg mass in Lepidoptera in general (Reavey 1992) because larger hatchlings can feed on tougher plants (Braby 1994; but see García-Barros & Munguira 1997).

Clearly egg size, and any other size, including adult size, is a highly complex trait with many ecological and evolutionary implications (García-Barros 2000c). Gradients in egg size could be caused by selection on adult size, by gradients in plant quality, or other factors that may directly affect adult size. This is because larger females often lay larger eggs, but often smaller relative to their own size (Fox & Czesak 2000).

Adult size

A large adult size may increase reproductive success, but size is subject to trade-offs with advantages of a short juvenile development time, as well as with some direct advantages of small size, such as lower food demands (for a review, see Nylin & Gotthard 1998).

The best-known rule relating to gradients in animal size is probably Bergmann's rule: animals should be larger in colder climates because a large body has a small surface-to-volume ratio and will thus retain warmth better. This idea applies most strongly to endothermic animals. Clines with larger size in colder climates do occur also in ectotherms (Van Voorhies 1996) and a strong possibility is that such patterns are due to direct effects of the environment and could sometimes be non-adaptive. Reaction norms where low temperatures experienced by juveniles lead to larger adults are common in ectotherms, suggesting to some that a constraint must be acting, because ectotherms should become larger rather than smaller in warmer temperatures (since growth is in general less costly when it's warm: Berrigan & Charnov, 1994). It has been suggested that the pattern could be an almost unavoidable consequence of constraints on ontogeny (Van Voorhies 1996). However, adaptive explanations have also been proposed. In seasonal environments, low temperatures possibly can signal that adding more generations the same season will be costly and that slow development to a larger size will instead result in higher overall fitness (see Atkinson 1995, Wedell et al. 1997).

If juvenile development time is correlated to adult size the opposite pattern to the one described by Bergmann's rule should be the norm within a particular pattern of voltinism (Fig. 15.1). In fact, the formal theory on sawtooth patterns was originally developed by Roff (1980, 1983) with adult size in mind. Roff envisioned that large size should correlate with high fecundity, but that this fitness advantage is opposed by increased mortality when development time becomes too long for the prevailing length of the season. The predicted pattern of increasing size in warmer climates is in fact very common in ectotherms (Mousseau 1997), and sawtooth patterns have been described in some insects (e.g. the cricket *Allonemobious socius*: Bradford & Roff 1995, Mousseau & Roff 1989).

What is observed in butterflies? In Britain, some species are smaller to the north (Dennis & Shreeve 1989; e.g. *Anthocaris cardamines*, *Hipparchia semele*, *Aphantopus hyperantus*) but others are larger (e.g. *Lycaena dispar*, *Polyommatus icarus*, *Pararge aegeria*). The species with clearest gradients of decreasing size to the north are univoltine, whereas those that are instead larger to the north are bivoltine and may be at least partially univoltine in their northernmost populations; if so these observations are consistent with sawtooth patterns (Fig. 15.1). In Sweden, Petersen (1947) studied size variation in a number of butterflies (Table 15.1). His summary was: 'when the wing-length of butterflies shows a geographic variation, it decreases to the north, and is in mountains shorter than in the lowlands of the same latitude. This variation may be partly due to selection, if the rate of development and size are correlated.' In *Pieris napi* Petersen found a more complex pattern consistent with predictions from sawtooth theory, and indeed Petersen's discussion of the phenomenon is similar to that of Roff (1980, 1983). His work was based mainly on specimens collected from the wild, but he also did some rearings suggesting that size variation among sites is at least partly genetic. The only observation in Petersen's material not entirely consistent with the theory is the lack of a sawtooth pattern in *Boloria euphrosyne*, which has a partial second generation in southern Sweden. However, its size cline is less steep than in the closely related *B. selene*, which is univoltine throughout Sweden.

Nylin & Svärd (1991) investigated size variation within Sweden in 16 species of satyrines and lycaenids, with similar results (Table 15.2). Thirteen of the species have constant patterns of voltinism in Sweden. Almost all of these species showed clear size clines with smaller size to the north (see for instance Fig. 15.2; they were also larger in continental Europe than in Sweden). Size clines were much less clear in the three species with shifting voltinism (*Pararge aegeria*, *Coenonympha pamphilus* and *Lycaena phlaeas*). In these

Table 15.2 *Butterfly species studied in Sweden and continental Europe by Nylin & Svärd (1991)*

Species	Size clines (to north in Sweden)	Larger in Europe	Voltinism
Lycaena hippothoe	Smaller (n.s.)	Yes	Univoltine in region
Lycaena virgaureae	Smaller	Yes	Univoltine
Lycaena helle	Smaller (n.s.)	No	Univoltine
Lycaena phlaeas	No	Yes (n.s.)	Shifting in region, 1–2
Coenonympha tullia	Smaller	Yes (n.s.)	Univoltine
Coenonympha arcania	Smaller	Yes (n.s.)	Univoltine
Coenonympha hero	Smaller	Yes (n.s.)	Univoltine
Coenonympha pamphilus	Smaller	Yes	Shifting in region, 1–2
Aphantopus hyperantus	Smaller	Yes	Univoltine
Maniola jurtina	No	Yes	Univoltine
Erebia ligea	Smaller	Yes	Semivoltine
Hipparchia semele	Smaller (n.s.)	Yes	Univoltine
Lasiommata megera	Smaller (n.s.)	Yes	Bivoltine
Lasiommata maera	Smaller	Yes	Univoltine in region
Lasiommata petropolitana	Larger (n.s.)	Yes	Univoltine
Pararge aegeria	Smaller	Yes	Shifting in region, 1–2

Correlations with latitude, or differences between Sweden and Europe, that were statistically significant at least in one of the sexes. Non-significant trends seen in both sexes noted, but depicted by n.s. 'No' indicates that sexes showed different (non-significant) trends.

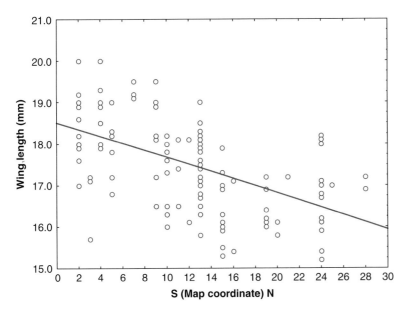

Figure 15.2. Example of a gradient with increasing butterfly size from north to south: male *Coenonympha tullia* in Sweden (data from Nylin & Svärd 1991). This species has a single generation per year all over Sweden. In *C. pamphilus*, where there is a shift in number of generations, the geographical pattern is not as strong, and more reminiscent of a sawtooth pattern (Fig. 15.1) (see Nylin & Svärd 1991).

species reversals of the trends in wing size with latitude seemed to coincide with areas where shifts in voltinism occur, although the sample sizes from critical areas were too small to permit documentation of sawtooth patterns. From America, the best example of a butterfly pattern compatible with sawtooth theory is the latitudinal size variation in the *Papilio glaucus/canadensis* species complex (Scriber 1994).

These studies are mainly based on collections of wild-caught specimens, which means that it is not possible to distinguish between size gradients caused by local genetic adaptation or by phenotypic plasticity. The potential for plasticity to affect size variation is evident already from the fact that generations of butterflies often vary in size. In *Pieris napi* the second generation is larger (Petersen 1947, Wiklund *et al.* 1991). This is despite a shorter development time, the explanation being that larvae of this generation grow faster (Wiklund *et al.* 1991). Because of this potential for plasticity to create size gradients, experiments where samples from several sites are reared side by side in common conditions are necessary to demonstrate local genetic adaptation. In one such study, of *Pararge aegeria* from different latitudes in Britain, Sibly *et al.* (1997) found clear evidence of genetic differences among sites, with butterflies from different sites showing consistent size differences at three different rearing temperatures. They found the opposite pattern to that predicted by sawtooth theory: northern individuals were larger, and field-collected specimens evidently also are larger to the north (Dennis & Shreeve 1989). The possibility that the northern populations are in fact partially univoltine or that there is a gradient in frequency of developmental pathways (this species is rather unique in having two hibernating stages – pupa or half-grown larva: Nylin *et al.* 1989) should be examined. Patterns of colonisation may also have contributed to the gradients, if colonising individuals are larger and northern populations were colonised most recently (Hill *et al.* 1999a).

In another developmental experiment, Fischer & Fiedler (2002) compared temperature reaction norms of different populations of *Lycaena hippothoe*. They reasoned (in line with sawtooth theory) that in multivoltine populations a short development time is important because of the possibility of extra generations, even at the expense of smaller size, which is not true in univoltine populations. Interestingly, they found that in univoltine populations temperature had little effect on pupal and adult weight, whereas in the multivoltine populations higher temperatures led to lower weights, especially in protandrous males, the category of individuals most stressed for time.

What is the situation further south? In his formulation of the sawtooth theory Roff (1980, 1983) noted that the fitness of insects in strongly seasonal environments is not well described by the intrinsic rate of increase (r) because of the constraint that diapause usually has to take place in a specific developmental stage. Generation length is an important component of r, implicitly assumed to be able to vary continuously. However, there can be no one-and-a-half generations in a seasonally constrained population, only mixtures of univoltine and bivoltine individuals. Under these circumstances, the *yearly* rate of increase is a more suitable fitness model.

As season length increases, going towards low latitudes and altitudes, the intrinsic rate of increase (r) or its counterpart (K) become progressively more realistic models of fitness in potentially multivoltine species as the non-seasonal situation implicitly assumed in r/K-theory is reached. This is of course even truer in areas with relatively weak seasonality, such as the island of Madeira (Nylin *et al.* 1993). In these situations, there could be r- or K-selection, in the sense of MacArthur & Wilson (1967) and Boyce (1984), depending on whether there are frequent population fluctuations due to density-independent mortality or more stable populations with density-dependent mortality.

Clearly, many different gradients in butterfly size are possible, even though most of the detailed studies on within-species patterns suggest that individuals from colder climates are smaller. There are many idiosyncrasies of particular species that could lead to other patterns. It should come as no surprise, therefore, that between-species patterns in size (averages for whole communities) are even less general, with size gradients differing among study areas (Hawkins & Lawton 1995) or among families (Hawkins & Devries 1996).

Protandry, sexual size dimorphism and lifespan

Protandry, the earlier entry of males into the population, is a very common pattern in butterflies and other insects (Wiklund & Fagerström 1977). There are two types of geographical gradients that may relate to protandry. Firstly, theoretical considerations suggest that, because of selection for protandry, males should be more prone to enter diapause than females (Wiklund *et al.* 1992). This has been confirmed by empirical findings from butterflies (Wiklund *et al.* 1992, Wedell *et al.* 1997). In other words, geographical gradients in seasonal sex ratios are expected. In areas of mixed voltinism there should be an overall female bias among directly developing individuals and a male bias among individuals that enter diapause. This has not been studied directly, but there is a documented male bias in some populations of overwintering *Danaus plexippus* (see Nylin *et al.* 1995a). Secondly, gradients in protandry itself can be expected, and also in sexual size dimorphism if the two are correlated. Singer (1982) suggested that protandry may be a major cause of the frequently observed pattern of males being smaller than females. If the sexes grow at the same rate, males would have to be smaller in order to be able to complete development earlier. For this reason Singer suggested patterns in sexual size dimorphism according to whether protandry should be expected. Less dimorphism is predicted in species or generations where diapause intrudes late in the life cycle and thus synchronises the sexes, so that protandry has to be achieved by breaking diapause earlier (Nylin 1992, Wiklund *et al.* 1996). More importantly, in the context of gradients, less protandry and dimorphism is expected in areas where generations partly overlap, so that a particular development time in males will not predictably lead to them emerging before the females. The extreme case occurs in relatively non-seasonal areas where butterflies fly continuously over the year, and it is not possible for males to enter the population before females.

In a test of the latter set of predictions, Nylin *et al.* (1993) studied populations of *Pararge aegeria* from two strongly seasonal habitats (Sweden and England), from a population where generations partly overlap (Spain) and from an almost non-seasonal habitat (Madeira). As predicted, protandry was weak in Spain and absent on Madeira. However, there was no corresponding variation in sexual size dimorphism, suggesting that selection pressures other than the trade-off with development time determine the sizes of the sexes. The lifespans of the sexes are also similar on Madeira whereas Swedish males live for a shorter time than females, perhaps because the Swedish population is more synchronised and protandrous, so that all females are soon mated and there is no selection on males to prolong their life (Gotthard *et al.* 2000b).

BEHAVIOUR

Host plant choice

A balance between the costs and benefits of specialisation should determine the width of the host plant range of a phytophagous insect. A species feeding on many different types of plants will be less likely to be constrained in its realised fecundity by lack of available host plants, and may gain additional advantages from risk-spreading in variable environments. However, some plants will inevitably be better in terms of overall offspring fitness than others, and if they are abundant enough and their quality predictable enough it may pay the female to only oviposit on these plants. Moreover, once specialisation on a particular plant has evolved, additional advantages can be gained from specific adaptations to utilising this particular plant. Most butterflies are oligophagous, feeding on several plant species in a single family or a few closely related families. A few are polyphagous and can feed on several relatively unrelated families (e.g. the nymphalids *Polygonia c-album*, *Nymphalis polychloros* and *Cynthia cardui*, the lycaenids *Callophrys rubi* and *Celastrina argiolus*), but some are even more specialised, and more or less dependent on a single host plant genus or even species. This will naturally limit distributions, causing gradients in abundance according to the occurrence of specific hosts.

What about genetic gradients in host plant utilisation, selected for because of environmental gradients? This is a possibility as soon as there is more than one potential host plant species. Often a clear hierarchy of host plants can be observed if females are given a choice under experimental conditions, or if the acceptance of each plant is measured in no-choice experiments. Such hierarchies may function to ensure that the best hosts are preferentially used when available, but that alternative plants are still used when needed (Wiklund 1981). There are gradients of hierarchical preferences over the geographical distribution of species, either in the order of preference ranking for different plants or in the range of plants used. There are many factors varying from site to site that could conceivably cause such gradients.

PLANT ABUNDANCE

By necessity, spatial variation in the availability of potential host plants will cause gradients in what host plants are actually used. Somewhat surprisingly, there are few examples of such spatial variation in plant occurrence causing genetic, rather than just ecological, changes in butterfly host plant utilisation. Evolutionarily conservative plant hierarchies, where females from different populations rank plants in the same order and even utilise them as hosts to similar degrees among populations in the laboratory, despite variation in the actual host plant use in the field (enforced by availability), have been found in the American *Papilio zeliacon* (Wehling & Thompson 1997). There are few European studies where females from different populations have been compared under controlled circumstances. Females of the lycaenid *Polyommatus icarus* from the Stockholm area in Sweden and the Baltic island of Öland do not differ in their relative preference for *Lotus corniculatus* and *Oxytropis campestris*, despite the fact that the latter is common on Öland but virtually absent elsewhere in Sweden (Bergström *et al.* 2004).

Such conservatism could be due to gene flow being strong enough to prevent local adaptation to plant abundance patterns even though selection maintains genetic variation (Bossart & Scriber 1995; for a contrast see Hagen 1990), or to specific constraints in the plant acceptance behaviour of females. *Polyommatus icarus* females could for instance perceive both *Lotus* and *Oxytropis* as acceptable plants based on the presence of the same chemical cue. It would still be possible for preference differences among populations to evolve by the use of chemical components that differ between the plant genera, but it is possible that the selection for this to happen is often weak. It may not be very costly to include rare or absent plants in the potential hierarchy, unless females actually spend time searching for them in the field and/or reject other potential plants and thereby reduce their fecundity (Bergström *et al.* 2004).

There are, however, examples of rapid evolution of butterfly host plant utilisation in response to changes in plant abundance, as in the Nearctic *Euphydryas editha* (Singer *et al.* 1993). In this species there are clear gradients in host plant utilisation, and evidence of frequent host plant shifts. Interestingly, the best European example is from the related butterfly, *Melitea cinxia*, which varies in its relative preference for *Plantago* and *Veronica* among sites in the Åland archipelago of Finland, according to local abundance (Kuussaari *et al.* 2000). This could relate to the strength of selection for adjusting preference to local plant abundance. Butterflies such as *E. editha* and *M. cinxia*, that lay their eggs in large clusters and where larvae frequently eat their entire native plant individual, may be under stronger selection to oviposit only on species abundant enough to provide more food nearby.

PLANT PHENOLOGY

Climatic and other environmental gradients can affect plant phenology and thereby the availability of suitable plant stages for oviposition. Plant phenology is especially important if only a particular plant stage is oviposited on, such as flower buds or young leaves. Rodríguez *et al.* (1994), working with the lycaenid *Cyaniris semiargus* in Spain highlight the common causal chain from physical gradients to plants

and then to herbivores. In the study area (Doñana National Park) the species is univoltine and monophagous on *Armeria velutina* (Plumbaginaceae) (elsewhere it feeds on *Trifolium* and other Fabaceae). Newly hatched larvae feed on flowers of this plant before entering an aestivation diapause, so the optimal butterfly phenology is constrained by that of the plants. The flowering phenology, in turn, is constrained by soil moisture in this dune area. Rodríguez and colleagues found that the plants flower later higher up on the hills, where the soil is drier, than at lower sites where groundwater is available. The scale is too small to permit local adaptation, and the authors found that synchronisation was good only at the lower, moister, sites. They also found that these low-elevation sites were sources for the population, with higher egg density and better survival of larvae than the sinks higher up.

SEASON LENGTH

In parallel with the reasoning regarding sawtooth patterns in development time and adult size, patterns in the degree of host plant specialisation may occur along climatic gradients. In areas where there is plenty of time for the number of generations that normally occur it could be advantageous to use several host plants, to gain the advantages of risk-spreading. However, in transitional areas where there is a shift between (especially) uni- and bivoltinism, only females ovipositing on host plants capable of supporting fast larval growth are likely to be successful in producing a second generation. By this reasoning, the partially bivoltine population of *Polygonia c-album* in the British Isles was predicted to be more specialised on *Urtica dioica* than the fully bivoltine Central Swedish population (Nylin 1988), and this was found to be the case (Nylin 1988, Janz & Nylin 1997), with evidence for a sex-linked genetic basis for this difference (Janz 1998) and for differences between the Swedish and other European populations (Nygren *et al.* 2006) (see also Fig. 15.3). A similar explanation was independently proposed to explain latitudinal variation in host plant specialisation in the American *Papilio glaucus* (Scriber & Lederhouse 1992).

Mobility and searching behaviour

Spatial variation in the availability of resources, including conspecific individuals for mating, is likely to cause corresponding gradients in mobility and other behaviours of insects, and sometimes in morphological features linked to these behavioural adaptations.

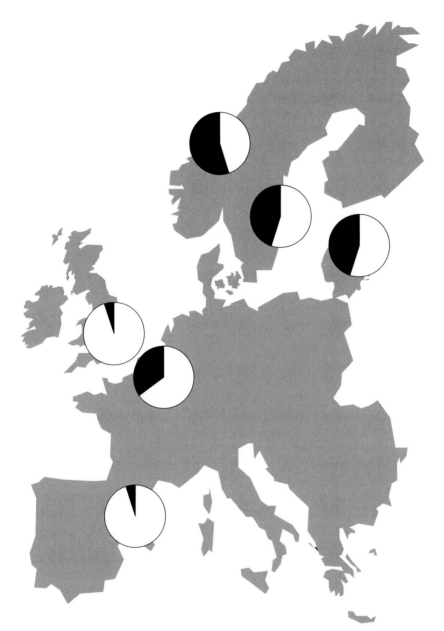

Figure 15.3. A schematic illustration of female host plant preferences in the spring generation of the comma butterfly *Polygonia c-album* across Europe (from Nygren *et al.* 2006, reproduced by permission of Blackwell Publishing). Black portions of pie charts correspond approximately to the percentage of eggs that a female would lay on *Salix caprea* when given a simultaneous choice of *Urtica dioica* or another urticalean rosid plant (white portion). Females in Sweden, Estonia, Norway and Belgium are less specialised on urticalean rosids than females from England and Spain.

Individuals should be more likely to accept a potential resource in areas where such resources are scarce and unpredictable. This selection pressure was suggested to explain the earlier onset of oviposition in *Pararge aegeria aegeria* from Madeira, as compared to *P. a. tircis* from Sweden (Gotthard *et al.* 1999a). The short and synchronised summer season in Sweden means that suitable grasses for oviposition are always abundantly available at the time when mated females are searching for oviposition sites. On Madeira, although suitable grasses are never absent, they are also never abundantly available at most sites. Consequently, females should accept them more or less whenever found (Gotthard *et al.* 1999a).

Interestingly, *P. aegeria* from Madeira are also much quicker to mate than butterflies from Sweden, suggesting that they are more likely to accept a potential mate (Gotthard *et al.* 1999a). Potential mates, even though never entirely absent, are more rare and unpredictable in a non-seasonal climate like the one on Madeira and should be accepted when found rather than risking that death comes before reproduction. Similarly, Dennis (1993) suggested that at high latitudes the reduced energy levels will often result in lower activity (despite counter-adaptations such as reduced thermal thresholds) and hence fewer chances of contact between conspecifics. This may explain the brighter coloration of northern *Polyommatus icarus* females in Britain (see also Chapter 13), and it could be predicted that butterflies at high latitudes and altitudes should be generally more prone to mate.

There is also a morphological component to the mate location behaviour in *Pararge aegeria*. Dark males are more likely to patrol, and pale males more likely to defend territories, probably because the former are better adapted for thermoregulation in closed forest and the latter in warm sunspots (Van Dyck & Matthysen 1998). Consequently behaviour as well as morphological gradients are expected. Berwaerts *et al.* (1998) studied populations of *P. aegeria* in Belgium differing in the amount of fragmentation of their forest landscapes. They found paler males in the least fragmented woodland. Males in the more fragmented woods were darker and also had a larger relative thorax size, indicating selection for increased mobility. At a somewhat larger geographical scale, gradients in thorax size (and wing morphology) have been found when comparing populations of *P. aegeria* differing in how long a time has passed since they were first colonised (Hill *et al.* 1999a). This suggests that new populations are likely to be founded by individuals that differ from the average in their ancestral populations.

CONCLUSIONS

The environment is replete with gradients that influence the lives of butterflies, directly or indirectly. Unless gene flow is too strong, butterflies can adapt to the local characteristics of such gradients, which creates gradients in their genetic and phenotypic features. When gene flow prevents local adaptation reaction norms that are adaptive over large areas, 'general-purpose genotypes' (Shapiro 1984b) will instead be favoured by natural selection.

Within species of butterflies, well-supported relationships exist between latitude and altitude, and features such as phenology, voltinism, wing colour and size. Such geographical patterns demonstrate that evolution is not something that happened a very long time ago, but is a continuous process contributing to the observed variation and to the present-day local ecology of butterflies. It should however be clear from this chapter not only that much knowledge exists, but also that most of the work still remains to be done. We are still far from a situation where we really understand how the environment shapes the features of butterflies and other organisms. There are many unresolved questions regarding for instance the relative roles of strictly genetic and plastic variation in life-history traits, why host plant hierarchies are sometimes very conservative and sometimes vary from site to site, what causes a given species to react to time stress in any particular way, or the role of ecological gradients in speciation processes. We are even further from a real understanding of how these features translate into population dynamics, an understanding that is necessary if we want to be able to predict the consequences of environmental change and changes in land management.

Part IV
Species in time and space: distribution and phylogeny

16 • Bad species

HENRI DESCIMON AND JAMES MALLET

SUMMARY

Taxonomists often added the term *bona species* after the Linnaean binomial. The implication is that there are also *malae species*. A 'bad species' is a taxonomic unit that does not conform to criteria used to delimit species. The advent of numerical taxonomy and cladistics has upset earlier taxonomic certainty and two different consensuses seem to be building among evolutionary biologists. The species concept either (a) takes the form of a minimal, Darwinian, definition which ignores evolutionary mechanisms to allow universal applicability or (b) attempts to combine a variety of species concepts together. Under both views, species may evolve or be maintained via multiple different routes. Whenever there is conflict between criteria, or whenever regular hybridization occurs, in spite of the fact that the taxa remain to some extent morphologically, ecologically or genetically distinct, or if populations are allopatric but seem at that stage of divergence at which species fusion is doubtful, one may speak of 'bad species'. The tools used in making a decision on the rank of taxa at this stage of divergence include morphological, chromosomal (karyological), molecular, and ecological characters.

Two main groups of questions are addressed. Firstly, do species exist as real entities in nature, or are they a construct of the human desire for categorization and classification? Secondly, what are species made of, how do they arise and how are they maintained? And, are species a homogeneous rank from this evolutionary point of view?

Around 16% of the 440 European butterfly species are known to hybridize in the wild. About half or more of these hybrids are fertile, and show evidence of backcrossing. Detailed accounts are given for (a) the genus *Hipparchia*, (b) *Polyommatus (Agrodiaetus) admetus* and the 'anomalous blue' group, (c) the sibling species *Leptidea sinapis* and *L. reali* – with a comparison to the situation in *Melitaea athalia*, (d) *Zerynthia rumina* and *Z. polyxena*, (e) for the frequent hybridizations and introgressions in sympatric Papilionidae (*Papilio machaon* and *P. hospiton*; *Parnassius apollo* and *P. phoebus*), (f) for *Polyommatus (Lysandra) coridon*, *L. hispana* and *L. albicans* with frequent hybridization everywhere (with species remaining distinguishable), (g) for the *Erebia tyndarus* group, (h) for *Erebia serotina* (a hybrid mistaken for a species) and (i) for some briefly mentioned further examples.

There is justification for reviving the rather neglected (and misused) rank of subspecies, with the trend among lepidopterists to consider only more strongly distinct forms (in morphology, ecology or genetics) as subspecies, and to lump dubious geographical forms as synonyms. These recommendations provide a useful compromise between descriptions of geographical variation, the needs of modern butterfly taxonomy, and Darwin's pragmatic use of the term species in evolutionary studies.

It is a Sisyphean task to devise a definitive, irrefutable definition of species, but species will continue to function as useful tools in biology for a long time. Studies of gene exchange in the many hierarchical layers of phenotype, genotype and genome in 'bad' species of butterflies will illuminate the nature of speciation and evolution at the species level more than discussions on the 'essence' of species.

INTRODUCTION: SPECIES CONCEPTS AND TAXONOMIC PRACTICE

Taxonomists, when describing a new species, often added the term *bona species* after the Linnaean binomial. The implication is that there are also *malae species*. A 'bad species' is a taxonomic unit that misbehaves with respect to criteria used to delimit species. There are a wide array of species definitions linked to theories of speciation and evolution (Harrison 1998, Coyne & Orr 2004) and there have been many debates, which often become abstruse and epistemological (Wilson 1999a, Hey 2006). The biological species concept (BSC), based on reproductive isolation and associated with the theory of allopatric speciation, prevailed for many years. More recently, the advent of numerical taxonomy (Sokal & Crovello 1970) and cladistics (Hennig 1968) has upset the earlier certainty. The establishment of a basis for conceiving (Maynard-Smith 1966) and observing (Bush 1969) sympatric speciation led to suspicions that species were more indefinite, even locally, than architects of the modern synthesis had imagined. Today, two different consensuses seem to be building among evolutionary biologists. The species concept either takes the form of a minimal, Darwinian, definition which is agnostic about evolutionary mechanisms to allow universal applicability (Mallet 1995, Feder 1998, Jiggins & Mallet 2000), or attempts to combine a variety of species concepts together (de Queiroz 1998, Templeton 1998a, Coyne & Orr 2004). Under both views, species may evolve or be maintained via multiple different routes.

Species concepts and criteria: speciation theory and systematic practice

When treating an actual fauna or flora, the central problem is of the purely taxonomic criteria for species status. For a long time, four kinds of criteria have been used to group members of a species: character-based or 'syndiagnostic' criteria (which may use morphological or genetic traits); phylogenetic or 'synepigonic' criteria; reproductive, 'mixiological', or 'syngamy' criteria; and finally geographical criteria, particularly 'sympatry', 'cohabitation', or geographical overlap (Poulton 1904b; see also Jordan 1905, Rothschild & Jordan 1906, Cuénot 1936). To be distinct at the level of species, taxa should provide at least some of these four kinds of evidence. With the advent of the BSC (Dobzhansky 1937, Mayr 1942), the main emphasis was put on reproductive isolation (i.e. mixiological) criteria. This caused something of a divorce between evolutionary theory and taxonomic practice. Although an overwhelming amount of work has been carried out on the genetics and evolution of species – studies of genetic structure within species, interspecific crosses in the laboratory and field studies on hybrid zones (Barton & Hewitt 1989, Berlocher 1998, Coyne & Orr 2004) – practising taxonomists often continue to use syndiagnostic methods based mainly on morphological characters.

Indeed, when taxonomists have a sample of specimens coming from an unexplored geographical area, they can find morphological differences with taxa already described, but it is difficult to determine whether they are due to a few pleiotropic gene changes (i.e. the new samples are merely morphs of described taxa), to intraspecific geographical variation (subspecies), or to differentiation at full species level. Sometimes, rare hybrids between well-known species have

even been mistaken for 'good' species. Since they are inaccessible, other criteria are simply ignored. Although they can reveal much about mixiological criteria, chromosomal and molecular characters are often used in much the same way as early taxonomists used morphological data; for instance, differences in chromosome numbers or the presence of diagnostic allozyme loci have been considered proof of distinct species, without consideration of geography or genetic relationships. We argue that these biological characteristics cannot be ignored.

Study of ecological niches is particularly important for associating morphological or genetic differences with different habitats (Sneath & Sokal 1973). Mayr, in later versions of his BSC (1982) argued that each species 'occupies a biological niche in nature'. Adaptive evolution is recognized as a primary means of both splitting and maintenance of separate lineages (Van Valen 1976, Templeton 1989, 1994, 1998a, Andersson 1990, Baum & Larson 1991, Schluter 2000). Sympatric speciation also involves ecological differentiation (Bush 1969, Feder 1998), and increasing evidence suggests that ecological divergence may directly cause reproductive isolation (Dodd 1989, Schluter 2001).

Nonetheless, mixiological criteria remain the most important within the BSC conceptual framework. They are reached through observation of the relations between the taxa either in sympatry, or in hybrid zones in the case of parapatry (O'Brien & Wolfluss 1991, Jiggins & Mallet 2000) – the latter are considered as 'natural laboratories for evolutionary studies' (Hewitt 1988) (see Chapter 19). Modelling as well as empirical studies suggest that hybrid zones can act as a barrier to gene flow (Barton & Hewitt 1989). Within them, the intensity of hybridization may vary. If hybrid genotypes predominate, the hybrid zone is considered 'unimodal', while, if genotypes are predominantly parental, with few intermediates, it appears phenotypically 'bimodal' (Harrison & Bogdanowicz 1997, Jiggins & Mallet 2000). Pairs of species that cohabit broadly and hybridize regularly can be studied genetically in the same way. In hybrid zones, the mixiological criterion of species depends on the fraction of genes that are actually exchanged between the taxa. Hybrids can be detected using morphological criteria, but this can be inaccurate, which makes it hard to estimate gene flow. Gene exchange, or introgression (Stebbins 1959), may transfer important genetic variation in some cases of adaptive evolution, especially in plants (Arnold 1992a, 1997, Mallet 2005). In birds and fish, hybridization is widespread (Grant & Grant 1992) and may be involved in rapid adaptive radiation and speciation (Grant & Grant

1998, Seehausen 2003). This also seems likely in *Heliconius* butterflies (Gilbert 2003, Bull *et al.* 2006). Introgression can affect the mitochondrial genome (Aubert & Solignac 1990) but, in Lepidoptera, where the Y-bearing sex is the female, Haldane's rule severely hinders mitochondrial introgression (see below and Sperling 1990, 1993, Aubert *et al.* 1997).

Based on the ideas of Mallet (1995) and Feder (1998), the separation of gene pools during speciation has been dubbed 'the genic view of speciation' by Wu (2001): speciation may not take place via separation of the whole gene pools, as postulated by the Dobzhansky–Mayr theory of speciation, but initially concerns only genes actively involved in reproductive isolation. The rest of the genome may still undergo sufficient gene flow to prevent differentiation, except in genomic regions tightly linked to 'speciation genes' (Ting *et al.* 2000). But what are speciation genes? Genes involved in divergent adaptation and mate choice should diverge first, and those causing hybrid sterility and inviability should be expected to diverge only after initial genetic separation. Complete separation should result from reinforcement of sexual isolation and further ecological differentiation (Noor 1999). Although Wu's genic view of speciation elicited an immediate rebuttal from the father of the BSC (Mayr 2001), it is clear that the proposed scheme is not that different from the 'classical' view of speciation according to Mayr. The most important distinction is that Wu's modification of Mayr's speciation scheme renders it compatible with a more substantial phase of gradual divergence in sympatry or parapatry.

An array of varied data obtained from difficult or 'bad' taxa can be used to support or refute the presence of additional species within a sample. The more concordant the data are, and the more bimodal the frequency distributions of phenotypes and genotypes, the more likely separate species status will be granted. These are methods termed 'genealogical concordance' or 'genotypic clustering' (Avise & Ball 1990, Mallet 1995). Similar syndiagnostic procedures were, in fact, being applied to morphological characters long before Darwinian times (Adanson 1763). As early as 1930, Nilsson (cited by Cuénot 1936) used the term 'genotypenkreis' to characterize species in *Salix*, a plant genus prone to hybridization.

This ideal procedure for species delimitation, careful study in zones of contact, is not always possible. In cases where concordance between criteria is imperfect, some argue for distinction at species level, and others against it. For instance, cryptic or sibling species (Dobzhansky 1937, Mayr 1963) fail to show diagnostic morphological characters;

species that are otherwise well characterized apparently share the same ecological niche; hybrid zones can be unimodal in some areas and bimodal in other parts of the range. Molecular markers may be strongly differentiated among populations within species; in other groups, species clearly distinct using other criteria can show little molecular differentiation, especially if speciation is recent compared with the rate of molecular divergence.

Cohabitation: the lumper's species criterion adopted here

The touchstone of all criteria for separate, biological species is the test of 'cohabitation': whether overlapping populations produce unimodal (in which case subspecies might be designated), or bimodal (in the case of separate species) morphological and genotypic frequency distributions. This procedure dates from the late nineteenth century, and was promoted particularly vigorously for the Lepidoptera by Karl Jordan (e.g. Jordan & Rothschild 1906). Other species criteria that do not depend on degree of hybridization or intermediacy in areas of overlap are also in use today. In particular, Cracraft's (1983, 1989) 'phylogenetic' or 'diagnostic' concept is contributing to taxonomic inflation of 'species' numbers in birds, primates, and other taxa (Isaac *et al.* 2004), even when no new populations have been discovered. In butterflies, the prohibitive diversity of morphologically or genetically diagnosable local populations, usually referred to in our literature as 'subspecies', has tended to prevent such rampant splitting (for the moment). Here, we adopt this traditional and more inclusive, polytypic or 'lumper's' criterion for species.

When sympatric taxa hybridize very rarely, they can be classified as separate species. But what can be concluded if the units to be compared are not in contact? Breeding and crossing experiments provide an apparent solution, but this can be misleading. In particular, viability of hybrids in the laboratory may appear normal while, in nature, hybrids could be severely disadvantaged. Pre-mating barriers to hybridization can also be reduced under artificial conditions. In both cases, the degree of mixiological separation estimated can be spurious.

Whenever there is conflict between criteria, or whenever regular hybridization occurs, in spite of the fact that the taxa remain to some extent morphologically, ecologically or genetically distinct, or if populations are allopatric but seem at that stage of divergence at which species fusion is doubtful, one may speak of 'bad species'. The tools used in making a decision on the rank of taxa at this stage of divergence include morphological, chromosomal (karyological), molecular and ecological characters. In addition, one may cross such taxa, to obtain criteria relevant to reproductive isolation and introgression, keeping in mind the caveat previously invoked. These tools are described in detail in the appendix.

As with any term, 'species' must have a definition that depends partly on theoretical considerations. At this point, one might ask two main groups of questions: (1) Do species exist as real entities in nature? Or are they a construct of the human desire for categorization and classification? (2) What are species made of? How do they arise? How are they maintained? And are species a homogeneous rank from this evolutionary point of view? To answer such questions, it is necessary to investigate actual problem cases in some depth, which is the main aim of the rest of this chapter.

HOW COMMON ARE BAD SPECIES IN EUROPEAN BUTTERFLIES?

It is often said that, although there are disagreements about species concepts, there are few cases where our ability to delimit species is severely challenged (e.g. Mayr 1963). However, hybridization and bad species are rather more common than field guides tend to mention. Taxonomists overlook 'dubious' individuals (which may often be hybrids) because they make species discrimination more difficult. Natural hybridization occurs between around 10% of all animal species, although there are many groups where hybridization rates are greater (Mallet 2005). Here we provide collated data on European species, one of the best-studied faunas in the world (Table 16.1). Overall, around 16% of the 440 butterfly species are known to hybridize with at least one other species in the wild. Of these perhaps half or more are fertile, and show evidence of backcrossing in nature.

CASE STUDIES: THE PRACTICE OF EUROPEAN BUTTERFLY TAXONOMISTS AT SPECIES LEVEL

European butterflies are taxonomically well known. In the first comprehensive work on European butterflies, Higgins & Riley (1970) enumerated 371 species (including the Hesperioidea); in a recent book of the same scope, Tolman & Lewington (1997) record 440 species, 69 more. Amongst the 'new' European species, hardly any are actually new finds; many

arise from 'taxonomic inflation', the upgrading of previously known subspecies to species level, or discoveries of known non-European species just inside the boundary (Dennis 1997, Isaac *et al.* 2004). In this section, we present an analysis of some decisions that illustrate how splitting and/ or lumping has been performed in particular cases.

The genus *Hipparchia*: splitters and lumpers at work

Some genera have undergone especially intense splitting, like the graylings (*Hipparchia* and *Neohipparchia*). According to Higgins & Riley (1970), there were only 10 species in Europe. Today, there are 19 (Tolman & Lewington 1997), to which one more, *H. genava*, can be added according to Leraut (1990). Mostly, this proliferation is due to elevation to species rank of forms inhabiting islands or other disjunct geographic regions (e.g. *H. azorina*, *H. caldeirense* and *H. miguelensis* in the Azores). However, this is not true for *H. alcyone* and *H. genava*, between which Leraut records a hybrid zone. In a revision of the genus (Kudrna 1977) elevation to species rank was based only on morphology. Morphometric analyses of multiple, well-replicated samples in the *semele* group based on genitalia, wing-pattern measurements and allozyme electrophoresis were later carried out by Cesaroni *et al.* (1994), who showed convincing congruence between the morphometric analysis of genitalia and allozymes, although wing patterns followed an obviously different evolutionary pathway. The number of taxa with specific status was reduced by Cesaroni *et al.* from eight to five. As the taxa were largely allopatric and often insular in distribution, cohabitation and hybrid zone criteria cannot be tested. Assignment to species level was therefore performed on the basis of 'sufficient' genetic distance (Nei's D between 0.07 and 0.26).

Later, Jutzeler *et al.* (1997) presented another treatment of the same group. Although devoted mainly to meticulous morphological description of certain taxa and their first instars, and lavishly illustrated with scanning electron microscope (SEM) pictures and excellent colour plates, the specific status of the various taxa was also discussed. The authors, it turns out, are extreme 'splitters', and even cite Cesaroni *et al.* (1994) to justify splitting – in complete contradiction to that paper. No morphometric analyses were performed while making these controversial decisions. More recently, even more 'insular splitting' has been carried out by Jutzeler *et al.* (2003a, b): taxa from the Tyrrhenian Islands were raised to species on the basis of morphological

Table 16.1A *Some examples of bad species in European butterflies, including all known records of interspecific hybridization in the wild*

Species 1	Species 2	Location	Hybrid frequency[a]	Characters, except morphology[b]	Taxonomic interpretation	Source
Papilio machaon	*P. hospiton*	Corsica, Sardinia	F	D, A, M, E, H(I), P!	Sibling species	See text
Iphiclides p. podalirius	*I. p. feisthamelii*	Languedoc	F	A, M	Parapatric sibling species	See text
Zerynthia polyxena	*Z. rumina*	Provence	R/E	P, G, C, A, M, E, H! I! S!	Parapatric species	See text, Plate 20b
Parnassius apollo	*P. phoebus*	Throughout the Alps	F	G, A, M, H(S)	Partially sympatric species	See text, Plate 19b
Artogeia napi	*A. bryoniae*	Alps	C/F	C,E,H^{-1}(I)	Parapatric sibling species	Bowden, 1996, Geiger & Shapiro, 1992; Porter & Geiger 1995
Artogeia napi	*A. balcana*	Balkans	?		?Parapatric subspecies	Tolman & Lewington, 1997
Artogeia napi	*A. rapae*	Britain, Germany	E	A	Sympatric species	Klemann, 1930; Heslop-Harrison, 1951
Pontia daplidice	*P. edusa*	Coastal S. France, Italy	F	A, H^0(I),G	Parapatric sibling species (narrow overlap)	Geiger *et al.*, 1988 ["semispecies"]; Wenger *et al.*, 1993 ["semispecies"]; Porter *et al.*, 1997 [regard as subspecies]
Euchloe crameri	*E. simplonia*	Alps, Pyrénées	Likely!	D, A Allozymes differ markedly (Geiger, pers. comm. to HD)	Parapatric, ecologically divergent forms	Lux, 1987; Descimon, unpub.
Anthocaris belia euphenoides	*A. cardamines*	S. France, Spain	R/E	H	Partially sympatric species	Legras in G&D, Plate 19c
Anthocaris damone	*A. gruneri*	Greece	E	A	Partially sympatric species	Rougeot, 1977
Colias crocea	*C. erate*	Greece, C. Europe	R/F	A	Partially sympatric species	Alberti, 1943
Colias hyale	*C. erate*	C. Europe	R	A	Partially sympatric species	Alberti, 1943
Colias crocea	*C. hyale*	Only in lab But likely to occur	E	Confused with aberrant *crocea?*	Sympatric species	Ryszka, 1949 5–7; Descimon in G&D
Colias myrmidone	*C. hyale*	E. Europe	E	A, M	Partially sympatric species	Mecke, 1923
Colias crocea	*C. phicomone*	Alps	R	E	Partially sympatric species	Descimon in G&D
Colias palaeno	*C. phicomone*	Alps	E	E	Partially sympatric species	Descimon in G&D G. Poluzzi, *fide* HD
Colias hecla	*C. tyche* (= *nastes*)	Norway, Sweden	F/R	"christiernsonni" Lampa	Sympatric species	Kaisila, 1950

Table 16.1A (cont.)

Species 1	Species 2	Location	Hybrid frequency[a]	Characters, except morphology[b]	Taxonomic interpretation	Source
Gonepteryx rhamni	G. cleopatra	S. Europe	E	E	Partially sympatric species	G&D; Descimon, unpub.
Leptidea sinapis	L. reali	Europe	None known	W−, P, G, A, M, E	Partially sympatric species	Descimon in G&D Bernardi, pers. comm. to HD
Lycaena tityrus subalpina	Lycaena hippothoe	French Alps	R	E!	Partially sympatric species	Higgins & Riley, 1970; Descimon, 1980
Lycaena tityrus tityrus	Lycaena t. subalpina		C	E, H(I)	Parapatric strong ssp.	D' Aldin, 1929
Cupido minimus	E. alcetas	W. France	E		Sympatric species	Wynne & Mallet, unpub.
Aricia agestis	A. artaxerxes	UK, possible elsewhere	E (ancient)	A, M, D, H(I)	Narrowly overlapping sympatric species	Schurian & Hoffmann, 1975
Agrodiaetus damon	A. rippartii	Balkans	R	A, M	Partially sympatric species	Rebel, 1920
Agrodiaetus damon	Polyommatus meleager	Alps	E	C, A, M, E	Intergeneric hybrid	Rebel, 1930b.
Agrodiaetus damon	Polyommatus icarus	Alps	E	A, M	Intergeneric hybrid	See text, Plate 19a
Lysandra coridon	L. bellargus	Europe	F	C,G,D, A polomus Zeller	Sympatric spp.	Cameron-Curry et al., 1980
Lysandra hispana	L. bellargus	S. France, Spain, Italy	R (rarer than polomus)	C, A, M =samsoni Verity?	Sympatric sp.	Gómez Bustillo & Fernandez-Rubio, 1974
Lysandra bellargus	L. albicans	S.W. Spain	R	D, C, A, M	Distant species	See text, Plate 19a
Lysandra coridon caelestissima	L. albicans	Central Spain	F	C, A, M caerulescens Vty	Partially sympatric species	Rebel, 1930a; Descimon, unpublished
Lysandra coridon	Agrodiaetus damon	Alps	E	G, C, A, M	Distant species	See text, Plate 19a
Lysandra coridon	Meleageria daphnis	Alps	R	C, A, M cormion Nabokov	Distant species	De Carpentrie, 1977
Lysandra albicans	Plebicula escheri	Spain	E	C, A, M, E	Intergeneric	Herrmann, 1926
Lysandra coridon	Polyommatus icarus	Germany	E	C, A, M, E	Intergeneric	Goodman et al., 1925
Lysandra coridon	Plebicula dorylas	France	E	C, A, M, E	Intergeneric	

Plebicula dorylas	Plebicula nivescens	Central Spain	R	C, A, M *caeruleonivescens* Verity	Partially sympatric related species	Verity in G&D; Descimon, unpub., Plate 19a
Polyommatus icarus	P. eros	Alps	E	A, M	Related species, sympatric on mountains	Descimon, unpub.
Polyommatus icarus	Plebejus argus	Germany	E	A, M, E	Intergeneric hybrid	Peter, 1928 140
Maculinea alcon	M. rebeli	All Europe	?	M, E	Good species or ecological races?	Wynhoff, 1998
Boloria pales	B. napaea	French Alps	F/R?	W-, G, A, M	Partially sympatric species	Descimon, unpub.
Euphydryas aurinia	E. desfontainii	Spain	R	G, A, M, E, HS	Partially sympatric species	De Lajonquière, 1966
Mellicta athalia athalia	M. athalia celadussa	Central France [?]	C	G, W-	Parapatric subspecies?	See text
Mellicta athalia	M. deione	Provence	E	G, A, M	Partially sympatric	Descimon, unpub.
Mellicta parthenoides	M. varia	Southern French Alps	F/R	C, H(I)!	Parapatric sibling species	Bernardi, pers. comm.; G&D
Melanargia russiae	M. lachesis	Eastern Pyrenees	E	A	Species	Tavoillot, 1967
Melanargia galathea	M. lachesis	France and Spain	F (only in some overlaps)		Parapatric sibling species	Higgins, 1969; Wagener, 1984; Essayan, 1990
Hipparchia semele	H. (senthes?) balletoi	Italy	R	G, A	Parapatric sibling species	Sbordoni, pers. comm.; but see text
Erebia flavofasciata	E. epiphron	Alps	R	G	Partially sympatric species	See text
Erebia pharte	E. epiphron	Alps	E	G	Sympatric species	Descimon in G&D
Erebia pronoe	E. epiphron	Pyrenees	R	H(I) ="serotina" Descimon & de Lesse	Sympatric species	See text, Plate 20a
Erebia pronoe	E. medusa	Carpathians	E		Distant species	See text
Erebia cassioides	E. hispania	Pyrenees	R, several zones in the Pyrenees	G, C, A, M	Parapatric sibling species	See text, Plate 20a
Erebia cassioides	E. tyndarus	Alps	F	A, M	Parapatric sibling species	See text
Erebia cassioides	E. nivalis	Alps	F	A, M, E	Parapatric sibling species	See text
Coenonympha arcania	C. hero	N. Europe	F *hero* nearly extinct		Partially sympatric species	Legras in G&D Gross, 1957
Coenonympha darwiniana	C. gardetta	Alps	F	A. darwiniana may be hybrid gardetta × arcania	Parapatric species	See text

Table 16.1B *Bad species supplementary data. Excluded from above because too doubtful or not studied enough; includes also some doubtful species/subspecies (these are included only if there is some cohabitation)*

Species 1	Species 2	Location	Hybrid frequency[a]	Characters, except morphology[b]	Taxonomic interpretation	Source
Pieris ergane	*P. napi*	S. Europe	L	W+/−, Lab	Partially sympatric species	Bred by Lorkovic
Euchloe crameri	*E. simplonia*	Maritime Alps	E to F? Not studied!	D, W+/−, A, E	Parapatric species (montane vs. lowland)	Lux, 1987
Euchloe simplonia	*Anthocaris cardamines*	Alps and Pyrenees	? (E)	A, HI, HS, Lab	Sympatric species	Obtained until pupa by HD
Lycaeides idas	*L. argyrognomon*	Central France	?, L+	W+/−	Sympatric species	HD's observations in Yonne
Lycaeides idas	*L. idas calliopis* Boisduval	S. French Alps	?	E, ?	Could be sibling species	Numerous observations since Boisduval, including HD's
Everes argiades	*E. alcetas*	S. Europe	L	W+/−	Partially sympatric	
Everes argiades	*E. decoloratus*	S. Europe	L	W+/− or W−	Partially sympatric	
Everes alcetas	*E. decoloratus*	S. Europe	L	W+/−	Partially sympatric	
Cupido lorquinii	*C. carsemelli*	S. Spain	L	W+/−, E	Partially sympatric	
Glaucopsyche alexis	*G. melanops*	S. W. Europe	L	W+/−	Partially sympatric	Possibly captured by HD
Maculinea teleius	*M. nausithous*	Europe	?		Partially sympatric	
Pseudophilotes baton	*P. panoptes*	Spain	?	W−	May be subspecies or parapatric sibling species	
Pseudophilotes baton	*P. abencerragus*	S. Spain	L	W+/−	Partially sympatric	
Aricia agestis	*A. morronensis*	Spain		W−	Partially sympatric	
Agriades glandon	*A. pyrenaica*	Pyrenees	?		Partially sympatric	
Agrodiaetus rippartii	*A. fabressei*	Central Spain		W−	Partially sympatric	Should be very difficult to detect
Agrodiaetus dolus	*A. damon*	S. Europe	L (E)		Partially sympatric	As likely as *L. coridon* × *A. damon*!
Agrodiaetus dolus	*Agrodiaetus* (brown sp.)	S. Europe	L (E)		Partially sympatric	As likely as *A. dolus* × *A. rippartii*!
Polyommatus icarus	*P. eroides*	S.E. Europe	L		Partially sympatric	
Polyommatus eros	*P. eroides*	S.E. Europe	L	W+/−	Partially sympatric	
Polyommatus icarus	*P. andronicus*	Greece		G+	Partially sympatric sibling species	

Apatura ilia	A. metis	S. E. Europe	L	W–E–	Parapatric species	
Argynnis adippe	A. niobe	Palaearctic		W+/–	Widespread species	
Brenthis hecate	B. daphne	W. Palaearctic		E, W+/–	Partially sympatric	Why not? The three *Brenthis* often fly in close vicinity, in spite of marked ecological differences.
Brenthis hecate	B. ino	W. Palaearctic		E, W+/–	Partially sympatric	
Brenthis daphne	B. ino	W. Palaearctic		E, W+/–	Partially sympatric	
Clossiana selene	C. euphrosyne	W. Palaearctic		E, W+/–	Partially sympatric	
Circumpolar *Clossiana*	(5 species)	Scandinavia		W+/–	Partially sympatric	
Melitaea parthenoides	M. aurelia	Europe	L	W+/–	Partially sympatric	Suspected to occur in Briançon region (French Southern Alps) by HD
Melitaea aurelia	M. britomartis	Central Europe	?	W–, G+	Partially sympatric	
Melitaea phoebe	M. aetherie	S. Spain		W+/–	Partially sympatric	
Melanargia occitanica	M. galathea or lachesis	S. Europe			Partially sympatric	
Melanargia occitanica	M. russiae	S. Europe		E	Partially sympatric	
Melanargia occitanica	M. ines	Spain			Partially sympatric	
Hipparchia fagi	H. alcyone	S. Europe	L	W–	Partially sympatric	
Hipparchia sp.				W+/–	Partially sympatric	Several opportunities within this complex genus
Chazara briseis	C. prieuri	Spain			Partially sympatric	Cf. *Hipparchia*
Pseudochazara sp.		S. E. Europe			Partially sympatric	Searched for by HD in Briançon region – in vain!
Satyrus actaea	S. ferula	S. Europe		W+/–	Partially sympatric	
Erebia ligea	E. euryale	European mountains	L		Partially sympatric	Possibly found by HD
Erebia pharte	E. melampus	Alps	L	W+/–	Partially sympatric	Possibly found by HD
Erebia aethiopella	E. mnestra	French Alps	F	W+/–	Parapatric species	Bimodal hybrid zone at Montgenèvre, French Southern Alps (HD's and Claude Herbulot's observations)
Erebia stirius	E. styx	Central Alps		W–	Partially sympatric	Cf. Lorković's works; could also hybridize with *E. montana*
Hyponephele lycaon	H. lupina	S. Europe		W+/–	Partially sympatric	Pairing rather often observed, hybrids never
Aphantopus hyperanthus	Maniola jurtina	Palaearctic			Largely sympatric	

Table 16.1B (cont.)

Species 1	Species 2	Location	Hybrid frequency[a]	Characters, except morphology[b]	Taxonomic interpretation	Source
Pyronia tithonus	P. bathseba	S. Europe		W+/−	Oceanic vs. Mediterranean	
Coenonympha sp.						Several candidates in the genus in addition to those observed
Lasiommata maera	L. megera	Europe		W+/−	Largely sympatric	Suspected around Marseilles by HD
Lasiommata maera and megera	L. petropolitana	Alps, Pyrenees		W+/−	Sympatric in Alps and Pyrenees	

[a] Hybrid frequency: C, Common (Hardy–Weinberg); F, Frequent >1%; R, Regular <1%; E, Exceptional <0.1%; L, Likely, but no data.

[b] Characters enabling detection of hybridization, apart from wing pattern (− means does NOT occur): W−, No wing pattern differences; W+/−, differences not striking enough to allow recognition without especial attention; P, Mate choice differences; D, Diapause; G, Genitalia; C, Chromosomes; A, Allozymes; M, Molecular (nuclear and mitochondrial DNA); E, Ecological; H, Haldane's rule; H^{-1}, Inverse Haldane's rule; H^0, Non-Haldane rule inviability; I, Inviability (e.g. H^{-1}(I)); S, Sterility; Lab, hybrids have been obtained in captivity.

[c] G&D, Guillaumin & Descimon, 1976.

and bionomic differences with continental relatives, again without any morphometric, karyological, mixiological or molecular justification. Most of these 'new' species are allopatric. We tend to side with the more conservative views of Cesaroni *et al.* (1994).

Polyommatus (*Agrodiaetus*) *admetus* and the 'anomalous blue' group: chromosome variation and allopatry

According to Lukhtanov *et al.* (2003), 'this complex is a real stumbling block in the taxonomy of the genus [*Agrodiaetus*]'. In a careful study using the 'classical' tools of typological taxonomy, Forster (1956) was uncertain about the taxonomic status of only a few forms or 'bad species'. Soon thereafter, de Lesse (1960a) used karyology to show that the picture was not simple but death prevented him from carrying his work further. The *admetus* group of *Agrodiaetus*, which included only three species in Higgins & Riley (1970), was raised to nine some 35 years later (Tolman & Lewington 1997, Wiemers 2003).

In *Agrodiaetus*, the males are generally blue, but the 'anomalous blues' all have similar, chocolate-brown uppersides in both sexes. In 1970, the species recognized were *A. admetus*, ranging from Eastern Europe to Asia Minor, *A. fabressei* known only from Spain and *A. ripartii* from scattered locations from Spain to Asia Minor. This treatment was supported by karyotyping: $n = 78–80$ for *admetus*, $n = 90$ with two large unequal chromosomes for *ripartii* and $n = 90$ with two large and two medium-sized chromosomes for *fabressei* (de Lesse 1960a). The taxa *fabressei* and *ripartii* cohabited without admixture in some Spanish localities (de Lesse 1961a).

The situation became more complex when wide karyotypic variation was found in Turkey and later in parts of Europe (Table 16.2).

More recently, allozyme studies have cast doubt on this multiplicity of species. *Agrodiaetus ripartii*, the most

Table 16.2 *Variation in chromosome number of described species within the subgenus* Agrodiaetus

Species of *Agrodiaetus* (according to Tolman & Lewington 1997, Wiemers 2003)[a]	Distribution	Chromosome number (*n*)
admetus Esper	Bulgaria	80
admetus Esper	Turkey	78–80
alcestis Zerny	Lebanon	20–21
aroaniensis Brown	Peloponnese	48
dantchenkoi Lukhtanov *et al.*	Turkey	42
demavendi Pfeffer	Iran, Turkey	68–71
eriwanensis Forster	Armenia	32–34
fabressei Oberthür	Spain	90 (86+2+2)
galloi Balletto & Toso	S. Italy	66
humedasae Toso & Balletto	N. Italy	38
interjectus de Lesse	Turkey	29–32
karacetinae Lukhtanov & Dantchenko	Turkey	19
nephohiptamenos Brown & Coutsis	N. Greece	8–11, or ~90[b]
ripartii Freyer	Spain–Turkey	90 (88 + 1 + 1)

[a] Taxa with no information on chromosome number are omitted, as are taxa of obviously subspecific rank.
[b] There are contradictory numbers counted by Brown & Coutsis (1978) and de Prins (unpublished); the $n = 90$ estimate seems most likely (Wiemers 2003).
Source: From Hesselbarth *et al.* (1995), Eckweiler & Häuser (1997), Häuser & Eckweiler (1997), Carbonell (2001), Lukhtanov & Dantchenko (2002a, b, 2003), Wiemers (2003) and Kandul *et al.* (2004).

widespread, proved as homogeneous genetically as in its karyotype; this is also true, to a lesser degree, for *A. admetus*. *Agrodiaetus fabressei* and the other taxa are poorly resolved and there is little correlation between allozymes and karyotype (Mensi *et al.* 1994). More recently, mitochondrial and nuclear DNA sequencing studies suggest that 'brown' *Agrodiaetus* are polyphyletic. The wing colour switch from the 'primitive' blue colour to brown in males seems to have occurred twice: once in the '*admetus*' group and once in *fabressei* (Wiemers 2003, Kandul *et al.* 2004). Most distinguishable entities are allopatric, and the only exceptions are the aforementioned *A. fabressei* and *A. ripartii*, and four species found close together in the Turkish Van province (Lukhtanov *et al.* 2003). In most other cases, nobody knows what would occur if these genetic entities flew together.

Clues are provided by the *fabressei–ripartii* case, which have the same chromosome number, but differ in details of the karyotype. They comply with the cohabitation criterion and are genetically distant (Lattes *et al.* 1994). Clearly, there is little doubt that these are good (albeit sibling) species. However, they are almost impossible to identify using morphology where they co-occur, since neither wing pattern nor skeletal morphology provide reliable criteria: karyotype and DNA sequencing are virtually the only ways to assure identification (Lukhtanov *et al.* 2003). Chromosomal information has also been used by Munguira *et al.* (1994), who merged the Spanish *agenjoi* Forster and *violetae* Gomez-Bustillo *et al.* into the known species: *fabressei* and *ripartii*. However, Gil-T & Gil-Uceda (2005) showed that these authors did not examine the 'true' *violetae* (rediscovered after more than 20 years) from Sierra de Almijara (its type locality), but populations coming from ca. 200 km to the northeast (Sierra de Cazorla). Both populations are morphologically well differentiated. New karyological and biochemical studies hopefully will determine its final taxonomic status (Lukhtanov *et al.* 2005).

Chromosome structure is unstable in *Agrodiaetus* and rearrangements are common even within populations, leading to the formation of multivalents during meiosis (Lukhtanov & Dantchenko 2002a, b, Lukhtanov *et al.* 2003). Limited abnormalities seem not to affect viability, although selection should eventually eliminate most rearrangement polymorphism. Why is chromosome structure so unstable in *Agrodiaetus*? Kandul *et al.* (2004) argue that tolerance of chromosomal polymorphism is related to centromere structure, and suggest that destabilization of chromosome numbers may be due to locally abundant transposons. In allopatric populations of *Agrodiaetus*, elimination of differences

will not take place and the karyotype diverges rapidly until a point of no return is reached, giving rise to a great deal of geographical variation, and ultimately speciation. Similarly, Wiemers (2003) boldly states that 'changes in the number of chromosomes do not lead to sympatric speciation, but instead appear as a by-product of allopatric speciation and such young species could only occur in sympatry after a sufficient differentiation in their phenotype to exclude erroneous matings'.

Leptidea sinapis and *L. reali*: sibling species and the almost 'perfect crime', with a comparison to the situation in *Melitaea athalia*

Until the end of the twentieth century, nobody suspected that two separate species lurked within the wood white, *Leptidea sinapis*. In 1962, Réal noticed that two different seasonal forms flew together in the French eastern Pyrénées, without considering the possibility that two species were involved (Réal 1962). By the late 1980s, after morphological studies on the genitalia, Lorković suggested to Réal that there were indeed two species. The latter described a new species under the name *lorkovicii* in 1988, an invalid name replaced by *reali* (Reissinger 1989). Further study confirmed that the two forms, characterized by male and female genitalia, were distinguishable and sympatric across much of Europe (Lorković 1994, Mazel & Leestmans 1996); in particular, the penis is short in *sinapis*, and long in *reali*. There are correlated differences in the females, with short vs. long ductus bursae. This strongly suggests a 'lock and key' mechanism is involved. Although other barriers may be present, it seems likely that these differences can explain reproductive isolation between the taxa. In contrast, earlier attempts to find reliable differences in wing pattern and ecology were in vain. *Leptidea sinapis* is present everywhere in Western Europe, while *reali*, if present, is always in sympatry with it.

Although the existence of two 'good' species is likely, it could be argued that there is merely a genitalic polymorphism, similar to that in *Melitaea athalia* and *M. celadussa* (see below). To address this point, a study based on multivariate morphometrics of genitalia, allozymes and mtDNA sequencing was undertaken by Martin *et al.* (2003) on six populations from southern France. A 728-bp fragment of the *ND1* gene showed a reliable and constant 3% divergence between the entities. Among 16 enzyme loci, none was completely diagnostic, but *Ak* and *Pgi* showed highly significant differentiation. Multivariate analysis demonstrated two well-separated 'genotypic clusters', with strong linkage disequilibria between loci. Furthermore, allozymes and the mtDNA were concordant. Morphometrics carried out on genitalia also yielded good concordance with molecular data, although there was some (<5%) overlap between the taxa. In 163 individuals of the two species, no hybrid was detected; the few individuals with doubtful genitalic measurements were clearly assigned to one or other taxon by molecular markers.

The necessity of dissecting individuals for identification makes ecological study difficult, and it was at first thought that the species fly together and share most foodplants. This should contradict Gause's principle but could explain the lack of consistent differences in wing pattern. However, the population genetic structure of the two species is somewhat different: *L. reali* is less polymorphic at allozymes (with heterozygosity $0.09 < H < 0.14$ in *sinapis* and $0.05 < H < 0.07$ in *reali*: Martin *et al.* 2003). Females and, to a lesser extent, males of both species discriminate between the species during mate choice, and only intraspecific matings occurred in captivity (Freese & Fiedler 2002). The two species are now known to differ in ecology: *L. sinapis* is a widespread generalist on various herbaceous Leguminosae from both wet and dry habitats, while *L. reali* specializes on *Lathyrus pratensis*, a plant confined to moist grasslands. In 347 localities in the Drôme department (southern France) where *L. sinapis* and/or *L. reali* were observed, *L. sinapis* was alone in 55% of the study sites, and *L. reali* in 22%, whereas both species were found together in 23% of them (Amiet 2004). There are also differences in phenology, response to temperature and habitat choice (Friberg *et al.* 2008). The situation seems to reverse in Eastern Europe, where *L. sinapis* becomes confined to warmer areas (Benes *et al.* 2003b). Freese & Fiedler (2002), in their mainly laboratory-based study, concluded that 'the two species are only weakly differentiated in ecological terms'; indeed, their egg-laying tests showed only a weak preference for *L. pratensis* in the females of *L. reali*; the larvae of both species prefer and perform better on another legume, *Lotus corniculatus*, a result rather discrepant with Amiet's (2004) field observations.

As in almost all 'perfect crimes', once the first clue was discovered, a cascade of confirmatory data was quickly revealed. At the end of the nineteenth century, the earliest dissectors of genitalia, such as Reverdin, could well have studied a series of *Leptidea* male genitalia and discovered the two species.

The latter did just this with *Melitaea athalia* (Reverdin 1920, 1922), where two types of male genitalia were

associated with two biogeographical entities, and he therefore split them into separate species. However, later study showed that the morphology of male genitalia was unimodal within a hybrid zone between the two taxa. The width of the hybrid zone varied from a few to several tens of kilometres (Bourgogne 1953). Since this differentiation is not associated with large and constant differences in allozymes or mtDNA, as in *Leptidea* (Zimmermann, unpublished), species separation in *Melitaea* was premature.

Zerynthia rumina and *Z. polyxena*: relativity of mixiological criteria

The genus *Zerynthia* contains two species, both recognized since the dawn of entomology: *Z. rumina*, a western Mediterranean species, and *Z. polyxena* from the eastern Mediterranean (Plate 20b). They overlap in southern France, where they display marked ecological differentiation, while in areas where only one species is found, both have a more extensive niche. Besides wing-pattern differences, there are diagnostic alleles between, with Nei's $D \approx 0.80$ (Braconnot, unpublished) and strong divergence in mitochondrial and nuclear gene sequences (Nazari *et al.* 2007). There is no doubt they are 'good' species. Both display marked intraspecific differentiation: wing patterns of the French subspecies *Z. rumina medesicaste* and *Z. polyxena cassandra* clearly differ from their respective nominal subspecies, but variation forms a wide cline within a continuous distribution.

Natural hybrids between the species are scarce (only five are known to HD), but interspecific pairing has been observed in the field (de Puységur 1947). A large series of crosses within and between species has been performed by HD, although only some have been published (Descimon & Michel 1989). When *Z. rumina medesicaste* was crossed with *Z. r. rumina*, remarkable hybrid vigour was observed in the F_1, followed by strong hybrid breakdown in the F_2 (i.e. $F_1 \times F_1$) with arrested embryonic development, larval weakness and difficulties of pupation. Fewer than 5% of ova reached the adult stage in about 10 parallel broods. The low viability persisted in further crosses; only backcrosses, with either parent subspecies (or, paradoxically, with *Z. polyxena*), restored viability. Crosses between *Z. p. polyxena* from Greece and *Z. p. cassandra* from southern France also produced F_1 hybrid vigour, and some F_2 hybrid breakdown. However, the F_2 viability was not too low (around 25%), and further crosses ($F_2 \times F_2$ and more) displayed markedly enhanced viability: incompatibility therefore seemed less marked than in the first case. Crosses between Austrian and French *Z. polyxena* produced no F$_2$ hybrid breakdown.

Mate choice was studied in cages containing 10 males and 10 females of each species. Only intraspecific matings were observed (including the aforementioned distinct subspecies), demonstrating strong prezygotic barriers between species. All females proved to have mated, and one female *polyxena* produced offspring consisting partly of *polyxena* and partly of hybrids. Clearly, she had mated twice, and with males of each species. The hybrids were viable, but while the F$_2$ resulted in no offspring, backcrosses with *polyxena* and *rumina* were successful. The backcross hybrids from either side could, however, be crossed with the more distant parental strains. Thus backcrossed individuals, which had 3/4 of their genes from one species and 1/4 from the other, gave symmetrical F$_3$ progeny with 3/8 *rumina* : 5/8 *polyxena* offspring and the reciprocal; the same scheme was applied in the F$_4$ and beyond. The possibilities for complex crosses increased with the rank of hybridization and some were practised (for a complete account, see Descimon & Michel 1989). The hybrids were viable provided they had at least one complete unrecombined genome from a parental strain. Much more surprisingly, two later hybrid × hybrid crosses (not many were tried) gave fairly viable offspring, with no significant departures from 1:1 sex ratio or diapause abnormalities. In spite of strong pre-mating isolation between the pure species, female hybrids were attractive to males of either species, and male hybrids were attracted to any female. Similar results on hybrid sexual attractiveness have been obtained in a number of other butterfly species (e.g. *Heliconius*: McMillan *et al.* 1997, Naisbit *et al.* 2001).

It was not possible to continue the crosses, but some clear facts emerge. Firstly, F$_2$ hybrid breakdown is not absolute in interspecific crosses. Secondly, it is not limited to interspecific crosses; it may take place between subspecies, as is known in other species (e.g. Oliver 1972, 1978, Jiggins *et al.* 2001). The latter is particularly paradoxical, since, within both species, broad, clinal, unimodal hybrid zones connect 'incompatible' populations. Careful field work could well disclose interesting features in these contacts. Hybrid inviability is therefore probably not a useful species criterion on its own in crosses between geographically distant taxa. The ease of playing ping-pong with the two species once initial barriers have been ruptured shows that there is no absolute threshold of postzygotic incompatibility at the species level.

Frequent hybridization and introgression in sympatric Papilionidae: *Papilio machaon* and *P. hospiton*; *Parnassius apollo* and *P. phoebus*

PAPILIO MACHAON AND P. HOSPITON

Hybridization is widespread in *Papilio* species, especially in North America (Sperling 1990). Hybrids between the Eurasian *Papilio machaon* and the endemic *P. hospiton* of Corsica and Sardinia have been known for a long time (e.g. Verity 1913). Although their habitats and distribution in Corsica are very different, there is a frequent overlap, and hybridization occurs regularly. Crosses revealed two especially important postzygotic barriers (Clarke & Sheppard 1953, 1955, 1956, Clarke & Larsen 1986). (1) An almost total inviability of F$_1$ × F$_1$ hybrid crosses, originally mistaken for F$_1$ sterility. However, non-hatching ova were not 'sterile'; instead embryos show arrested development at various stages between early segmented embryos and fully-developed larvae unable to break out of their egg shell. (2) Strong Haldane's rule F$_1$ hybrid effects. In *hospiton* male × *machaon* female crosses reared in Britain, female hybrid pupae became 'perpetual nymphs', that is pupae which are unable to resume development. However, in other *Papilio* interspecific hybrids with extended diapause, ecdysone and insulin injections can trigger development (Clarke *et al.* 1970, Arpagaus 1987). Descimon & Michel (in Aubert *et al.* 1997) showed that insulin could also trigger development in *machaon* × *hospiton* hybrids.

Both reciprocal F$_1$ crosses and various backcrosses proved possible. The experiments were carried out in the Paris region, in an oceanic climate, and in Marseilles, on the Mediterranean, but under long photoperiod summer in both cases (Aubert *et al.* 1997). In the case of *hospiton* male × *machaon* female crosses, results depended on rearing conditions. In Paris, growth and developmental time of males was normal, but the female pupae, which were markedly bigger than those of either parental species, became perpetual nymphs, as found by Clarke & Sheppard (1953). In Marseilles, females did not enter diapause and gave large, viable females. The other possible F$_1$ (*hospiton* female × *machaon* male) again gave healthy hybrid males, but females were small, with accelerated development and no diapause, in both climates. F$_1$ × F$_1$ crosses gave almost complete inviability at various stages of early development, as before. On the other hand, backcrosses were all viable. F$_1$ hybrid females, in particular, appeared not to be sterile, whether they had *hospiton* or *machaon* as mothers.

The results suggest that introgression is possible. Allozyme and restriction fragment length polymorphism (RFLP) analysis of mtDNA markers show strong differentiation between the two species, with diagnostic alleles at some loci and a rather high Nei's D and mtDNA sequence divergence (Aubert *et al.* 1997, Cianchi *et al.* 2003). Putative hybrids found in different localities in Corsica and Sardinia were most probably F_1s, and from both reciprocal crosses. No individuals were found with introgressed mtDNA RFLP types in a large sample, suggesting a lack of mitochondrial introgression. However, the same was not true for nuclear loci. Alleles from *hospiton* were found in Corsican *machaon*, but were always absent in continental *machaon* (Aubert *et al.* 1997, Cianchi *et al.* 2003). The frequency of hybrids was lower in the Italian than the French data set (approx. 1% vs. 5%), but this is probably because HD collected especially avidly in areas of cohabitation, whereas many samples obtained by the Italians contained only one species.

Classically, *hospiton* is considered single-brooded, while *machaon* is multi-brooded. However, broods reared from wild Corsican *hospiton* females give a proportion (5–100%) of non-diapausing pupae (Aubert *et al.* 1996a). Diapause control in *P. hospiton* (and in *P. machaon*) is highly heritable but not simple; temperature and photoperiod act in combination, with threshold effects which interact strongly with genetic factors. Multi-brooded individuals are particularly common where *hospiton* feeds on *Peucedanum paniculatum*, a perennial evergreen umbellifer endemic to northern Corsica; this plant is suitable throughout the warm season. Observations in July and August confirm the existence of the second brood (Aubert *et al.* 1996a, Guyot 2002, Manil & Diringer 2003). In most regions of Corsica and throughout Sardinia, the main food-plant, *Ferula communis*, withers down as early as May onwards. Even here, late larvae can be found when roadside mowing during late summer renders resprouting *Ferula* available (Descimon, pers. obs.).

Aubert *et al.* (1997) suggest that multivoltinism in *P. hospiton* may result from introgression from *P. machaon*. This hypothesis was criticized by Cianchi *et al.* (2003) because of doubt in the existence of the second brood of *P. hospiton* (this argument is not tenable, as we have seen). Of more weight is the difficulty of distinguishing ancestral from introgressed polymorphisms. Nonetheless, Cianchi *et al.* (2003) found up to 43% *hospiton* allozymes in *machaon* on the islands, though never present on the mainland, and they argued that this was due to introgression. Conversely, they found only a scattering of *machaon* alleles in *hospiton*.

They argued that this introgression was mostly ancient and that reinforcement of interspecific barriers took place early during the secondary contact. This conforms to the commonsense prediction that what we observe today is an equilibrium between gene flow and selection against introgression (Descimon *et al.* 1989).

PARNASSIUS APOLLO AND P. PHOEBUS

Parnassius apollo is a montane butterfly, widespread from Altai in central Asia to the Sierra Nevada in southern Spain. *Parnassius phoebus* has a more restricted, higher-elevation distribution; in Europe, it occurs and can hybridize with the *P. apollo* only in the Alps (Plate 19b). The species always occur in close proximity (dry, sunny slopes for *P. apollo* and banks of torrents and rills for *P. phoebus*), but this does not ensure hybridization. Not only are their preferred flight environments different, but *P. phoebus* also flies earlier in the year. Therefore, it is only in localities where the two kinds of habitats are closely interspersed and phenology is perturbed that hybridization takes place, often at rather high frequency (Descimon *et al.* 1989). In some localities, hybrids are observed almost yearly; in others, they occur only following a snowy winter, when avalanches accumulate in the bottom of thalwegs. Thus, rather 'soft' pre-mating barriers, such as habitat and phenology differences, prevent hybridization. In captivity, mating between male *apollo* and female *phoebus* is often observed, and hand-pairing easy. The reverse cross is more difficult, due to the small size of male *phoebus*. F_1 hybrids display typical vigour and females are not perturbed in diapause (which takes place in the first larval instar, inside the egg shell). Field observations on wild hybrids show a strikingly perturbed behaviour: males fly restlessly, constantly roaming between the types of habitat preferred by both parent species. In captivity, male hybrids backcross freely with females of both species and are highly fertile, but female hybrids are inevitably sterile, producing numerous small ova that never hatch.

Morphometric analyses of natural populations strongly suggested backcrossing as well as F_1 hybrids in the field (Descimon *et al.* 1989). Using four diagnostic allozymes and several other loci with different allele frequencies in the two species, F_1 hybrids and backcrosses were detected (Descimon & Geiger 1988). One individual with the pure *apollo* wing pattern was heterozygous at one of the diagnostic loci, suggesting that backcrossing continues beyond the F_2. Mitochondrial DNA analysis showed that hybridization took place in both reciprocal directions but also that backcrossing could involve hybrid females (Deschamps-Cottin *et al.* 2000). While this contradicts findings from some captive broods (Descimon *et al.* 1989), it conforms to others (Eisner 1966). Once again, introgression in nature seems possible and is demonstrated by the field results.

COMPARISONS BETWEEN THE TWO HYBRIDIZING PAIRS OF PAPILIONIDAE

It seems clear that most would regard the four swallowtails treated here as four distinct, if somewhat bad species. They are readily distinguishable on the basis of morphology, allozymes and mtDNA. Allozyme and mitochondrial divergences suggest an age of around 6 Myr for the *Papilio machaon–P. hospiton* pair (Aubert *et al.* 1999), and a similar age is probable for *Parnassius apollo* and *P. phoebus*. Regular hybridization is therefore not necessarily a sign of incomplete speciation, but rather of the inability of the taxa to erect complete pre-mating barriers.

In conclusion, species can remain stable in spite of frequent hybridization and introgression. While there has been significant progress in understanding this introgression, we still have little overall knowledge of the genomic distribution of introgressed and non-introgressed loci.

Polyommatus (Lysandra) coridon, L. hispana and *L. albicans*: frequent hybridization everywhere, strong gene flow and yet species remain distinguishable!

For a long time, the chalkhill blue was considered in Europe to be a single species, *L. coridon*. However, in *Polyommatus sensu lato*, species rarely show consistent differences in genitalia or wing pattern (Plate 19a). Because of this, complexity in the *coridon* group was recognized initially due to voltinism. In 1916, Verity observed three emergences of *Lysandra* in the hills around Florence, Italy and showed that this was due to the existence of two separate species: one single-brooded, *coridon sensu stricto*, one double-brooded, *hispana* H.-S. Later on, he recognized *L. caelestissima*, univoltine with a distinctive sky-blue colour, from Montes Universales, central Spain. In Spain, the situation is especially confusing: there are single- and double-brooded forms, and bimodal hybrid zones where they overlap. At one time, clear blue hybrids between *L. caelestissima* and *L. albicans* from Montes Universales were also considered a distinct species, *caerulescens*. For a while the number of species recognized varied from one to four; eventually three were recognized on the basis of chromosome number and voltinism (de Lesse 1960a, 1969). These are:

(1) *Lysandra coridon*: widespread, univoltine, with $n = 88–90$, with an isolate in central Spain, *caelestissima*, considered a subspecies with $n = 87$.
(2) *Lysandra albicans*, univoltine, southwestern Spain, $n = 82$.
(3) *Lysandra hispana*, central France and Italy to Northern Spain, bivoltine, $n = 84$.

De Lesse (1969) described ssp. *lucentina* (correctly: *semperi* Agenjo 1968) from the Alicante region, which he referred to *hispana* on the basis of chromosome number ($n = 84$); later it turned out to be univoltine like *albicans*. He also showed that *L. italaglauca*, described as a species from central Italy, was actually a rather abundant hybrid between *L. coridon* ($n = 88$) and *L. bellargus* ($n = 45$). This form, of intermediate colour between the greyish of *L. coridon* and the dazzling blue of *L. bellargus*, was identical to *L. × polonus* (Zeller 1845), formerly mistaken as a good species from Poland and later recognized as a hybrid (Tutt 1910). These hybrids occur wherever the parent species fly together, although their frequency varies widely. *Lysandra coridon* is univoltine and flies around August, while *L. bellargus* is bivoltine and flies in May and September; the hybrid flies in late June. The meiosis of these hybrids displays incoherent equatorial plates, strongly suggesting sterility (de Lesse 1960a). Ironically, a blue species, *L. syriaca*, from the Middle East was for a while mistaken for *polonus* (Lederer 1858). Tutt (1914), who had earlier deduced that *polonus* was a hybrid, also correctly interpreted *L. syriaca* as a 'good' species. By analogy, de Lesse interpreted *L. caerulescens* as a hybrid, but, in this case, karyotypes are similar and meiosis appears normal. Laboratory hybrids between *L. coridon* and *L. hispana* obtained by Beuret (1957) proved fertile and viable until the F_3 generation. Interestingly, individuals from the last generation had the most chromosomes, as in *Antheraea* moths (Nagaraju & Jolly 1986). Another 'blue' hybrid mistaken for a species, famous for the author who described it, '*Lysandra*' *cormion* (Nabokov 1941), turned out to be a *Lysandra coridon × Meleageria meleager* hybrid (Smelhaus 1947, 1948, Schurian 1991, 1997). Again, hybridization occurs regularly in some regions (Moulinet, Alpes Maritimes, France; Bavaria, Germany).

De Bast (1985) followed up de Lesse's work using morphometric analysis on imaginal morphology and wing pattern. He recognized five species, *L. coridon, L. caelestissima, L. albicans, L. hispana* and *L. semperi*. The latter could be referred either to *hispana* via karyotype and wing pattern or to *albicans* via voltinism. In 1989, Schurian, after breeding experiments, crosses and morphological studies of all instars from egg to imago, recognized only three species, *coridon*, *albicans* and *hispana* (*semperi* was included within *hispana*).

Based on a restricted sample of 15 populations, Mensi *et al.* (1988) separated *coridon* and *caelestissima* as species because of a diagnostic allozyme (*Pk-2–105*), absent in *caelestissima*. Lelièvre (1992) systematically sampled 75 populations, collected by himself and HD, in order to cover all known systematic units and to test for hybrid zones in France and Spain. Allozyme analysis showed that two main entities could be readily distinguished: *coridon* + *caelestissima*, and *hispana* + *albicans* + *semperi*, with Nei's $D \approx 0.05$ between the two groups. In contrast, *L. bellargus* was separated from the *coridon* group by a $D \approx 0.30$. No diagnostic alleles were found between *coridon* and *caelestissima*, contradicting Mensi *et al.* (1988). Therefore, there is little reason to consider them as separate species. The chief argument for separation is the colour of male imagines, but, in northern Spain, populations are often of intermediate colour (ssp. *manleyi* and *asturiensis*). A sex-limited morph, the blue '*syngrapha*' female, shared by *coridon* and *caelestissima* (Descimon 1989) also suggests conspecificity. Disjunct distributions of the two taxa prevent use of the cohabitation criterion. A conservative solution is thus to merge all the populations into a single species with some strong subspecies.

The tale of *L. coridon* in Tyrrhenian Islands is almost incredible. Its lime-loving foodplant, *Hippocrepis comosa*, is very scarce on the mainly acidic soil of these islands. The description in 1977 of ssp. *nufrellensis* from the remote granitic Corsican Muvrella massif by Schurian attracted scepticism, but was confirmed in 2006 by Schurian *et al.* – Muvrella granite is hyperalkaline and supports *H. comosa*! *L. coridon*, described as *gennargenti*, was also found in Sardinia on more easily accessed calcareous patches (Leigheb 1987). Both populations are well characterized by adult wing pattern (the males are vivid blue and females are always blue) and by preimaginal stages. Marchi *et al.* (1996), using allozyme analysis, left the form as a subspecies of *coridon*. However, Jutzeler *et al.* (2003a, b) did not lose an opportunity to raise yet another known form to species rank, based only on preimaginal morphology.

In the '*hispana–semperi–albicans*' complex, things are much more complicated. Populations assigned to one of these putative taxa by 'classical' criteria (namely, wing pattern, distribution and voltinism) are not distinguishable via allozymes. This is especially true for '*albicans*' and '*semperi*', which broadly overlap in their allozyme polymorphisms.

Hybrid zones between the taxa give rise to additional complexity. A hybrid zone exists between *caelestissima* and *albicans* in Montes Universales (central Spain); both are single-brooded and fly at the same time of year. The former flies at rather high elevation (1200–1800 m), the latter in lower zones (800–1400 m). They overlap at intermediate altitudes, where putative male hybrids ('*caerulescens*') can easily be detected by wing colour. We have studied three samples, each containing ~30 individuals: the first from a pure *caelestissima* locality (Paso del Portillo); the second from an *albicans* locality (Carpio del Tajo); and a third area of cohabitation, where hybrid *caerulescens* reach a frequency of 10% or more (Ciudad Encantada). Allozyme genotypes were concordant with colour pattern in 77% of the cohabiting sample. Discordant individuals were all '*caerulescens*', that is, presumably hybrids, and their allozyme genotypes were intermediate (Lelièvre 1992). The hybrid zone thus appears more or less bimodal, even though hybrids were rather abundant.

Two other hybrid zones were studied in northern Spain (at Ansó and Atarés in the Jaca region), where single-brooded *L. coridon manleyi* overlaps with double-brooded *L. hispana*. The former species again flies at a higher elevation, but the two overlap at intermediate altitudes. 'Pure' reference populations were again studied nearby: Aranqüite and Embalse de Oliana, respectively. In the hybrid zone at Ansó, the variously coloured butterflies were hard to separate genetically. Individuals were either genetically similar to those from one or other pure sample, or intermediates. In the second hybrid zone, at Atarés, two visually different categories of individuals were found, some with the obvious clear blue *coridon* phenotype, the others greyish-white and similar to *hispana*. Intermediate specimens were scarce and none was analysed genetically. Paradoxically, all genotypes from the cohabitation zone, including those classified as *hispana* by wing pattern, corresponded to *coridon* from Aranqüite, rather than to *hispana* from Oliana, so introgression is suspected (Lelièvre 1992).

More recently, bivoltine *Lysandra* populations flying in southern Slovakia were separated out as a species, *Polyommatus slovacus* (Vitaz *et al.* 1997), on the basis of subtle adult morphological differences (the bluish dorsal hue of male wing pattern and slight differentiation of male and female genitalia). A cohabitation criterion was used, since it apparently flies with univoltine *L. coridon* in some localities, although there is no mention of hybrids. There is no known genetic difference between *L. slovacus* and neighbouring populations of *L. coridon* (Schmitt *et al.* 2005). Voltinism remains the chief character.

In conclusion, there is one rather clear, homogeneous species, *L. coridon*, with strongly differentiated subspecies in Spain (*caelestissima*) and the Tyrrhenian Islands (*nufrellensis*); chromosome characters and phenology as well as allozyme data support the unity of this taxon. The geographically variable male wing colour pattern conforms to this diagnosis, since populations from northern Spain are intermediate. In contrast, the same criteria do not provide coherent evidence for splitting the *hispana* complex into several units. The forms *semperi* and *hispana* share the same karyotype ($n = 84$), but the former is univoltine like *albicans*, which, however, has a different chromosome number ($n = 82$). Allozymes have not yet proved very useful. HD has doggedly sought further contact zones between the three taxa of the *hispana* complex, but in vain. Lelièvre's (1992) work was extremely useful, but his premature death prevented a more complete analysis.

The *Erebia tyndarus* group: parapatry, hybrid zones and Gause's principle

This group (Plate 20c) illustrates the use of successively more sophisticated taxonomic criteria, and the difficulties of applying various species concepts; we therefore employ a historical approach. The *tyndarus* group is characterized by cryptic grey hind wing undersides, which provide good camouflage in rocky grasslands. Their distribution stretches from western North America, across the Pacific to Eurasia, and finally to the Asturias in Spain. Until the twentieth century, all were considered to belong to a single variable species. In 1898, Chapman piloted the use of male genitalia in *Erebia* and recognized *E. callias* Edwards from North America, and a submontane form from Asia Minor, *E. ottomana* H.-S., as separate species. In 1908, Reverdin studied wing pattern in Western European taxa, and showed that the Alpine forms could be arrayed in two groups, *E. tyndarus* Esper and *E. cassioides* Reiner & Hohenwarth. The latter can also be recognized in the Pyrénées, Apennines, Balkans and Carpathians. He further noted that the southernmost form, *hispania* Butler from the Sierra Nevada, could be grouped with others from the Pyrénées, *goya* Frühstörfer and *rondoui* Oberthür, without elevating them to species rank.

Warren (1936) recognized four species based on male genitalia: *tyndarus*, *cassioides*, *dromulus* Staudinger (from the mountains of Asia Minor) and *callias*, from North America, Central Asia, Elburz and the Caucasus. In 1949, he pointed out that *cassioides* and *rondoui* (previously included with *tyndarus*) overlapped in the Pyrénées and considered this

cohabitation evidence for separate species. In 1954, he extended this to *tyndarus sensu stricto* on the grounds of cohabition with *cassioides* in the Bernese Alps.

There is a striking feature in the *tyndarus* group: distributions of the taxa are typically parapatric and in a given region, there is only one form. Distributions overlap only in very narrow contact zones. Sometimes, hybrids are found in various proportions (see below); in other cases, hybridization is absent. Mutual exclusion can be attributed to Gause's (1934) principle: 'one species per ecological niche'. For the BSC, the *tyndarus* group was somewhat distressing: morphological criteria are weak, and ecological differences minimal, as shown by mutual geographical exclusion. Narrow cohabitation with little or no admixture therefore became the main distinguishing criterion within this group.

Warren never went beyond genitalic characters, but de Lesse and Lorković initiated a synthetic approach using karyotype, morphometrics of genitalia, wing-pattern variation, laboratory crosses, and detailed field studies on distribution and hybrid zones. There was great variation in chromosome number: *hispania*, with $n = 24$, stood out from *cassioides* and *tyndarus*, with $n = 10$ throughout their ranges (Lorković 1949, 1953, de Lesse 1953). Later, two cryptic species were discovered: *calcaria* Lrk. ($n = 8$), from the Julian Alps, and *nivalis* Lrk. & de Lesse ($n = 11$), limited to upper elevations of the Eastern Alps, where it flies above *cassioides* or *tyndarus* (Lorković 1949, Lorković & de Lesse 1954b). In addition, de Lesse (1955a, c) showed that *E. callias* from North America and *E. iranica* and *E. ottomana* from the Middle East displayed markedly different karyotypes ($n = 15$, 51, and 40, respectively). De Lesse (1960a) performed morphometric analyses of genitalia. He reinstated wing pattern as a valuable tool if concordant with other characters. In particular, he noticed that the dark hind-wing eyespots could be shifted distally in their fulvous surrounds, rather than being centred, enabling one to group the southernmost taxa, *hispania* and *iranica*, also characterized by high chromosome numbers ($n = 24–25$ and $51–52$). Recent studies have shown that satyrine eyespot variation often results from important developmental genetic shifts (Brakefield 2001). Locally adaptive camouflage wing patterns (see above), such as hind-wing underside colour, provided less useful criteria.

Lorković (1954) carried out crosses between several taxa (*calcaria* × *cassioides*, *calcaria* × *hispania* and *cassioides* × *ottomana*). All showed genetic and behavioural incompatibility: assortative mating, together with sterility of primary crosses and of F$_1$ hybrids (Lorković & de Lesse 1954a).

However, the taxa used were not the most significant: *otto-mana* is notoriously distant from the other members of the group (see below); *calcaria* and *hispania* differ in karyotype ($n = 8$ and 24 respectively) and their ranges are very distant. The most useful test is *calcaria* × *cassioides*: they have identical karyotypes ($n = 10$) and adjacent distributions, but clear incompatibilities were still found.

It was thus important to investigate contact zones and distribution in nature. A complex pattern of allopatric distribution of *hispania* and *cassioides* was found in the Pyrénées (de Lesse 1953, Descimon 1957), with very narrow zones of cohabitation. Only a single putative hybrid was captured by Descimon (de Lesse 1960a) among several hundred individuals in many zones of overlap. In the central Alps, *tyndarus* occurs as an outpost inserted between two disjunct populations of putative '*cassioides*'. In the absence of differences in chromosomes, genitalia and wing pattern provided the only useful criteria. Westwards, in Val Ferret, southwest Switzerland and in adjacent Italy, above Courmayeur, populations of *tyndarus* and '*cassioides*' are separated by narrow unoccupied regions (de Lesse 1952). Near Grindelwald, in the Bernese Oberland, a cohabitation site with phenotypically intermediate individuals was found. At the eastern end of the *cassioides–tyndarus* contact zone, in Niedertahl, Austria, a cohabitation site was found, but hybrids were not found, even though enhanced variability in genitalia suggested introgression (Lorković & de Lesse 1955).

Erebia nivalis Lrk. & de L., originally considered a smaller high-elevation form of *cassioides* (Lorković & de Lesse 1954b), was raised to species rank after discovery of its peculiar karyotype ($n = 11$). Cohabitation is often observed at the altitudinal boundary between the two, although hybrids are never found. Competitive exclusion is especially convincing: at Hohe Tauern, a different species occurs on each of two isolated massifs (*cassioides* on Weisseck and *nivalis* on Hochgolling); in both cases the entire span of alpine and subalpine zones (1800–2600 m) is occupied, suggesting competitive release (Lorković 1958). Similarly, in eastern parts of their distribution, *cassioides* and especially *tyndarus* reach higher elevations in the absence of *nivalis*. The distribution of *nivalis* is broadly fragmented into two parts: in the Austrian Alps and in a more restricted area in the Bernese Oberland. The gap between the two areas occupied by *nivalis* has been colonized by *tyndarus*. In the Grindelwald area, where all three taxa cohabit, *tyndarus* looks like the more aggressive competitor which has eliminated *nivalis* even from high-elevation habitats.

A rather clear picture emerges from these studies (Guillaumin & Descimon 1976): in Europe, the *tyndarus* group includes several well-defined species: *ottomana, hispania, calcaria* and *nivalis*. The *tyndarus–cassioides* pair is more puzzling. By now, a disjunct assemblage of seemingly subspecific forms were recognized as *cassioides*, including populations from the Asturias, the Pyrénées, Auvergne in French Massif Central, Western and Southern Alps, Eastern Alps, the Apennines and some Balkan massifs. The populations referable to *tyndarus* occurred in a continuous distribution inserted like a wedge between *cassioides* populations in the Central Alps. Lorković (1953) proposed that these taxa were examples of an intermediate category, 'semispecies' (Lorković 1953, Lorković & Kiriakoff 1958). However, in practice, *cassioides* and *tyndarus* were considered separate species by most lepidopterists (e.g. de Lesse 1960b).

In 1981, Warren published a supplement to his monograph of the genus *Erebia*. Arguing that chromosomes had little systematic value, he relied mainly on male genitalia and arranged the taxa in a somewhat confusing way. This was accentuated because he considered *cassioides* a *nomen nudum*, in spite of the lectotypification of the figure in Reiner & Hohenwarth by de Lesse (1955a) – he considered the figure was inaccurate. He recognized the following European species:

(1) *tyndarus* – Central Alps.
(2) *nivalis* – Austrian Alps and Bernese Oberland.
(3) *aquitania* Frhst. (= *cassioides pro parte*) – Southern Alps, Dolomites, Karawanken, Montenegro, Etruscan Apennines, Mont Blanc range and Pyrénées (part).
(4) *neleus* Frr. (= *cassioides pro parte*) – Transylvanian Alps, Austria, Rhodope, Macedonia, Central Alps, Pyrénées (part), Roman Apennines, Abruzzi, Auvergne.
(5) *calcarius* – Julian Alps.
(6) *hispania* – Sierra Nevada and Pyrénées.
(7) *ottomana* – considered very distinct from the other members of the group.

The species designated by Warren in the former *cassioides* group lacked zoogeographical coherence compared with those recognized by de Lesse & Lorković. The only serious (partial) support for Warren's theses was the suggestion that populations of *cassioides sensu lato* east of the *tyndarus* wedge could be called *neleus*, and the western ones *aquitania* (von Mentzer 1960). This prophetic suggestion, making zoogeographical sense, was largely overlooked at the time.

A much firmer position was adopted by Niculescu (1985): an extreme 'lumper', he used only morphological criteria to unite all of the group in a single polytypic species, *tyndarus*. Much earlier, de Lesse (1960a: 57), had warned about the exclusive use of morphology as criteria to delimit species, especially if already known to be labile and if the classification required illogical zoogeographical distributions. However, Gibeaux (1984) claimed he had discovered *E. calcaria* and *E. tyndarus* closely adjacent to *cassioides* in the Col Izoard region of the French Alps, on the base of wing pattern and genitalic morphology, without reference to karyotype, cohabitation and molecular criteria. Lorković (pers. comm. to HD) keenly argued that the genitalic characters used by Gibeaux could be explained by individual variation. Wing-pattern differences were confined to the strongly selected, taxonomically useless hindwing undersides.

Ten years later, a far more informative study, based on 17 allozyme loci, largely confirmed the common ground of previous authors: *ottomana*, the *hispania* complex and *nivalis* were very distinct from other members of the group, with Nei's $D > 0.20$ (Lattes *et al.* 1994). The single available sample of *tyndarus* differed by $D = 0.14$ from the cluster, while '*cassioides*' itself consisted of clearly differentiated 'western' and 'eastern' *cassioides* groups. Lattes *et al.* attempted to outflank Warren's rejection of the name *cassioides* by designating a neotype; an actual museum specimen from the Austrian Alps – *cassioides sensu stricto* therefore now refers specifically to the eastern taxon. Actually, the older valid name for western '*cassioides*' was *arvernensis* Oberthür (type locality: northern French Massif Central), and we use it instead of *neleus* below. The rather large genetic distance between *hispania sensu stricto* from Sierra Nevada and *rondoui* and *goya* from the Pyrénées (Nei's $D = 0.16$), added to slight differences in chromosome number ($n = 25$ vs. 24, respectively), led the authors to consider them different species. However, they did not do the same with two *ottomana* samples from the Italian Alps and southern French Massif Central, even though they were distant by a Nei's D of 0.18.

Most recently, a study using allozymes and sequence data from two mtDNA genes was carried out on a limited number of populations (Martin *et al.* 2002); eastern '*cassioides*', in particular, was lacking. There were large genetic distances between *ottomana* and *hispania sensu lato*, and their monophyly was confirmed; *tyndarus* (three populations) also proved monophyletic, while *nivalis* formed a strongly supported group together with *calcaria*; divergence at the mtDNA genes averaged 0.34%. The allozyme data showed a similar pattern to that found by Lattes *et al.* (1984): *nivalis* was located at the end of a long branch. In contrast to *tyndarus*, *arvernensis* did not group as a single cluster and appeared paraphyletic. The basal and terminal branches of these trees were well resolved, but the intermediate branches, which should define the phylogenetic relationships between *tyndarus*, *arvernensis*, *nivalis* and *calcaria*, remained unclear. The lack of eastern *cassioides sensu stricto* prevented accurate phylogenetic estimation, since we still do not know if this taxon clusters with *arvernensis*, *tyndarus*, or *nivalis* and *calcaria*.

A final and rather ludicrous episode of this tale occurred in the butterfly distribution atlases for France (Delmas *et al.* 1999) and Europe (Kudrna 2002). The former used the correct name *arvernensis* for 'western *cassioides*'. The resultant geographical distributions were correctly documented by Kudrna, but this author also reported older literature records from France (as well as from Spain, parts of Switzerland and Italy) as '*cassioides*'. Hence an extensive but entirely fictitious pseudo-sympatry of the two taxa was reported in the French Alps and Pyrénées, and even in the northern Massif Central.

Erebia serotina Descimon & de Lesse, 1953: a hybrid mistaken for a species

In September 1953, the 19-year-old HD captured two individuals of an unknown *Erebia* at 1000 m elevation in the Pyrenean valley of Cauterets and showed them to H. de Lesse. After careful examination, they concluded that the butterflies belonged to an unknown, late-flying species they named *E. serotina* (Descimon & de Lesse 1953) – a surprising finding in the mid twentieth century. Further individuals were captured regularly in the same region over a period of 10 years, always late in the season and at the same elevation (Descimon 1963) (Plate 20a). Chromosome study (Descimon & de Lesse 1954) disclosed a number of $n = 18$.

However, the absence of females in a sample of 18 individuals was intriguing; Bourgogne (1963) suggested that *E. serotina* was a hybrid between *E. epiphron* and *E. pronoe*, both also present in the region and having chromosome numbers of 17 and 19, respectively. This possibility had been rejected by Descimon & de Lesse, since the two species live at a higher elevation than *serotina* (over 1400 m and above the treeline). Moreover, de Lesse and later Lorković (pers. comm. to HD), who examined the histological preparations of *serotina* testes, considered chromosome pairing during meiosis to be normal. The debate was echoed by Riley (1975) and Perceval (1977), with no additional data. Higgins & Riley (1970) included *E. serotina* in their field guide, although the species was not mentioned in later editions or other guides.

A few other specimens were captured in the same valley (Lalanne-Cassou 1972, 1989) and 15 km to the west (Louis-Augustin 1985) and also in the Spanish Pyrénées, always late and at low elevation (Lantero & Jordana 1981). Warren (1981) was also inclined to the hypothesis of a hybrid, which he considered to be between *epiphron* and *manto*, another Pyrenean species, on the basis of morphology and against the chromosomal evidence – *manto* has $n = 29$, which should yield $n = 23$ for the hybrid. At this juncture, both 'hybrid' and 'good species' hypotheses seemed unlikely.

Forty years later, the retired HD again went in pursuit of *serotina* and found several individuals in September 2000 and 2002 close to Bagnères de Luchon, 60 km east of Cauterets (Descimon 2004). An analysed individual was heterozygous at all diagnostic allozyme loci between *epiphron* and *pronoe*, while mtDNA showed that *epiphron* was the mother (E. Meglécz *et al.* unpublished). Therefore, *serotina* is indeed a hybrid between *epiphron* and *pronoe*. Moreover, after a series of hand-pairing crosses, three hybrids similar to wild *serotina* were obtained by Chovet (1998). Bourgogne's hypothesis was therefore proved correct and the mystery of *Erebia serotina* solved; the absence of females may be due to arrested growth, while males undergo accelerated development and hatch before the cold season (see the *Papilio* case above). Now, the riddle has moved on towards other questions: why does *serotina* fly at altitudes where its parents do not? Why does it occur regularly in the Pyrénées, but not in other regions of parental contact?

Hybrids are scarce in *Erebia*: apart from the previously mentioned *arvernensis* × *hispania* hybrid, only two other cases have been recorded. The first, *intermedia* Schwnshs, is found in the Grisons, Switzerland; initially mistaken for a variety of *E. epiphron*, it was later shown to be a *flavofasciata* × *epiphron* hybrid (Warren 1981). The second has been collected only once, from the Carpathians, and was recognized immediately as a *pronoe* × *medusa* hybrid (Popescu-Gorj 1974). Taken in late September, like *serotina*, it was similar to it also in its genitalia. In all three cases, at least one of the parents of *serotina*, *E. epiphron* or *pronoe*, is involved.

Other cases of 'bad' species in European butterflies

Palaearctic butterflies demonstrate many other cases of uncertain or 'fuzzy' species (Tolman & Lewington 1997) (Table 16.1B). These cases suggest some general patterns of 'bad' species relations, often involving hybrid zones. Some such zones present ecological frontiers, in particular at boundaries between lowland and montane taxa: *Pieris napi* and *bryoniae*, *Euchloe crameri* and *simplonia*, *Lycaena tityrus* and *subalpina*, *Melitaea parthenoides* and *varia*, *Coenonympha arcania*, *gardetta* and *darwiniana*, *Pyrgus cirsii* and *carlinae*. *Coenonympha darwiniana* may actually be a stabilized hybrid between *arcania* and *gardetta*, since it is found at intermediate elevations between the areas where *arcania* and *gardetta* occur (Holloway 1980, Porter *et al.* 1995, Wiemers 1998). In most cases, the limit coincides with the elevation where two broods per year become impossible because of low mean temperature; a similar phenomenon in latitude is found in most areas where *Aricia agestis* meets its congener *artaxerxes*. Very often, there is a gap where neither form is regularly present, perhaps because in this area, a second brood can be triggered by photoperiod, but does not complete its growth before autumn, and fails. Here, a discrete biological response cannot easily track a continuous environmental change. Another striking feature is that differentiation between clearly distinct taxa is often observed in the Alps, while in the Pyrénées similar distribution gaps are observed, but with much weaker genetic differentiation between single- and double-brooded populations (e.g. *L. tityrus* and *M. parthenoides*). The case of *Maculinea alcon* and *M. rebeli* is so complex and the ecology of both taxa has given rise to so many papers that it deserves separate treatment. The case of these blues is the closest in butterflies to 'ecological races'. No differences were found at mtDNA or nuclear EF1-α gene sequences (Als *et al.* 2004). However, we know too little about gene exchange between the populations to locate them with precision on the bad species–good species spectrum (Wynhoff 1998, Als *et al.* 2004).

Other repeated patterns in contact zones suggest 'suture zones' (Remington 1968) caused by secondary contact of whole faunas from different Pleistocene or earlier refuges, especially the Iberian ('Atlanto-Mediterranean'), and Italian + Balkans refuges ('Ponto-Mediterranean': de Lattin 1957). *Iphiclides podalirius* and *feisthameli*, *Pontia edusa* and *daplidice*, *Colias hyale* and *alfacariensis*, *Lycaena alciphron* and *gordius*, *Melitaea athalia* and *celadussa*, and *Melanargia galathea* and *lachesis* appear to belong to this category. Desert species such as *Papilio saharae* and *Melitaea deserticola* meet with temperate counterparts in northern Africa, while montane species also provide examples of differentiation in various refuges followed by subsequent contact. A general feature of these contacts is Gausean exclusion and therefore parapatry; the cases of *Erebia pandrose* and *sthennyo*, *E. euryale* forms, *mnestra* and *aethiopellus* are comparable with the *tyndarus* group in this respect. Finally, Corsican and Sardinian endemics are somewhat different; they might be expected to provide parallels with *P. machaon* and *hospiton*, but they lack genetic differentiation or pre- and post-mating incompatibility; consequently, they are not able to cohabit.

GENERAL DISCUSSION

The examples studied here can serve as a testbed for theories and concepts of species and speciation, and of their use in answering questions such as: are there one, two, or more 'good' species involved, or is this an example of speciation in progress? Can we use the results to suggest a simple and unequivocal, or at least useful nomenclature? Is there a general procedure, using the tools and concepts already mentioned, to allow us to reach this goal?

The simplest case is *Erebia serotina*. Originally ranked as a species, it ended up as a mere hybrid: 1→0. Here the difficulty was technical: it was finally through the use of molecular markers that the parent species and the sexes involved in the cross were recognized. In the case of *Lysandra polonus* and *L. italaglauca*, the tools were cytological; in these cases, the sex of the parents involved remains unknown, although mtDNA analysis could easily solve the question. Among many other known hybrids (Table 16.1), the majority have been identified only via wing pattern. There is an opposite case, where a species, *Lysandra syriaca*, was recognized after being initially confused with the hybrid *polonus*: 0→1. Hybridization does not occur in all zones of cohabitation with the same frequency, as seen in all the cases studied here. The behaviour of hybrids can be not only different from either parent, but also not intermediate; this is especially striking with *serotina*, but is also observed with *Parnassius apollo* × *phoebus* hybrids (Descimon *et al.* 1989).

With *L. sinapis* and *reali*, we have an opposite, but equally clear case: 1→2. The data provide an unambiguous result under all species concepts: there are clear morphological differences; gene pools are completely isolated (to satisfy BSC adepts); the ecological niches are different and the two species form mutually monophyletic assemblages and thus raise no problem for phylogeneticists.

Things become more complex with *Zerynthia*. Few doubt that *Z. rumina* and *polyxena* are 'good' species. Again, there are obvious morphological differences, and there is a rather strong separation of gene pools – hybrids are scarce enough to satisfy BSC groupies, in spite of broad sympatry and character displacement in ecological preferences. Phylogeneticists

will be happy that each species constitutes a monophyletic assemblage. However, serious genomic incompatibilities were observed between distant populations within each of these species, especially within *rumina*. In fact, the level of incompatibility between the species was not markedly greater than within each. So does *Zerynthia* contain one, two, three, four or even more species? These findings occurred only as a result of crosses between forms which do not co-occur naturally; they are artefacts. Similar incompatibility effects have also recently been observed within the well-known tropical species *Heliconius melpomene* (Jiggins *et al.* 2001). It is wisest to conclude: 2→2.

The situation with *Papilio hospiton* and *P. machaon* is clearer, but fits less easily with theory. Obviously these two constitute 'good' species, conforming to morphological, biological and cladistic concepts. *Parnassius apollo* and *phoebus* are a similar case. However, the evidence for some mutual introgression corresponds more closely to the 'genic view' of speciation. Meanwhile, the asymmetrical character of introgression in *Papilio* fits less perfectly. It seems likely that these *Papilio* diverged beyond the point of no return in allopatry, and that introgression occurred only after *P. machaon* again became sympatric. The case of *Parnassius apollo* and *P. phoebus* is similar, but the two species seem likely to have been in close proximity for a long time. In this case, gene flow would have been progressively reduced. Yet, in spite of introgression, all four species remain 'good', in the sense of 'distinguishably different', wherever they overlap.

With the brown *Agrodiaetus*, the situation changes. Hybrids are morphologically undetectable. Karyotype becomes questionable, here, as a species criterion, unless one allows the concept of karyospecies (e.g. Wiemers 2003). Until recently, a karyotype markedly different, either in number or size of chromosomes, was taken as proof of species status because chromosomal differences directly provide mixiological incompatibility. On this basis, allopatric populations distinct in chromosome number were separated as 'good' species. However, frustratingly, Wiemers (2003) and Kandul *et al.* (2004) showed that karyotype variation in this group is sometimes associated with genetic and phylogenetic differentiation, and sometimes not. So how many 'species' are included in Western taxa of brown *Agrodiaetus*? Clearly, *A. ripartii* and *fabressei*, which occur in sympatry, must be distinct (ironically, they have the same chromosome number, but the karyotypes have different morphology). For the other populations, all allopatric and with very variable chromosome numbers, the question makes little sense. Nonetheless, in his excellent, exhaustive work on *Agrodiaetus* and related genera, Wiemers (2003) firmly comes down on the side of all of the other taxa being separate species.

In *Hipparchia*, it seems clear that the best solution is to ignore the more extreme splitters and adopt a moderate lumper approach (Cesaroni *et al.* 1994), but this remains somewhat arbitrary and, again, depends heavily on the status of allopatric units.

The situation observed today in the *Erebia tyndarus* group is typical of the present state of systematics. Taxonomic decisions made during the first half of the twentieth century lacked much biological insight, but the important contribution of genitalic morphology boosted knowledge. After Huxley's 'new systematics', even those specializing in morphology, like Warren, began to take the BSC into account, especially with respect to cohabitation, but also because genitalic differences were assumed to cause mechanical incompatibility during mating. The bulk of progress on the group was, however, made during the 1950s using karyology, in this case a highly efficient tool. Differences between chromosomal morphs are regularly associated with sterility and other deleterious side-effects of hybridization. However, morphometrics, research on contact zones and laboratory crosses were combined with chromosomal studies in a synthetic approach which continues to elicit admiration. It is worth noting the enormous contribution made by de Lesse & Lorković in this field. Access to most populations required ascending many hundreds of metres on foot. In his synthesis, de Lesse (1960a) provided impressive distribution maps. But while data on the most important contact zones and centres of distribution were published in detail, many distributional data accumulated by de Lesse remained unpublished, and were lost when he died.

Mostly, the polytypic or 'biological' species concept was employed. However, a number of pockets of resistance rebelled against any attempt at consensus. The *Erebia tyndarus* and the forms of the *cassioides–arvernensis* complex remain the most contentious. At present, it is clear that the Grindelwald contact forms a 'bimodal hybrid zone' (Jiggins & Mallet 2000). Gene flow might help to explain contradictions between allozyme and mtDNA sequence data elsewhere (Lattes *et al.* 1994, Martin *et al.* 2002). There are large allozyme distances between *nivalis* and the other taxa, and rather slight ones with mtDNA. Indeed, *nivalis* is more of a high-elevation species that must experience a markedly different thermal environment. Watt (2003) has demonstrated that 'differentiation or uniformity of polymorphic genotype frequencies over space may be driven by strong local

selection pressures'; allozyme divergence may not always yield results independent of selection.

What was the contribution of molecular markers to improve species delimitation in the *tyndarus* group? Lattes *et al.* (1994) used Nei's genetic distance to separate *cassioides* from *arvernensis* and *hispania* from *rondoui*, but ignored the larger differences between the two populations of *ottomana*, without any particular justification. The main problem of using genetic distance as a criterion of species is that the threshold level may differ in each group studied (Avise 1994). Finally attempts to determine the status of allopatric taxa (including experimental crosses) are rather like division by zero, the cohabitation criterion acting like the denominator that does not exist.

More significant was the much greater utility of molecular data for reconstructing phylogeny distinguishing monophyly from paraphyly. However, a phylogenetic species concept may be difficult to apply in this case. For example, in the tree published in Fig. 4 of Martin *et al.* (2002), *calcaria* and *nivalis* cluster within a group consisting of all the *arvernensis* samples, and together form the sister group to the monophyletic *tyndarus* assemblages. Yet *tyndarus* and *arvernensis* act as separate species, since they meet at a bimodal hybrid zone; this causes a logical anomaly for phylogenetic species, since more basal taxa do not seem to reach species rank, but form a paraphyletic group as far as sexual isolation is concerned (if sexual isolation is considered an apomorphy). Further research will perhaps help to resolve some of the tantalizing questions in this group, but, at present, we must confess an inability to answer precisely the question 'how many species are there?' One can propose a spectrum of solutions spanning two extremes: the 'lumper's' position, with *ottomana*, *hispania*, *tyndarus*; or the 'splitter's' position, with the various, very disjunct strains of *ottomana* as 'species', *hispania*, *rondoui*, *arvernensis*, *cassioides*, *tyndarus*, *calcaria* and *nivalis*. However, the precise decision along this spectrum will always be more or less arbitrary.

Although also complex, the *Lysandra coridon* group case is somewhat clearer. In particular, if the phylogenetic species concept is capable of wreaking havoc on the *Erebia tyndarus* group, Wu's (2001) 'genic view of species' aids in understanding puzzling features of the *coridon* group. We have mentioned the low level of allozyme differentiation within and between the species of this group, while habitus and ecological features yield stronger, better-supported patterns. One must keep in mind that chromosome number is very high in *Lysandra*. Therefore, each linkage group should be small and, hence, hitch-hiking will affect fewer loci during speciation. A majority of the genome might therefore be exchanged freely, while only regions linked to genes affecting sexual isolation and ecological specialization will be kept distinct by strong selection. Otherwise, in this group, the problem of characterizing species is relatively soluble, provided one cuts some Gordian knots. One example of such a unit is provided by *Lysandra coridon*, which displays a very 'open' population structure, with few if any genetic differences even between geographically distant populations (Lelièvre 1992, Schmitt *et al.* 2002). The main problems are the isolates at the southern periphery of its distribution: *caelestissima* in the mountains of central Spain and *nufrellensis–gennargenti* in Corsica and Sardinia. The stumbling block of the absence of cohabitation is again encountered. By far the simplest and most sensible solution based on such data would seem to be to merge all the forms into a single species, *coridon*, with some strong peripheral subspecies. Likewise, the *albicans–hispana–semperi* complex is best considered a single species with some variation in chromosome number (as in *coridon*) and adaptive features such as voltinism, in the absence of a clear indication from hybrid zones. On the contrary, the frequent occurrence of bimodal hybrid zones between populations of the *coridon* unit, as previously defined, and of members of the *albicans* complex precludes merging them into a single 'good' – or even 'bad' – species unit. This case, in common with the *Erebia tyndarus* group, demonstrates the phenomenon of local mutual exclusion due to similar ecological niches, especially foodplant choice. The criteria of voltinism and chromosome number, ranked highly by de Lesse, proved not much more reliable than other criteria. Therefore, to the question: 'how many species?', we finally answer 'two only' – a simple answer which unfortunately might fray the tempers of some lepidopterists.

CONCLUSIONS

'I have just been comparing definitions of species… It is really laughable to see what different ideas are prominent in various naturalists' minds, when they speak of "species". In some resemblance is everything & descent of little weight – in some resemblance seems to go for nothing & Creation the reigning idea – in some descent is the key – in some sterility an unfailing test, with others not worth a farthing. It all comes, I believe, from trying to define the undefinable' (Darwin 1856). Darwin would have found it even more laughable today: Mayden (1997) enumerated no fewer than 24 species concepts, most of them recent.

Whether species are material, 'real' objects, that exist in the absence of human observers as no other taxonomic rank does, or whether they are only a construction of our mind, is a philosophical problem beyond the scope of this chapter. Our aim is to use the totality of the existing evidence to suggest simple, practical solutions to taxonomic problems, and we attempt to avoid further adding to the vast slag-heap of useless concepts and definitions of the indefinable. Darwin used only a loose definition of species but he was an experienced taxonomist, knew a great deal about describing actual species, and it was sufficient to convince his readership of transpecific evolution. We believe that, even today, a pragmatic, taxonomic solution is more productive than attempting to decide whose concept is correct.

Two facts are undeniable:

(1) Taxonomic decisions based on biological or polytypic species concepts are still common. For instance, Kandul *et al.* (2004) use the term species to mean reproductively isolated populations. Many groups of organisms considered species are well behaved and obey not only the BSC, but also *most* definitions of species.
(2) However, a significant number of rakish taxa will probably always fail to conform to this species morality. They regularly conduct extramarital affairs and produce illegitimate offspring beyond the boundary of the species.

Rogue taxa such as these are the subject of the present chapter. Perhaps the most surprising conclusion we reach is that, in spite of increasing evidence from these well-known European taxa, in some cases flooding out of multiple laboratories using the most modern techniques, many 'bad' species stubbornly remain bad under a variety of species concepts. The existence of such rogues is of course a necessary outcome of gradual Darwinian evolution, and it shouldn't worry us. However, when it comes to placing specimens in drawers or data against a name, bad species are a problem. Unfortunately, constructing a perfect species definition that covers both well-behaved and bad species will almost certainly remain a matter of compromise.

Bernardi (1980) has shown that many a specialist in a given group has tinkered with his own special taxonomic categories to cover this kind of situation. An example is the 'semispecies' idea of Lorković & Mayr, but many other examples are scattered throughout the obscure or forgotten literature. Is the solution to house rogue taxa in a special fuzzy species ghetto? This might have been a good idea if bad species were a homogeneous group; however, as we have seen, the intermediate states are variable. In any case, there is no agreement today about the rank even of the supposedly most objective of taxa, the species itself (Isaac *et al.* 2004). We thus argue that classical taxonomic ranks – species and subspecies – are all we require, to avoid proliferation of ever more finely divided categories.

Returning to the actual bad species analysed above, let us ignore problem taxa that result from taxonomic error, such as the undetected 'good' species *Leptidea reali* or the hybrid *Erebia* 'serotina'. In the case of *Zerynthia*, there is intraspecies incompatibility, coupled with interspecies compatibility; this was discovered only through artificial crosses of geographically separate populations. Perhaps, therefore, we should proclaim the primacy of observations in natural contact or cohabitation over experimental tests, which can give an inaccurate impression of pre- and postzygotic compatibility (Mayr 1963, Mallet 1995). If geographically and genetically intermediate populations disappear, for some reason, we end up with the problem of allopatric entities (see below). Sometimes divergence is so great that it seems logical to classify allopatric taxa as species. But is it really necessary to consider continental and British strains of *Lasiommata megera* as different species because they display some genetic incompatibility (Oliver 1972)? We argue it is more informative not to do so.

In the three papilionids (*Zerynthia*, *Parnassius*, *Papilio*), most people looking at natural populations in zones of overlap would declare each pair of species to be 'good', even when hybridization occurs regularly, but sparsely, in at least some areas of cohabitation. We suggest that the same decision should apply to all other cases of bimodal phenotypic and genotypic distribution where hybrids occur (Jiggins & Mallet 2000), whether or not actual or potential gene flow (introgression) takes place. Similar decisions may be made without difficulty for parapatric species with a contact zone and limited or exceptional hybridization as in the *Erebia tyndarus* group. In the case of *Lysandra*, *Pontia daplidice* and *edusa*, and probably *Melanargia galathea* and *lachesis*, the presence of a bimodal hybrid zone allows us to consider the taxa in contact as species, but here we are near the boundary condition, because, if hybridization becomes much more frequent, hybrid swarms would result, and overlapping populations would become merged into a single, unimodal population. For *Pontia*, there are divergent opinions: Geiger *et al.* (1988) and Wenger *et al.* (1993) consider *daplidice* and *edusa* as (semi-)species, while Porter *et al.* (1997) grant them only subspecies rank.

Allopatric forms separated by major geographic discontinuities give rise to a virtually insoluble difficulty. Here, there is a Gordian knot to cut. Mayr (1942, 1963, 1982)

repeatedly justified the BSC as the only 'non-arbitrary definition of species', but even he (1982: 282) admits 'the decision whether to call such [allopatric] populations species is somewhat arbitrary'. Sperling (2003) likewise suggested that decisions should be made using information, such as genetic distance or karyotype, from closely related taxa that are in contact. This is essentially already implicit in the argument for the use of 'potential' gene flow in the BSC. An absolute threshold of similarity or distance is arbitrary, so no one should harbour illusions about the 'reality' of species delimited by this pragmatic approach. The most important objective is to preserve clarity, parsimony and stability in nomenclature. Therefore, endemics on Tyrrhenian or Atlantic islands might often be considered subspecies of mainland species if they are moderately differentiated, and we argue that this solution should be employed as far as possible on parsimony grounds. They should be considered species only if they present clear signs of very strong genetic, morphological and biological differentiation above that expected of related mainland species in contact with close relatives. When it comes to allopatric 'karyospecies', one might wish to follow Wiemers (2003), and give specific rank (especially if strongly divergent at other genetic markers). Even here, use of the same species name with chromosome number placed in parentheses would be as informative; this is typically applied, for example, in *Mus musculus*. In general, decisions about the species status of allopatric neighbours is always somewhat arbitrary, and a lot less interesting than obtaining field or genetic data from hybrid zones and parapatric contact zones, or from unimodal lines. Here, one deals with a concrete phenomenon, rather than an investigation into how many angels fit on the head of a pin.

We therefore argue for revival and a modern, scientific justification of the rather neglected and misused (and perhaps rightly, in many cases, much-maligned) rank of subspecies. Very often, subspecies have been used to describe geographical forms recognizable only to their author, which has led to disrepute. But today there is a refreshing trend among lepidopterists to consider only more strongly distinct forms (in morphology, ecology or genetics) as subspecies, and to lump more dubious geographical forms as synonyms. These general recommendations provide a useful compromise between description of geographical variation, the needs of modern butterfly taxonomy (for example, see Ehrlich & Murphy 1984, Sperling 2003), and Darwin's pragmatic use of the term species in evolutionary studies.

It is a Sisyphean task to try to give a definitive, irrefutable definition of species, but species will continue to function as useful tools in biology for a long time. To the question raised by the French population geneticist Le Guyader (2002): 'Must we give up on a species concept?' we answer: 'No!' We recommend that researchers of the future study gene exchange in the many hierarchical layers of phenotype, genotype and genome in 'bad' species of butterflies. This has been done in only a handful of species, such as the larch bud moth (Emelianov *et al.* 2004). Such studies will be surely much more illuminating about the nature of speciation and evolution at the species level than endless discussions on the 'essence' of species.

APPENDIX: TOOLS FOR TAXONOMIC PRACTICE AT SPECIES LEVEL IN BUTTERFLIES

The previous parts of this work presented first the theoretical background of taxonomic work on species, and then a series of analyses of peculiar real cases. To sum up, species are delimited by a series of criteria derived from the concept used and the speciation theory associated with it, with an accent on studies on populations in cohabitation or contact.

There are many different types of datasets that can be used. Wing colour morphology is perhaps the most obvious, and of course in butterflies is extremely important. Ecological, behavioural and distributional data are also important. Differences in genitalia have often been considered to be significant for reproductive isolation via a 'lock-and-key' hypothesis (Jordan 1896, Porter & Shapiro 1990). As already seen, genitalic data are useful in certain cases, but not always. Chromosomal data are more often reliable, but they can also be misleading. The same might also be true for pheromonal characters, which can be considered both as organismic and synepigonic, but there is little information on butterflies (but see Andersson *et al.* 2003).

We here present an overview of currently available methods for gathering and analysing taxonomic data and conducting biological and statistical studies to establish whether taxa might be species or taxa at some subspecific category. Nomenclatural aspects of species delimitation, however, do not form part of the remit of our chapter.

Morphological characters

Data acquisition
Empirical and intuitive, qualitative observations are still used, but biometrical methods have become more normal. Even with qualitative characters, records of a series of states

are often performed. In adults, the hard parts of the exoskeleton are most often studied, and genitalia have remained favourite characters since the late nineteenth century (Jordan 1896). Wing-pattern variation is used in butterflies predominantly because it is both evolutionarily labile and easy to detect and score, and provides useful data for identification in most cases. A still commonly used method in morphometrics consists of measuring anatomical structures under a microscope with a micrometer (see e.g. de Lesse 1960a, Cesaroni *et al.* 1994). Today, automated measurements employing digital imaging can also be used. Larval characters can also be useful: superficial features (pigmentation, pattern) are commonly used, but chaetotaxy of first-instar larvae sometimes provides very significant information. The microstructure of the eggs is a great favourite, especially using scanning electron microscopy (SEM). In using egg sculpturings, one must remember that it is actually an imaginal feature, since it results from the imprint of ovary follicles.

Data analysis
Analysis of morphological data may be performed character by character. It is also possible to integrate a dataset from a sample of individuals in multivariate, or reduced space, analyses (RSA). These methods have been great favourites for the French school of statisticians, long led by Benzecri. Systematists may sometimes be reluctant to use them, but they are powerful when correctly used. The reader should consult works such as Sneath & Sokal (1973) for details of clustering and ordination methods. In brief, there are three main categories of RSA: principal components analysis, using Euclidean distance, factorial correspondence analysis, using a chi-square-based distance, and factorial discriminant analysis (FDA). The latter seems to be the most appropriate to conduct a study on a dataset that may reasonably be supposed to include two (or more) different species. A frequent criticism of RSA is that these methods are descriptive, rather than inferential statistics. However, with some practice, they are excellent tools for exploring a dataset. Genetic data can also be analysed in the same way.

Chromosome characters
The study of chromosomes in butterflies was for a long time dominated by the work of Lorković (1941) and de Lesse (1960a). Since that time, interest has moved towards other types of genetic markers, but chromosome studies are still useful (e.g. Munguira *et al.* 1994, Wiemers 2003). Chromosome counting is typically practised on meiotic cells

in the testes during spermatogenesis. Generally rounded, small and numerous, lepidopteran chromosomes are not gratifying objects of study. In approximately 1000 species of Lepidoptera, the distribution of chromosome numbers proved markedly leptokurtic and asymmetrical, with a strong concentration around the modal number ($n = 31$), and an extreme scattering of frequencies for the higher numbers (Robinson 1971). Some members of *Polyommatus* (*Plebicula*) (Lycaenidae) display the highest chromosome numbers in metazoans (190–191 for *P. nivescens*, and for *P. atlantica*), while numbers less than 10 are observed in *Erebia* (de Lesse 1960a). Supernumerary chromosomes are often seen, especially in Satyridae and Hesperiidae, and may produce pronounced intraspecific variation, in particular in *Plebicula* (de Lesse 1960a).

The significance of chromosome number variation in butterflies has been widely debated (Lorković 1941, Robinson 1971, White 1973, Kandul *et al.* 2004). Polyploidy seems unlikely as a general mechanism for chromosome number variation in butterflies, despite Lorković's (1941) views. Centromeric fusion or fission seems a more probable cause of chromosomal number variation (Suomalainen 1965, White 1973, King 1993). This could be due to the structure of the lepidopteran centromere, which is reportedly 'diffuse' (Federley 1945, Suomalainen 1953; but see Gus *et al.* 1983). A diffuse centromere may allow some amelioration of damage suffered in chromosomal heterozygotes during meiosis. Another insect group with diffuse centromeres, scale insects, also show large variation in chromosome numbers (Cook 2000). On the other hand, the modality of chromosome number around 31 throughout the Lepidoptera is not easily accounted for under this scenario (White 1973). Kandul *et al.* (2004) suggest that instances of enhanced chromosome number variation could result from epidemics of transposable genetic elements.

In practice, chromosome study in butterflies is tedious because spermatogenesis often terminates early in adult life. Even in young males, meiotic metaphase equatorial plates in the spermatids, the most favourable stage for counting, are usually scarce. In addition, chromosomes are usually so highly condensed that little intrachromosomal structure is visible. However, particularly in *Polyommatus* (*Agrodiaetus*), differentiation of larger, so-called macro-chromosomes which vary in number and size has been found useful (de Lesse 1960b, Munguira *et al.* 1994, Lukhtanov & Dantchenko 2002b). Moreover, instead of producing conveniently visible giant polytene chromosomes as in Diptera, Lepidoptera appear to adopt polyploidy as a means of up-regulating gene expression in highly active somatic tissues – far less easy to use as a taxonomic or genetic marker.

Hesselbarth *et al.* (1995) put forward the hypothesis that chromosome fission and fusion could have an influence on adaptive abilities. Species with low chromosome numbers should be associated with greater genome stability and more supergenic association and therefore adapted to stable environments. Conversely, high chromosome numbers should ease recombination and generate many genotypes promoting adaptation to new or unstable environmental conditions. Wiemers (2003) found absolutely no evidence of such a phenomenon in *Agrodiaetus*, the genus displaying the largest variation in chromosome numbers in butterflies. We suggest another possible effect of high chromosome numbers: by increasing the average rate of recombination, they could limit hitch-hiking of genes causing incompatibility and could therefore ease introgression of 'neutral' genes in hybrid belts (e.g. in *Lysandra*).

Karyotypic differences between taxa are often taken as a proof of species-level distinction, and this argument can be legitimate. However, caution must be exercised. Supernumerary, genetically insignificant B-chromosomes are common (de Lesse 1960a, 1961b), and might sometimes be an indication of hybridization (Wiemers 1998); moreover, when morphologically and ecologically very similar groups of populations occurring in different areas display different karyotypes, it may be premature to base species separation on chromosomal number, in the absence of other evidence such as molecular studies. The term 'chromosome races' (Goldschmidt 1932) does not seem to have been used explicitly in butterflies, but de Lesse (1966) maintained, within a single species, allopatric populations of *Agrodiaetus dolus* from southern Europe with $n = 108$, 122 and 124; in contrast, Munguira *et al.* (1994) split the taxa into separate species with different karyotypes. Experiments carried out in moths of the genus *Antheraea* showed that two 'species', *A. roylei* and *pernyi* with $n = 18$ and 49 respectively, could be intercrossed for 32 generations with fertility and viability intact (Nagaraju & Jolly 1986).

Molecular characters

The history of molecular systematics can be divided into two major stages: a protein phase and a DNA phase. The former, based mainly on allozyme electrophoresis, became important at the end of the 1960s with studies on *Drosophila* and humans (Avise 1974, Richardson *et al.* 1986, Hillis *et al.* 1996), and played a major role in butterfly systematics from the 1970s onwards (Geiger 1990). The DNA phase really came into its own in the 1990s following the development of the polymerase chain reaction (PCR).

Protein data

Since the earliest days, electrophoretic study of protein polymorphism revealed a stunning amount of variation (Lewontin 1974). A bitter debate on the significance of these observations took place in the 1960s and 1970s: some championed selection as a cause for polymorphism, while others raised mathematical objections (Kimura 1968) and argued that it must be neutral. Current experimental (Watt 2003) and theoretical (Gillespie 1991) evidence suggests that both selection and neutral evolution may be important; consequently, when using protein variation to study taxonomic units, one must be careful that selected variation affecting ecological parameters, such as foodplants (Feder *et al.* 1997), does not obscure taxonomic conclusions.

Analysis of protein data

A classical method for analysing allozyme data is to reduce the multilocus data by means of a calculation of overall genetic distance (Hillis *et al.* 1996). This can be used in cluster analyses, and subsequently to phylogenetic inference, but there is no obvious level of genetic distance above which two samples can be confidently considered to be separate species. Nei's (1978) genetic identity (I) and distance ($D = -\ln I$) is regarded as particularly useful, because it corrects for small sample size and for multiple 'hits', and so should be proportional to time since divergence under a molecular clock. Closely related species of *Drosophila* may be in the range of Nei's D of 0.05–0.50 or so (Coyne & Orr 1997). In European butterflies, the genetic distances between species of the same genus range generally between 0.05 and 0.15 (Aubert *et al.* 1996b, Geiger 1990, Zimmermann *et al.* 1999). However, pairs of apparently closely related species may be more distant, and, more surprisingly, other pairs of species may coexist without hybridizing, but differ hardly at allozyme loci ($D < 0.01$). Diagnostic loci (fixed for different alleles in each population) are useful for studying hybridization and gene flow between taxa.

Allozyme studies within species have often attempted to estimate gene flow based on the neutral expectation of gene frequency variation. Firstly, one may estimate the variation of gene frequencies between populations via the use of F_{ST}, the standardized variance of gene frequencies, which measures the fraction of genetic diversity (heterozygosity, H_e) found between populations. If gene frequency variation can

be assumed to be a balance between homogenization via gene flow (m) and local divergence due to genetic drift (proportional to $1/2N_e$, where N_e is the effective population size), then $F_{ST} \approx 1/(1 + 4Nm)$. However, there are many problems with these methods, which allow the estimation only of the combined parameter $N_e m$. They should not be applied in any context other than under equilibrium between genetic drift and gene flow; it does not, for instance, apply in the case of gene flow and hybridization between two species, or between ecologically differentiated taxa (Mallet 2001), because here selection will be involved in the differentiation (*contra* Porter & Geiger 1995). In such cases, strong natural selection may lead to rapid equilibration of gene frequencies in the presence of gene flow. A much more useful method is available based on correlations (or linkage disequilibria) between loci diagnostic or with strong frequency differences between hybridizing taxa. Hybrid zones, in particular, allow estimation of selection and gene flow separately (Mallet *et al.* 1990, Porter *et al.* 1997, Mallet 2001, Blum 2002, Dasmahapatra *et al.* 2002).

A species criterion based on 'genotype clusters' (Mallet 1995) can be viewed as an extension of this multilocus method. Genotypes reach bimodality only when several characters or loci are in tight linkage disequilibrium. One may use 'assignment methods', likelihood or distance-based multivariate statistics (see above under 'Morphological characters') to cluster genotypes, to determine whether multilocus gaps between clusters are significant; if so, the clusters can be classified as separate species (Aubert *et al.* 1997, Feder *et al.* 1997, Deschamps-Cottin *et al.* 2000). Newer likelihood or Bayesian methods also allow estimation of the rates of hybridization in a sample of a pair of several, bimodally distributed taxa (Cianchi *et al.* 2003, Emelianov *et al.* 2003, 2004).

DNA data

DNA methods have outstripped allozyme electrophoresis, but are still in their infancy compared with what might be possible in a few years. The mitochondrial genome, with a mere 16 000 base pairs, has been far the most widely used in butterflies (Pashley & Ke 1992, Wahlberg & Zimmermann 2000), and elsewhere. Intraspecific mtDNA sequence polymorphism occurs in certain butterfly species but is absent in other cases. For instance, *Papilio machaon* displays polymorphism throughout its range (F. Michel, pers. comm.), as do *Euphydryas aurinia* and *Melitaea athalia* (Zimmermann *et al.* 2000). In contrast, no variation within *Euphydryas maturna* has been observed across a large range

(Zimmermann *et al.* 2000). The mitochondrial genome is very sensitive to genetic drift, since it has a N_e four times smaller than that of the nuclear genome. Comparison between closely related species usually shows 1–2% divergence, but strikingly low differences are observed in some instances: 0.2 % between *Euphydryas maturna* and *E. intermedia*. We therefore do not believe that any particular level of divergence can be used as a suitable benchmark or 'DNA barcode' for species status.

Nuclear gene sequences are beginning to be used with some success (e.g. Brower & Egan 1997, Beltrán *et al.* 2002), while microsatellite loci have proved disappointingly difficult to obtain in butterflies (Nève & Meglécz 2000, Meglécz *et al.* 2004). Amplified fragment length polymorphisms (AFLPs) can also be used as a very abundant source of 'fingerprint' markers in analyses of natural populations, including studies of hybridization in nature (Emelianov *et al.* 2004). Nonetheless, while useful in mapping, AFLPs are relatively untried as tools for studying populations.

In summary, marker data, whether morphological, cytological or molecular, have allowed us to search organisms for characters with increasing thoroughness, but are not fundamentally different from one another.

Ethological and ecological criteria

Treating ethological and ecological characters together seems hardly justified, since they are heterogeneous. However, they all play an active role both in cohesion *within* species and in maintaining separateness between species. They therefore give access to the very factor, reproductive isolation, important in speciation. We will consider the following most important categories. Firstly, there is the ecological niche and its main constituents: habitat and foodplant choice, phenology and diapause; secondly, sexual behaviour and pheromones; and thirdly, geographical distribution, particularly cohabitation.

According to Gause's principle (1934), if two species occupy the same niche, they will mutually exclude one another and will display parapatric distributions, with very limited cohabitation. These cohabitation zones may not necessarily imply hybridization and/or genetic proximity. Alternatively, if the two species share a large area of sympatry, they must be ecologically differentiated. The main difficulty in using such ecological information is circularity. Very often, field entomologists 'feel' that two putative species display subtle differences in habitat use but are unable to develop inferential tests to support their impression. Various parameters of the ecological niches occupied by butterflies frequently crop up in studies of butterfly species.

Larval foodplant choice

The host plant is perhaps the key niche dimension in the life of a phytophagous insect (Dethier 1954, Futuyma & Keese 1992, Feeny 1995, Berenbaum 1995), and feeding regime may play a major role in speciation, including in some Lepidoptera (Feder 1998, Drès & Mallet 2002). Butterflies are generally oligophagous and change in diet is likely to result in a selective regime that might lead to speciation and adaptive radiation (Ehrlich & Raven 1964). There is certainly evidence for rapid diet evolution in some taxa, such as the Papilionini (Aubert *et al.* 1999) or Melitaeini (Mazel 1982, Singer *et al.* 1992a). These changes may appear spectacular, with switching between plant families common (e.g. Rutaceae to Apiaceae in *Papilio*, Dipsacaceae to Caprifoliaceae and Valerianaceae in *Euphydryas*); however, these unrelated plants almost always have important chemical similarities (Bowers 1983, Berenbaum 1995). The evidence for host-related speciation in butterflies is thus somewhat weak (see e.g. Nice & Shapiro 2001, for a case in the Lycaenidae). In the North American *Euphydryas*, where rapid intraspecific diet evolution has been observed, new host adaptations normally evolve rapidly in local populations, and drive original preferences and adaptations to extinction, rather than causing speciation (Thomas & Singer 1998).

Diapause control and voltinism

A butterfly population is expected to have as many broods as climatic conditions and food availability allow. Intraspecific variation in voltinism is common in species that have a wide range. *Melitaea athalia*, for example, is univoltine in northern Europe, bivoltine in warm regions with a wet summer, and univoltine again in the Mediterranean and in mountains above 1000 m (see also the *Papilio* paragraph above). On the other hand, sometimes it forms a character presumed to differ at the species level: for example the species *Aricia artaxerxes* is univoltine and occurs in northern Europe, but is replaced by the bi- or multivoltine species *A. agestis* in southern Europe. Univoltinism can also be a constitutive character within a taxon: in the genus *Euphydryas*, for example, all species are single-brooded, even under conditions that could allow several broods.

The genetic determination of diapause in Lepidoptera has been studied in few cases, where it apparently involves a number of interacting loci (Held & Spieth 1999); in other cases, it appears to give a pattern suggesting sex-linked inheritance and few genetic factors. In some cases, crosses between related subspecies or species in the Papilionidae give classic 'Haldane's rule' asymmetry in diapause between males and females, suggesting the importance of Z-linkage (see the *Papilio* section in this chapter and below in this part).

Mixiological criteria

'Mixiological' is the term applied, especially in France, to phenotypic and behavioural traits which affect hybridization and introgression between pairs of taxa. In spite of a heated debate about the use of terms such as 'isolating mechanisms' (Lambert *et al.* 1987, Mallet 1995), all sides agree that a restriction of gene flow is the key process in speciation in sexual taxa. Since many factors may produce this result, it is normal to aggregate these heterogeneous traits under the same heading, 'reproductive isolation' (Mayr 1963); the two major kinds of reproductive isolation are prezygotic and postzygotic isolation.

Prezygotic barriers

These may involve spatial and temporal isolation (habitat choice and phenology), mating behaviour and courtship, pheromone differences, mechanical barriers to pairing, and physiological features of insemination before gametic fusion. Prevailing opinion about their origin is that prezygotic barriers are often formed as a by-product of intraspecific coevolution, with selection maintaining compatibility (as in the 'recognition' concept of Paterson 1985) while the system of mating or reproduction diverges. Another argument is that selection may cause divergence in pre-mating traits as a directly selected process ('reinforcement') to avoid the production of unfit hybrids. Reinforcement has been much debated (Paterson 1985, Lambert *et al.* 1987); however, the phenomenon has been demonstrated in some cases (e.g. Noor 1995, Lukhtanov *et al.* 2005), and is suspected in the tropical genus *Heliconius* (Jiggins *et al.* 2001).

Postzygotic barriers

These involve inviability or sterility acting on hybrids from the zygote stage onwards. Hybridization experiments show that hybrids between species are often inviable or sterile. Sterility was demonstrated, for example, using *Drosophila* as a research material (Dobzhansky 1937), but hybrid sterility had been recognized as early

as Buffon's time (Mayr 1982). However, hybrid sterility and inviability between taxa considered 'good species' is far from general (Darwin 1859, and several examples in the present chapter). Fitness is often reduced in hybrids (Rice & Hostert 1994), not only in physiology (intrinsic or endogenous selection) but also in ecological adaptations that allow individuals to exploit niches of parental taxa (extrinsic or exogenous selection) (Hatfield 1996, Jiggins & Mallet 2000). Crosses in captivity must be considered with utmost caution, since careful rearing and pampering can allow certain experimentally obtained hybrids to survive, while they would undoubtedly die under natural conditions. Conversely, the diseases associated with captivity and promiscuity, or unsuitable breeding conditions, can cause the loss of broods which could have thrived in the wild. This uncertainty allowed such wags as Loeliger & Karrer (2000) to cast doubt on earlier results of Clarke & Sheppard (1953, 1955, 1956) and Aubert *et al.* (1997), and to negate the existence of postzygotic incompatibilities between *Papilio machaon* and *P. hospiton* – an extraordinary assertion contradicted by all the evidence!

It has become de rigueur to refer to all kinds of hybrid inviability and sterility as Dobzhansky–Muller incompatibilities (Orr 1995), given that they rarely cause inviability or sterility within species, but only when transferred to another genetic background; in other words, their evil effects result from epistatic incompatibilities between genes. It is likely that hybrid inviability between populations can evolve, paradoxically, without producing fitness problems within populations at any time during its emergence.

Let us finish this work by looking more closely to a striking type of genomic incompatibility we have frequently evoked: Haldane's rule (Haldane 1922): in hybrids, the heterogametic sex (the one with heterogeneous sex chromosomes, e.g. XY) tends to be more sterile or unviable than the homogametic sex (e.g. XX). The heterogametic sex is the male in most insects, including *Drosophila*, as well as in mammals. The Lepidoptera and birds are notorious exceptions, having heterogametic females: their sex chromosome formula is ZW in the females and ZZ in males, yet obedience to Haldane's rule in Lepidoptera is as good as or better, in reversed form, as in the species with XX/XY sexdetermination (Presgraves 2002). It is surprising, perhaps, that agreement on the explanation, 'dominance theory', of the striking facts of Haldane's rule has been reached only recently: the earliest loci to diverge appear to cause incompatibilities only recessively; thus incompatibilities tend to affect the sex chromosome, and mainly in the heterogametic sex. In agreement with dominance theory, sex-linkage of incompatibilities holds for butterflies, where the female is most strongly affected (Grula & Taylor 1980, Sperling *et al.* 1990, Aubert *et al.* 1997, Jiggins *et al.* 2001, Naisbit *et al.* 2002) as well as for *Drosophila*, where it is the male (Coyne & Orr 1997). It is interesting that the general applicability of Haldane's rule in the Lepidoptera (Presgraves 2002) implies that maternally inherited markers, such as mitochondrial DNA or W-chromosomes, will rarely be transmitted between species (Sperling 1990). Thus, species identification based on mitochondrial 'DNA barcodes' may work better for Lepidoptera (Hebert *et al.* 2003) than in other taxa prone to hybridization and introgression.

17 • Faunal structures, phylogeography and historical inference

ROGER L. H. DENNIS AND THOMAS SCHMITT

SUMMARY

Every taxon has a unique evolutionary history and bio-geographical pattern, yet it is possible, using relationships between extant taxa and current environmental conditions, to reconstruct biogeographical histories. This search for pattern reveals similarities of the phylogeographies of several taxonomic groups and demonstrates how the current European butterfly fauna has a pattern of distribution and regional variation that reflects major environmental changes associated with glacial–interglacial cycles and European landform distribution. This pattern is dominated by glacial refugia for the majority of species in the Mediterranean, with accompanying isolation during long cold periods, with major range expansion during interglacials. Evidence of long-term evolution in isolation comes with the recognition of current contact zones between previously isolated units of particular taxa. By contrast, some montane species are isolated during the relatively shorter interglacial periods, with evidence of divergence through isolation during the current interglacial period. These contrasting patterns have led to different patterns of range-restricted taxa, those in the Mediterranean representing groups that became locked into narrow resource and condition sets during long cold periods and those in montane regions representing taxonomic groups currently isolated. They too may become locked into restricted resource sets over time. The work on species biogeographies has important implications for conservation effort and human impacts. The Mediterranean zone is perhaps more important than currently recognized; it contains many of Europe's endemic species, and most of the European butterfly species' evolutionary history. Molecular and morphological data, combined with knowledge of past climates and landforms, can be used to reconstruct biogeographical histories. Where adequate data exist, differing phylogeographical patterns emerge, though some patterns may be shared by different species groups. Whilst in its infancy, phylogeographical reconstruction can help focus conservation effort, not least by providing a mechanism to identify regional centres of genetic diversity.

THE LONG SEARCH FOR PATTERNS AND STRUCTURES

Recognition that organisms differ in distribution and geographical range has long encouraged insights into patterns of similarities and differences. The first worldwide distribution maps seem to be Bartholomew's *Atlas of Zoogeography* (Bartholomew *et al.* 1911), which illustrated maps of many genera showing for example that among European genera, *Charaxes* is mainly an African genus and *Neptis* mainly an Asian one. Particularly appealing has been the idea that groups of species exist that share similar distributions distinct from those of other groups of species. Scharff (1899) was one of the first to suggest the grouping of species according to their different 'original homes'. The terms describing these patterns developed from the concept of a fauna. Although the catchword 'fauna' is still widely used in biology, terms such as group, element, unit, region and type, prefixed with 'faunal', went out of fashion in the 1960s. These terms are conspicuously absent from recent biology texts and dictionaries (Wiley 1981, Lincoln *et al.* 1983, Collin 1988, Myers & Giller 1988, Cox & Moore 1993) but from the eighteenth century to the mid-twentieth century, they were powerful concepts, eagerly evoked in historical reconstructions of animal geography (Mayr 1965, de Lattin 1967). Current emphasis in biogeographical research places a premium on relating organism demography to geography and landscape, and seeking explanations in ecological rather than evolutionary time (Gaston 1994, Hanski & Gilpin 1997). The evolutionary focus has moved on from collective, faunal, concepts to phylogeography, tracing and delimiting cladistic links using molecular data (Avise 1994, 2000). Even so, as Mayr (1965) demonstrated, there is mileage in applying faunal concepts to taxa, particularly for groups lacking a fossil record. Associated with faunal concepts are techniques for determining historical components in current distribution patterns (Reinig 1937, de Lattin 1967, Udvardy 1969, Varga 1977) with Udvardy (1969) giving a useful summary.

A fauna is simply the animals found in a specified region. As such, fauna is a collective term for organisms

Ecology of Butterflies in Europe, eds. J. Settele, T. Shreeve, M. Konvička and H. Van Dyck. Published by Cambridge University Press. © Cambridge University Press 2009, pp. 250–280.

restricted to finite geographical areas, such as global biogeographical regions. For a long time it was taken for granted that floras and faunas of all regions were the products of these regions and that faunas owed their characteristics to the local physical environment (Mayr 1965). Darwin (1859) was the first to realize that this was not the case. Faunas comprise identifiable subdivisions, component parts, referred to as faunal groups and faunal elements, differing in origin and history (Dunn 1922, Simpson 1943, Mayr, 1946). Mayr (1965: 474) succinctly makes the point: 'As a result of various historical forces, a fauna is composed of unequal elements and no fauna can be fully understood until it is segregated into its elements and until one has succeeded in explaining the separate history of each of these elements.' It soon becomes evident that even the elements are not particulate (definitive) entities but are divisible into finer units based on age and origins of the component parts.

Within the Palaearctic region, faunal groups and faunal elements among butterflies have long been recognized (Pagenstecher 1909, de Lattin 1957b, 1967, Varga 1977), as has the clustered distributions of closely related taxa (Kostrowicki 1969). Much of this pattern recognition was achieved by eye from maps of superimposed distribution data and areal limits (Udvardy 1969). In this chapter, we first investigate the process of pattern recognition amongst European butterflies using multivariate statistical techniques. We then move on to interpret these patterns, to compare the contributions of influences in ecological and evolutionary time. Finally, we examine the fast developing approaches in phylogeography, the historical reconstructions of spatial patterns amongst closely related taxa. Historical reconstruction, dealing with lineage differentiation via vicariance events, requires accurate markers of relationship and time, now increasingly becoming available with the development of molecular studies tied into calibrated dating of genetic differentiation among clades; the techniques introduced in the last section provide the means for testing ideas about faunal relationships from the study of faunal elements and regions. Included in these are inferences and evidence of past events affecting species biogeographies and taxonomic differentiation drawn from a variety of sources for other taxonomic groups (for a recent review see Schmitt 2007). Simple universal patterns are elusive; some taxa share common patterns and others not. This chapter specifically summarises findings on the biogeography and phylogeography of West Palaearctic butterflies.

FAUNAL STRUCTURES: OUT OF FASHION OR USEFUL TOOL?

The faunal terminology

The definitions of faunal terms have varied widely over the years. Some terms are interchangeable (e.g. faunal unit with faunal group). Other terms make reference to causative agents or are operational definitions specific to a study. Compounding definitions with causation, however loosely, is self-defeating in biology; it presumes what is being sought: causation of pattern. The definitions below are purposely simple; their objective is to facilitate identification of pattern and structure that can be subsequently explained.

- A faunal group or unit is a collection of species that have similar distributions distinct from other collections of species of the same taxon.
- A faunal element is the spatial expression of the faunal group, as such it is defined by the distribution of members of the faunal group. Faunal elements can be defined on the basis of internal or external criteria to the study area. In the former, they are determined from data solely from within the study area; in the latter, taxa within an area or region are classified on some external regionalization, such as global scale biogeographical regions. Elements may be considered to have biological significance in terms of the numbers of replicates (species) they include. As each species has a unique distribution, it could be argued, not helpfully, that each species is a faunal element. Each species' unique geography, here, is simply described as its distribution or geographical range.
- A faunal region is a unique geographical area characterized by a set of taxa distinct from that of other geographical areas.
- A faunal type distinguishes taxa or groups of taxa by their geographical attributes, that is their location (i.e., longitude or latitude, or regional association) or geographical characteristics (i.e., montane, islandic).
- An endemic fauna is a taxon (taxa) unique to an area. These may be distinguished in terms of age as palaeoendemics or neoendemics, or in terms of origin as allochthonous endemics (relict endemic faunas) or autochthonous endemics (origin in situ).
- A relict fauna is a taxon (taxa) occurring in a fraction of its previous range and distribution. The areas in which relicts are found are generally referred to as refuges (refugia).
- A geographical distribution of a taxon (taxa) is the spatial pattern or arrangement of its members or population

units, described purely in terms of the location of individual members and population units.

- A geographical range of a species describes the spatial limits of its geographical distribution.
- A primary area is a spatial unit used for sampling a fauna or taxon.

Techniques for seeking structures

RESOLUTION OF THE SPECIES–AREA MATRIX

The simplest form in which distributional data exist is as a species × area matrix $X_{S \times A}$ where S are species, A are areas (or spatial sampling units, SUs) and x_{ij} are observations, typically presence (1) or absence (0) of a species for each spatial unit. Analysis of the $X_{S \times A}$ matrix has two forms:

- R mode or $X_{S \times S}$ (species × species matrix or simply here called the species matrix) for faunal elements of species
- Q mode or $X_{A \times A}$ (area × area matrix, here called the area matrix) for faunal regions of areas.

The conversion of $X_{S \times A}$ into $X_{S \times S}$ and $X_{A \times A}$ involves the calculation of resemblance functions for species and areas respectively. These take the form of similarity (distance) coefficients, overlap indices and correlation coefficients (Sneath & Sokal 1973, Ludwig & Reynolds 1988). Selection of different resemblance (proximity) functions for species and area has been advised on the basis that sampling units are independent samples whereas species are not (Orloci 1972, Legendre & Legendre 1983, Ludwig & Reynolds 1988). However, this is not always the case where SUs are contiguous spatial units as in a grid or bounded mapped units. In such cases, spatial autocorrelation typically occurs (Dennis *et al.* 2002). More importantly, the properties of resemblance functions affect the outcome of classifications (Sneath & Sokal 1973, Dennis *et al.* 1991, Hennig & Hausdorf 2006) as do the selection of network (trees) and ordination techniques used to explore and display resemblances. It is, therefore, recommended that results from a range of different measures and techniques are compared (Wakeham-Dawson *et al.* 2004). The objective of network techniques (e.g. SAHN clustering, additive clustering, divisive polythetic clustering) is to reduce the $X_{S \times A}$ matrix into 'homogenous' groups or clusters of areas or taxa. On the other hand, ordination attempts to reduce the dimensionality of the number of axes within which either S (R mode) or A (Q mode) are displayed. In effect, classifications are attempted, and the displays can be used to provide hypotheses about interrelationships in a reduced space.

HELPFUL STRANGERS: CLASSIFICATION USING EXTERNAL CRITERIA

Classifying fauna into elements based on geographical criteria external to a study region typically involves some predetermined regionalization. Categorization of species to zoogeographical regions is frequently employed (Larsen 1984). But, there is no reason why regions external to the study area, sufficiently extensive to account for the range of all species, should not be subdivided and for classification to proceed in the usual way on the species × area matrix. That there is value in searching for external geographical order on species within a smaller study region is exemplified by Larsen's (1986) examination of tropical affinities in the Mediterranean butterfly fauna. By first identifying the group and then examining the characteristics of the group – population dynamics, migration capacity, genetic variation and differentiation, endemism, ecology and vulnerability to environmental agents – he was able to determine their status in the conservation of Mediterranean fauna. The method expressing, as it does, biogeographical affinities, usefully generates historical and geographical inferences.

Standard texts generally specify the biogeographical regions to which European butterflies are believed to belong (Higgins & Hargreaves 1983, Higgins & Riley 1983). These are listed in Varga (1977) and Dennis *et al.* (1991) and more recently in Dennis (1997), the latter including the Atlantic islands (Azores, Canary Islands, Madeira). Within the boundaries of Europe (west of the Urals, north of the Caucasus), some 159 of 506 species (31.4%) are endemic to the region. The vast majority of the remainder occupy the Palaearctic, 322 species (63.6%), a number of which are also found throughout the Nearctic and thus have Holarctic ranges (34 of 322, 10.6%). A few Nearctic and Neotropical species occasionally cross the Atlantic Ocean (*Danaus plexippus*, *Vanessa virginiensis*). By comparison, very few species (24, 4.7%) are strictly tropical in origin, using the criteria established by Larsen (1986) and Perceval (1995); 10 are Afrotropical, 7 are Oriental, 1 is Neotropical, and a further 6 Palaeotropical with uncertain affinities within the tropical biogeographical regions. These mostly frequent the Mediterranean fringe of Europe. Two are specifically distinct from close relatives in the tropics (*Charaxes jasius*, *Libythea celtis*) and may be relicts in Europe (Larsen 1986).

Faunal regions and faunal elements: two tools for the understanding of biogeography

To date, the study of faunal structures among European butterflies has been the subject of six publications (Dennis *et al.* 1991, 1995a, b, 1998b, Dennis & Williams 1995, Dennis 1997). Due to incomplete mapping of European butterflies at the time (Kudrna 1996) the database is built up from maps in Higgins & Riley (1983) and comprises dichotomous scores (0, absence; 1, presence) on 393 species for 85 primary (natural) areas in the west Palaearctic (Fig. 17.1). Some additional areas were coded up in Eastern Europe and Turkey from data in Higgins & Hargreaves (1983). The region extends from northern Scandinavia to North Africa and from Britain and Portugal to the Urals and the Caspian Sea, but excludes the Atlantic islands. In many ways, this database of 393 species is deficient. Although some species have been demoted, the list of species continues to grow; in part owing to the inclusion of Turkey (see Kudrna 1986, 1996, 2002, Karsholt & Razowski 1996, Dennis 1997, van Swaay & Warren 1999, Lafranchis 2004). The quality of regional mapping is steadily improving towards the development of a continental atlas (Kudrna 2002); eventually the work can be revised using grid units. There is a valuable precedent for work on grid data (Dutreix 1988) which sought sub-regions within Bourgogne, France. Even so, the original database is sufficiently robust to be confident of the broad findings, though the regional pattern of small-scale endemics will change (Dennis 1997).

Analysis has been based on a selection of simple resemblance functions (proximities); for the area matrix (PHI coefficient and the Jaccard S_J coefficient), and for the species matrix (Jaccard S_J coefficient) (Sneath & Sokal 1973) (Fig. 17.2). A variety of spatial and network models have been used to explore the area and species matrices. Among network models are various hierarchical (SAHN) and clustering models (single, complete and average linkage, Ward) and a polythetic division algorithm, TWINSPAN (Hill 1979), which is based on reciprocal averaging and which divides ordinations in halves at successive steps. Among strict spatial models are principal components analysis, factor analysis (effectively principal co-ordinates (PCO) analysis: Gower 1969) and non-metric multidimensional scaling, which vary in their effectiveness in dealing with linearity and non-linearity in the data (Ludwig & Reynolds 1988).

FAUNAL REGIONS: GEOGRAPHY UNRAVELS BIOGEOGRAPHY

Applications of spatial and network techniques to area matrices produced good evidence for a regional structure in European butterflies. First, there is agreement for a number of regional

entities among SAHN clustering and eigenvector techniques (TWINSPAN, PCO, non-metric multidimensional scaling, SAHN clustering) (Fig. 17.3). This includes the division of north from south Europe, west from east Mediterranean, islands (British Isles, Corsica–Sardinia, Crete and Balearics) from continental regions and each other. In northern Europe, the arctic and Scandinavian regions are consistently reproduced. In the Mediterranean, Italy together with Sicily emerges as a separate entity although sharing characteristics with both the east and west Mediterranean. Second, contiguous areas making up faunal regions clearly emerge in SAHN and in eigenvector models. With few exceptions, discrepancies that exist between different methods involve geographically marginal (border) primary areas. The exceptions are montane units, especially the Alps and the Cantabrians. These tend to have substantial amounts of unique variance in eigenvector routines, indicative of a separate and unique identity. The Crimean peninsula, for which more recent data are now available, may also form a distinctive faunal region, as a number of endemic species have been found there (Nekrutenko 1985, Kolev & De Prins 1995, Kudrna 1996, Dennis 1997).

FAUNAL ELEMENTS: SPECIES UNRAVEL BIOGEOGRAPHY

There is an equal degree of consistency in the production of faunal elements for network and spatial models (Dennis *et al.* 1991, 1998b), with close agreement in classification of species (94% for three techniques). With one exception, a faunal element corresponds with a location of each faunal region. The exception is the British Isles, for which there is no faunal element equivalent for the faunal region. Typically, there is a multiplicity of faunal elements for each faunal region, that is, focusing on groups of primary areas within any one faunal region (Fig. 17.4). This is especially obvious for the west and east Mediterranean regions. Clustering and ordination of the species matrix also results in the clear emergence of faunal groups not well represented by faunal regions, particularly the numerous montane elements, the arctic–alpine element and a pan-Mediterranean element. Faunal groups overlap extensively but are clearly not random in location (Dennis & Williams 1995). Those south of the northern edge of the Alpine mountain chain tend to be relatively small closed systems, contrasting with those to the north, which are larger and clearly extend across the Palaearctic (Dennis *et al.* 1991). There are clear gradients in the number of faunal elements (groups). The peak numbers of faunal groups and of core

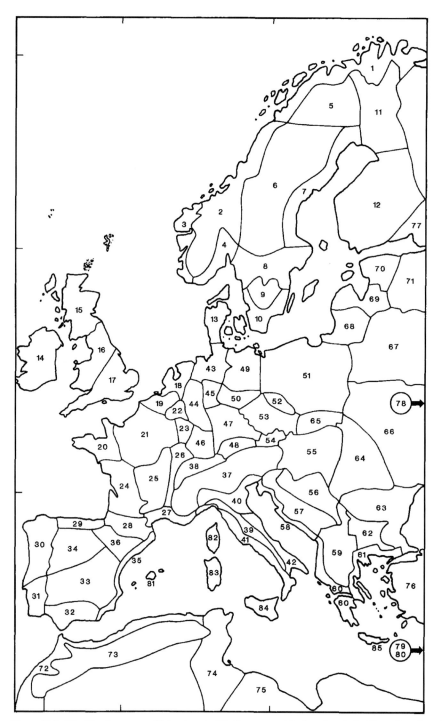

Figure 17.1. The 85 primary areas (defined a priori) in Europe used for faunal analysis. (From Dennis *et al.* 1991; courtesy of Blackwell Science.) Some of the following maps illustrate the centres for these areas, appropriately numbered. Some additional areas were used for regions outside this map for some figures (see Dennis *et al.* 1991).

OTU$_j$

		1	0	
OTU$_i$	1	a	b	$a+b$
	0	c	d	$c+d$
		$a+c$	$b+d$	$N=a+b+c+d$

Jaccard $S_{i,j} = a / (a+b+c)$

Kendall $\mathrm{PHI}_{i,j} = ad - bc / \sqrt{((a+b)(a+c)(b+d)(c+d))}$

$\qquad = \sqrt{\chi^2 / N}$

Figure 17.2. Format for the Jaccard similarity coefficient S_J and the Kendall PHI coefficient. PHI is the binary form of the Pearson product–moment correlation coefficient. 0 absence, and 1 presence, of an organism in a spatial unit (OTU, operational taxonomic unit);. a, b, c, d are frequencies of joint occurrence or absence.

primary areas for faunal elements occur in montane southern Europe, diminishing to the north and south (Fig. 17.5).

Faunal elements clearly have internal structure but, as with faunal regions, the form that this detailed structure takes differs from one faunal element to another. In the case of the largest faunal element, appropriately termed the extent element (= Siberian elements *sensu* de Lattin (1967)) as its full bounds extend beyond the region of Europe perhaps to include the entire Palaearctic, internal structure takes the form of a species richness gradient from east to west (Fig. 17.6). 'Siberian' element may be a misnomer as although a number of species belonging to it sweep across

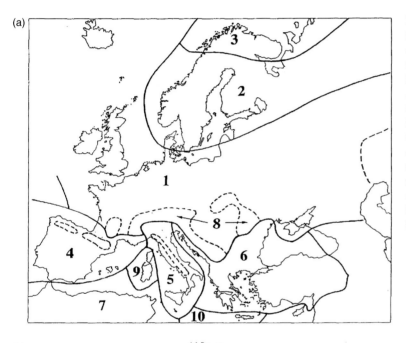

(a)

Figure 17.3. European faunal regions identified from ordination $\mathbf{X}_{A \times A}$ based on 85 primary areas and 393 species. (a) Factor analysis. 1, Extent region including Britain; 2, Scandinavian; 3, Arctic; 4, West Mediterranean; 5, Italian; 6, East Mediterranean; 7, North African; 8, Montane; 9, Sardinian–Corsican; 10, Cretan. (From Dennis *et al.* 1998b; courtesy of Blackwell Sciences.) (b) Non-metric multidimensional plot overlaid with minimal spanning tree from single linkage clustering (Kruskal stress = 0.13). (From Dennis *et al.* 1998b; courtesy of Blackwell Science.)

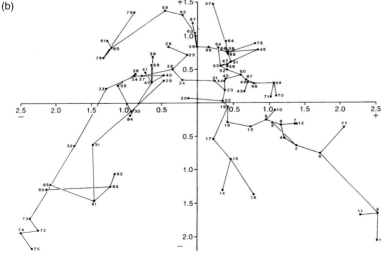

(b)

Figure 17.4. European faunal elements based on complete linkage cluster analysis of $X_{S \times S}$ based on 393 species (70% inclusion level used). Thick continuous lines, extent elements; thin continuous lines, southern elements including Mediterranean islands; dotted lines, northern elements; pecked lines montane elements. (From Dennis & Williams 1995b; courtesy of Springer.)

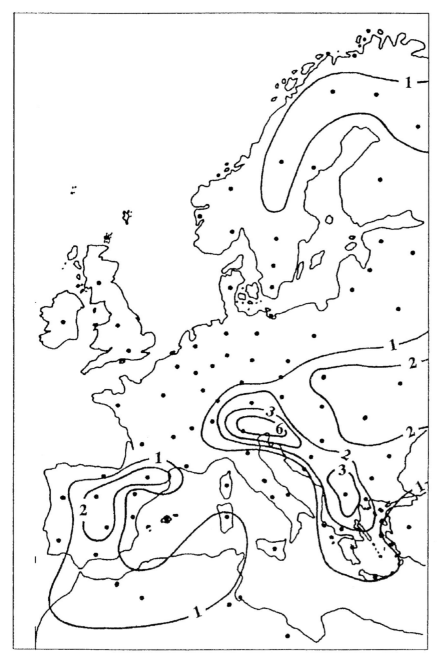

Figure 17.5. Mean frequency of key primary areas, effectively core areas, for faunal elements over Europe and North Africa. From four multivariate analyses each extracting 25 faunal groups. (From Dennis *et al.* 1998b; courtesy of Blackwell Science.)

the entirety of Europe and North Africa, these are not really Siberian taxa. Another important feature of this most important faunal element is that some species considered once to belong to it have subsequently been discovered to comprise sibling units with distinctive eastern and western ranges and which meet at tension zones (e.g. *Pontia daplidice* and *P. edusa*: Geiger *et al.* 1988; *Pyrgus malvae* and *P. malvoides*: Guillaumin & Descimon 1976; see Chapter 16).

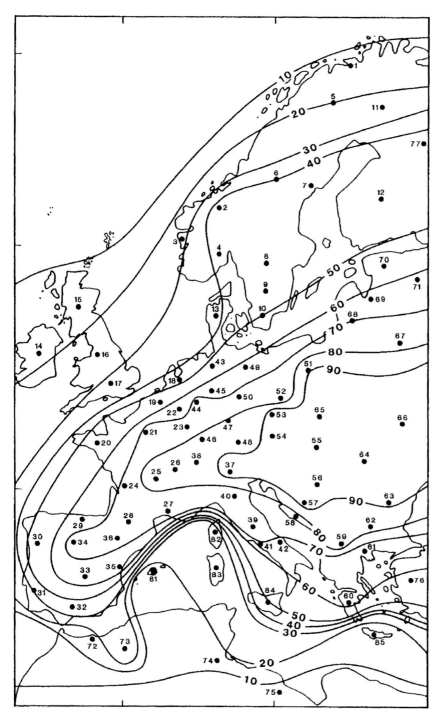

Figure 17.6. Isolines of relative species richness for the extent element (159 of 393 species analysed belong to the extent element). (From Dennis *et al.* 1991; courtesy of Blackwell Science.)

The same resemblance functions and range of techniques have been used in the determination of faunal elements and faunal regions. Even so, the strategies are very different. For faunal elements the idea is to uniquely group species (taxa) as a result of which the spatial units (elements) tend to overlap. For faunal regions the reverse holds; spatial units are grouped and there is taxonomic overlap for the regional boundaries. Despite these very different approaches, the faunal regions and faunal elements contain strikingly similar spatial structures: (i) latitudinal banding of units, (ii) tripartite division of the Mediterranean, (iii) large open units north of the Alps and generally smaller ones south of the Alps, (iv) montane and (v) island units. They also have other features in common; similar patterns of endemicity and of taxonomic bias (see below).

The differences between faunal regions and faunal elements are in themselves very informative. They differ in that: (i) there is a multiplicity of faunal elements for faunal regions, (ii) a lack of a British Isles faunal element for a faunal region and (iii) differences in the nature of internal diversity. Within faunal elements, species vary in range and distribution but are uniquely associated with that element. In faunal regions the taxa defining the region may be uniquely associated with that region but the region will also include species with very different distributions and ranges. A region may, as in the case of the British Isles, be defined purely on the basis of including an unusual combination of species, none of which is bounded within the region. In different ways then, both faunal elements and faunal regions have fuzzy boundaries. The greater number of elements than regions is caused largely by an excess of species over primary areas, allowing a finer filtering of spatial variation for faunal elements than faunal regions. The differences in composition between elements and regions reveal the shift in focus of the two entities. Faunal regions are derived by searching for areas with distinct combinations of species, whereas faunal elements are made up of species with similar, ideally identical, geographies. The British Isles faunal region comprises extreme examples of the extent element found across the whole of Europe. In Britain they very probably form an unusual combination (sample) of the extent group from the stochastics of Holocene colonization and extinction (Dennis 1977, 1993).

NORTHERN HOMOGENEITY VERSUS SOUTHERN HETEROGENEITY: AFFINITY GRADIENTS AMONG EUROPEAN BUTTERFLIES

Patterns of pair-wise resemblance functions describe the relative ordering and linkage among units (or taxa) but do not directly make use of the magnitude of resemblances for spatial sampling units. For this, resemblances (proximities) need to be converted into affinities and their dispersion parameters. Affinity is defined as mean proximity (here Jaccard's $\check{S}_J$) of any primary area with the remaining primary areas. Appropriate dispersion measures are the variance and coefficient of variation. The values obtained give a direct measure of the degree of resemblance for any primary area with every other primary area in a study region. The basic idea is that the pattern of means and variances should reflect on the uniqueness of faunas within primary areas, or lack of it. High affinity versus low affinity can be attributed to movement, reflecting on access and homogeneity as opposed to heterogeneity in biotopes and environments. High versus low variance for affinity indicates heterogeneity and homogeneity, respectively, in the levels of affinity for a specific primary area. A simple interpretation is that high values indicate similarity with many other primary areas and low values demonstrate uniqueness of the primary area.

Distinctive trends occur in affinity and its standardized dispersion (Dennis *et al.* 1995a). A zone of high affinity ($\check{S}_J > 0.5$) occupies a belt of central Europe north of the alpine fold mountain chain. Values decline north and south to minima ($\check{S}_J < 0.2$) in the arctic and in North Africa. The coefficient of variation has much of an inverse pattern to $\check{S}_J$. Peak values occur in peripheral regions of Europe, particularly the arctic and North Africa, and decline towards central Europe. However, the pattern is not identical and the correlation between $\check{S}_J$ and its variance is low ($r = -0.12$, NS). The lowest values occur on islands, such as Britain, Sicily and other islands in the Mediterranean Sea. This pattern does not take into account the relative isolation of primary areas, or the size of primary areas and their faunas. Affinity, as would be expected, is inversely related to isolation ($r = -0.73$, $P < 0.001$) and correlates directly with species richness ($r = 0.57$, $P = 0.001$). But, it does not correlate with the dimensions of primary areas ($r = 0.08$, NS). Some 66% of the variance in affinities is accounted for by species richness and geographical distances.

The pattern of residual affinity, controlled for isolation and species richness, is very different from the map of gross affinities (Fig. 17.7). Positive residuals dominate the

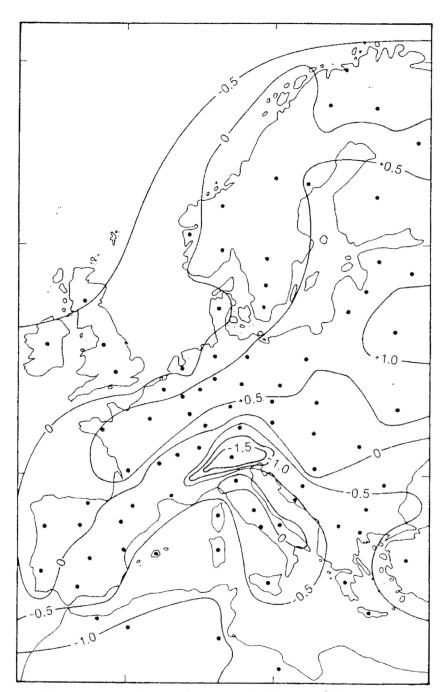

Figure 17.7. Standardized residuals from the regression of affinities (Jaccard $\check{S}_J$), between any one primary area and every other primary area, on geographical distances with the effects of affinities from a size stochastic function removed so as to eliminate the impact of primary area dimensions. The stochastic function is based on species numbers. (From Dennis *et al.* 1995a; courtesy of Pemberley Books.)

region north of the alpine mountain chain whereas negative residuals dominate the mountains, islands and the region to the south. Primary areas that have negative residuals also generally contain endemic species. Some exceptions do occur. In the northern region, the British Isles, the arctic zone, southwest Norway, Denmark and the Netherlands have negative scores. These tend to be islands or peninsulas and are peripheral areas. A few southern primary areas (e.g. western Iberia) have positive residuals. A noticeable feature is a west-to-east gradient in positive residuals in northern Europe. This matches the isolines for species richness of the extent faunal element.

The north-versus-south contrast for residual affinity, controlled for geographical distances and stochastic variation associated with species' richness, corresponds closely with other prominent features of faunal structure. It coincides with the north-versus-south distinction in: (i) the size and number of faunal units, and (ii) endemicity levels. The peak in positive residuals lies at the 'root' of the extent faunal element in northeastern Europe that covers all the West Palaearctic. In contrast, no elements having their 'core' areas in southern Europe are universal in the same way; most are very restricted. Numerous contact (potentially hybrid: see Chapter 16) zones, indicative of dynamic homeostasis, exist in northern Europe for closely related sister species separated in southern Europe. Finally, these patterns reflect physical geography and macroevolutionary dynamics, the open plain to the north of the alpine chain facilitating movement and integration and the isolated peninsulas locked off by sea and mountain to the south restricting movement and promoting isolation.

GRADIENTS OVER EUROPE: THE UNDERLYING TAXONOMIC DIFFERENTIATION OF BUTTERFLY FAUNAS

Part of the make-up of faunas, whether faunal regions or elements, is their taxonomic composition. This can be defined for different levels, but is perhaps most usefully explored at the family (or sub-family) level. Here, we distinguish Satyridae (= Satyrinae) from other Nymphalidae. Consideration is first given to the taxonomic composition of primary areas, species richness gradients, and then subsequently to that of faunal structures. Relative frequencies of species in taxa (families) within primary areas can be

calculated in two ways: (i) as a percentage of a particular taxon found within an entire study region (taxon fraction) which provides a suitable, global, base for comparisons throughout Europe, and (ii) as a percentage of the butterfly fauna found within any area or group (areal or group fraction) which is most useful for comparison between areas or groups (faunal structures) (Dennis et al. 1995b).

Species richness among Papilionoidea and Hesperioidea over Europe (Fig. 17.8) attains peak values (standard score $z > 2$) in the Alps and high values ($z > 1$) over montane southern and central Europe. From there, richness declines systematically northwards to arctic Scandinavia and southwards to peninsular Italy, Iberia and the Mediterranean islands. This spatial pattern is largely repeated for the six butterfly families (Dennis et al. 1995b). However, the proportion of species belonging to different families (areal fraction) is not constant for primary areas but varies over Europe. The pattern of species richness closely follows plant productivity in relation to annual actual evapotranspiration (Dennis 1993, Hawkins & Porter 2003).

Both the direction of gradients in relative frequencies and the location of peak or trough values contrast substantially for the six butterfly families (Fig. 17.9). In five of the six families (Papilionidae, Pieridae, Lycaenidae, Satyridae and Hesperiidae) relative abundance declines with latitude whereas in the Nymphalidae it increases northwards to arctic Scandinavia. Peak values in relative abundance (and thus importance for the respective primary areas) differ in location for these taxa. Peak values occur for: (i) the Papilionidae on Mediterranean islands; (ii) the Pieridae on Mediterranean islands but with a subsidiary mode in northwest Britain; (iii) the Lycaenidae in the Balkans and Iberia but also Denmark; (iv) the Nymphalidae in northern and central Scandinavia; (v) the Satyridae in montane south central Europe and most of the British Isles; and (vi) the Hesperiidae in southern Iberia and Turkey.

Global values in abundance (European taxonomic fraction) indicate where excess and deficit occur in regional occurrences. For five taxa (Papilionidae, Pieridae, Lycaenidae, Nymphalidae and Hesperidae) the means in relative frequency for primary areas (regional means) exceed the proportion (global fraction) for the taxon over the study area (Fig. 17.9). The difference is significant for the Nymphalidae ($\chi^2_{(1)} = 16.0$, $P < 0.001$). However, the situation is reversed for the Satyridae; in this case the global fraction exceeds the regional mean ($\chi^2_{(1)} = 22.1$, $P < 0.001$) indicative of many species having restricted

Figure 17.8. Species richness over Europe. Standard scores (z) of species richness based on 85 primary areas standardized by regression for 250-km^2 units. (From Dennis *et al*. 1995b; courtesy of Blackwell Science.)

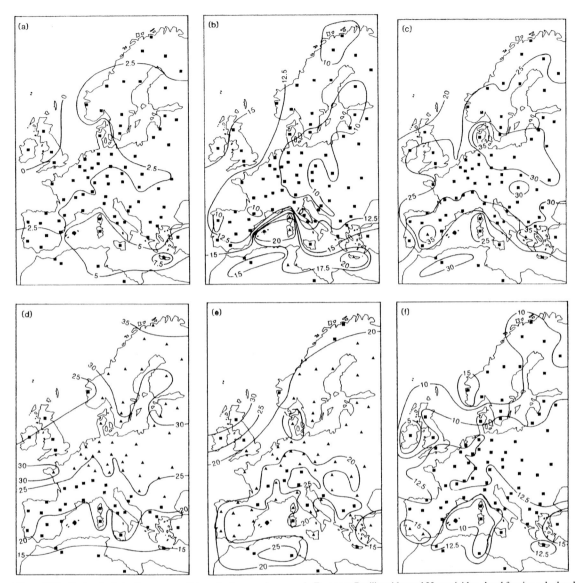

Figure 17.9. Regional relative importance of six major taxonomic groups in European Papilionoidea and Hesperioidea. Areal fraction calculated as the percentage of all species in each primary area. Black triangles, frequencies for primary areas differ significantly from global proportions at $P < 0.05$. (a) Papilionidae, global proportion = 2.8%, regional mean proportion = 3.5%, (b) Pieridae, global proportion = 10.7%, regional mean proportion = 11.8%, (c) Lycaenidae and Riodinidae, global proportion = 27.0%, regional mean proportion = 28.9%, (d) Nymphalidae and Libytheidae, global proportion = 17.6%, regional mean proportion = 24.9%, (e) Satyridae, global proportion = 30.8%, regional mean proportion = 19.8%, (f) Hesperiidae, global proportion = 11.2%, regional mean proportion = 11.8%. (From Dennis *et al.* 1995b; courtesy of Blackwell Science.)

distributions (e.g. narrow endemics in the genus *Erebia* in the European high mountain systems). Extremes of variation also occur in the Nymphalidae and Satyridae. In the Nymphalidae, most primary areas in northern Europe

have a significant relative surplus of species, whereas in the Satyridae they have largely a significant deficit of species. Significant variation is more limited and varies in location for other families, being completely absent in the

Table 17.1 *Frequencies of the number of species in different butterfly families of the main European faunal elements*

Faunal element		Family						Number of elements
		Papilionidae and Pieridae	Lycaenidae	Nymphalidae	Satyridae	Hesperiidae	Total	vs. expected number
Extent	Observed	24	50	46	30	22	172	–
	Expected	23	46	31	52	20		
Arctic/Scandinavian	Observed	2	2	8	8	3	23	–
	Expected	3	6	4	7	3		
West Mediterranean and Italian	Observed	6	19	4	14	6	49	+
	Expected	6	13	9	15	6		
East Mediterranean	Observed	12	16	2	18	5	53	–
	Expected	7	14	10	16	6		
North African	Observed	3	11	2	10	4	30	+
	Expected	4	8	5	9	4		
Mountains	Observed	4	5	7	37	4	57	+
	Expected	8	15	10	18	6		
Islands: Sardinia, Corsica and Crete[a]	Observed	2	2	1	4	0	9	
	Total	51	103	69	117	44	384	

[a] Expected values for islands are not calculated and these data are excluded from analysis because of small numbers. Expected frequencies are derived from the expectation of equal proportions of the six families in all faunal elements. Test of heterogeneity: $\chi^2_{(20)} = 77.0$, $P < 0.001$.

Source: From Dennis *et al.* (1998b), courtesy of Blackwell Science.

Papilionidae (Fig. 17.9). Patterns in relative abundance are most similar for the Lycaenidae and Hesperiidae and most disparate for the Hesperiidae and Nymphalidae (Dennis *et al.* 1995b).

Geographical bias for taxa extends to faunal groups. When species are classified to faunal elements and these elements, in turn, are classed into regions based on the geographical location of their core primary areas, then significant distinctions arise among faunal elements for the representation of taxa ($\chi^2_{(20)} = 77.0$, $P < 0.001$) (Table 17.1) (Dennis *et al.* 1998b). The largest contributions to the test statistic are for the montane and extent elements. There is a large surplus of Satyridae for the montane elements and a deficit of Lycaenidae. For the extent element, there is a considerable surplus of Nymphalidae and a deficit of Satyridae. For the Mediterranean elements, there are deficits of Nymphalidae and surpluses of Lycaenidae. These differences evoke distinctions in ecology beginning to emerge in studies of life history strategies (Shreeve *et al.* 2001, Dennis *et al.* 2004, Stefanescu *et al.* 2005b).

SMALL BUT CRUCIAL: ENDEMICS AND ENDEMISM

Endemism can be related to distinct geographical features (viz. islands, mountain tops) or to arbitrary regions. Below, endemism is discussed in relation to individual primary areas (primary area endemism) and to the entire study region of Europe (regional endemism).

If each taxon was equally susceptible to speciation and vulnerable to extinction, then the number of endemics in different families should be proportional to the number of species in the butterfly families. This is not the case ($\chi^2_{(5)} = 33.8$, $P < 0.001$), with a massive surplus of endemics among the Satyridae, mainly in the genus *Erebia*, and a substantial deficit in the Nymphalidae and Pieridae. Elsewhere, we have argued that these endemics are largely autochthonous, rather than allochthonous, in origin (Dennis *et al.* 1995b). Their geography coincides with isolated landscape features, where conditions for survival have persisted throughout the Pliocene and Quaternary, and to the east of the region their presence is replaced by congeneric taxa.

The vast majority of species endemic to Europe are restricted to areas south of the northern edge of the alpine mountain chain extending from the Cantabrian mountains in north Spain to the Carpathian mountains in Romania (Fig. 17.10) (Dennis *et al.* 1991, 1995b, Balletto 1995). Among islands only those in the Mediterranean Sea and the southern part of the Atlantic (Macaronesia) have endemic species, and these islands tend to have substantial areas at higher elevations (Dennis *et al.* 2000a, 2000b, 2001, 2008a, Dapporto & Dennis 2008a). The distribution of endemics corresponds largely with the distribution of species richness ($r_s = 0.74$, $P < 0.0005$). Only one European endemic species (*Pyrgus andromedae*, Hesperiidae) has a range that extends far into northern Europe. Peak endemicity values (relative frequencies) are distributed among the mountain ranges of southern Europe, e.g. in the Alps, Pyrenees, the Dinaric Alps and the Apennines. A number of species are endemic also to specific primary areas (Fig. 17.10); primary area endemism has a Mediterranean distribution, but is not just restricted to Mediterranean islands; it affects continental primary areas in northern Africa, Iberia, Italy and the Balkans.

The relative abundance of endemics in different regions differs substantially for the six different butterfly families. Spatial bias for endemicity in different taxa is highlighted in Fig. 17.11. For primary areas with >10 endemic species, spatial correlation in endemicity is most similar for the Papilionidae and Pieridae ($r_s = 0.51$, $P < 0.005$) and most divergent for the Lycaenidae and Satyridae ($r_s = -0.74$, $P < 0.005$) (Dennis *et al.* 1998b). Endemicity is disproportionately higher in genera with more than the mean number of species (i.e. replicates) than in those with less than the mean number. For all butterflies, the relationship is highly significant ($\chi^2_{(1)} = 41.6$, $P < 0.001$). It is also significant for the Pieridae (Fisher exact test, $P = 0.04$), Lycaenidae ($\chi^2_{(1)} = 6.0$, $P < 0.02$), Satyridae ($\chi^2_{(1)} = 12.9$, $P < 0.001$) and Hesperiidae (Fisher exact test, $P = 0.026$), but not the Papilionidae (Fisher exact test, $P = 0.65$) and Nymphalidae ($\chi^2_{(1)} = 3.6$, $P > 0.05$) (Dennis *et al.* 1995b).

There are clear distinctions among faunal elements for the frequency of endemic species ($\chi^2_{(6)} = 237.2$, $P < 0.0001$) (Table 17.2). When grouped by faunal regions, the extent element, the arctic and Scandinavian elements and the East Mediterranean elements are deficient in endemics. In the case of the extent element, the deficiency is contributing over a third of the value of the test statistic. Surpluses of endemic species are held by the West Mediterranean and Italian elements, montane elements and Mediterranean island elements. For the former two these surpluses are large. The slight deficiency for the East Mediterranean element is unexpected; this could be an artefact of delimiting the study region which truncates a number of distributions that extend further east into Turkey (e.g. *Heodes ottomanus*, *Agrodiaetus admetus*). A difference between the

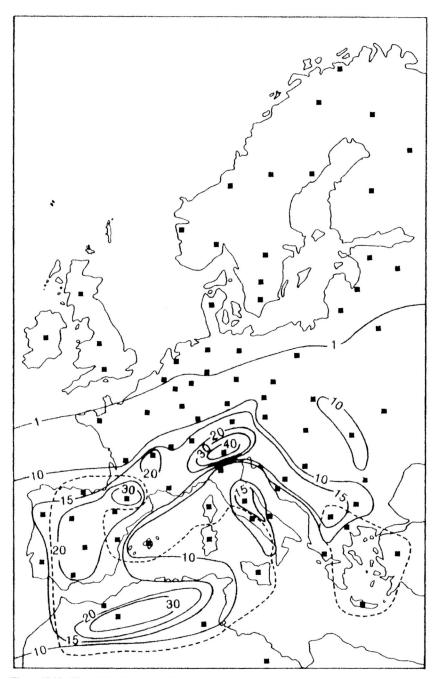

Figure 17.10. The relative frequency of European butterfly endemics as a percentage of the total number of endemics ($N = 138$ endemics). Pecked line indicates the region comprising species endemic to single primary areas. (From Dennis *et al.* 1995b; courtesy of Blackwell Science.)

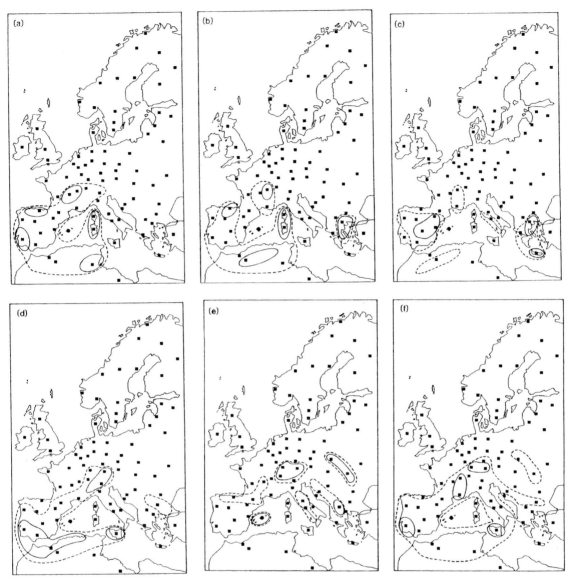

Figure 17.11. The distribution of endemics among European butterfly families. Pecked line, relative frequency of endemics exceeds the global proportion and/or the mean of regional relative frequencies for primary areas with ≥10 endemics; continuous line, the relative frequency of endemics exceeds the mean plus 1 SD of regional relative frequencies for primary areas with ≥10 endemics. (a) Papilionidae, global value 1.45 %, regional mean 3.8%; (b) Pieridae, global value 5.1 %, regional mean 6.0%; (c) Lycaenidae and Riodinidae, global value 25.4 %, regional mean 27.7%; (d) Nymphalidae and Libytheidae, global value 7.2 %, regional mean 8.6%; (e) Satyridae, global value 52.2 %, regional mean 47.9%; (f) Hesperiidae, global value 8.7 %, regional mean 12.1%. (From Dennis *et al.* 1995b; courtesy of Blackwell Science.)

Balkans on the one hand and Iberia and Italy on the other is produced by the large number of species belonging to the extent element in the Balkans and their restricted number in Italy and Iberia, thus making the latter faunas more distinct.

FUNCTIONAL SIGNIFICANCE OF FAUNAL STRUCTURES

Three questions are central to the study of faunal structures. First, are they real entities, with biological significance, or

Table 17.2 *Comparison between the observed and expected number of endemic species among main faunal elements of European butterflies*

Faunal element		Number of endemic species		
		Absent	Present	Total
Extent	Observed	170	2	172
	Expected	114	58	
Arctic/Scandinavian	Observed	22	1	23
	Expected	15	8	
West Mediterranean	Observed	14	35	49
and Italian	Expected	32	17	
East Mediterranean	Observed	40	13	53
	Expected	35	18	
North African	Observed	6	24	30
	Expected	20	10	
Mountains	Observed	9	48	57
	Expected	38	19	
Islands: Sardinia,	Observed	0	9	9
Corsica and Crete	Expected	6	3	
	Total	261	132	393

Notes: Expected frequencies are derived from the expectation of equal proportions of endemics and non-endemics across all faunal elements. Test of heterogeneity $\chi^2_{(6)} = 237.2$, $P < 0.001$.

Source: From Dennis *et al.* (1998b), courtesy of Blackwell Science.

just little more than random assemblages of species? Second, if real, do they relate at all to contemporary environmental agents or are they entirely the product of history? Finally, what is their significance for ongoing evolutionary dynamics?

Faunal structures: real entities or random assemblages?

There are clear indications that the faunal structures – faunal regions and elements – are real and not just artefacts. The spatial structures are reproduced by different multivariate analytic techniques from the species and area matrices. Their repeated appearance, even when widely different numbers of groups are extracted, arises because elements are fundamentally hierarchical in organization; distributions

of species are typically similar, but are rarely identical. Thus, groups of species are more or less compact depending on how many groups are extracted in ordination and on the level of thresholds for similarity coefficients adopted in cluster analysis. Consequently, extractions of different numbers of units or groups identify the same elements at different levels of resolution. There are other features that point to the faunal structures not being random in location and composition. Their margins largely coincide with physical (e.g. islands, mountains) and ecological (e.g. biomes such as tundra) boundaries (see below). They differ substantially in taxonomic composition, in the number of endemic species and in the representation of higher taxa.

It is clear that these structures have contemporary significance. The patterns exist and persist despite the known capacity of species to disperse and to respond rapidly to environmental changes. At very least, ecological, physiographic or historical limits are shared by species making up faunal structures, species that manifestly differ in other ways – in phylogenetic associations, resources (e.g. host-plants) and adaptations (e.g. life history). However, there is evidence that faunal elements extracted for the European butterflies also have historical and evolutionary significance. They are characterized by different sets of autochthonous endemic taxa that differ in age (Dennis *et al.* 1995b, 2001). They are characterized, too, by concentrations of different taxa (Dennis *et al.* 1998b), a predominance of different genera occurring in different elements (e.g. *Erebia* spp. in montane elements; *Agrodiaetus* spp. and Maniolini in East Mediterranean). Taxonomic differentiation extends to the existence of tension zones for sibling species partners in contiguous elements (e.g. *Pyrgus carlinae/cirsii*; *P. malvae/malvoides*; *S. sertorius/orbifer*; *P. baton/vicrama*: Dennis 1993). These components are end-products of long-term evolutionary processes.

Environmental correlates of faunal elements?

A distinctive feature of faunal structures is their delineation by, or association with, major landforms. Consequently, the elements and regions contrast for environmental parameters.

Two major physical structures, mountains and seas, separate elements and regions. The most prominent division faunistically, that between the northern and southern structures, is also the major divide geologically, physiographically and environmentally. The boundary line falls along the northern edge of the alpine fold mountain system from the Cantabrians (Iberia) to the Carpathians (Balkans). The elements north and south of this line comprise species with very different macroevolutionary dynamics. North of it, the species are evidently accustomed to migration, enforced by Pleistocene cyclic environmental changes, and limited autochthonous evolution; south of it, species need move less and evolve more, as evidenced in the dramatically higher levels of endemism (Dennis *et al.* 1991, 1995b, 1998b). Mountains also contribute to the isolation of the three Mediterranean blocks of Iberia, Italy and the Balkans, with their distinctive elements, cut off from one another largely by sea.

Elements are also located on physical structures. Some elements are limited to Mediterranean islands (e.g. Crete, Sardinia–Corsica), though not all large Mediterranean islands have distinct elements (e.g. Balearics). Islands, however, are also sites for faunal regions (e.g. Britain). A complex pattern of elements has developed in close association with adjoining blocks of southern European mountains and others occur separately on the plateaux of southern Europe (e.g. Iberia).

These divisions by physical barriers, and associations with physical structure, further segregate faunal elements and regions in terms of distinctive environments. Faunal structures differ with regard to latitudinal, longitudinal and altitudinal contrasts in climate and biome (vegetation type) (Dennis *et al.* 1991). The distributions and ranges of European butterflies have been shown to be sensitive to changes in climate and biotope (Dennis & Williams 1986, Turner *et al.* 1987, Dennis 1993, Parmesan *et al.* 1999, Dennis *et al.* 2000a, 2000b). The close link of faunal structure with biomes draws attention to 'communities' of associated fauna and flora (e.g. arctic elements with tundra; Mediterranean elements with sclerophyll forest), i.e. organisms having similar environmental limits, raising important questions about faunal structures over evolutionary time, particularly as regards the most recent geological episode of Pleistocene glaciations.

ICE AND CHANGE: THE FUNCTIONAL SIGNIFICANCE OF FAUNAL UNITS

The third question generated by faunal units is do faunal units (elements, regions) have functional significance for the species belonging to them? As species are historical entities, this is an issue of evolutionary time and must necessarily be considered against the backcloth of the Tertiary and Quaternary environmental disruptions.

Climatic oscillations of the Pleistocene, alternations between warm and cold phases (c. 46 cycles: see Ruddiman & Raymo 1988, Dennis 1993 for references), were responsible for dramatic changes in the distributions of organisms (Coope 1987, Huntley 1988) as surmised by earlier workers on butterflies (e.g. Reinig 1938, de Lattin 1949, 1967). During the first 1.6 or 1.7 million years of the Pleistocene, each of these cycles was some 41 000 years, changing to 100 000 year cycles in the middle Pleistocene (about 900 000 to 400 000 years BP) with conditions becoming particularly severe after 450 000 years BP (Ruddiman & Raymo 1988, West 1988). The severity of glacial stage climates is well known from studies of fossil organisms during the last cold stage, the Würm (southern Europe) or Weichselian (northern Europe) which began some 73 000 years BP and peaked at 18 000 years BP (Coope 1970, 1978, 1987, 1994, Prentice *et al.* 2000). Not only did temperature drop globally, but the amount of precipitation was dramatically reduced in most parts of the world (Prentice *et al.* 2000) including Central Europe (Tarasov *et al.* 2000); the Mediterranean region was not much drier than today and covered by steppe and steppe–forest (Elenga *et al.* 2000). Changes in conditions could also be rapid, as at the transition between the Late Glacial and the Holocene (Ashworth 1972, Dansgaard *et al.* 1989).

Cyclic climatic and environmental changes during the Pleistocene are considered to have impacted on faunal structures in two distinct ways: first by causing taxonomic changes in the (species) content of faunal elements between stages; second, through macroevolutionary processes and species turnover (speciation, extinction). Both have been generated by orthodox population processes (migration, colonization, population extinction) owing to ice cover, climate change and massive shifts in resources associated with spatial upheaval in biotopes and biomes (Dennis 1977, Dennis *et al.* 1991, Hewitt 1996). A crucial feature of these changes is the existence of refugia characteristic of particular elements during warm and cold stages (Dennis *et al.* 1991, Hewitt 1996, 2000, 2004); these have been integral to species' maintenance but also to speciation. Although each cold or warm stage is conventionally regarded as too short for macroevolution (Coope 1987, but see Dennis *et al.* 1991, 1995b), Pleistocene polyglaciation has potential for long-term evolutionary changes, as core populations are effectively isolated for the entire Quaternary, typically in southern Europe, whereas their range extensions become periodically erased (Dennis *et al.* 1991, Hewitt 1996). Examples are provided by temperate sibling taxa which meet at tension zones during interglacial stages (e.g. *Pontia daplidice* and *P. edusa*: Geiger *et al.* 1988), but which contract to localized refugia during glacial stages. Examples of the reversed pattern (i.e. interglacial refugia) are provided by species belonging to the arctic and montane elements (e.g. *Erebia* and *Oeneis* spp.: Douwes 1980, Mikkola & Kononenko 1989).

These two sets of processes have potential for generating changes in the content of faunal elements and faunal regions from one cycle to the next. Not only may new species appear and old ones disappear within the region but there will be a 'shuffling' of species among elements in response to stochastic environmental changes. A precedent and a prerequisite for this is that each cycle delivers unique patterns of vegetation cover, thus resources, despite superficial similarities in biomes and biotopes (Huntley 1988). Bearing in mind that species, within elements, differ in life histories, migration capacity, hostplants and adaptations, there is potential for interchange among elements for species. The overlapping, nested hierarchies of faunal elements support this observation (Dennis *et al.* 1991, 1998b).

The implications for the composition of faunal elements depend on species' attributes and location. A positive feedback between resource distributions and capacity for migration is expected to influence species' macroevolutionary dynamics (Dennis *et al.* 1995b); wider resource distributions encourage movement and colonization and narrow resource distributions the reverse, the former countering speciation and the latter generating it. As multiple clade formation initially requires expansion in ranges, the inference, backed by vegetation changes (Huntley 1988), is that resources undergo massive changes in spatial status driving macroevolution and extinction. The contrasting content of faunal elements, having different locations, in terms of taxa and endemics strongly suggests that potential for changes varies spatially. Different species in the same element or the same species in different elements are unlikely to experience the same potential for evolutionary change or persistence. Inasmuch as resources contrast for different taxa, the opportunities for different taxa are uniquely tied into specific landscapes (Dennis *et al.* 1995b).

The functional significance of elements for species raises evolutionary issues hinging on the conjoint occurrence of bias for taxa and endemicity. As taxa with more replicates contain more endemics, an argument can be made for the functional linkage of these two observations – biases for endemicity and taxa – particularly as these biases coexist within elements. The implication is that a species by belonging to one element, rather than to another, has a higher probability of being an endemic and belonging to a particular taxon. As faunal elements and higher taxa both differ substantially in the likelihood of containing endemics, a stronger inference can be made. The probability of a species becoming an endemic is directly influenced by its affiliation to particular elements and taxa. No doubt, the probability is affected by the *degree* of fidelity of a species to an element since elements are hierarchical structures and fidelity is not absolute. Thus, species belonging to northern elements (i.e. extent, arctic and Scandinavian elements) have a lower probability of being European endemics and a higher probability of being Nymphalidae. Species belonging to southern elements have a higher probability of being endemics and of being Lycaenidae (i.e. West Mediterranean) or Satyridae (i.e. montane elements).

The fidelity of a species to a faunal element will affect its evolutionary pathway since, by belonging to a specific faunal element, a species is subject to the distinctive processes linked to a unique landscape, its environmental conditions and the history of its biotopes (Dennis *et al.* 1998b). Species have the capacity of switching between faunal groups, through range changes and evolution, owing to environmental changes impinging on their resources and life conditions. In this process, transfer of species across different groups are not of equal likelihood, being greatest for species of contiguous continental elements and least for species belonging to island elements.

Reconstructing actual events is another matter. As species in any element can theoretically have 'transferred' from any other element, the composition of elements does not automatically convey information about origins. To determine and date the origin of species within elements, detailed phylogenetic analysis is required, the domain of phylogeography and biochemical markers.

PHYLOGEOGRAPHY

Taxonomy, distribution patterns, classical biogeography and ecology have been the predominant topics in the investigation of butterflies for a long time (see above). Only during the last few decades have genetic approaches become important; techniques that are considerably enhancing our understanding of biogeography, ecology and assessment of taxa for conservation. Here, we briefly review the state of the art in the scientific disciplines of phylogeography, address the different methods used for phylogeographical analysis and provide a synopsis on the phylogeographical studies performed on butterflies in Europe associated with different faunal elements. The following case studies present the evidence for historical patterns and processes, and illustrate how different taxonomic groups have responded uniquely to past events.

The importance of phylogeography for biogeography

Chorological analyses are useful tools to obtain an idea of past range changes in animals and plants (see above; e.g. Reinig 1938, de Lattin 1949, 1967, Holdhaus 1954, Varga 1977), although fossil remains are necessary to prove past distributions (Coope 1970, 1978, 1994, Gliemeroth 1995, Ponel 1997). However, actual distribution patterns can be misleading because they are not only subjected to processes in biogeography but to human and unpredictable stochastic influences. Even fossil evidence is sometimes problematic because good evidence is only available for a rather restricted collection of taxa and the absence of fossils causes misinterpretations of the existing data sets if only because it is almost impossible to prove an absence. These considerable weaknesses are largely overcome with the inclusion of genetic analyses in this field of science and the development of phylogeography (Avise *et al.* 1987). Furthermore, the real impact of glacial isolations and range changes on the species themselves can be demonstrated best by genetic analyses. This technique also offers the possibility of tracing detailed postglacial range change scenarios and enables precise assessments of contact zones and introgression between different genetic lineages (Hewitt 1996, 1999, 2000, 2001, 2004, Comes & Kadereit 1998, Taberlet *et al.* 1998).

Phylogeographical techniques

Since large-scale applications of DNA techniques became feasible, several techniques have become standard tools for the resolution of biogeographical questions. Individual molecular techniques have their advantages and disadvantages and may or may not be applicable to particular taxa (see Chapter 10 for a detailed discussion). However, the combination of techniques and increasingly sophisticated descriptive and analytical methods to distinguish between groups (e.g. hierarchical statistics, bootstrapping, various cluster algorithms, nested clade analysis) (e.g. Excoffier *et al.* 1992, Templeton 1998b) is beginning to reveal biogeographical patterns of taxonomic relationships, facilitating phylogeographical analyses.

Phylogeography of Mediterranean species

GLACIAL DIFFERENTIATION OF *POLYOMMATUS CORIDON/HISPANA*

Polyommatus coridon and its sibling species are probably, phylogeographically, the most intensively studied European butterfly species complex of Mediterranean origin, with more than 100 populations studied (Mensi *et al.* 1988, Lelièvre 1992, Marchi *et al.* 1996, Schmitt & Seitz 2001a, 2001b, 2002b, Schmitt *et al.* 2002, 2005a, Schmitt & Krauss 2004).

A study of 20 allozyme loci over most of *P. coridon*'s distribution range revealed a differentiation into two major genetic lineages with a mean genetic distance (Nei 1978) of 0.041 ($\pm$ 0.010 SD); 76.1 % of the total genetic variance among populations was between these two lineages (Schmitt & Seitz 2001a). This result is supported by differences in the chromosome numbers of these two lineages (western: 87 to 88 chromosomes; eastern: 90 to 92 chromosomes) (de Lesse 1969). The western lineage is of Adriatic–Mediterranean origin and the eastern one of Pontic–Mediterranean origin (Schmitt & Seitz 2001a). Both lineages represent monophyletic entities (Schmitt & Seitz 2001b) and differentiation of these two lineages probably took place in isolation in Italy and the Balkans during the last glaciation.

Interestingly, the differentiation between these two *P. coridon* lineages is of the same order of magnitude as in many of the following examples for butterflies. Therefore, we have to assume more or less similar events for the onset of these distinctions. The most plausible significant event, of suitable scale, underlying the intraspecific structure within these species is the onset of the last glaciation.

Spanish *P. coridon*, most probably, descend from a postglacial colonization originating from the Adriatic–Mediterranean region (Lelièvre 1992, Schmitt & Seitz 2001a). On the contrary, *P. coridon gennargenti* from Sardinia is genetically rather remote from *P. coridon* (genetic distance 0.39: Marchi *et al.* 1996). Therefore, this taxon must have been isolated for several glacial cycles from continental *P. coridon*. The bivoltine *P. hispana* was also disclosed to be a taxon in its own right by allozyme analyses (Lelièvre 1992, Schmitt *et al.* 2005a), most probably having differentiated from *P. coridon* in isolation in southeastern Iberia during the Würm or Riss glaciation (Schmitt *et al.* 2005a).

An overview on the Würm glacial differentiation centres of the taxa of the *P. coridon/hispana* complex is shown in Fig. 17.12a.

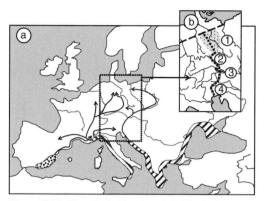

Figure 17.12. Hypothetical location of glacial refugia and possible postglacial expansion corridors of *Polyommatus coridon/hispana* (a) and reconstruction of the actual contact zone in Central Europe (inlaid graphic b) highlighting geographical structures that isolate the two major genetic lineages. (1, sandy area with acid soils in northeastern Germany; 2, mountain area with acid soils; 3, area along the Danube in the German–Austrian border region where phenotypically mixed populations are known; 4, western watersheds of eastern alpine river systems.) In the inlaid graphic, the northern distribution border of *P. coridon* (Tolman & Lewington 1997) is marked by a broken line, the border between the two genetic lineages after phenotypic differentiation by a dotted line. (From Schmitt & Seitz 2001a, 2002b, Schmitt *et al.* 2002, 2005a, Schmitt & Krauss 2004, Schmitt 2007 modified; courtesy of Blackwell Science and BioMed Central.)

POSTGLACIAL EXPANSION OF *POLYOMMATUS CORIDON*

Both major lineages of *P. coridon* come into contact in the eastern Alps of Austria and are separated further north by the mountains on the German–Czech border (Fichtelgebirge, Oberpfälzer Wald, Bayerischer Wald, Böhmerwald) and the acid sandy area of northeastern Germany (in particular in Sachsen-Anhalt and western Brandenburg) (Schmitt & Seitz 2001a) (Fig. 17.12b). Detailed allozyme studies within the western and the eastern lineage of *P. coridon* revealed further postglacial differentiation. Thus, both lineages showed northwards losses of genetic diversity, stepwise in the western (Schmitt *et al.* 2002) and continuously in the eastern (Schmitt & Seitz 2002b) lineage.

Postglacial range expansion of the western lineage very likely followed several corridors: (i) westwards through southern France to the Pyrenees, (ii) eastwards along the southern Alpine slopes and (iii) northwards along the Rhône and Saône river valleys to northeastern France (Schmitt *et al.* 2002). From northeastern France, Germany was colonized on three alternative routes: (i) western Germany along

the Moselle valley without major genetic modification of the populations; the other parts of Germany were colonized by a lineage experiencing a genetic bottleneck effecting approximately a 10% loss in the number of alleles by passing the Burgundian Gap (between the Vosges in the north and the Jura mountains in the south) into the southern Upper Rhine valley. Further colonization was split into (ii) a northwestern pathway expanding northwards along the Rhine valley and through Hesse to southern Lower Saxony and western Thuringia and (iii) a southeastern one along the southern German calcareous mountain systems through Württemberg and Bavaria to southern Thuringia, with both these groups differing genetically (Schmitt *et al.* 2002, Schmitt & Krauss 2004) (Fig. 17.12a).

Within the eastern lineage, postglacial expansion into eastern central Europe started in the northwestern Balkans and followed the calcareous mountains to western Hungary. From here on, two alternative pathways are reflected in two different genetic groups: (i) a western one through the Porta Hungarica to Moravia, westwards to Bohemia and northwards along the Vistula river and the Thorn-Eberswald glacial valley to eastern Brandenburg (northeastern Germany) and (ii) an eastern one along the Hungarian mountain ranges to eastern Slovakia (Schmitt & Seitz 2002b) (Fig. 17.12a).

Some preliminary allozyme data on 19 loci for the Adonis Blue *Polyommatus bellargus* show patterns which imply a similar biogeographical structure as in the *P. coridon/hispana* complex (Schmitt & Seitz 2001c).

MELANARGIA GALATHEA/LACHESIS

A rather similar genetic structure as in the *P. coridon/hispana* complex and in *P. bellargus* has been found in the *Melanargia galathea/lachesis* complex (Habel *et al.* 2005). A study of 19 allozyme loci of 13 populations of *M. galathea* from France, Germany, Austria, Slovenia and Montenegro and one population of *M. lachesis* from the Spanish Pyrenees revealed a moderately high genetic diversity of these populations compared to other butterfly species.

A cluster analysis has revealed a clear differentiation into three major genetic lineages. *Melanargia lachesis* and *M. galathea* are clearly differentiated from each other, with the latter being split into a western lineage of the populations from France and Germany and an eastern lineage of the populations from Austria, Slovenia and Montenegro. The average genetic distance (Nei 1978) between *M. galathea* and *M. lachesis* was 0.124 ($\pm$ 0.015 SD) and between the eastern and the western lineage of *M. galathea* 0.034 ($\pm$ 0.007 SD).

A hierarchical variance analysis distributed 91.1% of the among population variance (F_{ST} 13.1%) between these three taxa. Excluding the *M. lachesis* sample from the Spanish Pyrenees, F_{ST} dropped to 6.1%, and variance between these two taxa matched 80.4% of the total variance among populations.

For these reasons, the glacial differentiation centres of the *M. galathea/lachesis* complex, most probably, have been similar to those of the *P. coridon/hispana* complex, and also the contact zones between the taxa of these two species complexes are almost identical. However, no northwards decline in genetic diversity was found in the two *M. galathea* lineages, a clear contrast with *P. coridon*. This might be due to the more general distribution of *M. galathea* in the landscape and a broader use of possible resources. Therefore, *M. galathea*, most probably, has expanded northwards as a large phalanx without genetic impoverishment, and not in a stepping stone or leptokurtic fashion as in *P. coridon*. However, the genetic structure within the eastern lineage of *M. galathea* suggests three expansion corridors out of the Balkan differentiation centre (Schmitt *et al.* 2006a).

MANIOLA JURTINA

Two major allozyme studies researching the phylogeography of *Maniola jurtina* have been performed during the last two decades (Thomson 1987, 35 loci; Schmitt *et al.* 2005b, 21 loci). This rather widespread species, which exists in a great variety of different biotopes, has a much simpler genetic structure than, for example, the *P. coridon/hispana* complex. The combination of both *M. jurtina* data sets reveals a clear differentiation into two major genetic lineages. The western lineage includes individuals from Morocco, Portugal, Spain, France, the UK and Ireland; the eastern lineage is composed of individuals from Germany, Sweden, Switzerland, Austria, Italy, Slovakia, Hungary, Greece and Turkey. The genetic distance (Nei 1978) between both lineages (0.033) is identical with the one between the two *M. galathea* lineages (Habel *et al.* 2005), but somewhat less than in *Polyommatus coridon* (Schmitt & Seitz 2001a); 77% of the genetic differentiation among populations is contributed by differences between these two lineages (Schmitt *et al.* 2005b).

These two basic allozyme lineages are further supported by morphometric analyses (three out of four measured parameters revealed the same geographical structure) and by the structure of the male genitalia (Thomson 1987). The latter revealed a hybrid belt in eastern France, the Benelux countries and western Sweden, with its width increasing northwards.

Figure 17.13. Hypothetical distribution patterns of *Maniola jurtina* in Europe during the last glaciation maximum (dark shaded areas). Postglacial expansions are indicated by solid arrows. The postulated actual hybrid zone is shown by the hatched area. Question marks indicate lack of information concerning ancient distribution patterns in the south and present distribution of hybrid populations in the north. (From Schmitt *et al.* 2005b; courtesy of the Linnean Society of London.)

These data support the idea of two glacial differentiation centres: a western Atlanto-Mediterranean centre including Morocco and Iberia and an eastern one in the eastern and perhaps central Mediterranean (Ponto- or Adriato-Ponto-Mediterranean) (Fig. 17.13).

In contrast to other species of Mediterranean origin little genetic structure was observed within these two lineages. Thus, the genetic diversity was not reduced polewards and no remarkable genetic sublineages exist (Schmitt *et al.* 2005b). Therefore, postglacial northwards expansion, most probably, occurred along a wide phalanx without significant loss of genetic diversity (see Ibrahim *et al.* 1996).

SUMMARY MEDITERRANEAN SPECIES

Three main patterns of postglacial range expansion of Mediterranean taxa were formulated by Hewitt (1999, 2000): (i) expansion from all three Mediterranean peninsulas, (ii) expansion from Iberia and southeastern Europe, but not from Italy and (iii) expansion only from southeastern Europe, but not from Iberia and Italy. *Maniola jurtina* is the only species that matches, more or less, one of these three types, i.e. the second type. The data for the *Polyommatus coridon/hispana* complex (Lelièvre 1992, Schmitt & Seitz 2001a, 2002b, Schmitt *et al.* 2002, 2005a), *Polyommatus bellargus* (Schmitt & Seitz 2001c) and the *Melanargia galathea/lachesis* complex (Habel *et al.* 2005) disclose a pattern different from the three mentioned above: these taxa expanded from their Adriatic- and Pontic-Mediterranean

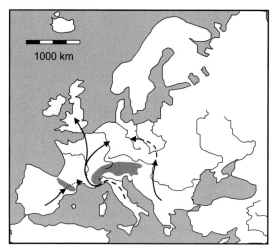

Figure 17.14. The fourth main pattern of postglacial range expansion of European terrestrial species as revealed from phylogeographical data from three butterfly species complexes. Solid lines, postglacial expansions proved for the *M. galathea/lachesis* complex; broken lines, postglacial expansions hypothesized for species following the fourth paradigm, but still in need of being tested in the *M. galathea/lachesis* complex; dotted lines, postglacial expansions proved for the *M. galathea/lachesis* complex, which might be followed only partly by other species of the fourth paradigm. Shaded areas represent mountain areas which might have acted as expansion obstacles. (From Habel *et al.* 2005; courtesy of Blackwell Science.)

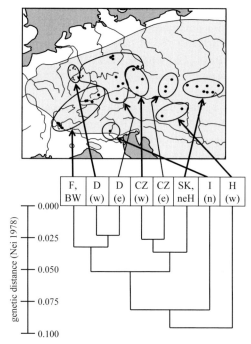

Figure 17.15. Geographical location of 53 populations of *Erebia medusa* and assignment of these populations towards the four major genetic lineages and their sub-groups. The UPGMA diagram is based on genetic distances (Nei 1978). The populations merged in the map represent the estimated minimum geographical extension of the respective genetic group. The distribution conforming Tolman & Lewington (1997) is shaded. (From Schmitt 2001). F, BW, French–Baden–Wüttemberg group; D (w), western German group; D (e), eastern German group (these three lineages represent the western lineage); CZ (w), Bohemian group; CZ (e), Moravian group; SK, neH, Slovakian–northeast Hungarian group (these three lineages represent the eastern lineage); I (n), southern Alps lineage; H (w), western Hungarian lineage.

differentiation centres but not, or if so only weakly, from Iberia. This pattern might represent a fourth general pattern of postglacial range expansion of Mediterranean taxa (Fig. 17.14). Indeed, it might be a typical feature in highly mobile invertebrates like butterflies.

Phylogeography of continental interior species

GLACIAL DIFFERENTIATION CENTRES OF *EREBIA MEDUSA*

Erebia medusa is the best studied of the continental interior species; investigation of 19 loci in more than 50 populations from France, Germany, Italy, Czech Republic, Slovakia and Hungary has revealed a strong differentiation within this species (Schmitt & Seitz 2001d) (Fig. 17.15). Over all these populations, an F_{ST} of 14.9% was calculated. A total of four major genetic entities were distinguished: a western lineage (populations from eastern France and Germany), an eastern lineage (populations from Czech Republic, Slovakia and northeastern Hungary), a western Hungarian lineage

(two populations from western Hungary) and a southern Alpine lineage (one sample from Monte Baldo, northern Italy). Most of the genetic differentiation among populations (68%) is attributed to these four lineages. In particular, the western Hungarian and the southern Alpine lineages were strongly differentiated from all the other lineages (mean genetic distance (Nei 1978) against other lineages 0.095, 0.082, respectively) (Fig. 17.15). Furthermore, the genetic diversity (e.g. mean number of alleles per population) was significantly higher in the eastern lineage than in the other three.

This strong intraspecific differentiation makes postglacial expansion from Siberia (as formerly proposed for this

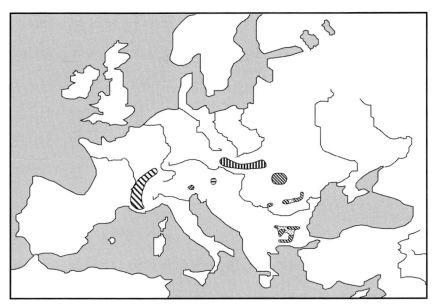

Figure 17.16. Possible extension of refugial areas (hatched) of *Erebia medusa* in Europe during the Würm glaciation. Some of the refugia at the southern slopes of the southern Carpathians and in Bulgaria were possibly (at least during interstadials) linked by gene flow (dotted). (From Schmitt & Seitz 2001d and Schmitt *et al.* 2007, modified.)

species: see de Lattin 1957b, Varga 1977) unlikely. On the contrary, survival in multiple extra-Mediterranean refugia in Europe appears a more reliable outcome. The three western lineages most probably evolved in three differentiation centres west, south and southeast of the glaciated Alps (Fig. 17.16). Work in progress provides evidence for a series of small differentiation centres along the southern Alpine margin. The proximity of these centres to the alpine glaciers might have provided the necessary humidity, which was lacking in the glacial steppes and very probably the most important factor excluding *E. medusa* from this area.

The eastern lineage was thought to have evolved in southeastern Europe (Schmitt & Seitz 2001d), but results of current research in southeastern Europe make survival in southern Moravia and/or south of the northern Carpathians more likely for this lineage (Schmitt *et al.* 2007) (Fig. 17.16). The reconstructed glacial distribution in southeastern Europe is highly complex and strongly distinguishes genetically rich populations in Bulgaria from genetically poor populations in Romania. This underlines glacial survival in both areas, but better survival conditions further south in Bulgaria. A strong genetic differentiation in Romania supports several strictly isolated areas of Würm glacial survival on the southern slopes of the southern Carpathians and most probably even in the eastern Carpathian Basin. The less

pronounced genetic differentiation in Bulgaria suggests a less scattered glacial distribution pattern in this area most probably with at least temporal gene flow among the centres of survival (Schmitt *et al.* 2007) (Fig. 17.16).

INFLUENCE OF THE CLIMATIC FLUCTUATIONS OF THE LATE WÜRM AND THE POSTGLACIAL ON *EREBIA MEDUSA*

The western lineage of *E. medusa* shows a further strong substructure. A first split separates the samples from eastern France and Baden-Württemberg (southwestern Germany) from all the other German samples (53% of the genetic differentiation among populations). Whereas the first group is rather homogeneous, the latter is further subdivided into a small western group (Saarland, western Rhineland-Palatinate) and a larger eastern group (eastern Rhineland-Palatinate, Thuringia, Bavaria), 47% of the genetic differentiation occurring among populations (Fig. 17.15).

This structure might be explained by climatic oscillations at the end of the Würm glaciation. A likely scenario is that during the first warming, *E. medusa* might have spread over a considerable area of western Central Europe. The climatic deterioration of the Older Dryas (14 300 – 14 000 years BP) might then have restricted the species to two core areas (i) in

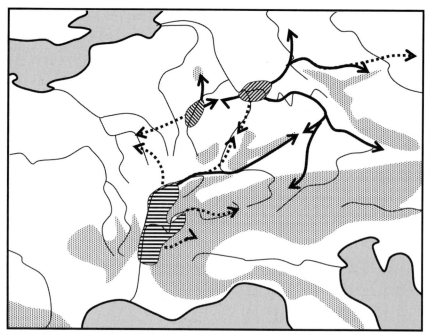

Figure 17.17. Hypothetical distribution of refugial areas of the western lineage of *Erebia medusa* during Younger Dryas and possible postglacial expansions. Refugia are hatched. Main expansion directions derived from our data are represented by solid arrows. Main expansion directions based on general biogeographical assumptions are represented by broken arrows. Mountain ranges are shaded. (From Schmitt & Seitz 2001d; courtesy of the Linnean Society of London.)

the western Alpine foothills and (ii) in a restricted area in western Central Europe, with further loss of genetic diversity in the latter group. After another expansion during the next warming (Alleröd, about 12 500 years BP), the species was restricted again, but two relicts survived in western Central Europe with further differentiation between them. During the postglacial, all these three centres expanded, contributing differently to the actual distribution (Fig. 17.17).

PARNASSIUS MNEMOSYNE

The population genetics of *Parnassius mnemosyne* has been intensively studied using allozymes in France (24 populations: Napolitano & Descimon 1994) and in Hungary (14 samples: Meglécz *et al.* 1997), the latter including one sample from western Romania. The French study included considerably more loci (24) than the Hungarian one (8). Both studies used horizontal starch gel electrophoresis and the differences between both regions are so pronounced that they should reflect different phylogeographical histories in France and Hungary. In general, genetic diversity of the populations is higher in Hungary than in France (mean number of alleles per locus: Hungary: 2.1, France: 1.4;

percentage of polymorphic loci: Hungary: 47%, France: 30%). Furthermore the French populations show a clear split into several hierarchically structured clades and a relatively high F_{ST} value (13.5%), whereas no clear structuring was obtained for the Hungarian samples and F_{ST} was about half of the French value (6.8%).

These data point to a similar phylogeographical structure as that in *Erebia medusa*. *Parnassius mnemosyne* may have had a large glacial refugium in southeastern Europe, with a high percentage of genetic diversity being preserved, and postglacial expansion over the eastern half of Europe. Glacial refugia in France might have been scattered along the Alps in geographically restricted areas with suitable climate and biotopes. Owing to relatively small population sizes, genetic erosion and differentiation would have taken place. As Napolitano & Descimon (1994) speculate, the expansion of *P. mnemosyne* to the Pyrenees and the Massif Central might have been a relatively recent (i.e. postglacial) phenomenon.

COENONYMPHA HERO

Coenonympha hero is a typical continental interior species, distributed in a relatively narrow band through central and northern Central Europe (Kudrna 2002), but occurring throughout temperate Asia (Tolman & Lewington 1997). Owing to this pattern of distribution, the species was often classified as a Siberian faunal element (e.g. Varga 1977).

Allozyme data of some 13 populations of *C. hero* from Sweden, Estonia and the Ural mountains are available (Cassel & Tammaru 2003). All analysed parameters of genetic diversity (mean number of alleles, percentage of polymorphic loci, heterozygosity) showed decreasing values westwards, e.g. the percentage of polymorphic loci decreased from a mean of 35% in the Ural mountains, to over 20% in Estonia to 17% in Sweden. Although the obtained F_{ST} value was relatively high (14.1%), the genetic distances were comparatively low and did not exceed 0.025 (Nei 1978). However, relatively high F_{ST} values are frequently observed in combination with low genetic distances in species with low genetic diversity within populations (see Schmitt & Seitz 2004).

The genetic patterns observed in *C. hero* suggest an eastern origin of the species. Due to the relative scarcity of data, an unambiguous centre of postglacial expansion cannot be proposed for the European populations of *C. hero*. However, the southern Urals might be a possible glacial refugium.

SUMMARY FOR CONTINENTAL INTERIOR SPECIES

This group includes many of the species that formerly were considered to be of Siberian origin (de Lattin 1967, Varga 1977). But, although some of them really might be of eastern origin, possibly *Coenonympha hero* (Cassel & Tammaru 2003) or *Maculinea teleius* (Figurny-Puchalska *et al.* 2000), the biogeography might be much more complex for many other species. The *Erebia medusa* pattern (Schmitt & Seitz 2001d, Schmitt *et al.* 2007) of major glacial differentiation centres in southeastern Europe and several small differentiation centres in the western half of Europe and especially around the glaciated Alps is, most probably, repeated in *Parnassius mnemosyne* (Napolitano & Descimon 1994, Meglécz *et al.* 1997). Future studies might reveal this pattern for more species of so-called Siberian origin.

Phylogeography of widespread species

POLYOMMATUS ICARUS

Polyommatus icarus is one of the most widely distributed butterflies of Europe (Kudrna 2002). An allozyme study of 19 loci including 29 populations from Portugal, France, Italy, Germany, Slovakia and Hungary (Schmitt *et al.*

2003) revealed a rather high genetic diversity within populations with a strikingly high mean number of alleles per locus (mean: 2.98). In contrast, differentiation among populations was very low (F_{ST} = 1.9%). Only a marginal trend of decline in genetic diversity from the south to the north was observed, but isolation by distance was strong on the European scale (r^2 = 0.69, significance tested by a Mantel test). No further genetic structure could be detected.

PIERIS NAPI

The allozymes of the Green-Veined White (including populations of the alpine sibling *P. bryoniae*) have been studied over a major part of Europe including samples from the UK, France, Switzerland, Germany, Italy, Croatia, Hungary, Denmark, Sweden and Norway (Porter & Geiger 1995). Apart from the two marginal races *P. napi britannica* (central and northern British Isles) and *P. napi adalwinda* (Scandinavian mountains), genetic differentiation was rather low on the European scale (F_{ST} = 2.6%). The genetic diversity is moderately high and no clines exist over Europe. Isolation by distance explaining 64% of the variance has been demonstrated for *P. napi meridionalis* (Geiger & Shapiro 1992).

AGLAIS URTICAE

Five allozyme loci of nine populations from the Netherlands, Belgium and France have been analysed for the Small Tortoiseshell (Vandewoestijne *et al.* 1999). The observed means of the parameters of genetic diversity were rather high (number of alleles per locus: 2.84; expected heterozygosity: 24.8%; polymorphic loci: 76%), whereas the genetic differentiation among populations was low (F_{ST} = 3.0%). This supports high gene flow between populations and the absence of major genetic bottlenecks even throughout the last glaciation or earlier.

SUMMARY FOR WIDESPREAD SPECIES

Although only three species of this group have been analysed on a major scale, the general patterns for this biogeographical group of species might be one of a high genetic diversity of individual populations, a lack of diversity gradients throughout Europe, but low genetic differentiation between populations, with some distance effects on a continental scale but a lack of deep phylogeographical structures. Therefore, it is probable that these species were widely distributed in the Mediterranean region during the last ice age and even during the Last Glacial Maximum (Dennis *et al.* 1991). Considerable gene flow among populations and regions, as well as large continuous distributions throughout the last glacial–interglacial cycle, are most probable. The northwards postglacial range expansion might have occurred along a large phalanx without major genetic erosion (see Ibrahim *et al.* 1996). The remarkable differentiation of the northern subspecies *P. napi britannica* is likely to have evolved during the postglacial, possibly due to selection of cold-adapted genotypes in these northern biotopes, as for example observed in *Colias* butterflies in North America (Watt *et al.* 1996). The subspecies *P. napi adalwinda* might be postglacial or somewhat older (e.g. Loch Lomond Readvance) (Dennis 1977).

Phylogeography of mountain species

ZYGAENA EXULANS

As no genetic data are available for a butterfly with typical arctic–alpine distribution, we refer here to a burnet moth species, *Zygaena exulans*. This species occurs in many of the Eurasian high mountain systems (e.g. Pyrenees, Alps, Carpathians) and in arctic regions (e.g. northern Fennoscandia, Scottish Highlands) (Naumann *et al.* 1999).

An allozyme study of 15 loci of two samples from the Pyrenees and five from the Alps (Schmitt & Hewitt 2004a) revealed a moderately high genetic diversity (e.g. mean number of alleles per locus 2.0). However, mean genetic diversities were generally higher in the Alps than in the Pyrenees. The F_{ST} was calculated as 5.4% and genetic distances (Nei 1978) among populations were generally low (mean: 0.022). However, 62% of the genetic variance among populations was between the Alps and Pyrenees. The two populations from the Pyrenees showed no internal differentiation, whereas significant differentiation was detected among the populations from the Alps ($F_{ST} = 2.8\%$).

This weak differentiation between Alps and Pyrenees implies that *Z. exulans* had a continuous distribution between these two high mountain systems during the last glaciation and that postglacial (current) isolation is contributing to differentiation.

EREBIA EPIPHRON

The Mountain Ringlet *Erebia epiphron* is one of the most widespread butterflies of the European high mountain systems (Cantabrian mountains, Pyrenees, Alps, Carpathians, high mountain systems of the western Balkans), but can also be found in the highest parts of some lower mountain ranges (French Massif Central, Vosges, Harz, Jesenik mountains); in the north, the species is restricted to two small areas

in northern England and Scotland (Kudrna 2002). In the high mountain systems of Bulgaria (Rila, Pirin and Stara Planina), the species is replaced by its sibling species *Erebia orientalis* (Tolman & Lewington 1997). The species is found in alpine and subalpine swards, but prefers patches of medium to higher-growing vegetation with some soil humidity (SBN 1987).

Sixteen populations of *E. epiphron* from the Alps, Pyrenees and the Jesenik mountains (northern Moravia) were studied for 18 allozyme loci (Schmitt *et al.* 2006b). The populations had moderately high mean genetic diversities for butterflies (e.g. mean number of alleles per locus 2.1, expected heterozygosity 15.4%). The differentiation among populations was rather strong ($F_{ST} = 29.1\%$) and mean genetic distances (Nei 1972) were higher than in most other studies based on butterflies. A neighbour-joining analysis distinguished two major genetic groups with a total of five subgroups: main group 1 included the Pyrenees and western Alps with the two subgroups (i) eastern Pyrenees and (ii) central Pyrenees and western Alps; and main group 2 included the rest of the Alps and the Jesenik mountains with three subgroups (iii) western southern Alps, (iv) eastern southern Alps and (v) northern Alps and Jesenik mountains. These five genetic lineages are supported by hierarchical variance analysis, which distributes 73% of the total variance among populations for these five groups.

The observed strong genetic differentiation of *E. epiphron* has to be explained by a long phase of isolation. The genetic structure of the Mountain Ringlet shows no congruence with the actual distribution of the species. Hence, the populations from the central Pyrenees and the western Alps, today isolated by several hundreds of kilometres, belong into the same genetic subgroup; the same features apply to the northern Alps and the Jesenik mountains. On the other hand, maximum differentiation has been observed for a transect north–south through the Alps from the Bernese Alps to the Aosta Valley. This structure strongly suggests disjunct glacial distribution patterns of *E. epiphron* with possibly five centres of differentiation: (i) southeast of the Pyrenees representing the source of the populations of the eastern Pyrenees, (ii) the hilly regions between the Pyrenees and western Alps, from where the species withdrew into the central Pyrenees and the western Alps after the melting of the glaciers, (iii, iv) two centres at the Alps, southern margin and (v) a long strip-like distribution north of the Alpine glaciers representing the source of the populations from the northern Alps and the Jesenik mountains (Fig. 17.18).

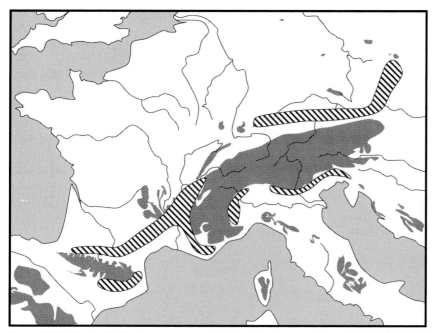

Figure 17.18. Hypothetical distribution patterns of *Erebia epiphron* in western Europe during the last glaciation. Refugia are hatched and grey areas show mountain areas above 1000 m asl. (From Schmitt *et al.* 2006b; courtesy of the European Society for Evolutionary Biology.)

This pattern raises the question why, in contrast to *Z. exulans*, a cold-resistant species like the Mountain Ringlet has not been fairly widespread in the periglacial steppes of Central Europe. A possible reason for this might have been the impact of aridity on resource distribution such as larval hostplants (SBN 1987). Therefore, it is very likely that the periglacial steppes of Central Europe over most of their range have been too dry and hence did not represent suitable biotope for the survival of *E. epiphron*.

PROCLOSSIANA EUNOMIA

The boreo-montane *Proclossiana eunomia* was studied over a major part of its native European distribution by Nève (1996). The allozymes of ten polymorphic loci were studied for 31 native populations. The highest number of alleles was found in Fennoscandia (25), followed by the Bulgarian Stara Planina (21) and the Pyrenees (20). Lower numbers of alleles were found in the Cantabrian mountains (17), the Bohemian forest (17) and the Ardennes (15). The populations from Fennoscandia are distinct from the populations from Central and Southern Europe. In the latter cluster, the Stara Planina is most differentiated, followed by the Cantabrian mountains. The Pyrenees, Ardennes and the Bohemian forest populations showed relatively close relationships.

These data imply postglacial colonization of Fennoscandia from an eastern refugium, maybe representing the Siberian differentiation centre *sensu* de Lattin (1967). Most parts of Western Europe were probably colonized from a refugium located in the vicinity of the Pyrenees, and populations lost genetic diversity due to stepping stone expansion (see Ibrahim *et al.* 1996). The populations of the Pyrenees, the Cantabrian mountains and the Stara Planina only performed altitudinal shifts and no major long-distance colonization because differentiation centres probably existed throughout the last glaciation in the foothills of these mountains. Therefore, this species has characteristics both of the group of mountain species and of the group of continental interior species.

PARNASSIUS APOLLO

Twenty populations of *Parnassius apollo* were studied over the entire distribution area of the species in France (Descimon 1995, Descimon *et al.* 2001). In general, genetic diversity was rather low and only six polymorphic loci could be analysed. However, they revealed a relatively high F_{ST} value (13.5%). A correspondence analysis clearly discriminated the Alpine populations (including the Jura mountains) from the ones of the Pyrenees and the Massif Central on the first axis, representing more than 90% of the variance. Only the second axis distinguished between the Pyrenees and the Massif Central. The genetic diversity of the Massif Central populations was considerably less than those of the Alps and Pyrenees.

Consequently, *P. apollo* populations from the Alps and Pyrenees could have been isolated from each other for a considerable time period, possibly for as long as the time since the Würm glaciation. Massif Central *P. apollo* populations are indicated to derive from Pyrenean populations most probably due to postglacial range fragmentation with subsequent genetic impoverishment of the former.

PARNASSIUS PHOEBUS

Twelve populations of *Parnassius phoebus* scattered over the French Alps have been analysed for 14 allozyme loci (Descimon 1995). In general, the populations have a moderately low genetic diversity within populations. The subspecies *P. phoebus gazeli* from the southernmost region of the Alps showed a strong genetic differentiation from the other populations. It was so strong that this taxon might be considered as a semispecies. However, most of the differentiation was due to genetic impoverishment. All the other populations were only moderately differentiated from one another, but did display an east–west genetic cline. This might be the result of isolation, east and west, of the glaciated French Alps. A scenario envisaged is that owing to postglacial range translocations up-mountain, both lineages met and mixed in the central French Alps. The subspecies *P. phoebus gazeli* might be a young taxon too, possibly evolving in some ice-free, but small, pocket at the southernmost margin of the Alps during the last glaciation.

ARICIA ARTAXERXES

Mitochondrial DNA (cytochrome *b*) and allozymes of nine populations of *Aricia artaxerxes* were analysed in the northern UK and in Scandinavia (Aagaard *et al.* 2002). Three haplotypes were found, each distinguished by one single mutation. Haplotype 5 was the only one found in Scandinavia; haplotype 6 was largely dominant in the northern UK and haplotype 7 was found only twice in one UK sample. Allozymes did not distinguish between the northern UK and Scandinavian populations. This close relatedness between British and Scandinavian *A. artaxerxes* implies a short period of divergence (i.e. the postglacial) and a single source for the postglacial colonization of these two regions.

SUMMARY MOUNTAIN SPECIES

Two different patterns seem to be found in the group of mountain species. (i) Some species were widely distributed throughout the glacial steppes of the Pleistocene, so that large-scale gene flow was only interrupted by the postglacial disjunction into different high mountain systems. No, or only limited, genetic differentiation evolved during separation as in the case of *Zygaena exulans* (Schmitt & Hewitt 2004a). (ii) Other species have probably been subject to disjunct distribution patterns during glacial and interglacial phases, thus accumulating much greater differentiation than species of the first group. Typical examples for this group are *Erebia epiphron* (Schmitt *et al.* 2006b), *Proclossiana eunomia* (Nève 1996) as well as, most probably, *Parnassius apollo* (Descimon 1995) and *Parnassius phoebus* (Descimon 1995). The deglaciation of the high mountain systems and their subsequent colonization has given rise to secondary contacts between different genetic lineages and eventual intermixing and hybridization, as in the case of *Parnassius phoebus*.

The importance of phylogeography for nature conservation

As it has become evident that maintenance of genetic diversity within populations and species is vital for their long time conservation (Frankham *et al.* 2002, Hansson & Westerberg 2002, Reed & Frankham 2003), the question arises as to how the above-mentioned different phylogeographical structures, and thus different distribution patterns of genetic diversity, influence the regional viability and adaptability of species. That they strongly do so is shown in a meta-analysis correlating *Red Data Book* data (van Swaay & Warren 1999) and data of genetic diversity of various butterfly species (Schmitt & Hewitt 2004b). In general, the highest loss rates of populations over the last 25 years have occurred in the countries with the lowest genetic diversities, and greater stability experienced in countries with high genetic diversities of the populations. Due to the remarkable differences between the different phylogeographical groups, several patterns of endangerment were found.

This finding of a strong interaction between phylogeographical structures and the regional development of populations emphasizes that detailed knowledge of the molecular biogeography of species is essential for undertaking successful conservation measures.

ACKNOWLEDGEMENTS

Our thanks, for helpful comments, to Dr Gabriel Nève, Professor Henri Descimon and Dr Tim Shreeve.

18 • Butterfly richness patterns and gradients

DAVID GUTIÉRREZ

SUMMARY

Species richness gradients within Europe vary with latitude, altitude and climate and current richness gradients reflect historical and ecological phenomena. Richness gradients are not just determined by habitat management and successional change. The richest area of Europe is within the latitudes 36° to 42° N, with high vegetation heterogeneity and which has been the least affected by the Quaternary glaciation. The latitudinal range of species is greatest at 55° N, a finding that does not conform to Rapoport's rule. There are competing theories to explain species richness gradients, based on either historical frameworks and glacial–interglacial cycles, with specialised species located within southern parts of Europe and (fewer) generalist species at higher latitudes, or richness gradients being explained by physiology and current climate regimes. In areas where there are elevational gradients with altitude, both tend to peak at mid-altitude. No single factor can explain these patterns. They may represent the product of temperature and moisture gradients, with peaks occurring in favourable conditions, they may represent the area where low- and high-elevation species coincide or the elevational pattern of richness might be influenced by a loss of species at low elevations due to human pressures. Historical and current ecological processes and area and isolation can explain patterns of species richness on island groups, but the relative importance of these differs with the location and stability of ecological processes on the island groups. Contemporary area and isolation dominate islands that have been relatively stable over glacial–interglacial periods (Aegean), whilst species production may only be evident on islands that have been stable over the longest time periods (Canaries and Madeira). The incidence of species on islands in northwest Europe is dominated by latitudinal gradients of species richness on the neighbouring mainland, and island populations may be dependent on current colonisation events.

INTRODUCTION

Ecologists have provided many answers to the question 'What determines the number of species in a given site?' by exploring relationships between species richness and a variety of predictors, such as latitude, climate, sizes and isolation of islands, productivity, habitat heterogeneity and disturbance (Schluter & Ricklefs 1993). The importance of ecological phenomena in determining species richness, however, is a relatively recent perspective, and for nearly a century the study of species richness was assessed from a historical point of view. Only recently have the connections been recognised between ecological and historical phenomena in determining species richness over a variety of spatial scales (Ricklefs & Schluter 1993, Dynesius & Jansson 2000, Jansson 2003).

Like most other taxa, butterfly species richness varies with many environmental gradients on continental mainlands (e.g. Kerr 2001, Kerr et al. 2001, Choi 2004). In a European context, the patterns of butterfly species richness along gradients have been largely discussed, among other factors, in terms of habitat management and successional change (e.g. Erhardt 1985, Erhardt & Thomas 1991, Porter et al. 1992, Väisänen 1992, Feber et al. 1996, Steffan-Dewenter & Tscharntke 1997, Wettsein & Schmid 1999, Balmer & Erhard 2000, Grill & Cleary 2003). In this chapter, I concentrate on two apparently universal, but still intensively debated gradients in species richness: the latitudinal gradient and the elevational gradient (Brown & Lomolino 1998, Willig et al. 2003). For a wide range of organisms, it has been shown that the number of species decreases from the equator to the poles, and from lowlands to high mountains, although the underlying causes of the pattern are still poorly understood.

If we also ask whether the kinds of species vary from one site to another along both environmental gradients, we could look at the gradient in species' range size. It has been shown for a wide variety of taxa that species tend to have larger latitudinal or elevational ranges at, respectively, higher latitudes or elevations. These patterns, called Rapoport's

Ecology of Butterflies in Europe, eds. J. Settele, T. Shreeve, M. Konvička and H. Van Dyck. Published by Cambridge University Press. © Cambridge University Press 2009, pp. 281–295.

latitudinal and elevational rules respectively (Stevens 1989, 1992), are suggested to have important implications for richness through the Rapoport-rescue hypothesis. Basically, both rules suggest that abiotic conditions are more variable in space and time at higher latitudes and elevations; consequently, species inhabiting those habitats would tend to have relatively larger climatic tolerances because they must be capable of tolerating the wider breadth of climatic conditions encountered (Addo-Bediako et al. 2000). Emigrants from higher latitudes or elevations would 'rescue' populations from lower-elevation assemblages from local extinction, and consequently would inflate species richness. Because of the impact and discussion of this hypothesis (Rhode et al. 1993, Rhode & Heap 1996, Gaston et al. 1998), I will present data on latitudinal and elevational ranges along with species richness data.

MacArthur & Wilson (1967) developed their seminal work called the equilibrium theory of island biogeography to explain the richness patterns in insular systems. They modelled mathematically the changing numbers of species with island area and isolation, based on the idea that the number of species on an island is the consequence of a dynamic equilibrium between immigration and extinction. Given the importance of insular systems in general for ecological studies and in particular for European butterflies, they will be considered in this chapter separately from mainland studies.

GENERAL MAINLAND PATTERNS

Latitudinal gradients

SPECIES RICHNESS

Although there are some exceptions, most taxa reach their greatest species richness at low latitudes. These patterns have been examined by counting the number of species in regions of approximately equal size, in natural regions, or in different latitudinal bands (Brown & Lomolino 1998).

The species richness patterns and composition of European butterflies have been thoroughly examined by Dennis et al. (1991, 1995b, 1998b), based on distributional data for 393 species (Higgins & Riley 1983) in 82 'primary areas' or natural regions. After correcting for the size of primary areas, Dennis et al. (1995b) showed that regions in montane southern central Europe had higher species richness than expected. Using equal area cells (220 km × 220 km) and two different sources of data (Tolman & Lewington 1997, Kudrna 2002), Hawkins & Porter (2003) found a similar

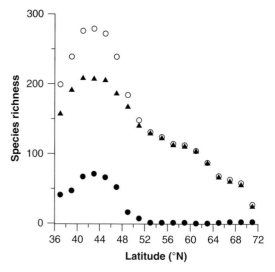

Figure 18.1. Relationship between species richness of European butterflies and latitude (in degrees N). Data extracted from distribution maps in Kudrna (2002) using 2° latitudinal bands. Open circles: total species richness. Solid circles: number of species endemic to the Western Palaearctic following Dennis et al. (1998b). Solid triangles: species richness excluding endemics.

pattern of species richness. More recently, and based on country data from the *Red Data Book* of European butterflies (Van Swaay & Warren 1999), Konvička et al. (2006) have shown that species richness is higher in those states with high habitat heterogeneity, located on the European mainland and that were less affected by Quaternary glaciation (see also Ulrich & Buszko 2003, 2005).

To examine in a simpler manner the global latitudinal trend in species richness across Europe, I counted the number of butterfly species for continental Europe at each 2° band of latitude (from Kudrna 2002). I excluded northern Africa, Turkey and European islands to minimise the effects of sea barriers on the pattern. The pattern is clearly hump-shaped (Fig. 18.1), increasing from 36° to 42° N, and then decreasing up to the northernmost latitudes in Scandinavia.

A substantial proportion (30%) of the European fauna is endemic to the Western Palaearctic (west of the Urals, north of the Caucasus mountains and Turkey: Dennis et al. 1998b). Endemics in Europe are thought to be autochthonous rather than allochthonous though they may vary in age, and except for *Pyrgus andromedae*, their ranges do not extend above 52° N (Dennis et al. 1995b; Fig. 18.1). Because patterns in species richness could depend on net species production (i.e. endemics), I also plotted the number of butterflies in each latitudinal band excluding endemics. Figure 18.1 shows that the pattern of decreasing species richness with latitude is maintained, but is not so steep below 50° N. Latitudinal bands vary considerably in area and this should be taken into account, because low latitudes could show highest richness if they are larger in area (Rhode 1997, 1998, Rosenzweig & Sandlin 1997). To examine in more detail the relative contributions of latitude and area, I performed a stepwise multiple regression of total species richness against both independent variables, and an additional third, maximum elevation, as an indicator of climatic diversity and habitat heterogeneity within the latitudinal band (Kerr et al. 2001, Konvička et al. 2006a). The results show that most variation in richness is accounted for by latitude (86%), with area and maximum elevation explaining only a limited proportion of variance (Table 18.1). In the case of endemics, latitude was also the main predictor of species richness.

RANGE SIZE

As intriguing as the pattern in richness is the latitudinal pattern in species distributions, and particularly, in the extent of their latitudinal ranges. This relationship was given particular attention when observed for the geographical ranges of subspecies of North American mammals by Rapoport (1982), and accordingly was named Rapoport's rule (see Introduction).

I examined to what extent the European butterflies conform to the expectations of Rapoport's rule. To calculate the mean latitudinal range of species at each latitudinal band, I used the midpoint method (Rhode et al. 1993). It basically consists of averaging the latitudinal ranges of all species with their latitudinal midpoint in the same latitudinal band. The results suggest that mean latitudinal ranges clearly peak at mid-latitudes (55° N) and there is no pattern conforming to Rapoport's rule at all (Fig. 18.2).

MECHANISMS

Species richness
There are a great number of hypotheses that attempt to explain the latitudinal gradient in species richness (Rhode 1992, 1999). However, Rhode (1992) suggests that the most plausible hypotheses to explain the latitudinal gradient are those based on time or direct solar effects, including those concerning evolutionary and ecological time, temperature dependence of chemical reactions, solar energy and effective

evolutionary time. For European butterflies, the stepwise regression analysis presented in Table 18.1 shows a strong influence of latitude and a limited effect of maximum elevation on species richness. Here, I shall concentrate on the two attempts to explain the patterns of butterfly species richness in Europe.

Table 18.1 *Summary of the stepwise multiple regression analysis performed to test the effects of latitude (° N), area (km²) and maximum elevation (m) on total and endemic species richness for eighteen 2° latitudinal bands in the European mainland*

Independent variable	Parameter	Cumulative r^2
All species		
Latitude	− 5.324	0.86
Area	0.0001	0.92
Maximum elevation	0.012	0.94
Endemic species		
Latitude	− 1.385	0.64
Maximum elevation	0.010	0.81

Note: Only significant variables in the model are shown. Both regression models: $P < 0.001$.

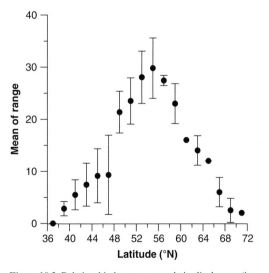

Figure 18.2. Relationship between mean latitudinal range (in degrees) and latitude (North). Mean latitudinal range is calculated from only those species with their midpoint in a given latitudinal band (midpoint method: $r_s = 0.11$, $P = 0.669$). Bars represent SDs. Data extracted from Kudrna (2002).

The thorough study by Dennis et al. (1995b), in which they paid particular attention to the patterns of taxonomic bias in species richness and endemicity across different European regions, suggests that evolutionary and ecological time hypotheses could partly explain the species richness patterns of European butterflies. The evolutionary time hypothesis claims that species within certain sites have not yet had time to evolve, whereas the ecological time hypothesis suggests that species that would be able to survive in certain sites exist, but they have not yet had time to disperse into them (Brown & Lomolino 1998).

In European butterflies, endemics are restricted to latitudes below the 52° N latitudinal band (Fig. 18.1), coinciding with the southern limits of the ice cover extending from Scandinavia during the last glacial maximum, 18 000 years BP. Although the lack of fossil evidence in European butterflies makes any evolutionary hypothesis speculative, there are at least three lines of evidence that suggest some potential evolutionary change during glacial–interglacial cycles (Dennis et al. 1991, 1995b).

Firstly, habitats for butterflies in southern Europe have been available for longer than in northern latitudes. Summer and winter temperatures during glacial periods in northern Europe probably were beyond the tolerance of adult and diapausing stages of most species, giving rise to periodic mass extinctions over this area, which should be subsequently recolonised during the interglacial periods (Dennis 1993). This point is supported by the fact that there is lower genetic diversity in butterfly populations from areas supposedly colonised after glaciation (Schmitt & Hewitt 2004b).

Secondly, populations currently isolated in southern Europe were probably more isolated during the glacial phases than now. If species distributions contracted southwards during glacial periods, population isolation should have been more effective because the European Mediterranean basin consists of three separate blocks of land (Iberian peninsula, Italian peninsula and the Balkans) bounded by physical barriers (Pyrenees and Alps to the north and the Mediterranean sea to the south: Dennis et al. 1991).

Thirdly, endemicity occurs more frequently in families containing a substantial proportion of poorly mobile habitat specialists, supporting the idea that potential evolutionary change could depend on dispersal ability. Overall, the evidence suggests that southern Europe acted as a generator of species during the glacial–interglacial cycles, whereas northern Europe's fauna has depended on successful migration from glacial refuges, and therefore, the current European butterfly fauna could partly be the result of non-equilibrium processes.

The second attempt to explain the gradient of butterfly species richness in Europe has been carried out by Hawkins & Porter (2003). Using two sources of data (Tolman & Lewington 1997, Kudrna 2002) they examined the potential influence of 11 environmental variables on the species richness of European butterflies (Table 18.2). A measure of water-energy balance, actual evapotranspiration, accounted for 72–79% of variance in species richness, depending on the data source. The effects of this variable can operate in two ways, directly via the physiological effects of heat/cold stress and water availability, and indirectly as trophic effects via its influence on productivity (Hawkins & Porter 2003). Contrastingly, in their analyses, the pattern of glacial retreat during the last 20 000 years contributed little to the richness gradient (Table 18.2). These results suggest that the richness gradient is in equilibrium with the current climate in Europe, as found for tree species in the same area (Adams & Woodward 1989).

Though the analysis by Hawkins & Porter (2003) is in some way more refined than that by Dennis et al. (1995b) because it explores a wider range of hypotheses, it may not be the definitive explanation for the butterfly richness gradient in Europe. Probably, many of the potential explanatory variables are collinear, hampering the detection of key environmental factors underlying the diversity–environment relationships identified by regression analyses. Alternative statistical methods (such as variation and hierarchical partitioning, and structural equation modelling) could provide deeper insights into the relationship of butterfly richness with environmental variables (Heikkinen et al. 2004, Menéndez et al. 2007). Alternatively, molecular phylogenies with estimated dates for diversification events could be an important source of information for testing hypotheses concerning the significance of historical events (Gaston 2000).

Range size

As stated above, Rapoport's rule is one of the most currently debated explanations for gradients in species richness (Gaston et al. 1998). In European butterflies, the rule is not supported because latitudinal ranges for high latitudes are fairly similar to those of low-latitude species. Nevertheless it is worth taking into account that some

Table 18.2 *Significant (P < 0.05) variables from the stepwise multiple regression models of butterfly species richness performed by Hawkins & Porter (2003); data are from range maps from Tolman & Lewington (1997) and Kudrna (2002)*

Variable	r^2
Species richness from Tolman & Lewington (1997)	
Actual evapotranspiration	0.788
Log range in elevation	0.025
Annual precipitation	0.010
Summer Global Vegetation Index – measure of summer greenness	0.016
Annual Global Vegetation Index – measure of annual greenness	0.008
Area age – years before present when the area was exposed for colonisation	0.020
January temperature – mean daily temperature in the coldest month	0.005
(Potential evapotranspiration)^2	0.040
Annual mean temperature	0.005
Species richness from Kudrna (2002)	
Actual evapotranspiration	0.721
Plant species richness	0.037
January temperature – mean daily temperature in the coldest month	0.030
Annual precipitation	0.011
Log range in elevation	0.008
Summer Global Vegetation Index – measure of summer greenness	0.017

southern species have large ranges in Africa that are not included in Kudrna's (2002) atlas, which would make less likely Rapoport's effect. On the other hand, the wider latitudinal ranges of intermediate-latitude species could reflect a geometrical constraint (mid-domain effect: Colwell & Lees 2000, Colwell *et al.* 2004), rather than a biological phenomenon. Because latitudinal extents are estimated within a bounded area (whole Europe) and many species abut the land ends (Mediterranean and North Sea coastlines), species with wide latitudinal ranges can only have midpoints close to the latitudinal centre of Europe (Colwell & Hurtt 1994, Colwell & Lees 2000). Consequently, available evidence suggests that there is no support to Rapoport's rule from European butterflies and that it cannot explain the latitudinal gradient in species richness.

Elevational gradients

SPECIES RICHNESS

The elevational gradient of species richness has often been claimed to mirror the latitudinal gradient, in that species richness declines with increasing elevation. This apparently intuitive pattern is not so generalised as previously thought. Rahbek (1995) thoroughly reviewed the relevant literature and found that, for several taxa and biomes, 66 (73%) out of 90 data sets showed either a hump-shaped pattern, or a horizontal species richness curve up to certain elevation before declining. Only 21% of studies fitted a monotonic decrease of species richness with elevation. Still, all these data suggest that species richness peaks at some point along the elevational gradient.

Apparently, the elevational gradient of species richness has not been considered by many studies on European butterflies. There are some detailed surveys concerning the distribution of butterfly assemblages across relatively wide elevational gradients carried out in Austria (Ryrholm & Huemer 1995, Hauser 1996), the Czech Republic (Beneš *et al.* 2000), Italy (Balletto *et al.* 1977, 1982, 1988), Spain (Baz 1987, Molina 1988, Viejo & Martín 1988, Sánchez-Rodríguez & Baz 1995, Gutiérrez 1997, Stefanescu *et al.* 2004) and Switzerland (Gonseth 1993, 1994a, b). To discuss the elevational

Table 18.3 *Published studies on European butterfly species richness and composition along an elevational gradient. All studies attempted to standardise sample size and to span at least one complete peak flight season of butterflies*

Region	Country	Geographical location	Area (km^2)	Elevational range (m)	No. sampling sites	No. species	No. individuals	Reference
Sierra Norte de Sevilla[a]	Spain	38° 00′ N, 5° 50′ W	248	120–800	6	61	na	1
Sierra de Javalambre	Spain	40° 05′ N, 1° 00′ W	22	1100–2000	10	101	2120	2
Sierra de Gredos	Spain	40° 17′ N, 5° 10′ W	867	400–1900	10	92	2221	3
Sistema Ibérico	Spain	40° 30′ N, 2° 00′ W	4536	1060–1630	20	89	736	4
Picos de Europa	Spain	43° 15′ N, 5° 00′ W	169	400–2150	17	70	2517	5
Dolomites	Italy	46° 00′ N, 11° 00′ E	na	940–2250	15	56	3137	6
Sengsengebirge[a]	Austria	47° 00′ N, 10° 00′ E	na	730–1905	10	47	na	7
Jura of Neuchatel (woodland clearings)[b]	Switzerland	47° 00′ N, 7° 00′ E	na	640–1470	22	74	2104	8

[a] Presence–absence data; [b] abundance rank data; na, data not available.
References: (1) Molina (1988); (2); Sánchez-Rodríguez & Baz (1995); (3) Viejo & Martín (1988); (4) Baz (1987); (5) Gutiérrez (1997); (6) Balletto *et al.* (1988); (7) Hauser (1996); (8) Gonseth (1993).

patterns on a relatively homogeneous set of studies, I reanalysed the raw data, selecting studies that attempted some standardisation of the data (catch-effort and transect method) and were based on sampling at local sites (i.e. point sampling) (Table 18.3). Therefore, species richness is comparable between sampling sites within a region, but not between different regions. The selected studies also attempted to span the whole or at least the core flying season of butterflies at each sampling site. This is an important point because the sampling regime is a potential factor determining the pattern of insect species richness along elevational transects (Wolda 1987, McCoy 1990). As the butterfly flying season may be longer at low elevations (e.g. Gutiérrez & Menéndez 1998), sampling over a short time period would tend to underestimate the richness of low elevation assemblages relative to high elevation ones.

As shown in Fig. 18.3 and Table 18.4, species richness significantly decreases with increasing elevation in five regions, but, in all these cases and the Sierra de Gredos, there was a trend of species richness peaking at mid-elevations. In the Sierra Norte, species richness increases with elevation and peaks at the highest points sampled across the gradient (c. 600–800 m). It is worth noting that this is the most southern and lowest altitude region

sampled. To examine this relationship more thoroughly, I fitted the species richness data to a second-order polynomial of elevation (Li *et al.* 2003). Four of the second-order polynomial regressions were globally significant, but the regression coefficients of elevation and (elevation)^2 were non-significant. There is no butterfly species endemic to any of the regions included in Table 18.3.

Limiting the analysis to species richness does not take into account which species occur in each site, and whether those species show an apparent pattern of variation of abundance along the gradient. One possible way to examine species assemblages along ecological gradients is by multivariate ordination techniques, such as detrended correspondence analysis (DCA: Gauch 1982, ter Braak & Prentice 1988). Resulting ordination axes can be tested by correlation against quantifiable environmental factors, in the present case, elevation. In six out of eight studies, there was a significant correlation between DCA axis 1 scores of sites and elevation (Table 18.4). In the case of the Sierra Norte, no relationship was found between elevation and axis 1, but was found between elevation and axis 2, suggesting that there is a more influential variable determining species' distributions (Table 18.4). The pattern is illustrated for the Sierra de Javalambre data set out in Fig. 18.4.

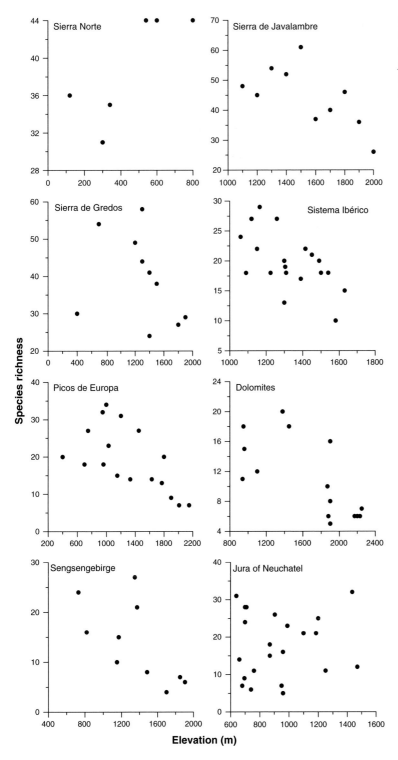

Figure 18.3. Relationship between species richness and elevation in eight European mountainous regions. Note that both *x*- and *y*-axes are differently scaled. Data from the studies included in Table 18.3. Regression statistics in Table 18.4.

Table 18.4 *Coefficients of determination (r^2) and regression of the first- and second-order least squares polynomial regressions of species richness of butterflies against elevation (in m). Spearman's rank correlation coefficients (r_s) between elevation and the first axis scores from detrended correspondence analyses (DCAs) on species composition are also included*

| Site | First order | | Second order | | | Richness peak (m) | DCA axis 1 |
	r^2	elevation	r^2	elevation	elevation^2		
Sierra Norte de Sevilla	0.64[a]	0.019[a]	0.64	0.012	0.000006	Plateau	0.60[e]
Sierra de Javalambre	0.43[b]	−0.022[b]	0.64[b]	0.168	−0.00006	1500	−0.99[d]
Sierra de Gredos	0.10	−0.008	0.44	0.062	−0.00003	1300	−0.88[c]
Sistema Ibérico	0.34[c]	−0.016[c]	0.34[b]	0.012	−0.00001	1165	0.46[b]
Picos de Europa	0.38[c]	−0.010[c]	0.54[c]	0.026	−0.00001[b]	1000	0.93[d]
Dolomites	0.48[c]	−0.007[c]	0.56[c]	0.021	−0.000009	1380	0.69[c]
Sengsengebirge	0.42[b]	−0.013[b]	0.44	0.012	−0.000009	1350	0.56[a]
Jura of Neuchatel (woodland clearings)	0.01	0.003	0.036	−0.045	0.00002	no pattern	−0.77[d]

[a] $P < 0.1$;   [b] $P < 0.05$;   [c] $P < 0.01$;   [d] $P < 0.001$;   [e] relationship between elevation and DCA axis 2: $r_s = 0.83$[b].

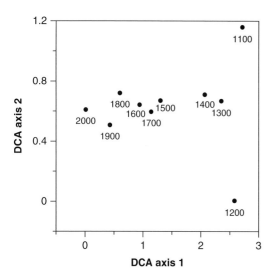

Figure 18.4. Detrended correspondence analysis (DCA) ordination of the butterfly fauna of the Sierra de Javalambre. Figures next to dots are the elevation of sites in metres. Axis 1 is ecologically meaningful, expressing the elevational gradient (see Table 18.4). Axis scales are in units of average standard deviations of species turnover (Gauch 1982). Eigenvalues (and percentage variance of species composition) of axis 1 and 2 were 0.455 (34.7%) and 0.071 (5.4%), respectively. From data in Sánchez-Rodríguez & Baz (1995).

RANGE SIZE

In an extension of Rapoport's latitudinal rule, Stevens (1992) hypothesised that an increase in the elevational range of species with increasing elevation is also expected (see Introduction). We should then expect a net gradual variation in species' range sizes (measured as elevational range) across an elevational gradient. The pattern has been supported by butterfly assemblages of the Great Basin in North America (Fleishman *et al.* 1998), but, as far as I know, Rapoport's rule has not been explicitly tested for European butterfly assemblages (but see Baz, 1987, Sánchez-Rodríguez & Baz 1995).

As with latitude, I used the midpoint method to calculate mean elevational ranges (Rhode *et al.* 1993). Because the data sets included in Table 18.3 are based on point sampling and not on elevational bands, I subdivided the elevational range sampled in each region into as many elevational bands as the number of sampling sites in that region (Table 18.5). Elevational band width was consequently different between regions and not all bands contained at least one species (Table 18.5). There was no significant correlation between mean elevational range and elevation (Table 18.5). Instead, mean elevational ranges peaked at mid-elevations, as illustrated for the Sierra de Javalambre (Fig. 18.5).

Table 18.5 *Spearman rank correlation coefficients between mean elevational range of species at a given site and elevation for the eight regions included in Table 18.3. Mean elevational range (in m) is calculated from only those species with their midpoint in a particular elevational band. Elevational band width differs among regions and was a function of the number of sampling sites and elevational range examined*

Region	Number of elevational bands	r_s
Sierra Norte de Sevilla	5	−0.30
Sierra de Javalambre	9	0.50
Sierra de Gredos	10	−0.15
Sistema Ibérico	16	0.18
Picos de Europa	15	0.05
Dolomites	12	0.16
Sengsengebirge	9	0.13
Jura of Neuchatel (woodland clearings)	13	0.17

No correlation was significant at $P < 0.05$.

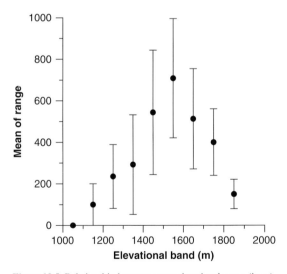

Figure 18.5. Relationship between mean elevational range (in m) and elevation for the butterflies of the Sierra de Javalambre. The elevational gradient was subdivided into equal-width elevational bands (100 m), and for each band, means and SDs were calculated for those species with their midpoint in that band. Bars represent SDs. Spearman rank correlation statistics in Table 18.5. From data in Sánchez-Rodríguez & Baz (1995).

MECHANISMS

Species richness

The clearest pattern along elevational gradients is the marked species turnover shown by the multivariate analyses (DCA). Elevation is closely correlated with climate, particularly temperature, and that variable is often correlated also with precipitation. There are many direct and indirect ways in which climate, and particularly temperature, could determine the abundance of a particular butterfly species (summarised in Dennis 1993, Boggs & Murphy 1997). Nevertheless, as far as I know, there is no detailed study concerning the elevational distributions of particular butterfly species in relation to those of their host plants and other potential factors (but see Randall 1982 and Alonso 1999 for excellent cases of moth species).

However, the fact that climate has a significant impact on butterfly distributions through a wide variety of mechanisms does not explain why species richness along an elevational gradient shows a particular pattern or peaks at a specific elevation. Butterfly species composition could be different at different elevations and yet not show any pattern in species richness, only reflecting the turnover due to the different ecological requirements or physiological restrictions of species. In six of eight mountain ranges detailed in Table 18.4 species richness peaks at mid-elevations ranging from 800 to 1500 m, and in no case at the lowest elevation sampled. Does it mean that there are some 'intermediate' conditions that favour more butterfly species and consequently increase richness? On Mediterranean mountains (Sierra de Javalambre, Sierra de Gredos and Sistema Ibérico), where there is a marked drought period which is a potentially limiting factor, the pattern could actually reflect two different environmental gradients. Temperature decreases with increasing elevation, whereas moisture availability increases. Both gradients have important influences on butterfly biology as well as on host plant growth and condition. Thus, the physical conditions would be 'more favourable' for butterflies at the intermediate elevations and could explain the peak in species richness (Stefanescu *et al.* 2004). However, the peak in species richness also occurs at intermediate elevations outside the Mediterranean area, as in the Picos de Europa, the Dolomites and the mountain range Sengsengebirge, in which there is no apparently marked summer drought period (Tables 18.3 and 18.4). Some authors have attributed the lower species richness at low elevations to increased human pressure on habitats

(McCoy 1990, Sánchez-Rodríguez & Baz 1995). However, the available data presented here do not include the details to test this hypothesis.

Range size

As with latitude, I found the typical hump-shaped pattern illustrated in Fig. 18.5, and no Rapoport's effect for any of the regions. Given that no region showed an apparently monotonic decrease in species richness with elevation either, European butterfly assemblages do not follow Rapoport's elevational rule.

THE PARTICULAR CASE OF ISLANDS

A summary of island biogeography theory

The island biogeography theory by MacArthur & Wilson (1967) predicts that the number of species increases with area and decreases with isolation, and there is a continual temporal turnover of species on an island. Immigration rates depend on distance from the species source, and extinction rates are inversely related to population sizes, which are in turn directly related to island area. Other and probably complementary explanations to species richness patterns in insular systems, particularly with respect to area effects, include the habitat-diversity hypothesis and the sampling hypothesis. The habitat-diversity hypothesis emphasises the role of habitat diversity in determining species richness, proposing that the amount of different habitats increases with increasing area. The sampling hypothesis suggests that passive sampling from the species pool determines that larger areas receive more species from the source simply because they are larger targets for immigrants, generating a positive species–area relationship regardless of habitat differences and population processes (Connor & McCoy 1979).

Richness patterns in insular systems

In a European context, insular systems are particularly important in the North, the Baltic, the Irish, the Adriatic and the Aegean seas. There is a dearth of studies concerning the patterns of species richness in European islands, which probably results from the limited availability of complete data on the butterfly faunas on individual islands. However, a few surveys concerning butterfly faunas on

European islands are already available, providing an excellent opportunity to examine the patterns of species richness within different insular systems showing markedly contrasting characteristics, such as origin, degree of isolation, size and geographical position.

This section of the chapter will focus on discussing the relative importance of ecological and evolutionary factors in determining species richness, as well as the relative roles of geographical attributes of the islands (size, isolation, geographical position and extent). Species endemic to one or more islands are evidence of evolutionary events within an archipelago and they are expected to occur more frequently in more isolated and large islands and archipelagos, circumstances that favour intra- or inter-archipelago speciation (Adler & Dudley 1994, Adler et al. 1995, Losos 1996). The archipelagos examined are the British and Irish offshore islands (Dennis & Shreeve 1996, 1997), the Aegean archipelago (Dennis et al. 2000a, 2001) and the Canary Islands and Madeira (Jones et al. 1987, Fernández-Palacios & Andersson 1993, Wiemers 1995, Báez 1998). Because the number of archipelagos considered here is small, the focus here is to provide some guidelines to explain European island biogeography, which hopefully will soon be tested with more empirical evidence on Palaearctic island butterfly faunas.

British and Irish mainland and offshore islands are continental in origin. It is thought that the extreme temperatures and the ice cover in Britain during the last maximum glaciation (c. 20 000 years BP) made most of the land unsuitable for all current butterfly species. The current fauna appears to have begun to enter since the late glacial (c. 10 000 years BP: Dennis & Shreeve 1996). Thus, connections of islands with the British and Irish mainland could be of relevance for colonisation during the Holocene. The British mainland had a long postglacial (from c. 12 000 years BP) connection with the European continent (to c. 7800 years BP), but this period was much shorter for the connection of Ireland with Britain (to c. 9500 years BP). Many British and Irish islands were connected to their corresponding mainland in postglacial times, but some of the furthest ones, such as the Outer Hebrides, apparently have had no land bridge during that period and must have been colonised by sea crossing (Dennis 1977, 1992b).

The patterns in species richness in the British Isles have been assessed on several occasions, with some disparity between the results obtained by the different surveys (Hockin 1981, Reed 1985, Dennis 1992b, Dennis & Shreeve 1996, 1997). I will concentrate here on the most complete and recent

Table 18.6 *Ecological factors and their corresponding variables included to assess species richness on British and Irish offshore islands, the Aegean archipelago, and the Canary Islands and Madeira. The mean and range of values are included in brackets*

Ecological factor	British islands	Aegean archipelago	Canary Islands and Madeira
Island size	Island area (162; 0.06–2051 km^2)	Island area (3535; 9–92545 km^2)	Island area (1030; 278–2058 km^2)
		Island maximum elevation (862; 150–2456 m)	Island maximum elevation (1790; 609–3718 m)
Isolation	Distance from the nearest mainland of France, Britain or Ireland (29; 0.1–230 km)	Distance from the nearest mainland of Greece or Turkey (42; 1.6–161 km)	Distance from the nearest mainland of Africa (294; 90–630 km)
	Distance from the nearest source of equivalent size (7; 0.1–50 km)	Distance from nearest source of equivalent size (18; 1.6–96 km)	Distance from the nearest source (85; 11–420 km)
		Distance from the nearest point of the Turkish mainland (102; 1.8–355 km)	
		Distance from the nearest point on the Greek mainland (181; 1.6–790 km)	
		Isolation based on distances to all the other islands in the archipelago weighted by the area of islands	
Size of faunal source	Species richness within 50 km of the nearest mainland source (28; 13–67)		
	Species richness at the nearest equivalent area (23; 5–67)		
Geographical position	Island latitude (54° 27′ N; 49° 09′ N – 60° 29′ N)	Island latitude (37° 22′ N; 35° 00′ N – 40° 39′ N)	Island latitude (28° 52′ N; 27° 45′ N – 32° 48′ N)
	Latitude of the mainland source (54° 26′ N; 49° 23′ N – 58° 40′ N)	Island longitude (26° 00′ E.; 22° 59′ E – 33° 00′ E)	Island longitude (16° 13′ W; 13° 36′ W – 18° 01′ W)

From data in Jones *et al.* (1997), Fernández-Palacios & Andersson (1993), Dennis & Shreeve (1996, 1997), Báez (1998), and Dennis *et al.* (2000a, 2001).

study by Dennis & Shreeve (1996, 1997; see also Dennis *et al.* 1998b, 2000b). They used a subset of 73 islands that are relatively well surveyed from a set of 219 islands containing at least one butterfly record (Dennis & Shreeve 1996). In total, they included 57 species in the analyses. Endemic components are completely lacking from the British butterfly fauna.

Dennis & Shreeve (1996, 1997) identified four important ecological factors determining the patterns of species richness on islands: area, isolation, size of faunal source and geographical position (Table 18.6). Using simple correlations, they showed that the size of faunal source, measured as the number of species on nearby islands and at proximate areas on adjacent mainland, was the factor accounting for most variation in species richness (Fig. 18.6). Among the measures of area and isolation, only distance from the nearest mainland source was significantly negatively correlated with species richness. Because some independent variables were cross-correlated, Dennis & Shreeve (1996, 1997) performed several stepwise

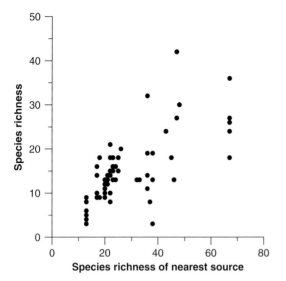

Figure 18.6. Relationship between species richness on British and Irish islands and species richness of the nearest mainland source. Data are for 73 sufficiently well-surveyed islands from Dennis & Shreeve (1996, 1997). The relationship is highly significant after log-transformation of data ($r^2 = 0.41$, $P < 0.001$).

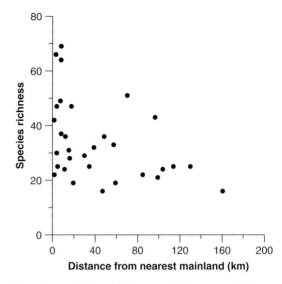

Figure 18.7. Relationship between species richness on the Aegean archipelago and distance from the nearest mainland of Greece or Turkey. Data are for 31 sufficiently well-surveyed islands from Dennis et al. (2000a, 2001). The relationship is highly significant and shows a decline in species richness as islands are further from mainland sources ($r^2 = 0.45$, $P < 0.001$, after log-transformation of data).

regression analyses in which the steps were predetermined. Their results showed that after controlling for the size of faunal source, area accounted for more variation in species richness than isolation.

A study by Dennis et al. (2000a, 2001) has examined the patterns of species richness for the butterflies within the Aegean archipelago, including Megisti and Cyprus islands. Most islands in the archipelago have been in contact with the surrounding landmasses and with nearby islands at different times and for different periods. Only the southern arch of the Cyclades has been formed by recent volcanic activity, and some of those islands have remained isolated from the nearby landmasses since their formation (Dennis et al. 2000a).

Dennis et al. (2000a, 2001) selected a subset of 31 well-surveyed islands from a larger data file of 64 islands. The study deals with 127 butterfly species, of which eight are endemic to the Aegean archipelago (6%). Selected ecological factors which potentially influence species richness patterns were island size, degree of isolation and geographical position (Table 18.6). For the Aegean islands, species richness closely correlates with distance from the nearest potential source on the mainland of Greece or Turkey and island area, together accounting for 80% of variation

(Fig. 18.7). Endemic species are restricted to large isolated islands in the south part of the archipelago (e.g. Kriti), indicating that autochthonous evolutionary events have been limited.

The Canary Islands and Madeira are located in the eastern part of the Atlantic Ocean, at approximately 90 km from the nearest point on the Saharan coast in Africa. They are all of volcanic origin, and are well separated from the African mainland and from each other, indicating that there has been no land connection in geological times. Only Lanzarote and Fuerteventura are separated by shallower waters, and probably formed a larger island in the past. Madeira is the oldest island (30 million years BP), and El Hierro the youngest (1 million years BP) (Fernández-Palacios & Andersson 1993, Brunton & Hurst 1998, Juan et al. 2000).

The butterfly fauna of the Canary Islands and Madeira consists of 34 species, of which 11 (32%) are endemic to one or more islands (Jones et al. 1987, Wiemers 1995, Báez 1998, Brunton & Hurst 1998). I performed an assessment of species richness of the Canary Islands and Madeira using the same geographical variables as for the Aegean archipelago (Table 18.6). There was a closely positive relationship

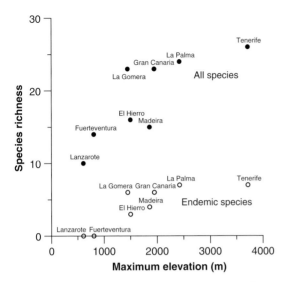

Figure 18.8. Relationship between species richness on the Canary Islands and Madeira and maximum elevation of the island for all species (solid circles) and for endemics to the archipelago (open circles). The plot shows that species richness for both groups increases as maximum elevation of the island increases. Data are from Jones *et al.* (1987) and Báez (1998). The relationships are significant in a stepwise multiple regression analysis (all species: $r^2 = 0.62$, $P = 0.016$; endemics: $r^2 = 0.67$, $P = 0.013$), and no other geographical variables included in Table 18.6 accounted for residual variance in any of both groups of species.

between species richness and maximum elevation of the island (Fig. 18.8), but no other geographical variables entered in the analysis. Surprisingly, island area appears to have no significant effect on species richness, as large islands such as Fuerteventura and Lanzarote are the poorest in number of species. The number of endemic butterflies follows a similar pattern, related closely to maximum elevation of the island (Fig. 18.8). Finally, the number of non-endemic species is also positively related to elevation, but with more scatter in the data ($r^2 = 0.51$, $P = 0.048$).

MECHANISMS

The patterns of species richness for three Palaearctic archipelagos suggest that they are determined by strongly differing geographical variables. Only the species richness of butterflies of the Aegean archipelago appears to be dominated by the influences of area and isolation, as expected by island biogeography theory. Different

contributions of geographical variables, however, do not reflect contradictory results, but probably differences in the range of values of the geographical variables of each archipelago. Also, the contribution of evolutionary processes to species richness is likely to differ strongly between archipelagos, as a result of differences in spatial and temporal isolations experienced by the insular systems.

The British Isles are located along a relatively wide latitudinal gradient in northern Europe (>10°) (Table 18.6), where the rate of decrease in species richness towards the north is particularly strong on the European mainland (see Fig. 18.1). In particular, it has been shown that butterfly richness declines with latitude along the British mainland as a result of a gradient of summer sunshine and temperature (Dennis & Williams 1986, Turner 1986, Turner *et al.* 1987, Dennis 1992b, 1993). For most islands, isolation lies within the usual distances moved by British butterflies, and island areas are large relative to habitat patches sustaining populations on the British mainland. Nevertheless, at least for a small set of well-surveyed islands, it has been shown that short-term maintenance of island populations depends in part on immigration from mainland sources (Dennis 2000). Islands at high latitudes have a smaller potential number of colonist species from the mainland than islands off the southern coast of Britain, suggesting that species richness on British islands is the result of the sharp diversity gradient at the faunal source. On the other hand, the British mainland and islands have been exposed to strong glacial–interglacial cycles probably driving populations to frequent extinction. This phenomenon along with the low isolation of the British offshore islands probably prevents the generation of endemic species.

The Aegean archipelago is located along a narrower latitudinal gradient than the British islands (c. 5°), but has a much larger range of island areas (Table 18.6). The range of distances from the nearest mainland source is roughly similar to the British islands, but the distances tend to be longer on average: 68% of islands are further than 10 km from the nearest mainland in the Aegean archipelago vs. 51% in the British islands. Dennis *et al.* (2000a) suggested that contemporary geography is the dominant influence on the butterfly fauna of the Aegean archipelago. Apart from the reported direct effects of area and isolation on richness, their hypothesis is supported by additional evidence concerning individual species distributions not reported here (details in Dennis *et al.* 2000a).

The Canary Islands and Madeira is the southernmost island system of the three presented here. This archipelago is located along a 5° latitudinal gradient, and is more isolated from the nearest mainland source than the British and the Aegean islands. Three lines of evidence support the hypothesis that evolutionary changes appear to have had an important role on the present butterfly fauna. Firstly, no effects of area or isolation on species richness were found, but species richness is strongly correlated with maximum elevation. Secondly, there is a substantial proportion of endemic species on the archipelago (32%). Notably, many of the non-endemic species are long-distance migrants originally from remote regions (e.g., *Cynthia virginiensis* and *Danaus plexippus* from North America: Báez 1998); this migration propensity has probably maintained sufficient gene flow between mainland and island populations to prevent subsequent speciation. And thirdly, the number of endemic butterflies is closely related to elevation and matches the variation of total species richness (Fig. 18.8).

The extensive area of sea surrounding the archipelago was (and still is) probably a strong barrier for most species coming from the nearest potential sources, limiting the total number of species within the island system. Following colonisation, differentiation from ancestral colonists was the dominant mechanism in at least some of the genera (the less mobile ones) of Canarian butterflies (Brunton & Hurst 1998). However, although speciation has been a noteworthy process in the Canary Islands, colonisation and extinction still should have played a determinant role in establishing the species richness patterns. Each individual island has no more than one endemic species of the same genus, and if there is more than one endemic species from the same genus, they consistently occupy different islands. Thus, there has been substantial net species production within the archipelago, but it has been subject to successful between-island colonisation and within-island population persistence. This is supported by the positive relationship between elevation of the island and number of endemic species: wider elevational gradients can provide higher heterogeneity of habitats and climate, which could promote greater population lifetime and stability (Adler & Dudley 1994, Adler *et al.* 1995).

CONCLUDING REMARKS

In this chapter, attention has focused on the latitudinal and elevational gradients in species richness, and island biogeography theory. Some support has been demonstrated for the latitudinal and elevational gradients, but evidence shows that hump-shaped patterns are more the rule than the exception in butterflies rather than monotonic patterns that have been generally accepted for other taxa. Nevertheless, the mechanisms involved are likely to be different for latitude and elevation. Evolutionary processes have apparently had some influence on the latitudinal pattern of European butterfly species richness, whereas their effects appear to be negligible on the regional elevational gradients. This may be partly due to the greatly different spatial scales at which each pattern is examined. One potential aspect of importance in geographical gradients is the mid-domain effect or geographical constraint (Colwell & Lees 2000), which provides the basis for generating a null model of gradients bounded by geographical barriers. Its relative contribution in European butterfly species richness still remains unanswered. Another important issue is that no support has been found for Rapoport's rule, even though the European mainland is located in the northern hemisphere at the latitudes at which this pattern has been frequently shown. All these patterns occur at regional and geographical scales in which there are practical constraints on conducting experiments for explicitly testing hypotheses. However, more refined studies of the factors determining the warm and cold limits to species ranges could shed light on the mechanisms involved in generating butterfly richness gradients. Finally, it is worth mentioning here that Europe is a small part of a larger landmass (the Palaearctic region), which has been practically omitted in all studies of latitudinal richness patterns. For instance, it remains to be examined whether Europe as a whole shows the peninsular effect (i.e. decrease of species richness towards the tips of peninsulas), as found for smaller areas in Europe and elsewhere (e.g. Martín & Gurrea 1990, Choi 2004).

The effects of area and isolation expected from island biogeography theory are partly supported by European island butterfly assemblages. Species richness in more isolated systems could be dominated by evolutionary processes, whereas in systems close enough to the mainland to allow frequent immigration events species richness could be dominated by the effects of environmental gradients. This hypothesis, however, should be evaluated in a greater number of European archipelagos taking into account the geographical location of the system, particularly latitude. A major concern in island surveys is the lack of knowledge about basic habitat availability of species to control for the habitat diversity hypothesis. Data on island floras could be a feasible approach to partly overcome this concern, and could be a fundamental topic to be included in future work.

ACKNOWLEDGEMENTS

The first draft of this chapter was written while holding a contract funded by the Spanish Ministry of Science and Education. I sincerely thank Emilio Balletto, Roger Dennis, Martin Konvička, Tomáš Kuras and Marcos Méndez for kindly providing some references. Markus Kiefer and Rosa Viejo helped with the translation of the references in German. Rosa Viejo, Rob Wilson, Martin Konvička, Dirk Maes and Chris Van Swaay provided interesting comments on earlier drafts of the manuscript.

19 • Ecological genetics and evolutionary ecology in hybrid zones

ADAM H. PORTER

SUMMARY

Natural hybridization offers opportunities to study evolutionary pressures acting on the morphological, behavioral, physiological and life-history differences that hybridizing taxa often exhibit. Depending on the nature of the contact regions between hybridizing species, very powerful insights can be obtained using rather straightforward sampling schemes in natural populations. Lepidopterists have been slow to adopt established analytical methods because they are often presented in a mathematically challenging framework. The methods are extremely powerful, and not terribly difficult to use once the central concepts are apparent. The central ideas of hybrid zone analysis are presented here in a largely graphical framework, providing an accessible introduction to the topic. The methods should belong in every field biologist's toolbox, as considerable advances can be made in understanding European butterfly biology in particular and evolutionary biology in general, when these methods are applied. The strategy for ecological geneticists working with hybrid zones is to use microevolutionary models as an analytical framework and fit data to these models to estimate microevolutionary parameters. The range of situations that lend themselves to this approach is laid out and an overview of possible sources of correlations among traits used to explain how the ecological setting determines analytical tractability and how narrow hybrid zones work. Single-locus clines under selection and correlations among loci in hybrid zones are explored. Instructions are given for how to work with data, how to apply hypothesis testing in clines, and in particular how to devise an adequate sampling pattern.

While cline analysis is a very powerful tool, it is not a panacea. Measures of dispersal rates and selection coefficients provided by cline analysis represent a suite of hypotheses about evolutionary processes. In running these analyses, assumptions are made that the underlying models to which the data are fitted are in fact appropriate. Some of the analyses involve model selection, so this concern is somewhat ameliorated. To be sure, common-garden experiments are needed, as well as transferring individuals of known genotype across the cline and following their fates. Finally,
habitat selection has to be ruled out as a contributor to the clinal patterns – it is not built into the models, which assume random dispersal. Cline shape is sensitive to selection coefficients small enough to be missed using field experiments alone but that are strong enough to rapidly influence evolution in the traits they affect.

There are numerous cases where subspecies or closely related species meet and hybridize. These taxa differ in traits ranging from minor elements in their wing patterns through to major differences in host use and the timing of their life cycles. Tools are available to analyze the selection and dispersal processes that act upon these traits. With better understanding of these microevolutionary forces, we will not only have a much better understanding of the roles of ecology and genetics in shaping the traits that define these taxa, but be in a much better position to reach informed conclusions about their taxonomic relationships.

INTRODUCTION

Hybridization continues to retain its place of primary and longest-enduring interest to lepidopterists. Butterflies of hybrid ancestry often show remarkable and bizarre wing patterns highly sought by collectors, and historically their discovery has generated sometimes emotional exchanges about the taxonomic status of the populations involved. In this chapter, I touch on this but emphasize a feature of natural hybridization that has been much less exploited by lepidopterists, which is the opportunities it presents to study evolutionary pressures acting on the morphological, behavioral, physiological and life-history differences that hybridizing taxa often exhibit. Depending on the nature of the contact regions between hybridizing species, very powerful insights can be obtained using rather straightforward sampling schemes in natural populations. The analytical methods are now over 25 years old, and I believe lepidopterists have been slow to adopt them (with notable exceptions among *Heliconius* biologists working in the New World tropics) because they are presented in a mathematically challenging framework. The methods are extremely

Ecology of Butterflies in Europe, eds. J. Settele, T. Shreeve, M. Konvička and H. Van Dyck. Published by Cambridge University Press. © Cambridge University Press 2009, pp. 296–311.

powerful, and not terribly difficult to use once the central concepts are apparent.

Broadly, hybrid zones are areas where populations of related taxa meet and hybridize. In this chapter, I provide a primer on how evolutionary ecologists, ecological geneticists, and even species-level systematists can profitably use hybrid zones to address a great diversity of questions. I present the central ideas of hybrid zone analysis in a largely graphical framework, which is an accessible introduction to the topic. I include equations only where their use will help readers relate to the broader literature or see how the methods capture central ideas in evolutionary biology. I believe these methods belong in every field biologist's toolbox and that considerable advances can be made in our understanding of European butterfly biology in particular, and evolutionary biology in general, as more people apply these methods.

The great strength of hybrid zone analysis is that it can provide direct empirical estimates of extremely important processes in population biology. These measures include the dispersal rate of genes through the habitat, the strengths of natural selection acting separately on individual traits in the vicinity of the zone, and on the hybrid genome as a whole in the center of a hybrid zone. One may also measure the extent of genetic isolation between hybridizing species, the rates of introgression of individual traits past the hybrid zone, and even the extent that selection acting in one set of traits can block introgression in other traits in the genome (space does not allow elaboration on these issues here). These are all central questions in evolutionary biology, addressed time and again for diverse traits in diverse species but rarely measured. Being able to measure them in a straightforward way is of great value.

The strategy for ecological geneticists working with hybrid zones is to use microevolutionary models of how hybrid zones work as an analytical framework. By fitting data to these models, it is possible to then estimate the microevolutionary parameters of interest. These models do not pertain to all situations where hybridization is present, so before delving into the models, I will lay out the range of situations that lend themselves to this approach. But first, some often-cherished tacit assumptions about interspecific hybridization need to be challenged.

THE ECOLOGICAL GENETICIST'S PERSPECTIVE

Natural historians have long thought of natural hybridization from an almost typological perspective, with populations comprised of "pure" taxa and their hybrids; this oversimplification continues to dominate most biological subdisciplines. We often use language that tacitly presumes natural hybrids contain equal proportions of the genes from the two parental species, as if all hybrids were derived from pure crosses. And we may presume that a specimen is genetically "pure" if all its diagnostic traits conform with norms from one of the species. When we study microevolutionary forces acting on the hybridization rate or the fate of hybrids, this typological perspective can force us to adopt some appalling tacit assumptions. For example, we must presume that natural selection that acts on a single trait, favoring, perhaps, one species in dry habitats, extends magically to the rest of the genome and blocks the introgression of all traits from the other species. We may inadvertently presume that all hybrids are average, and neglect the wide phenotypic and genotypic variation that can arise when different genomes interact. And most fundamentally, we may presume that the diagnostic traits we use to distinguish the "parental" forms are somehow useful for interpreting the extent that the remaining genes introgress and are shared between the species. Of course, if any hybrids are viable and fertile, then the "parental" genomes can recombine in diverse ways following simple Mendelian rules, and this typological way of thinking is not only inadequate, but actively misleading. We need a more sophisticated way to represent hybridization if we are to understand its implications. We need to account for the possibility that individuals can have, in principle, a pedigree that includes any conceivable proportion of ancestry from the two "parental" species. At the same time, suites of diagnostic parental traits sometimes *do* co-occur predictably in individuals of hybrid ancestry, and our system must be flexible enough to accommodate this possibility.

An appropriate perspective is to consider hybridizing populations one gene at a time. We can then study how microevolutionary forces act on these single genes to determine their frequencies in different populations. From there, we can extend the analysis to multiple genes at once, to study how and why frequencies of different genes are correlated, spatially and within individuals. This perspective lies squarely within the realm of ecological genetics. We may ultimately find that all genes in the genome are similarly correlated, and even verify empirically the existence of discrete classes of genetically "pure" and "hybrid" constitutions. But more likely, we will find degrees of differentiation and blending that vary not just geographically through the region of contact, but also from region to region within the genome. This puts us into a position to ask much more insightful and sophisticated questions about the causes and consequences of natural hybridization than any typological perspective could give us.

THE POSSIBLE SOURCES OF CORRELATIONS AMONG TRAITS

A second central assumption about hybridizing taxa that needs to be revisited involves how correlations among traits might be created and maintained. It is very tempting to presume that the traits we use to differentiate and diagnose "pure" taxa that hybridize are maintained by some sort of genetic isolation, and evolutionary biologists since at least the Neo-Darwinian Synthesis have made this assumption (e.g. Mayr 1942). Notwithstanding, the potential reasons for maintaining such correlations are diverse (Table 19.1).

(I)   Correlated genes might be especially closely linked on the same chromosome, so that recombination breaking up allelic combinations is rare. Even neutral genes can experience the selective forces acting on linked loci, but the linkage has to be very tight not to be broken eventually, and leakage will gradually erode initial patterns of differentiation. Chromosomal rearrangements can decrease the rate of recombination, but even this will gradually erode due to rare recombination wherever selection is weak.

(II) Correlations can arise because two genes interact to generate a single phenotype, such as when genes are part of the same regulatory or metabolic pathway, or when different wing pattern elements contribute to an overall pattern of crypsis or warning coloration. Alleles of such genes are coadapted and can be maintained as suites by natural selection, regardless of their linkage.

(III) Different genes can contribute to physiologically independent traits that are themselves influenced by the same environmental conditions. For example, larval growth rate, adult body size, and the propensity to enter diapause may covary geographically, switching together at the ecological boundary where two generations are favored over one (Roff 1983).

(IV) Correlations can arise when genes are completely unlinked, functionally independent, and influenced by different environmental conditions, if those environmental conditions happen to be correlated spatially. For example, in montane regions, host-plant species often change with elevation, as does temperature and growing season, so genes for surmounting respective host defenses may be found to co-occur with those influencing diapause, even though they are entirely unrelated genetically and physiologically (e.g., Hagen 1990).

(V)   Correlations can arise in phenotypically plastic traits that are responding to the same or correlated environmental cues, even when there are no genetic differences between the taxa. This may occur, for example, in larval performance on different hosts in the *Pieris napi–bryoniae* complex (Porter 1997), where the species use different hosts that occur at different elevations.

(VI) And, of course, genes for diagnostic traits may be correlated because the taxa truly are genetically isolated, unable to produce fertile hybrids or backcrosses.

Given these potential causes, we cannot simply invoke genetic isolation to explain correlations of diagnostic traits.

Table 19.1 *Reasons why correlations may be observed between traits, and the spatial scale at which the correlation may be observed*

	Reason	Brief description	Spatial scale of the correlation[a]
I.	Linkage	On the same chromosome or in the same inversion	Local and global
II.	Epistasis	Traits that influence the same fitness component	Local and global
III.	Shared selection	Under selection by the same environmental condition	Usually global
IV.	Correlated selection	Traits under selection by independent, but spatially correlated, environmental conditions	Usually global
V.	Correlated plasticity	Phenotypic plasticity in both traits, affected by the same or correlated environments	Usually global
VI.	Genetic isolation	Barrier to introgression blocks homogenization	Local or global

[a] Local: among individuals in local populations; global: among individuals in larger geographic areas, as when local populations are pooled before the correlation coefficient is calculated.

THE ECOLOGICAL SETTING DETERMINES ANALYTICAL TRACTABILITY

Hybrid zones may be sorted into geographic categories that correspond to different underlying fitness patterns for the traits involved. Three common patterns are shown in Fig. 19.1. These are:

(1) *Broad clinal contact*: broad geographic regions of gradual blending, often many times broader than the average

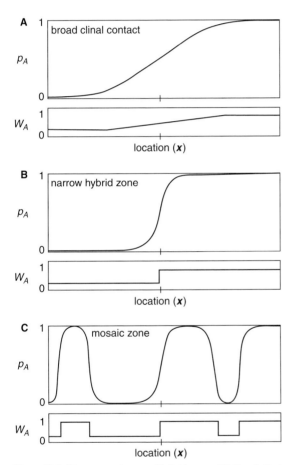

Figure 19.1. Three general types of hybrid zones. (A) Broad clinal contact. A gradual transition in the frequency of a trait (p_A), corresponding to a gradual change in the underlying fitness (W_A). (B) Narrow hybrid zone. A relatively abrupt change in the frequency of a trait (p_A), corresponding to an abrupt change in the underlying fitness (W_A). (C) Mosaic hybrid zone. A series of narrow hybrid zones between populations adapted to different habitat types, often found in regions where one broadly distributed habitat type meets another.

lifetime dispersal distance, corresponding to gradual changes in fitness of the underlying traits. This pattern is not analytically tractable using the methods reviewed in this chapter.

(2) *Narrow hybrid zones*: narrow geographic areas of blending, typically on the order of a few lifetime dispersal distances, corresponding to abrupt changes in fitness of the underlying traits.

(3) *Mosaic contact*: broad geographic areas of patchy co-mingling of populations with blending at the patch boundaries, regionally usually much broader than the dispersal distance. However, where the patches meet locally, widths are often on the order of a few dispersal distances, generated by sharp fitness differences between patch types. If sampling is sparse, mosaic contact may seem to be a broad cline.

A helpful feature to distinguish broad clinal contact is that the frequencies of diagnostic traits change gradually and over a much broader geographic area than the lifetime dispersal distance. This pattern can arise in three ways. (1) It can be caused by local adaptation to a broad gradient in environmental conditions (Endler 1977), as in Fig. 19.1A. Selection can be quite strong or it can be weak, but as long as the environmental gradient is much wider than the dispersal distance, the cline shape will closely track the gradient. Inasmuch as this adaptation represents a stable, equilibrium situation, it is not possible to tell without additional data whether the pattern arose from primary or secondary contact. However, natural selection is not strictly necessary to maintain broad clines. (2) Broad clinal contact may also result when the diagnostic traits are neutral or nearly so, so that dispersal tends to homogenize them. This clinal pattern can be a result of secondary contact in the not-too-distant past, such that blending has not reached completion. In this case, it is possible to calculate the time since contact if the dispersal rate is known or vice versa. (3) Clinal variation can even be caused in neutral traits by primary contact: genetic drift can create isolation by distance over large spatial scales (Endler 1977). Indeed, Endler showed that surprisingly steep clines in single traits, on the order of several dispersal distances, can arise de novo and be maintained for thousands of generations in neutral traits, even steeper when a stretch of inhospitable habitat imposes a partial barrier. When multiple traits are considered, overlapping clines can arise and persist just by chance even when the traits are entirely neutral. Without appropriate experiments, there is really no way to tell whether traits are blending because they are neutral or

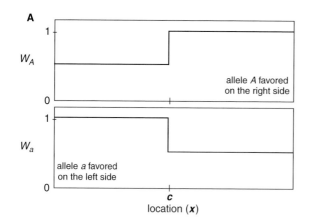

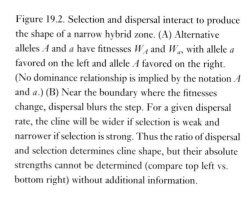

Figure 19.2. Selection and dispersal interact to produce the shape of a narrow hybrid zone. (A) Alternative alleles A and a have fitnesses W_A and W_a, with allele a favored on the left and allele A favored on the right. (No dominance relationship is implied by the notation A and a.) (B) Near the boundary where the fitnesses change, dispersal blurs the step. For a given dispersal rate, the cline will be wider if selection is weak and narrower if selection is strong. Thus the ratio of dispersal and selection determines cline shape, but their absolute strengths cannot be determined (compare top left vs. bottom right) without additional information.

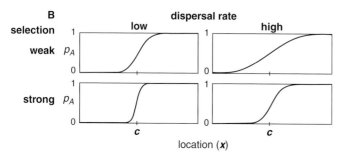

because they are under some type of selection. None of these types of broad clines presents any sort of a barrier to gene flow at other loci. Unfortunately, none of these types of broad clines can be analyzed using methods in this chapter.

Narrow hybrid zones (Fig. 19.1B) are characterized by a sharp step in fitness at the boundary (Fig. 19.2A). (If the fitness difference is small at the step, these clines may not be particularly narrow, and it may be difficult to distinguish them from broad clines without some information on the dispersal rate. But the term "narrow hybrid zone" has some inertia in the literature and is not challenged here.) This fitness step may be produced by either extrinsic and intrinsic underlying environmental conditions, which are not mutually excusive. Extrinsic conditions refer to the usual abiotic and biotic conditions that ecologists study – temperature regimes, host availability, parasitoid density, etc. Intrinsic conditions refer to the genetic background – the suite of alleles at other loci that may interact physiologically or developmentally with the gene in question. It can be thought of as the physiological environment that the gene finds itself in, and it may operate regardless of the extrinsic environment that the organism occupies. Intrinsic selection

is therefore frequency dependent, determined by loci in the remainder of the genome. From the perspective of the shapes of the clines that are produced, this distinction is not especially profound, but from the point of view of the ecological geneticist interested in the mechanisms and consequences of differential adaptation, it is.

Mosaic hybrid zones (Fig. 19.1C), however broad they may appear at larger spatial scales, can be parsed into a series of smaller, narrow hybrid zones with sharp habitat boundaries. It is therefore possible to treat them as special cases of narrow hybrid zones, making them tractable for analysis.

HOW NARROW HYBRID ZONES WORK

Narrow hybrid zones are generated by opposing processes: natural selection across the central step increases differentiation and narrows the cline, and dispersal blurs these differences and broadens the cline (Fig. 19.2B). If conditions persist for a sufficient time, as determined by the strengths of dispersal and selection and the initial conditions of contact (a few tens of generations or less for moderate selection), then genetic differentiation settles into equilibrium with a characteristic sigmoid shape (Fig. 19.2B). Most narrow hybrid zones involve more than one trait, and the fine details of the shape vary for each trait depending on their correlation to other traits and how selection acts on each one.

Fortunately, given this potential for complexity, commonalities remain among the shapes that help rather than hinder a quantitative analysis. To learn how to exploit narrow hybrid zones for evolutionary insights on separate traits as well as the genome as a whole, it is helpful first to see how patterns arise concerning clines involving independent loci, then to see how clines at these loci interact to generate narrow hybrid zones. From there, we can see how data might be fit to these models, and microevolutionary parameters estimated. Finally, it is useful to think further about what assumptions have been made to permit quantitative analysis, how conclusions may be inadvertently shaped by these assumptions, and what we can do to mitigate it.

Throughout this section, I will rely on diagrams to present major concepts visually rather than mathematically. However, behind each diagram is a mathematical formula that relates the graphical shape to underlying microevolutionary causes, and I will highlight these relationships. For clarity, I will sometimes present these concepts in different terms than seen in the original mathematical literature. My treatment is still equivalent to the original, and I will point out where a little algebraic manipulation can bring them into agreement. In adopting this strategy, I hope to help make hybrid zone dynamics and analysis more intuitively accessible, and easier to incorporate into research programs.

SINGLE-LOCUS CLINES UNDER SELECTION

I will start with a single locus A, with allele a favored on the left side, and allele A on the right side of the cline (think of A as representing *Any* locus). (Throughout, locus names are capitalized and allele names italicized, and no dominance is implied by the capitalization.) If we assume the lifetime dispersal pattern follows a normal (bell-shaped) distribution, then once the equilibrium genetic structure has been reached, the frequency of the allele takes a simple sigmoid shape (Fig. 19.2B) that depends on dispersal, selection, and to a lesser extent, the mode of inheritance. The equilibrium shape in Fig. 19.2B is mathematically described by a hyperbolic tangent function, $p_A = (1 + \tanh[2(x - c)/w])/2$. Figure 19.3A describes the variables and how they influence the shape of the cline. This formulation assumes that alleles are fixed on different sides of the cline, but this need not be

the case. As shown in Fig. 19.3B, we can scale the shape so that the cline runs between frequencies p_L on the left and p_R on the right.

The width turns out to be a key parameter in this model, because it holds all the microevolutionary information of interest. At equilibrium, the width stabilizes at $w = k\sigma/\sqrt{s}$, where s is the strength of selection favoring alleles on different sides of the zone, and σ is the standard deviation of the dispersal distribution (= the dispersal "rate"). This makes intuitive sense too: as dispersal increases in the numerator, the cline gets wider, and as selection increases in the denominator, the cline gets narrower. The constant k scales the shape to the appropriate mode of inheritance. For codominant loci and extrinsic selection (as in Fig. 19.2A), $k = \sqrt{3}$; for dominant loci, $k = 1.782$; for codominant loci and selection against heterozygotes, $k = 2$; and for quantitative traits, $k = (\bar{p}_R - \bar{p}_L)/(2\sqrt{V_G})$ where V_G is the genetic variance in the trait and the $\bar{p}$'s are the trait means on the two sides of the cline. It follows that when the locus or trait is neutral, then $s = 0$ in the denominator and the equilibrium width is infinite. In fact, the slope $(1/w)$ is 0 and the frequencies will be equal everywhere in the cline (ignoring the random fluctuations of genetic drift).

The most appealing aspect of this approach is that these shape equations were derived from first principles of how selection, gene flow and the mode of inheritance should interact to generate clines (Barton & Gale 1993), so the shape holds microevolutionary information that can be extracted when data are fit to the mathematical models. There is one caveat: because the width depends on the ratio of dispersal and selection, the width only tells us their strengths relative to one another (Fig. 19.2B). But, if we can get information on dispersal independently, then we can use the width to measure selection, or vice versa. Such information is indeed available in multilocus clines, and we harness it in the analyses below.

CORRELATIONS AMONG LOCI IN HYBRID ZONES

The single-locus perspective forms the basis of the analysis of clines involving multiple loci. However, correlations can occur among loci, with broad implications not only for the microevolutionary dynamics of multilocus clines, but also for our ability to exploit narrow hybrid zones to study the ecological genetics of the underlying traits. These genetic correlations, called gametic or linkage disequilibria when they involve individual genetic loci, can arise and be destroyed

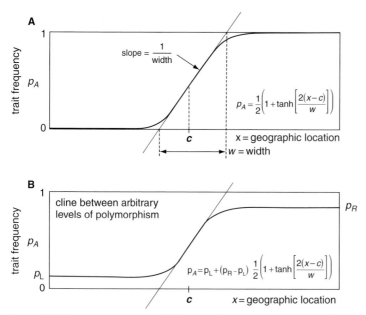

Figure 19.3. Translating between graphical and mathematical forms of simple cline models. (A) The cline in allele A is described mathematically as the frequency of A (p_A) at any geographic location x. The overall sigmoid shape is described by a hyperbolic tangent function (tanh). The width has units of distance along the x-axis per unit of change in marker frequency, so it is the inverse of the slope in the center. The center is at location c, so the term ($x - c$) gives the distance and direction of a given location x from the center. The terms $1/2$ and 1 shift the tanh function, which normally spans the y-axis from -1 to 1, into the range from 0 to 1. The tanh function is not assigned by convenience, rather it comes from a series of assumptions about the underlying biology. These include assuming that the underlying fitnesses are stepped as in Figs. 19.1B and 19.2A, that selection isn't so strong that it dramatically changes the phenotypic distribution, and that dispersal follows a normal distribution with standard deviation σ. Under those conditions, the width becomes $w = k\sigma/\sqrt{s}$ (see text). Simulations show that the tanh function is a good approximation even when selection is strong (Barton & Gale 1993). (B) As in (A), but when allele frequencies are not fixed on either side.

by several microevolutionary processes. It will be expedient first to examine the biology behind these correlations before considering their effects on multilocus cline shape.

Disequilibrium (D) measures the statistical dependence among alleles at different loci within a population. For example, when $D > 0$, it tells us that when an individual carries an A allele, that it has an increased chance of carrying a B allele as well – it implies an excess of AB and ab genotypes over Ab and aB, compared to proportions expected by chance alone. D is therefore a type of correlation, and loci are completely independent when $D = 0$. But more importantly, D is a numerical way to represent an idea that is fundamental to evolutionary ecologists and species-level systematists alike, namely the extent to which the genomes of the interacting taxa cohere or "hang together," perhaps due to suites of coadapted gene complexes or some degree of genetic isolation. To gain insight into these elemental questions in evolutionary biology, it is clearly

very important to think about how disequilibrium is created and destroyed in hybrid zones. Although disequilibrium can vary between 0.25 (say, when A always occurs with B) and –0.25 (when A always occurs with b), whether the value is positive or negative also depends on how we label the alleles. By convention, we label alleles A and B as the ones at higher frequency on one side of the zone, and alleles a and b on the other, so that $D > 0$ implies a positive correlation among diagnostic markers. When there are multiple markers, then D is usually presented as some type of an average of pairwise disequilibria, and higher-order (3-way, etc.) correlations are ignored.

When relating disequilibrium in hybrid zones as a measure of genetic coadaptedness or isolation, it is especially important to consider the spatial scale over which D is being calculated. The appropriate spatial scale is that of the local population or, in a continuously distributed species, the genetic neighborhood. When we think of genetic isolation or coadaptedness, we may intuitively consider "populations" at larger spatial scales, say across the entire hybrid zone or even of entire species complexes (e.g., Table 19.1). As spatial scale increases beyond the panmictic unit to include geographically variable populations, D increases because it captures spatial covariation. This spatial covariation could arise for any of the "global" reasons in Table 19.1, and is not necessarily evidence for coadapted gene complexes or genetic isolation. In hybrid zone analysis, we restrict the calculation of D to the local geographic scale, so that it does not include any contribution from spatial covariation.

Four processes can potentially act to generate D within local populations: gene flow, genetic drift, natural selection and non-random mating. Of these, gene flow is the most important because it applies to all loci in the genome. The effect of gene flow on disequilibrium is strongest when adjacent populations have different allele frequencies at multiple loci, such as near the center of hybrid zones where clines are steepest. Near the center, individuals on the right side have more A and B alleles, and those on the left have more a and b alleles. An individual dispersing from right to left will therefore tend to carry both A and B genes, automatically creating disequilibrium (more AB genotypes than expected by chance) in the population where it arrives. The same holds for a and b alleles in individuals dispersing the opposite way. The more different the allele frequencies in nearby populations, the stronger the increase in D after a dispersal episode. The result is that D tends to be high near the centers of hybrid zones. Genetic drift can create disequilibrium in small populations, but the effect is transitory and potentially strong only for very small populations. Natural selection can also create D for particular combinations of loci by favoring pairs of alleles, say AB and ab combinations over Ab and aB combinations. Non-random mating can create disequilibrium, too, particularly for genes influencing mating behaviors. The change in D is always positive for gene flow, but it can be positive or negative for the others, depending on the nature of the underlying association. Nevertheless, hybrid zones tend to have high D near their centers, so the contribution is small at best from factors that would promote negative D.

By far the most important process breaking down disequilibrium is recombination. If we define r as the probability that two loci will recombine, then a standard result in population genetics is that after one generation, $D_{t+1} = (1-r)$ D_t. If loci A and B are on different chromosomes, then they will randomly recombine each generation and $r = 1/2$, so D will be reduced towards 0 by 50%. At this rate populations approach $D = 0$ within about 6 generations and effectively reach $D = 0$ within 10 generations. Closely linked loci will recombine less and D will persist longer, hundreds of generations for extremely tightly linked loci and even longer for loci on a chromosomal inversion. But without any process creating disequilibrium, recombination eventually breaks down all disequilibrium in the genome. The dynamics differ when disequilibrium involves sex-linked or cytoplasmic genes, but the general principles are the same. Technically, natural selection, non-random mating and even genetic drift can sometimes break down disequilibria for particular allelic combinations, but the scenarios are transitory and very restrictive.

These competing processes generate a stable pattern of disequilibrium in the hybrid zone (an equilibrium of disequilibrium?) that varies both geographically and, because different loci experience different conditions of natural selection and non-random mating, across the genome. Geographically, D is highest in the hybrid zone center where allele frequencies are most different and generating forces are strongest, and lower towards the tails where generating forces are weak and recombination has had longer to operate (Fig. 19.4A). This is an important point: the further an alien allele is found past the cline center, the more generations of dispersal have likely occurred since its ancestor immigrated across the cline. Thus, the more recombination events that allele has likely experienced, the less of that ancestor's genetic background is likely to be present in the individual carrying that allele. Geographic distance is therefore a proxy for the history of recombination, and therefore, disequilibrium drops off away from the zone center. (There are statistical reasons that D drops off, too: D is maximized only when the allele frequencies of the loci involved are at 50%. Some of this effect can be removed by scaling, however.) Across the genome, D will tend to remain highest in chromosomal regions that are near loci under the strongest differential selection or involved in assortative mating, and will become weakest in neutral chromosomal segments. D will also remain high in regions with inversion polymorphisms or other regions where recombination is reduced, such as near centromeres.

For the purposes of hybrid zone analysis, it is useful to consider that D is created by dispersal and destroyed by recombination, and to treat the roles of selection and non-random mating as secondary. In this perspective, we can think of selection and non-random mating as operating mostly to maintain the *spatial* allele frequency differences that permit the dispersal to generate disequilibrium across

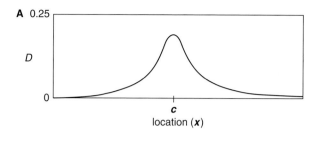

A 0.25

D

0

c

location (x)

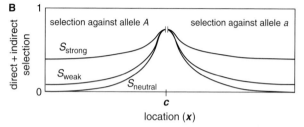

B 1

direct + indirect selection

selection against allele A selection against allele a

S_{strong}

S_{weak}

$S_{neutral}$

0

c

location (x)

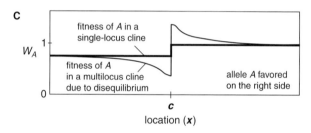

C

fitness of A in a single-locus cline

1

W_A

fitness of A in a multilocus cline due to disequilibrium

allele A favored on the right side

0

c

location (x)

Figure 19.4. Disequilibrium (D) and its effects in a narrow hybrid zone. D describes the overrepresentation of AB and ab genotypes compared to random expectation. (A) Disequilibrium is highest in the center of a hybrid zone where its generating forces are strongest. It weakens with distance from the center, as generating forces weaken and more generations of recombination accumulate in dispersers. (B) This affects the realized (direct + indirect) selection acting on a trait. At the left and right edges, realized selection against the unfavored allele includes only direct selection. In the center, all traits experience a peak in realized selection where D with other selected traits is strong; even neutral traits experience this indirect selection. In this figure, the left side diagrams selection against A; on the right it represents selection against a. (C) This translates into location-specific fitness effects for particular alleles. On the unfavorable side (left), realized fitness declines towards the center; on the favored side (right), it increases toward the center.

the genome. Selection favouring coadapted AB and ab combinations is presumed to be weaker, and to act only in a subset of loci that remain differentiated across hybrid zones.

DISPERSAL AND DISEQUILIBRIUM IN HYBRID ZONES

For evolutionary geneticists, dispersal is a synonym of gene flow and measures the distribution of distances between where parents are born and where their offspring are born. Often these distributions are summarized using the standard deviation. There are two general approaches to measuring dispersal – ecological and genetic. Each has drawbacks and neither is precise. Ecological methods generally follow individuals for some fraction of their lives, but for butterflies as for most organisms, researchers typically do not know where both parents were born or where each of the eggs was laid. These measures typically underestimate dispersal (Endler 1979). Genetic methods, on the other hand, rely on models of the relationship between dispersal and the level of variation in allele or genotype frequencies. This variation often contains information from multiple generations, so

the dispersal estimate is some sort of a multigenerational average. Regardless of the method used, the measure of dispersal that results is necessarily a standard deviation, not a mean. Standard deviations have larger margins of error than means, and asymmetrical confidence intervals. Unless sample sizes are astronomical, any study of dispersal will necessarily have large error clouds, and it will often be necessary to arrive at relatively qualitative conclusions once all the calculations are made. I will present genetic measures of dispersal here, because they fall conveniently out from multilocus marker data often collected in hybrid zones.

When we adopt the perspective of dispersal as the primary generating force for disequilibrium, then straightforward mathematical modeling can be used to find their quantitative relationship in the cline center: $D = (r + 1)\sigma^2/rw^2$ (Barton & Gale 1993). This immediately rearranges: $\sigma^2 = Dw^2r/(r + 1)$ – and since D, width w, and recombination rate r are easy to measure empirically, we can use this formula to estimate the dispersal rate σ. This method is applied in Boxes 19.1 and 19.2. Using this method does require some assumptions. We could inadvertently overestimate dispersal if D in the population were generated in part

Box 19.1   Dispersal and selection in the *Pontia daplidice–edusa* hybrid zone in Italy

The pierids *Pontia daplidice* and *P. edusa*, distinguished primarily by allozyme markers and subtle differences in genital morphology, meet in a hybrid zone in northern Italy (Porter *et al.* 1997). The allozyme clines in this hybrid zone have a small step in the center indicating a standardized width of $w = 28$ km, flanked by broad, somewhat asymmetrical tails of introgression on either side (see figure). Disequilibrium of $D = 0.148$ in the center of the zone yields a gene-dispersal rate

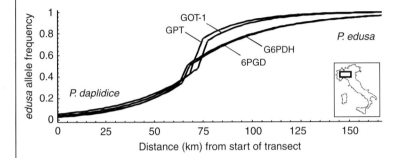

of $\sigma = 4.5$ km/gen, the standard deviation of distances between birthplaces of parents and offspring. The shapes of the tails extending into each taxon imply that weak selection (ca. 1.5–3%) maintains the diagnostic differences between them. The locus 6PGD is sex-linked, but it has the same cline shape as the autosomal locus G6PDH. Interestingly, these loci perform nearby steps in the pentose-phosphate pathway, suggesting that synergy between the diagnostic alleles at these loci might affect metabolic efficiency and thereby provide a target for natural selection independently of the remainder of the genome. Similarly, loci GOT-1 and GPT, which also have indistinguishable cline shapes, perform adjacent steps linking glycolysis and amino-acid synthesis. The shapes of the tails on the *P. edusa* side suggest that selection there is stronger on the GPT/GOT-1 pair than on the G6PDH/6PGD pair, but no such difference appears on the *P. daplidice* side. Regardless of how selection operates on these loci, the broad tails of introgression demonstrate that selection is far too weak to enforce genome-wide isolation between the taxa.

by epistatic selection or non-random mating (Table 19.1) with respect to the markers. However, when molecular markers are used to calculate D, that may not be so serious a risk. More subtly, we could underestimate dispersal if individuals tend to notice habitat cues at the cline center and are reluctant to cross to the opposite side. In that case, we would still measure dispersal in the vicinity of the boundary, but we would need more information to apply that dispersal estimate to areas outside the cline. This dispersal estimate is also necessarily an average of some sort over the sexes, which may themselves disperse at different rates, because D includes correlations left over from previous generations in addition to newly generated correlations.

It is also possible to use this logic to estimate dispersal when the only traits available are quantitative traits (Barton & Gale 1993). The idea is to decompose the phenotypic variances and covariances of traits into their genetic and environmental components. Disequilibrium in the

underlying genes responsible for the traits contributes to their genetic covariances, and dispersal in the center of the cline should maximize disequilibrium there. So, by comparing genetic covariances in the center of the cline with covariances far out in the tails, we get a disequilibrium measure. We then convert this to a dispersal measure as above. To do all this, we need to make further assumptions about the underlying genetic structure of the traits as they vary across the hybrid zone. Especially, we have to assume that the environmental variance does not increase in the hybrid zone center, and that the identities of the genes contributing to the traits do not change across the hybrid zone. (These assumptions seem hopeful to me, so my preference is to treat such dispersal estimates as working hypotheses at best, and certainly not as defensible measures.)

Before we apply this dispersal estimate to decompose cline width w and estimate selection strength s, we need to understand the effects of disequilibrium on multilocus cline shape.

Box 19.2   Dispersal and selection in a *Heliconius erato* hybrid zone in Peru

Heliconius butterflies are warningly colored and involved in Müllerian mimicry rings, yet they often have wing patterns that change dramatically from region to region (Turner 1981). Where geographic races (often named as subspecies) meet, they form hybrid zones maintained largely by frequency-dependent selection for different color patterns. The genetic control of wing pattern in *Heliconius erato* is relatively simple (Mallet 1989), and hybrids of various genotypes are usually recognizable. Mallet & Barton (1989a) fit clines to wing pattern alleles at three loci that differ between *H. erato favonius* and *H. erato emma* (see figure), and measured dispersal and the strength of selection on loci contributing to warning coloration. Even though these loci are unlinked, selection against the nonconformist hybrid color patterns is expected to produce clines that vary together geographically. Indeed, the congruence is high (see figure). The cline in locus D^{Ry} is

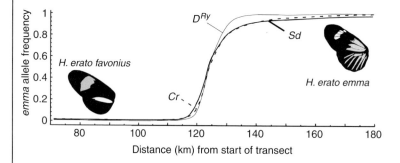

steeper on the *H. e. emma* side, suggesting stronger selection; the D^{Ry}/D^{Ry} genotype shows the yellow forewing patch and red rays of *H. e. emma*, producing a striking change. Clines in the loci *Cr* and *Sd* are virtually perfectly congruent; recessive alleles at these loci control elements of the yellow hindwing band in *H. e. favonius*. They have a relatively small effect on wing pattern in the *H. e. emma* background, and the *H. e. emma* alleles are dominant, so the shallower cline there indicates that the *H. e. favonius* alleles introgress further into the *H. e. emma* side. Disequilibrium among these loci averages $D = \sim0.35$ in hybrid populations, and the average cline width is $w = \sim9.6$ km, so these methods produce a dispersal estimate of $\sigma = \sim 2.7$ km/gen. Selection acting on the mimicry genes is therefore on the order of $s = 10\%$ against regionally inappropriate D^{Ry} wing patterns, and $s = \sim 9\%$ against inappropriate *Cr* and *Sd* genotypes (accounting for the steeper D^{Ry} cline and the shallower *Cr* and *Sd* = $\sim$ 2.6 km/gen) and much stronger selection ($s = \sim23\%$) estimates, agreeing better with field measures (Mallet & Barton 1989b). Mimicry in *Heliconius* is clearly maintained by quite strong selection.

MULTILOCUS CLINE SHAPE

We have seen that the shape of a cline in a single locus is determined by the competing processes of selection and dispersal. Now consider the effects of a cline in a second locus, with alleles *b* and *B*, also under selection. If it existed alone, locus B would have a sigmoid equilibrium shape as does locus A, but when loci A and B are together, they interact and two things happen to their shapes (Fig. 19.5).

First, both clines get steeper in the center. To see why, imagine selection as the probability that individuals carrying the wrong allele will die (this argument extends to other forms of selection). Just left of the cline center, where

alleles *a* and *b* are favored, consider an individual carrying allele *A*. Selection will kill it off with probability s_A. If it survives, then it still has a problem: disequilibrium is high near the cline center, so this individual has an increased chance that it will also be carrying allele *B*. If it is a carrier, its probability of being killed is s_B. In more general and quantitative terms, selection on allele *A* is composed of two quantities: direct selection s_A and indirectly, selection acting on *B* that is transferred through its correlation (disequilibrium) with allele *A*: $s^*_A = s_A + Ds_B$. (Technically, these fitness effects are multiplicative rather than additive, but when *s* is small, the additive approximation is very close.)

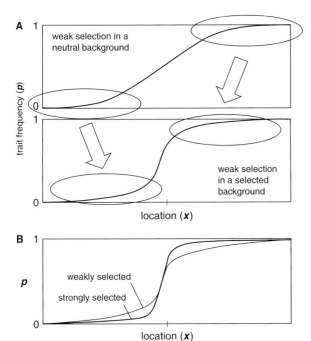

Figure 19.5. The effects of disequilibrium on cline shape. (A) In a weakly selected trait, disequilibrium in a selected genetic background steepens the center while permitting the tails to retain their flatter shapes. This gives the appearance of shifting the tails together (bottom), relative to a neutral genetic background (top). The extent of this shift represents the barrier to introgression imposed by selection on correlated traits in the genetic background. (B) Traits dissociate in the tails of the cline, with weakly selected traits having shallower tails (Box 19.1). The most strongly selected traits may show no evidence of shallower tails (Box 19.2). The shapes may be asymmetrical if relative fitnesses differ on opposite sides of the center (Box 19.1), or if the inheritance pattern differs (Box 19.2).

In general, selection on locus A includes the direct component, plus indirect components acting through any loci it is in disequilibrium with. Near the center of narrow hybrid zones, disequilibrium is likely to occur throughout the entire genome (Fig. 19.4A), so even neutral genes will experience selection there ($s^*_A = 0 + Ds_B$) (Fig. 19.4B). Indeed, evolutionary pressures acting through indirect selection of this sort tend to cause clines near enough to one another to converge toward a common center position (Endler 1977). Thus, if we were to consider the average of all shapes in the center of the zone, it would be very close to the shape of each individual locus. This is a useful point later, because it means that we can increase our statistical power to estimate cline shape parameters near the center of the cline by averaging over loci, and over all multilocus traits. This commonality among shapes is what gives robust statistical power to cline analysis.

The second thing that happens in multilocus clines is that the tails of the clines for individual loci or traits begin to dissociate from one another (Fig. 19.5B). The reason is that disequilibrium dissipates in the tails of the cline (Fig. 19.4A), so indirect selection dissipates with it (Fig. 19.4B, C) and the clines follow their evolutionary fates increasingly independently of one another. The tails of the clines therefore approach the shapes they would

have in a neutral or randomized genetic background, with weakly selected traits taking on much shallower tails than strongly selected traits (Fig. 19.5B). Weakly selected traits thereby introgress further into the genetic background on the opposite side, before being eliminated by selection. This feature gives ecological geneticists considerable analytical leverage – we use these shape differences to measure microevolutionary forces acting on different traits independently.

Taken together, these effects of disequilibrium on cline shape mean that separate traits should have clines with very similar shapes in the center region and increasingly independent shapes in the tails (Fig. 19.5B). During analysis, it is statistically much more powerful to use all the traits at once for estimating the shape in the center region, rather than estimating shapes for each cline separately. But in the tails, estimating a common shape over all traits eliminates any insight into evolutionary forces that act on traits individually. Analysts therefore break narrow hybrid zones into three distinct regions. In the center, a common sigmoid shape is estimated and in each tail, shapes are measured separately for different traits (Fig. 19.5). The models for the tails have shallower widths and (so that these tails join smoothly to the steeper central cline) shifted centers, which we could represent for the left and right sides respectively as $p = (1 + \tanh[2(x - c_L)/w_L])/2$ and $p = (1 + \tanh[2(x - c_R)/w_R])/2$. However, since these two tail regions and the steeper central region are sections of a single multilocus cline and therefore merge into one another, it is common to join them together into a single statistical framework. To do this, we can represent the widths of the tail sections as proportions of the width in the center. Dispersal is the same for all traits, so $\sigma = \frac{1}{k}w\sqrt{s} = \frac{1}{k_L}w_L\sqrt{s_L}$, which rearranges to $w_L = w\frac{k_L}{k}\sqrt{\theta_L}$ where $\theta_L = s_L / s$; for each tail region. That is, the tail shape differs from the center shape by the fraction θ, which captures the ratio between weaker selection in the tail and stronger selection in the middle. The fraction k_L/k scales clines with different inheritance patterns, and it is traditional to use $k = 2$ for the shared central shape and $k = \sqrt{3}$ for dominant markers (see Barton & Gale 1993). For quantitative traits, $k = (\bar{p}_R - \bar{p}_L)/(2\sqrt{V_G})$ and it is necessary to measure broad-sense heritability separately to obtain V_G.

Similarly, the shifted centers c_L and c_R in the tails can be represented as distances from the overall center, $z_L = c_L - c$ and $z_R = c - c_R$. And lastly, a tanh function is composed of two inverted exponential decay functions, asymptoting downward to the left and upward to the right (Fig. 19.1). Since we are only interested in the respective tails, we can use the exponential forms. With these manipulations, we rephrase the tail shapes as $p = \exp\left[4(x - c - z_L)\sqrt{\theta_L}/w\right]$ on the left and $p = 1 - \exp\left[-4(x - c + z_R)\sqrt{\theta_R}/w\right]$ on the right. When $z = 0$ and $\theta = 1$, the exponential tail equations fall on top of the central tanh equation. This rephrasing is therefore really only a heuristic device, since the interconversions can be done before or after the parameters are estimated, but the exponential form is widely used in analyses (Szymura & Barton 1986, 1991, Barton & Baird 1999, Porter 2005). These exponential formulas assume that frequencies range from 0 to 1, but they can readily be scaled to range between any upper and lower limits as in Fig. 19.3B. They therefore apply well to clines between arbitrarily different levels of polymorphism, and clines in quantitative traits.

We now have the basis for relating the shape of a cline to the underlying microevolutionary processes that generate its shape. We can see that marker frequency varies across clines in proportion to selection strength, dispersal rate, background selection and the inheritance pattern. If we have data on more than one marker, then we can estimate dispersal rates from the genetic correlations that arise. Then, from simple measures of the shape, we can measure selection. The selection measures apply to the hybrid genome as a whole in the middle region of the cline,

using $s = (k\sigma / w)^2$, and they apply to individual traits in the left and right tails of the cline using $s_L = s\theta_L$ and $s_R = s\theta_R$. Once a disequilibrium measure is available in the cline center so that dispersal can be estimated, this method of measuring selection applies to any trait that can be mapped across a hybrid zone. It is therefore an important tool for ecological geneticists.

WORKING WITH DATA

Here we look at how to fit data to cline models, so that we can arrive at estimates of the evolutionary forces acting on both the underlying traits and the entire taxa. The statistical approach, based in maximum likelihood, is rather simple at its core: we try to minimize the differences between observed and expected values (Fig. 19.6). Maximum likelihood uses probability models to do this, under the reasonable assumption that random sampling error is responsible for discrepancies between observed and expected values. A

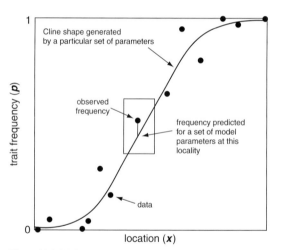

Figure 19.6. Fitting a cline to data using likelihood. The parameters of the cline (center, width, etc.) determine its shape, and different parameter values give different shapes. At each location where a sample is taken, we use the shape equation to find its predicted frequency. We then ask, given the sample size at that location, what is the probability that the predicted frequency could have produced the observed frequency? That is the likelihood of those parameter values for that locality, and we can accumulate these likelihoods across all localities in the data set. The maximum likelihood estimate of the cline shape is the set of parameters that, overall, would have the highest probability of producing the data. We use a computer to vary parameters until we find the most likely combination.

maximum likelihood estimate is simply the one of many "expected values" that would have the highest probability of yielding the observed values. In terms of cline analysis, the "observed values" are traits measured from individuals found at sampled locations along the cline – their individual genotypes or measures of any quantitative traits they exhibit. The "expected values" are predictions that we get by assigning numerical values to each cline parameter (center, width, disequilibrium, etc.) and solving the cline-model equation to get the predicted measures at the points where the samples were taken. The maximum likelihood estimate of cline shape is therefore the set of parameter values that together have the highest probability of yielding the observed data along the cline. Using the same logic, we can also find confidence limits on these parameter values. Likelihoods can take very small values when sample sizes are large, so it is normal to calculate them using natural logs. For a thorough introduction to maximum likelihood, see Edwards (1992). Because of the potential for interdependencies among some of the shape parameters, it is best to find these maxima using numerical methods, which are highly computer intensive, rather than using partial derivatives (which assume independence or known relationships). Software is available (Analyse: Barton & Baird 1999; ClineFit: Porter 2005) that generates estimates and confidence limits from multilocus data.

HYPOTHESIS TESTING IN CLINES

Since likelihood methods allow models with different numbers of parameters to be fit to clines, what do the different models tell us? The answer depends on which patterns emerge. Intrinsic selection (due to genetic background) tends to produce clines that lie atop one another (Barton & Gale 1993), so traits with different cline centers are probably influenced by different extrinsic (ecological) conditions. Very commonly, however, clines for different traits have similar centers and widths (although polygenic traits are more susceptible to environmental variation [V_E] and tend to be wider). Thus, the most valuable information often comes from determining whether there are significant cline tails in the traits of interest. If so, then direct selection on each trait can be measured from the slope of that cline's tail, and those traits are only partially responsible for overall genomic differentiation. For a trait where no significant tail is found, selection on it is likely to be highly correlated to the major fitness difference across the zone. Caution is warranted in reaching this conclusion, however, because greater sampling around a cline's center may be able to distinguish a

tail that the current sample missed. The shapes of the cline tails may also be asymmetrical, and a tail may even be found on one side but not the other. These correspond to asymmetrical patterns of selection and introgression in the hybrid zone. The corollaries – whether underlying evolutionary forces differ among traits – are fundamental to our understanding of the ecological genetics of individual traits and our interpretation of the degree of genome-wide segregation in hybrid zones. When the function of the trait is known (e.g., host preference, diapause susceptibility, disease resistance), then these selection measures provide fundamental insights about not only the importance of the trait to the organisms themselves, but the role of that trait in contributing to genetic isolation across the zone.

To choose among alternative cline shapes, statistical comparison is needed. There are two general kinds of questions: (i) What cline shape fits best? For example, is a simple sigmoid (tanh) shape appropriate for the whole cline, or does the cline have significantly flatter tails? (ii) Do shapes differ significantly among traits? If so, do they differ for all parameters, or only some? For any such statistical comparison, we always need to use the same data in order to rule out sampling error. That is, we have to construct our questions to fit the form, "For these data, which shape is best?" The details otherwise differ among the comparisons.

For question (i), which shape fits best, the problem is that the more parameters added to a model, the better the fit should be. At some stage, the improvement will be negligible and we ought to use Occam's Razor to eliminate overelaborate explanations. We take advantage of a simple statistical property of likelihood: twice the ratio of any two log-likelihoods follows a χ^2 distribution, and the degrees of freedom are the number of parameters that differ between the models. So, say the log-likelihood of a sigmoid model with two parameters, center and width, is $\ln L_1$. The log-likelihood model for a flat-tailed model with 6 parameters (center, width, and the two θs and zs in the tails) is $\ln L_2$. The difference in parameters is $6 - 2 = 4$. We calculate $2 * \ln L_1 / \ln L_2$, and conclude that the 6-parameter model is better only if our calculated value exceeds the 95% threshold in the χ^2 table with df $= 4$.

For question (ii), whether two traits have different cline shapes, there are two approaches. In both, we first calculate separate maximum-likelihood shapes for each data set. Say we are comparing simple sigmoid (2-parameter) models; we therefore solve for $\ln L_1(data_1)$ and $\ln L_2(data_2)$ with a total of 4 parameters (center and width for each trait). The most widely used method (Hilborn & Mangel 1997) is to then calculate a new best-fitting shape for the combined data, $\ln L_{1\&2}(data_1 \ \& \ data_2)$, which has 2 parameters (i.e., the same center and width for both data sets). Now, sum the single-trait likelihoods to get a likelihood of a combined, 4-parameter model, $\ln L_1(data_1) + \ln L_2(data_2)$. Finally, compare the likelihoods as in the paragraph above, calculating $2*$ $(\ln L_1(data_1) + \ln L_2(data_2))/\ln L_{1\&2}(data_1 \ \& \ data_2)$, and conclude the shapes are significantly different only if this calculated value exceeds the 95% threshold in the χ^2 table with $df = 4 - 2 = 2$. An alternative method is to take the best-fit parameters from trait 1 and calculate their likelihood using data from trait 2, and vice versa. The traits have different shapes if the likelihood falls outside the 95% confidence intervals (Porter *et al.* 1997). In this approach, if the traits have different sample sizes or error distributions, the reciprocal analyses may show different significance levels. These methods can be adapted for any suite of trait comparisons, by varying parameters of interest and holding the others constant.

THE SAMPLING PATTERN

To maximize the information that can be gleaned from a hybrid zone analysis, there are some important principles to consider when collecting the data. These are:

(1) Choose a region where there is a sufficiently narrow hybrid zone that the underlying selection gradient is likely to be step-shaped. Further, choose a region where populations are rather continuously distributed, without gaps in the habitat that might affect local dispersal patterns.

(2) Sample a transect that runs perpendicular to the hybrid zone. Off the perpendicular, the cline will appear wider than it really is, while the disequilibrium measure will stay the same. The result will be to overestimate dispersal and underestimate selection. Moreover, the tails may become more difficult to differentiate from the center region because the step will seem less pronounced. However, the accidental bias is proportional to the cosine of the angle away from the perpendicular, so errors of up to $\sim 10°$ are not likely to produce detectable biases, given the often broad confidence limits on these estimates.

(3) Samples at each locality have to be independent – for example, caterpillars from the same larval mass aren't random samples from the population.

(4) Sampling should to be most intense across geographic localities near the center of the zone. If samples are sparse, it will be hard to distinguish whether the cline has a steep step on one or both sides. The presence and steepness of this step greatly influence the measurement of selection on individual traits in the cline.

(5) Large samples of populations are needed at the center of the cline, so that disequilibrium can be accurately measured. D can be averaged over central populations, but there are significant constraints. First, D is a correlation and is thereby subject to high statistical uncertainty. This translates directly into uncertainty in the dispersal measure, which translates into uncertainty in measures of selection strength. Many of the relationships among these parameters involve ratios and sometimes squared values, and these have the effect of compounding initial uncertainty in the data. Second, D is expected to drop away from the cline center for biological reasons (Fig. 19.3A), so combining samples over too wide an area will dilute its value and underestimate dispersal.

(6) For measuring selection on individual traits, the most valuable information is in the tails of the cline, where traits are also rarest. Here large samples are needed in order to get accurate estimates of the trait distributions in each location. But if the tail is sufficiently flat, it is often possible to spread the sampling over a broader area when accumulating sample size.

(7) In all these analyses, it is best if traits are measured in the same individuals, because disequilibrium and phenotypic correlations are measured by comparing data from the same individuals. Once a dispersal estimate is available, this is less of an issue and studies of additional traits measured in later years need not entail re-collecting the entire data set. But, unless there is some way to account for the uncertainty around the dispersal estimate, parameter estimates that rely on dispersal may underreport sampling error by a significant proportion.

(8) To measure disequilibrium, and therefore dispersal and selection coefficients, it is extremely valuable to obtain genotypes of two or more codominant, autosomal nuclear genes in each individual. It is possible to measure D using phenotypic correlations among quantitative traits, if you have independent information on heritabilities, and if you are willing to assume that laboratory-based measures extend to the field. Given that environments often change near the centers of hybrid zones, that assumption may be hard to defend.

(9) For quantitative traits, it is necessary to measure broad-sense heritabilities to obtain an estimate of selection, because the scaling constants k_L and k_R include the genetic variance V_G. This can be done by rearing full-sib families in a common garden on one or both sides of the cline, using a split-brood design to block potential microenvironmental effects. The variance components for the within-family effects estimate the genetic variance (Lynch & Walsh 1998).

While cline analysis is a very powerful tool, it is also not a panacea. The measures of dispersal rates and selection coefficients it provides represent a suite of hypotheses about these evolutionary processes. In running these analyses, you have significantly assumed that the underlying class of models to which the data were fit is in fact appropriate. Some of the analysis involves model selection – comparison of models with different sets of shape parameters – so this concern is ameliorated somewhat. But to be sure, common-garden experiments are needed, transferring individuals of known genotype across the cline and following their fates (see Mallet & Barton 1989a). Finally, you need to rule out habitat selection as a contributor to the clinal patterns – it is not built into the models, which assume random dispersal. This requires good fieldwork. But, cline shape is sensitive to selection coefficients small enough to be missed using field experiments alone, but strong enough to rapidly influence evolution in the traits they affect.

One need only look in any butterfly field guide or faunal treatment to find cases where subspecies or closely related species meet and hybridize. These taxa differ in traits ranging from minor elements in their wing patterns through to major differences in host use and the timing of their life cycles. Tools are available to analyze the selection and dispersal processes that act upon these traits, and the data collection is not especially difficult for any contemporary research laboratory (and increasingly, well-funded secondary schools). With better understanding of these microevolutionary forces, we will not only have a much better understanding of the roles of ecology and genetics in shaping the traits that define these taxa, we will be in a much better position to reach informed conclusions about their taxonomic relationships.

ACKNOWLEDGEMENTS

I thank H. Descimon, E. Jakob, M. Kovička, N. Johnson, G. Neve, and S. Tulchinsky for their helpful comments. W. Watt and C. Boggs first noticed the functional relationships of the *Pontia* allozyme loci.

Part V
Global change and conservation

20 • Climate warming and distribution changes in butterflies

JANE K. HILL, RALF OHLEMÜLLER, RICHARD FOX
AND CHRIS D. THOMAS

SUMMARY

There is evidence that many species are responding to climate warming by shifting their distributions polewards and/or to higher latitudes. While at very large spatial scales, climate is important in limiting distributions, biotic interactions become more important in determining species distributions at smaller scales. Given the rapidity of current warming, it is important that we determine general patterns underlying species' responses to climate change. Climate response surfaces and bioclimatic distribution models are useful for interpreting how climate limits butterfly distributions. Analyses of current range expansions at species' northern margins demonstrate that expansions are confined to widespread, generalist species. Rates of range expansions are influenced by the role of habitat in determining patterns of expansion, the evolution of dispersal at range margins, changes in genetic diversity associated with expansion and changes in larval hostplant use. These factors are likely to determine future species richness patterns. Distribution changes that are taking place appear to be reasonably predictable; generalist and mobile species are tracking climate change. The rate at which (and whether) species expand can be understood in terms of the amount of habitat available to a species. As a parallel to these among-species patterns, some increases in niche breadth and mobility can be observed in species that are successfully expanding; indicative of traits that predispose species to respond favourably to climate change being under selection. Meanwhile species are starting to retreat in the parts of their ranges where the climate is deteriorating.

The major threats to butterflies in northern Europe seems to be loss of suitable climatic conditions for the northern species, and the inability of many other species to spread northwards across heavily human-modified landscapes. Habitat loss and fragmentation, which have already led to the declines of most butterfly species, are now impeding their ability to adjust their distributions to keep track of climate change.

INTRODUCTION

Global climates warmed by approximately 0.74 °C during the twentieth century, and are predicted to increase by >1 to >6 °C this century (IPCC 2007). There is evidence that many species are responding to this warming by shifting their distributions polewards and/or to higher latitudes (Walther *et al.* 2002, 2005, Parmesan & Yohe 2003, Root *et al.* 2003, Hickling *et al.* 2006, Parmesan 2006). Most evidence comes from observations of species' expansions northwards at high-latitude northern range boundaries, and many species have shown rates of range expansion consistent with observed temperature changes (5 km yr^{-1} northwards, 5 m yr^{-1} uphill: Parmesan *et al.* 1999). Nonetheless, there is continuing debate over the relative importance of biotic factors (e.g. competitive interactions: Davis *et al.* 1998a, b; host plant distributions: Araújo & Luoto 2007) versus abiotic factors (e.g. climate: Hodkinson 1999, Huntley *et al.* 2004) in limiting species distributions, and thus over the reliability of predictions of future range changes that are based primarily on assuming climatic limits to species' distributions. Recent research indicates that the distributions of species spanning a range of trophic levels (primary producers, herbivores and predators) produce equally good fits to climate variables, indicating that, at very large spatial scales, climate is important in limiting distributions of a wide range of different types of organisms (Huntley *et al.* 2004). At small spatial scales, biotic interactions become more important in determining species distributions (Pearson & Dawson 2003). However it is clear that not all species are expanding their distributions polewards during recent climate warming (Parmesan *et al.* 1999, Warren *et al.* 2001, Parmesan & Yohe 2003); given the rapidity of current warming, it is important that we determine general patterns underling species' responses to current climate change.

Insects comprise more than 60% of global biodiversity and play essential roles in the functioning of terrestrial ecosystems. Insects include species that are pests and vectors of disease, as well as pollinators and other beneficial species,

Ecology of Butterflies in Europe, eds. J. Settele, T. Shreeve, M. Konvička and H. Van Dyck. Published by Cambridge University Press. © Cambridge University Press 2009, pp. 315–321.

and species of conservation importance (Speight *et al.* 1999); understanding the responses of insects to climate change is thus important for economists and conservationists. Insects are poikilotherms and many aspects of their growth and survival are temperature dependent. Thus, insects are likely to be particularly responsive to climate and changes in butterfly abundance and phenology are related to climate (Pollard 1988, Roy & Sparks 2000, Roy *et al.* 2001). At continental scales, the distributions of many insects, including butterflies, appear to be constrained by climate (Uvarov 1931, Coope 1978, Pollard 1979, Turner *et al.* 1987, Dennis & Shreeve 1991, Dennis 1993, Parmesan 1996, Hill *et al.* 1999c, Parmesan *et al.* 1999, Bryant *et al.* 2000, Hawkins & Porter 2003, Konvička *et al.* 2006a). Evidence from the palaeoecological record indicates that insects would be expected to respond to climate changes by shifting their distributions, rather than to adapt to climate changes *in situ*, or go extinct (Coope 1978). However, insects have to respond to climate across heavily fragmented, human-modified landscapes, and future climatic conditions may lie outside those experienced for several million years, so Quaternary responses of species may not be replicated in modern landscapes (Warren *et al.* 2001).

In this chapter, we use climate response surfaces and bioclimatic distribution models to investigate the role of climate in limiting butterfly distributions in Europe. Butterflies are an excellent group on which to carry out this sort of study; they are highly visible and popular with both the general public and ecologists and so they have been well studied. As a result, good records exist for current and historical distributions, particularly in Britain. Butterflies also contain species with a range of dispersal abilities and habitat requirements and so the importance of climate in limiting distributions can be investigated across species with contrasting ecologies. After a short review of our research investigating distribution changes in British butterflies we examine rates of range expansion at species' northern margin and demonstrate that expansions are confined to widespread, generalist species. We then focus on these expanding species and examine factors associated with expansion; specifically, we examine (1) the role of habitat in determining patterns of expansion, (2) the evolution of dispersal at range margins, (3) changes in genetic diversity associated with expansion, and (4) changes in larval host-plant use in affecting distribution changes. We describe how distributions are likely to change in the future as the climate continues to warm, and then examine how these changes may affect species richness patterns in the future.

HISTORICAL EVIDENCE FOR THE ROLE OF CLIMATE IN LIMITING SPECIES' RANGES

Over the past 200+ years, many butterflies species have undergone marked changes in their British distributions (Heath *et al.* 1984, Emmet & Heath 1990, Asher *et al.* 2001, Fox *et al.* 2006). For example, approximately 60% of non-migratory British butterflies had more extensive distributions in the early twentieth century than they do currently (Emmet & Heath 1990). For many of these species, changes in land-use and the loss of breeding habitats have been responsible for their decline (Pollard & Eversham 1995), but for others, changes in distribution are related to changes in climate (Pollard 1979, Hill *et al.* 1999c, Parmesan *et al.* 1999). For example, the satyrines *Pararge aegeria* and *Pyronia tithonus* occurred throughout most of Britain during the nineteenth century, probably occurring as far north as central Scotland (Thomson 1980). However, these species underwent marked contraction of their ranges in the nineteenth century at a time when the British climate appears to have been cooler. The distributions of several non-satyrine butterflies also contracted at this time (Thomson 1980, Pratt 1986–87). By 1930, *P. tithonus* had disappeared from Scotland and *P. aegeria* was generally restricted to southwest England and Wales (Emmet & Heath 1990). However, from the 1930s onwards, the climate in the UK has been warming, and at least 15 British butterfly species are currently expanding their distributions northwards (Asher *et al.* 2001, Fox *et al.* 2006). Additional evidence for range limits to be climate driven is that UK range expansions temporarily contracted or halted during the cooler decades of the 1950s and 1960s, before resuming expansion from the 1970s onwards (Jackson 1980).

INVESTIGATING THE ROLE OF CLIMATE IN LIMITING RANGE MARGINS AT CONTINENTAL SCALES IN EUROPE

We have used 'climate response surfaces' to investigate the potential role of climate in limiting the distributions of butterflies in Europe (Huntley *et al.* 1995, Hill *et al.* 1999c, 2002). These models fit each species' European distribution (Tolman & Lewington 1997) to three bioclimate variables that reflect principal limitations on butterfly growth and survival. These variables are; (1) annual temperature sum above 5 °C (GDD5: developmental threshold for larvae);

(2) mean temperature of the coldest month (MTCO: related to over-wintering survival); (3) moisture availability (AET/PET: related to host-plant quality and expressed as an estimate of the ratio of actual to potential evapotranspiration – Huntley *et al.* 1995). For all but one species (*Erebia epiphron*), there were excellent fits between observed and simulated butterfly distributions (Hill *et al.* 2002), indicating that climate is important in limiting species' distributions at continental scales. The distributions of species with different ecologies and mobilities showed equally good fits to climate.

THE ROLE OF HABITAT IN AFFECTING RANGE EXPANSION

Many previously extensive tracts of natural habitat are now highly fragmented, requiring species to move from one isolated habitat fragment to another if they are to keep track of current climate changes successfully. We determined the degree to which butterflies are tracking climate by examining existing long-term distribution records (from Butterfly Conservation and the Biological Records Centre). We used data from two time periods (1970–82: Heath *et al.* 1984; 1995–99: Asher *et al.* 2001), to examine the degree to which species have expanded their ranges northwards during a period of anthropogenic climate warming. We selected only those species that reach their northern range limits in Britain and quantified range changes as the difference in the number of 10-km grid squares occupied during the two periods. We predict that all species should have expanded northwards if they are responding solely to climate, but our analysis showed that 34 of the 46 study species had declined over time, and only 12 species had expanded northwards as predicted. In addition, expansions were generally confined to mobile, generalist species (*sensu* Pollard & Yates 1993), and most specialists declined. Thus most species are failing to track recent climate warming (Warren *et al.* 2001).

For those generalist species that are expanding, habitat is also important in determining their rate of range expansion. For example, the satyrine butterfly *Pararge aegeria* is currently expanding its distribution in Britain and we examined expansion rates in two regions that differ in their degree of habitat availability. Our comparison of rates of range expansion between England and Scotland showed that range expansion over the past 30 years was approximately twice as fast in Scotland where there is approximately 25% more habitat, a difference which is predicted successfully when the expansion of *P. aegeria* is simulated using a colonisation model (MIGRATE) applied to the two landscapes (Hill *et al.* 2001).

FACTORS ASSOCIATED WITH CLIMATE-DRIVEN RANGE EXPANSIONS IN BUTTERFLIES

Species traits

The fossil record provides little evidence for evolutionary adaptations of species to Quaternary climate changes. Nonetheless, evidence from our studies indicates that evolutionary changes are occurring in populations that are currently expanding during recent climate warming. These include evolutionary changes in dispersal, fecundity and larval host-plant use, which may affect species' responses to future climate change. For example, any changes in dispersal that increase the ability of species to colonise fragmented habitats would increase the likelihood that species could track climate changes. Increased dispersal is often documented in newly established populations (e.g. Niemela & Spence 1991, Thomas *et al.* 2001a, Hanski *et al.* 2002, Simmons & Thomas 2004). This is because colonists are not a random selection of individuals from source populations but share a suite of traits associated with dispersal. Our data examining populations of *P. aegeria* at expanding range margins also showed that increased dispersal was evident at expanding range margins. We measured dispersal indirectly through measures of adult flight morphology (increased flight ability in butterflies is associated with relatively larger, broader thoraxes and smaller abdomens: Dempster *et al.* 1976, Srygley & Chai 1990a, Chai & Srygley 1990). We showed that increased dispersal was evident at considerable distances from the range boundary, indicating benefits to increased dispersal for many generations after colonisation, and especially in more fragmented landscapes (Hill *et al.* 2006). However, there are physiological trade-offs between flight and reproduction such that increased dispersal in expanding populations is balanced by reduced fecundity (Hughes *et al.* 2003, 2007) and so population growth rates in newly colonised sites may be lower than would otherwise be expected. Comparable patterns have also been observed in bush crickets (Simmons & Thomas 2004).

Genetic structure

Reduced genetic diversity is often evident in populations as a consequence of postglacial expansion (Schmitt & Seitz 2002b) as well as in populations in fragmented habitats (e.g. Berwaerts *et al.* 1998). These genetic changes arise as a consequence of repeated founder events, population bottlenecks and/or genetic drift in small populations. We compared allozyme diversity in three species of butterfly in Britain: two expanding species (*Pyronia tithonus*, *Pararge aegeria*) and a control (non-expanding, related) species (*Maniola jurtina*) (Hill *et al.* 2006). Overall levels of genetic diversity were lower in the expanding species compared with the control species. There was also a latitudinal cline in diversity in species with the most fragmented habitat (*P. aegeria*) and significantly reduced diversity in populations at the expanding range margin. These data indicate that anthropogenic habitat loss may reduce genetic diversity during range expansion, and thus species' ability to adapt to novel environmental changes, which could affect their long-term persistence (Schmitt & Hewitt 2004a, Hill *et al.* 2006).

Environmental tolerances

Many species have more restricted realised niches towards the boundaries of their ranges (Hengeveld & Haeck 1982, Brown 1984a). Butterflies, for example, are often restricted to particularly warm micro-habitats at their cool, northern range margins but are able to occupy a wider variety of habitats towards the core of their range. Thus we might expect climate warming to benefit some species through increasing niche breadth, availability of habitat and/or abundances at range margins (Wilson *et al.* 2002, Davies *et al.* 2005), thus assisting species in tracking climate warming. Some butterflies have shown very rapid rates of expansion. *Polygonia c-album* has expanded its range margin northwards in Britain by approximately 200 km during the past 60 years (Hill *et al.* 2002). Larvae of *P. c-album* are polyphagous and rapid range expansion in this species may be associated with changes in host-plant preferences over time. There is some evidence that historically *P. c-album* fed primarily on hops (*Humulus lupulus*) but the species is now more commonly reported feeding on common nettle (*Urtica dioica*) and wych elm (*Ulmus glabra*) (Asher *et al.* 2001, Pratt 1986–87). We have investigated larval performance on these different larval host plants (Braschler & Hill 2007). Not only are *Urtica dioica* and *Ulmus glabra* more widely available than *H. lupulus* in northern Britain, but larval performance is better, suggesting higher population growth rates on these hosts. Compared with hops, larvae have higher survival on *Urtica dioica* and *Ulmus glabra* and adults are less likely to enter diapause and thus *P. c-album* is likely to develop through two generations rather than one

generation per year (Braschler & Hill 2007). The recent rapid range expansion of *Aricia agestis* in Britain may also be associated with changes in larval host-plant preferences which increase the availability of habitat in the landscape (Thomas *et al.* 2001a). However, it should be noted that most British butterfly species have not broadened their host ranges in response to climate change; most specialist species have remained specialised, and confined to their traditional (and dwindling) habitats.

IMPACTS OF CLIMATE WARMING ON SPECIES RICHNESS

Most studies have focused on individual species' responses to current climate warming, and it is clear that the different species respond in different ways. The question is whether there are also predictable community or species assemblage responses to climate change. In terms of understanding recent changes and making future projections, should we model individual species separately and then derive a composite community measure (e.g. species richness) from the individual predictions, or can species richness be accurately predicted directly? We test this by fitting a model of UK butterfly species distribution, richness and climate in time period 1 (1970–82) and predicting butterfly species distribution and richness in time period 2 (1995–1999). This is one of the few data sets worldwide where species distributions for two time periods are sufficiently well known to enable us to compare predicted with observed species richness in the later (1995–99) period (1970–82: Heath *et al.* 1984; 1995–99: Asher *et al.* 2001). We modelled patterns of butterfly species richness in two ways:

(I) Species richness per grid cell was directly modelled as the response variable.
(II) Individual species' distributions were modelled separately and all probability-of-occurrence values were summed for each grid cell.

In order to account for uncertainties arising from the modelling method used (Thuiller 2004), all analyses were carried out using three commonly used methods: parametric generalised linear models (GLM), semi-parametric generalised additive models (GAM) and non-parametric regression trees (RT). Distributions of individual species and species richness were fitted to four climate predictor variables known to influence the growth, development and activity of butterflies: growing degree-days above 5 °C, mean temperature of the coldest month, mean temperature of the warmest month and total annual sunshine hours.

Table 20.1 *Accuracy of UK butterfly species richness predictions in 1995–99 at the 10 × 10 km grid cell resolution for two different prediction methods (I: species richness per grid cell was directly modelled as a response variable, II: individual species distributions were modelled and values of probability of occurrence were summed up as a measure of species richness) and three modelling methods (GLM, generalised linear models; GAM, generalised additive models; RT, regression trees)*

| | | Proportion of grid cells for which the predicted 1995–99 species richness was: | | | |
Prediction method	Modelling method	more than 25% **lower** than the actual richness	**between ±25%** of the actual richness	more than 25% **higher** than the actual richness	Total
I	GLM	42%	39%	20%	100%
	GAM	57%	27%	16%	100%
	RT	55%	31%	14%	100%
II	GLM	40%	42%	17%	100%
	GAM	55%	29%	17%	100%
	RT	49%	38%	14%	100%

The predictive accuracy of our models was evaluated by calculating the root mean square error (RMSE) between our predicted species richness values and the observed species richness for 1995–99 (Asher *et al.* 2001). An RMSE of zero indicates 100% predictive accuracy and increasing RMSE values indicate decreasing predictive accuracy. Models included 49 resident British butterflies that occurred in more than 50 10-km grid squares in the earlier time period. Our models suggest that the accuracy of species richness predictions was very similar for both the two prediction methods (I and II above: Fig. 20.1). All methods over- or underestimated the observed 1995–99 species richness by, on average, 7–9 species (Fig. 20.1).

Depending on the modelling method used, species richness was predicted with an accuracy of ±25 % of the actual richness for 27–39% of all grid cells using richness as a predictor variable (I) and for 29–42% of the grid cells using individual species predictions (II) (Table 20.1). For ca. 50% of the grid cells, the actual species richness was underestimated by all methods and for ca. 20% it was overestimated (Table 20.1). Species-rich areas in southern Britain were generally underestimated (i.e. predicted with lower than observed richness) by all methods, whereas areas of low richness in northern Britain were generally overestimated (predicted with higher than observed richness) by all methods (Fig. 20.2). We found that regional differences in species richness estimations were influenced more by the choice of modelling method (GLM, GAM, RT) than by the

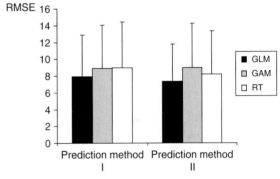

Figure 20.1. Predictive accuracy of 1995–99 butterfly species richness using the two prediction methods (I: species richness per grid cell was directly modelled as a response variable, II: individual species distributions were modelled and values of probability of occurrence were summed up as a measure of species richness) and three modelling methods (GLM, generalised linear models; GAM, generalised additive models; RT, regression trees). The root mean square error (RMSE) indicates the average number of species by which the actual species richness was over- or underestimated. Error bars indicate standard deviations.

choice of individualistic vs. assemblage-based predictions (Fig. 20.2). Generalised linear models (GLM) generally performed slightly better than GAMs and RTs as indicated by the lower RMSE (Fig. 20.1) and there were larger areas in Britain for which the predicted species richness was within ±25% of the actual richness (Table 20.1). These results

GLM   GAM   RT

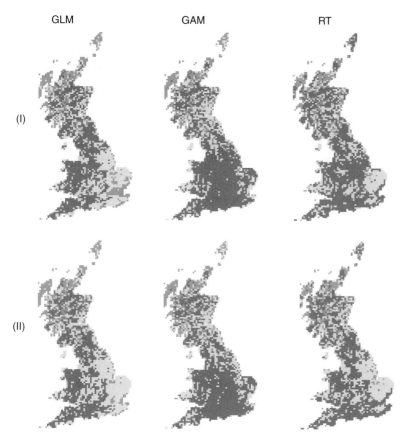

(I)

(II)

Figure 20.2. Deviance in predicted 1995–99 from observed 1995–99 butterfly species richness, given as proportion of observed 1995–99 richness for three modelling methods (GLM, GAM, RT – see Fig. 20.1) and community (I) vs. individualistic (II) species richness predictions. Dark grey, 1995–99 species richness underestimated by more than 25% of actual richness; medium grey, 1995–99 species richness overestimated by more than 25% of actual richness; light grey, 1995–99 species richness estimated within ± 25% of actual richness. Only grid cells with >0 species in the 1995–99 period are shown.

suggest that, for large-scale biodiversity predictions under different climate scenarios and for certain areas, the composite measures of richness may work just as well as using individual species projections. It may therefore be feasible to model future shifts in biodiversity using data based on complete species richness inventories at selected locations (which are comparatively cheap and quick to obtain) rather than using data based on extensive species surveys (which are comparatively expensive, time-consuming and generally limited to a few species). However, for other areas, factors other than climate might be responsible for changes in species richness and neither individual nor composite predictions of richness may reliably predict species richness.

On this basis, we modelled species richness directly to evaluate the extent to which current species richness patterns were, or were not, keeping up with climate change (Menéndez *et al.* 2006). Average species richness did increase between 1970–82 and 1995–99, but increased more slowly than would have been expected if species richness had been tracking climate change perfectly. Only about one-third of

the predicted increase in richness had taken place, and this was almost entirely attributable to the expansion of relatively mobile southern habitat generalists, suggesting that a combination of limited habitat availability and low dispersal capacity have constrained the ability of the fauna to respond to climate change. Butterfly species richness has lagged behind climate change and assemblages have increasingly become dominated by habitat generalists (Menéndez *et al.* 2006).

MODELLING POTENTIAL FUTURE EUROPEAN DISTRIBUTIONS

Our climate response surface models predict that species' European distributions have the potential to shift polewards in the future (2070–99: Hill *et al.* 2002). For many southerly distributed species, our research suggests that the overall sizes of species ranges will change relatively little, although the location will change. Northern and montane species have little opportunity to shift further northwards and will disappear from locations in the south resulting in greatly reduced distribution size in the future. These estimates of future range size assume that species' southern and northern range boundaries are equally sensitive to climate. There is some evidence that southern range boundaries may be more stable and less sensitive to climate warming (Parmesan & Yohe 2003), although long-term, fine-resolution data show that southern range margins are retracting at a rate expected if they were responding solely to climate (Wilson *et al.* 2005, Franco *et al.* 2006). Our studies suggest that the majority of southern species have failed to track climate (see above) and so we revised our predictions for future distributions accordingly for these species, and we estimated 65% and 24% declines for northerly and southerly distributed species, respectively (Hill *et al.* 2002). We consider these more reliable estimates of butterfly range sizes in the future. Our models of simulated potential future distributions give an indication of the availability of suitable climate in the future and of the magnitude of the potential impact of climate change on species distributions, but they are not forecasts of future butterfly distributions. The degree to which species do or do not achieve these future distributions remains to be seen, but we predict that lags between species distributions and suitable climate will increase in the future. In our analysis of UK butterfly species richness changes between 1970–82 and 1995–99 (Table 20.1), our climate models predicted butterfly species richness to be >25% lower than the actual species richness for ca. 50% of all grid cells in the later time period. This indicates that for approximately half of the grid cells in the UK, we find more species than we would expect from 1970–82 richness/climate relationships; however, a bias in sampling intensity between the two time periods and poor model fits may have to be considered when interpreting these results in terms of potential lags. Our initial results suggest that analyses of community-level responses could complement species-specific projections, and they may be more robust because they will not be greatly affected by the individualistic responses of some species that do not respond as projected. However, this requires further testing, because the results may vary from region to region (e.g. depending on whether temperature or moisture availability is the main limit to species richness).

Despite considerable uncertainties, not least in what future conditions will be like, the distribution changes that are taking place appear to be reasonably predictable; generalist and mobile species are tracking climate change, even if they lag somewhat behind the climatic changes themselves. The rate at which (and whether) species expand can be understood in terms of the amount of habitat available to a species. As a parallel to these among-species patterns, we are seeing some increases in niche breadth and mobility in those species that do expand successfully; in some cases suggesting that the traits that predispose species to respond favourably to climate change are also under selection within species. Meanwhile species are starting to retreat in the parts of their ranges where the climate is deteriorating. The major threats to butterflies in northern Europe seem to be loss of suitable climatic conditions for the northern species, and the inability of many other species to spread northwards, across heavily human-modified landscapes. Habitat loss and fragmentation, which have already led to the declines of most butterfly species, are now impeding their ability to adjust their distributions to keep track of climate change.

ACKNOWLEDGEMENTS

We thank the huge number of volunteers whose records have contributed to the data sets held at the Biological Records Centre and Butterfly Conservation. The research was funded by NERC and by the EU FP6 project "ALARM" (Settele *et al.* 2005).

21 • Conservation status of European butterflies

CHRIS A. M. VAN SWAAY, DIRK MAES AND MARTIN S. WARREN

SUMMARY

The status of butterflies in Europe was published by the Council of Europe in 1999. The methods and criteria used in the *Red Data Book* are presented and the status of the 576 species of butterfly known to occur in Europe is assessed. For almost all of the 19 endemic species, a decline in distribution has been apparent for the last 25 years, making them a major conservation priority in Europe. In the *Red Data Book* almost 40 species are considered to be *Near Threatened*. These are most likely to move up to the category *Vulnerable* if no actions for conservation are taken. Following the concept used for birds, the assessments also identified butterflies termed Species of European Conservation Concern (SPECs), divided into four categories depending on their global conservation status, their European Threat Status and the proportion of their world range in Europe. Countries with the greatest responsibilities for the conservation of butterflies in Europe are identified, as well as the most species-rich, and the most threatened biotopes in Europe. The biggest threats to European butterflies are from agricultural improvements as well as abandonment. Resulting from massive direct loss of suitable areas, isolation and fragmentation of population units emerge as a growing threat. Within the conservation strategies the following elements are discussed: (1) improved legislation, (2) habitat protection, (3) habitat management, (4) research and monitoring, (5) fragmentation of habitats and the wider countryside, (6) climate change, (7) the development of a European Butterfly Indicator, and (8) the growing urgency for coordination of an overall action plan for European butterflies. The last is envisaged under the umbrella of Butterfly Conservation Europe, an organisation formed in 2004, with the aims of conserving butterflies, moths and their habitats across Europe (www.bc-europe.eu).

INTRODUCTION

The first overview of threatened butterflies in Europe was published in 1981 by John Heath, based on information from around 23 countries. Since that time a large amount of new information has become available on the status of butterflies following detailed surveys and mapping schemes in many countries. By the late 1990s it was clear that there was an urgent need to conduct a more comprehensive and up-to-date review in order to identify priorities and plan an effective conservation strategy for this high profile group of insects.

In 1997 the Council of Europe asked Chris van Swaay from Dutch Butterfly Conservation and Martin Warren from Butterfly Conservation (UK) to produce a new and revised overview on the present status of butterflies in Europe (all countries belonging to the Council of Europe, including Madeira, the Azores, the Canary Islands, Cyprus, the whole of Turkey and Russia east to the Urals). The aims of the review were to provide an up-to-date assessment of the threat status of all resident species and a consistent continental conservation framework for butterflies in Europe. The project was supported financially by the Council of Europe, English Nature, the Dutch Ministry of Agriculture, Nature Conservation and Fisheries, and 'in kind' contributions from Dutch Butterfly Conservation and Butterfly Conservation (UK). In this chapter we summarise the main results of the review, which was published by the Council of Europe in 1999 (Van Swaay & Warren 1999), and provide additional information on the main habitats for butterflies in Europe.

METHODS AND CRITERIA USED IN THE *RED DATA BOOK*

Data were primarily collated by distributing questionnaires to expert national compilers in all European countries. These data are ultimately based on the field work carried out by hundreds or even thousands of amateur lepidopterists over many years, often drawing on detailed distribution data.

Using these questionnaires, data were collected on all native species within each country covering:

- Present distribution
- Trend over the last 25 years
- Main habitat used by the species (CORINE classification)

Ecology of Butterflies in Europe, eds. J. Settele, T. Shreeve, M. Konvička and H. Van Dyck. Published by Cambridge University Press. © Cambridge University Press 2009, pp. 322–338.

Species just reaching their natural boundaries in Europe, perhaps having established only temporary populations, are considered marginal to Europe and were excluded from the review. For all remaining species we calculated their European distribution class and trend over the whole continent, giving weight to the size of each country. Species names were according to Karsholt & Razowski (1996).

To identify the threat status it was necessary to distinguish between species restricted to Europe and species that can also be found outside Europe:

- For species restricted to Europe the 1994-IUCN criteria (IUCN 1994) can be applied directly since the European status reflects their global status. The main differences are that data on trends in butterfly populations are mostly available over the last 25 years, rather than the 10-year period used by IUCN. Another problem is that the IUCN criteria refer to trends in population size whereas the data available for butterflies in Europe is nearly always based on coarse distribution data. However, for many colonial butterflies, range declines assessed from distribution data based on grid cells (e.g. UTM) have been shown to seriously underestimate population decline (e.g. by an average of 32% at mapping scales of 10×10 km: Thomas & Abery 1995). Furthermore, not only rare species are underestimated using grid cells as units of distribution; more 'common' species also often show strongly declining trends when actual habitat areas are used to calculate trends in changes in distribution area (Cowley et al. 1999, León-Cortés et al. 1999, 2000). Thus we have adjusted the IUCN criteria which are based on population decline over a 10-year period to a roughly equivalent distribution decline over a 25-year period for which data are available on European butterflies. Nevertheless for most species our data were not precise enough. Therefore local specialists were consulted for more detailed information if necessary.
- For species that can also be found outside Europe, the 1994-IUCN criteria for assessing global threat cannot be directly used since no information is available on the trend and abundance outside Europe. We therefore assessed European threat by adapting the 1994-IUCN criteria for use at a European level. With the data gathered, it was relatively easy to apply the IUCN criteria for rates of decline but far more difficult to interpret the other criteria which relate to rarity. We therefore chose to concentrate on trend and incorporate rarity only for very rare species that occur in less than 1% of Europe.

We believe this follows the IUCN criteria as closely as possible with the data currently available.

- Since the publication of the European *Red Data Book*, IUCN have published new criteria (IUCN 2001) and guidelines for regional *Red Lists* (IUCN 2003). So far these new criteria have only been used at a national or local scale (e.g. Van Swaay 2006), not at a European scale.

Participants have been asked to indicate the quality of the estimation of present abundance and trend. If more than half of the present European distribution of a species is in countries where the quality of trend estimation is considered 'poor' by the national participants or where the trend is unknown, then this is indicated in the tables with the results.

Because the 1994-IUCN criteria are based on estimates of rates of decline and extinction risk as well as rarity, they produce a different, but more useful, assessment compared to the older criteria which are based on more subjective and less explicit criteria. One result of these criteria is that widespread but rapidly declining species are included for the first time, highlighting large-scale changes that might otherwise have been ignored until species reached critical levels. The 1994-IUCN criteria are consequently felt to be a far preferable method for assessing conservation priorities amongst European butterflies and identifying species requiring conservation action than earlier versions. They have also been adapted successfully for use at the national level in Britain (Warren et al. 1997), Flanders (north Belgium) and the Netherlands (Maes & van Swaay 1997).

THREATENED BUTTERFLIES IN EUROPE

Species restricted to Europe

The study assessed the status of the 576 species of butterfly known to occur in Europe. Table 21.1 shows the 19 threatened butterfly species whose global range is restricted to Europe. Most of the species have a very restricted distribution, at present often less than 1% of the area of Europe. The two exceptions are *Pyrgus cirsii* and *Maculinea rebeli*. The first species is very difficult to distinguish from close relatives and is mainly distributed in southern France, Spain and Portugal. The taxonomic status of *M. rebeli* has been disputed for a long time and recent publications suggest that together with *M. alcon* it is a single ecologically differentiated species (Als et al. 2004), but at the time of the production of the *Red Data Book* it was treated as a separate species.

Table 21.1 *Globally threatened butterflies restricted to Europe*

Species	Global threat status[a]	Distribution class[b]	Trend class[c]	Conservation status (SPEC)[d]
Pyrgus cirsii	VU	1–5%	decrease 20–50%	1
Zerynthia caucasica	VU	<1%	decrease 20–50%	1
Pieris wollastoni	CR	<1%	decrease 80–100%	1
Pieris cheiranthi	VU	<1%	decrease 20–50%	1
Gonepteryx maderensis	EN	<1%	unknown	1
Lycaena ottomanus	VU	<1%	decrease 20–50%	1
Maculinea rebeli	VU	1–5%	decrease 20–50%	1
Plebejus trappi	VU	<1%	unknown	1
Plebejus hesperica	VU	<1%	decrease 20–50%	1
Polyommatus humedasae	EN	<1%	unknown	1
Polyommatus dama	EN	<1%	decrease 50–80%	1
Erebia christi	VU	<1%	decrease 20–50%	1
Erebia sudetica	VU	<1%	decrease 20–50%	1
Erebia epistygne	VU	<1%	decrease 20–50%	1
Hipparchia maderensis	VU	<1%	decrease 20–50%	1
Hipparchia azorina	VU	<1%	decrease 20–50%	1
Hipparchia occidentalis	VU	<1%	decrease 20–50%	1
Hipparchia miguelensis	VU	<1%	decrease 20–50%	1
Pseudochazara euxina	VU	<1%	decrease 20–50%	1

[a] Global threat status: CR, Critically Endangered; EN, Endangered; VU, Vulnerable.

[b] Distribution class: the percentage European landcover where the species is reported after 1980, divided into five classes.

[c] Trend class: change in species distribution between 1970 and 1995.

[d] Conservation status (SPEC): depending on threat and proportion of world range in Europe. SPEC 1, species of global conservation concern because they are restricted to Europe and considered threatened in Europe.

Furthermore many of these species are typical of mountain areas in Europe (e.g. the *Plebejus* and *Erebia* species) or Atlantic archipelagos (e.g. the *Hipparchia* and *Pieris* species in the list).

Yet, for almost all of these endemic species a decline in distribution has been apparent for the last 25 years, making them a major conservation priority in Europe. *Gonepteryx maderensis*, *Plebejus trappi* and *Polyommatus humedasae* are species with a very limited distribution in the world. Although the exact trend is unknown, local butterfly specialists report a severe decrease during the last 25 years.

Four European endemics (*Pyrgus cinarae*, *Pararge xiphia*, *Erebia melas* and *Hipparchia mersina*), are classified as *Near Threatened*. They have a present distribution area of less than 1% of Europe and show a decrease of 15–20% during

the last 25 years. These will be the first species to move up to the category *Vulnerable* if their status deteriorates.

Species also found outside Europe

Table 21.2 lists the species threatened in Europe, but which can also be found outside Europe. One species is considered regionally extinct in the investigated part of Europe. There is a chance that *Polyommatus caeruleus* is still present in one of the Caucasian Republics, but no recent information was available. In the *Red Data Book* almost 40 species are considered to be *Near Threatened*. They are listed in Table 21.3. These will be the ones most likely to move up to the category *Vulnerable* if no actions for conservation are taken.

Table 21.2 *Butterflies threatened in Europe found both within and outside Europe*

| Species | European threat status | | European distribution class[b] | Trend class in Europe[c] | European conservation status (SPEC)[d] |
	Red Data Book	Global criteria[a]			
Spialia osthelderi	CR	EN	<1%	decrease 50–80%	3
Muschampia proteides	EN	VU	<1%	decrease 20–50%	3
Pyrgus centaureae	VU	VU	5–15%	decrease 20–50%	3
Thymelicus acteon	VU	VU	5–15%	decrease 20–50%	2
Archon apollinus	EN	VU	<1%	decrease 20–50%	3
Archon apollinaris	EN	VU	<1%	decrease 20–50%	3
Parnassius phoebus	VU	LR(nt)	<1%	decrease 15–20%	3
Parnassius apollo	VU	VU	5–15%	decrease 20–50%	3
Leptidea morsei	CR	EN	<1%	decrease 50–80%	3
Anthocharis damone	VU	LR(nt)	<1%	decrease 15–20%	3
Euchloe simplonia	EN	VU	<1%	decrease 20–50%	3
Colias nastes	VU	VU	5–15%	decrease 20–50%	3
Colias hecla	VU	VU	1–5%	decrease 20–50%	3
Colias myrmidone	VU	VU	5–15%	decrease 20–50%	2
Colias chrysotheme	VU	VU	5–15%	decrease 20–50%	3
Lycaena helle	VU	VU	5–15%	decrease 20–50%	3
Tomares ballus	VU	VU	1–5%	decrease 20–50%	2
Tomares nogelii	EN	VU	<1%	decrease 20–50%	2
Tomares callimachus	EN	VU	<1%	decrease 20–50%	2
Neolycaena rhymnus	EN	VU	<1%	decrease 20–50%	3
Pseudophilotes vicrama	VU	VU	5–15%	decrease 20–50%	3
Pseudophilotes bavius	EN	VU	<1%	decrease 20–50%	3
Scolitantides orion	VU	VU	>15%	decrease 20–50%	3
Glaucopsyche alexis	VU	VU	>15%	decrease 20–50%	3
Maculinea arion	EN	EN	5–15%	decrease 50–80%	3
Maculinea teleius	VU	VU	5–15%	decrease 20–50%	3
Maculinea nausithous	VU	VU	5–15%	decrease 20–50%	3
Maculinea alcon	VU	VU	5–15%	decrease 20–50%	3
Polyommatus eroides	CR	EN	<1%	decrease 50–80%	3
Polyommatus poseidon	EN	VU	<1%	decrease 20–50%	3
Polyommatus caeruleus	RE	RE	0%	Regionally extinct	3
Polyommatus damone	VU	VU	1–5%	decrease 20–50%	3
Boloria titania	VU	VU	1–5%	decrease 20–50%	3
Boloria thore	VU	VU	5–15%	decrease 20–50%	3
Boloria frigga	VU	VU	5–15%	decrease 20–50%	3
Nymphalis xanthomelas	VU	VU	1–5%	decrease 20–50%	3
Nymphalis vaualbum	EN	EN	1–5%	decrease 50–80%	3
Euphydryas intermedia	EN	VU	<1%	decrease 20–50%	3
Euphydryas maturna	VU	VU	1–5%	decrease 20–50%	3

Table 21.2 (*cont.*)

Species	European threat status		European distribution class[b]	Trend class in Europe[c]	European conservation status (SPEC)[d]
	Red Data Book	Global criteria[a]			
Euphydryas aurinia	VU	VU	5–15%	decrease 20–50%	3
Euphydryas orientalis	CR	CR	<1%	decrease 80–100%	3
Melitaea aetherie	EN	VU	<1%	decrease 20–50%	3
Melitaea aurelia	VU	VU	5–15%	decrease 20–50%	3
Melitaea britomartis	VU	VU	5–15%	decrease 20–50%	3
Lopinga achine	VU	VU	>15%	decrease 20–50%	3
Coenonympha tullia	VU	VU	5–15%	decrease 20–50%	3
Coenonympha oedippus	CR	CR	1–5%	decrease 80–100%	3
Coenonympha hero	VU	VU	>15%	decrease 20–50%	3
Triphysa phryne	CR	CR	<1%	decrease 80–100%	3
Erebia embla	VU	VU	5–15%	decrease 20–50%	3
Erebia medusa	VU	VU	5–15%	decrease 20–50%	3
Melanargia titea	EN	VU	<1	decrease 20–50%	3

[a] Threat status: CR, Critically Endangered; EN, Endangered; VU, Vulnerable; LR(nt), Low Risk near threatened.

[b] Distribution class: the percentage European landcover where the species is reported after 1980, divided into five classes.

[c] Trend class: change in species distribution between 1970 and 1995.

[d] Conservation status (SPEC): depending on threat and proportion of world range in Europe. SPEC 2, species is concentrated in Europe and considered threatened in Europe; SPEC 3, species has its headquarters within and outside Europe and considered threatened in Europe.

Table 21.3 *Thirty-nine butterfly species also found outside Europe with the threat status Near Threatened according to the* Red Data Book

Erynnis marloyi	*Colias palaeno*	*Polyommatus damon*
Spialia phlomidis	*Hamearis lucina*	*Boloria chariclea*
Muschampia poggei	*Lycaena virgaureae*	*Boloria improba*
Muschampia cribrellum	*Lycaena hippothoe*	*Neptis sappho*
Pyrgus onopordi	*Lycaena candens*	*Apatura metis*
Thymelicus novus	*Satyrium ledereri*	*Erebia aethiops*
Gegenes pumilio	*Tarucus theophrastus*	*Erebia polaris*
Pelopidas thrax	*Tarucus balkanica*	*Erebia ottomana*
Zerynthia cerisy	*Zizeeria knysna*	*Melanargia hylata*
Euchloe belemia	*Cupido lorquinii*	*Hipparchia pellucida*
Euchloe charlonia	*Pseudophilotes abencerragus*	*Pseudochazara geyeri*
Euchloe penia	*Plebejus argyrognomon*	*Oeneis bore*
Pieris krueperi	*Polyommatus eros*	*Oeneis jutta*

Effect of IUCN regional guidelines

If we apply the new IUCN regional guidelines (IUCN 2003) to the data gathered for the *Red Data Book*, and use the global criteria unaltered for species found both inside and outside, the effect is generally to reduce the threat status of species listed (Table 21.2). In this assessment we did not perform the downgrading by one class that IUCN (2003) advise if there are conspecific populations outside the region that may affect the European extinction risk. If these were applied in Table 21.2, columns CR would change to EN, EN to VU and VU to LR(nt) leaving only nine species as threatened in Europe. However, if the global criteria are applied unaltered only two species (*Parnassius phoebus* and *Anthocharis damone*, both originally classed as *Vulnerable*) would become *Near Threatened*. In contrast, the effect on the *Near Threatened* species is far more drastic as none of them qualifies for this status under the new draft guidelines.

In the case of European butterflies, we strongly believe that downgrading is not appropriate because there is no knowledge on the effect of immigration from outside Europe and no evidence that this could provide any rescue effect. Also, since there is no information on the status and trend of butterflies in Northern Africa or Western Asia, downgrading is not appropriate according to figure 2 in IUCN (2003). Moreover, there is no reason to suppose that the situation is different outside Europe and we consider that the data obtained for a large region such as Europe is likely to represent the global range for most species. We therefore believe that it is sensible to take a precautionary approach until other information becomes available in non-European parts of the species' range.

CONSERVATION STATUS

The aim of this assessment is to identify species that are of conservation concern at a European scale, following the concept used for birds by Tucker & Heath (1994). These butterflies are termed Species of European Conservation Concern (SPECs) and are divided into four categories depending on their global conservation status, their European threat status and the proportion of their world range in Europe:

SPEC 1 Species of global conservation concern because they are restricted to Europe and considered globally threatened (Critically Endangered, Endangered or Vulnerable).

SPEC 2 Species whose global distribution is concentrated in Europe and that are considered threatened in Europe (Critically Endangered, Endangered or Vulnerable).

SPEC 3 Species whose global distribution is not concentrated in Europe, but that are considered threatened in Europe (Critically Endangered, Endangered or Vulnerable).

SPEC 4 4a Species whose global distribution is restricted to Europe, but that are not considered threatened either globally or in Europe.

4b Species whose global distribution is concentrated in Europe, but that are not considered threatened either globally or in Europe.

The 19 SPEC 1 species are the same as the globally threatened butterflies listed in Table 21.1. From the 52 species in Table 21.2, five are regarded as SPEC 2 species since their global distribution is concentrated in Europe and they are considered threatened in Europe: *Thymelicus acteon*, *Colias myrmidone*, *Tomares ballus*, *Tomares nogelii* and *Tomares callimachus*. The remaining species in Table 21.2 are regarded as SPEC 3 species, since they have their distribution headquarters both within and outside Europe.

The conservation status of a further 170 species is considered SPEC 4a (European endemics, not threatened) and of 33 species SPEC 4b (species largely concentrated in Europe, but not considered threatened in Europe). Only one of the Lower Risk (near threatened) species in Table 21.3. is considered a SPEC 4b species: *Polyommatus eros*.

IMPORTANT COUNTRIES AND HABITATS

By counting the total number of species and SPEC species per country and per CORINE biotope type as indicated by the national compilers of the different countries that participated in the compilation of the *Red Data Book* (van Swaay & Warren 1999), we can determine which countries have the greatest responsibilities for the conservation of butterflies in Europe.

Important countries

The most species-rich countries (with more than 200 species) are situated in the Mediterranean region (Italy, France, Greece, Spain, Balkan States) and in the East (Turkey (Asian part), Russia, Ukraine and Bulgaria) (Fig. 21.1). Most Central European countries, some of the remaining

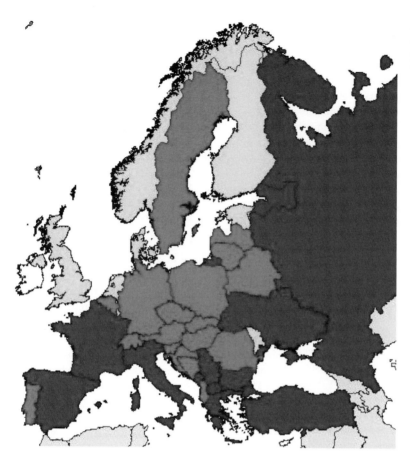

Figure 21.1. Number of butterfly species per country in Europe: light grey = 1–50, mid grey = 51–100, dark grey = 101–200, black >200 species.

Mediterranean countries and Sweden have between 100 and 200 species while the other countries (mostly situated in Western and Northern Europe) hold fewer than 100 species (Dennis *et al.* 1991, Dennis & Williams 1995). Island groups like the Azores and Madeira have small species numbers but a relatively high number of them are of European conservation concern (Fig. 21.1).

The absolute number of SPEC 1–3 species in each country is given in Table 21.4 and shown in Fig. 21.2. All European countries (except Malta) have SPEC 1–3 species and therefore all have responsibilities for the conservation of these species. The number of threatened species generally decreases from the east to the west and from the centre of Europe towards the north and the south. The highest number of threatened species (SPEC 1–3) are found in Russia (40), Ukraine (34) and the Asian part of Turkey (31). Lower numbers (21–30) of SPEC 1–3 species can be found in the central part of Europe (from Belarus and Romania to France and Italy). Two larger regions (the Caucasian republics and

Greece and the Scandinavian and Baltic countries) together with Spain and Belgium hold between 10 and 20 SPEC 1–3 species. All other countries (almost all northwest European plus Portugal, Moldova and Cyprus) have fewer than 10 SPEC 1–3 species.

These results underline the high importance of the Asian part of Turkey, being very rich in habitats and species. Obviously the large number of SPEC 1–3 species in European Russia, Ukraine and the Asian part of Turkey is due partly to the great size of the region and the associated high diversity of habitats. Although low numbers of SPEC 1–3 species occur on the Azores and Madeira, these islands are of considerable importance for several endemic SPEC 1 species.

Important habitats

The most species-rich biotopes in Europe are dry grassland biotopes like dry calcareous grasslands and steppes

Table 21.4 *Total number of species and number of SPECs (Species of European Conservation Concern) per country; in total 576 species are mentioned*

Country	SPEC category			Number of SPECs 1–3 (% total no. of species)	SPEC category		Number of SPECs 1–4	Total number of species
	1	2	3		4a	4b		
Albania	1	1	12	14 (8.0%)	20	13	47	176
Andorra	2	1	4	7 (5.2%)	22	15	44	135
Austria	2	2	25	29 (14.9%)	36	11	76	195
Belarus	–	1	25	26 (20.3%)	2	5	33	128
Belgium	1	1	12	14 (13.2%)	3	7	24	106
Bosnia	1	2	14	17 (9.1%)	23	14	54	187
Bulgaria	2	2	15	19 (9.3%)	28	13	60	205
Croatia	1	2	18	21 (11.4%)	16	14	51	184
Cyprus	–	1	1	2 (4.7%)	2	1	5	43
Czech Republic	2	2	22	26 (17.2%)	6	9	41	151
Denmark	–	–	5	5 (7.4%)	1	2	8	68
Estonia	–	–	14	14 (14.1%)	1	2	17	99
FYR of Macedonia	1	1	12	14 (7.0%)	26	14	54	201
Finland	–	–	17	17 (17.2%)	4	–	21	99
France	4	2	20	26 (11.1%)	53	22	101	235
Germany	2	2	23	27 (15.2%)	25	11	63	178
Greece	1	1	14	16 (7.0%)	37	13	66	229
Hungary	2	2	20	24 (15.3%)	4	12	40	157
Ireland	–	–	2	2 (6.9%)	2	1	5	29
Italy	5	1	22	28 (10.9%)	65	19	112	257
Latvia	–	–	16	16 (14.7%)	1	3	20	109
Liechtenstein	1	–	12	13 (11.6%)	21	5	39	112
Lithuania	–	–	14	14 (12.4%)	2	4	20	113
Luxemburg	–	1	9	10 (10.5%)	3	6	19	95
Malta	–	–	–	0 (0%)	–	–	0	18
Moldova	–	1	6	7 (8.6%)	1	5	13	81
Montenegro and Serbia	1	2	16	19 (9.3%)	26	14	59	205
Netherlands	–	1	7	8 (11.4%)	1	3	12	70
Norway	–	–	12	12 (12.9%)	3	1	16	93
Poland	2	2	21	25 (16.7%)	8	8	41	150
Portugal	1	2	3	6 (5.2%)	9	13	28	116
Azores	3	–	–	3 (33.3%)	–	–	3	9
Madeira	3	–	–	3 (17.6%)	1	–	4	17
Romania	1	3	22	26 (14.5%)	13	12	51	179
Russia (European part)	2	3	35	40 (16.6%)	13	10	63	241
Slovakia	1	2	24	27 (16.2%)	9	12	48	167
Slovenia	2	2	22	26 (14.4%)	19	12	57	180
Spain	4	2	10	16 (7.3%)	40	23	79	218
Canary Islands	1	1	–	2 (7.1%)	7	1	10	28
Sweden	–	–	18	18 (16.8%)	4	2	24	107

Table 21.4 (*cont.*)

Country	SPEC category			Number of SPECs 1–3 (% total no. of species)	SPEC category		Number of SPECs 1–4	Total number of species
	1	2	3		4a	4b		
Switzerland	5	1	21	27 (14.1%)	35	13	75	192
Turkey (Asian part)	4	3	24	31 (9.3%)	62	21	114	334
Turkey (European part)	–	1	9	10 (7.7%)	6	10	26	130
Ukraine	1	4	29	34 (16.4%)	21	11	66	207
United Kingdom	–	1	3	4 (6.8%)	3	3	10	59

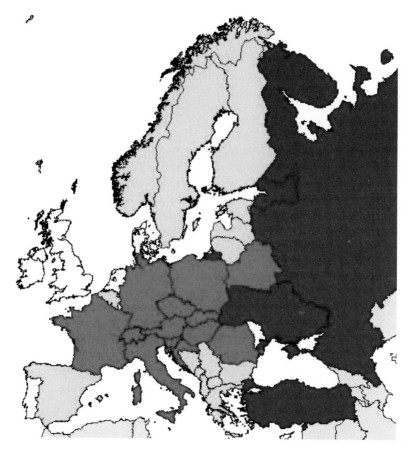

Figure 21.2. Number of SPEC 1–3 species per country: light grey = 1–10; mid grey = 11–20; dark grey = 21–30; black >30.

(274 species), alpine and subalpine grasslands (261), mesophile grasslands (223), dry siliceous grasslands (220 species), followed by heath and (sclerophyllous) scrub (189 species) and different types of woodlands like mixed woodland (187 species), broadleaved deciduous forests (186 species), coniferous woodland (156 species) and humid grasslands and tall herb communities (171 species) (Table 21.5).

The habitats with the largest absolute numbers of SPEC 1–3 species are mainly grasslands: mesophile grasslands (39 SPEC 1–3 species), dry calcareous grasslands and

Table 21.5 *Total number of species, number of SPEC 1–3 species and percentage SPEC 1–3 species per CORINE biotope*

CORINE habitat	Total number of species	Number of SPEC 1–3 species	Percentage SPEC 1–3 species
Blanket bogs	45	14	31.1
Raised bogs	48	13	27.1
Fens, transition mires and springs	59	15	25.4
Water-fringe vegetation	75	15	20.0
Mesophile grasslands	223	39	17.5
Humid grasslands and tall herb communities	171	27	15.8
Mixed woodland	187	29	15.5
Alluvial and very wet forests and brush	100	15	15.0
Coniferous woodland	156	23	14.7
Dry calcareous grasslands and steppes	274	37	13.5
Broadleaved deciduous forests	186	25	13.4
Heath and scrub	189	25	13.2
Alpine and subalpine grasslands	261	34	13.0
Dry siliceous grasslands	220	27	12.3
Inland sand dunes	43	5	11.6
Broadleaved evergreen woodland	67	6	9.0
Inland cliffs and exposed rocks	70	6	8.6
Tree lines, hedges, small woods, bocage, parkland dehesa	128	11	8.6
Phrygana	137	11	8.0
Screes	88	7	8.0
Fallow land, waste places	104	8	7.7
Orchards, groves and tree plantations	95	6	6.3
Cliffs and rocky shores	17	1	5.9
Sclerophyllous scrub	202	12	5.9
Urban parks and large gardens	96	5	5.2
Coastal sand dunes and sand beaches	40	2	5.0
Scrub and grassland	28	1	3.6
Towns, villages and industrial sites	66	2	3.0
Improved grasslands	74	1	1.4

Note: Three SPEC 1–3 species on the Azores (*Hipparchia miguelensis*, *H. occidentalis* and *H. azorina*) are mentioned for agricultural land and artificial landscapes but are not given in the table.

steppes (37), alpine and subalpine grasslands (34) and humid grasslands and tall herb communities and dry siliceous grasslands (27). Different types of woodlands hold lower numbers of SPEC 1–3 species: mixed woodland (29), broadleaved deciduous forests (25) and coniferous woodland (23) while heath and scrub have 25 SPEC 1–3 species.

Relative to the total number of species per CORINE habitat, the most threatened habitats are bogs and marshes (blanket bogs, raised bogs, fens, transition mires and springs, water-fringe vegetation, humid grasslands and tall herb communities, mesophile grasslands), followed by different types of woodlands (mixed woodland, coniferous woodland, broadleaved deciduous forests) and dry habitats (dry calcareous grasslands and steppes, alpine and subalpine grasslands, dry siliceous grasslands). Very specific biotopes like vulcanic features, islets and rock stacks or inland rocks, screes and sands are not discussed due to their low total number of species mentioned but can be of great importance locally (e.g. volcanic features with *Hipparchia maderensis* on Madeira and *Scolitantides orion* in Eastern Europe, islets and

Table 21.6 *Main threats for the butterflies considered threatened in Europe*

Threat	Number of species	Average grade of threat[a]
Agricultural improvements	63	2.1
Isolation and fragmentation of habitat	62	2.1
Built development (inc. roads, housing and mining)	58	1.8
Chemical pollution (inc. herbicides and pesticides)	55	1.8
Afforestation on non-woodland habitats	53	1.9
Recreational pressure and disturbance	48	1.8
Agricultural abandonment and changing management	46	2.1
Collecting (killing or taking)	46	1.4
Felling/destruction of woodland	45	2.1
Abandonment and change of woodland management	45	1.9
Climatic change	45	1.7
Land reclamation/coastal development	41	2.1
Natural ecological change (e.g. myxomatosis effect on rabbits)	37	1.8
Land drainage	33	2.2

[a] Average grade of threat: 1 = low, 2 = medium, 3 = high.

rock stacks with *Parnassius apollo* or inland rocks, screes and sands with *Glaucopsyche alexis*). An overview of the habitat preferences of all European butterflies is published in Van Swaay *et al.* (2006).

MAIN THREATS

Data on suspected threats have been gathered for the threatened species (SPEC 1–3 species) only. The results are shown in Table 21.6. The biggest threat is from agricultural improvement, which affects almost 90% of the threatened species. This threat comprises a wide range of activities from conversion of unimproved grasslands to arable crops, to fertilisation of pastureland. The increasing use of herbicides and pesticides on farmland is also a serious problem for butterflies (affecting 80% of the threatened species), especially in some eastern countries where economic pressures are more severe and regulations are less strict. Building developments such as roads, quarries and housing are also important (affecting 80% of the threatened species). As a result of this massive direct loss of breeding areas, a growing threat arises from the subsequent isolation and fragmentation of habitats, which now affect 87% of the threatened species.

Perhaps the second main threat constitutes the abandonment of agricultural land and changing habitat management. This is thought to affect 65% of the threatened species and

is symptomatic of the widespread cessation of traditional farming systems, which is known to have a negative impact on a variety of other wildlife groups (Tucker & Heath 1994, Poole *et al.* 1998). Examples of changing management include the cessation of cutting of damp hay meadows (affecting species like *Maculinea nausithous*, *M. teleius* and *Lycaena helle*) and abandonment of pasture land (affecting species such as *Euphydryas aurinia* and *Maculinea alcon*). Drainage of wetlands is also a serious problem for many of these species as it is for species restricted to bogs and wet heathland (e.g. *Pyrgus centaureae*, *Boloria frigga* and *Coenonympha tullia*).

Similar problems of abandonment and changing management were also reported in woodland habitats, affecting 63% of the threatened species. This has been recognised as a major problem in western countries for many years (e.g. Warren & Key 1991) but is obviously becoming a widespread European problem. Afforestation of non-woodland habitats is also a major threat to many species, especially those occurring in small breeding areas such as *Parnassius apollo*.

Contrary to many people's views of threats to butterflies, collecting was reported to be of very minor or local importance. However, there were some important exceptions of species which are possibly quite seriously threatened by collecting, notably *Parnassius apollo*, *Polyommatus humedasae*,

Polyommatus poseidon, *Polyommatus damone*, *Euphydryas maturna* and *Coenonympha oedippus*. Nevertheless, all these species are suffering far more seriously from problems such as habitat loss or changing habitat management.

Climatic change is also mentioned as a potential threat to several species, notably highly restricted montane endemics which have co-evolved with specific vulnerable habitats and which have a very limited possibility of adapting to climate change (see Dennis 1993). Yet more recent research (e.g. Warren *et al.* 2001) shows that the negative effects of climate change might be larger than expected and not restricted to a relatively small group of butterflies.

When considering threats, it is worth stressing that Europe is a large and diverse region, and it is therefore clear that the types of threat vary considerably among countries. This partly reflects the fact that the types of habitat used by each species vary naturally in different climatic areas, but also reflect the wide variation of economic and political situations. For example, Romania reported major problems for butterflies resulting from recent political changes which have led to the subdivision of land, the break-up of former nature reserves and increasing use of persistent chemicals on farmland. The increased demand for livestock production also leads to overgrazing of pastureland while many habitats suffer from development pressure. Thus, while there are obviously common threats that operate throughout Europe, each country has its individual set of problems which need to be addressed in a locally adequate conservation strategy.

It is likely that most major threats identified for butterflies will continue to operate in the foreseeable future, and may grow even worse in some countries. The accession of many new countries in Central and Eastern Europe to the European Union has already had large implications on their butterfly fauna, especially reducing the area with semi-natural, extensively managed grasslands. This poses a particularly serious potential threat as these countries hold a disproportionately large number of threatened butterflies. It is doubtful whether new initiatives, like the formation of High Nature Value Farmland areas (European Environmental Agency 2004), will be able to stop this, unless there will also be substantial financial support.

On the plus side, there is a growing move to reform EU agricultural and forestry policies to encourage more environmentally sustainable systems, for example within mechanisms such as the Agri-environment Regulation (EU Regulation 2078/92). Although schemes currently being funded under such regulations comprise a very small proportion of the agricultural budget, they have the potential to slow down some of the trends reported. However, much wider reforms of agricultural policies are also urgently needed (e.g. see Tucker & Heath 1994, Poole *et al.* 1998). Policies such as the EU Habitats and Species Directive may also help to slow declining trends but many countries have been slow to implement this Directive (e.g. Flanders: Maes & Van Dyck 2001) and its likely impact on butterflies remains uncertain.

CONSERVATION STRATEGY

In the past, the conservation of threatened butterflies has mainly been conducted at a national level and there has been no overall co-ordination across Europe. The only relevant pan-European legislation concerning butterflies is the Council of Europe *Convention on the Conservation of European Wildlife and Natural Habitats*, known as the Bern Convention. This has four appendices listing species for which signatories are obligated to take appropriate measures for protection of their habitats or to give strict protection to adults and life stages. Twenty-one butterfly species are listed in the Bern Convention and many countries have used these to direct their own domestic legislation. The lists were also used to identify species covered by the EU Habitats and Species Directive, which is leading to greatly improved protection of some sites. However, the implementation of the Bern Convention amongst Member States is patchy and several countries have failed to ratify the Convention. Also, it is now widely recognised that the species selected for these appendices did not reflect the threat of butterflies across Europe and that a more up-to-date list is needed, based on more objective criteria. The European *Red Data Book* provides the basis for such a revision.

The continuing rapid decline of butterflies highlighted in the *Red Data Book*, and the shortcomings of existing legislation have demonstrated that a new and more comprehensive conservation strategy for butterflies is needed urgently. Some of the specific conservation measures needed are described below. They apply mostly to the countries of the EU, not only because most of the information available comes from these countries, but also because there is still little attention given to butterfly conservation in the remaining non-EU countries in Europe, with the exception of Switzerland and Norway.

Improved legislation

Legislation can play a crucial role in the conservation of Europe's butterflies, provided that it is directed primarily towards the protection and proper management of important butterfly habitats. It must be stressed that a simple ban on collecting is not an effective way of conserving butterflies and can even be counter-productive since it hinders butterfly research by amateurs.

The various lists and annexes of existing legislation should be updated at the earliest opportunity, using the new priorities identified in the *Red Data Book*. Specifically, the threatened species endemic to Europe (SPEC 1) and all *Extinct, Critically Endangered* and *Endangered* species found within and outside Europe (SPEC 2 and 3) should be added to the Bern Convention and the relevant species listed in any revision of Annexe II of the EU Habitats and Species Directive. Individual European countries should also incorporate European threatened species when amending their domestic legislation, so that they address international priorities as well as national and regional ones.

Habitat protection

It is clear that important wildlife habitats are still being destroyed across Europe and that a vastly improved system of habitat protection is needed both within individual countries and at a pan-European level. The EU Habitats and Species Directive is undoubtedly improving the situation within the EU, but progress in implementing the Directive has been very slow in some Member States, and inadequate in others. It is also unclear to what extent key butterfly habitats are being protected under the new measures; this would be a very useful topic for future study.

The improved protection of habitats needs to occur both at the local, site level but also at a larger landscape level. Experience within the UK, the Netherlands and Flanders (north Belgium) has shown that species continue to disappear from nature reserves often because they are too small and isolated (Thomas 1984b, 1995c, Wynhoff & van Swaay 1995, Maes & Van Dyck 2001). Far more emphasis needs to be placed on the protection of whole landscapes as well as individual sites (Warren 1992c, Thomas 1995c).

To help identify priorities for action, a follow-up project was co-ordinated by Dutch Butterfly Conservation and Butterfly Conservation (UK) to identify an initial selection of Prime Butterfly Areas of Europe, similar to the Important Bird Areas identified by Grimmett & Jones (1989). Because resources were limited, the selection was confined to identifying sites for 34 target species, selected either because they were threatened or protected under EU legislation (van Swaay & Warren 2003). Two types of area were included:

(1) discrete sites that support one or more rare or threatened species or (2) wider areas (such as mountain ranges or valley systems) where a target species occurs in scattered populations that may or may not be considered as a metapopulation.

A total of 433 Prime Butterfly Areas (PBAs) were identified among 37 countries and three island archipelagos, covering more than 21 million ha, equivalent to 1.8% of the land area of Europe (Fig. 21.3). Information on trends shows that many target species are declining within PBAs, even within protected areas. This is extremely alarming and indicates that breeding habitats are deteriorating rapidly on most PBAs and that conservation measures are needed urgently. Very few species have undergone a recent increase on PBAs, but trends of target species are not known on many PBAs, indicating the general need for increased monitoring of populations.

Habitat management

Butterflies have very specific habitat requirements and occupy very specific and narrow ecological niches. Many are restricted not only to just one or two food-plants and to particular types of vegetation, but also to particular successional stages. For example, a large number of species rely on grassland biotopes that would normally succeed to woodland unless regularly managed. For centuries, such grasslands have been maintained by traditional systems of livestock grazing or hay-cutting and many studies have shown that butterfly losses are directly caused not by habitat destruction but by the breakdown of some traditional farming or forestry system (e.g. van Swaay 1990, Erhardt 1995, Maes & Van Dyck 2001). The future of many butterflies in Europe will thus depend on the continuation of such traditional regimes, or some near equivalent that will produce a similar range of habitat conditions (Thomas 1993). This presents a major challenge to conservationists in the face of increasing pressure to intensify and modernise agricultural and timber production. It is important to recognise that this issue is vital to a wide range of other wildlife groups and has been recognised by many other conservationists, for example WWF International and the European Forum on Nature Conservation and Pastoralism (Baldock *et al.* 1994, Tucker & Heath 1994, Poole *et al.* 1998).

Far more attention needs to be paid to maintaining appropriate management systems within protected areas and on semi-natural habitats throughout Europe. This will require changes at a strategic policy level as well as producing management plans for individual sites or areas

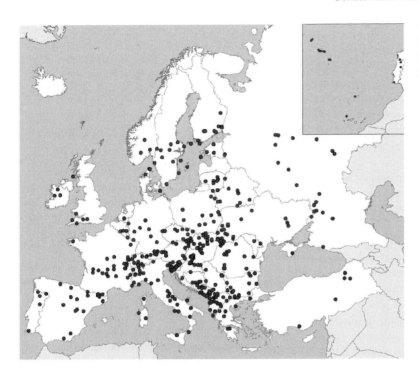

Figure 21.3. The locations of the 431 Prime Butterfly Areas of Europe, identified for 34 target species (after van Swaay & Warren 2003).

(e.g. Bergman & Kindvall 2004, Maes *et al.* 2004, WallisDeVries 2004). For example the EU Common Agricultural Policy, which has been a major engine of agricultural intensification and habitat loss, but also has a huge potential to enhance semi-natural habitats through its Agri-environment Regulation. As a first step we urge a rapid expansion of this programme and a better integration of environmental objectives into all aspects of European agricultural and forestry policy.

Research and monitoring

The ability to identify priorities and major conservation issues is limited by the quality of the data that is available on European butterflies. While a few countries have extremely good recording and monitoring schemes, others have very poor data and species assessment has to be far more subjective. It is thus a priority to establish adequate recording and monitoring in every country as soon as possible, not only to allow a more accurate assessment in the future, but also to allow continuous updates on species status and progress in conservation. An initial attempt to produce distribution maps for all European butterflies has recently been completed (Kudrna 2002), which should help refine

our knowledge of threatened species and target conservation action.

There is currently a responsibility for all signatories to the 1992 Convention on Biological Diversity (which includes most European countries and the European Union itself) to adequately monitor the impact of land-use activities on the environment, including its biodiversity. Butterflies have great potentials as a key indicator of the biological impact of such policies and the European *Red Data Book* has highlighted the major gaps in our knowledge of butterfly populations and where future improvements need to be targeted. It would also be extremely valuable to develop a European-wide butterfly monitoring strategy, combining many of the schemes that are now operating successfully in individual countries (e.g. Pollard & Yates 1993, van Swaay *et al.* 1997, Stefanescu 2000a).

Because each butterfly species has specific requirements, conservation action is only likely to be successful if it is based on thorough knowledge of both the species and its habitats (e.g. Thomas 1984b, Warren 1992c, Dennis *et al.* 2003). Ecological research is thus vital to underpin any conservation strategy and must take a priority in any new programme. While much is known about the ecology of butterflies in general, the requirements of many threatened species remain poorly known.

Fragmentation of habitats and the wider countryside

The dramatic loss of habitats in recent decades has brought about additional problems to butterflies due to the fragmentation and isolation of remaining patches. Recent studies have shown that the persistence of many species depends on metapopulations which consist of networks of small habitat patches. Within these patches, there is periodic extinction and recolonisation and the regional survival of the species probably depends on the maintenance of a whole network of nearby habitats, not all of which may be occupied at any moment in time (Hanski & Gilpin 1991, Thomas 1995c, Hanski & Ovaskainen 2000).

The main implication of studies on threatened species is that we need to consider the conservation of whole landscapes, as well as the individual components of that landscape (e.g. Thomas 1993, Warren 1994, Hanski et al. 1995a, Thomas 1995c, Asher et al. 2001). This will require the integration of all land-use policies, including the Common Agricultural Policy and the planning system to ensure sustainable development and the maintenance of biodiversity. The potential of new positive measures under the EU Agri-environmental Regulation has been mentioned earlier and could play an important role in wider landscape conservation. This has been recognised in the UK and schemes such as Environmentally Sensitive Areas are now being refined to include specific biodiversity objectives, including the needs of several threatened butterflies (e.g. Warren & Bourn 1997). Recent research has made clear that declines are not restricted any more to rare and localised specialists, but more and more common and widespread species are affected as well (Van Dyck et al. 2009).

Climate change

Butterflies may also be adversely affected by other widespread environmental changes, including pollution and global climate change. Data on the likely impact of the latter is patchy at present but has been considered by several authors (e.g. Dennis 1993, Elmes & Free 1994, Parmesan 2006). There is also evidence that climatic warming has led to a northward shift in the ranges of several species (Parmesan et al. 1999), though the implications for threatened species are unclear. Climate change is a particular threat to highly restricted montane endemics in Europe because their habitats are vulnerable and because they are intimately adapted to specific conditions (Wilson et al. 2005). Clearly further research is needed on the impact of such factors on all wildlife,

but studies on butterflies should be included due to their sensitivity to environmental change (Thomas et al. 2004). In 2008 the *Climatic Risk Atlas of European Butterflies* (Settele et al. 2008) was published giving an overview of the possible effects of climate change on the distribution of future climatic spaces of European butterflies.

Developing a European Butterfly Indicator

Recent years have seen political agreements on halting or significantly reducing the current rate of loss of biodiversity by 2010 (the 2010 target). This is accompanied by a growing consensus on the need for structured European co-ordination of biodiversity monitoring, indicators, assessment and reporting efforts, with a long-term perspective and sound funding basis.

Butterflies are good indicators of changes in insects, which represent over 50% of the terrestrial biodiversity in Europe (Thomas 2005). To test the possibilities to use butterfly monitoring data of ten regions in Europe for a European Butterfly Indicator, a grassland butterfly indicator has been developed, containing seven widespread grassland species and ten grassland specialists (Van Swaay & Van Strien 2005).

Data was available for:

- Ukraine (Transcarpathia only) since 1983 (only for *Erynnis tages*)
- Pfalz region (Germany) since 1989 (only for *Maculinea nausithous*)
- The Netherlands since 1990
- Flanders (Belgium) since 1991
- Catalonia (Spain) since 1994
- Aargau (Switzerland) since 1998
- Finland since 1999
- Nordrhein Westfalen (Germany) since 2001
- Doubs and Dordogne (France) since 2001

In the meantime new butterfly monitoring schemes were started in Jersey (Channel Islands), Germany, France and Slovenia. These data could not be used in this assessment.

The method closely follows the one used for the birds (Gregory et al. 2005).

(1) National level. The indices for each species are produced for each country, using TRIM (Pannekoek & Van Strien 2005). TRIM is a programme to analyse time-series of counts with missing observations using Poisson regression.

(2) Supranational level. To generate European trends, the difference in national population size of each species in each country has to be taken into account. This weighting

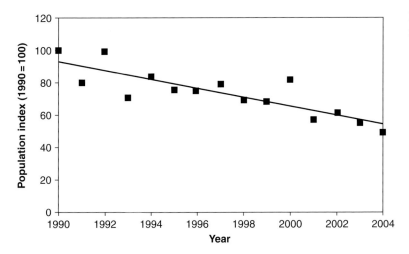

Figure 21.4. European Butterfly Indicator for 17 grassland butterflies.

allows for the fact that different countries hold different proportions of a species' European population (Gregory *et al.* 2005). A weighting factor is established as the proportion of the country (or region) in the European distribution (Van Swaay & Warren 1999). The missing year totals are estimated by TRIM in a way equivalent to imputing missing counts for particular sites within countries (Gregory *et al.* 2005).

(3) Multi-species level. For each species the geometric mean of the supranational indices is calculated.

The resulting graph (Fig. 21.4) of the European Butterfly Indicator for 17 grassland butterflies since 1990 shows a very severe, significant decline of almost 50% over this 15-year period.

An extension of this method to new indicators (e.g. a Wetland Butterfly Indicator or an Open Woodland Butterfly Indicator) will add new information and knowledge and help us to focus conservation (Van Swaay & Van Strien 2005, European Environment Agency 2007). The European Grassland Indicator has been updated successfully in 2008 (Van Swaay & Van Strien 2008). The same trend as shown in Figure 21.4 persisted.

The co-ordination of an overall action plan for European butterflies

The implementation of such a wide variety of measures, across such a wide range of countries and organisations, will require considerable co-ordination. This would best be done under an overall action plan that combines both habitat and species approaches. All species identified as threatened in Europe (SPEC 1–3) should receive specific attention, concentrating on the most threatened European species, but many of the remainder could be combined into habitat plans.

Within more and more European countries, there now exist rapidly growing national societies dedicated to the conservation of butterflies (Butterfly Conservation (UK), Dutch Butterfly Conservation/De Vlinderstichting, Flemish Butterfly Working Group and others), and these organisations are now drawing up strategic Species Action Plans for threatened butterflies (Maes & Van Dyck 1999, Van Swaay 1999, Bourn 2002, Warren 2002). We believe these can act as models for developing European wide plans, but would require a hitherto unprecedented level of pan-European co-operation and commitment. There is now considerable interest in the conservation of butterflies throughout Europe and many countries have active research groups, and specialists within national conservation organisations.

In 2004, a new umbrella organisation, Butterfly Conservation Europe was formed with the aim of conserving butterflies, moths and their habitats across Europe. The new organisation operates through Network Partners in each country and aims to co-ordinate existing activities and stimulate action on priority projects. A website has been created to provide information on European conservation initiatives and to act as a focal point for activities (www.bc-europe.eu). Funds are now being sought as a matter of urgency to run the new organisation and bring together existing groups with the common aim of conserving Europe's rich but threatened butterfly heritage.

In 2007 Butterfly Conservation Europe held its inaugural meeting for partners in Laufen (Germany). The main aims were:

(1) To identify Butterfly Conservation Europe's priorities for reversing the decline of butterflies and moths in Europe and contributing to the 2010 target of halting biodiversity loss.
(2) To identify ways of working together to achieve our goals of a countryside rich in butterflies and moths.
(3) To provide an opportunity for networking amongst partners and board members.

The meeting was a great success, with partners from more than 30 countries participating and keynote speakers from the EU Environment Directorate, IUCN and Birdlife International. Details of the proceedings are available from www.bc-europe.eu

ACKNOWLEDGEMENTS

We are extremely grateful to the national compilers who provided data for the European *Red Data Book* and especially to Eladio Fernandez Galiano of the Council of Europe and David Sheppard of Natural England (formerly English Nature) for helping establish the project.

Note: In 2009 the *Red Data Book* will be updated into a new *Red List of European Butterflies*, which will be published by Butterfly Conservation Europe and the IUCN.

22 • (Meta)population viability analysis: a crystal ball for the conservation of endangered butterflies?

NICOLAS SCHTICKZELLE AND MICHEL BAGUETTE

SUMMARY

Butterflies are popular study organisms and their conservation status is relatively well known across Europe. Several species are considered endangered with conservation targets at the regional, national or pan-European or even global level. Management decisions about endangered species may benefit from guidelines based on the quantitative analysis of the predicted future of the considered (meta)population. Population viability analysis (PVA) is currently the main tool conservation biologists can use to determine whether a (meta)population is at risk of extinction, according to the various threats it faces. It made its debut in the 1970s, and has since then been developed and applied to several other endangered organisms, including butterflies. Environmental stochasticity is considered to be the dominant random force in large populations; demographic stochasticity has a real impact only in very small populations, or populations where the effective population size is limited (e.g. highly skewed sex-ratio). There is a controversy about the importance of genetics in the persistence of populations and the necessity to incorporate it into PVA. According to the way they incorporate spatial dynamics, there are three main categories of current PVA models for metapopulations: stochastic patch occupancy models (SPOM), structured population models (SPM) and individual-based models (IBM). Differences among the models are illustrated by an overview of butterfly PVA studies. Studies were rarely based on a complete set of estimated parameters; most used surrogate parameters or even parameters fixed arbitrarily to some specific value. This illustrates that one major problem of PVA is the availability of information to parameterise the models. The chapter also discusses a series of potential problems and how they could be solved. PVA should be considered as a useful tool that can be helpful in the selection of concurrent management scenarios, provided that the basic rules as highlighted in this chapter are respected.

INTRODUCTION

Butterflies are declining all over Europe (see Chapter 21) and both species richness and abundance are decreasing owing to habitat destruction or change in habitat quality. The abundance of local populations of both generalist and specialist species has been strongly decreasing since ca. 1950, following drastic changes in agricultural practices and land use (e.g. Van Swaay & Warren 1999, Van Swaay & van Strien 2005). Consequently, the distributions of many butterfly species are becoming more and more fragmented at the landscape scale. This results in a continuous process of extinction at the margin of these species' distribution area, generating a shrinking distribution range. This rarefaction process seems faster in butterflies than in other living organisms (Thomas *et al.* 2004, Thomas 2005). In Europe, many butterfly species currently occur as scattered, relict (meta)populations, sometimes quite far away from the more continuous part of their distribution range. The successful management of such (meta)populations is a conservation priority: in the context of the implementation of EU conservation policies like the Natura 2000 network or agri-environment schemes, relict (meta)populations may act as cores for the successful recolonisation of restored (semi-) natural habitats.

Making management decisions about endangered species demands a tool, like a crystal ball, that facilitates 'predicting' the future in order to know what kind of actions need to be taken to ensure that species will persist (Beissinger & McCullough 2002). Population viability analysis (PVA) is such a tool. It has progressively been developed by conservation biologists since the 1970s using the biology of population extinction as theoretical framework. The foundation of this framework may be traced back to Darwin's prediction (1859, p. 301): 'rarity precedes extinction'. Factors increasing population rarity can be divided into two categories: (1) regular and rather continuous environmental pressures, potentially leading to deterministic extinction if they persist,

Ecology of Butterflies in Europe, eds. J. Settele, T. Shreeve, M. Konvička and H. Van Dyck. Published by Cambridge University Press. © Cambridge University Press 2009, pp. 339–352.

and (2) random perturbations, potentially leading to stochastic extinction when population size is small. These two categories correspond to what Caughley (1994) has called the 'declining population' and the 'small population' paradigms, respectively. Most often, extinction is due to the joint action of both categories: a deterministic event (e.g. habitat destruction) pushes the population under a size threshold where stochastic events become a threat. Nevertheless, stochasticity itself also presents a cumulative deterministic tendency to decrease population sizes, sometimes driving them to extinction (Lande 2002). Therefore, the dichotomy between these two paradigms is rather artificial. It is mainly a by-product of research history, with the small population paradigm being driven by theoretical interests whereas the declining population paradigm was documented mainly by empirical studies. This separation is currently disappearing (Hedrick et al. 1996, Boyce 2002).

Whilst deterministic pressures, such as habitat loss and climate change, are highly variable in nature and intensity, stochastic perturbations responsible for natural fluctuations of populations can be classified in four categories (Shaffer 1981, Gilpin & Soulé 1986, Morris & Doak 2002):

- *Demographic stochasticity*: random variation coming from the rounding effect due to the discrete nature of birth and death events: whatever the demographic rates, each individual either dies or survives, and gives birth to a number of offspring that is an integer number. Demographic stochasticity is best understood through a parallel with coin flipping: flipping three coins can obviously not produce the expected average result (1.5 heads and 1.5 tails); when the number of coins flipped increases, the rounding effect decreases and the proportion of heads and tails is closer to the 50%/50% average. This variation is uncorrelated between individuals and its magnitude is inversely proportional to population size.
- *Environmental stochasticity*: variation in the external environment (both temporal and spatial) affecting demographic properties of an entire population. This variation is correlated between individuals (i.e. it affects all individuals), and its magnitude is usually independent of population size.
- *Genetic stochasticity*: changes in gene frequencies due to founder effects, random drift or inbreeding. It may lead to loss of adaptive potential and reduced fitness and finally to extinction, especially when habitat destruction and fragmentation have reduced the ability of populations/species to alter their distribution and range in response to major environmental changes.

- *Natural catastrophes*: extreme and rare events that may occur at random intervals through time and cause sudden collapses in population size. The distinction between normal temporal variation (environmental stochasticity) and catastrophes is obviously arbitrary.

Each endangered species in a given landscape follows its own 'extinction vortex', a positive feedback loop leading to extinction because one or several of those pressures decrease the sizes of populations, which then become more vulnerable to these pressures, leading to further decrease ... (Gilpin & Soulé 1986). Environmental stochasticity is considered to be the dominant random force in populations with more than about 100 individuals (Leigh 1981, Soulé 1987, Pimm et al. 1988; but see Belovsky et al. 1999, 2002). Demographic stochasticity has a real impact only in very small populations, or populations where the effective population size is limited by, for example, a highly skewed sex-ratio (Gerber 2006). There is a controversy about the importance of genetics in the persistence of populations and the necessity to incorporate it into PVA. Some authors argue that population genetics and genetic stochasticity have been demonstrated to have important effects on extinction risk, but only on populations so small that they already risk immediate extinction due to demographic and environmental stochasticities; therefore genetic problems are likely to be neither major threats for natural populations nor the ultimate cause of extinction for the vast majority of species (Berry 1971, Lande 1988, Caughley 1994, Belovsky et al. 1999, 2002). Others have suggested the contrary: inbreeding depression can decrease persistence of populations of several hundreds of individuals, mainly for species with low intrinsic growth rates and stable social systems (Frankham 1995, Lande 1995, Saccheri et al. 1998, Lacy 2000a, Brook et al. 2002b). Still others have suggested that we have insufficient information about the effects of inbreeding in most wild populations to be able to incorporate genetics into PVA (Boyce 1992, Beissinger & Westphal 1998, Morris & Doak 2002). Anyway, the importance of inbreeding depression and other genetic effects for persistence of populations is likely to differ between organisms, for example frequent and important for large vertebrates which frequently form effective populations with only a few individuals, but rare and anecdotal for insects (Ehrlich 1992; but see Saccheri et al. 1998, Nieminen et al. 2001, Vandewoestijne et al. 2008). Nevertheless, demographic and genetic stochasticities can play major roles in the success or failure of colonisation due to the usually low number of propagules (Grapputo et al. 2005, Lavergne & Molofsky 2007).

Plate 1a: *Thymelicus sylvestris* visiting a flower of *Dianthus carthusianorum* in the female phase. Note labial palps with pollen touching the stigma (picture: Andreas Erhardt).

Plate 1b: *Gonepteryx rhamni* female visiting the orchid *Gymnadenia conopsea*. Note proboscis inserted into the floral spur (picture: Andreas Erhardt).

Plate 2: A hilltopping *Iphichlides podalirius* male
(picture: Per–Olof Wickman).

Plate 3: *Leptidea sinapis* males repeatedly eject their proboscis during courtship (cf. Wiklund, 1977) (picture: Per-Olof Wickman).

Plate 4: Males of *Hipparchia semele* have androconial scales on their front wings. During courtship males place themselves in front of females flipping their wings and tilting the body in jerky movements toward the female. These movements probably direct the scent from the androconial scales over the female antennae (see Tinbergen *et al*. 1942). The unwilling female (left) flutters her open wings (picture: Per-Olof Wickman).

Plate 5a: An unwilling *Pieris napi* female elevates the tip of her abdomen when emitting the olfactory substance that functions as an anti-aphrodisiac to the courting male (picture: Per-Olof Wickman).

Plate 5b: During cool weather in the spring *Aglais urticae* basks dorsally and presses wing margins to the substrate (picture: Per-Olof Wickman).

Plate 6a: Reflectance basking in a male *Pieris napi*. During copulation the male carries the female in flight (picture: Per-Olof Wickman).

Plate 6b: During cool weather some lateral baskers like *Hipparchia semele* tilt the body and wing plane to increase the amount of intercepted solar radiation (picture: Per-Olof Wickman).

Plate 7a: *Mesochorus* sp. (Ichneumonidae: Mesochorinae), an obligatory true hyperparasitoid, ovipositing through the body of a small *Melitaea cinxia* larva and into the primary parasitoid larva within (picture: S. van Nouhuys).

Plate 7b: A pseudohyperparasitoid, *Lysibia nanus* (Gravenhorst) (Ichneumonidae: Cryptinae), ovipositing into cocoons of *Cotesia glomerata* (picture: P Mabbott).

Plate 8a: *Hyposoter horticola* (Ichneumonidae: Campopleginae)
with an egg cluster of *Melitaea cinxia*. Unusually, this is an egg–
larval parasitoid: the female visits egg clutches of all ages but
oviposits only when the eggs are close to hatching (picture: Saskya
van Nouhuys).

Plate 8b: *Itoplectis aterrima* Jussila (Ichneumonidae: Pimplinae)
host-feeding on a Campopleginae (Ichneumonidae) cocoon.
Instead of laying an egg in it, this facultative
pseudohyperparasitoid has wounded the inhabitant of the cocoon
with its ovipositor and is now sucking up the exuding fluids
(picture: M. R. Shaw).

Plate 9a: *Trogus lapidator* (Ichneumonidae: Ichneumoninae), a specialist larva–pupal parasitoid of *Papilio machaon*. Emergence through the host's wing case is characteristic of *Trogus*, *Psilomastax* and very close relatives (picture: G. Nobes).

Plate 9b: *Ichneumon eumerus* (Ichneumonidae: Ichneumoninae) ovipositing into a final instar larva of *Maculinea rebeli* (picture: J. A. Thomas).

Plate 10a: *Pimpla rufipes* (Ichneumonidae: Pimplinae), a common idiobiont endoparasitoid of butterfly pupae (picture: S. Edwards).

Plate 10b: An adult *Cotesia melitaearum* (Braconidae: Microgastrinae) from the small broods that emerge quickly from post-hibernation larvae of *Euphydryas aurinia*. These attack the same host generation to produce large broods later in the year (picture: J. Planas).

Plate 11: The tough solitary cocoon of *Cotesia gonepterygis*
(Braconidae: Microgastrinae) persists through the summer and
following winter, only hatching in spring when early instar larvae
of its host, *Gonepteryx* spp., are again available (picture: J. R. Salas).

Plate 12a: A pupa of *Aglais urticae* from which about 40 *Pteromalus puparum* (Chalcidoidea: Pteromalidae) adults have emerged, all using the same exit hole (picture: M. R. Shaw).

Plate 12b: *Pteromalus puparum* (Chalcidoidea: Pteromalidae), a gregarious endoparasitic idiobiont, ovipositing into a freshly formed *Euphydryas desfontainii* pupa (picture: J. Planas).

Plate 13a: Fully fed larvae of the gregarious ectoparasitic
koinobiont *Euplectrus ?flavipes* (Chalcidoidea: Eulophidae) leaving
a larva of *Charaxes jasius* (picture: J. Planas).

Plate 13b: *Trichogramma brassicae* (Chalcidoidea:
Trichogammatidae) ovipositing into eggs of *Pieris brassicae*
(picture: N. Fatouros).

Plate 14: A phoretic female *Trichogramma brassicae* (Chalcidoidea:
Trichogrammatidae) on the leg of a mated female *Pieris brassicae*.
When the butterfly lays her eggs the parasitoid will dismount and
parasitise them (picture: N. Fatouros).

Plate 15: *Sturmia bella* (Tachinidae), a larva–pupal parasitoid of Nymphalini that has recently expanded its range to Britain. It has been speculated that simultaneous declines in, especially, *Aglais urticae* there have been due to this now common species (picture: C. Raper).

Plate 16: *Aplomya confinis* (Tachinidae) ovipositing onto a
second–instar larva of *Polyommatus icarus* (picture: K. Fiedler).

Plate 17: *Melanargia galathea* as perceived within the human visual range (left) and in the UV range (right). Under UV the normally bright and obvious coloration becomes cryptic against UV absorbing vegetation.

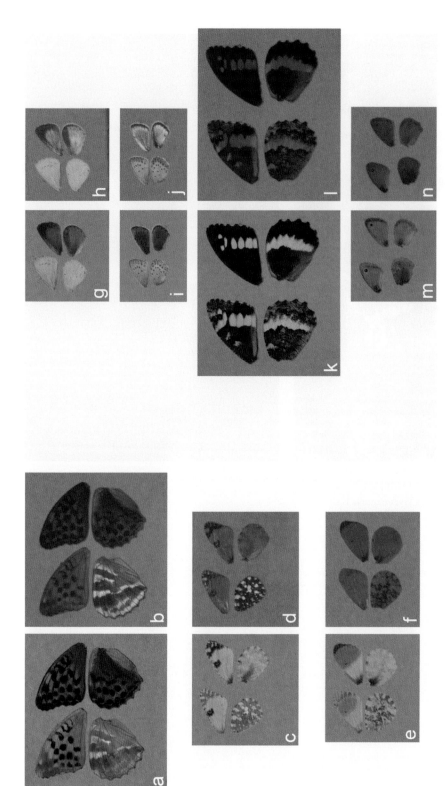

Plate 18: Paired human visible (left) and UV (right) images of wings of (a,b) *Argynnis paphia* (female), (c,d) *Euchloe ausonia* (female), (e,f) *Anthocharis cardamines* (male), (g,h) *Celastrina argiolus* (female), (i,j) *Plebeius argus* (male), (k,l) *Brintesia circe* (male), and (m,n) *Coenonympha pamphilus* (male), illustrating the variability of UV reflection within Palaearctic butterflies.

Plate 19a: *Polyommatus sensu lato* species and their hybrids (*Lysandra + Meleageria + Plebicula*) (all males). Top row: *L. coridon*, Cauterets, Hautes Pyrénées, France, 12/08/1954, H. Descimon leg.; "polonus" (*L. coridon × L. bellargus*), Quès, Pyrénées Orientales, France, 15/07/1969, H. Descimon leg.; *L. bellargus*, Cauterets, Hautes Pyrénées, France, 03/09/1955, H. Descimon leg.; *P. dorylas*, Vallecillo, Teruel, Spain, 28/07/1988, H. Descimon leg. Second row: "cormion" (*L. coridon × M daphnis*), 20/06/1989, bred by Klaus Schurian; "samsoni" (*L. hispana × L. bellargus*), Digne, Alpes de Haute Provence, France, 06/08/1923, Dr. Glais leg.; "caerulescens" (*L. coridon caelestissima × L. albicans*), Ciudad Encantada, Cuenca, Spain, 08/08/1989, H. Descimon leg.; "caeruleonivescens" (*P. dorylas × P. nivescens*), Vallecillo, Teruel, Spain, 28/07/1988, H. Descimon leg. Third row: *M daphnis*, Le Fournel, l'Argentière-la Bessée, Hautes-Alpes, France, 25/07/1966, H. Descimon leg.; *L. hispana* Digne, Alpes de Haute Provence, France, 15/05/1980, H. Descimon leg.; *L. coridon caelestissima*, Ciudad Encantada, Cuenca, Spain, 08/08/1989, H. Descimon leg.; *P. nivescens*, Vallecillo, Teruel, Spain, 28/07/1988, H. Descimon leg.

Plate 19b. *Parnassius* species and hybrid (all males).
Top: *P. apollo*, Ravin de Roche Noire, Monêtier les Bains, Hautes Alpes, France, 30/07/1974, H. Descimon leg.; middle: *P. apollo × phoebus* hybrid, same locality, date and collector; bottom: *P. phoebus*. Same locality, date and collector.

Plate 19c. *Anthocaris* species and hybrid (all males).
Top: *A. belia euphenoides*, Auriol, Bouches du Rhône, France, 02/04/1981; middle: *A. belia euphenoides × cardamines* hybrid, bred by Z. Lorkovič, 1938; bottom, *A. cardamines*, Châteauvert, Var, France, 04/04/1981.

Plate 20a: *Erebia* species and hybrid (all males).
Top: E. *epiphron*, Barèges, Hautes-Pyrénées, France, 18/07/1958, H. Descimon leg.; middle: "serotina" (hybrid offspring of *E. pronoe* male × *E. epiphron* female), Pont de la Raillère, Cauterets, Hautes Pyrénées, France, 07/09/1954, H. Descimon leg.; bottom: *E. pronoe* Le Campbasque, Cauterets, Hautes Pyrénées, France, 25/08/1958, H. Descimon leg.

Plate 20b: *Zerynthia* species and hybrid (all males).
Top: *Z. polyxena*, Châteauvert, Var, France, 04/04/1981, H. Descimon leg.; middle: *Z. polyxena* × *Z. rumina* hybrid, bred by H. Descimon, 15/04/82; bottom: *Z. rumina*, Châteauvert, Var, France, 15/04/1980, H. Descimon leg.

Plate 20c: Distribution of brassy ringlets (*Erebia tyndarus* group) in Europe. Red = E. *hispania* (*hispania* from Sierra Nevada not distinguished from *rondoui* in the Pyrénées). Sky blue = E. *arvernensis*. Dark blue = E. *cassioides*. Green = E. *tyndarus*. Purple (not very distinctive) = E. *calcaria*. Orange = E. *nivalis*. Dark regions between colour patches cohabitation zones. Populations from the Italian peninsula are considered to be *E. arvernensis* on the basis of allozymes, while those from the Balkans are assigned to *E. cassioides* somewhat arbitrarily. Crosses represent intersections between

latitude and longitude at five degree intervals.
Note 1: The distribution of *hispania* and *cassioides* in the Pyrénées is strongly interdigitated. (The distribution in Spain is not well known; expect more surprises!)

Note 2: *E. nivalis* is fragmented into two nuclei by the distribution of *tyndarus*, which seems more competitive than *cassioides* or *arvernensis*. Competitive exclusion probably explains patchwork distributions and scarcity of overlap.

Plate 21a: Recently coppiced forest in Germany. The dynamics of coppicing – which unfortunately in the course of modern seminatural forest management hardly exists any longer – makes this an especially valuable habitat (picture: Matthias Dolek).

Plate 21b: Typical potential habitat for both *Lopinga achine* and *Euphydryas maturna* is the Laelatu wooded meadow in Estonia, known for its extremely high richness of higher plants (July 2008; picture: Doris Vetterlein).

Plate 22a: Bog at the edge of the Bavarian Alps. The locality is inhabited by a number of typical butterfly species, e.g. *Colias palaeno, Coenonympha tullia, Boloria aquilonaris* and *Euphydryas aurinia* (June 2007; picture: Matthias Dolek).

Plate 22b: The Estonian bog area of Viru in the Lahemaa National Park is a typical example of the long-term dynamics of oligotrophic bogs. They have developed from a lake which due to succession turned into a bog area with clear signs of forest developments (July 2008; picture: Josef Settele).

Plate 23a: Calcareous grassland with limestone rocks in the southern Franconian Jura. The locality is managed by grazing and holds almost 60 butterfly species. Nevertheless, the strong growth of trees and shrubs is clearly visible (picture: Matthias Dolek).

Plate 23b: Calcareous grasslands are home to many butterfly species even with contrasting habitat needs. This locality was until recently inhabited by a population of *Chazara briseis*; the last remnants of stony slopes – an important element of the species' habitats – are visible in the lower centre (picture: Matthias Dolek).

Plate 24a: Horse pasture in the Apennines, Italy. The gentian, *Gentiana cruciata* (with eggs of *Maculinea rebeli*), is not eaten by the horses and stands out from the short turf (Sept. 2003; picture: Matthias Dolek).

Plate 24b: Farm track delineated by hedges within an arable landscape in northeast Croatia (Korenici, Istria), with a high abundance of species like *Melitaea cinxia*, *Euphydryas aurinia* and *Limenitis reducta* (May 2008; picture: Josef Settele).

Population viability analysis is currently the main tool conservation biologists can use to determine whether a (meta)population is at risk of extinction, according to the various threats it faces. 'It is the process of predicting the risk of extinction from the combined effects of the deterministic threats – such as habitat loss, overexploitation, pollution and species introductions – and stochastic threats, including demographic, environmental and genetic fluctuations and catastrophes' (Frankham 2002, p. 18). It helps formulating management guidelines for effective conservation. Population viability analysis is a rapidly growing research area and therefore many different approaches and opinions emerge, broadening the scope of what is called PVA. In this chapter, along with Ralls *et al.* (2002), Reed *et al.* (2002) and others, we restrict the use of PVA to quantitative modelling of the risk of extinction (or a closely related measure of population viability): a stochastic simulation model is parameterised from available biological and landscape data and used to project (meta)population dynamics over time. Through various estimates of persistence (extinction risk, time to extinction, distribution of (meta)population sizes, frequency and duration of local extinctions, etc.), viability is estimated and compared under different scenarios of landscape or population management to design effective conservation guidelines. For a broader view of PVA, including alternatives to PVA *sensu stricto* for conservation management, see Akçakaya & Sjögren-Gulve (2000) and Beissinger & McCullough (2002).

A BRIEF HISTORY OF POPULATION VIABILITY ANALYSIS

The history of PVA is short: it made its debut in the 1970s, with controversy over the Yellowstone National Park population of grizzly bears (*Ursus arctos*), for which different deterministic models led to different predictions (see Beissinger 2002 and references therein). Then a new direction was launched by Shaffer with a stochastic simulation model incorporating chance events (demographic and environmental stochasticities) and producing extinction probabilities and an estimate of minimum viable population size (MVP: smallest population size with a certain probability chance – 95%, 99%,... – of remaining extant for a certain period – 100, 1000, ... years) (Shaffer 1981). Soon, several papers estimating MVPs began to appear (e.g. Shaffer & Samson 1985). Population viability analysis was more firmly established in 1986–87 with the release of two books (Gilpin & Soulé 1986, Soulé 1987). Applications

of PVA grew rapidly in the late 1980s and early 1990s with the large application of MVP to captive populations (notably in zoos), the necessity to implement laws concerning the preservation of animals and plants (e.g. U.S. Endangered Species Act of 1973) through quantification of extinction risk, and also through the development of computer power and user-friendly packages specially designed to run complex stochastic PVA simulations. A well-known example of the use of PVA as part of the process of designing guidelines to protect an endangered species is the case of the northern spotted owl *Strix occidentalis caurina* in the USA (Noon & McKelvey 1996).

The way PVA is performed has evolved since its onset (Beissinger 2002 and references therein): from exactly estimating MVP and extinction risk to ranking different management scenarios, from models emphasising genetics over demography to models that consider demography but not genetics, and from single populations to metapopulations. Currently, ever more complicated PVA models are designed, following the growth of geographic information systems (GIS).

CURRENT FORMS OF (META) POPULATION VIABILITY ANALYSIS MODELS

Three main categories of current PVA models for metapopulations exist, according to the way they incorporate spatial dynamics (Akçakaya & Sjögren-Gulve 2000, Akçakaya 2000b). Only categories 2 and 3 are suitable for modelling single population viability.

(1) *Stochastic patch occupancy models (SPOM)* (e.g. incidence function models: Hanski 1994b, Sjögren-Gulve & Hanski 2000) describe each population as full (occupied) or empty (extinct). The basis of these models is the balance between stochastic local extinctions and dispersal-mediated recolonisations in each generation (Fig. 22.1). They are suitable for classical (Levins 1968) metapopulations in highly fragmented landscapes, where the habitat patches are numerous, relatively small and represent only a tiny fraction of the landscape. In such situations regional dynamics (extinctions and colonisations) dominate local dynamics (changes in population size) and these latter may be ignored: only the presence or the absence of local populations in habitat patches is considered and the metapopulation is assumed to approach the extinction–recolonisation equilibrium

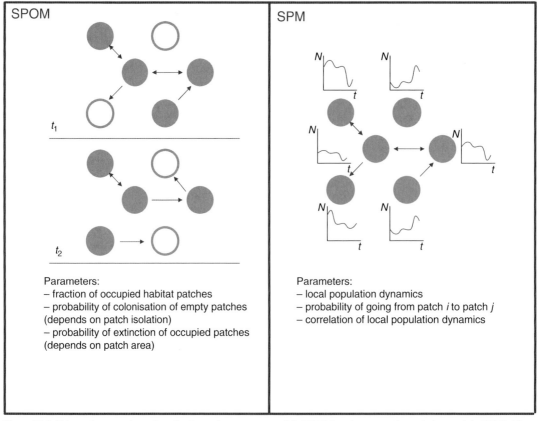

Figure 22.1. Schematic comparison of stochastic patch occupancy models (SPOM) and structured population models (SPM). The SPOM refers to the incidence function model, which relates colonisation and extinction processes to patch isolation and area respectively. Arrows indicate dispersal of individuals between populations.

(Hanski 2002). Due to these assumptions, SPOM can be parameterised with relatively small amounts of data (e.g. only one snapshot of presence/absence; but see Clinchy *et al.* 2002). But the same assumptions limit their applicability to specific cases where the system consists of many habitat patches, i.e. usually not the threatened species for which PVA are urgently needed (Ralls *et al.* 2002, Baguette 2004; but see Lopez & Pfister 2001, Hanski 2004). Nevertheless, SPOM could be used to extend PVA application beyond purely threatened species, to ' "ordinary" metapopulations living in "ordinary" landscapes' as advocated by Hanski (2002, pp. 101–2) because 'large numbers of currently not-so-endangered species will become highly endangered in the future if current trends in habitat loss and fragmentation continue'. Stochastic patch occupancy models can

be implemented with META-X (Grimm *et al.* 2004) and SPOMSIM (Moilanen 2004) packages.

(2) *Structured population models (SPM)* (Akçakaya 2000a) describe the dynamics of each local population with discrete-time structured demographic models (i.e. Leslie matrix models: Caswell 2001) and incorporate spatial dynamics by modelling dispersal and correlation between each pair of patches according to their location, making these models spatially realistic (Fig. 22.1). Structured population models require more data to be parameterised than SPOM but are more realistic and can be used for cases where modelling regional dynamics is not sufficient and local dynamics need to be modelled too (i.e. a lot of endangered species with few local populations). Finally, SPM are based on a number of common techniques (e.g. matrix algebra, generation of

random variables) that allow their implementation as generic platforms (e.g. RAMAS/GIS: Akçakaya 2002).

(3) *Individual-based models (IBM)* (e.g. Lacy 2000a, Grimm & Railsback 2005) describe the metapopulation through modelling the behaviour and fate of each individual, following various rules of dispersal, reproduction, etc., according to the aim of the model designed. They are extremely flexible but require a large quantity of data in order to be parameterised correctly and are computationally intensive and sometimes difficult to analyse because of their complexity. Whereas many IBM are implemented using completely unique programming, packages (e.g. VORTEX: Lacy 2000b; SELES: Fall & Fall 2001) offer a framework for their construction.

APPLICATION TO BUTTERFLIES

Table 22.1 presents an overview of published PVA on butterflies. Mainly stochastic patch occupancy models (SPOM) and structured population models (SPM) have been used so far; the large amount of precise data required by individual-based models seems to have precluded their use in modelling the fate of butterfly (meta)populations, until very recently (study 18). The basic differences between SPOM and SPM are well illustrated in Table 22.1, particularly concerning three main aspects of metapopulation dynamics: extinction probability (SPOM) / demography (SPM), dispersal, and correlation between populations (called regional stochasticity in the context of SPOM). It appears that only three studies, one SPOM (study 11), one SPM (study 12) and one IBM (study 18), were based on a complete set of estimated parameters. All the other studies used, to varying extents, surrogate parameters from other species and/or metapopulation systems, or even parameters fixed arbitrarily to some specific value. This illustrates that one major problem of PVA is the availability of information to parameterise the models.

STOCHASTIC PATCH OCCUPANCY MODELS

Stochastic patch occupancy models (SPOM) are relatively well suited for butterflies for several reasons. Most butterfly species in temperate regions show discrete generations, which is an essential, and often neglected, feature that makes the SPOM approach possible. Several species present metapopulations with a high number of habitat patches, both empty and occupied, and occupying only a small fraction

of the landscape. For the incidence function model (IFM), the most popular and developed SPOM so far, the total number of habitat patches in the landscape must be at least 30, with at least 10 occupied and 10 empty patches, and habitat must constitute only a small fraction of the landscape (ca. 20%) (Hanski 1999, pp. 233–4).

The generality of species distribution in the form of a long-term equilibrium between stochastic extinction and recolonisation has been criticised by Baguette (2004). Even in highly fragmented landscapes, most specialist butterfly species saturate all available habitat patches, provided that (1) habitat quality is sufficient, (2) area of suitable patches exceeds a minimum threshold and (3) distance between patches is in the range of the species' dispersal distance. A striking example of this situation is provided by two metapopulations of *Boloria aquilonaris* living in landscapes differing in suitable habitat size and distance between local populations (Schtickzelle *et al.* 2005b). Total landscape areas amount to 220 km² and 400 km², with 14 and 42 patches of suitable habitat totalling to 26 ha and 3 ha, respectively. Proportions of occupied patches were 100% and 33% in each landscape. This empirical example, which is not anecdotal in butterflies, illustrates that the stochastic 'blinking' of local populations (extinction–recolonisation dynamics) begins in the landscape when the total area of suitable habitat and/or its connectivity goes below a certain threshold. A model based on turnover of local populations might be suitable to predict the fate of metapopulations in highly fragmented and degraded habitat (i.e. below these thresholds), but current SPOM approaches require the existence of several tens of habitat patches. Anyway, measuring such thresholds is certainly as important for species conservation as is developing sophisticated models of population viability.

The interest of SPOM lies in the small number of parameters required. For the most complete version of IFM (Moilanen 2004 and references therein), the number of parameters that need to be defined are, in addition to landscape configuration (patch sizes and interpatch distances): one or two for dispersal, two for connectivity, one for colonisation probability, two for extinction probability, and at least one for correlation of extinction (i.e. regional stochasticity). Depending on the complexity of the IFM model version used, and on the fact that metapopulation dynamics are at a quasi-equilibrium steady state (i.e. the local extinction and recolonisation events are balanced at each generation), one or a few snapshots of presence/absence of local populations in suitable habitat patches are enough to estimate these parameters. However, the quality of model

Table 22.1 *Overview of the butterfly PVAs published in scientific journals. We made all efforts to produce an exhaustive list, but other butterfly PVAs may exist, especially in the grey literature*

Reference	Species	Region	Study area	Type of model[a]	Number of patches	Percent vacant sites	Extinction probability (SPOM) or demography (SPM)[b]	Dispersal[b]	Correlation[b]
1. Hanski & Thomas 1994	Melitaea cinxia	Åland, Finland	15 km²	SPM	50	16%	Density-dependent growth (untested); surrogate, evaluated or estimated	Arbitrary, evaluated or estimated	Not included
2. Hanski & Thomas 1994	Plebejus argus	North Wales, UK	110 km²	SPM	70	56%	Density-dependent growth (untested); evaluated or estimated	Arbitrary or evaluated	Not included
3. Hanski & Thomas 1994	Hesperia comma	South England, UK	2800 km² 1500 km²	SPM	94 75	25%	Density-dependent growth (untested); evaluated or estimated	Arbitrary or evaluated	Not included
4. Hanski 1994a	Melitaea cinxia	Åland, Finland	20–100 km² (same as 1)	SPOM	50	16%	Evaluated	Estimated	Not included
5. Hanski 1994a	Hesperia comma	South England, UK	20–100 km² (part of 3)	SPOM	64	20%	Evaluated	Surrogate from *Melitaea cinxia*	Not included
6. Hanski 1994a	Scolitantides orion	South Finland	20–100 km²	SPOM	70	37%	Evaluated	Surrogate from *Melitaea cinxia*	Not included
7. Wahlberg et al. 1996	Melitaea diamina	Tampere, Finland	600 km²	SPOM	94	63%	Evaluated	Surrogate from *Melitaea cinxia*	Not included
8. Sawchik et al. 2002	Proclossiana eunomia	Ardenne, Belgium	300 km²	SPM	32	0%	Density-dependent growth (untested); arbitrary	Estimated	Arbitrary
9. Wahlberg et al. 2002a	Euphydryas aurinia	Southeastern Finland	150 km²	SPOM	114	52%	Evaluated	Evaluated	Not included
10. Schultz & Hammond 2003	Icaricia icarioides fenderi	Oregon, USA Area not specified		SPM	12	N.a. (modelled separately)	Density-independent growth (untested); estimated	N.a.	Estimated
11. Bergman & Kindvall 2004	Lopinga achine	Östergötland, Sweden	60 km²	SPOM	158	49%	Evaluated	Estimated	Estimated

Reference	Species	Location	Area	Model[a]		%	[b]		
12. Schtickzelle & Baguette 2004	*Proclossiana eunomia*	Ardenne, Belgium	10 km²	SPM	20	0%	Density-dependent growth (tested); estimated	Estimated	Estimated
13. Gutiérrez 2005	*Erynnis tages*	North Wales, UK	35 km²	SPOM	34	57%	Evaluated	Arbitrary or evaluated	Arbitrary
14. Schtickzelle et al. 2005a	*Euphydryas aurinia*	Ardenne, Belgium	0.28 km²	SPM	11	0%	Density-dependent growth (tested); estimated	Estimated	Surrogate from *P. eunomia*
15. Schtickzelle et al. 2005b	*Boloria aquilonaris*	Ardenne, Belgium	220 km²	SPM	14	0%	Density-dependent growth (tested); estimated	Estimated	Surrogate from *P. eunomia*
16. Schtickzelle et al. 2005b	*Boloria aquilonaris*	Drenthe, The Netherlands	400 km²	SPM	45	67%	Surrogates from *B. aquilonaris* in Ardenne, Belgium		
17. Bulman et al. 2007	*Euphydryas aurinia*	Dorset, UK	625 km²	SPOM	123	11%–16%	Evaluated	Surrogate from *Meliteaea cinxia*	arbitrary
18. McIntire et al. 2007	*Icaricia icarioides fenderi*	Oregon, USA		IBM	48	Unspecified	Density-dependent growth (tested); estimated	Estimated	Estimated

[a] SPOM, stochastic patch occupancy model; SPM, structured population model; IBM, individual-based model.

[b] The way parameter values have been determined has been classified in four types. Arbitrary, fixed to a specific value not supported by data or information; Surrogate, generalised from information available on other species or study system; Evaluated, evaluated by fitting the SPOM PVA model to the observed patch occupancy pattern; Estimated, estimated from external data (such as capture–mark–recapture data).

predictions is likely to be improved when some parameters can be estimated from independent data.

A major input in SPOM is the spatial scale of dispersal, i.e. the distance at which empty habitat patches may be (re)colonised by individuals emigrating from extant local populations, quantified by the α parameter. Classically, the number of dispersing individuals travelling a certain distance decreases with the distance from the starting point. Several probability distributions are used to describe this process; the negative exponential and the inverse power function are the most frequently used in butterflies. All SPOM used so far in butterfly PVA used the negative exponential function as the dispersal kernel. Hanski (1999) advocated that the α parameter should be estimated from capture–mark–recapture (CMR) data; this can be done for example with the virtual migration model (VM: Hanski et al. 2000). When such data are not available, the parameter must be numerically estimated (Hanski 1999). Out of the seven SPOM PVA realised on butterflies (Table 22.1), only two used CMR data to estimate α values (study 4 for *Lopinga achine* and study 11 for *Melitaea cinxia*). Two studies (9 and 13) evaluated α by best fitting the occupancy patterns and the three remaining studies (5, 6 and 7) used surrogate values coming from *Melitaea cinxia*, but surprisingly studies 5, 6 and 17 used $\alpha = 2$ while study 7 used $\alpha = 1$. Such a focus on α values is not purely academic: this parameter scales the geographic range of metapopulation dynamics and should therefore reliably capture the field situation. When the negative exponential function is used as the dispersal kernel, some minor changes in α values have dramatic consequences on the probability that a butterfly flies a certain distance and therefore founds a local population elsewhere in the landscape. For instance, Hanski (1999) reported that the α value of 2 used to parameterise the *M. cinxia* model was estimated from a CMR experiment. The maximum dispersal distance observed during this experiment was 3.1 km. A negative exponential dispersal kernel provides an associate probability of 0.002 and 0.0450 for $\alpha = 2$ and $\alpha = 1$ respectively, that is a 22.5-fold increase in the chance of travelling this particular distance. Moreover, recorded variations in the α parameter according to sex, space and time (Petit et al. 2001, Wahlberg et al. 2002a, 2002b, Schtickzelle et al. 2006) are extra difficulties that preclude its use with a general surrogate value in different PVAs, as sometimes done (Table 22.1). More generally, the use of a single, monotonically decreasing dispersal kernel is probably an oversimplification: dispersal is far from a fixed trait, and there is mounting evidence showing that both resident and 'super-

dispersers' should coexist in populations subject to selective pressures associated with habitat fragmentation (see review in Baguette & Van Dyck 2007).

Another important aspect of metapopulation dynamics, namely the correlation of extinction–recolonisation among local populations due to common environmental conditions, is not routinely implemented in existing SPOM PVA. Completely absent from the basic IFM model (Hanski 1994b), regional stochasticity can be implemented in more recent developments of the model (Moilanen 2004). However, it seems to be the least developed aspect of SPOM, despite being likely to act significantly on metapopulation viability. Except for study 11, we are not aware of butterfly PVA with SPOM taking this important parameter into account. In this study, environmental variation was modelled by allocating common, 'regional' conditions to a fraction of the local populations, whereas others experienced localised conditions. In other words, this modelling procedure was spatially implicit because the correlation between local dynamics depends on the distance, and is therefore clearly spatially explicit. Local populations close to each other experience more similar environmental conditions than those that are located further away, and this is likely to synchronise their dynamics. The spatial scale at which environmental factors generate synchrony in local population dynamics is controversial. Some studies on aphids and moths report correlations at distances up to 800 km (Hanski & Woiwod 1993), whereas other studies restrict this distance to 0.5 km (Sutcliffe et al. 1997b) or 1–2 km (Schtickzelle & Baguette 2004). This is an important issue for conservation that certainly deserves future research efforts.

Raw outputs of SPOM are occupancy patterns, i.e. the proportion of suitable habitats occupied by local populations in the landscape, which can be simulated generation after generation on a given time period with a high number of replications (usually >1000). The extinction probability of the species in the landscape can be computed as the proportion of replications when extinction occurred, and can be matched to the spatial configuration of the metapopulation in the landscape to identify which local populations are the most important for the persistence of the species in the landscape (Gutiérrez 2005).

STRUCTURED POPULATION MODELS

In structured population models (SPM), the dynamics of each local population are taken into account, which should improve precision because trends in population size are more informative than the two states (presence vs. absence) issued from SPOM. As an exception, this statement might not be true for metapopulations with a very high number of patches and a substantial turnover of populations, but such situations are rarely applicable to threatened species. However, SPM require a larger amount of information to be parameterised, and all this information must come from independent data.

Accurate time series of population size are required to capture the dynamics of the system. Contrary to other taxa, there are few long-term monitoring data of butterfly meta-populations. However, a review of several time series of population dynamics revealed that density dependence appears common in temperate-zone butterflies (Baguette & Schtickzelle 2006). Accordingly, a deterministic density-dependent function of population growth may be used as core of the model. In the following example from our study on the bog fritillary *Proclossiana eunomia* (study 12), the population growth rate R_t was assumed to depend on the density the year before N_{t-1} and on a series of n weather parameters W_{it}:

$$\ln(R_t) = \beta_0 + \beta_1 \cdot N_{t-1} + \sum_{i=2}^{n+1} \beta_i \cdot \ln(W_{it}).$$

Relationships between growth rate and relevant environmental variables were investigated using regression analyses. To estimate the impact of weather conditions on population growth rate, the year was divided in four periods corresponding to the successive developmental stages of the butterfly: imagos and eggs (June), first larval instars (July–September), larval diapause (October–March) and last larval instars (April–May). Relevant weather parameters chosen a priori were collected for each of those periods (mean diurnal temperature, sunshine duration and rainfall). Demography was then implemented in the PVA model by sampling the population growth rate from a log-normal distribution with the mean growth rate provided by the density-dependent function and the standard deviation mimicking the environmental stochasticity. Assuming that the mean growth rate is due to density dependence while environmental stochasticity is due to variations in weather conditions, the standard deviation was computed from series of N_t estimates obtained from the equation above using environmental conditions of the year but considering that the populations were at carrying capacity each year. This procedure requires high-quality data and parameter estimates. Although this is logistically and computationally intensive, the combination of density

dependence and environmental variation provided a predictive envelope of the population demographic trajectory that matched the observed trajectory very well.

Data on dispersal movements can be obtained from multi-site CMR experiments, but their analysis to extract probabilities of dispersal is not so straightforward. Simple dispersal kernels such as negative exponential or power functions give an oversimplified picture of dispersal by relating it to distance between patches only. In particular, three basic features of dispersal are neglected. Firstly, dispersal depends on the willingness of an individual to leave a given habitat patch, i.e. its dispersal propensity. Comparative studies of several metapopulations showed that dispersal propensity markedly decreased with increased levels of habitat fragmentation (Schtickzelle *et al.* 2006). As in SPM, local dynamics are taken into account, and it is worth determining how many individuals emigrate from the population at each generation. Secondly, dispersal is usually not cost-less: according to the configuration of the landscape (i.e. the relative position of suitable habitat patches) and its structure (i.e. the nature of the matrix and the presence of barriers), dispersing individuals incur variable risks of mortality. Thirdly, dispersal is not isotropic; the use of real dispersal rates takes into account the asymmetry of individual fluxes among local populations. Some modelling approaches allow circumvention of these major problems by the use of simpler kernels relating dispersal to distance. As an example, the virtual migration (VM) model, frequently used for butterflies (e.g. Petit *et al.* 2001, Wahlberg *et al.* 2002a, 2002b, Schtickzelle & Baguette 2003, Mennechez *et al.* 2004, Schtickzelle *et al.* 2006), is based on assumptions relating dispersal to features of landscape configuration: emigration depends negatively on patch area, survival during dispersal depends on patch connectivity, which decreases with distances between patches but increases with patch areas. Dispersal in studies 12, 14, 15 and 16 was modelled as a matrix of the probability of going from every patch i to every patch j, computed from parameters estimated using the VM model.

Correlation of local dynamics of the various local populations can be estimated from time series of local demography provided they are available for several populations. In the study of *Proclossiana eunomia* (12), we used the number of (re)captures of adult butterflies in each patch summed over the whole flight period as a surrogate of local demography, taking advantage that the sampling effort was similar in each patch, and therefore the capture probability too (Schtickzelle *et al.* 2002). The numbers of (re)captures (CR) in each local population were then transformed in series of yearly change rates (CR_t/CR_{t-1}), Spearman correlation coefficients between pairs of local populations were computed, and non-linear regression used to model a negative exponential decay of correlation with distance between populations. This function, predicting a steep decline with distance (0.91 at 0 m, 0.17 at 1000 m and 0.03 at 2000 m), was used to correlate local population growth rate R_t values sampled during PVA simulation according to the distance between pairs of patches.

Outputs of SPM are distributions of local population sizes along the time horizon of the simulations. From those distributions, patterns of local population occupancy or probabilities of metapopulation extinction can be easily computed. However, the most commonly used output of SPM is the interval quasi-extinction risk, i.e. the probability that the metapopulation size falls under a given value at least once during the time course of the simulation. Quasi-extinction means that the experimenter can decide that extinction occurs at any given small population size greater than zero, to take into account the absence in the PVA model of processes occurring in very small populations only (e.g. Allee effect: Allee *et al.* 1949). In reality, extinction is likely to happen deterministically once the population goes below a certain abundance threshold. Usually internal quasi-extinction risk is a sigmoid function of metapopulation size, and allows an easy and visual comparison of the various scenarios, either in the frame of sensitivity analysis or for the ranking of different management measures.

PRACTICAL USE OF BUTTERFLY POPULATION VIABILITY ANALYSIS

A recurrent criticism of PVA is their lack of applicability, such as 'PVA are essentially games played with guesses' (Caughley 1994). However, at least in butterflies, several PVAs seem to have contributed to the design of practical conservation measures and to their implementation. Authors of some studies listed in Table 22.1 were asked whether to their knowledge (1) their PVA produced conservation guidelines and (2) whether those guidelines were used to implement practical conservation measures. On 10 studies listed in Table 22.1 (studies 7 to 12, 14 to 16, and 18), nine produced conservation guidelines and all those nine have led – or will soon lead – to the practical implementation of management measures. We also questioned these authors about their confidence in PVA: all of them recommend the use of this procedure, even if most of them pleaded for more detailed data as basic input to parameterise the models (see below).

Several authors emphasised that quantitative output produced by PVA had the potential to convince more efficiently a non-expert reserve manager than a verbal statement of the obvious.

PROBLEMS AND ADVICE FOR POPULATION VIABILITY ANALYSIS

Our brief survey of butterfly PVA provided promising results, but also a series of problems and dangers associated with the building and use of a PVA model. Most PVA do not surface above the level of the grey literature (Caughley 1994). However, some PVA for butterflies are already published in the literature (Table 22.1) even though fair and rigorous PVA models are clearly rare for butterflies, and for that reason should have priority for publication in a peer-reviewed journal.

Because of all these potential problems and associated misuses, the efficiency of PVA in conservation planning is frequently and continuously questioned. Especially, the easy availability of software packages makes it easy (or too easy?) to construct a model that is presented as a PVA. Despite all its drawbacks, PVA is currently one of the best decision-support tools available in many circumstances for conservation planning (Brook *et al.* 2002a). Considering that conservation biology is a crisis discipline (Soulé 1985) and that decisions must be made quickly and in the face of incomplete data, the use of PVA is clearly justified despite all its drawbacks.

It is of prime importance that conservation biologists and managers are aware of the limitations, disadvantages and potential sources of error of PVA. Here, we synthesise several important general lines of problems and advice (Boyce 1992, Beissinger & Westphal 1998, Reed *et al.* 1998, 2002, Akçakaya & Sjögren-Gulve 2000, Burgman & Possingham 2000, Morris & Doak 2002, Ralls *et al.* 2002); some of these are particularly well underlined in the META-X PVA package (Grimm *et al.* 2004), which therefore offers a high heuristic value.

Problem: Subjectivity and uncertainty in parameter estimates and choices about model assumptions and structure

Advice: Adapt the model structure to the system modelled, the aim of the PVA, and to what the data at hand can support

Obtaining the data to satisfactorily parameterise a PVA model is surely the main problem with PVA, especially for metapopulations: *garbage in, garbage out*. Long-term and high-quality *data*, along with good *analysis methods*, are usually needed to obtain estimates of *parameters* and their *variance* as accurate as possible because any error is propagated into the viability results. Furthermore, some assumptions about the biology of the species (e.g. concerning mating and social systems) or the model structure (e.g. type of model or software used, correlation among parameters, probability distribution of parameters) can greatly affect the results. There is no type of PVA model with a consistent best performance; the choice of the model type suitable for a particular PVA depends on the features of the species and study system, the questions addressed, and also on the data available. It is therefore of prime importance to carefully evaluate whether choices and assumptions are valid for the system modelled and whether parameter specification is consistent with software implementation.

The parameters to be correctly estimated depend on the situation modelled, but generally important factors include: (1) fitness-related rates (survival, fecundity, etc.), (2) strength and functional form of density dependence (if any), (3) dispersal rates and mortality during dispersal (i.e. connectivity), (4) correlation between population dynamics (regional stochasticity), (5) variance in temporal series of parameters meaningful to quantify environmental stochasticity, (6) frequency and impact of catastrophes, and (7) classification of the landscape elements (e.g. habitat vs. matrix, or different habitat quality levels). When there are no data available, professional judgement-based estimates may be used instead. Although some of the parameter generalisations applied in butterfly PVA models (Table 22.1) are justified by the biological information available, some are totally arbitrary, and are usually round values that clearly reflect the lack of relevant parameter estimate. It is of the utmost important to avoid using such arbitrary values or default values given by some software packages (e.g. VORTEX: Lacy 2000b), because this leads to pseudo-realistic models giving a strong, but false, impression of realism.

More generally, there is a trade-off between generality, precision and realism of the model. On the one hand, it is not reasonable to construct a detailed model when basic information on crucial parameters is lacking. On the other hand, oversimplified models that fail to incorporate key processes are not likely to give a sufficiently realistic, and therefore useful, picture of the system. Incomplete information does not necessarily mean that meaningful results are out of reach for the PVA model. The model could at least direct further research.

Problem: Difficulties in incorporating uncertainties about parameter estimates

Advice: Perform sensitivity analysis on all model parameters as an integrant part of the *PVA*, and report how uncertainties in the model create uncertainties in the predictions

'Uncertainty is just about the only certainty in PVA' (Beissinger 2002, p. 11). The simplest approach for incorporating uncertainty on some parameter estimates is to build best-case and worst-case models. This can be extended to a wider range of parameter values to make a sensitivity analysis. Complex alternatives have been recently proposed, but have not been used extensively so far. Alternatives include focusing on time to extinction (as a stochastic population prediction interval (PPI): Saether & Engen 2002) instead of extinction probability, and using PVA based on Bayesian statistical theory (Wade 2002). When uncertainties in terms of viability exist, it is important to identify them clearly, especially when overestimation of the viability is suspected. Otherwise the credibility of the PVA is simply unknown. Hence, some kind of sensitivity analysis is essential, particularly for parameters or assumptions that are not underpinned with real data.

Problem: Weak ability to validate model predictions

Advice: Validate model components individually as far as possible, and evaluate relative rather than absolute extinction risks

Once a PVA model has been developed, the next critical step is its validation, i.e. the confrontation of model predictions to what happens in the real, natural world. Validation of stochastic results is difficult, especially when these results deal with events that we try to prevent, such as extinction. Some results can nevertheless be validated by comparing predicted values with values observed in the field. It is better to test this on an independent data set. Furthermore, components of the PVA model (e.g. density-dependence function) or some secondary model predictions (i.e. not the ones relating directly to persistence, e.g. frequency of extinctions, movement pattern of individuals) can often be more easily validated than the whole model.

'The future is much more easily knowable in a qualitative sense than it is predictable in a quantitative sense' (Shaffer *et al.* 2002, p. 133). As a general rule advocated by many PVA scientists, it is safer to compare viability under different scenarios than to estimate absolute extinction risk. This is so because the relative errors of the assessments of the scenarios are likely to be more or less the same. Qualitative results of PVA have been shown to be valid by controlled experiments, even if some important quantitative differences appeared (Belovsky *et al.* 2002). Because estimates are obtained in specific (and current) conditions, a PVA is usually more useful in projecting (i.e. predicting what will happen if conditions do not change) than in forecasting (i.e. predicting what will really happen in the future); this is a subtle but not a trivial distinction.

Problem: Absence of universally recognised measures and benchmarks for quantifying viability

Advice: Clearly detail the quantity used to assess viability, how it is computed and what are the underlying assumptions; give results in terms of complete distributions rather than point estimates

A general measure to quantify viability, i.e. the probability of persistence, seems to be lacking (Beissinger & Westphal 1998) despite some attempts to unify the field (e.g. Grimm *et al.* 2004). Various measures have been used according to the species and system studied, the kind of model used, and author's preferences: mean time to extinction, (quasi-)extinction risk, metapopulation occupancy pattern (e.g. Burgman *et al.* 1993, Akçakaya 2002). There is no universally recognised benchmark for a viable (meta)population, or for an acceptable level of extinction risk. Is for example a 5% risk of extinction within 50 years low enough to state that the system is viable? This is not just an ecological issue since such choices also have social aspects (Committee of Scientists 1999). Therefore, PVA studies must explicitly report on how viability had been quantified and the threshold of persistence defined.

In particular, it is often unjustified to fix a reference point of population size that defines extinction, such as 0. Indeed, in the case when no information is available to correctly model problems specific to very small populations, such as inbreeding depression or Allee effect, extinction might well already be certain before the population collapsed to 0 individuals. Then it is more valuable to give the complete probability distribution of risk to go below a certain population size (quasi-extinction risk: Ginzburg *et al.* 1982) than any point estimate such as the mean or the median (Ludwig 1996, Akçakaya 2000a, 2002).

Problem: End-users of the PVA for policy or management decision-making are rarely the scientists who made them

Advice: Communicate clearly the ins and outs

The aim of a PVA is to help to make choices for complex policy or management questions about species conservation. But applying scientific tools outside the academic world involves issues other than the science itself. One of the main bottlenecks to translating PVA modelling results to the field is the fact that there is a considerable gap between the scientists developing PVA models and the end-users applying (or not) the results for policy-making or management decisions. This is actually a broader problem for conservation biology (Pullin & Knight 2001).

Scientists should expend effort to ensure that the end-users receive and perceive the ins and outs of the PVA model and what it means for conservation. This is not an easy task because it is often difficult to clearly communicate results to non-ecologists, especially when the output is expressed in terms of probabilities instead of absolute facts or numbers, which causes confusion and suspicion. Our role is therefore to communicate as clearly as possible the facts as we see them, our conclusions with their uncertainties but also with their certainties: a (meta)population predicted to rapidly go extinct is highly likely to do so, the uncertain element is when rather than whether! This is the precautionary principle (International Union for Conservation of Nature (IUCN) 1994, sec. II.7): 'in cases with uncertainties in the estimate of the extinction risk, it is legitimate to use the credible estimate that gives the highest risk of extinction' (Saether & Engen 2002, p. 192).

Ecologists also need to better understand how economic, social, political and ethical factors interweave with ecological science (Ludwig *et al.* 2001) in order to improve their communication with decision-makers. This aspect is well beyond the scope of the present text, however.

INTERACTIONS BETWEEN POPULATIONS AND LANDSCAPE: CAUTION TO THE USE OF SURROGATE DATA IN POPULATION VIABILITY ANALYSIS

The use of surrogate data in PVA is widespread, because of the many difficulties in gathering the whole set of parameters needed. However, we have to warn against the frequent,

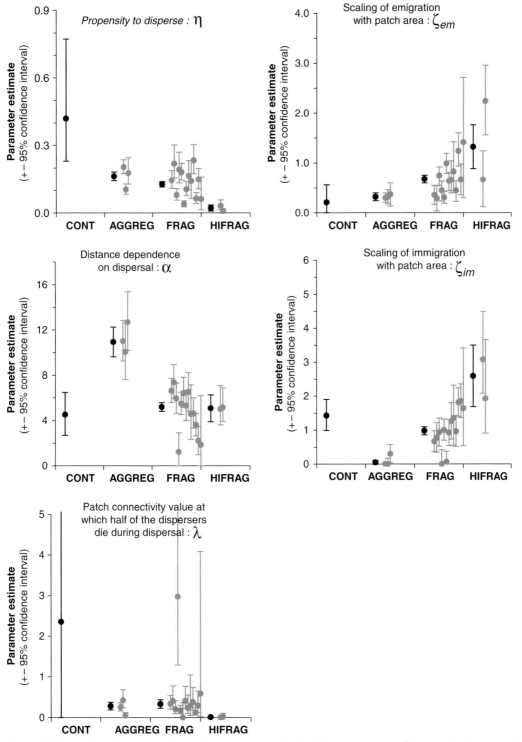

Figure 22.2. Illustration of the potential danger of using surrogate value for PVA input parameters. The case of spatio–temporal variability in dispersal, a key PVA parameter, for the bog fritillary *Proclossiana eunomia* in four landscapes in increasing order of habitat fragmentation. Parameter estimates were obtained from capture–make–recapture data using the virtual migration model (Hanski *et al.* 2000). Grey: yearly estimates (1 year = 1 generation), illustrating temporal variation; black: mean estimate per landscape, illustrating spatial variation. (Reprinted with permission from Schtickzelle *et al.* 2006.)

or even automatic, use of surrogate values. Indeed, populations and landscapes are in interaction: life-history traits are variable between species, even between those which are closely related, and even among different (meta)populations within the same species.

Dispersal provides a good example of this intricacy. Firstly, comparative studies of butterfly dispersal provided evidence that dispersal behaviour is affected by landscape fragmentation in a complex way, affecting emigration rate (Schultz & Crone 2001, Crone & Schultz 2003, Merck *et al.* 2003, Schtickzelle & Baguette 2003), movement strategy (Van Dyck & Baguette 2005), dispersal mortality (Schtickzelle *et al.* 2006, 2007) and perceptual range (i.e. the distance at which dispersing individuals perceive a suitable habitat) (Merckx & Van Dyck 2007). Secondly, dispersal also varies between individuals, because of differing dispersal tendencies (Hanski *et al.* 2004), dispersal metabolism (Haag *et al.* 2005) and covariation with fecundity (Hanski *et al.* 2006). These variations lead to differences in local population growth rates according to the kind of individuals that founded populations (Hanski & Saccheri 2006). Thirdly, life-history traits may change with the passing of time, affecting the location of suitable habitat patches, their relative suitability, and the landscape context (i.e. the nature of the matrix, including the presence of barriers or corridors), all of which may alter dispersal rates. Few empirical studies document such temporal variation of dispersal (but see Petit *et al.* 2001, Schtickzelle *et al.* 2006). Such spatio-temporal variations in dispersal parameters are particularly well illustrated in the bog fritillary *Proclossiana eunomia* (Fig. 22.2).

The emergence of a behavioural component to dispersal in fragmented landscape suggests that selective pressures associated with landscape configuration generate adaptive answers in butterfly movements (Mennechez *et al.* 2003, 2004, Schtickzelle *et al.* 2006). The evidence that the physiological attributes of dispersing females are associated with particular fitness traits, themselves leading to differential local population growth rates, implies that selective pressures on butterfly movements have consequences far beyond changes in dispersal capacity. As landscape configuration affects dispersal and local population demography, we advocate that input data used to parameterise PVA should be as far as possible collected in the populations interacting with the particular landscape under investigation.

CONCLUSIONS

We have shown that life-history traits and landscape configuration are not independent, and we advocate the use of demographic and dispersal parameters collected in the landscape under investigation. Should this definitely preclude the use of surrogate data in PVA models? The answer is no, provided that extreme care is taken to investigate the effect of parameters based on surrogate data on PVA outputs. The use of surrogate data thus implies detailed sensitivity analysis to reveal the potential bias due to their use. Our view is that such an approach is not yet clearly perceived nor applied for butterfly PVA. We encourage all butterfly researchers engaged in conservation to consider PVA as a useful tool that can be helpful in the selection of concurrent management scenarios, provided that the basic rules we highlighted here are respected.

ACKNOWLEDGEMENTS

We thank the editors for the opportunity to advertise a tool that could be helpful for the conservation of many butterfly species in Europe, and all the authors of PVA who kindly and quickly answered to our enquiry: K.-O. Bergmann, M. Wallis de Vries, M. Dufrêne, Ph. Goffart and N. Wahlberg. Comments by H. Van Dyck and M. Wallis de Vries greatly improved the manuscript. We are grateful to our funding agencies (Belgian Science Policy Office, Fund for Scientific Research-FNRS). The Ministère de la Région Wallonne gave us nature reserve access, capture licences, and the opportunity to test PVA predictions by introducing too many rustic cows in one of our favourite study sites. NS is Research Associate of the Fund for Scientific Research-FNRS. Virginie M. Stevens provided moral support and inspiration to MB.

23 • Butterflies of European ecosystems: impact of land use and options for conservation management

JOSEF SETTELE, JOHN DOVER, MATTHIAS DOLEK
AND MARTIN KONVIČKA

SUMMARY

Land use has a profound effect on the presence and abundance of butterfly species. Butterflies rely on the presence and quality of a series of ecological resources for all life stages, all of which can be strongly influenced by the management of vegetation. This chapter gives an overview of current knowledge and experience with management aspects of different European biotopes from the viewpoint of butterfly conservation. Traditional management practices have been the driving force for the recent evolution of plant and animal communities of European ecosystems. In modern landscapes, most butterflies cannot survive without active management of their habitats. Biotopes reviewed include woodlands, alpine, subalpine and arctic ecosystems, heathlands, wetlands, grasslands, arable landscapes, and disturbed and urban landscapes. There is a recurrent pattern of dependency on early succession sites. This is also true for many woodland species; management has to be based on maintaining low tree densities with either permanent or dynamically managed clearings. A critical factor for butterfly survival in several biotopes is the continuation of natural disturbances (e.g. such as landslides, or animal grazing). Many sites of conservation interest are too small to sustain viable populations of threatened specialist species, consequently increased connectivity between remaining habitats (e.g. creation of stepping stones) can be crucial to the long-term survival of many species. Unified prescriptions applied across entire regions can be deleterious and possibly even disastrous for some species. Management strategies need to be regionally or locally adapted based on adequate data and knowledge. Agri-environmental schemes may offer the potential to conserve species of the wider countryside, but they should be appropriately targeted to improve significantly the resource base for wildlife. Urban areas also offer opportunities for butterfly survival; a small shift in aesthetic perception may turn city parks and gardens into enjoyable biodiversity spots.

INTRODUCTION

This chapter deals with European ecosystems as butterfly biotopes. We focus on the impact of land use on butterfly populations, and discuss the major European ecosystems from rarely disturbed or managed ecosystems like woodlands, up to urban ecosystems. However, as it is surely not always straight forward to talk about typical butterflies of European ecosystems (compare Chapter 1 by Roger Dennis), we present some additional aspects on this question in Box 23.1.

CONSERVING BUTTERFLIES IN EUROPEAN ECOSYSTEMS

Woodlands

Most naturalists see butterflies as sun-loving creatures that avoid shady conditions of closed-canopy forests. The few species tolerating dense forests (e.g. *Pararge aegeria*; Shreeve 1984) or the handful of European canopy dwellers (e.g. *Neozephyrus quercus*, *Apatura* spp. or *Limenitis populi*) are rather the exception. As a rule, European woodland butterflies utilise sunny habitats *within* woodlands, such as sparse stands, bogs, streamsides, clearings, rides or edges.

About a third of the European species may be found in such open situations and for about one-fifth they represent the major biotope. Many of the latter are seriously threatened. Nineteen woodland species are threatened continentally (out of 71 classified in Van Swaay & Warren 1999). At the level of individual states, proportions of threatened species are considerably larger, and declines of species such as *Euphydryas maturna* have become apparent even on maps with a coarse resolution covering the whole continent (Kudrna 2002). For the conservation of woodland habitats it is good that under Annex II of the 1992 EU Habitats Directive the following woodland species were listed as requiring designation of special areas for conservation: *Parnassius mnemosyne*, *Leptidea morsei*, *Apatura metis*,

Box 23.1   Are there typical butterflies of European ecosystems?

Characterising a particular species as typical for a specific ecosystem is difficult if not impossible at a European level. Butterflies tend to tolerate increasingly shady conditions in warmer areas. For example, *Parnassius mnemosyne* inhabits hay meadows in Scandinavia (Luoto *et al.* 2001), woodland edges in Estonia (Meier *et al.* 2005), clearings and sparse woods in Central Europe (Kudrna & Seufert 1991, Konvička & Kuras 1999 Meglécz *et al.* 1999), and mountain meadows in southern Europe (Descimon & Napolitano 1993, Napolitano & Descimon 1994). A similar gradient exists in *Leptidea sinapis* (Benes *et al.* 2003b, Amiet 2004, see also Chapter 16). *Limenitis reducta* used to be restricted to sparse coppices in the Czech Republic and Slovakia, where it is largely extinct now, but is found in high forests along narrow streamsides in the Mediterranean. The breadth of biotope use also decreases towards range edges (Thomas 1993, Shreeve *et al.* 1996). The best evidence concerns grassland species (e.g. Wilson *et al.* 2002, Davies *et al.* 2006). A woodland example is *Melitaea athalia*, which is restricted to coppiced woodlands and heath in Britain (Asher *et al.* 2001) but occupies grasslands, woodlands and even bogs in Central Europe (Schwarzwalder *et al.* 1997, Settele *et al.* 2000). Our current view on biotope use may also be obscured by recent losses of populations. In Germany, *Hamearis lucina* deserted woodlands after they became too shaded without traditional management; now it only occurs in mosaics of grasslands and scrub (Fartmann 2006).

Nymphalis vaualbum, Fabriciana elisa, Euphydryas maturna, Coenonympha hero and *Lopinga achine*. Under a broader definition of woodland, *Zerynthia polyxena, Colias myrmidone, Coenonympha oedippus* and *Erebia sudetica* are also listed under Annex II.

Managing woodlands for many threatened butterflies consists of maintaining relatively low tree density and/or permanent or dynamically managed clearings. However, the conjecture that forest closure would lead to species losses has generally not been reflected in conservation policies until very recently. It was believed that since woodlands represent the climax vegetation of most of Europe, it would be sufficient for associated wildlife to preserve the natural composition of the vegetation (e.g. preventing coniferisation, or leaving some areas to develop naturally), without regard to structure. This vision and policy caused losses of species and populations, even from reserves.

We argue, following Vera (2000), that:

(i) woodlands in ancient landscapes, in the absence of human influence, would have had more open structure than woodlands operated by modern forestry;

(ii) in modern landscapes, woodland butterflies cannot survive without active management of the sites;

(iii) in many situations, traditional management practices have been and are the driving force for the evolution of plant and animal communities of European woodlands – in contrast to many modern forestry operations.

The role of active woodland management for butterflies was first demonstrated for British populations of *Melitaea athalia*. The species hung at the brink of extinction, although

its localities retained both a natural composition of vegetation and high densities of host plants (Warren *et al.* 1984). Detailed research revealed that:

(a) both larval development and adult activity were restricted to woodland clearings regenerated after cuts;

(b) dispersal among clearings was limited since adults hesitated to cross shady forest;

(c) each clearing would become unsuitable after about ten years following a cut; and

(d) adults were able to colonise new clearings established up to ca. 600 m from existing ones (Warren 1987a, b, c).

The reason behind the decline of *M. athalia* in Britain was the demise of coppicing, a practice in which wood is cut on short rotations (4–30 years) and which allows regrowth from stumps or roots (Rackham 2003). Regular coppicing creates a fine mosaic of successional stages (see Plate 21.a and Fig. 23. 1), from open ground to closed canopy, and in such close proximity that a fast turnover of local colonies is possible (new habitat is created within the dispersal range of *M. athalia*). Re-establishment of coppicing resulted in a population recovery (Warren 1991) and the butterfly is now relatively safe, although still highly restricted (Fox *et al.* 2006).

Similar situations apply for other species. In woodland, British populations of *Boloria euphrosyne* and *B. adippe* require sunny patches with *Viola* spp. (Greatorex-Davies *et al.* 1992, Clarke and Robertson 1993, Barnett & Warren 1995a, Warren 1995) and larvae of *Hamearis lucina* feed on sun-exposed *Primula* plants at sheltered sites (Leon-Cortes *et al.* 2003b, Sparks *et al.* 2005). British populations of *Satyrium pruni* develop on *Prunus spinosa* which regenerates

Figure 23.1. Coppice area in the Steigerwald, northern Bavaria (August 2002). These 'forests' host numerous endangered forest species, among them the butterflies *Euphydryas maturna*, *Lopinga achine* and *Coenonympha hero* (picture: M. Dolek).

after periodic cutting in sheltered locations (e.g. often along woodland rides: Bourn & Warren 1998). The case of *Parnassius mnemosyne* in Central Europe (e.g. Kudrna & Seufert 1991, Meglécz *et al.* 1999, Kuras *et al.* 2000b) parallels that of British *Melitaea athalia*. Its host plants (*Corydalis* spp.) tolerate varying light levels including shade, but females of *P. mnemosyne* oviposit only on sun-exposed plants (Konvička & Kuras 1999, Välimäki & Itämies 2003, Bergstrom 2005). *Lopinga achine* requires a delicate balance between sunlight and humidity, as demonstrated in Swedish studies. It develops on *Carex montana* growing under 60–80% of canopy cover, as traditionally maintained by cattle grazing in woodland (Bergman 2001). A sparser canopy causes desiccation of the host plants, whereas shadier conditions become too cool for the larvae (Bergman 1996 (2000), Bergman 1999). Pre-diapause larvae of Central European populations of *Euphydryas maturna* develop on saplings and low-hanging branches (1.5 to 3 m) of *Fraxinus* spp. located in sunny but damp conditions (Konvička *et al.* 2005, Freese *et al.* 2006). This species also displays regional variation in habitat use. Finnish populations develop on herbs (Wahlberg 1998) and use younger successional stages than they do in Central Europe (typical habitats for both *Lopinga achine* and *Euphydryas maturna* are shown on Plate 21.b and Fig. 23.1.)

Less intensively studied examples include *Satyrium ilicis* relying on shrubby *Quercus* spp. (Hermann & Steiner 2000, Koschuh & Fauster 2005); *Coenonympha hero* depending on waterlogged cleanings (Kulfan 1993, Steiner & Hermann 1999) and *Neptis sappho* and *Leptidea morsei* historically

reported from 'sparse woods' in Central Europe (e.g. Gartner 1861, Moucha 1951, Lorkovic 1993). Even canopy-dwelling butterflies (e.g. *Apatura* spp. or *Limenitis populi*) use open structures, either because they prefer sun-exposed host plant parts or to locate resources like salt or carrion (e.g. Ebert & Rennwald 1991a, Weidemann 1995, Dell *et al.* 2005, Füldner 2006). Many species occur in mosaics of sparse forest, meadows and bogs in Central Europe, but inhabit sparse woodlands in the north. Examples include *Colias palaeno*, *Lasiommata petropolitana* or *Boloria titania* (cf. SBN 1987).

Although a wide range of mechanisms drives the association of woodland butterflies with open and sparse forests, there is a recurrent pattern of dependency on early successional sites. Such sites, given enough time, will eventually succumb to successional regrowth. Successional processes were incorporated into metapopulation models dynamics of *Lopinga achine* in Sweden (Bergman & Kindvall 2004) and *Euphydryas aurinia* in Finland (Wahlberg *et al.* 2002a), and extinctions were predicted in the absence of active management to keep woodlands open.

Butterflies lend support to the idea that European woodlands had rather open structures in their 'natural' state (see Box 23.2). Many woodland species are strictly confined to the nemoral latitudinal belt, not extending to Boreal taiga in the North or to Mediterranean/Pontic zones in the South both in Europe and Asia (see Dubatolov & Kosterin 2000 for Siberia). They would have nowhere to go, had the woodlands been more closed. The degree of openness and the prevailing factors securing it would differ with respect to

---

Box 23.2   The past of European wooded landscapes?

It is critical to understand what 'primeval' woodland would have looked like in absence of human activity. How open would European forests be in a pristine state? This debate has revived since Vera (2000) presented a theory that the primeval vegetation of most of temperate Europe was not a closed forest, but rather a 'wood pasture'. This includes mosaics of grasslands, scrubland and denser groves maintained by large ungulates, such as aurochs and European bison. After early farmers had invaded such landscapes, domestic animals replaced wild herbivores, securing a continuation of essentially 'natural' grazing processes. Vera supported his theory with observations of germination patterns of oaks, with vegetation dynamics of woodland reserves without herbivores, and with a reinterpretation of historical records. Further support comes from modelling (Kirby 2004) and experiments (Bakker *et al.* 2004); contradictory evidence stems from studying pollen records in areas that historically lacked ungulate herbivores (Ireland) (Mitchell 2005). The issue is still debated, but with a focus on the relative importance of ungulates versus other processes such as fire, floods, waterlogging or outbreaks of herbivorous insects. Siberian woodlands also demonstrate a more diversified landscape (i.e. areas of sparse tree cover, broad ecotones and even woodland meadows) than most of the current European woodlands. Habitat descriptions of woodland butterflies by Russian authors do refer to 'steppefied forest' or burned woodlands (e.g. Korshunov & Gorbunov 1995, Tuzov *et al.* 1997, 2000).

---

geography and woodland type. Large ungulates were probably more important in more southerly regions, where grazed savannahs rather than forests would have developed (Rackham 1998). Factors such as fire were likely more important in Boreal regions (Lindbladh *et al.* 2003). The activity of ungulates would presumably create finer mosaics of open (i.e. sunny) and closed (i.e. shady) conditions, whereas catastrophic disturbances like fires would create larger blocks in different successional stages. We might predict that the former landscapes should select for long-term persistence without great demands for long-range dispersal, in contrast to the latter.

A scale factor is also important since long-range dispersal is a rare event (see Chapter 9). When (open) woodlands covered most of Europe, the chances of a disturbance followed by a fast build-up of a butterfly population, and subsequent long-range dispersal, were much higher than in present fragmented landscapes. This may explain why large woods host more species than small woods (Shreeve & Mason 1980). Scale considerations shed some light on the importance of traditional woodland management techniques. Historical management used to proceed by small 'coupes' with a fast turnover. Multiple use, such as combining pannage, coppicing and bee-keeping, was the rule. Resulting habitat diversity had contributed to the survival of a rich 'primeval' diversity of European woodlands even after the woodlands became fragmented by human activity.

An early first woodland management technique was (as already mentioned) coppicing. It provides large areas of

open habitats in high forests of equal size and reduces the cost of butterfly dispersal when coppiced zones occur in close proximity (Warren 1987b, Cizek & Konvička 2005). A special case of coppicing is shredding, i.e. periodic harvesting of lateral tree branches as livestock fodder (Rackham 2003). Nowadays it is only practised regionally in southern Europe. Effects on butterflies should be similar to coppicing as shredding also opens up stand structure. Another technique historically associated with wood pasture is pollarding, in which trees are cut above the browsing height of grazing animals (Rackham 2003) allowing a continuous supply of woody regrowth. Rather than a fine-grained mosaic, woodland pasture produces relatively homogeneous but sparse growth of trees and shrubs. Depending on tree density, it may favour species requiring half-shaded conditions (e.g. *Lopinga achine* in Sweden), or species thriving under full sun. High grazing intensity results in savannah-like biotopes in which grassland conditions gradually prevail over woodland conditions. Forest pasture is also important for mountain butterflies by maintaining an open stand structure (e.g. below the timberline in the Alps). Litter mowing and raking were once widespread practices supplying rural communities with animal bedding (Glatzel 1991). Repeated exports of nutrients affected the composition of the ground flora, slowed down tree regeneration and gradually opened the stands. These practices – for long despised by foresters – gradually lost their economic importance (Farrell *et al.* 2000). However, experimental re-establishment of these practices may restore sensitive plant and fungal communities

(Baar & Kuyper 1998, Dzwonko & Gawronski 2002). We hypothesise that their effects would be similar to that of forest pasture in promoting vegetation of nutrient-poor soils and butterflies (see Pollard *et al.* 1998, Dennis *et al.* 2004, Tudor *et al.* 2004, Öckinger *et al.* 2006).

Fire is rarely thought of as a management practice in Europe, but it was historically used to improve pasture quality in savannah-like systems, especially in the Mediterranean (Grove & Rackham 2001) but also elsewhere (e.g. Carpathian wooded meadows: Kralicek & Gottwald 1984). Bee-keepers in some Mediterranean regions sometimes burn woody vegetation encroaching nectar-rich heaths. The effects on butterflies are largely unknown from Europe, but in North America, prescribed burning has been gradually accepted as a tool for conserving butterflies of open woodlands (Huntzinger 2003, Kleintjes *et al.* 2004, Waltz & Convington 2004).

All the above practices moulded European woodlands for millennia, maintaining their open and highly diversified structure. Some are still in use in some regions (e.g. coppicing in Italy and France, woodland pasture in northern Greece, shredding of small groups of trees in northern Spain), but many have been entirely abandoned. There are now large regions where modern forestry has completely replaced open woodlands with even-aged, structurally uniform stands. In the Czech Republic, where >6% of woodlands were still coppiced in the 1920s (i.e. after most of the conversions already occurred), only 0.1% is still under such a regime. The demise of coppicing was brought about by a political decision in the 1950s. Consequently, coppices within large regions were either cut and replanted, or left to mature. This led to the extirpation of woodland butterflies from most of the country, including reserves (e.g. Vrabec 1994, Konvička *et al.* 2006b).

In Europe there are few realistic opportunities to revert to vast woodlands with disturbance dynamics mediated by catastrophes and ungulate herbivores. Hence, preserving several threatened butterflies is impossible without maintaining or restoring some of the traditional management practices (Bergman 2001, Konvička *et al.* 2005, Liegl & Dolek 2008, Häusler *et al.* 2008). However this is often hindered by legal, technical, cultural and economic obstacles, such as legal bans on short-rotation cycles, lack of traditional practical skills, cultural preferences for high forest and poor markets for coppice products (Freese *et al.* 2006). Some of the obstacles may be overcome (e.g. increasing energy costs increase demands for fuelwood), but in the near future, coppicing or forest pasture will probably be confined to reserves. This puts special emphasis on

politically and economically acceptable alternatives, suitable even for commercial high forests, including plantations. Examples include maintaining wide margins along roads and glades (Warren 1985, Greatorex-Davies *et al.* 1992, Füldner 2006), creating box junctions at ride intersections and scalloped bays along rides (Carter & Anderson 1987, Konvička *et al.* 2006b) or managing structures such as power lines for butterflies (Smallidge & Leopold 1997, Forrester *et al.* 2005). Another option might be presented by deer parks, which maintain ungulate densitites comparable to those of pasture woodlands, but have both positive and negative effects on butterflies (Feber *et al.* 2001, Benes *et al.* 2006). A first proof of the viability of creating dynamic landscape mosaics for nature conservation purposes on a small scale is presented by Kleyer *et al.* (2007).

Alpine, subalpine and arctic ecosystems

Ecological studies of European alpine butterflies are rather scarce, perhaps due to the combination of difficult access to localities and obstacles to studying butterflies in harsh climates. This contrasts with North America, where several classical studies of butterfly ecology originated from alpine environments (Watt *et al.* 1977, Gall 1984) and alpine butterflies continue to be used as model species (Boggs & Murphy 1997, Roland *et al.* 2000, Fleishman *et al.* 2003, Matter *et al.* 2004, Boggs & Freeman 2005). The dearth of studies from Europe may also stem from the belief that alpine environments are still relatively intact, with the more threatened lowlands being prioritised for research. High Arctic butterflies are studied even less. On the other hand, mountain and arctic environments have always attracted flocks of visiting lepidopterists and an impressive amount of ecological information has been accumulated as a result (SBN 1987, Varga 1997, Sonderegger 2005).

Based on the information available, it seems that alpine butterflies are faring better than butterflies of other habitats. The *Red Data Book* of European butterflies (Van Swaay & Warren 1999) lists seven species as continentally threatened, but some (e.g. *Erebia christi*, *Erebia sudetica*) are listed mainly on the basis of their restricted distribution. European mountains contain the oldest, largest and most famous national parks. Modern agriculture and forestry have not affected the biotopes above the timberline. Infrastructure development is limited to sport and recreation facilities, mostly with locally restricted environmental impacts. Traditional land use, especially grazing, has persisted to the present in the Alps, Carpathians, Pyrenees and the mountains of southern

Europe. All the above aspects are likely to contribute to relatively few butterfly losses in topographically heterogeneous mountain areas (Konvička *et al.* 2006b).

Still, there are reasons for concern. European mountains host disproportionately high numbers of endemic species with narrow distributions (Thomas 1995b, Dennis *et al.* 1998b) and extremely specialised requirements for altitude, slope, aspect or humidity (e.g. Sonderegger 2005; see also Chapters 16 and 17). For many of them, even basic data on population status are missing, while others are threatened by tourism development (e.g. *Lycaena golgus* or *Agriades zullichi* in Spain: Munguira & Martin 1993). Many species have retained their strongholds in mountains despite losses elsewhere (Deschamps-Cottin *et al.* 1999, Van Swaay & Warren 2003). The most serious threat is climate change threatening the existence of entire alpine (and arctic) biota (Wilson *et al.* 2005, Luoto *et al.* 2006, IPCC 2007, Notaro *et al.* 2007). Retreats of species at low-altitude range margins are already occurring (see Chapter 20). Mountain populations in small areas of alpine biotopes are particularly imperilled (e.g. the Hercynian mountains of Central Europe: Schmitt *et al.* 2005c).

Periods of warm climate occurred during the Holocene (e.g. the Atlantic era, 8000–6000 BP). Apparently, alpine butterflies of the middle high mountains, such as *Erebia epiphron* in the Jesenik Mountains. (Czech Republic) survived these periods *in situ* (Schmitt *et al.* 2006b). As this species never enters closed forest (Cizek *et al.* 2003), it is likely that a critical factor for its survival is a continuation of natural disturbances, such as landslides, avalanches, outbreaks of tree-killing insects, and animal grazing, preventing or slowing down an upward shift of the timberline. The most species-rich alpine biotopes (e.g. moraines, glacial cirques, valley headwalls and stream canyons) are disturbance dependent (see Box 23.3).

We clearly need a more thorough understanding of the disturbance dynamics of alpine and arctic biotopes and their effects on butterfly populations if we are to counter the effects of a warming climate. At present, it is not even known which 'natural' factors would maintain a high diversity of alpine biotopes currently grazed by domestic animals. Even greater doubt relates to our understanding of habitats and populations of butterflies occurring in High Arctic conditions (Luoto *et al.* 2006) where grazing by reindeer and other mammals represents the main disturbance factor (Olofsson *et al.* 2001, 2004). Moderate levels of grazing intensity maintain high plant species diversity, which in turn has been suggested to support a rich insect fauna

---

Box 23.3   Butterflies, mountains and disturbance

In the Jesenik Mountains (Czech Republic), *Erebia sudetica sudetica* occurs in several large colonies on avalanche slopes just at the timberline (Kuras *et al.* 2001, 2003). Additional colonies may be found below the timberline at streamsides and clearings, but it is not known for how long such small colonies persist. Still, the ability to survive at disturbed sites within forests may become critical if avalanche frequency declines or if the timberline rises due to warming climate. In the same mountains, *Erebia epiphron* forms a large population of tens of thousands of individuals in (sub)alpine grasslands. Despite their seemingly natural character, the grasslands were grazed until 1939. Cessation of grazing allowed the slow spread of a non-native shrub (i.e. the dwarf pine *Pinus mugo*) which was planted by foresters a century ago to stop avalanches. Although *E. epiphron* is not imminently threatened, its population is genetically subdivided (Schmitt *et al.* 2005c) and local extinctions of several other alpine invertebrates due to the encroachment by dwarf pine have been detected (Kuras & Tuf 2005).

Similar developments may be affecting other mountain areas. Brčák (1952) conducted a pioneering quantitative study of Lepidoptera in the Slovakian High Tatras, using transect counts. Five species of *Erebia* were abundant and widely distributed. A recent survey by Kulfan *et al.* (2004) detected decreased abundances and distribution ranges. Climatic effects are possible, but another factor is the absence of grazing since the High Tatras National Park was established in 1949.

Disturbance dynamics are important even for species occurring within the timber belt. In Norway, *Parnassius mnemosyne* is only found in two fjord valleys where avalanches keep woody vegetation open. The persistence of the butterfly does not require human intervention as suitable conditions are created by natural events (Aagaard & Hanssen 1992). In contrast, populations of *Parnassius apollo* in the Pieniny Mountains, southern Poland, depend on calcareous grasslands and scree, kept open by animal grazing and landslides. Afforestation by spruce almost drove the butterfly to extinction; it required costly spruce removal to restore the population (Witkowski *et al.* 1997).

(Suominen & Olofsson 2000) including butterflies (Välimäki 2005).

The arctic and alpine areas represent an unconquered frontier of European butterfly ecology, and at the same time may be the most severely imperilled by changing climate.

Heathlands

In Europe, heathlands mostly arose from intensive agricultural activity on thinly wooded sites with poor, typically acid, soils that were cleared about 4000 years ago and used for a range of agricultural activities from stock grazing to fuel cutting and fodder collection. Ericaceous dwarf shrubs such as *Calluna* spp. are typical of heathlands, and may have been present as an understorey prior to woodland clearance (Webb 1998). Heathlands range in composition from those completely dominated by dwarf shrubs to those with shrubs interspersed between grassy patches; the hydrological state of the area is important in determining vegetational composition. Grazing, harvesting of materials and rotational burning kept the areas open and continued nutrient depletion. In the past, heathland would have covered several millions of hectares of Europe, but now only a highly fragmented fraction remains (estimated at 350 000 ha: Webb 1998, Webb & Vermaat 1990). Loss of heathlands has many causal factors: sand and gravel extraction, afforestation, scrubbing-up due to lack of traditional management, over-grazing (mostly in the uplands), housing and industrial development, and infrastructure work such as road building (Fry & Lonsdale 1991).

A typical butterfly resident in the heathlands of northwestern Europe is *Plebejus argus* which has been extensively studied in the UK. Temperature is a determinant of niche occupation for this species. In north Wales it is restricted to recently burnt swards of heathland on south- and west-facing slopes (Dennis & Bardell 1996) which are 8–10 °C warmer than adjacent tall vegetation and 7 °C warmer than nearby north-facing slopes (Thomas 1985). In Devon, however, which has a much warmer climate, the butterfly can exploit the cooler northern slopes, and uses much older heather stands. Hence, in north Wales a habitat patch can only last for a maximum of 5 years, whereas in Devon it can last up to 15 years with a wider range of heather morphologies (Erhardt & Thomas 1991). Interestingly, this work has also demonstrated that the habitat (as opposed to biotope: Webb 1994) can vary over relatively short distances. This species' biotope has been extensively fragmented, and habitat conditions within the remaining fragments are not all optimal; some fragments are probably too small to permit the establishment of stable populations. Internal fragmentation with large stretches of unsuitable biotope (and scrub barriers) between habitat patches and external fragmentation between heathlands is frequently too great for dispersal with inter-patch distances too wide to maintain a metapopulation-type structure (Thomas 1991b).

Wetlands

Butterflies associated with wetland tend to be highly specialised, and many of them are threatened due to habitat loss (e.g. Fischer *et al.* 1999, L'Honoré & Lagarde 1999, Stefanescu *et al.* 2005b). Most continental wetlands have been lost due to canalisation, land drainage and conversion to arable crops. On the positive side, the importance of wetland conservation is increasingly accepted by the general public, and many of the remaining European wetlands have some legal protection, and generous restoration schemes are being applied throughout the continent.

Two broad wetland types with distinct butterfly assemblages can be identified: oligotrophic, *Sphagnum*-dominated bogs of cooler or high-altitude regions (see Plates 22a and 22b) and eutrophic wetlands dominated by sedges and reeds (with tyrphophilous/hygrophilous butterflies). Due to their rather extreme environments, butterfly assemblages inhabiting wetlands tend to be well differentiated from those of surrounding landscapes (Väisänen 1992, Nekola 1998, Wettstein & Schmid 1999, Dapkus 2004, Sawchik *et al.* 2005). The casual factors behind this pattern are straightforward in the case of peat bog specialist species: the larvae develop on the dominant constituents of bog vegetation (e.g. *Vaccinium oxycoccus*, *Vaccinium uliginosum*, *Eriophorum* spp.) (Spitzer & Danks 2006). Other mechanisms relate to climate: Mikkola & Spitzer (1983) pointed out that the composition of wetland assemblages varies regionally, with a tendency that widespread northern species perform as peat bog specialists in more southerly latitudes. As requirements of particular species might be fulfilled in different ecosystems across its distribution, ecosystem-related classifications are sometimes quite difficult (see Box 23.1 for general and Box 23.4 for wetland-specific considerations).

The dominant factor affecting wetland insects is water. In wetlands, depending on surface run-off, partial or complete submergence of entire habitats is not uncommon. How butterflies respond to flooding is important as wetland restoration often employs the use of sluices and dams to artificially manipulate water levels. Larvae of *Lycaena dispar batavus* from regularly flooded sites survive up to 28 days

---

Box 23.4   Wetland species or not?

The complexity of patterns of biotope associations (see also Box 23.1) can be illustrated with species of eutrophic wetlands using different host plants across different regions.

Northwestern subspecies of *Lycaena dispar* (extinct ssp. *dispar* and extant ssp. *batavus*) are strictly associated with wetland-growing 'large' species of docks (e.g. *Rumex hydrolapathum*). In Central and Eastern Europe, the bivoltine subspecies *L. dispar rutilus* uses a wider array of dock species, including the common weedy species *Rumex obtusifolius* and *R. crispus* (Martin & Pullin 2004a, b). As a consequence, the *rutilus* race is not endangered and is actually expanding (Pullin *et al.* 1998, Konvička *et al.* 2003).

The British subspecies of *Papilio machaon* is strictly monophagous on *Peucedanum palustre*, a marshland member of the Apiaceae, whereas continental populations feed on a wide range of Apiaceae (Dempster 1995), including those grown as vegetables. *Brenthis ino* can develop on a wide range of Rosaceae plants in a wide range of habitats. One of those plants (*Filipendula ulmaria*) forms tall and nearly monospecific swards at wet areas. This plant is increasing in abundance due to abandonment of wet meadows; *B. ino* tracks this development, but this makes a false impression that it 'depends' on wetlands (Zimmerman *et al.* 2005).

Brenthis ino represents the species that are not restricted to wetlands, but their abundance on wetlands exceeds those of other habitats. Spitzer & Danks (2006) described this phenomenon for peat bogs, coining such species 'tyrphophilous' (as opposed to 'tyrphobiontic'); they did not offer any interpretation apart from that they retreated to wetlands from other, human-altered habitats. One reason for such retreats might be the pervasive loss of open woodlands. Mainland European bogs tend to retain a high architectural diversity with alternating open and partially shaded sections, for longer than commercially managed woods. Examples of the tyrphophilous species with alternative optima in open woodlands include, among others, *Callophrys rubi*, *Melitaea athalia* and *Boloria titania*. Even such textbook tyrphobionts as *Colias palaeno* may occur away from bogs, provided that its host plant and an open canopy are present, and some of the sites can be quite arid indeed (e.g. serpentinite heaths in the Slavkovskyles Mountains, Czech Republic).

---

of submergence, but longer floods increase mortality (Webb & Pullin 1998). However, the larvae actively escape floods if their plants are only partially flooded (Nicholls & Pullin 2003). Active escape is also shown by larvae of *Coenonympha tullia*, whose habitats are normally not flooded and for whom submergence as short as three days significantly increases mortality (Joy & Pullin 1997, 1999). However, an American subspecies adapted to salt marshes, *C. tullia nipisiquit*, survives submergence in brackish water, illustrating a wide range of existing adaptations (Sei & Porter 2003, Sei 2004). A wetland-inhabiting population of *Neptis rivularis* represents an intermediate situation, surviving under water for about one week and compensating the mortality caused by occasionally prolonged floods by colonisation from adjacent upland populations (Konvička *et al.* 2002b).

Again, responses to the environment are context dependent. The same applies for demography. Based on the assumption that peat bogs of temperate Europe represent relatively stable remnants of cooler periods of the Holocene (Spitzer & Danks 2006), it is often assumed that specialised insects should exhibit low dispersal ability and numerical

stability of populations. Recent work on two peat bog specialists, *Proclossiana eunomia* and *Boloria aquilonaris* allows us to scrutinise these assumptions (see Chapter 22). In the Belgian Ardennes, both species exist as metapopulations with fluctuating local densities (Schtickzelle *et al.* 2002, Baguette & Schtickzelle 2003). Although the habitats are highly fragmented, *Boloria aquilonaris* is capable of relatively long-range dispersal (Baguette 2003). This must also have been the case historically, as many bog areas in the Palatine Forest (southern Germany), which have been inhabited by *B. aquilonaris* and *C. tullia* for several hundreds of years, are of human origin (these bogs have developed at the edges of dystrophic ponds, created for milling and other purposes) and must have been colonised from primary habitat complexes over several dozens of kilometres (Settele *et al.* 1992). Adults of *P. eunomia* spend much time outside of habitat patches (which contain both adult and larval resources) (Baguette & Neve 1994, Petit *et al.* 2001) but leave the habitat more readily in landscapes with high habitat connectivity (Mennechez *et al.* 2003), while dispersal mortality is lower in more fragmented landscapes, suggesting an evolutionary trade-off (Schtickzelle *et al.* 2006). Therefore, their demographic traits do not substantially differ from species inhabiting more transient environments, such as woodland openings. Indeed, there are two known instances of *P. eunomia* expanding into a network of suitable empty habitats, one from the River Morvan in France (an introduced population: Néve *et al.* 1996a) and one from the Sumava Mountains in Czech Republic (spontaneous expansion from refuge sites: Pavlíčko 1996).

Preserving hydrological conditions that led to the formation of a particular locality is a necessary precondition for wetland conservation. However, even wetlands are dynamic systems, subject to successional changes. This is obvious for eutrophic sites (e.g. marshes formed in oxbow lakes), but sometimes unacknowledged for oligotrophic bogs with much slower dynamics. Even then succession towards closed forests becomes inevitable, once the peat layer grows out of contact with the water table (i.e. terrestrification; compare Plate 22b). On the other hand, bog-like conditions may arise de novo via invasion of waterlogged sites by *Sphagnum* and other bog-forming plants (i.e. palludification: Klinger 1996). The examples of *B. aquilonaris* and *P. eunomia* illustrate that within certain limits, specialised bog butterflies can track habitat dynamics.

The complexity of wetland management is apparent in the case of litter meadows and the fen habitats of pre-alpine slopes in Germany. These fens used to be mown in autumn to obtain bedding for domestic animals, and the management is considered as optimal for such species as *Maculinea alcon*, *M. nausithous*, *M. teleius*, *Euphydryas aurinia* and *Proclossiana eunomia* (Bräu *et al.* 2006, Völkl *et al.* 2008). Although continuation of this management is prohibitively costly, it was long thought that other types of management, such as grazing, would be detrimental for local biota (Blab 1993). A comparison of grazed and mown fens showed that specialised butterflies similarly use both types (Dolek & Geyer 1997). It was also found that the 'historical' mowing management is actually quite new, not more than two centuries old, as these habitats were used as pastures in the more distant past (Radlmair *et al.* 1999). Obviously, grazing has to be extremely light, and it is difficult to prepare and implement precise management guidelines, as there is a high occurrence of chance effects compared to mowing. Impacts of specific actions may also differ among years (high vs. low water table), and among focal species. From a practical point of view, regional and local conservation authorities strive for flexible programmes to support farmers under conditions adapted to locally appropriate frameworks. This is often hindered by higher-level authorities, including the European Union, who strive to create standardised support programmes.

To summarise the major problems of wetland butterflies:

(a) many remaining wetland sites are too small to sustain viable populations of specialised species (Schtickzelle *et al.* 2005a, b). For instance, two out of three tyrphobiont butterflies had been lost during the last decade from the Cervene Blato, a bog in the Czech Republic (Spitzer & Jaros 1993)

(b) many wetlands, including peat bogs in mountain areas, are threatened by accelerating terrestrification caused by drainages in wider aquifers

(c) river channelisation and forest drainage prevent the spontaneous formation of wetland conditions.

Comprehensive data are missing, but it is likely that these factors contribute to deterioration of both habitat quality and quantity of bog biotopes across Europe. However, quality can be manipulated by preventing successional changes within existing bogs (Dennis & Eales 1999), whereas quantity and connectivity may be enhanced by initiating bog-formation processes, such as restoration of hydrological conditions in wider aquifers, or by restoring bogs that have been milled out (i.e. when commercial extraction has exhausted the peatland resource). Several successful projects illustrate that this can be done (e.g. Campeau *et al.* 2004, Lanta *et al.* 2006), although effects on butterflies are rarely reflected in mainstream literature. One such positive example concerns the formerly milled out Borkovicka Blata bog in the Czech Republic which was restored by blocking drainage channels. This led to spectacular increase of a population of the nationally threatened *Coenonympha tullia*.

Grasslands

INTRODUCTION

About 50% of European butterflies inhabit semi-natural grasslands created and maintained by human activities: grazing by domestic animals and hay production (Erhardt & Thomas 1991). Van Swaay (2002) calculated that 48% of the 576 native European butterflies occur on calcareous grasslands alone. Land use changes threaten grassland biotopes in two opposite ways: abandonment and intensification (Bignal & McCracken 1992, Settele & Henle 2003); the latter in productive areas, especially the lowlands, the former in marginal areas such as the uplands.

Abandonment leads first to dominance by a few competitive herbs, usually grasses, and then to scrub and tree invasion

(Erhardt & Thomas 1991). Grassland areas shrink, remnants become increasingly fragmented and species losses inevitably follow (Wenzel *et al.* 2006, Polus *et al.* 2007). Cessation of traditional grazing threatens the biodiversity of the foothills of the Carpathians in Romania (Cremene *et al.* 2005, Schmitt & Rakosy 2007), the Ukraine (Elligsen *et al.* 1997, 1998), the Swabian Jura in Germany (Beinlich & Plachter 1995) and elsewhere. An extreme form of change, which has aspects of both abandonment and intensification, is intentional afforestation, sometimes opted for by owners to extract at least some profit from otherwise useless lands, but too often promoted by national or European agencies under the guise of helping the environment. It represents a particularly grave threat in areas with a high diversity of species with restricted ranges, such as on the Mediterranean island of Sardinia (Grill *et al.* 2005) or in pre-alpine grasslands (Nunner 2006).

Intensification affects both pastures and hay meadows and can be caused by fertilising, reseeding, drainage, overstocking of pasture and employment of new technologies, such as silage and haylage (early baling of part-dry hay), on hay meadows. When land use intensifies, the landscape may not obviously change with the same physical areas of grassland remaining, but habitat quality deteriorates. Thomas (1984) observed in the UK that improved grasslands support three butterfly species, whereas unimproved pastures have 28 species. Traditionally grazed grassland in the subalpine Tavetsch Valley in Switzerland hosted 41 species whereas fertilised and mown grasslands in the same valley hosted only eight (Erhardt 1992). The nature of the butterfly assemblage also changes, with very few xenotopic species (migrants/visitors) in traditionally managed grasslands compared with over a third in heavily fertilised ones (Erhardt & Thomas 1991). Increased nutrients through fertilisation suppress competitively inferior plants, reducing butterfly resources (Ockinger *et al.* 2006). Effects of fertilisation may even increase by atmospheric deposition of nitrogen pollutants (Fischer & Fiedler 2000a). Effects of overstocking of pastures are quite similar, although the mechanism here is physical destruction of resources. Overstocking represents a particular threat to some Mediterranean areas (Balletto 1992, Balletto & Casale 1991).

It might seem that both threats can be reversed by incentives to farmers to relax the intensity of farming in favour of more traditional techniques (de-intensification) and to keep marginal grasslands in use (prevention of abandonment). Insights from recent studies reveal that the whole situation is more complex. Traditional family farms were smaller than most farms today, producing diverse outputs consumed either by farmers' families or at local markets, and because farming relied on manual or animal labour, any impact on biotopes was distributed across longer timespans. Each farmer applied slightly different methods in different years (e.g. timing of management). The resulting landscapes were finely grained mosaics of differently managed sites, offering diverse resources in close proximity. For butterflies, traditional farming allowed (i) easy localisation of resources within short-range movements (Ouin *et al.* 2004), (ii) swift responses to temporarily changing habitat quality (Dennis & Sparks 2006), and (iii) fast recolonisation following local depletions of populations (Dover and Settele 2009). If we add dense networks of field banks, hedgerows, orchards or forest edges, we can see how traditional landscapes supported the long-term coexistence of multiple species, even with contrasting requirements (Settele 1998, 2005).

Below we present some case studies of effects of hay mowing, grazing and fallow land. It should be kept in mind that under traditional land use, those methods often occurred close to one another, and even varied within a single site.

HAY MEADOWS

Mowing affects butterflies via direct mortality of adults and pre-imaginal stages and removing nectar sources. The effects depend on the timing, frequency and technology used. Mowing in late summer destroys larval nests of *Euphydryas aurinia* in the Czech Republic (Hula *et al.* 2004), although the harm may be mitigated if a longer sward (15–20 cm) is left above dense growths of the host plant *Succisa pratensis*. Abandonment is not an option here, as ovipositing females prefer plants growing in low-sward conditions (Anthes *et al.* 2003b, Konvička *et al.* 2003). Late-summer mowing also decimates local populations of *Maculinea teleius* and *M. nausithous* in Central Europe (Johst *et al.* 2006, Völkl *et al.* 2008), as it happens precisely at the time when females oviposit on flowerheads of *Sanguisorba officinalis*. Hence, both species prefer less intensively mown structures, such as banks of channels and road verges, in intensively farmed areas (Settele 1998, Settele unpubl.). In Switzerland *Lycaena hippothoe eurydame* – a specialist on moist meadows with *Polygonum bistorta* – appears to be adapted to hay cutting. Adults disperse to nearby nectar sources in unmown meadows before returning to already mown meadows to lay eggs (Erhardt 1992). Schwarzwalder *et al.* (1997) show a similar case for a Swiss population of *Melitaea athalia*.

These examples illustrate the importance of finely grained, mosaic-like management, with small patches of grasslands periodically uncut allowing for the persistence of butterfly

Figure 23.2. Sheep grazing on calcareous grasslands in the Franconian Jura (August 1997) is the main method used to preserve diversity. The locality holds 54 butterfly species, including *Maculinea arion*, *Melitaea didyma*, *Melitaea aurelia*, and *Carcharodus alceae* (picture: M. Dolek).

populations. Likewise, mowing that does not promote patchiness, or is too frequent is detrimental. This is surprisingly poorly documented, but immensely important, because mowing is a favoured practice of landscape management in many parts of Europe, richly subsidised, among others, by the Common Agricultural Policy (CAP) of the EU. With modern machinery, vast stretches of land may be cut almost instantly, which would never have happened in the past. Balmer & Erhardt (2000) warned that uniform mowing of reserves, although beneficial for rare plants, can be fatal for insects.

An option to combat negative developments in mowing regimes has been intensively investigated for hay meadows in southwestern Germany, inhabited by *Maculinea teleius* and *M. nausithous*. Based on long-term ecological data (Settele 1998, Settele unpubl.) the species ecology was integrated into ecological (Johst *et al.* 2006) and economic (Wätzold *et al.* 2006) models, and acceptance surveys were conducted among the local residents (Lienhoop *et al.* 2005). All studies combined have shown that public awareness was very limited but can be improved, leading to a much more positive attitude by farmers as well as administrative officials to land management for butterflies. The willingness to pay for improved biodiversity conservation is also very high, as long as people in the region are part of the conservation strategy (Lienhoop *et al.* 2005).

PASTURES

The effects of grazing (and browsing) can mainly be related to grazing intensity and type (or species) of grazer. Sheep,

for example, create uniform short swards with multiple horizontal 'tracks' on hilly terrains (compare Plate 23b and Fig. 23.2), while goats create longer, 'notched' swards while keeping scrub regeneration in check. A peculiar aspect of grazing is the creation of multiple small gaps ('hoofprints') that may provide shelter and microclimate for the development of temperature-sensitive species. Such gaps can be selected for oviposition sites (e.g. *Hesperia comma* in Britain: Thomas 1983). On substrates like sand, animal trampling leads to the formation of patches of bare ground, facilitating, for example, guilds of thermophilous species at Dutch sand dunes (WallisDeVries & Raemakers 2001). A well-known species that depends, at least in the UK, on warm microclimates maintained by heavy grazing by sheep is *Maculinea arion*. In England it became extinct after exclusion of grazing from its last sites, but it has been reintroduced (Barnett & Warren 1995b, Mouquet *et al.* 2005, Thomas 1995).

In the southeast of England, *Hesperia comma* lives on closely grazed chalk grasslands containing *Festuca ovina*. This skipper was once widespread, but post-war reduction of grazing and the subsequent introduction of myxomatosis led to changes in grassland structure; only 46 populations of the butterfly persisted in 1982 (Hill *et al.* 1996). Later on, re-establishment of management and resurgence in rabbit populations restored several habitat patches. Sixty-nine potential patches were identified in the study area by Hill *et al.* (1996) in 1982 and 1991; 48 patches were occupied in 1982 and 53 in 1991. A recent survey by Davies *et al.* (2005)

revealed that the 'metapopulation success story' had continued. In 2000, the butterfly occupied 109 2 × 2 km grid squares compared to 30 in 1982, forming no less then 257 breeding colonies. The factors behind the increase were further recovery of rabbits, successful restoration of grazing under an agri-environment scheme, and, importantly, more suitable climatic conditions (Davies *et al.* 2006).

However, some species require higher turf heights. Bourn & Thomas (2002) describe a situation in southwest England, where *Polyommatus bellargus* and *Thymelicus acteon* co-occur on calcareous grasslands. The former requires short, intensively grazed turf, the latter longer swards of the grass *Brachypodium pinnatum*. A decrease of grazing intensity in the 1980s combined with myxomatosis suppressed populations of *P. bellargus* to a point that extinction seemed unavoidable, but allowed for a recovery of *T. acteon*. Later on, the prospects for *P. bellargus* improved as rabbits recovered and restoration of grazing returned some sites to a more hospitable state. Roy & Thomas (2003) also discovered that spring and summer broods of *P. bellargus* differ in microsite requirements, the latter using cooler parts of habitats with taller swards for oviposition.

Dolek & Geyer (2002) reported another example of co-occurring species differing in preferred grazing intensity. Calcareous grasslands of the Franconian Jura – traditionally grazed by sheep – host two butterflies critically endangered in Central Europe, *Chazara briseis* and *Polyommatus damon*. While *C. briseis* needs sparse and short vegetation produced by extremely heavy grazing (see Plate 23b), *P. damon* tolerates only weak and patchy grazing in autumn or, according to ongoing studies, carefully selected patchy grazing in summer (Geyer & Dolek unpubl.). Instant extinction following unrestricted (large-scale) sheep grazing was reported from the Rhön Mountains in Germany (Kudrna 1998). Seufert & Grosser (1996) studied *C. briseis* in central Germany (Halle/Saale). They found that short-sward grasslands were used for oviposition, but adults required substantial areas of tall, flower-rich vegetation. The short grasslands were limited to rocky outcrops. A landscape-scale genetic study by Johannesen *et al.* (1997) found that the hills were interconnected by high rates of gene flow. The entire system of hills and meadows is needed to sustain a large structured population (Veith *et al.* 1999).

A similar pattern applies to Finnish populations of *Parnassius apollo*. They occur in a landscape with scattered rocky outcrops that support its succulent host plant *Sedum telephium*. Because these patches cannot support vital nectar plants, adult butterflies need to move substantial distances

(up to 1840 m) to reach nectar (Brommer & Fred 1999). Importantly, because this incurs some cost for the population, the next-year generation depends mainly on females that eclosed at host plant patches situated close to nectar supplies (Fred *et al.* 2006).

Broadening the focus to butterfly communities, we should be aware that traditional pastoral landscapes may not host high butterfly densities, as any visitor to the vast traditional rangelands of the Romanian or Ukrainian Carpathians can testify (see Elligsen *et al.* 2007). Because grazing exerts constant pressure on both larval and adult resources, local butterfly densities are low and it is the sheer size of such landscapes that allows the existence of viable populations. In most of Europe, formerly vast traditional pastures have already been greatly reduced in extent. Therefore, managers willing to use grazing for conservation purposes must pay utmost attention to grazing intensity, frequency and timing, and the choice of grazing animals (see Plates 23b and 24a; Figs. 23.2 and 23.3). Several studies have addressed these issues at the level of butterfly communities. Soderstrom *et al.* (2001) in Sweden and Kruess & Tscharntke (2002) in Germany found that relaxing grazing intensity on intensively managed pastures *increased* the species richness of butterflies, bees and wasps over a single year, and the diversities remained higher even 5–10 years after cessation of grazing. Pöyry *et al.* (2004) found an essentially identical pattern in Finland. Clearly, pastures are best for butterflies just shortly after abandonment, when there is a high supply of resources and successional deterioration is not yet apparent. However, the Finnish group also detected a distinct group of 'old pasture species' consisting of butterflies requiring a continuous grazing pressure. Those specialists failed to colonise formerly abandoned pastures even 5 years after re-establishment of grazing (Pöyry *et al.* 2005).

ABANDONED LAND AND FALLOW

The above German and Finnish examples illustrate that after cessation of active management, temporary increases in butterfly richness and abundance regularly follow. The increase in butterfly abundance and particularly in wild flower abundance is higly visible and led to *bans* on grazing in many reserves established until about the 1980s. However, the increases were followed by serious butterfly losses as the succession proceeded (e.g. Erhardt 1985, Balmer & Erhardt 2000).

Some species actually prefer taller-sward conditions found at pasture margins or (temporarily) abandoned sites. Two cases of butterflies profiting from abandonment,

Figure 23.3. Horse grazing in the Rhön Mountains, northern Bavaria. Horses seem to replace cattle in some parts of Western Europe as the main reason for the maintenance of pastures and hay meadows (picture: M. Dolek).

Brenthis ino and *Proclossiana eunomia*, were discussed in the wetland section. The latter underwent a spectacular expansion in the Sumava Mountains, Czech Republic, where formerly managed wet meadows went out of cultivation after a forced exodus of local inhabitants of German nationality following the end of European hostilities in 1945. The meadows got damper as old drainage systems deteriorated and the butterfly's host, *Polygonum bistorta*, became a locally dominant plant. Another mechanism leading to a preference for fallow may stem from use of taller plants, including scattered trees, for perching and roosting. *Lycaena helle* shares its host plant with *P. eunomia*, but requires sheltered corners for establishing territorial perches and taller trees for roosting (Fischer *et al.* 1999, Nunner 2006). In a pre-alpine area (Bavaria, Germany), where the two species co-occurred, *L. helle* was considerably scarcer, presumably due to more exacting requirements combined with lower mobility (Nunner 2006). The list of examples may be further expanded by species inhabiting shrubs, which obviously profit from scrub encroachment (e.g. *Satyrium pruni* or *S. acaciae*).

GRASSLAND SUMMARY

Although the key to the conservation of butterflies of semi-natural grasslands is securing the continuation of traditional management techniques, a blind application of a few simple prescriptions may not suffice in modern, fragmented landscapes, in which the entire context (landscape grain, landholding patterns, mowing/grazing methods, etc.) differs from the past. In cases where only a small area of a grassland remains such as in reserves, management must accommodate a much higher diversity of conditions and resources within limited areas than in the past. Unless there is a concern for a single priority species (as in the case of *Maculinea arion* reintroduction sites in Britain: Barnett & Warren 1995b, Mouquet *et al.* 2005), the aim should be to create as wide a diversity of vegetation structure as possible, with some short and tall areas and occasional patches of bare ground. In landscapes where larger areas of grasslands are still available, the management goal should focus on networks of differently managed grasslands with varying mowing and/or grazing intensities. It should always be kept in mind that at least portions of both mown and grazed lands should be left fallow each year. In contrast, unified prescriptions applied across entire regions (e.g. that all meadows must be mown by a certain date) are obviously deleterious and possibly disastrous for some species.

Arable landscapes

The principal factors affecting butterflies in arable landscapes are the extremely reduced area of habitat and the impacts of farming. Intensively farmed landscapes are, at the extreme, exemplified by large fields separated by thin boundary structures. At the other extreme, small patchwork fields separated by lush hedgerows are exemplified by the bocage landscape of Brittany in France.

Because of intense farming pressure, the principal remaining biotope for butterflies in the arable landscapes are field boundaries. These may be as simple as a farm track, fence-post-and-wire, or stream, river and road corridor. Within most farm landscapes there will also be areas of rough, uncultivated, waste ground between buildings or in inaccessible areas, farm woodlands (often abandoned deciduous coppice or coppice now managed for game conservation), commercial conifer plantings or game cover crops (e.g. Dover *et al.* 1992). Field boundary composition is very variable; perhaps the best for butterflies is a vigorous multi-species hedgerow on a wide free-draining bank with species-rich grassland or woodland flora (see for example plots of butterfly abundance in Dover 1996; compare Plate 24b). Other common field boundaries may be wood edges, flat rough grass strips or raised banks, or simply dusty farm tracks.

Generally, butterflies of arable landscapes tend to be restricted by the host plants that remain in the field boundaries. These are typically ubiquitous species such as the Gramineae or species that favour nutrient-rich conditions, such as *Urtica dioica*, but may also include other species like *Lotus corniculatus*. Of the 39 butterfly species that have been recorded in British hedgerows only 26 were considered to breed there. Of those, 10 use Gramineae as host plants, 4 use *U. dioica*, 2 use common hedgerow Cruciferae, 2 commercial Brassicaceae, 1 uses thistles (*Cirsium* spp.), 1 *L. corniculatus*, 1 *Rumex* spp., and 5 use shrub species of the hedge (Dover & Sparks 2000).

Butterfly densities in arable landscapes are usually low (Dover *et al.* 1990). This is probably due to their small habitat area (for breeding) and the direct and indirect impact of farming, including insecticides (Sinha *et al.* 1990, Davis *et al.* 1991). Because of the extremely narrow field boundaries acting as habitat for these species, they are exposed to strong variation in environmental conditions. Dover (1996) and Dover *et al.* (1997) found insolation and shelter were two abiotic factors determining the incidence of individuals in farmland. Green lanes (typically double lines of hedges separated by an unmetalled farm track) which have enhanced shelter and reduce exposure to farming operations are significant as butterfly habitat in arable landscapes (Dover *et al.* 2000, Dover & Sparks 2001). The absence of shelter may also inhibit movement through the landscape, as butterflies are known to be reluctant to cross arable fields (Dover 1990, Fry & Main 1993, Dover & Fry 2001).

Experimental manipulations in arable fields in Norway have demonstrated that physical and visual aspects of hedges can influence butterfly flight direction (Dover & Fry 2001). Nectar plant abundance is also important in determining butterfly abundance in field boundaries (Dover 1996); flowers appear to be particularly scarce during the spring in arable boundaries (Dover 1989, 1997). Pesticide and fertiliser drift (Rew *et al.* 1992) cause changes in the floral composition of field boundaries from a rich perennial flora to a species-poor annual flora (Boatman 1992, Snoo & Poll 1999). This has consequences for host plant presence, nectar plant availability (Marshall & Birnie 1985), flowering (Marrs *et al.* 1989) and nectar quality (Corbet 1995, 2000), and ultimately for butterfly longevity and fecundity (Watt *et al.* 1974, Murphy *et al.* 1983, Fischer & Fiedler 2001a). Increasing nectar availability increased butterfly abundance in field margins through aggregation and also probably through changes in longevity and fecundity (Dover *et al.* 1990, Dover 1991).

Much of the recent interest in the ecology of butterflies on farmland is connected to new initiatives to improve conditions for biodiversity such as the game conservation technique known as 'conservation headlands' where the pesticide regime of the outer 6 m of cereal fields is modified to create a more diverse crop flora (Rands & Sotherton 1986, de Snoo 1994, Dover 1997) and within EU agri-environmental schemes which aim to buffer the effects of intensification by supplementing and/or restoring habitat. There is good evidence that techniques such as restoring hedgerows (Dover & Sparks 2000), sowing grass margins alongside field boundaries (Field *et al.* 2005), pollen–nectar seed mixtures (Pywell *et al.* 2004) or a combination of grass and wildflowers (Feber *et al.* 1996; Meek *et al.* 2002, Critchley *et al.* 2006) can improve farmland quality for butterflies, at relatively low cost. On the other hand, some authors express scepticism of the potential of such schemes to reverse alarming biodiveristy loss. It is not the general philosophy that is disputed, but rather the efficiency of the schemes (Warren *et al.* 2005), given the fact that they probably represent the largest-ever investment to biodiversity conservation (e.g. Kleijn *et al.* 2006).

Part of such doubt relates to the efficiency of the schemes in maintaining landscape heterogeneity. As regularly disturbed systems, agricultural biotopes necessarily select for species with a good dispersal ability and fast reproduction compared to the duration of farming cycles. Ouin *et al.* (2004) showed that farmland butterflies used short-range movements among differently managed patches to complement for locally missing resources. All other factors being equal, heterogeneous farmlands with smaller fields intersected by field banks, hedges and roads should be more hospitable than homogeneous ones (Dover & Sparks 2000, Schneider & Fry 2001, Bergman *et al.* 2004, Croxton *et al.* 2005). This has recently been supported by strong evidence from Finland; Kivinen *et al.* (2006) isolated the factors affecting butterfly species richness in agricultural areas and identified extent of forest and areas with low-intensity management as positive factors while the area of intensively managed crop fields was negative. In Sweden, farms with identical and supposedly benign management (organic farming) contained more species when they were composed of finer mosaics and were surrounded by a heterogeneous landscape (Weibull *et al.* 2000, Weibull & Ostman 2003, Rundlöf & Smith 2006). Interestingly, Westergaard (2006) found no difference between organic and conventional farms in a study of landscape characteristics in Denmark, and Jeanneret *et al.* (2003) showed that butterflies were sensitive to habitat and plant species richness but not to landscape features in Switzerland. Ockinger & Smith (2006, 2007) demonstrated that both quantity of habitat (grasslands embedded within arable fields) and their connectivity within a landscape contribute to butterfly diversity and abundance. Weibull *et al.* (2003) showed that landscape characters at the scale of the individual farm were important for butterflies rather than at the broader landscape scale. Accordingly, they reasoned that the decisions of individual farmers were critical in the context of farmland wildlife. Recent interest in the behaviour of butterflies within the seemingly inhospitable 'matrix' of farmland (i.e. crop fields) has shown that even such apparently barren structures can contain, albeit temporarily, resources for some species (Dennis 2004), and Dover *et al.* (1992) showed that non-production sowings of crops for game cover were beneficial to butterflies. As yet, the relative importance of landscape structure/composition, agricultural practices, individual farmer's motivations, etc. on butterflies remains unclear (e.g. Hole *et al.* 2005). However, the point of agri-environment schemes within arable systems is to reintroduce some heterogeneity or expand habitat in areas where a fine-grained mosaic is absent and unlikely to be (re)introduced through standard agricultural practice. Rundlöf & Smith (2006) effectively showed that organic farming in heterogeneous landscapes was no better for butterflies than conventional farming, but was significantly better in homogeneous ones. Agri-environmental schemes appear, at present, to be the only mechanism with the potential to safeguard species of the 'wider countryside' although the targeting of organic production in homogeneous

landscapes coupled with habitat creation as part of the organic regulations could make an important contribution (Rundlöf & Smith 2006). Pywell *et al.* (2004) rightly conclude that agri-environmental schemes should be appropriately targeted so that they improve the resource base for wildlife, and that monitoring should be an integral part of these generally desirable policies.

Disturbed and urban landscapes

There are two main drivers for an interest in landscapes heavily disturbed by humans, including urban areas. The first is an increased understanding that disturbances are common in 'pristine' nature and that many species have evolved to track and exploit disturbance-dependent resources. Humans, however, abhor barren and early successional conditions; the entire profession of agriculture and forestry aims to put landscapes into some use, maintaining middle successional stages instead of the earliest (or the oldest) ones. Disturbed biotopes, such as gravel banks of unregulated rivers, continental sand dunes, exposed landslides or burned-out forests have almost disappeared from much of Europe, falling victim to regulation, cultivation or afforestation. Thomas *et al.* (1994b) showed that specialists of early successional stages represent a disproportionately large proportion of invertebrate species that have become extinct in Britain during the twentieth century. They are closely matched by the other extreme specialists of old wood, whereas mid-succession species hold their grounds. A cursory overview of *Red Data Books* for other regions of Europe reveals a similar picture.

The second reason is more philosophical. As humans are taking over the greater part of the Earth's surface, it becomes impossible, due to fundamental limitations of available space, to save all species in set-aside wilderness reserves. To fulfil its mission, conservation must accommodate a proportion of original biodiversity, as high as possible, in landscapes where people live and work, to reconcile wild species and human enterprise (Rosenzweig 2003, 2005, Hanski 2005). This is not impossible. In rich societies such as those in Europe, supporting factors include agricultural and industrial innovation, and abandonment of economically marginal areas. These processes may imperil the well-being of species, as we illustrate in many of the examples above. However, they also enhance the opportunities for restoring natural habitats (Young 2000).

The list of butterflies that exploit natural disturbances is long. We only recall here that some species require quite extensive patches of bare ground, e.g. several *Pyrgus* skippers exploiting gravel beds of unregulated rivers (SBN 1987). *Hipparchia statilinus* specialises on near-barren sands (Steiner & Trusch 2000), larvae of *Pseudophilotes baton* develop on *Thymus* plants colonising exposed gravels. In some regions, the last large areas with sufficiently intensive disturbance levels, absence of fertilisation and minimum emphasis on modern forestry are represented by military training ranges, functioning as regional butterfly host spots (Väisänen *et al.* 1994b, Maes & Van Dyck 2001, Pavlíčko 2001). Even industry, mining, traffic and related activities may support butterflies. *Erynnis tages* readily colonises sub-urban wastelands, railway tracks and derelict (brownfield) land in Britain (Gutiérrez *et al.* 1999, Ellis 2005, Gutiérrez 2005). Kulfan (1989) studied a large population of *Cupido decoloratus* thriving at waste dumps of cement mills. Disused limestone quarries, if not 'reclaimed' for forestry or filled with waste, often harbour butterflies of calcareous grasslands and rocks. Even *Parnassius apollo*, an icon of butterfly conservation in some countries, can thrive in disused quarries (Kudrna *et al.* 1994) (Fig. 23.4). Benes at al. (2003a) compared 21 operating and closed quarries in the Czech Republic and found that even operating quarries hosted species of conservation interest, provided that quarrying operations have preserved heterogeneity of successional conditions. The conservation potential of disused quarries, sand or gravel pits is increasingly recognised by both biologists (Novak & Prach 2003) and technicians (Tränkle & Hehmann 2002, Tränkle *et al.* 2003). A pioneering study by Munguira & Thomas (1992) highlighted the potential of

roadside vegetation for grassland butterflies. Cars cause some mortality, but in intensively farmed landscapes, the net gain from the existence of unproductive grasslands outweighs the losses. Ries *et al.* (2001) evaluated a program in Iowa (USA) that used seed taken from roadsides to recreate native prairie vegetation instead of the more usual practice of using commercial grass mixtures. A narrow strip of restored prairie contained twice as many specialist butterfly species and five times as many individuals than ordinary roadsides. Although the grassland strips along roads have long been appreciated as corridors rather than as habitats proper, recent results from Finland challenge the view. Saarinen *et al.* (2005b) showed that richness and abundance of grassland butterflies is highest along the largest roads, something expected if they function as habitats rather than corridors (see also Valtonen & Saarinen 2005). However, roads are susceptible to rapid colonisation by invasive plant species with potentially negative impacts on butterflies (Valtonen *et al.* 2006).

Towns and cities represent largely untapped opportunities for butterfly conservation. Historical cities tend to be located in naturally diverse areas, probably because such areas offered a wider array of resources for their early settlers (Kühn *et al.* 2004), and many have retained considerable species richness. For instance, Prague alone hosts 60 % of butterflies occurring in the Czech Republic (Cila & Skyva 1993), and similar figures exist for other cities (e.g. Vienna: Höttinger 2002). Many cities have dense reserve systems (e.g. 84 small to middle-sized reserves within the city limits of Prague), or alternatives such as green belts set aside for recreation opportunities. Most importantly, intensive

Figure 23.4. Secondary habitat in the southern Franconian Jura (Bavaria) shortly after creation. The owner signed the locality: 'Do not enter. Habitat creation for apollo butterfly.' The picture was taken in 2001; today *Parnassius apollo* is inhabiting the locality (picture: M. Dolek).

agriculture and forestry are unimportant for the urban economy, whereas enjoyment of nature is highly valued (and paid for) by urban dwellers.

In an analysis of atlas data from Greater Manchester in England, Hardy & Dennis (1999) found a surprisingly low decrease in species richness with increasing urban cover (0.81 species lost for every 10% increase in urbanisation). Kadlec et al. (2008) did not find a decrease in species richness in 48 Prague reserves and parks between the 1980s and the present, which staggeringly contrasts with high losses reported in studies focusing on reserves surrounded by farmland (Wenzel et al. 2006). Although the two cities studied differ (Manchester is species poor, Prague is a Czech butterfly hotspot), and although both studies dealt with already impoverished faunas (major losses occurred a century ago in both regions), and the Czech study demonstrated changes of species composition post 1980, their counterintuitive results demonstrate two points. First, impacts of urbanisation, up to a certain point, are not necessarily fatal for butterflies, and second, the urban landscapes surrounding urban reserves may be more hospitable for butterflies than, for example, intensively farmed arable fields. Compared to arable land, cities are more heterogeneous and harbour more 'unproductive' biotopes such as brownfield sites. They also contain networks of linear structures (e.g. railway tracks), which supply essential connectivity among habitat islands.

Accepting that habitat quantity can hardly be increased in cities, there is much room to improve habitat quality. In both Manchester and Prague, butterflies dwelling in city centres were those associated either with ruderal weeds (e.g. *Pieris rapae*) or with shrubs and trees (*Celastrina argiolus*, *Thecla betulae*); grassland species, even common species such as *Maniola jurtina*, were missing. The canopy of city parks evidently provides a 'natural' habitat, whereas repeatedly cut lawns form hostile deserts. On the other hand, stony walls and rock gardens near the centre of Prague supported *Scolitandites orion*, a continentally declining species associated with calcareous rocks. It illustrates that much more might be achieved with scientifically informed promotion of urban habitat restoration. As shown by Gaston et al. (2005) many urban conservation campaigns are abortive due to poor consideration of the ecology of focal species. (The authors surveyed nettle patches left uncut in gardens to attract nymphalid larvae, and found that none of 60 such patches was occupied.) However, efficient methods, such as more patchy mowing of urban lawns, are available and compatible with gardening at the individual habitation and park level (Comba et al. 1999a, b). As noted by Hanski

(2005), a small shift in aesthetic perception may turn city parks into biodiversity havens. This may not be easy: people like freshly cut lawns and may oppose plans to 'litter' them with weedy strips. But they also love butterflies, and this may make the difference.

MANAGEMENT OF EUROPEAN ECOSYSTEMS AND BUTTERFLY CONSERVATION: SUMMARY CONSIDERATIONS

Despite the high diversity of European ecosystems and their significance for butterflies, there are some common features which should guide conservation in nearly all of them:

- in modern landscapes, butterflies cannot survive without active management of their habitats;
- in many situations, traditional management practices have been (and still are) the driving force for the evolution of plant and animal communities of European ecosystems – in contrast to many modern agriculture and forestry operations (including the mostly negative or at least inefficient but fashionable 'natural development' strategies in forests);
- although a wide range of mechanisms drive the survival of butterflies, there is a recurrent pattern of dependency on early successional sites (even or especially for woodland inhabiting species, where management for butterflies has to be based on maintaining low tree densities with either permanent or dynamically managed openings);
- a critical landscape factor for butterfly survival is a continuation of natural disturbances, such as landslides, avalanches, outbreaks of tree-killing insects, or animal grazing (in alpine areas this particularly helps to prevent or slow down an upward shift of the timberline; see Box 23.3);
- many remaining sites of high relevance are too small to sustain viable populations of specialised species; consequently increased connectivity between remaining habitats (which might include habitat restoration to create stepping stones) might be crucial to the long-term survival of many species;
- in cases where remnant habitats remain small and isolated in the long term, as in many reserves, management must accommodate much higher diversity of conditions and resources within a limited area; unless there is a concern for a single priority species, this means small-scaled and patchy application of diverse management techniques;

- in landscapes where larger habitat areas are still represented, the management goal should be networks of differently managed sites with varying land use regimes and intensities;
- while the impact of direct destruction of sites is obvious, indirect effects also have to be taken into account; many wetlands (including peat bogs) are threatened by accelerating terrestrification caused by drainage in wider aquifers; river channelisation, forest drainage, etc. prevent the spontaneous formation of wetland conditions (e.g. via the palludification processes);
- agri-environmental schemes appear, at present, to be the only mechanism with the potential to safeguard species of the 'wider countryside'. However, in order to be efficient, they should be appropriately targeted so that they improve the resource base for wildlife (thus, monitoring should be an integral part of these policies);
- support programmes as the agri-environmental schemes are at present far from being optimal as they are designed to be easy to oversee and control, while the actual needs of biodiversity are of second priority for the final implementation;

- urban areas offer some important opportunities for butterfly survival; a small shift in aesthetic perception may turn city parks into biodiversity havens. This may not be easy: people like freshly cut lawns and may oppose plans to litter them with weedy strips. But they also love butterflies, which could make the difference.
- AND/BUT: unified prescriptions applied across entire regions (e.g., that all meadows must be mown by a certain date) are surely deleterious and possibly disastrous for some species.

ACKNOWLEDGEMENTS

Risto Heikkinen and Juha Pöyry from SYKE (Finland) gave important hints to publications on arctic systems. Martin Warren and one anonymous referee kindly commented on earlier drafts of the manuscript and also provided stylistic improvements. JS was partly funded by the EU under the FP projects COCONUT (www.coconut-project.net; SSPI-CT-2006-044343), MacMan (EVK2-CT-2001-00126; Settele & Kühn 2009) and ALARM (www.alarmproject.net; GOCE-CT-2003-506675; Settele *et al.* 2005).

References

We intentionally have arranged the references in a strict user-friendly order and not as we would have to do for use the references of a chapter on its own. Thus, there seem to be some inconsistencies in the alphabetical order (if one includes the initials of the first names of authors). In the present book there is an unique alphabetical reference to each citation without using initials and while keeping a strict temporal order within the three categories used: 'one author', 'two authors', and 'more than two authors'.

Aagaard K, Hanssen O (1992) Population studies of *Parnassius mnemosyne* (Lepidoptera) in Sunndalen, Norway. In: Pavlicek-van Beek T, Ovaa AH, van der Made JG (eds.) *Future of Butterflies in Europe: Strategies for Survival*, Wageningen, Netherlands: Agricultural University of Wageningen, pp. 160–166.

Aagaard K, Hindar K, Pullin AS, James CH, Hammarstedt O, Balstad T, Hanssen O (2002) Phylogenetic relationships in brown argus butterflies (Lepidoptera: Lycaenidae: *Aricia*) from north-western Europe. *Biological Journal of the Linnean Society* **75**: 27–37.

Abrams PA, Leimar O, Nylin S, Wiklund C (1996) The effect of flexible growth rates on optimal sizes and development times in a seasonal environment. *American Naturalist* **147**: 381–395.

ACIA (2005) *Arctic Climate Impact Assessment*. Cambridge: Cambridge University Press.

Adams JM, Woodward FI (1989) Patterns in tree species richness as a test of the glacial extinction hypothesis. *Nature* **339**: 699–701.

Adanson M (1763) *Familles des plantes*. Paris: Vincent.

Addo-Bediako A, Chown SL, Gaston KJ (2000) Thermal tolerance, climatic variability and latitude. *Proceedings of the Royal Society of London B* **267**: 739–745.

Adler GH, Dudley R (1994) Butterfly biogeography and endemism on tropical Pacific islands. *Biological Journal of the Linnean Society* **51**: 151–162.

Adler PH, Pearson DL (1982) Why do male butterflies visit puddles? *Canadian Journal of Zoology* **60**: 322–325.

Adler GH, Austin CC, Dudley R (1995) Dispersal and speciation of skinks among archipelagos in the tropical Pacific Ocean. *Evolutionary Ecology* **9**: 529–541.

Agnew K, Singer MC (2000) Does fecundity drive the evolution of insect diet? *Oikos* **88**: 533–538.

Akaike H (1974) A new look at statistical model identification. *IEEE Transactions on Automatic Control* **19**: 716–723.

Akçakaya HR (2000a) Population viability analysis with demographically and spatially structured models. *Ecological Bulletins* **48**: 23–38.

Akçakaya HR (2000b) Viability analyses with habitat-based metapopulation models. *Population Ecology* **42**: 45–53.

Akçakaya HR (2002) *RAMAS/GIS: Linking Spatial Data with Population Viability Analysis (version 4.0)*. Setauket, NY: Applied Biomathematics.

Akçakaya HR, Sjögren-Gulve P (2000) Population viability analyses in conservation planning: an overview. *Ecological Bulletins* **48**: 9–21.

Akçakaya HR, McCarthy MA, Pearce JL (1995) Linking landscape data with population viability analysis: management options for the helmeted honeyeater *Lichenostomus melanops cassidix*. *Biological Conservation* **73**: 169–176.

Akino T, Knapp JJ, Thomas JA, Elmes GW (1999) Chemical mimicry and host specificity in the butterfly *Maculinea rebeli*, a social parasite of *Myrmica* ant colonies. *Proceedings of the Royal Society of London B* **266**: 1419–1426.

Alberti B (1943) Zur Frage der Hybridisation zwischen *Colias erate* Esp., *hyale* L. und *edusa* F. und über die Umgrenzung der 3 Arten. *Mitteilungen der Münchner Entomologischen Gesellschaft* **33**: 606–625.

Alcock J (1984) Convergent evolution in perching and patrolling site preferences of some hilltopping insects of the Sonoran Desert. *Southwestern Naturalist* **29**: 475–480.

Alcock J (1987) Leks and hilltopping in insects. *Journal of Natural History* **21**: 319–328.

Alcock J (1989) *Animal Behavior*, 4th edn. Sunderland, MA: Sinauer Associates.

Alcock J (1994) Alternative mate-locating tactics in *Chlosyne californica* (Lepidoptera, Nymphalidae). *Ethology* **97**: 103–118.

Alcock J, Gwynne D (1988) The mating system of *Vanessa kershawi*: males defend landmark territories as mate encounter sites. *Journal of Research on the Lepidoptera* **26**: 116–124.

Alcock J, O'Neill KM (1986) Density-dependent mating tactics in the grey hairstreak, *Strymon melinus* (Lepidoptera: Lycaenidae). *Journal of Zoology* **209**: 105–113.

Alexandrino J, Froufe E, Arntzen JW, Ferrand N (2000) Genetic subdivision, glacial refugia and postglacial recolonization in the golden-striped salamander, *Chioglossa lusitanica* (Amphibia: Urodela). *Molecular Ecology* **9**: 771–781.

Allee WC, Emerson AE, Park O, Park T, Schmidt KP (1949) *Principles of Animal Ecology*. Philadelphia, PA: W. B. Saunders.

Alm MJ, Ohmeiss TE, Lanza J, Vriesenga L (1990) Preference of cabbage white butterflies and honey-bees for nectar that contains amino-acids. *Oecologia* **84**: 53–57.

Alonso C (1999) Variation in herbivory by *Yponomeuta mahalabella* on its only host plant *Prunus mahaleb* along an elevational gradient. *Ecological Entomology* **24**: 371–379.

Als TD, Vila R, Kandul NP, Nash DR, Yen S, Hsu Y, Mignault AA, Boomsma JJ, Pierce NE (2004) The evolution of alternative parasitic life histories in large blue butterflies. *Nature* **432**: 386–390.

Althoff DM, Segraves KA, Pellmyr O (2005) Community context of an obligate mutualism: pollinator and florivore effects on *Yucca filamentosa*. *Ecology* **86**: 905–913.

Altizer SM, Oberhauser KS, Brower LP (2000) Associations between host migration and the prevalence of a protozoan parasite in natural populations of adult monarch butterflies. *Ecological Entomology* **25**: 125–139.

Amiet JL (2004) Séparation des niches écologiques chez deux espèces jumelles sympatriques de *Leptidea* (Lepidoptera, Pieridae). *Revue d'Ecologie (Terre et Vie)* **59**: 433–452.

Andersson M (1990) The driving force: species concepts and ecology. *Taxon* **39**: 375–382.

Andersson S (2003) Antennal responses to floral scents in the butterflies *Inachis io*, *Aglais urticae* (Nymphalidae), and *Gonepteryx rhamni* (Pieridae). *Chemoecology* **13**: 13–20.

Andersson S, Dobson HEM (2003) Behavioral foraging responses by the butterfly *Heliconius melpomene* to *Lantana camara* floral scent. *Journal of Chemical Ecology* **29**: 2303–2318.

Andersson M, Simmons LW (2006) Sexual selection and mate choice. *Trends in Ecology and Evolution* **21**: 296–302.

Andersson J, Borg-Karlson AK, Wiklund C (2000) Sexual cooperation and conflict in butterflies: a male-transferred anti-aphrodisiac reduces harassment of recently mated females. *Proceedings of the Royal Society of London B* **267**: 1271–1275.

Andersson J, Borg-Karlson AK, Wiklund C (2003) Antiaphrodisiacs in pierid butterflies: a theme with variation! *Journal of Chemical Ecology* **29**: 1489–1499.

Andersson J, Borg-Karlson AK, Wiklund C (2004) Sexual conflict and anti-aphrodisiac titre in a polyandrous butterfly: male ejaculate tailoring and absence of female control. *Proceedings of the Royal Society of London B* **271**: 1765–1770.

Angiosperm Phylogeny Group (1998) An ordinal classification for the families of flowering plants. *Annals of the Missouri Botanical Garden* **85**: 531–553.

Angold PG, Sadler JP, Hill MO, Pullin A, Rushton S, Austin K, Small E, Wood B, Wadsworth R, Sanderson R, Thompson K (2006) Biodiversity in urban habitat patches. *Science of the Total Environnement* **360**: 196–204.

Anholt BR (1991) Measuring selection on a population of damselflies with a manipulated phenotype. *Evolution* **45**: 1091–1106.

Anthes N, Fartmann T, Hermann G (2003a) Wie lässt sich der Rückgang des Goldenen Scheckenfalters (*Euphydryas aurinia*) in Mitteleuropa stoppen? Erkenntnisse aus populationsökologischen Studien in voralpinen Niedermoorgebieten und der Arealentwicklung in Deutschland. *Naturschutz und Landschaftsplanung* **35**: 279–287.

Anthes N, Fartmann T, Hermann G, Kaule G (2003b) Combining larval habitat quality and metapopulation structure: the key for successful management of pre-alpine *Euphydryas aurinia* colonies. *Journal of Insect Conservation* **7**: 175–186.

Anthes N, Fartmann T, Hermann G (2008) The Duke of Burgundy butterfly and its dukedom: larval niche variation in *Hamearis lucina* across Central Europe. *Journal of Insect Conservation* **12**: 3–14.

Anton C, Zeisset I, Musche M, Durka W, Boomsma JJ, Settele J (2007) Population structure of a large blue butterfly and its specialist parasitoid in a fragmented landscape. *Molecular Ecology* **16**: 3828–3838.

Aplin RT, d'Arcy Ward R, Rothschild M (1975) Examination of the large white and small white butterflies (*Pieris* spp.) for the presence of mustard oils and mustard oil glycosides. *Journal of Entomology A* **50**: 73–78.

Araújo MB, Luoto M (2007) The importance of biotic interactions for modelling species distributions under climate change. *Global Ecology and Biogeography* **16**: 743–753.

Araújo MB, New M (2007) Ensemble forecasting of species distributions. *Trends in Ecology and Evolution* **22**: 42–47.

Arendt J (2007) Ecological correlates of body size in relation to cell size and cell number: patterns in flies, fish, fruits and foliage. *Biological Reviews* **82**: 241–256.

Arikawa K (2003) Spectral organization of the eye of a butterfly, *Papilio xuthus*. *Journal of Comparative Physiology*

A – Neuroethology Sensory Neural and Behavioural Physiology
189: 791–800.

Arikawa K, Takagi N (2001) Genital photoreceptors have crucial role in oviposition in Japanese yellow swallowtail butterfly, *Papilio xuthus. Zoological Science* **18**: 175–179.

Armbruster P, Bradshaw WE, Ruegg K, Holzapfel CM (2001) Geographic variation and the evolution of reproductive allocation in the pitcher-plant mosquito, *Wyeomyia smithii. Evolution* **55**: 439–444.

Arms K, Feeny P, Lederhouse RC (1974) Sodium: stimulus for puddling behaviour by tiger swallowtail butterflies, *P. glaucus. Science* **185**: 372–374.

Arnold RA (1983a) Ecological studies of six endangered butterflies (Lepidoptera, Lycaenidae): island biogeography, patch dynamics, and the design of habitat preserves. *Entomology* **99**: 1–161.

Arnold SJ (1983b) Morphology, performance and fitness. *American Zoology* **23**: 347–361.

Arnold ML (1992a) Natural hybridization as an evolutionary process. *Annual Review of Ecology and Systematics* **23**: 237–261.

Arnold SJ (1992b) Constraints on phenotypic evolution. *American Naturalist* **140**: 85–107.

Arnold ML (1997) *Natural Hybridization and Evolution.* Oxford: Oxford University Press.

Arnqvist G, Nilsson T (2000) The evolution of polyandry: multiple mating and female fitness in insects. *Animal Behaviour* **60**: 145–164.

Arnyas E, Bereczki J, Toth A, Pecsenye K, Varga Z (2006) Egg-laying preferences of the xerophilous ecotype of *Maculinea alcon. European Journal of Entomology* **103**: 587–595.

Arpagaus M (1987) Vertebrate insulin induces diapause termination in *Pieris brassicae* pupae. *Roux Archive for Developmental Biology* **196**: 527–530.

Artemyeva EA (2005) Clinal variation in populations of the common blue butterfly *Polyommatus icarus* Rott. (Lepidoptera, Lycaenidae). *Russian Journal of Genetics* **41**: 859–870.

Arthur AP (1962) *Thymelicus lineola* (Ochs.) (Lepidoptera: Hesperiidae) and its parasites in Ontario. *Canadian Entomologist* **94**: 1082–1089.

Arthur AP (1966) The present status of the introduced skipper, *Thymelicus lineola* (Ochs.) (Lepidoptera: Hesperiidae), in North America and possible methods of control. *Canadian Entomologist* **98**: 622–626.

Asher J, Warren MS, Fox R, Harding P, Jeffcoate G, Jeffcoate S (2001) *The Millennium Atlas of Butterflies in Britain and Ireland.* Oxford: Oxford University Press.

Ashworth AC (1972) A late-glacial insect fauna from Red Moss, Lancashire, England. *Entomologica Scandinavica* **3**: 211–224.

Askew RR (1971) *Parasitic Insects.* London: Heinemann.

Askew RR (1994) Parasitoids of leaf-mining Lepidoptera: what determines their host ranges? In: Hawkins BA, Sheehan W (eds.) *Parasitoid Community Ecology*, Oxford: Oxford University Press, pp. 177–202.

Askew RR, Shaw MR (1974) An account of the Chalcidoidea (Hymenoptera) parasitising leaf-mining insects of deciduous trees in Britain. *Biological Journal of the Linnean Society* **6**: 289–335.

Askew RR, Shaw MR (1986) Parasitoid communities: their size, structure and development. In: Waage J, Greathead D (eds.) *Insect Parasitoids*, London: Academic Press, pp. 225–264.

Askew RR, Shaw MR (1997) *Pteromalus apum* (Retzius) and other pteromalid (Hym.) primary parasitoids of butterfly pupae in Western Europe, with a key. *Entomologist's Monthly Magazine* **133**: 67–72.

Askew RR, Shaw MR (2005) Observations on the biology of *Baryscapus* (Hymenoptera: Eulophidae: Tetrastichinae) with description of a new koinobiont hyperparasitoid with delayed development. *Acta Societatis Zoologicae Bohemicae* **69**: 11–14.

Asmussen MA, Arnold J, Avise JC (1989) The effects of assortative mating and migration on cytonuclear associations in hybrid zones. *Genetics* **122**: 923–934.

Atkinson D (1994) Temperature and organism size: a biological law for ectotherms? *Advances in Ecological Research* **25**: 1–58.

Atkinson D (1995) Effects of temperature on the size of aquatic ectotherms: exceptions to the general rule. *Journal of Thermal Biology* **20**: 61–74.

Aubert J, Solignac M (1990) Experimental evidence for mitochondrial DNA introgression between *Drosophila* species. *Evolution* **44**: 1272–1282.

Aubert J, Barascud B, Descimon H, Michel F (1996a) Systématique moléculaire des Argynnes (Lepidoptera: Nymphalidae). *Comptes Rendus de l'Académie des Sciences de Paris, Série III, Sciences de la Vie* **319**: 647–651.

Aubert J, Descimon H, Michel F (1996b) Population biology and conservation of the Corsican swallowtail butterfly *Papilio hospiton* Géné. *Biological Conservation* **78**: 247–255.

Aubert J, Barascud B, Descimon H, Michel F (1997) Ecology and genetics of interspecific hybridization in the swallowtails, *Papilio hospiton* Géné and *P. machaon* L., in Corsica (Lepidoptera: Papilionidae). *Biological Journal of the Linnean Society* **60**: 467–492.

Aubert J, Legal L, Descimon H, Michel F (1999) Molecular phylogeny of swallowtail butterflies of the tribe Papilionini

(Papilionidae, Lepidoptera). *Molecular Phylogenetics and Evolution* 12: 156–167.

Augustin NH, Mugglestone MA, Buckland ST (1996) An autologistic model for the spatial distribution of wildlife. *Journal of Applied Ecology* 33: 339–347.

Austin MP (1985) Continuum concept, ordination methods and niche theory. *Annual Review of Ecology and Systematics* 16: 39–61.

Austin MP (2002) Spatial prediction of species distribution: an interface between ecological theory and statistical modelling. *Ecological Modelling* 157: 101–118.

Austin MP (2007) Species distribution models and ecological theory: a critical assessment and some possible new approaches. *Ecological Modelling* 200: 1–19.

Avise JC (1974) Systematic value of electrophoretic data. *Systematic Zoology* 23: 465–481.

Avise JC (1994) *Molecular Markers, Natural History and Evolution.* London: Chapman and Hall.

Avise JC (2000) *Phylogeography: The History and Formation of Species.* Cambridge, MA: Harvard University Press.

Avise JC (2004) *Molecular Markers, Natural History and Evolution,* 2nd edn. Sunderland, MA: Sinauer Associates.

Avise JC, Ball RM (1990) Principles of genealogical concordance in species concepts and biological taxonomy. In: Futuyma DJ, Antonovics J (eds.) *Oxford Surveys in Evolutionary Biology,* Vol. 7, Oxford: Oxford University Press, pp. 45–67.

Avise JC, Arnold J, Ball RM, Bermingham E, Lamb T, Neigel JE, Reeb CA, Sanders NC (1987) Intraspecific phylogeography: the mitochondrial DNA bridge between population genetics and systematics. *Annual Review of Ecology and Systematics* 18: 489–522.

Awmack CS, Leather SR (2002) Host plant quality and fecundity in herbivorous insects. *Annual Review of Entomology* 47: 817–844.

Ayala FJ, Powell JR (1972) Allozymes as diagnostic characters of sibling species of *Drosophila. Proceedings of the National Academy of Sciences, USA* 69: 1094–1096.

Ayres MP, Scriber JM (1994) Local adaptation to regional climates in *Papilio canadensis* (Lepidoptera: Papilionidae). *Ecological Monographs* 64: 465–482.

Azevedo RBR, French V, Partridge L (1996) Thermal evolution of egg size in *Drosophila melanogaster. Evolution* 50: 2338–2345.

Baar J, Kuyper TW (1998) Restoration of aboveground ectomycorrhizal flora in stands of *Pinus sylvestris* (Scots pine) in The Netherlands by removal of litter and humus. *Restoration Ecology* 6: 227–237.

Babendreier D, Kuske S, Bigler F (2003a) Parasitism of non-target butterflies by *Trichogramma brassicae* Bezdenko (Hymenoptera: Trichogrammatidae) under field cage and field conditions. *Biological Control* 26: 139–145.

Babendreier D, Kuske S, Bigler F (2003b) Non-target host acceptance and parasitism by *Trichogramma brassicae* Bezdenko (Hymenoptera: Trichogrammatidae) in the laboratory. *Biological Control* 26: 128–138.

Báez M (1998) *Mariposas de Canarias.* Madrid: Editorial Rueda.

Baguette M (2003) Long distance dispersal and landscape occupancy in a metapopulation of the cranberry fritillary butterfly. *Ecography* 26: 153–160.

Baguette M (2004) The classical metapopulation theory and the real, natural world: a critical appraisal. *Basic and Applied Ecology* 5: 213–224.

Baguette M, Mennechez G (2004) Resource and habitat patches, landscape ecology and metapopulation biology: a consensual viewpoint. *Oikos* 106: 399–403.

Baguette M, Nève G (1994) Adult movements between populations in the specialist butterfly *Proclossiana eunomia* (Lepidoptera, Nymphalidae). *Ecological Entomology* 19: 1–5.

Baguette M, Schtickzelle N (2003) Local population dynamics are important to the conservation of metapopulations in highly fragmented landscapes. *Journal of Applied Ecology* 40: 404–412.

Baguette M, Schtickzelle N (2006) Negative relationship between dispersal distance and demography in butterfly metapopulations. *Ecology* 87: 648–654.

Baguette M, Van Dyck H (2007) Landscape connectivity and animal behavior: functional grain as a key determinant for dispersal. *Landscape Ecology* 22: 1117–1129.

Baguette M, Convié I, Nève G (1996) Male density affects female spatial behaviour in the butterfly *Proclossiana eunomia. Acta Oecologica* 17: 225–232.

Baguette M, Vansteenwegen C, Convié I, Nève G (1998) Sex-biased density-dependent migration in a metapopulation of the butterfly *Proclossiana eunomia. Acta Oecologica* 19: 17–24.

Baguette M, Petit S, Quéva F (2000) Population spatial structure and migration of three butterfly species within the same habitat network: consequences for conservation. *Journal of Applied Ecology* 37: 100–108.

Baguette M, Mennechez G, Petit S, Schtickzelle N (2003) Effect of habitat fragmentation on dispersal in the butterfly *Proclossiana eunomia. Comptes Rendus Biologies* 326: 200–209.

Baker RR (1969) The evolution of the migratory habit in butterflies. *Philosophical Transactions of the Royal Society of London* 253: 310–341.

Baker RR (1970) Bird predation as a selective pressure of the immature stages of the cabbage butterflies, *Pieris rapae* and *P. brassicae. Journal of Zoology* **162**: 43–59.

Baker RR (1972) Territorial behaviour of the nymphalid butterflies, *Aglais urticae* (L.) and *Inachis io* (L.). *Journal of Animal Ecology* **41**: 453–469.

Baker HG (1986) Yuccas and the Yucca moths: a historical commentary. *Annals of the Missouri Botanical Garden* **73**: 556–564.

Baker HG, Baker I (1973) Amino acids in nectar and their evolutionary significance. *Nature* **241**: 543–545.

Baker HG, Baker I (1975) Studies of nectar-constitution and pollinator-plant coevolution In: Gilbert LE, Raven PH (eds.) *Coevolution of Animals and Plants*, Austin, TX: University of Texas Press, pp. 100–140.

Baker HG, Baker I (1977) Interspecific constancy of floral nectar amino acid complements. *Botanical Gazette* **138**: 183–191.

Baker HG, Baker I (1982) Chemical constituents of nectar in relation to pollination mechanisms and phylogeny. In: Nitecki HM (ed.) *Biochemical Aspects of Evolutionary Biology*, Chicago, IL: University of Chicago Press, pp. 131–171.

Baker HG, Baker I (1983) Floral nectar sugar constituents in relation to pollinator type. In: Jones CE, Little RJ (eds.) *Handbook of Experimental Pollination Biology*, New York: Scientific and Academic Editions, pp. 117–141.

Bakker ES, Olff H, Vandenberghe C, De Maeyer K, Smit R, Gleichman JM, Vera FWM (2004) Ecological anachronisms in the recruitment of temperate light-demanding tree species in wooded pastures. *Journal of Applied Ecology* **41**: 571–582.

Balachowsky AS (ed.) (1962) *Entomologie appliquée à l'agriculture*, tome 2 Lépidoptères, vols. 1, 2. Paris: Masson.

Baldock D, Beaufoy G, Clark J (1994) *The Nature of Farming: Low Intensity Farming Systems in Nine European Countries*. London: Institute for European Environmental Policy.

Balletto E (1992) Butterflies in Italy: status, problems and prospects. In: Pavlicek-van Beek T, Ovaa AH, van der Made JG (eds.) *Future of Butterflies in Europe: Strategies for Survival*, Wageningen, Netherlands: Agricultural University of Wageningen, pp. 53–64.

Balletto E (1995) Endemism, areas of endemism, biodiversity and butterfly conservation in the Euro-Mediterranean area. *Bollettino del Museo Regionale di Scienze Naturali – Torino* **13**: 445–491.

Balletto E, Toso GG (1979) On a new species of *Agrodiaetus* (Lycaenidae) from Southern Italy. *Nota Divulg. Ist. Agr. Univ. Napoli Portici* **2**: 13–22.

Balletto E, Toso G, Barberis G, Rossaro B (1977) Aspetti dell'ecologia dei lepidotteri ropalocei nei consorzi erbacei alto Appenninici. *Animalia* **4**: 277–343.

Balletto E, Toso GG, Barberis G (1982) Le comunita' di lepidotteri ropaloceri nei consorzi erbacei dell'Appenino. *Quaderni sulla struttura delle zoocenosi terrestri. 2. La montagna. II.1. I pascoli altomontani.*

Balletto E, Lattes A, Cassulo L, Toso G (1988) Studi sull'ecologia dei Lepidotteri Ropaloceri in alcuni ambienti delle Dolomiti. *Studi Trentini di Scienze Naturali* **64** (Suppl. Acta Biologica): 87–123.

Balloux, F (2001) EASYPOP (version 1.7): a computer program for population genetics simulations. *Journal of Heredity* **92**: 301–302.

Balmer O, Erhardt A (2000) Consequences of succession on extensively grazed grasslands for central European butterfly communities: rethinking conservation practices. *Conservation Biology* **14**: 746–757.

Banno H (1990) Plasticity of size and relative fecundity in the aphidophagous lycaenid butterfly, *Taraka hamada. Ecological Entomology* **15**: 111–113.

Barascud B, Martin JF, Baguette M, Descimon H (1999) Genetic consequences of an introduction–colonization process in an endangered butterfly species. *Journal of Evolutionary Biology* **12**: 697–709.

Barbosa P (1988) Some thoughts on "The evolution of host range". *Ecology* **69**: 912–915.

Barnett LK, Warren MS (1995a) *Pearl-Bordered Fritillary*, Boloria euphrosyne: *Species Action Plan*. Wareham, Dorset: Butterfly Conservation.

Barnett LK, Warren MS (1995b) *Large Blue*, Maculinea arion: *Species Action Plan*. Wareham, Dorset: Butterfly Conservation.

Barron AB (2001) The life and death of Hopkins' Host-Selection Principle. *Journal of Insect Behavior* **14**: 725–737.

Bartholomew JG, Clarke WE, Grimshaw PH (1911) *Atlas of Zoogeography*, Vol. 5 of *Bartholomew's Physical Atlas*. Edinburgh: Bartholomew.

Barton NH, Baird SJE (1999) *Analyse: Software for Analysis of Geographic Variation and Hybrid Zones*. Edinburgh: University of Edinburgh.

Barton NH, Gale KS (1993) Genetic analysis of hybrid zones. In: Harrison RG (ed.) *Hybrid Zones and the Evolutionary Process*, Oxford: Oxford University Press, pp. 13–45.

Barton NH, Hewitt GM (1989) Adaptation, speciation and hybrid zones. *Nature* **341**: 497–503.

Bauer E, Kohavi R (1999) An empirical comparison of voting classification algorithms: bagging, boosting, and variants. *Machine Learning* **37**: 105–139.

Bauerfeind SS, Fischer K (2005a) Effects of adult-derived carbohydrates, amino acids and micronutrients on female reproduction in a fruit-feeding butterfly. *Journal of Insect Physiology* 51: 545–554.

Bauerfeind SS, Fischer K (2005b) Effects of food stress and density in different life stages on reproduction in a butterfly. *Oikos* 111: 514–524.

Baughman JF, Murphy DD, Ehrlich PR (1990) A reexamination of hilltopping in *Euphydryas editha*. *Oecologia* 83: 259–260.

Baum DA, Larson A (1991) Adaptation reviewed: a phylogenetic methodology for studying character macroevolution. *Systematic Zoology* 40: 1–18.

Baumgarten HT, Fiedler K (1998) Parasitoids of lycaenid butterfly caterpillars: different patterns in resource use and their impact on the hosts' symbiosis with ants. *Zoologischer Anzeiger* 236: 167–180.

Baylis M, Pierce NE (1991) The effect of host-plant quality on the survival of larvae and oviposition by adults of an ant tended lycaenid butterfly, *Jalmenus evagoras*. *Ecological Entomology* 16: 1–10.

Bayliss JL, Simonite V, Thompson S (2005) The use of probabilistic habitat suitability models for Biodiversity Action Planning. *Agriculture, Ecosystems and Environment* 108: 228–250.

Baz A (1987) Abundancia y riqueza de las comunidades forestales de mariposas (Lepidoptera: Rhopalocera) y su relación con la altitud en el Sistema Ibérico meridional. *Graellsia* 43: 179–192.

Bazin A, Goverde M, Erhardt A, Shykoff JA (2002) Influence of atmospheric carbon dioxide enrichment on induced response and growth compensation after herbivore damage in *Lotus corniculatus*, *Ecological Entomology* 27: 271–278.

Beaumont LJ, Hughes L (2002) Potential changes in the distributions of latitudinally restricted Australian butterfly species in response to climate change. *Global Change Biology* 8: 954–971.

Beaumont LJ, Hughes L, Poulsen M (2005) Predicting species distributions: use of climatic parameters in BIOCLIM and its impact on predictions of species' current and future distributions. *Ecological Modelling* 186: 251–270.

Beck JR, Shultz EK (1986) The use of ROC curves in test performance evaluation. *Archives of Pathology and Laboratory Medicine* 110: 13–20.

Beck J, Muhlenberg E, Fiedler K (1999) Mud-puddling behaviour in tropical butterflies: in search of proteins or minerals? *Oecologia* 119: 140–148.

Begon M, Parker GA (1986) Should egg size and clutch size decrease with age? *Oikos* 47: 293–302.

Begon M, Harper JL, Townsend CR (1990) *Ecology: Individuals, Populations and Communities*. Oxford: Blackwell Scientific Publications.

Behura SK (2006) Molecular marker systems in insects: current trends and future avenues. *Molecular Ecology* 15: 3087–3113.

Beinlich B, Plachter H (eds.) (1995) Schutz und Entwicklung der Kalkmagerrasen der Schwäbischen Alb.*Beihefte zu den Veröffentlichungen für Naturschutz und Landschaftspflege in Baden-Württemberg* 83.

Beissinger SR (2002) Population Viability Analysis: past, present, future. In: Beissinger SR, McCullough DR (eds.) *Population Viability Analysis*, Chicago, IL: University of Chicago Press, pp. 5–17.

Beissinger SR, McCullough DR (2002) *Population Viability Analysis*. Chicago, IL: University of Chicago Press.

Beissinger SR, Westphal MI (1998) On the use of demographic models of population viability in endangered species management. *Journal of Wildlife Management* 62: 821–841.

Beldade P, Brakefield PM (2002) The genetics and evo-devo of butterfly wing patterns. *Nature Review Genetics* 3: 442–452.

Beldade P, Koops K, Brakefield PM (2002) Developmental constraints versus flexibility in morphological evolution. *Nature* 416: 844–847.

Belovsky GE, Mellison C, Larson C, Van Zandt PA (1999) Experimental studies of extinction dynamics. *Science* 286: 1175–1177.

Belovsky GE, Mellison C, Larson C, Van Zandt PA (2002) How good are PVA models? Testing their predictions with experimental data on the brine shrimp. In: Beissinger SR, McCullough DR (eds.) *Population Viability Analysis*, Chicago, IL: University of Chicago Press, pp. 257–283.

Belshaw R (1993) Tachinid flies. Diptera: Tachinidae. *Handbooks for the Identification of British Insects* 10(4a(i)): 1–169.

Belshaw R (1994) Life history characteristics of Tachinidae (Diptera) and their effect on polyphagy. In: Hawkins BA, Sheehan W (eds.) *Parasitoid Community Ecology*, Oxford: Oxford University Press, pp. 145–162.

Beltrán MS, Jiggins CD, Bull V, Linares M, Mallet J, McMillan WO, Bermingham E (2002) Phylogenetic discordance at the species boundary: comparative gene genealogies among rapidly radiating *Heliconius* butterflies. *Molecular Biology and Evolution* 19: 2176–2190.

Benedick S, White TA, Searle JB, Hamer KC, Mustaffa N, Vun Khen C, Mohamed M, Schilthuizen M, Hill JK (2007) Impacts of habitat fragmentation on genetic diversity in a tropical forest butterfly on Borneo. *Journal of Tropical Ecology* 23: 623–634.

Beneš J, Kuras T, Konvička M (2000) Assemblages of montainous day-active Lepidoptera in the Hruby Jesenik Mts, Czech Republic. *Biologia, Bratislava* 55: 159–167.

Beneš J, Kepka P, Konvička M (2003a) Limestone quarries as refuges for European xerophilous butterflies. *Conservation Biology* 17: 1058–1069.

Beneš J, Konvička M, Vrabec V, Zamecnik J (2003b) Do the sibling species of small whites, *Leptidea sinapis* and *L. reali* (Lepidoptera, Pieridae) differ in habitat preferences? *Biologia, Bratislava* 58: 943–951.

Beneš J, Cizek O, Dovala J, Konvička M (2006) Intensive game keeping, coppicing and butterflies: the story of Milovicky Wood, Czech Republic. *Forest Ecology and Management* 237: 353–365.

Bennett ATD, Cuthill IC (1994) Ultraviolet vision in birds: what is its function? *Vision Research* 34: 1471–1478.

Bennett ATD, Cuthill IC, Norris KJ (1994) Sexual selection and the mismeasure of color. *American Naturalist* 144: 848–860.

Bennetts RE, Nichols JD, Lebreton JD, Pradel R, Hines JE, Kitchens WM (2001) Methods for estimating dispersal probabilities and related parameters using marked animals. In: Clobert J, Danchin E, Dhondt AA, Nichols JD (eds.) *Dispersal*, Oxford: Oxford University Press, pp. 3–17.

Benson WW, Brown KS Jr, Gilbert LE (1975) Coevolution of plants and herbivores: passion flower butterflies. *Evolution* 29: 659–680.

Benson J, van Driesche RG, Pasquale A, Elkinton J (2003) Introduced braconid parasitoids and range reduction of a native butterfly in New England. *Biological Control* 28: 197–213.

Benyamini D (1999) The biology and conservation of *Iolana alfierii* Wiltshire, 1948: The Burning Bush Blue (Lepidoptera: Lycaenidae). *Linneana Belgica* 17: 119–134.

Berembaum M (1980) Adaptive significance of midgut pH in larval Lepidoptera. *American Naturalist* 115: 138–146.

Berenbaum M (1995) Chemistry and oligophagy in the Papilionidae. In: Scriber JM, Tsubaki Y, Lederhouse RC (eds.) *Swallowtail Butterflies: Their Ecology and Evolutionary Biology*, Gainesville, FL: Scientific Publishers, pp. 27–38.

Bergman KO (1996) [2000] Oviposition, host plant choice and survival of a grass feeding butterfly, the Woodland Brown (*Lopinga achine*) (Nymphalidae: Satyrinae). *Journal of Research on the Lepidoptera* 35: 9–21.

Bergman KO (1999) Habitat utilization by *Lopinga achine* (Nymphalidae: Satyrinae) larvae and ovipositing females: implications for conservation. *Biological Conservation* 88: 69–74.

Bergman KO (2001) Population dynamics and the importance of habitat management for conservation of the butterfly *Lopinga achine*. *Journal of Applied Ecology* 38: 1303–1313.

Bergman KO, Kindvall O (2004) Population viability analysis of the butterfly *Lopinga achine* in a changing landscape in Sweden. *Ecography* 27: 49–58.

Bergman KO, Landin J (2002) Population structure and movements of a threatened butterfly (*Lopinga achine*) in a fragmented landscape in Sweden. *Biological Conservation* 108: 361–369.

Bergman KO, Askling J, Ekberg O, Ignell H, Wahlman H, Milberg P (2004) Landscape effects on butterfly assemblages in an agricultural region. *Ecography* 27: 619–628.

Bergström A (2005) Oviposition site preferences of the threatened butterfly *Parnassius mnemosyne*: implications for conservation. *Journal of Insect Conservation* 9: 21–27.

Bergström J, Wiklund C (2002) Effects of size and nuptial gifts on butterfly reproduction: can females compensate for a smaller size through male-derived nutrients? *Behavioral Ecology and Sociobiology* 52: 296–302.

Bergström J, Wiklund C (2005) No effect of male courtship intensity on female remating in the butterfly *Pieris napi*. *Journal of Insect Behavior* 18: 479–489.

Bergström J, Wiklund C, Kaitala A (2002) Natural variation in female mating frequency in a polyandrous butterfly: effects of size and age. *Animal Behaviour* 64: 49–54.

Bergström J, Nylin S, Nygren GH (2004) Conservative resource utilization in the common blue butterfly: evidence for low costs of accepting absent host plants? *Oikos* 107: 345–351.

Bergström A, Janz N, Nylin S (2006) Putting more eggs in the best basket: clutch size regulation in the comma butterfly. *Ecological Entomology* 31: 255–260.

Berlocher SH (1998) Origins: a brief history of research on speciation. In: Howard DJ, Berlocher SH (eds.) *Endless Forms: Species and Speciation*, New York: Oxford University Press, pp. 3–15.

Bernard GD (1986) Butterfly color-vision: spectral properties of photoreceptors and wing patterns. *Journal of the Optical Society of America A – Optics Image Science and Vision* 3: 44–88.

Bernardi G (1980) Les catégories taxonomiques de la systématique évolutive. *Mémoires de la Société Zoologique de France* 40: 373–425.

Bernays EA, Barbehenn R (1987) Nutritional ecology of grass foliage-chewing insects. In: Slansky FJ, Rodriguez JG (eds.) *Nutritional Ecology of Insects, Mites, Spiders, and Related Invertebrates*, New York: John Wiley, pp. 147–175.

Bernays EA, Chapman RF (1976) Antifeedant properties of seedling grasses. *Symposia Biologica Hungarica* 16: 41–46.

Bernays EA, Chapman RF (1994) *Host-Plant Selection by Phytophagous Insects*. New York: Chapman and Hall.

Bernays EA, Janzen D (1988) Saturnid and sphingid caterpillars: two ways to eat leaves. *Ecology* **69**: 1153–1160.

Berrigan D, Charnov EL (1994) Reaction norms for age and size at maturity in response to temperature: a puzzle for life historians. *Oikos* **70**: 474–478.

Berry RJ (1971) Conservation aspects of the genetical constitution of populations. In: Duffey E, Watt AS (eds.) *The Scientific Management of Animal and Plant Communities for Conservation*, Oxford: Blackwell Scientific Publications, pp. 177–206.

Berry PM, Dawson TP, Harrison PA, Pearson RG (2002) Modelling potential impacts of climate change on the bioclimatic envelope of species in Britain and Ireland. *Global Ecology and Biogeography* **11**: 453–462.

Berwaerts K, Van Dyck H (2004) Take-off performance under optimal and suboptimal thermal conditions in the butterfly *Pararge aegeria*. *Oecologia* **141**: 536–545.

Berwaerts K, Van Dyck H, van Dongen S, Matthysen E (1998) Morphological and genetic variation in the speckled wood butterfly (*Pararge aegeria* L.) among differently fragmented landscapes. *Netherlands Journal of Zoology* **48**: 241–253.

Berwaerts K, Van Dyck H, Vints E, Matthysen E (2001) Effect of manipulated wing characteristics and basking posture on thermal properties of the butterfly *Pararge aegeria* (L.). *Journal of Zoology* **255**: 261–267.

Berwaerts K, Van Dyck H, Aerts P (2002) Does flight morphology relate to flight performance? An experimental test with the butterfly *Pararge aegeria*. *Functional Ecology* **16**: 484–491.

Berwaerts K, Aerts P, Van Dyck H (2006) On the specific mechanisms of butterfly flight: flight performance relative to flight morphology, wing kinematics, and sex in *Pararge aegeria*. *Biological Journal of the Linnean Society* **89**: 675–687.

Besold J, Huck S, Schmitt T (2008) Allozyme polymorphisms in the small heath, *Coenonympha pamphilus*: recent ecological selection or old biogeographical signal? *Annales Zoologici Fennici* **45**: 217–228.

Betts CR, Wootton RJ (1988) Wing shape and flight behavior in butterflies (Lepidoptera, Papilionoidea and Hesperioidea): a preliminary analysis. *Journal of Experimental Biology* **138**: 271–288.

Beuret H (1957) Studien über den Formenkreis *Lysandra coridon-hispana-albicans*: Ein Beitrag zum Problem der Artbildung (2 Studie). *Mitteilungen der Entomologischen Gesellschaft Basel, N. F.* **7**: 17–36, 37–59.

Biedermann R (2004) Modelling the spatial dynamics and persistence of the leaf beetle *Gonioctena olivacea* in dynamic habitats. *Oikos* **107**: 645–653.

Bierne BP (1955) Natural fluctuations in abundance of British Lepidoptera. *Entomologist's Gazette* **6**: 21–52.

Biesmeijer JC, Roberts SPM, Reemer M, Ohlemüller R, Edwards M, Peeters T, Schaffers AP, Potts SG, Kleukers R, Thomas CD, Settele J, Kunin WE (2006) Parallel declines in pollinators and insect-pollinated plants in Britain and the Netherlands. *Science* **313**: 351–354.

Bignal EM, McCracken DI (1992) *Prospects for Nature Conservation in European Pastoral Farming Systems: A Discussion Document*. Peterborough, Cambs.: JNCC.

Bink FA (1970) Parasites of *Thersamonia dispar* Haw. and *Lycaena helle* Den., Schiff. (Lep., Lycaenidae). *Entomologische Berichten* **30**: 30–34.

Bink FA (1985) Host plant preference of some grass feeding butterflies. *Proceedings of the 3rd Congress of European Lepidopterology*, Cambridge 1982: 23–29.

Bink FA (1992) *Ecologische Atlas van de Dagvlinders van Noordwest-Europa*. Haarlem, Netherlands: Schuyt.

Bink FA, Siepel H (1996) Nitrogen and phosphorus in *Molinia caerulea* (Gramineae) and its impact on the larval development in the butterfly species *Lasiommata megera* (Lepidoptera: Satyridae). *Entomologia Generalis* **20**: 271–280.

Binzenhöfer B, Schröder B, Biedermann R, Strauss B, Settele J (2005) Habitat models and habitat connectivity analysis for butterflies and burnet moths: the example of *Zygaena carniolica* and *Coenonympha arcania*. *Biological Conservation* **126**: 247–259.

Binzenhöfer B, Biedermann R, Settele J, Schröder B (2008) Connectivity compensates for low habitat quality and small patch size in the butterfly *Cupido minimus*. *Ecological Research* **23**: 259–269.

Bio AMF, De Becker P, De Bie E, Huybrechts W, Wassen M (2002) Prediction of plant species distribution in lowland river valleys in Belgium: modelling species response to site conditions. *Biodiversity and Conservation* **11**: 2189–2216.

Bird CD, Hilchie GJ, Kondla NG, Pike EM, Sperling FAH (1995) *Alberta Butterflies*. Edmonton, Alberta: Provincial Museum of Alberta.

Biró LP, Balint Z, Kertesz K, Vertesy Z, Mark GI, Horvath ZE, Balazs J, Mehn D, Kiricsi I, Lousse V, Vigneron JP (2003) Role of photonic-crystal-type structures in the thermal regulation of a lycaenid butterfly sister-species pair. *Physical Review E* **67**: Art. No. 021907.

Bisset GA (1938) Larvae and pupae of tachinids parasitizing *Pieris rapae* L. and *P. brassicae* L. *Parasitology* **30**: 111–122.

Bissoondath CJ, Wiklund C (1995) Protein content of spermatophores in relation to monandry/polyandry in butteflies. *Behavioral Ecology and Sociobiology* **37**: 365–371.

Bissoondath CJ, Wiklund C (1996a) Effect of male mating history and body size on ejaculate size and quality in two polyandrous butterflies, *Pieris napi* and *Pieris rapae* (Lepidoptera: Pieridae). *Functional Ecology* **10**: 457–464.

Bissoondath CJ, Wiklund C (1996b) Male butterfly investment in successive ejaculates in relation to mating system. *Behavioral Ecology and Sociobiology* **39**: 285–292.

Bissoondath CJ, Wiklund C (1997) Effect of male body size on sperm precedence in the polyandrous butterfly *Pieris napi*. *Behavioral Ecology* **8**: 518–523.

Bitzer RJ, Shaw KC (1979) Territorial behavior of the red admiral, *Vanessa atalanta* (L.) (Lepidoptera: Nymphalidae). *Journal of Research on the Lepidoptera* **18**: 36–49.

Bitzer RJ, Shaw KC (1983) Territorial behavior of *Nymphalis antiopa* and *Polygonia comma* (Nymphalidae). *Journal of the Lepidopterists' Society* **37**: 1–13.

Bitzer RJ, Shaw KC (1995) Territorial behavior of the red admiral, *Vanessa atalanta* (Lepidoptera: Nymphalidae). I. The role of climatic factors and early interaction frequency on territorial start time. *Journal of Insect Behavior* **8**: 47–66.

Blab J (1993) *Grundlagen des Biotopschutzes für Tiere*, 4th edn. Greven, Germany: Kilda Verlag.

Blanckenhorn WU, Fairbairn DJ (1995) Life history adaptation along a latitudinal cline in the water strider *Aquarius remigis* (Heteroptera: Gerridae). *Journal of Evolutionary Biology* **8**: 21–41.

Blau WS (1981) Life history variation in the black swallowtail butterfly. *Oecologia* **48**: 116–122.

Bloch D, Erhardt A (2008) Selection towards shorter flowers by butterflies whose probosces are shorter than floral tubes. *Ecology* **89**: 2453–2460.

Bloch D, Werdenberg N, Erhardt A (2006) Pollination crisis in the butterfly-pollinated wild carnation *Dianthus carthusianorum*? *New Phytologist* **169**: 699–706.

Blum MJ (2002) Rapid movement of a *Heliconius* hybrid zone: evidence for phase III of Wright's shifting balance theory? *Evolution* **56**: 1992–1998.

Boatman ND (1992) Herbicides and the management of field boundary vegetation. *Pesticide Outlook* **3**: 30–34.

Boggs CL (1981a) Nutritional and life history determinants of resource allocation in holometabolous insects. *American Naturalist* **117**: 692–709.

Boggs CL (1981b) Selection pressures affecting male nutrient investment at mating in Heliconiinae butterflies. *Evolution* **35**: 931–940.

Boggs CL (1986) Reproductive strategies of female butterflies: variation in and constraints on fecundity. *Ecological Entomology* **11**: 7–15.

Boggs CL (1987) Ecology of nectar and pollen feeding in Lepidoptera. In: Slansky F Jr, Rodriguez JG (eds.) *Nutritional Ecology of Insects, Mites, Spiders, and Related Invertebrates*, New York: John Wiley, pp. 369–391.

Boggs CL (1988) Rates of nectar feeding in butterflies: effects of sex, size, age and nectar concentration. *Functional Ecology* **2**: 289–295.

Boggs CL (1990) A general model of the role of male-donated nutrients in female insects' reproduction. *American Naturalist* **136**: 598–617.

Boggs CL (1992) Resource allocation: exploring connections between foraging and life history. *Functional Ecology* **6**: 508–518.

Boggs CL (1995) Male nuptial gifts: phenotypic consequences and evolutionary implications. In: Leather SR, Hardie J (eds.) *Insect Reproduction*, Boca Raton, FL: CRC Press, pp. 215–242.

Boggs CL (1997a) Dynamics of reproductive allocation from juvenile and adult feeding: radiotracer studies. *Ecology* **78**: 192–202.

Boggs CL (1997b) Reproductive allocation from reserves and income in butterfly species with differing adult diets. *Ecology* **78**: 181–191.

Boggs CL (2003) Environmental variation, life histories and allocation. In: Boggs CL, Watt WB, Ehrlich PR (eds.) *Butterflies: Ecology and Evolution Taking Flight*, Chicago, IL: University of Chicago Press, pp. 185–206.

Boggs CL, Dau B (2004) Resource specialization in puddling Lepidoptera. *Environmental Entomology* **33**: 1020–1024.

Boggs CL, Freeman KD (2005) Larval food limitation in butterflies: effects on adult resource allocation and fitness. *Oecologia* **144**: 353–361.

Boggs CL, Gilbert LE (1979) Male contribution to egg production in butterflies: evidence for transfer of nutrients at mating. *Science* **206**: 83–84.

Boggs CL, Murphy DD (1997) Community composition in mountain ecosystems: climatic determinants of montane butterfly distribution. *Global Ecology and Biogeography Letters* **6**: 39–48.

Boggs CL, Nieminen M (2004) Checkerspot reproductive biology. In: Ehrlich PR, Hanski I (eds.) *On the Wings of Checkerspots: A Model System for Population Biology*, Oxford: Oxford University Press, pp. 92–111.

Boggs CL, Ross CL (1993) The effect of adult food limitation on life-history traits in *Speyeria mormonia* (Lepidoptera, Nymphalidae). *Ecology* **74**: 433–441.

Boggs CL, Watt WB (1981) Population structure of pierid butterflies. IV. Genetic and physiological investment in offspring by male *Colias. Oecologia* **50**: 320–324.

Boggs CL, Watt WB, Ehrlich PR (eds.) (2003) *Butterflies: Ecology and Evolution Taking Flight.* Chicago, IL: University of Chicago Press.

Bohonak AJ (1999) Dispersal, gene flow, and population structure. *Quarterly Review of Biology* **74**: 21–45.

Bonn A, Schröder B (2001) Habitat models and their transfer for single- and multi-species groups: a case study of carabids in an alluvial forest. *Ecography* **24**: 483–496.

Boppré M (1983) Leaf-scratching: a specialized behavior of Danaine butterflies (Lepidoptera) for gathering secondary plant-substances. *Oecologia* **59**: 414–416.

Boppré M (1984) Chemically mediated interactions between butterflies. In: Ackery PR, Vane-Wright RI (eds.) *The Biology of Butterflies*, Symposium of the Royal Entomological Society of London, London: Academic Press, pp. 259–276.

Boppré M (1990) Lepidoptera and pyrrolizidine alkaloids: exemplification of complexity in chemical ecology. *Journal of Chemical Ecology* **16**: 165–185.

Boppré M, Vane-Wright RI (1989) Androconial systems in Danainae (Lepidoptera): functional morphology of *Amauris, Danaus, Tirumala* and *Euploea. Zoological Journal of the Linnean Society* **97**: 101–133.

Borcard D, Legendre P, Drapeau P (1992) Partialling out the spatial component of ecological variation. *Ecology* **73**: 1045–1055.

Bossart JL, Prowell DP (1998) Genetic estimates of population structure and gene flow: limitations, lessons and new directions. *Trends in Ecology and Evolution* **13**: 202–206.

Bossart JL, Scriber JM (1995) Maintenance of ecologically significant genetic variation in the tiger swallowtail butterfly through differential selection and gene flow. *Evolution* **49**: 1163–1171.

Boughton DA (1999) Empirical evidence for complex source–sink dynamics with alternative states in a butterfly metapopulation. *Ecology* **80**: 2727–2739.

Boughton DA (2000) The dispersal system of a butterfly: a test of source–sink theory suggests the intermediate-scale hypothesis. *American Naturalist* **156**: 131–144.

Boulinier T, Danchin E (1997) The use of conspecific reproductive success for breeding patch selection in terrestrial migratory species. *Evolutionary Ecology* **11**: 505–517.

Bourgogne J (1953) *Melitaea athalia athalia* Rott. et *M. athalia helvetica* Rühl (*pseudathalia* Rev.) en France: étude biogéographique (Lep. Nymphalidae). *Annales de la Société Entomologique de France* **72**: 131–176.

Bourgogne J (1963) Réflexions au sujet d'une espèce singulière, *Erebia serotina* (Nymphalidae: Satyrinae). *Alexanor* **3**: 363–368.

Bourgogne J (1979) Ordre des Lépidoptères. In: Grassé PP (ed.) *Traité de Zoologie*, Vol. 10, 2nd edn, Paris: Masson, pp. 174–448.

Bourn NAD (2002) Habitats, the landscape and butterflies. In: Stone D, Tither J, Lacey P (eds.) *English Nature: The Species Recovery Programme*, Proceedings of the 10th Anniversary Conference, Peterborough: English Nature, pp. 72–75.

Bourn NAD, Thomas JA (1993) The ecology and conservation of the Brown Argus butterfly *Aricia agestis* in Britain. *Biological Conservation* **63**: 67–74.

Bourn NAD, Thomas JA (2002) The challenge of conserving grassland insects at the margins of their range in Europe. *Biological Conservation* **104**: 285–292.

Bourn NAD, Warren MS (1998) *Black Hairstreak*, Satyrium pruni: *Species Action Plan.* Wareham, Dorset: Butterfly Conservation.

Bowden SR. (1966) 'Sex-ratio' in *Pieris* hybrids. *Journal of the Lepidopterists' Society* **20**: 189–196.

Bowden SR (1990) Experimental breeding of butterflies. In: Kudrna O (ed.) *Butterflies of Europe 2: Introduction to Lepidopterology*, Wiesbaden, Germany: Aula, pp. 437–448.

Bowers MD (1983) The role of iridoid glycosides in host-plant specificity of checkerspot butterflies. *Journal of Chemical Ecology* **9**: 475–493.

Bowers MD (1993) Aposematic caterpillars: life styles of the warningly colored and unpalatable. In: Stamp NE, Casey TM (eds.) *Caterpillars: Ecological and Evolutionary Constraints on Foraging*, New York: Chapman and Hall, pp. 331–371.

Bowler DE, Benton TG (2005) Causes and consequences of animal dispersal strategies: relating individual behaviour to spatial dynamics. *Biological Reviews* **80**: 205–225.

Boyce MS (1984) Restitution of *r*- and *K*-selection as a model of density-dependent natural selection. *Annual Review of Ecology and Systematics* **15**: 427–447.

Boyce MS (1992) Population viability analysis. *Annual Review of Ecology and Systematics* **23**: 481–506.

Boyce MS (2002) Reconciling the small-population and declining-population paradigms. In: Beissinger SR, McCullough DR (eds.) *Population Viability Analysis*, Chicago, IL: University of Chicago Press, pp. 41–49.

Boyce MS, Vernier PR, Nielsen SE, Schmiegelow FKA (2002) Evaluating resource selection functions. *Ecological Modelling* **157**: 281–300.

Braby MF (1994) The significance of egg size variation in butterflies in relation to hostplant quality. *Oikos* **71**: 119–129.

Braby MF (2001) *Butterflies of Australia*, 2 vols. Melbourne, Australia: CSIRO.

Braby MF (2002) Life history strategies and habitat templets of tropical butterflies in North-Eastern Australia. *Evolutionary Ecology* 16: 399–413.

Braby MF, Jones RE (1995) Reproductive patterns and resource allocation in tropical butterflies: influence of adult diet and seasonal phenotype on fecundity, longevity and egg size. *Oikos* 72: 189–204.

Bradford MJ, Roff DA (1995) Genetic and phenotypic sources of life history variation along a cline in voltinism in the cricket *Allonemobius socius*. *Oecologia* 103: 319–326.

Bradshaw WE (1976) Geography of photoperiodic response in a diapausing mosquito. *Nature* 262: 384–385.

Bradshaw WE, Holzapfel CM (1983) Life cycle strategies in *Wyeomyia smithii*: seasonal and geographic adaptations. In: Brown VK, Hodek I (eds.) *Diapause and Life Cycle Strategies in Insects*, The Hague: Junk Publishers, pp. 169–187.

Brakefield PM (1982) Ecological studies on the butterfly *Maniola jurtina* in Britain. I. Adult behaviour, microdistribution and dispersal. *Journal of Animal Ecology* 51: 713–726.

Brakefield PM (1984) The ecological genetics of quantitative characters of *Maniola jurtina* and other butterflies. In: Vane-Wright RI, Ackery PR (eds.) *Biology of Butterflies*, London: Academic Press, pp. 167–190.

Brakefield P (1990) Case studies in ecological genetics. In: Kudrna O (ed.) *Butterflies of Europe*, Vol. 2, Wiesbaden, Germany: Aula, pp. 307–331.

Brakefield PM (1994) Egg size declines with female age in the fruit-feeding tropical butterfly *Bicyclus anyana* (Satyrinae). *Proceedings of Experimental and Applied Entomology* 5: 53–54.

Brakefield PM (1996) Seasonal polyphenism in butterflies and natural selection. *Trends in Ecology and Evolution* 11: 275–277.

Brakefield P (2001) Structure of a character and the evolution of butterfly eyespot patterns. *Journal of Experimental Zoology* 291: 93–104.

Brakefield PM (2003a) Artificial selection and the development of ecologically relevant phenotypes. *Ecology* 84: 1661–1671.

Brakefield PM (2003b) The power of evo-devo to explore evolutionary constraints: experiments with butterfly eyespots. *Zoology* 106: 283–290.

Brakefield PM (2006) Evo-devo and constraints on selection. *Trends in Ecology and Evolution* 21: 362–368.

Brakefield PM, Frankino WA (2009) Polyphenisms in Lepidoptera: multidisciplinary approaches to studies of evolution and development. In: Whitman D, Ananthakrishnan TN (eds.) *Phenotypic Plasticity in Insects:*

Mechanisms and Consequences, Enfield NH: Science Publishers, pp. 121–152.

Brakefield PM, French V (1999) Butterfly wings: the evolution of development of colour patterns. *BioEssays* 21: 391–401.

Brakefield PM, Larsen TB (1984) The evolutionary significance of dry and wet season forms in some tropical butterflies. *Biological Journal of the Linnean Society* 22: 1–12.

Brakefield PM, Mazzotta V (1995) Matching field and laboratory environments: effects of neglecting daily temperature variation on insect reaction norms. *Journal of Evolutionary Biology* 8: 559–573.

Brakefield PM, Shreeve TG (1992a) Case studies in evolution. In: Dennis RLH (ed.) *The Ecology of Butterflies in Britain*, Oxford: Oxford University Press, pp. 197–216.

Brakefield PM, Shreeve TG (1992b) Diversity within populations. In: Dennis RLH (ed.) *The Ecology of Butterflies in Britain*, Oxford: Oxford University Press, pp. 178–196.

Brakefield PM, van Noordwijk AJ (1985) The genetics of spot pattern characters in the meadow brown butterfly *Maniola jurtina* (Lepidoptera, Satyrinae). *Heredity* 54: 275–284.

Brakefield PM, Shreeve TG, Thomas JA (1992) Avoidance, concealment, and defence. In: Dennis RLH (ed.) *The Ecology of Butterflies in Britain*, Oxford: Oxford University Press, pp. 93–119.

Brakefield PM, Kesbeke F, Koch B (1998) The regulation of phenotyic plasticity of eyespots in the butterfly *Bicyclus anynana*. *American Naturalist* 152: 853–860.

Brändle M, Öhlschläger S, Brandl R (2002) Range sizes in butterflies: correlation across scales. *Evolutionary Ecology Research* 4: 993–1004.

Braschler B, Hill JK (2007) Role of larval host plants in the climate-driven range expansion of the butterfly *Polygonia c-album*. *Journal of Animal Ecology* 76: 415–423.

Brattsen LB (1988) Enzymatic adaptations in leaf-feeding insects to host-plant allellochemicals. *Journal of Chemical Ecology* 14: 1919–1939.

Bräu M, Gros P, Nunner A, Stettmer C, Settele J (2006) Der verlustreiche Weg in die Sicherheit eines Wirtsameisen-Nestes: neue Daten zur Entwicklungsbiologie und zur Mortalität der Präimaginalstadien von *Maculinea alcon* sowie zum Einfluss der Mahd. In: Fartmann T, Hermann G (eds.) Larvalökologie von Tagfaltern und Widderchen in Mitteleuropa. *Abhandlungen aus dem westfälischen Museum für Naturkunde* 68: 197–219.

Brčák J (1952) Biocenologie lepidopter Temnosmrečinové doliny ve Vysokých Tatrách. *Biologický sborník SAVU* 7: 113–131.

Breiman L (1996) Bagging predictors. *Machine Learning* 24: 123–140.

Breiman L (2001) Random forests. *Machine Learning* **45**: 5–32.

Brennan LA, Block WM, Guttiérrez RJ (1986) The use of multivariate statistics for developing habitat suitability index models. In: Verner J, Morrison ML, Ralph CJ (eds.) *Wildlife 2000: Modeling Habitat Relationships of Terrestrial Vertebrates*, Madison, WI: University of Wisconsin Press, pp. 177–182.

Breuker CJ, Brakefield PM (2002) Female choice depends on size but not symmetry of dorsal eyespots in the butterfly *Bicyclus anynana. Proceedings of the Royal Society of London B* **269**: 1233–1239.

Breuker CJ, Brakefield PM, Gibbs M (2007) The association between wing morphology and dispersal is sex-specific in the Glanville fritillary butterfly *Melitaea cinxia* (Lepidoptera: Nymphalidae). *European Journal of Entomology* **104**: 445–452.

Bridle JR, Vines TH (2007) Limits to evolution at range margins: when and why does adaptation fail? *Trends in Ecology and Evolution* **22**: 140–147.

Briers RA, Warren PH (2000) Population turnover and habitat dynamics in *Notonecta* (Hemiptera: Notonectidae) metapopulations. *Oecologia* **123**: 216–222.

Briscoe AD, Bernard GD, Szeto AS, Nagy LM, White RH (2003) Not all butterfly eyes are created equal: rhodopsin absorption spectra, molecular identification and localization of UV-, blue- and green-sensitive opsins in the retina of *Vanessa cardui. Journal of Comparative Neurology* **458**: 334–349.

Britten HB, Brussard PF, Murphy DD, Ehrlich PR (1995) A test for isolation-by-distance in central Rocky Mountain and Great Basin populations of Edith's checkerspot butterfly (*Euphydryas editha*). *Journal of Heredity* **86**: 204–210.

Brock JP (1990) Origins and phylogeny of butterflies. In: Kudrna O (ed.) *Butterflies of Europe*, Vol. 2, Wiesbaden, Germany: Aula, pp. 209–233.

Brodeur J (1992) Caterpillars are more than a nutrient reservoir for parasitic wasps. *Proceedings of the Section Experimental and Applied Entomology, Netherlands Entomological Society* **3**: 57–61.

Brommer JE, Fred MS (1999) Movement of the Apollo butterfly *Parnassius apollo* related to host plant and nectar plant patches. *Ecological Entomology* **24**: 125–131.

Brook BW, Burgman MA, Akçakaya HR, O'Grady JJ, Frankham R (2002a) Critiques of PVA ask the wrong questions: throwing the heuristic baby out with the numerical bath water. *Conservation Biology* **16**: 262–263.

Brook BW, Tonkyn DW, O'Grady JJ, Frankham R (2002b) Contribution of inbreeding to extinction risk in threatened species. *Conservation Ecology* **6**: 16 (online: http://www.consecol.org/vol6/iss1/art16).

Brookes MI, Graneau YA, King P, Rose OC, Thomas CD, Mallet JLB (1997) Genetic analysis of founder bottlenecks in the rare British butterfly *Plebejus argus. Conservation Biology* **11**: 648–661.

Brooks JS, Feeny P (2004) Seasonal variation in *Daucus carota* leaf-surface and leaf-tissue chemical profiles. *Biochemical Systematics and Ecology* **32**: 769–782.

Brooks DR, McLennan DA (1991) *Phylogeny, Ecology and Behavior: A Research Program in Comparative Biology*. Chicago, IL: University of Chicago Press.

Brotons L, Thuiller W, Araújo MB, Hirzel AH (2004) Presence–absence versus presence-only modelling methods for predicting bird habitat suitability. *Ecography* **27**: 437–448.

Brower LP (1963) The evolution of sex-limited mimicry in butterflies. *Proceedings of the 16th International Congress on Zoology, Washington* **4**: 173–179.

Brower LP (1984) Chemical defence in butterflies. In: Vane-Wright RI, Ackery PR (eds.) *The Biology of Butterflies*, London: Academic Press, pp. 109–134.

Brower LP, Calvert WH (1985) Foraging dynamics of bird predators on overwintering monarch butterflies in Mexico. *Evolution* **39**: 340–349.

Brower AVZ, Egan MG (1997) Cladistic analysis of *Heliconius* butterflies and relatives (Nymphalidae: Heliconiiti): a revised phylogenetic position for *Eueides* based on sequences from mtDNA and a nuclear gene. *Proceedings of the Royal Society of London B* **264**: 969–977.

Brower LP, Brower JVZ, Cranston FP (1965) Courtship behavior of the queen butterfly *Danaus gilippus berenice* (Cramer). *Zoologica* **50**: 1–39.

Brown J (1976a) Notes regarding two previously described European taxa of the genera *Agrodiaetus* Hübner 1822 and *Polyommatus* Kluck 1801 (Lep.: Lycaenidae). *Entomologist's Gazette* **27**: 77–84.

Brown J (1976b) On two previously undescribed species of Lycaenidae (Lepidoptera) from Greece. *Entomologische Berichten, Amsterdam* **36**: 46–47.

Brown KS (1981) The biology of *Heliconius* and related genera. *Annual Review of Entomology* **26**: 427–456.

Brown JH (1984a) On the relationship between abundance and distribution of species. *American Naturalist* **124**: 255–279.

Brown KS (1984b) Adult obtained pyrrolizidine alkaloids defend ithomiine butterflies against a spider predator. *Nature* **309**: 707–709.

Brown WD, Alcock J (1991) Hilltopping by the red admiral butterfly: mate searching alongside congeners. *Journal of Research on the Lepidoptera* **29**: 1–10.

Brown J, Coutsis JG (1978) Two newly discovered lycaenid butterflies (Lepidoptera: Lycaenidae) from Greece, with notes on allied species. *Entomologist's Gazette* **29**: 201–213.

Brown IL, Ehrlich PR (1980) Population biology of the Checkerspot Butterfly, *Euphydryas chalcedona*: structure of the Jasper Ridge Colony. *Oecologia* **47**: 239–251.

Brown JH, Lomolino MV (1998) *Biogeography*, 2nd edn. Sunderland, MA: Sinauer Associates.

Bruinsma M, Van Dam NM, Van Loon JJA, Dicke M (2007) Jasmonic acid-induced changes in *Brassica oleracea* affect oviposition preference of two specialist herbivores. *Journal of Chemical Ecology* **33**: 655–668.

Brunton CFA, Hurst GDD (1998) Mitochondrial DNA phylogeny of Brimstone butterflies (genus *Gonepteryx*) from the Canary Islands and Madeira. *Biological Journal of the Linnean Society* **63**: 69–79.

Brunton CFA, Baxter JD, Quartson JAS, Panchen AL (1991) Altitude-dependent variation in wing pattern in the Corsican butterfly *Coenonympha corinna* Hubner (Satyridae). *Biological Journal of the Linnean Society* **42**: 367–378.

Brunzel S (2002) Experimental density-related emigration in the cranberry fritillary *Boloria aquilonaris*. *Journal of Insect Behavior* **15**: 739–750.

Brussard PF, Ehrlich PR (1970) Population structure of *Erebia epipsodea* (Lepidoptera – Satyrinae). *Ecology* **51**: 119–129.

Bryant SR, Thomas CD, Bale JS (1997) Nettle-feeding nymphalid butterflies: temperature, development and distribution. *Ecological Entomology* **22**: 390–398.

Bryant SR, Bale JS, Thomas CD (1998) Modification of the triangle method of degree-day accumulation to allow for behavioural thermoregulation in insects. *Journal of Applied Ecology* **35**: 921–927.

Bryant SR, Thomas CD, Bale JS (2000) Thermal ecology of gregarious and solitary nettle-feeding nymphalid butterfly larvae. *Oecologia* **122**: 1–10.

Bryant SR, Thomas CD, Bale JS (2002) The influence of thermal ecology on the distribution of three nymphalid butterflies. *Journal of Applied Ecology* **39**: 43–55.

Buckland ST, Burnham KP, Augustin NH (1997) Model selection: an integral part of inference. *Biometrics (Journal of the Biometric Society)* **53**: 603–618.

Bukovinszky T, Potting RPJ, Clough Y, Lenteren JC, Vet LEM (2005) The role of pre- and post-alighting detection mechanisms in the responses to patch size by specialist herbivores. *Oikos* **109**: 435–446.

Bull V, Beltrán M, Jiggins CD, McMillan WO, Bermingham E, Mallet J (2006) Polyphyly and gene flow between non-sibling *Heliconius* species. *BMC Biology* **4**: 11.

Bullock JM, Kenward RE, Hails RS (2002) *Dispersal Ecology*. Oxford: Blackwell.

Bulman CR, Wilson RJ, Holt AR, Bravo LG, Early RI, Warren MS, Thomas CD (2007) Minimum viable metapopulation size, extinction debt, and the conservation of a declining species. *Ecological Applications* **17**: 1460–1473.

Burghardt F, Fiedler K, Proksch P (1997) Uptake of flavonoids from *Visca villosa* by the lycaenid butterfly, *Polyommatus icarus*. *Biochemical Systematics and Ecology* **25**: 527–536.

Burgman MA, Possingham HP (2000) Population viability analysis for conservation: the good, the bad, and the undescribed. In: Young AG, Clark GM (eds.) *Genetics, Demography, and Viability of Fragmented Populations*, Cambridge: Cambridge University Press, pp. 97–112.

Burgman MA, Ferson S, Akçakaya HR (1993) *Risk Assessment in Conservation Biology*. London: Chapman and Hall.

Burke S, Pullin AS, Wilson RJ, Thomas CD (2005) Selection for discontinuous life-history traits along a continuous thermal gradient in the butterfly *Aricia agestis*. *Ecological Entomology* **30**: 613–619.

Burkhardt D (1996) Ultraviolet perception by bird eyes and some implications. *Naturwissenschaften* **83**: 492–497.

Burnham KP, Anderson DR (1998) *Model Selection and Inference: A Practical Information–Theoretic Approach*. Berlin: Springer-Verlag.

Burns JM (1968) Mating frequency in natural populations of skippers and butterflies as determined by spermatophore counts. *Proceedings of the National Academy of of Sciences, USA* **61**: 852–859.

Bush GL (1969) Sympatric host race formation and speciation in frugivorous flies of the genus *Rhagoletis* (Diptera, Tephritidae). *Evolution* **23**: 237–251.

Caballero VE (1996) Biología y ecología del género *Euchloe* (Lepidoptera: Pieridae) en el sur de la Península Ibérica. Unpublished Ph.D. thesis, Universidad de Córdoba, Córdoba.

Cabeza M, Araújo MB, Wilson RJ, Thomas CD, Cowley MJR, Moilanen A (2004) Combining probabilities of occurrence with spatial reserve design. *Journal of Applied Ecology* **41**: 252–262.

Cain ML, Ecclestone J, Kareiva PM (1985) The influence of food plant dispersion on caterpillar searching success. *Ecological Entomology* **10**: 1–7.

Cameron-Curry V, Leigheb G, Cameron-Curry P. (1980) Due ibridi di *Lysandra bellargus* Rott. (Lepidoptera, Lycaenidae). *Bollettino della Società Entomologica Italiana* **112**: 41–42.

Campbell IM (1962) Reproductive capacity in genus *Choristoneura* Led. (Lepidoptera: Tortricidae).

I. Quantitative inheritance and genes as controllers of rates. *Canadian Journal of Genetics and Cytology* **4**: 272–288.

Campeau S, Rochefort L, Price JS (2004) On the use of shallow basins to restore cutover peatlands: plant establishment. *Restoration Ecology* **12**: 471–482.

Cant ET, Smith AD, Reynolds DR, Osborne JL (2005) Tracking butterfly flight paths across the landscape with harmonic radar. *Proceedings of the Royal Society of London B* **272**: 785–790.

Capdeville P (1978) *Les races géographiques de* Parnassius apollo. Compiègne, France: Sciences Naturelles.

Cappuccino N, Kareiva P (1985) Coping with a capricious environment: a population study of a rare pierid butterfly. *Ecology* **66**: 152–161.

Carbonell F (2001) Contribution à la connaissance du genre *Agrodiaetus* Hübner (1822) *A. ahmadi* et *A. khorasanensis*, nouvelles espèces dans le nord de l'Iran. *Linneana Belgica* **18**: 105–110.

Carey DB (1994a) Diapause and the host plant affiliations of lycaenid butterflies. *Oikos* **69**: 259–266.

Carey DB (1994b) Patch dynamics of *Glaucopsyche lygdamus* (Lycaenidae): correlations between butterfly density and host species diversity. *Oecologia* **99**: 337–342.

Carl KP (1968) *Thymelicus lineola* (Lepidoptera: Hesperiidae) and its parasites in Europe. *Canadian Entomologist* **100**: 785–801.

Carter DJ (1984) *Pest Lepidoptera of Europe, with special Reference to the British Isles*. Dordrecht, Netherlands: Kluwer.

Carter CI, Anderson MA (1987) *Enhancement of Lowland Forest Ridesides and Roadsides to Benefit Wild Plants and Butterflies*. Farnham, Hants.: Forestry Commission.

Casey TM (1993) Effects of temperature on foraging caterpillars. In: Stamp NE, Casey TM (eds.) *Caterpillars: Evolutionary Constraints on Foraging*, New York: Chapman and Hall, pp. 5–28.

Cassel A, Tammaru T (2003) Allozyme variability in central, peripheral and isolated populations of the scarce heath (*Coenonympha hero*: Lepidoptera, Nymphalidae): implications for conservation. *Conservation Genetics* **4**: 83–93.

Cassel A, Windig J, Nylin S, Wiklund C (2001) Effects of population size and food stress on fitness-related characters in the scarce heath, a rare butterfly in western Europe. *Conservation Biology* **15**: 1667–1673.

Caswell H (2001) *Matrix Population Models*. Sunderland, MA: Sinauer Associates.

Cates RG (1980) Feeding patterns of monophagous, oligophagous and polyphagous insect herbivores: the effect of resource abundance and plant chemistry. *Oecologia* **46**: 22–31.

Caughley G (1994) Directions in conservation biology. *Journal of Animal Ecology* **63**: 215–244.

Céréghino R, Giraudel JL, Compin A (2001) Spatial analysis of stream invertebrates distribution in the Adour–Garonne drainage basin (France), using Kohonen self-organizing maps. *Ecological Modelling* **146**: 167–180.

Céréghino R, Santoul F, Compin A, Mastrorillo S (2005) Using self-organizing maps to investigate spatial patterns of non-native species. *Biological Conservation* **125**: 459–465.

Cesaroni D, Lucarelli M, Allori P, Russo F, Sbordoni V (1994) Patterns of evolution and multidimensional systematics in graylings (Lepidoptera: *Hipparchia*). *Biological Journal of the Linnean Society* **52**: 101–119.

Chai P, Srygley RB (1990) Predation and the flight, morphology, and temperature of neotropical rain-forest butterflies. *American Naturalist* **135**: 748–765.

Chaplin SB, Wells PH (1982) Energy reserves and metabolic expenditures of monarch butterflies overwintering in southern California. *Ecological Entomology* **7**: 249–256.

Chapman TA (1898) A review of the genus *Erebia* based on an examination of the male appendages. *Transactions of the Entomological Society of London* **3**: 209–240.

Chardon JP, Adriaensen F, Matthysen E (2003) Incorporating landscape elements into a connectivity measure: a case study for the speckled wood butterfly (*Pararge aegeria* L.). *Landscape Ecology* **18**: 561–574.

Chatfield C (1995) Model uncertainty, data mining and statistical inference (with discussion). *Journal of the Royal Statistical Society A* **158**: 419–466.

Chesmore D, Monkman G (1994) Automated analysis of variation in Lepidoptera. *Entomologist* **113**: 171–182.

Chevan A, Sutherland M (1991) Hierarchical partitioning. *American Statistician* **45**: 90–96.

Chew FS (1977) Coevolution of pierid butterflies and their cruciferous food plants. II. The distribution of eggs on potential food plants. *Evolution* **31**: 568–579.

Chew FS, Robbins RK (1984) Egg-laying in butterflies. In: Vane-Wright RI, Ackery PR (eds.) *The Biology of Butterflies*, London: Academic Press, pp. 65–80.

Choi SW (2004) Trends in butterfly species richness in response to the peninsular effect in South Korea. *Journal of Biogeography* **31**: 587–592.

Chovet G (1998) L'accouplement chez les Lépidoptères Rhopalocères: divers aspects éthologiques, morphologiques et physiologiques et leurs implications biosystématiques. Unpublished Ph.D. thesis, Université Pierre et Marie Curie, Paris.

Chown SL, Gaston KJ (1999) Exploring links between physiology and ecology at macro-scales: the role of respiratory metabolism in insects. *Biological Reviews of the Cambridge Philosophical Society* **74**: 87–120.

Chown SL, Nicolson SW (2004) *Insect Physiological Ecology: Mechanisms and Patterns*. Oxford: Oxford University Press.

Church NS (1960) Heat loss and the body temperature of flying insects. II. Heat conduction within the body and its loss by radiation and convection. *Journal of Experimental Biology* **37**: 186–212.

Church SC, Bennett ATD, Cuthill IC, Partridge JC (1998) Ultraviolet cues affect the foraging behaviour of blue tits. *Proceedings of the Royal Society of London B* **265**: 1509–1514.

Cianchi R, Ungaro A, Marini M, Bullini L (2003) Differential patterns of hybridization and introgression between the swallowtails *Papilio machaon* and *P. hospiton* from Sardinia and Corsica islands (Lepidoptera, Papilionidae). *Molecular Ecology* **12**: 1461–1471.

Cila P, Skyva J (1993) Výsledky faunistického průzkumu motýlů (Lepidoptera) na území Prahy – 1. část. [Results of the faunistic research of Lepidoptera in Prague area – Part 1]. *Klapalekiana* **29**: 71–86.

Cizek O, Konvička M (2005) What is a patch in a dynamic metapopulation? Mobility of an endangered woodland butterfly, *Euphydryas maturna*. *Ecography* **28**: 791–800.

Cizek O, Bakesova A, Kuras T, Beneš J, Konvička M (2003) Vacant niche in alpine habitat: the case of an introduced population of the butterfly *Erebia epiphron* in the Krkonose Mountains. *Acta Oecologica* **24**: 15–23.

Cizek L, Fric Z, Konvička M (2006) Host plant defences and voltinism in European butterflies. *Ecological Entomology* **31**: 337–344.

Clark FS, Faeth SH (1997) The consequences of larval aggregation in the butterfly *Chlosyne lacinia*. *Ecological Entomology* **22**: 408–415.

Clark BR, Faeth SH (1998) The evolution of egg clustering in butterflies: a test of the egg desiccation hypothesis. *Evolutionary Ecology* **12**: 543–552.

Clarke CA, Larsen TB (1986) Speciation problems in the *Papilio machaon* group of butterflies (Lepidoptera, Papilionidae). *Systematic Entomology* **11**: 175–181.

Clarke SA, Robertson PA (1993) The relative effects of woodland management and pheasant *Phasianus colchicus* predation on the survival of the pearl-bordered and small pearl-bordered fritillaries *Boloria euphrosyne* and *B. selene* in the south of England. *Biological Conservation* **65**: 199–203.

Clarke CA, Sheppard PM (1953) Further observations on hybrid swallowtails. *Entomologist's Record and Journal of Variation* (Suppl.) **65**: 1–12.

Clarke CA, Sheppard PM (1955) The breeding in captivity of the hybrid swallowtail *Papilio machaon gorganus* Frühstorfer (F) × *Papilio hospiton* Géné (M). *Entomologist* **88**: 265–270.

Clarke CA, Sheppard PM (1956) A further report on the genetics of the *machaon* group of swallowtail butterflies. *Evolution* **10**: 66–73.

Clarke CA, Sheppard PM, Willig A (1970) The use of ecdysone to break a two and a half year pupal diapause in *Papilio glaucus* female × *Papilio rutulus* male hybrids. *Entomologist* **105**: 137–138.

Clarke RT, Thomas JA, Elmes GW, Hochberg ME (1997) The effects of spatial patterns in habitat quality on community dynamics within a site. *Proceedings of the Royal Society of London B* **264**: 347–354.

Clausen CP (1940) *Entomophagous Insects*. New York: McGraw-Hill. (Reprinted 1972, New York: Hafner.)

Cleary DFR, Descimon H, Menken SBJ (2002) Genetic and ecological differentiation between the butterfly sister species *Colias alfacariensis* and *Colias hyale*. *Contributions to Zoology* **71**: 131–139.

Clench HK (1966) Behavioral thermoregulation in butterflies. *Ecology* **47**: 1021–1034.

Clinchy M, Haydon DT, Smith AT (2002) Pattern does not equal process: what does patch occupancy really tell us about metapopulation dynamics? *American Naturalist* **159**: 351–362.

Clobert J, Danchin E, Dhont AA, Nichols JD (2001a) *Dispersal*. Oxford: Oxford University Press.

Clobert J, Wolff JO, Nichols JD, Danchin E, Dhont AA (2001b) Introduction. In: Clobert J, Wolff JO, Nichols JD, Danchin E, Dhont AA (eds.) *Dispersal*, Oxford: Oxford University Press, pp. xvii–xxi.

Cohen J (1960) A coefficient of agreement for nominal scales. *Educational and Psychological Measurement* **20**: 37–46.

Cohen D (1970) A theoretical model for the optimal timing of diapause. *American Naturalist* **104**: 389–400.

Cole LR (1959) On the defences of lepidopterous pupae in relation to the oviposition behaviour of certain Ichneumonidae. *Journal of the Lepidopterists' Society* **13**: 1–10.

Cole LR (1967) A study of the life-cycles and hosts of some Ichneumonidae attacking pupae of the green oak-leaf roller moth, *Tortrix viridana* (L.) (Lepidoptera: Tortricidae) in England. *Transactions of the Royal Entomological Society of London* **119**: 267–281.

Collin PH (1988) *Dictionary of Ecology and the Environment*. Teddington, Middlesex: Peter Collin Publishing.

Colwell RK, Hurtt GC (1994) Nonbiological gradients in species richness and a spurious Rapoport effect. *American Naturalist* 144: 570–595.

Colwell RK, Lees DC (2000) The mid-domain effect: geometric constraints on the geography of species richness. *Trends in Ecology and Evolution* 15: 70–76.

Colwell RK, Rahbek C, Gotelli NJ (2004) The mid-domain effect and species richness patterns: what have we learned so far? *American Naturalist* 163: E1–E23.

Comba L, Corbet SA, Barron A, Bird A, Collinge S, Miyazaki N, Powell M (1999a) Garden flowers: insect visits and the floral reward of horticulturally modified variants. *Annals of Botany* 83: 73–86.

Comba L, Corbet SA, Hunt L, Warren B (1999b) Flowers, nectar and insect visits: evaluating British plant species for pollinator friendly gardens. *Annals of Botany* 83: 369–383.

Comes HP, Abbott RJ (1998) The relative importance of historical events and gene flow on the population structure of a Mediterranean ragwort, *Senecio gallicus* (Asteraceae). *Evolution* 52: 355–367.

Comes HP, Kadereit JW (1998) The effect of Quarternary climatic changes on plant distribution and evolution. *Trends in Plant Science* 3: 432–438.

Comins HN, Hamilton WD, May RM (1980) Evolutionarily stable dispersal strategies. *Journal of Theoretical Biology* 82: 205–230.

Committee of Scientists (1999) *Sustaining the People's Lands: Recommendations for Stewardship of the National Forests and Grasslands into the Next Century*. Washington, DC: US Department of Agriculture.

Conner JK, Davis R, Rush S (1995) The effect of wild radish floral morphology on pollination efficiency by four taxa of pollinators. *Oecologia* 104: 234–245.

Connor EF, McCoy ED (1979) The statistics and biology of the species–area relationship. *American Naturalist* 113: 791–833.

Conover DO, Schultz ET (1995) Phenotypic similarity and the evolutionary significance of countergradient variation. *Trends in Ecology and Evolution* 10: 248–252.

Conradt L, Roper TJ (2006) Nonrandom movement behavior at habitat boundaries in two butterfly species: implications for dispersal. *Ecology* 87: 125–132.

Conradt L, Bodsworth EJ, Roper TJ, Thomas CD (2000) Non-random dispersal in the butterfly *Maniola jurtina*: implications for metapopulation models. *Proceedings of the Royal Society of London B* 267: 1505–1510.

Conradt L, Roper TJ, Thomas CD (2001) Dispersal behaviour of individuals in metapopulations of two British butterflies. *Oikos* 95: 416–424.

Cook LG (2000) Extraordinary and extensive karyotype variation: a 48-fold range in chromosome number in the gall-inducing scale insect *Apiomorpha* (Hemiptera: Coccoidea: Eriococcidae). *Genome* 43: 255–263.

Coope GR (1970) Interpretation of Quaternary insect fossils. *Annual Review of Entomology* 15: 97–120.

Coope GR (1978) Constancy of insect species versus inconstancy of Quaternary environments. In: Mound LA, Waloff N (eds.) *Diversity of Insect Faunas*, Symposium of the Royal Entomological Society No. 9, Oxford: Blackwell Scientific Publications, pp. 176–187.

Coope GR (1987) The response of Late Quaternary insect communities to sudden climatic changes. In: Gee JHR, Giller PS (eds.) *Organization of Communities, Past and Present*. Oxford: Blackwell Scientific Publications, pp. 421–438.

Coope GR (1994) The response of insect faunas to glacial–interglacial climatic fluctuations. *Philosophical Transactions of the Royal Society of London* 344: 19–26.

Corander J, Waldmann P, Marttinen P, Sillanpaa MJ (2004) BAPS 2: enhanced possibilities for the analysis of genetic population structure. *Bioinformatics* 20: 2363–2369.

Corbet D (1978) Bee visits and the nectar of *Echium vulgare* L. and *Sinapis alba* L. *Ecological Entomology* 3: 25–37.

Corbet SA (1995) Insects, plants and succession: advantages of long-term set-aside. *Agriculture, Ecosystems and Environment* 53: 201–217.

Corbet SA (2000) Butterfly nectaring flowers: butterfly morphology and flower form. *Entomologia Experimentalis et Applicata* 96: 289–298.

Cordero C (2000a) Is spermatophore number a good measure of mating frequency in female *Callophrys xami* (Lycaenidae)? *Journal of the Lepidopterists' Society* 53: 169–170.

Cordero C (2000b) The number of copulations of territorial males of the butterfly *Callophrys xami* (Lycaenidae). *Journal of Research on the Lepidoptera* 35: 78–89.

Cordero CR, Soberón J (1990) Non-resource based territoriality in males of the butterfly *Xamia xamia* (Lepidoptera: Lycaenidae). *Journal of Insect Behavior* 3: 719–732.

Corke D (1999) Are honeydew/sap-feeding butterflies (Lepidoptera: Rhopalocera) affected by particulate air pollution? *Journal of Insect Conservation* 3: 5–14.

Courtnet SP, Forsberg J (1988) Host use by two pierid butterflies varies with host density. *Functional Biology* 2: 67–75.

Courtney SP (1981) Coevolution of pierid butterflies and their cruciferous plants. III. *Anthocharis cardamines* (L.) survival, development and oviposition on different host plants. *Oecologia* 51: 91–96.

Courtney SP (1982a) Coevolution of pierid butterflies and their cruciferous host plants. IV. Host apparency and *Anthocharis cardamines* oviposition. *Oecologia* **52**: 258–265.

Courtney SP (1982b) Notes on the biology of *Zegris eupheme* (Pieridae). *Journal of the Lepidopterists' Society* **36**: 132–135.

Courtney SP (1983) The ecology of movement in pierid butterflies (Lep., Pieridae). *Atalanta* **14**: 110–121.

Courtney SP (1984a) The evolution of egg clustering by butterflies and other insects. *American Naturalist* **123**: 276–281.

Courtney SP (1984b) Habitat versus foodplant selection. In: Vane-Wright RI, Ackery PR (eds.) *The Biology of Butterflies*, London: Academic Press, pp. 89–90.

Courtney SP (1986) The ecology of pierid butterflies: dynamics and interactions. *Advances in Ecological Research* **15**: 51–131.

Courtney SP, Anderson K (1986) Behaviour around encounter sites. *Behavioral Ecology and Sociobiology* **19**: 241–248.

Courtney SP, Courtney S (1982) The 'edge-effect' in butterfly oviposition: causality in *Anthocharis cardamines* and related species. *Ecological Entomology* **7**: 131–137.

Courtney SP, Duggan AE (1983) The population biology of the orange tip butterfly, *Anthocharis cardamines* in Britain. *Ecological Entomology* **8**: 271–281.

Courtney SP, Forsberg J (1988) Host use by two pierid butterflies varies with host density. *Functional Ecology* **2**: 65–75.

Courtney SP, Parker GA (1985) Mating behaviour of the tiger blue butterfly (*Tarucus theophrastus*): competitive mate-searching when not all females are captured. *Behavioral Ecology and Sociobiology* **17**: 213–221.

Courtney SP, Hill CJ, Westermann A (1982) Pollen carried for long periods by butterflies. *Oikos* **38**: 260–263.

Courtney SP, Chen GK, Gardner A (1989) A general model for individual host selection. *Oikos* **55**: 55–65.

Courvoisier L (1907) Über Zeichnungs-Aberrationen bei Lycaeniden. *Zeitschrift für wissenschaftliche Insektenbiologie* **3**: 8–11, 33–39, 73–78.

Cowley MJR, Thomas CD, Thomas JA, Warren MS (1999) Flight areas of British butterflies: assessing species status and decline. *Proceedings of the Royal Society of London B* **266**: 1587–1592.

Cowley MJR, Wilson RJ, León-Cortés JL, Gutiérrez D, Bulman CR, Thomas CD (2000) Habitat-based statistical models for predicting the spatial distribution of butterflies and day-flying moths in a fragmented landscape. *Journal of Applied Ecology* **37**: 60–72.

Cowley MJR, Thomas CD, Roy DB, Wilson RJ, León-Cortés JL, Gutiérrez D, Bulman CR, Quinn RM, Moss D, Gaston KJ

(2001a) Density-distribution relationships in British butterflies. I. The effect of mobility and spatial scale. *Journal of Animal Ecology* **70**: 410–425.

Cowley MJR, Thomas CD, Wilson RJ, León-Cortés JL, Gutiérrez D, Bulman CR (2001b) The density and distribution of British butterflies. II. An assessment of mechanisms. *Journal of Animal Ecology* **70**: 426–441.

Cox CB, Moore PD (1993) *Biogeography: An Ecological and Evolutionary Approach*. Oxford: Blackwell Science.

Coyne JA, Orr HA (1997) "Patterns of speciation in *Drosophila*" revisited. *Evolution* **51**: 295–303.

Coyne JA, Orr HA (2004) *Speciation*. Sunderland, MA: Sinauer Associates.

Cracraft J (1983) Species concepts and speciation analysis. *Current Ornithology* **1**: 159–187.

Cracraft J (1989) Speciation and its ontology: the empirical consequences of alternative species concepts for understanding patterns and processes of differentiation. In: Otte D, Endler JA (eds.) *Speciation and its Consequences*, Sunderland, MA: Sinauer Associates, pp. 28–59.

Crawley MJ (2002) *Statistical Computing: An Introduction to Data Analysis using S-Plus*. New York: John Wiley.

Crawley MJ (2005) *Statistics: An Introduction using R*. New York: John Wiley.

Cremene C, Groza G, Rakosy L, Schileyko AA, Baur A, Erhardt A, Baur B (2005) Alterations of steppe-like grasslands in Eastern Europe: a threat to regional biodiversity hotspots. *Conservation Biology* **19**: 1606–1618.

Critchley CNR, Fowbert JA, Sherwood AJ, Pywell RF (2006) Vegetation development of sown grass margins in arable fields under a countryside agri-environment scheme. *Biological Conservation* **132**: 1–11.

Crone EE, Schultz CB (2003) Movement behaviour and minimum patch size for butterfly population persistence. In: Boggs CL, Watt WB, Ehrlich PR (eds.) *Butterflies: Ecology and Evolution Taking Flight*, Chicago, IL: University of Chicago Press, pp. 561–576.

Croxton PJ, Hann JP, Greatorex-Davies JN, Sparks TH (2005) Linear hotspots? The floral and butterfly diversity of green lanes. *Biological Conservation* **121**: 579–584.

Crozier LG (2003) Winter warming facilitates range expansion: cold tolerance of the butterfly *Atalopedes campestris*. *Oecologia* **135**: 648–656.

Crozier LG (2004) Warmer winters drive butterfly range expansion by increasing survivorship. *Ecology* **85**: 231–241.

Cruden RW, Hermann-Parker SM (1979) Butterfly pollination of *Caesalpinia pulcherrima*, with observations on psychophilous syndrome. *Ecology* **67**: 155–168.

Cuénot L (1936) *L'Espèce*. Paris: Doin.

Cullenward MJ, Ehrlich PR, White RR, Holdren CE (1979) The ecology and population genetics of an alpine checkerspot butterfly, *Euphydryas anicia*. *Oecologia* 38: 1–12.

Cushman JH, Boggs CL, Weiss SB, Murphy DD, Harvey AW, Ehrlich PR (1994) Estimating female reproductive success of a threatened butterfly: influence of emergence time and host plant phenology. *Oecologia* 99: 194–200.

D'Aldin A. (1929) Un hybride de *Lycaena* en Dordogne. *Procès-Verbaux. Actes de la Société Linnéenne de Bordeaux* 85: 142–143.

Daily GC, Ehrlich PR, Wheye D (1991) Determinants of spatial distribution in a population of the subalpine butterfly *Oeneis chryxus*. *Oecologia* 88: 587–596.

Dal B (1978) *Fjärilar i naturen*, Vol. 1. Stockholm: Wahlström and Widstrand.

Dal B (1981) *Fjärilar i naturen*, Vol. 2. Stockholm: Wahlström and Widstrand.

Dall SRX, Giraldeau LA, Olsson O, McNamara JM, Stephens DW (2005) Information and its use by animals in evolutionary ecology. *Trends in Ecology and Evolution* 20: 187–193.

Danilevskii AS (1965) *Photoperiodism and Seasonal Development of Insects*, English edn. Edinburgh: Oliver and Boyd.

Danks HV (ed.) (1994) *Insect Life-Cycle Polymorphism: Theory, Evolution and Ecological Consequences for Seasonality and Diapause Control*. Dordrecht, Netherlands: Kluwer.

Dansgaard W, White JWC, Johnsen SJ (1989) The abrupt termination of the Younger Dryas event. *Nature* 339: 532–533.

Dapkus D (2004) Lepidoptera of a raised bog and adjacent forest in Lithuania. *European Journal of Entomology* 101: 63–67.

Dapporto L, Dennis RLH (2008a) Species' richness, rarity and endemicity on Italian offshore islands: complementary signals from island-focused and species-focused analyses. *Journal of Biogeography* 35: 664–674.

Dapporto L, Dennis RLH (2008b) Size is not the only consideration: identifying priorities for the conservation of butterflies on Italian offshore islands. *Journal of Insect Conservation* 12: 237–249.

Darwin C (1856) Letter to J. D. Hooker.

Darwin C (1859) *On the Origin of Species by Means of Natural Selection or the Preservation of Favored Races in the Struggle for Life*, 5th edn (1872). New York: Appleton and Co.

Darwin C (1871) *The Descent of Man, and Selection in Relation to Sex*. London: John Murray.

Dasmahapatra KK, Blum M, Aiello A, Hackwell S, Davies N, Bermingham EP, Mallet J (2002) Inferences from a rapidly moving hybrid zone. *Evolution* 56: 741–753.

David WAL, Gardiner BOC (1962) Oviposition and the hatching of eggs of *Pieris brassicae* in a laboratory culture. *Bulletin of Entomological Research* 53: 91–109.

Davies NB (1978) Territorial defence in the speckled wood butterfly (*Pararge aegeria*): the resident always wins. *Animal Behaviour* 26: 138–147.

Davies CR, Gilbert N (1985) A comparative study of the egg laying behaviour and larval development of *Pieris rapae* L and *P. brassicae* on the same host plants. *Oecologia* 67: 278–281.

Davies ZG, Wilson RJ, Brereton TM, Thomas CD (2005) The re-expansion and improving status of the silver-spotted skipper butterfly (*Hesperia comma*) in Britain: a metapopulation success story. *Biological Conservation* 124: 189–198.

Davies ZG, Wilson RJ, Coles S, Thomas CD (2006) Changing habitat associations of a thermally constrained species, the silver-spotted skipper butterfly, in response to climate warming. *Journal of Animal Ecology* 75: 247–256.

Davis BNK, Lakhani KH, Yates TJ (1991) The hazards of insecticides to butterflies of field margins. *Agriculture, Ecosystems and Environment* 36: 151–161.

Davis AJ, Jenkinson LS, Lawton JH, Shorrocks B, Wood S (1998a) Making mistakes when predicting shifts in species range in response to global warming. *Nature* 391: 783–786.

Davis AJ, Lawton JH, Shorrocks B, Jenkinson LS (1998b) Individualistic species responses invalidate simple physiological models of community dynamics under global environmental change. *Journal of Animal Ecology* 67: 600–612.

Day KR, Leather SR (1997) Threats to forestry by insects in Europe. In: Watt AD, Stork NE, Hunter MD (eds.) *Forests and Insects*, London: Chapman and Hall, pp. 177–205.

De Bast B (1985) La notion d'espèce dans le genre *Lysandra* Hemming, 1933 (Lepidoptera, Lycaenidae). *Linneana Belgica* 10: 98–110, 175–190, 194–208.

DeCarpentrie M. (1977) Communication [sur l'accouplement de *Lysandra albicans* m. x *Plebicula escheri* f.]. *Bulletin du Cercle des Lépidoptéristes de Belgique* 6: 86–87.

de Lajonquière Y. (1966) À propos de l'*Euphydryas desfontainii* Godart et de ses sous-espèces. *Alexanor* 4: 347–353.

de Lattin G (1949) Beiträge zur Zoogeographie des Mittelmeergebietes. *Verhandlungen der deutschen Zoologischen Gesellschaft, Kiel* 1948: 143–151.

de Lattin G (1957a) Die Ausbreitungszentren der holarktischen Landtierwelt. *Verhandlungen der deutschen Zoologischen Gesellschaft Hamburg* 1956: 380–410.

de Lattin G (1957b) Die Lepidopteren-Fauna der Pfalz. I. Teil A. *Mitteilungen der Pollichia, Pfälzer Verein für Naturkunde und Naturschutz, III Reihe,* 4: 51–167.

de Lattin G (1967) *Grundriss der Zoogeographie.* Stuttgart, Germany: Gustav Fischer.

de Lesse H (1952) Contribution à l'étude du genre *Erebia*: répartition de *E. tyndarus* Esp. et *E. cassioides* R et H.; dans la partie occidentale du Valais. *L'Entomologiste* 4: 68–70.

de Lesse H (1953) Formules chromosomiques nouvelles du genre *Erebia* (Lépid.: Rhopal.) et séparation d'une espèce méconnue. *Comptes Rendus de l'Académie des Sciences de Paris, Série III, Sciences de la Vie* 236: 630–632.

de Lesse H (1955a) Nouvelles formules chromosomiques dans le groupe d'*Erebia tyndarus* Esp. (Lépidoptères: Satyrinae). *Comptes Rendus de l'Académie des Sciences de Paris, Série III, Sciences de la Vie* 240: 347–349.

de Lesse H (1955b) Distribution holarctique d'un groupe d'espèces du genre *Erebia* (Lépidoptères) récemment séparées d'après leurs formules chromosomiques. *Comptes Rendus de la Société de Biogéographie* 276: 12–18.

de Lesse H (1955c) Une nouvelle formule chromosomique dans le groupe d'*Erebia tyndarus* Esp. (Lépidoptères: Satyrinae). *Comptes Rendus de l'Académie des Sciences de Paris, Série III, Sciences de la Vie* 241: 1505–1507.

de Lesse H (1960a) Les nombres de chromosomes dans la classification du groupe d'*Agrodiaetus rippartii* Freyer (Lepidoptera: Lycaenidae). *Revue française d'Entomologie* 27: 240–262.

de Lesse H (1960b) Spéciation et variation chromosomique chez les Lépidoptères Rhopalocères. *Annales de Sciences Naturelles, Zoologie, 12è série*, 2: 1–233.

de Lesse H (1961a) Cohabitation en Espagne d'*Agrodiaetus rippartii* Freyer et *A. fabressei* Oberthür. *Revue française d'Entomologie* 28: 50–53.

de Lesse H (1961b) Les nombres de chromosomes chez *Agrodiaetus dolus* et les espèces voisines. *Alexanor* 2: 57–63.

de Lesse H (1966) Variation chromosomique chez *Agrodiaetus dolus* Hübner (Lepidoptera: Lycaenidae). *Annales de la Société Entomologique de France* 5: 67–136.

de Lesse H (1969) Le nombre de chromosomes dans le groupe de *Lysandra coridon* Poda. *Annales de la Société Entomologique de France* 5: 469–522.

De Liñán C (ed.) (1998) *Entomología agroforestal: insectos y ácaros que dañan a los montes, cultivos, jardines e invernaderos.* Madrid: Ediciones Agrotécnicas.

de Puységur K (1947) Note sur un accouplement entre *Zerynthia polyxena-creusa* Meig. et *Z. rumina-medesicaste* Ill. *Revue française de Lépidoptérologie* 11: 10–15.

de Queiroz K (1998) The general lineage concept of species, species criteria, and the process of speciation: a conceptual unification and terminological recommendations. In:

Howard DJ, Berlocher SH (eds.) *Endless Forms: Species and Speciation.* New York: Oxford University Press, pp. 57–75.

de Queiroz K, Donoghue MJ (1988) Phylogenetic systematics and the species problem. *Cladistics* 4: 317–338.

de Snoo GR (1994) Unsprayed field margins on arable land. *Med. Fac. Landbouww. Univ. Gent*, 59/2b: 549–559.

de Snoo, GR, van der Poll RJ (1999) Effect of herbicide drift on adjacent boundary vegetation. *Agriculture, Ecosystems and Environment* 73: 1–6.

De Vries PJ, Kitching IJ, Vane-Wright RI (1985) The systematic position of *Antirrhea* and *Caerois*, with comments on the classification of the Nymphalidae (Lepidoptera). *Systematic Entomology* 10: 11–32.

De'ath G, Fabricius KE (2000) Classification and regression trees: a powerful yet simple technique for ecological data analysis. *Ecology* 81: 3178–3192.

Debat V, David P (2001) Mapping phenotypes: canalization, plasticity and developmental stability. *Trends in Ecology and Evolution* 16: 555–561.

DeChaine EG, Martin AP (2004) Historic cycles of fragmentation and expansion in *Parnassius smintheus* (Papilionidae) inferred using mitochondrial DNA. *Evolution* 58: 113–127.

Deinert EI (2003) Mate location and competition for mates in a pupal mating butterfly. In: Boggs CL, Watt WB, Ehrlich PR (eds.) *Butterflies: Ecology and Evolution Taking Flight*, Chicago, IL.: University of Chicago Press, pp. 91–108.

Deinert EI, Longino JT, Gilbert LE (1994) Mate competition in butterflies. *Nature* 370: 23–24.

Dell D, Burckhardt D (2004) *Psilomastax pyramidalis* (Hymenoptera, Ichneumonidae), ein Parasitoid von *Apatura iris* (Lepidoptera, Nymphalidae): Beobachtungen aus der Region Basel (CH, F) aus den Jahren 1982–2002. *Mitteilungen der entomologischen Gesellschaft, Basel* 54: 83–87.

Dell D, Sparks TH, Dennis RLH (2005) Climate change and the effect of increasing spring temperatures on emergence dates of the butterfly *Apatura iris* (Lepidoptera: Nymphalidae). *European Journal of Entomology* 102: 161–167.

Delmas S, Maechler J, Sibert JM (1999) Catalogue permanent de l'entomofaune française, *Vol. 2*, Lepidoptera: Rhopalocera. Dijon, France: Union de l'Entomologie Française.

Dempster JP (1967) The control of *Pieris rapae* with DDT. I. The natural mortality of the young stages of *Pieris. Journal of Applied Ecology* 4: 485–500.

Dempster JP (1968) The control of *Pieris rapae* with DDT. II. Survival of the young stages of *Pieris* after spraying. *Journal of Applied Ecology* 5: 451–462.

Dempster JP (1983) The natural control of populations of butterflies and moths. *Biological Reviews* **58**: 461–481.

Dempster JP (1984) The natural enemies of butterflies. In: Vane-Wright RI, Ackery PR (eds.) *The Biology of Butterflies*, London: Academic Press, pp. 97–104.

Dempster JP (1991) Fragmentation, isolation and mobility of insect populations. In: Collins NM, Thomas JA (eds.) *Conservation of Insects and their Habitats*, London: Academic Press, pp. 143–154.

Dempster JP (1992) Evidence of an oviposition-deterring pheromone in the orange-tip butterfly, *Anthocharis cardamines* (L.). *Ecological Entomology* **17**: 83–85.

Dempster JP (1995) The ecology and conservation of *Papilio machaon* in Britain. In: Pullin AS (ed.) *Ecology and Conservation of Butterflies*, London: Chapman and Hall, pp. 137–149.

Dempster JP (1997) The role of larval food resources and adult movement in the population dynamics of the orange-tip butterfly (*Anthocharis cardamines*). *Oecologia* **111**: 549–556.

Dempster JP, Hall ML (1980) An attempt at re-establishing the swallowtail butterfly at Wicken Fen. *Ecological Entomology* **5**: 327–334.

Dempster JP, Pollard E (1981) Fluctuations in resource availability and insect populations. *Oecologia* **50**: 412–416.

Dempster JP, King ML, Lakhani KH (1976) The status of the swallowtail butterfly in Britain. *Ecological Entomology* **1**: 71–84.

Dempster JP, McLean IFG (eds.) (1998) *Insect Populations in Theory and Practice*, 19th Symposium of the Royal Entomological Society. Dordrecht, Netherlands: Kluwer.

Den Otter CJ, Behan M, Maes FW (1980) Single-cell responses in female *Pieris brassicae* (Lepidoptera: Pieridae) to plant volatiles and conspecific egg odours. *Journal of Insect Physiology* **26**: 465–472.

Dennis RLH (1977) *The British Butterflies: Their Origin and Establishment*. Faringdon, Oxon.: EW Classey.

Dennis RLH (1982a) Mate location strategies in the wall brown butterfly, *Lasiommata megera* (L.) (Lepidoptera: Satyridae): wait or seek? *Entomological Record and Journal of Variation* **94–95**: 209–214, 7–10.

Dennis RLH (1982b) Patrolling behaviour in orange tip butterflies within the Bollin Valley in north Cheshire, and a comparison with other pierids. *Vasculum* **67**: 17–25.

Dennis RLH (1983a) Hierarchy and pattern in the spatial responses of ovipositing *Anthocharis cardamines* (Lep.). *Vasculum* **68**: 27–43.

Dennis, RLH (1983b) The 'edge effect' in butterfly oviposition: a simple calculus for an insect in a hurry. *Entomologist's Gazette* **34**: 5–8.

Dennis RLH (1983c) Egg-laying cues in the wall brown butterfly, *Lasiommata megera* (L.) (Lepidoptera: Satyridae). *Entomologist's Gazette* **34**: 89–95.

Dennis RLH (1984) The edge effect in butterfly oviposition: batch siting in *Aglais urticae* (L.) (Lepidotera: Nymphalidae). *Entomologist's Gazette* **35**: 157–173.

Dennis RLH (1985) Small plants attract attention! Choice of egg-laying sites in the green-veined white butterfly (*Artogeia napi*) (L.) (Lep.: Pieridae). *Bulletin of the Amateur Entomologists Society* **44**: 77–82.

Dennis RLH (ed.) (1992a) *The Ecology of Butterflies in Britain*. Oxford: Oxford University Press.

Dennis RLH (1992b) Islands, ranges, and gradients. In: Dennis RHL (ed.) *The Ecology of Butterflies in Britain*, Oxford: Oxford University Press, pp. 1–21.

Dennis RLH (1993) *Butterflies and Climate Change*. Manchester: Manchester University Press.

Dennis RLH (1995) *Euchloe ausonia* (Hübner) (Lepidoptera: Pieridae) oviposition on *Brassica nigra* (L.) Koch (Cruciferae): big immature plants are preferred. *Entomologist's Gazette* **46**: 253–255.

Dennis RLH (1997) An inflated conservation load for European butterflies: increases in rarity and endemism accompany increases in species richness. *Journal of Insect Conservation* **1**: 43–62.

Dennis RLH (2000) The comparative influence of source population size and migration capacity on the persistence of butterfly species on a small offshore island. *Entomologist's Gazette* **51**: 39–52.

Dennis RLH (2004) Butterfly habitats, broad scale biotope affiliations and structural exploitation of vegetation at finer scales: the matrix revisited. *Ecological Entomology* **29**: 744–752.

Dennis RLH, Bardell P (1996) The impact of extreme weather events on local populations of *Hipparchia semele* (L.) (Nymphalidae) and *Plebejus argus* (L.) (Lycaenidae): hindsight, inference and lost opportunities. *Entomologist's Gazette* **47**: 211–225.

Dennis RLH, Bramley MJ (1985) The influence of man and climate on dispersion patterns within a population of adult *Lasiommata megera* (L.) (Satyridae) at Brereton Heath, Cheshire (U.K.). *Nota lepidopterologica* **8**: 309–324.

Dennis RLH, Eales HT (1997) Patch occupancy in *Coenonympha tullia* (Müller, 1764) (Lepidoptera: Satyrinae): habitat quality matters as much as patch size and isolation. *Journal of Insect Conservaton* **1**: 167–176.

Dennis RLH, Eales HT (1999) Probability of site occupancy in the large heath butterfly *Coenonympha tullia* determined from geographical and ecological data. *Biological Conservation* **87**: 295–302.

Dennis RLH, Shreeve TG (1988) Hostplant-habitat structure and the evolution of butterfly mate-locating behaviour. *Zoological Journal of the Linnean Society* **94**: 301–318.

Dennis RLH, Shreeve TG (1989) Butterfly wing morphology variation in the British Isles: the influence of climate, behavioral posture and the hostplant-habitat. *Biological Journal of the Linnean Society* **38**: 323–348.

Dennis RLH, Shreeve TG (1991) Climatic change and the British butterfly fauna: opportunities and constraints. *Biological Conservation* **55**: 1–16.

Dennis RLH, Shreeve TG (1996) *Butterflies on British and Irish Offshore Islands: Ecology and Biogeography*. Wallingford, Oxon.: Gem Publishing Co.

Dennis RLH, Shreeve TG (1997) Diversity of butterflies on British islands: ecological influences underlying the roles of area, isolation and the size of faunal source. *Biological Journal of the Linnean Society* **60**: 257–275.

Dennis RLH, Shreeve TG (2003) Gains and losses of French butterflies: tests of predictions, under-recording and regional extinction from data in a new atlas. *Biological Conservation* **110**: 131–139.

Dennis RLH, Sparks TH (2005) Landscape resources for the territorial nymphlid butterfly *Inachis io*: microsite landform selection and behavioral responses to environmental conditions. *Journal of Insect Behavior* **18**: 725–742.

Dennis RLH, Sparks TH (2006) When is a habitat not a habitat? Dramatic resource use changes under differing weather conditions for the butterfly *Plebejus argus*. *Biological Conservation* **129**: 291–301.

Dennis RLH, Williams WR (1986) Butterfly 'diversity': regressing and a little latitude. *Antenna* **10**: 108–111.

Dennis RLH, Williams WR (1987) Mate location behaviour of the large skipper butterfly *Ochlodes venata*: flexible strategies and spatial components. *Journal of the Lepidopterists' Society* **41**: 45–64.

Dennis RLH, Williams WR (1995) Implications of biogeographical structures for the conservation of European butterflies. In: Pullin AS (ed.) *Ecology and Conservation of Butterflies*, London: Chapman and Hall, pp. 213–247.

Dennis RLH, Porter K, Williams WR (1984) Ocellation in *Coenonympha tullia* (Müller) (Lepidoptera: Satyridae). I. Structures in correlation matrices. *Nota lepidopterologica* **7**: 199–219.

Dennis RLH, Porter K, Williams WR (1986) Ocellation in *Coenonympha tullia* (Muller) (Lepidoptera. Satryridae). II. Population differentiation and clinal variation in the

context of climatically-induced anti-predator defence strategies. *Entomologist's Gazette* **37**: 133–172.

Dennis RLH, Williams WR, Shreeve TG (1991) A multivariate approach to the determination of faunal structures among European butterfly species (Lepidoptera: Rhopalocera). *Zoological Journal of the Linnean Society* **101**: 1–49.

Dennis RLH, Shreeve TG, Williams WR (1995a) Affinity gradients among European butterflies: evidence for an historical component to species distributions. *Entomologist's Gazette* **46**: 141–153.

Dennis RLH, Shreeve TG, Williams WR (1995b) Taxonomic differentiation in species richness gradients among European butterflies (Papilionoidea, Hesperioidea): contribution of macroevolutionary dynamics. *Ecography* **18**: 27–40.

Dennis RLH, Shreeve TG, Sparks TH (1998a) The effects of island area, isolation and the source population size on the presence of the grayling butterfly *Hipparchia semele* (L.) (Lepidoptera: Satyrinae) on British and Irish offshore islands. *Biodiversity and Conservation* **7**: 765–776.

Dennis RLH, Williams WR, Shreeve TG (1998b) Faunal structures among European butterflies: evolutionary implications of bias for geography, endemism and taxonomic affiliation. *Ecography* **21**: 181–203.

Dennis RLH, Shreeve TG, Olivier A, Coutsis J (2000a) Contemporary geography dominates butterfly diversity gradients within the Aegean archipelago (Lepidoptera: Papilionoidea, Hesperioidea). *Journal of Biogeography* **27**: 1365–1383.

Dennis RLH, Donato B, Sparks TH, Pollard E (2000b) Ecological correlates of island incidence and geographic range among British butterflies. *Biodiversity and Conservation* **9**: 343–359.

Dennis RLH, Olivier A, Coutsis JG, Shreeve TG (2001) Butterflies on islands in the Aegean archipelago: predicting numbers of species and incidence of species using geographical variables. *Entomologist's Gazette* **52**: 3–39.

Dennis RLH, Shreeve TG, Sparks TH, Lhonore JE (2002) A comparison of geographical and neighbourhood models for improving atlas databases: the case of the French butterfly atlas. *Biological Conservation* **108**: 143–159.

Dennis RLH, Shreeve TG, Van Dyck H (2003) Towards a resource-based concept for habitat: a butterfly biology viewpoint. *Oikos* **102**: 417–426.

Dennis RLH, Hodgson JG, Grenyer R, Shreeve TG, Roy DB (2004) Host plants and butterfly biology: do host-plant strategies drive butterfly status? *Ecological Entomology* **29**: 12–26.

Dennis RLH, Shreeve TG, Van Dyck H (2006) Habitats and resources: the need for a resource-based definition to conserve butterflies. *Biodiversity and Conservation* **15**: 1943–1966.

Dennis RLH, Shreeve TG, Sheppard DA (2007) Species conservation and landscape management: a habitat perspective. In: New T, Stewart A, Lewis O (eds.) *Insect Conservation Biology*, RES symposium volume, London: Royal Entomological Society, pp. 92–126.

Dennis RLH, Dapporto L, Shreeve TG, John E, Coutsis JG, Kudrna O, Saarinen K, Ryrholm N, Williams WR (2008a) Butterflies of European islands: the implications of the geography and ecology of rarity and endemicity for conservation. *Journal of Insect Conservation* **12**: 205–236.

Dennis RLH, Hardy PB, Shreeve TG (2008b) The importance of resource databanks for conserving insects: a butterfly biology perspective. *Journal of Insect Conservation* **12**: 711–719.

Deschamps-Cottin M, Lemperière G, Descimon H (1999) Bilan sur le suivi des populations de *Parnassius apollo* L. en France (Lepidoptera: Papilionidae). *Annales de la Société Entomologique de France* **35** (Suppl.): 281–289.

Deschamps-Cottin M, Aubert J, Barascud B, Descimon H (2000) Hybridation et introgression entre "bonnes espèces": le cas de *Parnassius apollo* et *P. phoebus*. *Comptes Rendus de l'Académie des Sciences de Paris, Série III, Sciences de la Vie* **323**: 287–297.

Descimon H (1957) Nouvelles données sur la répartition d'*Erebia cassioides* Reiner et Hohenwarth et d'*E. hispania* Butler dans les Pyrénées centrales et occidentales. *Revue française de Lépidoptérologie* **16**: 48–59.

Descimon H (1963) Nouvelles captures d'*Erebia serotina*. *Alexanor* **3**: 71–80.

Descimon H (1980) *Heodes tityrus tityrus* Poda et *H. tityrus subalpina* Speyer (Lycaenidae): un problème de spéciation en milieu alpin. *Nota lepidopterologica* **2**: 123–125.

Descimon H (1989) La variation géographique du polymorphisme chez les Lépidoptères: femelles bleues et femelles brunes chez *Lysandra coridon* Poda dans le sud-ouest de l'Europe (Lepidoptera: Lycaenidae). *Alexanor* **16**: 23–41.

Descimon H (1995) La conservation des *Parnassius* en France: aspects zoogéographiques, écologiques, démographiques et génétiques. *Rapports d'études de l'OPIE* **1**: 1–54.

Descimon H (2004) *Erebia serotina*: retour aux sources. *Alexanor* **22**: 451–465.

Descimon H, de Lesse H (1953) Découverte d'un nouvel *Erebia* dans les Hautes Pyrénées. *Revue française de Lépidoptérologie* **14**: 119–122.

Descimon H, de Lesse H (1954) Nouvelle note sur *Erebia serotina*. *Revue française de Lépidoptérologie* **14**: 237–241.

Descimon H, Geiger H (1988) Electrophoretic detection of interspecific hybrids in *Parnassius* (Lepidoptera: Papilionidae). *Génétique, Sélection et Evolution* **20**: 435–440.

Descimon H, Michel F (1989) Expériences d'hybridation intra- et interspécifiques dans le genre *Zerynthia* (Papilionidae): relativité des critères mixiologiques de l'espèce. *Nota lepidopterologica* **12** (Suppl. 1): 28–31.

Descimon H, Napolitano M (1993) Enzyme polymorphism, wing pattern variability, and geographical isolation in an endangered butterfly species. *Biological Conservation* **66**: 117–123.

Descimon H, Renon C (1975) Mélanisme et facteurs climatiques. II. Corrélation entre la mélanisation et certains facteurs climatiques chez *Melanargia galathea* (Linné) (Lepidoptera: Satyridae) en France. *Archives de Zoologie Expérimentale et Générale* **116**: 437–468.

Descimon H, Genty F, Vesco JP (1989) L'hybridation naturelle entre *Parnassius apollo* (L.) et *P. phoebus* (F.) dans les Alpes du sud (Lepidoptera: Papilionidae). *Annales de la Société Entomologique de France, N.S.*, **25**: 209–234.

Descimon H, Zimmermann M, Cosson E, Barascud B, Nève G (2001) Diversité génétique, variation géographique et flux géniques chez quelques Lépidoptères Rhopalocères français. *Genetics Selection Evolution* **33** (Suppl. 1): 223–249.

Dethier VG (1947) *Chemical Insect Attractants and Repellents*. New York: McGraw-Hill.

Dethier VG (1954) Evolution of feeding preference in phytophagous insects. *Evolution* **8**: 33–54.

Dethier VG (1959) Food-plant distribution and density and larval dispersal as factors affecting insect populations. *Canadian Entomologist* **91**: 581–596.

Dethier VG, MacArthur RH (1964) Field capacity to support butterfly population. *Nature* **201**: 728–729.

DeVries P (1979) Pollen-feeding rainforest *Parides* and *Battus* butterflies in Costa Rica. *Biotropica* **11**: 237–238.

DeVries P (1987) *The Butterflies of Costa Rica and Their Natural History. Papilionidae, Pieridae, Nymphalidae*. Princeton, NJ: Princeton University Press.

DeVries PJ, Penz CM (2000) Entomophagy, behavior, and elongated thoracic legs in the myrmecophilous neotropical butterfly *Alesa amesis* (Riodinidae). *Biotropica* **32**: 712–721.

Dickinson JL, Rutowski RL (1989) The function of the mating plug in the chalcedon checkerspot butterfly. *Animal Behaviour* **38**: 154–162.

Digby PSB (1955) Factors affecting the temperature excess of insects in sunshine. *Journal of Experimental Biology* **32**: 279–298.

Dingle H (1996) *Migration: The Biology of Life on the Move*. New York: Oxford University Press.

Dingle H (2001) The evolution of migratory syndromes in insects. In: Woiwod IP, Reynolds DR, Thomas CD (eds.) *Insect Movement: Mechanisms and Consequences*, Proceedings of the Royal Entomological Society's 20th Symposium, Wallingford, Oxon.: CABI Publishing, pp. 159–181.

Diniz-Filho JAF, Bini LM, Hawkins BA (2003) Spatial autocorrelation and red herrings in geographical ecology. *Global Ecology and Biogeography* 12: 53–64.

Doak P, Kareiva P, Kingsolver JG (2006) Fitness consequences of choosy oviposition for a time-limited butterfly. *Ecology* 87: 395–408.

Dobkin DS, Olivieri I, Ehrlich PR (1987) Rainfall and the interaction of microclimate with larval resources in the population dynamics of checkerspot butterflies (*Euphydryas editha*) inhabiting serpentine grassland. *Oecologia* 71: 161–166.

Dobzhansky T (1937) *Genetics and the Origin of Species.* New York: Columbia University Press.

Dobzhansky T (1950) Mendelian populations and their evolution. *American Naturalist* 84: 401–418.

Dodd DMB (1989) Reproductive isolation as a consequence of divergence in *Drosophila pseudoobscura*. *Evolution* 43: 1308–1311.

Dolek M, Geyer A (1997) Influence of management on butterflies of rare grassland ecosystems in Germany. *Journal of Insect Conservation* 1: 125–130.

Dolek M, Geyer A (2001) Der Violette Feuerfalter (*Lycaena alciphron* Rottemburg, 1775): Artenhilfsprogramm für einen wenig bekannten Tagfalter. *Schriftenreihe des bayerischen Landesamtes für Umweltschutz* 156: 341–354.

Dolek M, Geyer A (2002) Conserving biodiversity on calcareous grasslands in the Franconian Jura by grazing: a comprehensive approach. *Biological Conservation* 104: 351–360.

Dolek M, Geyer A, Bolz R (1998) Distribution of *Maculinea rebeli* and hostplant use on sites along the river Danube. *Journal of Insect Conservation* 2: 85–89.

Doncaster CP (2000) Extension of ideal free resource use to breeding populations and metapopulations. *Oikos* 89: 24–36.

Dormann CF, McPherson JM, Araújo MB, Bivand R, Bolliger J, Carl G, Davies R, Hirzel A, Jetz W, Kissling WD, Kühn I, Ohlemüller R, Peres-Neto PR, Reineking B, Schröder B, Schurr FM, Wilson R (2007) Incorporating spatial autocorrelation in the analysis of ecological species distribution data: a user's guide. *Ecography* 30: 609–628.

Douwes P (1975) Territorial behaviour in *Heodes virgaureae* L. (Lep., Lycaenidae) with particular reference to visual stimuli. *Norwegian Journal of Entomology* 22: 143–154.

Douwes P (1976a) Activity in *Heodes virgaureae* (Lep., Lycaenidae) in relation to air temperature, solar radiation, and time of day. *Oecologia* 22: 287–298.

Douwes P (1976b) Distribution of a population of the butterfly *Heodes virgaureae*. *Oikos* 26: 332–340.

Douwes P (1978) Adult feeding in the scarce copper, *Heodes virgaureae* L. (Lep., Lycaenidae). *Entomologisk Tidskrift* 99: 1–10.

Douwes P (1980) Periodical appearance of species in the genera *Oeneis* and *Erebia* in Fennoscandia (Lep: Satyridae). *Entomologia Generalis* 6: 151–157.

Dover JW (1989) The use of flowers by butterflies foraging in cereal field margins. *Entomologist's Gazette* 40: 283–291.

Dover JW (1990) Butterflies and wildlife corridors. In: Nodder C (ed.) *The Game Conservancy Review of 1989.* Fordingbridge, Hants.: The Game Conservancy, pp. 62–64.

Dover JW (1991) The conservation of insects on arable farmland. In: Collins NM, Thomas JA (eds.) *The Conservation of Insects and their Habitats*, London: Academic Press, pp. 294–318.

Dover JW (1996) Factors affecting the distribution of satyrid butterflies on arable farmland. *Journal of Applied Ecology* 33: 723–734.

Dover JW (1997) Conservation headlands: effects on butterfly distribution and behaviour. *Agriculture, Ecosystems and Environment* 63: 31–49.

Dover JW, Fry GLA (2001) Experimental simulation of some visual and physical components of a hedge and the effects on butterfly behaviour in an agricultural landscape. *Entomologia Experimentalis et Applicata* 100: 221–233.

Dover J, Settele J (2009) The influences of landscape structure on butterfly distribution and movement: a review. *Journal of Insect Conservation* 13: 3–27.

Dover J, Sparks T (2000) A review of the ecology of butterflies in British hedgerows. *Journal of Environmental Management* 60: 51–63.

Dover JW, Sparks TH (2001) Green lanes: biodiversity reservoirs in farmland? In: Barr CJ, Petit S (eds.) *Hedgerows of the World: Their Ecological Functions in Different Landscapes.* Cheshire: Lymm, pp. 241–250.

Dover J, Sotherton N, Gobbett K (1990) Reduced pesticide inputs on cereal field margins: the effects on butterfly abundance. *Ecological Entomology* 15: 17–24.

Dover JW, Clarke SA, Rew L (1992) Habitats and movement patterns of satyrid butterflies (Lepidoptera: Satyridae) on arable farmland. *Entomologist's Gazette* 43: 29–44.

Dover JW, Sparks TH, Greatorex-Davies JN (1997) The importance of shelter for butterflies in open landscapes. *Journal of Insect Conservation* 1: 89–97.

Dover JW, Sparks T, Clarke S, Gobbett K, Glossop S (2000) Linear features and butterflies: the importance of green lanes. *Agriculture, Ecosystems and Environment* **80**: 227–242.

Dowd PF, Smith CM, Sparks TC (1983) Detoxification of plant toxins by insects. *Insect Biochemistry* **13**: 453–468.

Dowdeswell WH (1981) *The Life of the Meadow Brown*. London: Heinemann.

Dowdeswell WH, Ford EB (1953) The influence of isolation on variability in the butterfly *Maniolia jurtina* L. *Symposia of the Society for Experimental Biology* **7**: 253–273.

Drake VA, Gatehouse AG (1995) *Insect Migration: Tracking Resources through Space and Time*. Cambridge: Cambridge University Press.

Dreisig H (1995) Thermoregulation and flight activity in territorial male graylings, *Hipparchia semele* (Satyridae), and large skippers, *Ochlodes venata* (Hesperiidae). *Oecologia* **101**: 169–176.

Drès M, Mallet J (2002) Host races in plant-feeding insects and their importance in sympatric speciation. *Philosophical Transactions of the Royal Society of London B* **357**: 471–492.

Dronamraju KR (1960) Selective visits of butterflies to flowers: a possible factor of sympatric speciation. *Nature* **186**: 178.

Drummond BA (1984) Multiple mating and sperm competition in the Lepidoptera. In: Smith RL (ed.) *Sperm Competition and the Evolution of Animal Mating Systems*, Orlando, FL: Academic Press, pp. 291–370.

Dubatolov VV, Kosterin OE (2000) Nemoral species of Lepidoptera (Insecta) in Siberia: a novel view on their history and the timing of their range disjunctions. *Entomologica Fennica* **11**: 141–166.

Dudley R (1991) Thermoregulation in unpalatable danaine butterflies. *Functional Ecology* **5**: 503–506.

Du Merle P (1964) Cycle biologique d'un Diptère du genre *Villa*. *Comptes Rendus Hebdomadaires des Séances de L'Académie des Sciences* **259**: 1657–1659.

Du Merle P (1979a) Biologie de la larve planidium de *Villa brunnea* Beck., Diptère Bombyliide parasite de la processionnaire du pin. I. Recherche et découverte de l'hôte. *Annales de Zoologie – Ecologie Animale* **11**: 289–304.

Du Merle P (1979b) Biologie de la larve planidium de *Villa brunnea* Beck., Diptère Bombyliide parasite de la processionnaire du pin. II. Infestation de l'hôte. *Annales de Zoologie – Ecologie Animale* **11**: 305–326.

Du Merle P (1979c) Biologie de la larve planidium de *Villa brunnea*, Diptère Bombyliidae parasite de la processionnaire du pin. III. Le début de la vie endoparasitaire du planidium, les réactions de l'hôte et les échecs du parasitisme. *Annales de la Société de Entomologique de France* **15**: 525–551.

Dunlap-Pianka HL (1979) Ovarian dynamics in *Heliconius* butterflies: correlations among daily oviposition rates, egg weights, and quantitative aspects of oögenesis. *Journal of Insect Physiology* **25**: 741–749.

Dunlap-Pianka H, Boggs CL, Gilbert LE (1977) Ovarian dynamics in heliconiine butterflies: programmed senescence versus eternal youth. *Science* **197**: 487–490.

Dunn ER (1922) A suggestion to zoogeographers. *Science* **56**: 336–338.

Dussourd DE (1993) Foraging with finesse: caterpillar adaptations for circumventing plant defenses. In: Stamp NE, Casey TM (eds.) *Caterpillars: Ecological and Evolutionary Constraints on Foraging*, New York: Chapman and Hall, pp. 92–131.

Dutreix C (1988) *Le peuplement des Lépidoptères de la Bourgogne*. Autun, France: Société d'Histoire Naturelle d'Autun.

Dynesius M, Jansson R (2000) Evolutionary consequences of changes in species' geographical distributions driven by Milankovitch climate oscilations. *Proceedings of the National Academy of Sciences, USA* **97**: 9115–9120.

Dzwonko Z, Gawronski S (2002) Effect of litter removal on species richness and acidification of a mixed oak–pine woodland. *Biological Conservartion* **106**: 389–398.

Eaton MD, Lanyon SM (2003) The ubiquity of avian ultraviolet plumage reflectance. *Proceedings of the Royal Society of London B* **270**: 1721–1726.

Eberhard WG (1996) *Female Control: Sexual Selection by Cryptic Female Choice*. Princeton, NJ: Princeton University Press.

Eberhard WG (1997) Sexual selection by cryptic female choice in insects and arachnids. In: Choe JC, Crespi BJ (eds.) *Mating Systems in Insects and Arachnids*, Cambridge: Cambridge University Press, pp. 32–57.

Eberhard WG, Cordero C (1995) Sexual selection by cryptic female choice on male seminal products: a new bridge between sexual selection and reproductive physiology. *Trends in Ecology and Evolution* **10**: 493–496.

Ebert G, Rennwald E (1991a) *Die Schmetterlinge Baden-Württembergs*, Vol. 1, *Tagfalter I*. Stuttgart, Germany: Eugen Ulmer.

Ebert G, Rennwald E (1991b) *Die Schmetterlinge Baden-Württembergs*, Vol. 2, *Tagfalter II*. Stuttgart, Germany: Eugen Ulmer.

Eckweiler W, Häuser CL (1997) An illustrated checklist of *Agrodiaetus* Hübner, 1822, a subgenus of *Polyommatus* Latreille, 1804 (Lepidoptera: Lycaenidae). *Nachrichten des entomologischen Vereins Apollo* **16** (Suppl.): 113–168.

Edwards A (1992) *Likelihood*. Baltimore, MD: Johns Hopkins University Press.

Efron B, Tibshirani RJ (1991) Statistical data analysis in the computer age. *Science* **253**: 390–395.

Efron B, Tibshirani RJ (1993) *An Introduction to the Bootstrap*. New York: Chapman and Hall.

Eggleton P, Gaston KJ (1990) 'Parasitoid' species and assemblages: convenient definitions or misleading compromises? *Oikos* **59**: 417–421.

Eggleton P, Gaston KJ (1992) Tachinid host ranges: a reappraisal (Diptera: Tachinidae). *Entomologist's Gazette* **43**: 139–143.

Ehrlich PR (1961) Has the biological species concept outlived its usefulness? *Systematic Zoology* **10**: 167–176.

Ehrlich PR (1984) The structure and dynamics of butterfly populations. In: Vane-Wright RI, Ackery PR (eds.) *The Biology of Butterflies*, Princeton, NJ: Princeton University Press, pp. 25–40.

Ehrlich PR (1992) Population biology of checkerspot butterflies and the preservation of global biodiversity. *Oikos* **63**: 6–12.

Ehrlich AH, Ehrlich PR (1978) Reproductive strategies in the butterflies. I. Mating frequency, plugging, and egg number. *Journal of the Kansas Entomological Society* **51**: 666–697.

Ehrlich PR, Hanski I (eds.) (2004) *On the Wings of Checkerspots: A Model System for Population Biology*. Oxford: Oxford University Press.

Ehrlich PR, Murphy DD (1981) The population biology of checkerspot butterflies (*Euphydryas*). *Biologisches Zentralblatt* **100**: 613–629.

Ehrlich PR, Murphy DD (1983) Butterflies and biospecies. *Journal of Research on the Lepidoptera* **21**: 219–225.

Ehrlich PR, Murphy DD (1984) On butterfly taxonomy. *Journal of Research on the Lepidoptera* **23**: 19–34.

Ehrlich PR, Murphy DD (1987) Conservation lessons from long-term studies of checkerspot butterflies. *Conservation Biology* **1**: 122–131.

Ehrlich PR, Murphy DD (1988) Plant chemistry and host range in insect herbivores. *Ecology* **69**: 908–909.

Ehrlich PR, Raven PH (1964) Butterflies and plants: a study in coevolution. *Evolution* **18**: 586–608.

Ehrlich PR, Raven PH (1969) Differentiation of populations. *Science* **165**: 1228–1231.

Ehrlich PR, Wheye D (1986) 'Nonadaptive' hilltopping behavior in male checkerspot butterflies (*Euphydryas editha*). *American Naturalist* **127**: 477–483.

Ehrlich PR, Wheye D (1988) Hilltopping butterflies revisited. *American Naturalist* **132**: 460–461.

Ehrlich PR, White RR, Singer MC, McKechnie SW, Gilbert LE (1975) Checkerspot butterflies: a historical perspective. *Science* **188**: 221–228.

Ehrlich PR, Murphy DD, Singer MC, Sherwood CB, White RR, Brown IL (1980) Extinction, reduction, stability and increase: the responses of checkerspot butterfly (*Euphydryas*) populations to the California drought. *Oecologia* **46**: 101–105.

Eichel S, Fartmann T (2008) Management of calcareous grasslands for Nickerl's fritillary (*Melitaea aurelia*) has to consider habitat requirements of the immature stages, isolation, and patch area. *Journal of Insect Conservation* **12**: 677–688.

Eisner C (1966) Parnassiidae-Typen in der Sammlung JC Eisner. *Zoologische Verhandelingen* **81**: 1–190.

Elenga H, Peyron O, Bonnefille R, Jolly D, Cheddadi R, Guiot J, Andrieu V, Bottema S, Buchet G, de Beaulieu J-L, Hamilton AC, Maley J, Marchant R, Perez-Obiol R, Reille M, Riollet G, Scott L, Straka H, Taylor D, Van Campo E, Vincens A, Laarif F, Jonson H (2000) Pollen-based biome reconstruction for southern Europe and Africa 18 000 yr BP. *Journal of Biogeography* **27**: 621–634.

Elgar MA, Pierce NE (1988) Mating success and fecundity in an ant-tended lycaenid butterfly. In: Clutton-Brock TH (ed.) *Reproductive Success*, Chicago, IL: University of Chicago Press, pp. 59–75.

Eliasson CU, Shaw MR (2003) Prolonged life cycles, oviposition sites, foodplants and *Cotesia* parasitoids of Melitaeini butterflies in Sweden. *Oedippus* **21**: 1–52.

Eliasson CU, Ryrholm N, Gardenfors U (2005) *Nationalnyckeln till Sveriges Flora och Fauna*. Fjarilar, Sweden: Dagfjarilar, Pensoft and Stenstrup.

Elith J, Graham CH, Anderson RP, Dudik M, Ferrier S, Guisan A, Hijmans RJ, Huettmann F, Leathwick JR, Lehmann A, Li J, Lohmann LG, Loiselle BA, Manion G, Moritz C, Nakamura M, Nakazawa Y, Overton JMM, Peterson AT, Phillips SJ, Richardson K, Scachetti-Pereira R, Schapire RE, Soberon J, Williams S, Wisz MS, Zimmermann NE (2006) Novel methods improve prediction of species' distributions from occurrence data. *Ecography* **29**: 129–151.

Ellegren H (2008) Sequencing goes 454 and takes large-scale genomics into the wild. *Molecular Ecology* **17**: 1636–1647.

Ellers J, Boggs CL (2002) The evolution of wing color in *Colias* butterflies: heritability, sex linkage, and population divergence. *Evolution* **56**: 836–840.

Ellers J, Boggs CL (2003) The evolution of wing color: male mate choice opposes adaptive wing color divergence in *Colias* butterflies. *Evolution* **57**: 1100–1106.

Ellers J, Boggs CL (2004a) Functional ecological implications of intraspecific differences in wing melanization in *Colias* butterflies. *Biological Journal of the Linnean Society* **82**: 79–87.

Ellers J, Boggs CL (2004b) Evolutionary genetics of dorsal wing colour in *Colias* butterflies. *Journal of Evolutionary Biology* **17**: 752–758.

Elligsen H, Beinlich B, Plachter H (1997) Effects of large-scale cattle grazing on populations of *Coenonympha glycerion* and *Lasiommata megera* (Lepidoptera: Satyridae). *Journal of Insect Conservation* **1**: 13–23(11).

Elligsen H, Beinlich B, Plachter H (1998) Large-scale grazing systems and species protection in the eastern Carpathians of Ukraine. *La Cañada* **9**: 10–12.

Ellis J (ed.) (2005) *Register of West Midlands Brownfield Sites with High Lepidopteran Interest*. Wareham, Dorset: Butterfly Conservation.

Elmes GW, Free A (1994) *Climate Change and Rare Species*. London: HMSO.

Elmes GW, Thomas JA, Wardlaw JC (1991a) Larvae of *Maculinea rebeli*, a large-blue butterfly, and their *Myrmica* host ants: wild adoption and behaviour in ant-nests. *Journal of Zoology* **223**: 447–460.

Elmes GW, Wardlaw JC, Thomas JA (1991b) Larvae of *Maculinea rebeli*, a large-blue butterfly, and their *Myrmica* host ants: patterns of caterpillar growth and survival. *Journal of Zoology* **224**: 79–92.

Elmes GW, Clarke RT, Thomas JA, Hochberg ME (1996) Empirical tests of specific predictions made from a spatial model of the population dynamics of *Maculinea rebeli*, a parasitic butterfly of red ant colonies. *Acta Oecologica* **17**: 61–80.

Emelianov I, Simpson F, Narang P, Mallet J (2003) Host choice promotes reproductive isolation between host races of the larch budmoth *Zeiraphera diniana*. *Journal of Evolutionary Biology* **16**: 208–218.

Emelianov I, Marec F, Mallet J (2004) Genomic evidence for divergence with gene flow in host races of the larch bud moth. *Proceedings of the Royal Society of London B* **271**: 97–105.

Emmet AM, Heath J (eds.) (1989) *The Moths and Butterflies of Great Britain and Ireland*, Vol. 7, Part 1, *Hesperiidae to Nymphalidae*. Colchester, Essex: Harley Books.

Emmet AM, Heath J (1990) *The Butterflies of Great Britain and Ireland*. Colchester, Essex: Harley Books.

Endler JA (1977) *Geographic Variation, Speciation and Clines*. Princeton, NJ: Princeton University Press.

Endler JA (1979) Gene flow and life history patterns. *Genetics* **93**: 263–284.

Endler JA (1984) Progressive background matching in moths, and a quantitative measure of crypsis. *Biological Journal of the Linnean Society* **22**: 187–231.

Endler JA (1986) *Natural Selection in the Wild*. Princeton, NJ: Princeton University Press.

Enfjall K, Leimar O (2005) Density-dependent dispersal in the Glanville fritillary, *Melitaea cinxia*. *Oikos* **108**: 465–472.

Epperson BK (2003) *Geographical Genetics*. Princeton, NJ: Princeton University Press.

Erhardt A (1985) Diurnal Lepidoptera: sensitive indicators of cultivated and abandoned grassland. *Journal of Applied Ecology* **22**: 849–861.

Erhardt A (1988) Pollination and reproduction in *Dianthus silvester* Wulf. In: Cresti M, Gori P, Pacini E (eds.) *Sexual Reproduction in Higher Plants*, Proceedings of the 10th Internation Symposium on the Sexual Reproduction of Higher Plants, Berlin, Germany: Springer-Verlag, pp. 351–356.

Erhardt A (1991) Flower preferences, nectar preferences and pollination effect of Lepidoptera. *Advances in Ecology* **1**: 239–260.

Erhardt A (1992) Impact of grassland management on diurnal Lepidoptera in the Swiss Central Alps. In: Pavlicek-van Beek T, Ovaa AH, van der Made JG (eds.) *Future of Butterflies in Europe: Strategies for Survival*, Wageningen, Netherlands: Agricultural University of Wageningen, pp. 146–155.

Erhardt A (1995) Ecology and conservation of alpine Lepidoptera. In: Pullin, AS (ed.) *Ecology and Conservation of Butterflies*, London: Chapman and Hall, pp. 258–276.

Erhardt A, Baker I (1990) Pollen amino acids: an additional diet for a nectar feeding butterfly? *Plant Systematics and Evolution* **169**: 111–121.

Erhardt A, Rusterholz HP (1998) Do peacock butterflies (*Inachis io* L.) detect and prefer amino acids and other nitrogenous compounds? *Oecologia* **117**: 536–542.

Erhardt A, Thomas JA (1991) Lepidoptera as indicators of change in the semi-natural grasslands of lowland and upland Europe. In: Collins NM, Thomas JA (eds.) *The Conservation of Insects and their Habitats*, London: Academic Press, pp. 213–236.

Erhardt A, Rusterholz HP, Stöcklin J (2005) Elevated carbon dioxide increases nectar production in *Epilobium angustifolium* L. *Oecologia* **146**: 311–317.

Espeland M, Aagaard K, Balstad T, Hindar K (2007) Ecomorphological and genetic divergence between lowland and montane forms of the *Pieris napi* species complex (Pieridae, Lepidoptera). *Biological Journal of the Linnean Society* **92**: 727–745.

Essayan R. (1990) Contribution lépidoptérique à la cartographie des invertébrés européens. XVII. La cartographie des satyrines de France (*Erebia* non compris) (Lep., Nymphalidae, Satyrinae). *Alexanor* **16**: 291–328.

Estoup A, Cornuet JM (1999) Microsatellite evolution: inferences from population data. In: Goldstein DB,

Schlötterer C (eds.) *Microsatellites: Evolution and Applications*, Oxford: Oxford University Press, pp. 49–65.

European Environmental Agency (2004) *High Nature Value Farmland: Characteristics, Trends and Policy Challenges*, 2 EEA report No. 1/2004. Luxemburg: Office for Official Publications of the European Communities.

European Environment Agency (2007) *Halting the Loss of Biodiversity by 2010: Proposal for a First Set of Indicators to Monitor Progress in Europe.* Luxemburg: Office for Official Publications of the European Communities.

Excoffier L, Smouse PE, Quattro JM (1992) Analysis of molecular variance inferred from metric distances among DNA haplotypes: application to human mitochondrial DNA restriction data. *Genetics* 131: 479–491.

Faegri K, van der Pijl L (1979) *The Principles of Pollination Biology*, 3rd edn. Oxford: Pergamon Press.

Falconer DS, Mackay T (1996) *Introduction to Quantitative Genetics.* Harlow, Essex: Longman.

Fall A, Fall J (2001) A domain-specific language for models of landscape dynamics. *Ecological Modelling* 141: 1–18.

Farrell EP, Fuhrer E, Ryan D, Andersson F, Huttl R, Piussi P (2000) European forest ecosystems: building the future on the legacy of the past. *Forest Ecology and Management* 132: 5–20.

Fartmann T (2004) Die Schmetterlingsgemeinschaften der Halbtrockenrasen-Komplexe des Diemeltales: Biozönologie von Tagfaltern und Widderchen in einer alten Hudelandschaft. *Abhandlungen aus dem Westfälischen Museum für Naturkunde* 66: 1–256.

Fartmann T (2005) Quendel-Ameisenbläuling *Glaucopsyche arion* (Linnaeus, 1758). *Naturschutz und Biologische Vielfalt* 20: 175–180.

Fartmann T (2006) Oviposition preferences, adjacency of old woodland and isolation explain the distribution of the Duke of Burgundy butterfly (*Hamearis lucina*) in calcareous grasslands in central Germany. *Annales Zoologici Fennici* 43: 335–347.

Fartmann T, Hermann G (2006) Larvalökologie von Tagfaltern und Widderchen in Mitteleuropa, von den Anfängen bis heute. *Abhandlungen aus dem Westfälischen Museum für Naturkunde* 68: 11–57.

Fartmann T, Mattes H (2003) Störungen als ökologischer Schlüsselfaktor beim Komma-Dickkopffalter (*Hesperia comma*). *Abhandlungen aus dem Westfälischen Museum für Naturkunde* 65: 131–148.

Fatouros NE, Huigens ME, van Loon JJA, Dicke M, Hilker M (2005) Butterfly anti-aphrodisiac lures parasitic wasps. *Nature* 433: 704.

Fauvelot C, Cleary DFR, Menken SBJ (2006) Short-term impact of disturbance on genetic diversity and structure of Indonesian populations of the butterfly *Drupadia theda* in East Kalimantan. *Molecular Ecology* 15: 2069–2081.

Feber RE, Smith H (1995) Butterfly conservation on arable farmland. In: Pullin AS (ed.) *Ecology and Conservation of Butterflies*, London: Chapman and Hall, pp. 84–97.

Feber RE, Smith H, McDonald DW (1996) The effect on butterfly abundance of the management of uncropped edges of arable fields. *Journal of Applied Ecology* 33: 1191–1205.

Feber RE, Brereton TM, Warren MS, Oates M (2001) The impacts of deer on woodland butterflies: the good, the bad and the complex. *Forestry* 74: 271–276.

Feder JL (1998) The apple maggot fly, *Rhagoletis pomonella*: flies in the face of conventional wisdom. In: Howard DJ, Berlocher SH (eds.) *Endless Forms: Species and Speciation*, New York: Oxford University Press, pp. 130–144.

Feder JL, Roethele JB, Wlazlo B, Berlocher SH (1997) Selective maintenance of allozyme differences among sympatric host races of the apple maggot fly. *Proceedings of the National Academy of Sciences, USA* 94: 11 417–11 421.

Feder ME, Bennett AF, Huey RB (2000) Evolutionary physiology. *Annual Review of Ecology and Systematics* 31: 315–341.

Federley H (1945) Polyploidie und non-disjonction in der Gametogenese einiger Lepidopteran. *Commentationes Biologicae, Societas Scientiarum Fennica* 9: 1–9.

Feeny P (1995) Ecological opportunism and chemical constraints on the host associations of swallowtail butterflies. In: Scriber JM, Tsubaki Y, Lederhouse RC (eds.) *Swallowtail Butterflies: Their Ecology and Evolutionary Biology*, Gainesville, FL: Scientific Publishers, pp. 9–15.

Feeny PP, Rosenberry L, Carter M (1983) Chemical aspects of oviposition behavior in butterflies. In: Ahmad S (ed.) *Herbivorous Insects: Host-Seeking Behavior and Mechanisms*, New York: Academic Press, pp. 27–76.

Feeny PP, Stadler E, Ahman I, Carter M (1989) Effects of plant odor on oviposition by the black swallowtail butterfly, *Papilio polyxenes* (Lepidoptera: Papilionidae). *Journal of Insect Behavior* 2: 803–827.

Feltwell J (1982) *Large White Butterfly: The Biology, Biochemistry, and Physiology of* Pieris brassicae *(Linnaeus).* The Hague: Junk Publishers.

Fernández-Palacios JM, Andersson C (1993) Species composition and within archipelago co-occurrence patterns in the Canary Islands. *Ecography* 16: 31–36.

Fiedler K (1991) Systematic, evolutionary, and ecological implications of myrmecophily within the Lycaenidae

(Insecta: Lepidoptera: Papilionoidea). *Bonner zoologische Monographien* **31**: 1–210.

Fiedler K (1998) Diet breadth and host plant diversity of tropical- vs. temperate-zone herbivores: South-East Asian and West Palaearctic butterflies as a case study. *Ecological Entomology* **23**: 285–297.

Fiedler K, Seufert P, Pierce NE, Pearson JG, Baumgarten H-T (1995) Exploitation of lycaenid–ant mutualisms by braconid parasitoids. *Journal of Research on the Lepidoptera* **31**(1992): 153–168.

Field SA, Hardy ICW (2000) Butterfly contests: contradictory but not paradoxical. *Animal Behaviour* **59**: F1–F3.

Field RG, Gardiner T, Mason CF, Hill J (2005) Agri-environment schemes and butterflies: the utilisation of 6 m grass margins. *Biodiversity and Conservation* **14**: 1969–1976.

Fielding AH, Bell JF (1997) A review of methods for the assessment of prediction errors in conservation presence–absence models. *Environmental Conservation* **24**: 38–49.

Fielding AH, Haworth PF (1995) Testing the generality of bird-habitat models. *Conservation Biology* **9**: 1466–1481.

Figurny E, Woyciechowski M (1998) Flowerhead selection for oviposition by females of the sympatric butterfly species *Maculinea teleius* and *Maculinea nausithous* (Lepidoptera: Lycaenidae). *Entomologia Generalis* **23**: 215–222.

Figurny-Puchalska E, Gadeberg RME, Boomsma JJ (2000) Comparison of genetic structure of the large blue butterflies *Maculinea nausithous* and *M. teleius*. *Biodiversity and Conservation* **9**: 419–432.

Findlay R, Young MR, Findlay JA (1983) Orientation behaviour in the grayling butterfly: thermoregulation or crypsis? *Ecological Entomology* **8**: 145–153.

Fiori G (1956) *Strymon ilicis* Esp. (Lepidoptera, Lycaenidae). *Bollettino di Entomologia di Bologna*, **22**: 205–256.

Fischer K (1998) Population structure, mobility and habitat selection of the butterfly *Lycaena hippothoe* (Lycaenidae: Lycaenini) in Western Germany. *Nota lepidopterologica* **21**: 14–30.

Fischer K (2006) Reduced mating vigor in selection lines of the butterfly *Bicyclus anynana*. *Journal of Insect Behavior* **19**: 657–668.

Fischer K, Fiedler K (2000a) Response of the copper butterfly *Lycaena tityrus* to increased leaf nitrogen in natural food plants: evidence against the nitrogen limitation hypothesis. *Oecologia* **124**: 235–241.

Fischer K, Fiedler K (2000b) Sex-related differences in reaction norms in the butterfly *Lycaena tityrus* (Lepidoptera: Lycaenidae). *Oikos* **90**: 372–380.

Fischer K, Fiedler K (2001a) Effects of larval starvation on adult life-history traits in the butterfly species *Lycaena tityrus* (Lepidoptera: Lycaenidae). *Entomologia Generalis* **25**: 249–254.

Fischer K, Fiedler K (2001b) Egg weight variation in the butterfly *Lycaena hippothoe*: more small or fewer large eggs? *Population Ecology* **43**: 105–109.

Fischer K, Fiedler K (2001c) Resource-based territoriality in the butterfly *Lycaena hippothoe* and environmentally induced behavioural shifts. *Animal Behaviour* **61**: 723–732.

Fischer K, Fiedler K (2002) Reaction norms for age and size at maturity in response to temperature: a test of the compound interest hypothesis. *Evolutionary Ecology* **16**: 333–349.

Fischer K, Beinlich B, Plachter H (1999) Population structure, mobility and habitat preferences of the violet copper *Lycaena helle* (Lepidoptera, Lycaenidae) in Western Germany: implications for conservation. *Journal of Insect Conservation* **3**: 43–52.

Fischer K, Zwaan BJ, Brakefield PM (2002) How does egg size relate to body size in butterflies? *Oecologia* **31**: 375–379.

Fischer K, Brakefield PM, Zwaan BJ (2003) Plasticity in butterfly egg size: why larger offspring at lower temperatures? *Ecology* **84**: 3138–3147.

Fischer K, Bot ANM, Zwaan BJ, Brakefield PM (2004a) Genetic and environmental sources of egg size variation in the butterfly *Bicyclus anynana*. *Heredity* **92**: 163–169.

Fischer K, O'Brien DM, Boggs CL (2004b) Allocation of larval and adult resources to reproduction in a fruit-feeding butterfly. *Functional Ecology* **18**: 656–663.

Fischer K, Bot ANM, Brakefield PM, Zwaan BJ (2006) Do mothers producing large offspring have to sacrifice fecundity? *Journal of Evolutionary Biology* **19**: 380–391.

Fitton MG, Shaw MR, Gauld ID (1988) Pimpline ichneumon-flies: Hymenoptera, Ichneumonidae (Pimplinae). *Handbooks for the Identification of British Insects* **7**(1): 1–110.

Fitzgerald TD (1993) Sociality in caterpillars. In: Stamp NE, Casey TM (eds.) *Caterpillars: Ecological and Evolutionary Constraints on Foraging*, New York: Chapman and Hall, pp. 372–403.

Fleishman E, Austin GT, Weiss AD (1998) An empirical test of Rapoport's elevational rule: elevational gradients in montane butterfly communities. *Ecology* **79**: 2482–2493.

Fleishman E, Nally RM, Fay JP, Murphy DD (2001) Modeling and predicting species occurrence using broad-scale environmental variables: an example with butterflies of the Great Basin. *Conservation Biology* **15**: 1674–1685.

Fleishman E, Ray C, Sjörgen-Gulve P, Boggs CL, Murphy DD (2002) Assessing the roles of patch quality, area and isolation in predicting metapopulation dynamics. *Conservation Biology* **16**: 706–716.

Fleishman E, Nally RM, Fay JP (2003) Validation tests of predictive models of butterfly occurrence based on environmental variables. *Conservation Biology* **17**: 806–817.

Fleishman E, Thomson JR, Mac Nally R, Murphy DD, Fay JP (2005) Using indicator species to predict species richness of multiple taxonomic groups. *Conservation Biology* **19**: 1125–1137.

Fletcher NH (1978) Acoustical response of hair receptors in insects. *Journal of Comparative Physiology* **127**: 185–189.

Ford EB (1953) The genetics of polymorphism in the Lepidoptera. *Advances in Genetics* **5**: 43–87.

Ford EB (1964) *Ecological Genetics*. London: Methuen.

Ford EB (1975) *Ecological Genetics*, 2nd edn. London: Chapman and Hall.

Ford EB (1977) *Butterflies*, 3rd edn. London: Collins.

Ford EH, Ford EB (1930) Fluctuation in numbers and its influence on variation in *Melitaea aurinia* Rott. (Lepidoptera). *Transactions of the Entomological Society of London* **78**: 345–351.

Fordyce JA, Nice CC (2003) Variation in butterfly egg adhesion: adaptation to local host plant senescence characteristics? *Ecology Letters* **6**: 23–27.

Fordyce JA, Shapiro AM (2003) Another perspective on the slow-growth/high-mortality hypothesis: chilling effects on swallowtail larvae. *Ecology* **84**: 263–268.

Fordyce JA, Nice CC, Forister ML, Shapiro AM (2002) The significance of wing pattern diversity in the Lycaenidae: mate discrimination by two recently diverged species. *Journal of Evolutionary Biology* **15**: 871–879.

Forister ML (2004) Oviposition preference and larval performance within a diverging lineage of lycaenid butterflies. *Ecological Entomology* **29**: 264–272.

Forrester JA, Leopold DJ, Hafner SD (2005) Maintaining critical habitat in a heavily managed landscape: effects of power line corridor management on Karner blue butterfly (*Lycaeides melissa samuelis*) habitat. *Restoration Ecology* **13**: 488–498.

Forsberg J (1987) Size discrimination among conspecific hostplants in two pierid butterflies, *Pieris napi* L. and *Pontia daplidice* L. *Oecologia* **72**: 52–57.

Forsberg J, Wiklund C (1989) Mating in the afternoon: time-saving in courtship and remating by females of a polyandrous butterfly *Pieris napi*. *Behavioral Ecology and Sociobiology* **25**: 349–356.

Forster W (1956) Bausteine zur Kenntnis der Gattung *Agrodiaetus* Scudd. (Lep.: Lycaen.). *Internationaler Zeitschrift der Wiener Entomologische Geselsschaft* **41**: 42–61, 70–89, 116–212.

Fox J (2002) *An R and S-Plus Companion to Applied Regression.* Thousand Oaks, CA: Sage Publications. Online: http://socserv.mcmaster.ca/jfox/Books/Companion/

Fox CW, Czesak ME (2000) Evolutionary ecology of progeny size in arthropods. *Annual Review of Entomology* **45**: 341–369.

Fox R, Asher J, Brereton T, Roy D, Warren MS (2006) *The State of Butterflies in Britain and Ireland*. Newbury, Berks: Pisces Publications.

Franco AMA, Hill JK, Kitschke C, Collingham YC, Roy DB, Fox R, Huntley B, Thomas CD (2006) Impacts of climate warming and habitat loss on extinctions at species' low-latitude range boundaries. *Global Change Biology* **12**: 1545–1553.

Frankham R (1995) Inbreeding and extinction: a threshold effect. *Conservation Biology* **9**: 792–799.

Frankham R (2002) Predicting extinction risk. *Nature* **419**: 18–19.

Frankham R, Ballou JD, Briscoe DA (2002) *Introduction to Conservation Genetics*. Cambridge: Cambridge University Press.

Frankino WA, Zwaan BJ, Stern DL, Brakefield PM (2005) Natural selection and developmental constraints in the evolution of allometries. *Science* **307**: 718–720.

Franklin J (1998) Predicting the distribution of shrub species in southern California from climate and terrain-derived variables. *Journal of Vegetation Science* **9**: 733–748.

Franzén M, Ranius T (2004) Occurrence patterns of butterflies (Rhopalocera) in semi-natural pastures in southeastern Sweden. *Journal for Nature Conservation* **12**: 121–135.

Fred MS, Brommer JE (2003) Influence of habitat quality and patch size on occupancy and persistence in two populations of the Apollo Butterfly (*Parnassius apollo*). *Journal of Insect Conservation* **7**: 85–98.

Fred MS, O'Hara RB, Brommer JE (2006) Consequences of the spatial configuration of resources for the distribution and dynamics of the endangered *Parnassius apollo* butterfly. *Biological Conservation* **130**: 183–192.

Freese A, Fiedler K (2002) Experimental evidence for specific distinctness of the two wood white butterfly taxa, *Leptidea sinapis* and *L. reali* (Pieridae). *Nota lepidopterologica* **25**: 39–59.

Freese A, Beneš J, Bolz R, Cizek O, Dolek M, Geyer A, Gros P, Konvička M, Liegl A, Stettmer C (2006) Habitat use of the endangered butterfly *Euphydryas maturna* and forestry in Central Europe. *Animal Conservation* **9**: 388–397.

Friberg M, Bergman M, Kullberg J, Wahlberg N, Wiklund C (2008) Niche separation in space and time between two sympatric sister species: a case of ecological pleiotropy. *Journal of Animal Ecology* **22**: 1–18.

Fric Z, Konvička M (2000) Adult population structure and behaviour of two seasonal generations of the European Map Butterfly, *Araschnia levana*, species with seasonal polyphenism (Nymphalidae). *Nota lepidopterologica* 23: 2–25.

Fric Z, Konvička M (2002) Generations of the polyphenic butterfly *Araschnia levana* differ in body design. *Evolutionary Ecology Research* 4: 1017–1033.

Fric Z, Konvička M, Zrzavy J (2004) Red, black or black, white? Phylogeny of the *Araschnia* butterflies (Lepidoptera: Nymphalidae) and evolution of seasonal polyphenism. *Journal of Evolutionary Biology* 17: 265–278.

Fric Z, Klimova M, Konvička M (2006) Mechanical design indicates mobility differences among butterfly generations. *Evolutionary Ecology Research* 8: 1511–1522.

Friedlander TP (1985) Egg mass design relative to surface-parasitizing parasitoids, with notes on *Asterocampa clyton* (Lepidoptera: Nymphalidae). *Journal of Research on the Lepidoptera* 24: 250–257.

Friedman JH (2002) Stochastic gradient boosting. *Computational Statistics and Data Analysis* 38: 367–378.

Friedrich E (1986) *Breeding Butterflies and Moths: A Practical Handbook for British and European Species*. Colchester, Essex: Harley Books.

Frohawk FW (1912) Completion of the life-history of *Melanargia japygia* subsp. *sumarovius*. *Entomologist* 14: 1–5.

Fry R, Lonsdale D (eds.) (1991) *Habitat Conservation for Insects: A Neglected Green Issue*. London: Amateur Entomologists' Society.

Fry GLA, Main AR (1993) Restoring seemingly natural communities on agricultural land. In: Saunders DA, Hobbs RJ, Ehrlich PR (eds.) *Nature Conservation*, Vol. 3, *Reconstruction of Fragmented Ecosystems*, Chipping Norton, Australia: Surrey Beatty and Sons, pp. 225–241.

Füldner K (2006) Die Tagfalterarten der Wälder und ihre Beeinflussung durch das Waldmanagement. *Oedippus* 24: 1–28.

Futuyma DJ (1986) *Evolutionary Biology*. Sunderland, MA: Sinauer Associates.

Futuyma DJ, Keese MC (1992) Evolution and coevolution of plants and phytophagous arthropods. In: Rosenthal GA, Berenbaum MR (eds.) *Herbivores: Their Interactions with Secondary Plant Metabolites*, Vol. 2, San Diego, CA: Academic Press, pp. 439–475.

Futuyma DJ, Slatkin M (eds.) (1983) *Coevolution*. Sunderland, MA: Sinauer Associates.

Gadeberg RME, Boomsma JJ (1997) Genetic population structure of the large blue butterfly *Maculinea alcon* in Denmark. *Journal of Insect Conservation* 1: 99–111.

Gage MJG (1994) Associations between body size, mating pattern, testes size and sperm lengths across butterflies. *Proceedings of the Royal Society of London B* 258: 247–254.

Gaggiotti OE (2003) Genetic threats to population persistence. *Annales Zoologici Fennici* 40: 155–168.

Gall LF (1984) The effect of capturing and marking on subsequent activity of *Boloria acrocnema* (Lepidoptera: Nymphalidae), with a comparison of different numerical models that estimate population size. *Biological Conservation* 28: 139–154.

Gamisans J (1991) *La végétation de la Corse*. Geneva, Switzerland: Conservatoire et Jardin Botanique de la Ville de Genève.

Gandon S, Michalakis Y (2001) Multiple causes of the evolution of dispersal. In: Clobert J, Danchin E, Dhondt AA, Nichols JD (eds.) *Dispersal*, Oxford: Oxford University Press, pp. 155–167.

García-Barros E (1986) Notas sobre el comportamiento de las hembras de *B. circe* (Fabricius) (Lep., Nymphalidae-Satyrinae): estridulación y agresividad. Actas VIII Jornadas Asociación española de Entomología, Universidad de Sevilla, pp. 849–854.

García-Barros E (1987) Observaciones sobre la biología de *Maniola jurtina* (L., 1758) en el centro de la Península Ibérica: fenología general del ciclo biológico, duración del periodo de prepuesta y fecundidad potencial de las hembras (Lep., Nymphalidae). *Boletín de la Asociación española de Entomología* 11: 235–247.

García-Barros E (1989a) Estudio comparativo de los caracteres biológicos de dos satirinos, *Hipparchia statilinus* (Hufnagel, 1766) e *Hipparchia semele* (L., 1758) (Lepidoptera, Nymphalidae, Satyrinae). *Miscellánia Zoológica* 13: 85–96.

García-Barros E (1989b) Morfología larvaria de *Hipparchia (Pseudotergumia) fidia* (L., 1767) (Lepidoptera, Nymphalidae). *Nouvelle Revue d'Entomologie, N.S.*, 6: 71–83.

García-Barros E (1992) Evidence for geographic variation of egg size and fecundity in a satyrine butterfly, *Hipparchia semele* (L.) (Lepidoptera, Nymphalidae-Satyrinae). *Graellsia* 48: 45–52.

García-Barros E (1994) Egg size variation in European satyrine butterflies (Nymphalidae, Satyrinae). *Biological Journal of the Linnean Society* 51: 309–324.

García-Barros E (1998) Delayed ovarian maturation in the butterfly *Hipparchia semele* as a possible response to summer drought. *Ecological Entomology* 13: 391–398.

García-Barros E (2000a) Comparative data on the adult biology, ecology and behaviour of species belonging to the genera *Hipparchia*, *Chazara* and *Kanetisa* in central Spain (Nymphalidae: Satyridae). *Nota lepidopterologica* 23: 119–140.

García-Barros E (2000b) Notas sobre la biología de los adultos de *Pandoriana pandora* (Denis, Schiffermüller, 1775) en la España central. *Shilap, Revista de Lepidopterología* **28**: 97–102.

García-Barros E (2000c) Body size, egg size, and their specific relationships with ecological and life history traits in butterflies. *Biological Journal of the Linnean Society* **70**: 251–284.

García-Barros E (2006) Within and between species scaling in the weight, water, carbon and nitrogen contents of eggs and neonate larvae of twelve satyrine butterflies (Lepidoptera: Nymphalidae). *European Journal of Entomology* **103**: 559–568.

García-Barros E, Munguira ML (1997) Uncertain branch lengths, taxonomic sampling error, and the egg to body size allometry in temperate butterflies (Lepidoptera). *Biological Journal of the Linnean Society* **61**: 201–221.

Gardener M, Gillman M (2001) The effects of soil fertilization on amino acids in the floral nectar of corncockle, *Agrostema githago* (Caryophyllaceae). *Oikos* **92**: 101–106.

Garland T Jr, Dickerman AW, Janis CM, Jones JA (1993) Phylogenetic analysis of covariance by computer simulation. *Systematic Biology* **42**: 265–292.

Garraway E, Bailey AJA (1992) Parasitoid induced mortality in the eggs of the endangered giant swallowtail butterfly *Papilio homerus* (Papilionidae). *Journal of the Lepidopterists' Society* **46**: 233–234.

Gartner A (1861) *Limenitis aceris* Fabricius und ihre ersten Stände. *Jahresheft Naturwiss. Sect. K.K. Mahr. Shles. Ges. Ackerbau, Natur-, Landeskunde* **1860**: 7–19.

Gaston KJ (1994) *Rarity*. London: Chapman and Hall.

Gaston KJ (2000) Global patterns in biodiversity. *Nature* **405**: 220–227.

Gaston KJ, Blackburn TM (2000) *Pattern and Process in Macroecology*. Oxford: Blackwell Science.

Gaston KJ, Blackburn TM, Spicer JI (1998) Rapoport's rule: time for an epitaph? *Trends in Ecology and Evolution* **13**: 70–74.

Gaston KJ, Smith RM, Thompson K, Warren PH (2005) Urban domestic gardens. II. Experimental tests of methods for increasing biodiversity. *Biodiversity and Conservation* **14**: 395–413.

Gatto M, Ghezzi LL (1996) Optimal life strategies in organisms exposed to recurrent critical events. *Journal of Optimization Theory and Applications* **90**: 79–94.

Gauch HG Jr (1982) *Multivariate Analysis in Community Ecology*. Cambridge: Cambridge University Press.

Gauld ID (1988) Evolutionary patterns of host utilization by ichneumonoid parasitoids (Hymenoptera: Ichneumonidae and Braconidae). *Biological Journal of the Linnean Society* **35**: 351–377.

Gauld I, Bolton B (eds.) (1988) *The Hymenoptera*. London: British Museum (Natural History) and Oxford: Oxford University Press.

Gause GF (1934) *The Struggle for Existence*. Baltimore, MD: Williams and Wilkins.

Gavrilets S, Scheiner SM (1993) The genetics of phenotypic plasticity. V. Evolution of reaction norm shape. *Journal of Evolutionary Biology* **6**: 31–48.

Geiger R (1965) *The Climate Near the Ground*. Cambridge, MA: Harvard University Press.

Geiger W (ed.) (1987) *Les Papillons de jour et leurs biotopes*. Basel, Switzerland: Ligue Suisse pour la Protection de la Nature.

Geiger H (1990) Enzyme electrophoretic methods in studies of systematics and evolutionary biology of butterflies. In: Kudrna O (ed.) *Butterflies of Europe*, Vol. 2, Wiesbaden, Germany: Aula, pp. 397–436.

Geiger H, Shapiro AM (1992) Genetics, systematics and evolution of holarctic *Pieris napi* species group populations (Lepidoptera, Pieridae). *Zeitschrift für Zoologie, Systematik und Evolutionsforschung* **30**: 100–122.

Geiger H, Descimon H, Scholl A (1988) Evidence for speciation within nominal *Pontia daplidice* (Linnaeus, 1758) in southern Europe (Lepidoptera: Pieridae). *Nota lepidopterologica* **11**: 7–20.

Gerber LR (2006) Including behavioral data in demographic models improves estimates of population viability. *Frontiers in Ecology and the Environment* **4**: 419–427.

Geuder M, Wray V, Fiedler K, Proksch P (1997) Sequestration and metabolism of host-plant flavonoids by the lycaenid butterfly *Polyommatus bellargus*. *Journal of Chemical Ecology* **23**: 1361–1372.

Gibbs M, Lace LA, Jones MJ, Moore AJ (2005) Egg size-number trade-off and a decline in oviposition site choice quality: female *Pararge aegeria* butterflies pay a cost of having males present at oviposition. *Journal of Insect Science* **5**: 39.

Gibbs M, Lace LA, Jones MJ, Moore AJ (2006) Multiple host-plant use may arise from gender-specific fitness effects. *Journal of Insect Science* **6**: 04. Online: insectscience.org/6.04/

Gibeaux C (1984) *Erebia tyndarus* Esper et *E. calcaria* Lorković capturés en France (Lep.: Nymphalidae). *Entomologica Gallica* **1**: 49–60.

Gil-T F (2001) Estudio sobre la influencia de parasitoides (Hymenoptera: Ichneumonoidea) en poblaciones del raro lepidóptero *Iolana iolas* Ochsenheimer, 1816 (Lepidoptera: Lycaenidae). *Boletín de la Sociedad Entomológica Aragonesa* **29**: 85–88.

Gil-T F (2004) Nuevos datos sobre la biología de *Iolana iolas* Ochsenheimer (Lepidoptera: Lycaenidae) y su interacción con himenópteros mirmecófilos, fitófagos y parasitoides (Hymenoptera, Formicidae, Eurytomidae, Ichneumonoidea). *Boletín de la Sociedad Entomológica Aragonesa* **34**: 139–145.

Gil-T F, Gil-Uceda T (2005) *Agrodiaetus violetae* (Gómez-Bustillo, Expósito, Martínez, 1979): morfología comparada y descripción de *Agrodiaetus fabressei subbaeticus* ssp. nov. del sureste de la Península Ibérica (Lepidoptera, Lycaenidae). *Boletín de la Sociedad Entomológica Aragonesa* **36**: 357–364.

Gilbert LE (1972) Pollen feeding and reproductive biology of *Heliconius* butterflies. *Proceedings of the National Academy of Sciences, USA* **69**: 1403–1407.

Gilbert LE (1975) Ecological consequences of a coevolved mutualism between butterflies and plants. In: Gilbert LE, Raven PH (eds.) *Coevolution of Animals and Plants*, Austin, TX: University of Texas Press, pp. 210–240.

Gilbert LE (1976) Postmating female odor in *Heliconius* butterflies: a male contributed antiaphrodisiac. *Science* **193**: 419–420.

Gilbert LE (1980) Ecological consequences of a coevolved mutualism between butterflies and plants. In: Gilbert LE, Raven PH (eds.) *Coevolution of Animals and Plants*, 2nd edn, Austin, TX: University of Texas Press, pp. 210–240, 253–255.

Gilbert N (1984) Control of fecundity in *Pieris rapae*. I. The problem. *Journal of Animal Ecology* **53**: 581–588.

Gilbert LE (2003) Adaptive novelty through introgression in *Heliconius* wing patterns: evidence for a shared genetic 'toolbox' from synthetic hybrid zones and a theory of diversification. In: Boggs CL, Watt WB, Ehrlich P (eds.) *Butterflies: Ecology and Evolution Taking Flight*. Chicago, IL: University of Chicago Press, pp. 281–318.

Gilbert LE, Singer MC (1973) Dispersal and gene flow in a butterfly species. *American Naturalist* **107**: 58–72

Gilbert MC, Singer MC (1975) Butterfly ecology: annual review. *Ecological Systematics* **6**: 365–397.

Gilchrist GW (1990) The consequences of sexual dimorphism in body size for butterfly flight and thermoregulation. *Functional Ecology* **4**: 475–487.

Gilchrist GW, Rutowski RL (1986) Adaptive and incidental consequences of the alba polymorphism in an agricultural population of *Colias* butterflies: female size, fecundity, and differential dispersion. *Oecologia* **68**: 235–240.

Gillespie JH (1991) *The Causes of Molecular Evolution*. Oxford: Oxford University Press.

Gilpin ME, Soulé ME (1986) Minimum viable populations: processes of species extinction. In: Soulé ME (ed.) *Conservation Biology: The Science of Scarcity and Diversity*, Sunderland, MA: Sinauer Associates, pp. 19–34.

Ginzburg L, Slobodkin LB, Johnson K, Bindman AG (1982) Quasiextinction probabilities as a measure of impact on population growth. *Risk Analysis* **21**: 171–181.

Glassl H (2005) Parnassius apollo: *Seine Unterarten*. Möhrendorf, Germany: Helmut Glassl.

Glatzel G (1991) The impact of historic land use and modern forestry on nutrient relations of central European forest ecosystems. *Fertilizer Research* **27**: 1–8.

Gliemeroth AK (1995) *Paläoökologische Untersuchungen über die letzten 22.000 Jahre in Europa*. Stuttgart, Germany: Gustav Fischer.

Godfray HCJ (1994) *Parasitoids: Behavioral and Evolutionary Ecology*. Princeton, NJ: Princeton University Press.

Godfrey GL, Miller JS, Carter DJ (1989) Two mouth part modifications in larval Notodontidae (Lepidoptera): their taxonomic distributions and putative functions. *Journal of the New York Entomological Society* **97**: 455–470.

Goedart JB (1662) *Metamorphosis et Historia Naturalis Insectorum*. Middlesbrough.

Goldschmidt RB (1932) Prä oder Postreduktion der Chromosomen? Die Lösung eines alten Problems. *Naturwissenschaften* **20**: 358–362.

Goldstein DB, Schlötterer C (eds.) (1999) *Microsatellites: Evolution and Applications*. Oxford: Oxford University Press.

Gómez Bustillo MR, Fernandez Rubio F (1974) *Mariposas de la Peninsula Ibérica: Ropaloceros (II)*. Madrid: Ministerio de Agricultura.

Gómez Bustillo MR, Exposito Hermosa A, Martínez Borrego P (1979) Una nueva especie por la ciencia: *Agrodiaetus violetae* (Lep.: Lyc.). *Sociedad Hispano-Luso-Americana de Lepidopterología, Revista de Lepidopterología* **7**: 47–54.

Gómez-Zurita J, Vogler AP (2003) Incongruent nuclear and mitochondrial phylogeographic patterns in the *Timarcha goettingensis* species complex (Coleoptera, Chrysomelidae). *Journal of Evolutionary Biology* **16**: 833–843.

Gompert Z, Nice CC, Fordyce JA, Forister ML, Shapiro AM (2006) Identifying units for conservation using molecular systematics: the cautionary tale of the Karner blue butterfly. *Molecular Ecology* **15**: 1759–1768.

Gonseth Y (1993) Les Lépidoptères diurnes (Lep. Rhopalocera) des clairières et des chemins forestiers du Jura neuchâtelois. *Bulletin de la Société Entomologique Suisse* **66**: 283–302.

Gonseth Y (1994a) La faune des Lépidoptères diurnes (Rhopalocera) des pâturages, des pelouses sèches et des prairies de fauche du Jura neuchâtelois. *Bulletin de la Société Entomologique Suisse* **67**: 17–36.

Gonseth Y (1994b) La faune des Lépidoptères diurnes (Rhopalocera) des milieux humides du canton de Neuchâtel II.

Tourbières, prés à litière, mégaphorbiées. *Bulletin de la Société Neuchâteloise des Sciences Naturelles* **117**: 33–57.

Goodman OR, Turner HJ, Pierce FN (1925) A supposed hybrid *Polyommatus* (*coridon-hylas*). *Entomological Record and Journal of Variation* **37**: 21–25.

Gossard TW, Jones RE (1977) The effects of age and weather on egg-laying in *Pieris rapae*. *Journal of Applied Ecology* **14**: 65–71.

Gotthard K (1998) Life history plasticity in the satyrine butterfly *Lasiommata petropolitana*: investigating an adaptive reaction norm. *Journal of Evolutionary Biology* **11**: 21–39.

Gotthard K (2000) Increased risk of predation as a cost of high growth rate: an experimental test in a butterfly. *Journal of Animal Ecology* **69**: 896–902.

Gotthard K (2004) Growth strategies and optimal body size in temperate Pararginii butterflies. *Integrative and Comparative Biology* **44**: 471–479.

Gotthard K, Nylin S (1995) Adaptive plasticity and plasticity as an adaptation: a selective review of plasticity in animal morphology and life history. *Oikos* **74**: 3–17.

Gotthard K, Nylin S, Wiklund C (1999a) Mating system evolution in response to search costs in the speckled wood butterfly, *Pararge aegeria*. *Behavioral Ecology and Sociobiology* **45**: 424–429.

Gotthard K, Nylin S, Wiklund C (1999b) Seasonal plasticity in two satyrine butterflies: state-dependent decision making in relation to daylength. *Oikos* **83**: 453–462.

Gotthard K, Nylin S, Wiklund C (2000a) Individual state controls temperature dependence in a butterfly (*Lasiommata maera*). *Proceedings of the Royal Society of London B* **267**: 589–593.

Gotthard K, Nylin S, Wiklund C (2000b) Mating opportunity and the evolution of sex-specific mortality rates in a butterfly. *Oecologia* **122**: 36–43.

Gottsberger G (1996) Floral ecology, report on the years 1992 (1991) to 1994 (1995). *Progress in Botany* **57**: 368–415.

Gottsberger G, Arnold T, Linskens HF (1989) Intraspecific variation in the amino acid content of floral nectar. *Botanica Acta* **102**: 141–144.

Gottsberger G, Arnold T, Linskens HF (1990) Variation in floral nectar amino acids with aging of flowers, pollen contamination and flower damage. *Israel Journal of Botany* **39**: 167–176.

Goudet J (2005) Hierfstat, a package for R to compute and test hierarchical *F*-statistics. *Molecular Ecology Notes* **5**: 184–186.

Goulson D (1993) Allozyme variation in the butterfly *Maniola jurtina* (Lepidoptera: Satyrinae) (L.): evidence for selection. *Heredity* **71**: 386–393.

Goulson D (1994) Determination of larval melanization in the moth, *Mamestra brassicae*, and the role of melanin in thermoregulation. *Heredity* **73**: 471–479.

Goverde M, van der Heiden M, Wiemken A, Sanders I, Erhardt A (2000) Arbuscular mycorrhizal fungi influence life history traits of a lepidopteran herbivore. *Oecologia* **125**: 362–369.

Goverde M, Erhardt A, Niklaus PA (2002) In situ development of a satyrid butterfly on calcareous grassland exposed to elevated carbon dioxide. *Ecology* **83**: 1399–1411.

Gower JC (1969) Multivariate analysis and multidimensional geometry. *Statistician* **17**: 13–28.

Graham MH (2003) Confronting multicollinearity in ecological multiple regression. *Ecology* **84**: 2809–2915.

Graham SM, Watt WB, Gall LF (1980) Metabolic resource allocation vs. mating attractiveness: adaptive pressures on the 'alba' polymorphism of *Colias* butterflies. *Proceedings of the National Academy of Sciences, USA* **77**: 3615–3619.

Grant PR, Grant BR (1992) Hybridization of bird species. *Science* **256**: 193–197.

Grant BR, Grant PR (1998) Hybridization and speciation in Darwin's finches: the role of sexual imprinting on a culturally transmitted trait. In: Howard DJ, Berlocher SH (eds.) *Endless Forms: Species and Speciation*, New York: Oxford Univerity Press, pp. 404–422.

Grapputo A, Boman S, Lindstrom L, Lyytinen A, Mappes J (2005) The voyage of an invasive species across continents: genetic diversity of North American and European Colorado potato beetle populations. *Molecular Ecology* **14**: 4207–4219.

Graur D (1985) Gene diversity in Hymenoptera. *Evolution* **39**: 190–199.

Greatorex-Davies JN, Roy D (2005) *The Butterfly Monitoring Scheme: Report to Recorders, 2004*. Monks Wood, Huntingdon: Centre for Ecology and Hydrology.

Greatorex-Davies JN, Hall ML, Marrs RH (1992) The conservation of pearl-bordered fritillary butterfly (*Boloria euphrosyne* L.): preliminary studies on the creation and management of glades in conifer plantations. *Forest Ecology and Management* **53**: 1–14.

Gregory RD, Van Strien AJ, Vorisek P, Gmelig Meyling AW, Noble DG, Foppen RPB, Gibbons DW (2005) Developing indicators for European birds. *Philosophical Transactions of the Royal Society of London B* **360**: 269–288.

Grevstad FS, Herzig AL (1997) Quantifying the effects of distance and conspecifics on colonization: experiments and models using the loosestrife leaf beetle, *Galerucella calmariensis*. *Oecologia* **110**: 60–68.

Grill A, Cleary DFR (2003) Diversity patterns in butterfly communities of the Greek nature reserve Dadia. *Biological Conservation* **114**: 427–436.

Grill A, Knoflach B, Cleary DFR, Kati V (2005) Butterfly, spider, and plant communities in different land-use types in Sardinia, Italy. *Biodiversity and Conservation* **14**: 1281–1300.

Grill A, Raijmann LEL, Van Ginkel W, Gkioka E, Menken SBJ (2007) Genetic differentiation and natural hybridization between the Sardinian endemic *Maniola nurag* and the European *Maniola jurtina*. *Journal of Evolutionary Biology* **20**: 1255–1270.

Grimes LR, Neunzig HH (1986a) Morphological survey of the maxillae in last stage larvae of the suborder Ditrysia (Lepidoptera): Palpi. *Annals of the Entomological Society of America* **79**: 491–509.

Grimes LR, Neunzig HH (1986b) Morphological survey of the maxillae in last stage larvae of the suborder Ditrysia (Lepidoptera): mesal lobes (Laciniogaleae). *Annals of the Entomological Society of America* **79**: 510–526.

Grimm V, Railsback SF (2005) *Individual-Based Modeling and Ecology*. Princeton, NJ: Princeton University Press.

Grimm V, Lorek H, Finke J, Koester F, Malachinski M, Sonnenschein M, Moilanen A, Storch I, Singer A, Wissel C, Frank K (2004) META-X: generic software for metapopulation viability analysis. *Biodiversity and Conservation* **13**: 165–188.

Grimmet RFA, Jones TA (1989) *Important Bird Areas in Europe*, Technical Publication No. 9. Cambridge: International Council for Bird Preservation.

Grodnitsky DL (1993) Preliminary data on body movement of freely flying butterflies. *Zoologichesky Zhurnal* **72**: 84–94.

Groß FJ. (1957) Ein vermutlicher Bastard zwischen *Coenonympha arcania* L. und *C. hero* L. *Entomologische Zeitschrift* **66**: 169–170.

Grossmueller DW, Lederhouse RC (1985) Oviposition site selection: an aid to rapid growth and development in the tiger swallowtail butterfly, *Papilio glaucus*. *Oecologia* **66**: 68–73.

Grossmueller DW, Lederhouse RC (1988) The role of nectar source distribution in habitat use and oviposition by the tiger swallowtail butterfly. *Journal of the Lepidopteran Society* **41**: 159–165.

Grove AT, Rackham O (2001) *The Nature of Mediterranean Europe: An Ecological History*. New Haven, CT: Yale University Press.

Grula JW, Taylor OR (1980) Some characteristics of hybrids derived from the sulfur butterflies, *Colias eurytheme* and *C. philodice*: phenotypic effects of the X-chromosome. *Evolution* **34**: 673–687.

Grula JW, McChesney JD, Taylor OR (1980) Aphrodisiac pheromones of the sulfur butterflies *Colias eurytheme* and *C. philodice*. *Journal of Chemical Ecology* **6**: 241–256.

Grundel R, Pavlovic NB, Sulzman CL (2000) Nectar plant selection by the Karner blue butterfly (*Lycaeides melissa samuelis*) at the Indian Dunes National Lakeshore. *American Midland Naturalist* **144**: 1–10.

Gu H, Hughes J, Dorn S (2006) Trade-off between mobility and fitness in *Cydia pomonella* L. (Lepidoptera: Tortricidae). *Ecological Entomology* **31**: 68–74.

Guillaumin M, Descimon H (1976) La notion d'espèce chez les Lépidoptères. *Mémoires de la Société Zoologique de France* **38**: 129–201.

Guillot G, Mortier F, Estoup A (2005) GENELAND: a computer package for landscape genetics. *Molecular Ecology Notes* **5**: 712–715.

Guisan A, Thuiller W (2005) Predicting species distribution: offering more than simple habitat models. *Ecology Letters* **8**: 993–1009.

Guisan A, Zimmermann NE (2000) Predictive habitat distribution models in ecology. *Ecological Modelling* **135**: 147–186.

Guisan A, Theurillat J-P, Kienast F (1998) Predicting the potential distribution of plant species in an alpine environment. *Journal of Vegetation Science* **9**: 65–74.

Guisan A, Weiss SB, Weiss AD (1999) GLM versus CCA spatial modeling of plant species distribution. *Plant Ecology* **143**: 107–122.

Guisan A, Edwards TC, Hastie T (2002) Generalized linear and generalized additive models in studies of species distributions: setting the scene. *Ecological Modelling* **157**: 89–100.

Guisan A, Lehmann A, Ferrier S, Austin M, Overton JM, Aspinall R, Hastie T (2006) Making better biogeographical predictions of species' distributions. *Journal of Applied Ecology* **43**: 386–392.

Gumpertz ML, Graham JM, Ristaino JB (1997) Autologistic model of spatial pattern of *Phytophthora* epidemic in bell pepper: effects of soil variables on disease presence. *Journal of Agricultural, Biological, and Environmental Statistics* **2**: 131–156.

Gunn A, Gatehouse AG (1987) The influence of larval phase on metabolic reserves, fecundity and lifespan of the African armyworm moth, *Spodoptera exempta* (Walker) (Lepidoptera: Noctuidae). *Bulletin of Entomological Research* **77**: 651–660.

Guppy CS (1986). The adaptive significance of alpine melanism in the butterfly *Parnassius phoebus* F. (Lepidoptera, Papilionidae). *Oecologia* **70**: 205–213.

Gus R, Schifino MT, Araújo AM (1983) Occurrence of localized centromeres in Lepidoptera chromosomes. *Revista Brasileira de Genética* 6: 769–774.

Gutiérrez D (1997) Importance of historical factors on species richness and composition of butterfly assemblages (Lepidoptera: Rhopalocera) in a northern Iberian mountain range. *Journal of Biogeography* 24: 77–88.

Gutiérrez D (2005) Effectiveness of existing reserves in the long-term protection of a regionally rare butterfly. *Conservation Biology* 19: 1586–1597.

Gutiérrez D, Menéndez R (1995) Distribution and abundance of butterflies in a mountain area in the northern Iberian peninsula. *Ecography* 18: 209–216.

Gutiérrez D, Menéndez R (1998) Phenology of butterflies along an altitudinal gradient in northern Spain. *Journal of Zoology* 244: 249–264.

Gutiérrez D, Thomas CD (2000) Marginal range expansion in a host-limited butterfly species *Gonopteryx rhamni*. *Ecological Entomology* 27: 165–170.

Gutiérrez D, Thomas CD, León-Cortés JL (1999) Dispersal, distribution, patch network and metapopulation dynamics of the dingy skipper butterfly (*Erynnis tages*). *Oecologia* 121: 506–517.

Gutiérrez D, Fernandez P, Seymour AS, Jordano D (2005) Habitat distribution models: are mutualist distributions good predictors of their associates? *Ecological Applications* 15: 3–18.

Guyot H (2002) Découverte d'une nouvelle plante-hôte de *Papilio hospiton* en Corse (Lepidoptera: Papilionidae). *Alexanor* 21: 285–287.

Gwynne DT (1989) Does copulation increase the risk of predation? *Trends in Ecology and Evolution* 4: 54–56.

Haag KL, Araújo AM (1994) Inbreeding, genetic load and morphometric variation in natural populations of *Dryas iulia* (Lepidoptera, Nymphalidae). *Revista Brasileira de Genética* 17: 35–39.

Haag CR, Saastamoinen M, Marden JH, Hanski I (2005) A candidate locus for variation in dispersal rate in a butterfly metapopulation. *Proceedings of the Royal Society of London B* 272: 2449–2456.

Habel JC, Schmitt T, Müller P (2005) The fourth paradigm pattern of postglacial range expansion of European terrestrial species: the phylogeography of the Marbled White butterfly (Satyrinae, Lepidoptera). *Journal of Biogeography* 32: 1489–1497.

Habel JC, Meyer M, El Moussadik A, Schmitt T (2008) Africa goes Europe: the complete phylogeography of the Marbled White butterfly species complex *Melanargia galathea/*

M. lachesis (Lepidoptera: Satyridae). *Organisms, Diversity and Evolution* 8: 121–129.

Haddad NM (1999a) Corridor and distance effects on interpatch movements: a landscape experiment with butterflies. *Ecological Applications* 9: 612–622.

Haddad NM (1999b) Corridor use predicted from behaviors at habitat boundaries. *American Naturalist* 153: 215–227.

Haddad N (2000) Corridor length and patch colonization by a butterfly, *Junonia coenia*. *Conservation Biology* 14: 738–745.

Haddad NM, Baum KA (1999) An experimental test of corridor effects on butterfly densities. *Ecological Applications* 9: 623–633.

Haeselbarth E (1979) Zur Parasitierung der Puppen von Forleule (*Panolis flammea* [Schiff.]), Kiefernspanner (*Bupaulus pinarius* [L.]) und Heidelbeerspanner (*Boarmia bistortana* [Goeze]) in bayerischen Keifernwäldern. *Zeitschrift für angewandte Entomologie* 87: 186–202, 311–322.

Hagen RH (1990) Population structure and host use in hybridizing subspecies of *Papilio glaucus* (Lepidoptera: Papilionidae). *Evolution* 44: 1914–1930.

Haikola S (2003) Effects of inbreeding in the Glanville fritillary butterfly (*Melitaea cinxia*). *Annales Zoologici Fennici* 40: 483–493.

Haikola S, Fortelius W, O'Hara RB, Kuussaari M, Wahlberg N, Saccheri IJ, Singer MC, Hanski I (2001) Inbreeding depression and the maintenance of genetic load in *Melitea cinxia* metapopulations. *Conservation Genetics* 2: 325–335.

Haikola S, Singer MC, Pen I (2004) Has inbreeding depression led to avoidance of sib mating in the Glanville fritillary butterfly (*Melitaea cinxia*)? *Evolutionary Ecology* 18: 113–120.

Hainsworth FR (1989) 'Fast food' vs 'haute cuisine': painted ladies, *Vanessa cardui* (L.), select food to maximize net meal energy. *Functional Ecology* 3: 701–707.

Hairston NG Jr, Ellner SP, Geber MA, Yoshida T, Fox JA (2005) Rapid evolution and the convergence of ecological and evolutionary time. *Ecology Letters* 8: 1114–1127.

Haldane JBS (1922) Sex ratio and unisexual sterility in hybrid animals. *Journal of Genetics* 12: 101–109.

Hall JPW, Willmott KR (2000) Patterns of feeding behaviour in adult male riodinid butterflies and their relationship to morphology and ecology. *Biological Journal of the Linnean Society* 69: 1–23.

Hamback PA, Summerville KS, Steffan-Dewenter I, Krauss J, Englund G, Crist TO (2007) Habitat specialization, body size, and family identity explain lepidopteran density–area relationships in a cross-continental comparison. *Proceedings of the National Academy of Sciences, USA* 104: 8368–8373.

Hamilton WD, May RM (1977) Dispersal in stable habitats. *Nature* 269: 578–581.

Handford PT (1973a) Patterns of variation in a number of genetic systems in *Maniola jurtina*: the boundary region. *Proceedings of the Royal Society of London B* **183**: 265–284.

Handford PT (1973b) Patterns of variation in a number of genetic systems in *Maniola jurtina*: the Isles of Scilly. *Proceedings of the Royal Society of London B* **183**: 285–300.

Hanski I (1994a) A practical model of metapopulation dynamics. *Journal of Animal Ecology* **63**: 151–162.

Hanski I (1994b) Patch-occupancy dynamics in fragmented landscapes. *Trends in Ecology and Evolution* **9**: 131–135.

Hanski I (1998) Metapopulation dynamics. *Nature* **396**: 41–49.

Hanski I (1999) *Metapopulation Ecology*. Oxford: Oxford University Press.

Hanski I (2002) Metapopulations of animals in highly fragmented landscapes and population viability analysis. In: Beissinger SR, McCullough DR (eds.) *Population Viability Analysis*, Chicago, IL: University of Chicago Press, pp. 86–108.

Hanski I (2003) Biology of extinctions in butterfly metapopulations. In: Boggs CL, Watt WB, Ehrlich PR (eds.) *Butterflies: Ecology and Evolution Taking Flight*, Chicago, IL: University of Chicago Press, pp. 577–602.

Hanski I (2004) Metapopulation theory, its use and misuse. *Basic and Applied Ecology* **5**: 225–229.

Hanski I (2005) *The Shrinking World: Ecological Consequences of Habitat Loss*. Oldendorf/Luhe, Germany: International Ecology Institute.

Hanski I, Gaggiotti OE (2004) *Ecology, Genetics, and Evolution of Metapopulations*. Heidelberg, Germany: Elsevier.

Hanski I, Gilpin ME (1991) Metapopulation dynamics: brief history and conceptual domain. *Biological Journal of the Linnean Society* **42**: 3–16.

Hanki I, Gilpin ME (eds.) (1997) *Metapopulation Biology: Ecology, Genetics, and Evolution*. London: Academic Press.

Hanski I, Meyke E (2005) Large-scale dynamics of the Glanville fritillary butterfly: landscape structure, population processes, and weather. *Annales Zoologici Fennici* **42**: 379–395.

Hanski I, Ovaskainen O (2000) The metapopulation capacity of a fragmented landscape. *Nature* **404**: 755–758.

Hanski I, Saccheri I (2006) Molecular-level variation affects population growth in a butterfly metapopulation. *PLoS Biology* **4**: 129.

Hanski I, Singer MC (2001) Extinction–colonization dynamics and host-plant choice in butterfly metapopulations. *American Naturalist* **158**: 341–353.

Hanski I, Thomas CD (1994) Metapopulation dynamics and conservation: a spatially explicit model applied to butterflies. *Biological Conservation* **68**: 167–180.

Hanski I, Woiwod IP (1993) Spatial synchrony in the dynamics of moth and aphid populations. *Journal of Animal Ecology* **62**: 656–668.

Hanski I, Kouki J, Halkka A (1993) Three explanations of the positive relationship between distribution and abundance of species. In: Ricklefs RE, Schluter D (eds.) *Species Diversity in Ecological Communities: Historical and Geographical Perspectives*, Chicago, IL: University of Chicago Press, pp. 108–116.

Hanski I, Kuussaari M, Nieminen M (1994) Metapopulation structure and migration in the butterfly *Melitaea cinxia*. *Ecology* **7**: 747–762.

Hanski I, Pakkala T, Kuussaari M, Lei G (1995a) Metapopulation persistence of an endangered butterfly in a fragmented landscape. *Oikos* **72**: 21–28.

Hanski I, Pöyry J, Kuussaari M, Pakkala T (1995b) Multiple equilibria in metapopulation dynamics. *Nature* **377**: 618–621.

Hanski I, Alho J, Moilanen A (2000) Estimating the parameters of survival and migration of individuals in metapopulations. *Ecology* **81**: 239–251.

Hanski I, Breuker CJ, Schops K, Setchfield R, Nieminen M (2002) Population history and life history influence the migration rate of female Glanville fritillary butterflies. *Oikos* **98**: 87–97.

Hanski I, Erälahti C, Kankare M, Ovaskainen O, Siren H (2004) Variation in migration propensity among individuals maintained by landscape structure. *Ecology Letters* **7**: 958–966.

Hanski I, Saastamoinen M, Ovaskainen O (2006) Dispersal-related life-history trade-offs in a butterfly metapopulation. *Journal of Animal Ecology* **75**: 91–100.

Hansson B, Westerberg L (2002) On the correlation between heterozygosity and fitness in natural populations. *Molecular Ecology* **11**: 2467–2474.

Hardy ICW (1998) Butterfly battles: on conventional contests and hot property. *Trends in Ecology and Evolution* **13**: 385–386.

Hardy PB, Dennis RLH (1999) The impact of urban development on butterflies within a city region. *Biodiversity and Conservation* **8**: 1261–1279.

Hardy PB, Dennis RLH (2007) Seasonal and daily shifts in substrate use by settling butterflies: conserving resources for invertebrates has a behavioral dimension. *Journal of Insect Behavior* **20**: 181–199.

Haribal M, Feeny P (2003) Combined roles of contact stimulant and deterrents in assessment of host-plant quality by ovipositing zebra swallowtail butterflies. *Journal of Chemical Ecology* **29**: 653–670.

Harmer AS (2000) *Variation in British Butterflies*. Lymington, Hants.: Paphia Publishing.

Harper GL, Maclean N, Goulson D (2003) Microsatellite markers to assess the influence of population size, isolation and demographic change on the genetic structure of the UK butterfly *Polyommatus bellargus*. *Molecular Ecology* **12**: 3349–3357.

Harper GL, Maclean N, Goulson D (2006) Analysis of museum specimens suggests extreme genetic drift in the adonis blue butterfly (*Polyommatus bellargus*). *Biological Journal of the Linnean Society* **88**: 447–452.

Harper GL, MacLean N, Goulson D (2008) Molecular evidence for a recent founder event in the UK populations of the Adonis blue butterfly (*Polyommatus bellargus*). *Journal of Insect Conservation* **12**: 147–153.

Harrell FE Jr (2001) *Regression Modeling Strategies: With Applications to Linear Models, Logistic Regression, and Survival Analysis*. Berlin: Springer-Verlag.

Harrell FE Jr, Lee KL, Mark DB (1996) Multivariable prognostic models: issues in developing models, evaluating assumptions and adequacy, and measuring and reducing errors. *Statistics in Medicine* **15**: 361–388.

Harrison RG (1991) Molecular changes in speciation. *Annual Review of Ecology and Systematics* **22**: 281–308.

Harrison RG (1998) Linking evolutionary pattern and process: the relevance of species concepts for the study of speciation. In: Howard DJ, Berlocher SH (eds.) *Endless Forms: Species and Speciation*, New York: Oxford University Press, pp. 19–31.

Harrison RG, Bogdanowicz SM (1997) Patterns of variation and linkage disequilibrium in a field cricket hybrid zone. *Evolution* **51**: 493–505.

Hartl DL, Clark AG (1989) *Principles of Population Genetics*, 2nd edn. Sunderland, MA: Sinauer Associates.

Harvey GT (1983) A geographical cline in egg weights in *Choristoneura fumiferana* (Lepidoptera: Tortricidae) and its significance in population dynamics. *Canadian Entomologist* **115**: 1103–1108.

Harvey PH, Pagel MD (1991) *The Comparative Method in Evolutionary Biology*. Oxford: Oxford University Press.

Hassell MP (1998) The regulation of populations by density-dependent processes. In: Dempster JP, McLean IFG (eds.) *Insect Populations in Theory and in Practice*. Dordrecht, Netherlands: Kluwer, pp. 29–51.

Hassell MP (2000) Host–parasitoid population dynamics. *Journal of Animal Ecology* **69**: 543–566.

Hassell MP, Southwood TRE (1978) Foraging strategies of insects. *Annual Review of Ecology and Systematics* **9**: 75–98.

Hastie TJ, Tibshirani RJ (1990) *Generalized Additive Models*. London: Chapman and Hall.

Hastie T, Tibshirani R, Friedman JH (2001) *The Elements of Statistical Learning: Data Mining, Inference, and Prediction*. New York: Springer-Verlag.

Hastings A (1983) Can spatial variation alone lead to selection for dispersal? *Theoretical Population Biology* **24**: 244–251.

Hatfield T (1996) Genetic divergence in adaptive characters between sympatric species of sticklebacks. *American Naturalist* **149**: 1009–1029.

Haughton AJ, Champion GT, Hawes C, Heard MS, Brooks DR, Bohan DA, Clark SJ, Dewar AM, Firbank LG, Osborne JL, Perry JN, Rothery P, Roy DB, Scott RJ, Woiwod IP, Birchall C, Skellern MP, Walker JH, Baker P, Browne EL, Dewar AJG, Garner BH, Haylock LA, Horne SL, Mason NS, Sands RJN, Walker MJ (2003) Invertebrate responses to the management of genetically modified herbicide-tolerant and conventional spring crops. II. Within-field epigeal and aerial arthropods. *Philosophical Transactions of the Royal Society of London B* **358**: 1863–1877.

Hauser E (1996) Vergleichende Analyse der Zönosen tagaktiver Schmetterlinge im Sengsengebirge (Oberösterreich) (Lepidoptera). *Nota lepidopterologica* **18**: 247–265.

Häuser CL, Eckweilerm W (1997) A catalogue of the species-group taxa in *Agrodiaetus* Hübner, 1822, a subgenus of *Polyommatus* Latreille, 1804 (Lepidoptera: Lycaenidae). *Nachrichten des entomologischen Vereins Apollo* **16** (Suppl.): 53–112.

Häusler A, Dolek M, Güthler W, Market R (2008) Conservation efforts and strategies for forest canopies in Germany: a review of conservation programmes. In: Floren A, Schmidl J (eds.) *Canopy Arthropod Research in Central Europe: Basic and Applied Studies from the High Frontier*, Nuremberg, Germany: Bioform, pp. 563–572.

Hawkins BA, Devries PJ (1996) Altitudinal gradients in the body sizes of Costa Rican butterflies. *Acta Oecologica* **17**: 185–194.

Hawkins BA, Lawton JH (1995) Latitudinal gradients in butterfly body sizes: is there a general pattern? *Oecologia* **102**: 31–36.

Hawkins BA, Porter EE (2003) Water-energy balance and the geographic pattern of species richness of western Palearctic butterflies. *Ecological Entomology* **28**: 678–686.

Hayes JL (1982) A study of the relationships of diapause phenomena and other life history characters in temperate butterflies. *American Naturalist* **120**: 160–170.

Hazel WN, Smock R, Johnson MD (1990) A polygenic model for the evolution and maintenance of conditional strategies. *Proceedings of the Royal Society of London B* **242**: 181–187.

Heath J (1981) *Threatened Rhopalocera (butterflies) in Europe*. Strasbourg, France: Council of Europe.

Heath J, Pollard E, Thomas JA (1984) *Atlas of Butterflies in Britain and Ireland*. Harmondsworth, Middx.: Viking.

Hebert PDN (1983) Egg dispersal and adult feeding behaviour in the Lepidoptera. *Canadian Entomologist* 115: 1477–1481.

Hebert PDN, Beaton MJ (1993) *Methodologies for Allozyme Analysis Using Cellulose Acetate Electrophoresis*. Beaumont, TX: Helena Laboratories.

Hebert PDN, Cywinska A, Ball SL, deWaard JR (2003) Biological identifications through DNA barcodes. *Proceedings of the Royal Society of London B* 270: 313–321.

Hedrick PW (2000) *Genetics of Populations*. Sudbury, MA: Jones and Bartlett.

Hedrick PW, Lacy RC, Allendorf FW, Soulé ME (1996) Directions in conservation biology: comments on Caughley. *Conservation Biology* 10: 1312–1320.

Heikkinen RK, Luoto M, Virkkala R, Rainio K (2004) Effects of habitat cover, landscape structure and spatial variables on the abundance of birds in an agricultural–forest mosaic. *Journal of Applied Ecology* 41: 824–835.

Heikkinen RK, Luoto M, Kuussaari M, Pöyry J (2005) New insights into butterfly–environment relationships using partitioning methods. *Proceedings of the Royal Society of London B* 272: 2203–2210.

Hein S, Binzenhöfer B, Poethke HJ, Biedermann R, Settele J, Schröder B (2007a) The generality of habitat suitability models: a practical test with two insect groups. *Basic and Applied Ecology* 8: 310–320.

Hein S, Voss J, Poethke HJ, Schröder B (2007b) Habitat suitability models for the conservation of thermophilic grasshoppers and bush crickets. *Journal of Insect Conservation* 11: 221–240.

Heinrich B (1981) *Insect Thermoregulation*. New York: John Wiley.

Heinrich B (1986a) Comparative thermoregulation of four montane butterflies of different mass. *Physiological Zoology* 59: 616–626.

Heinrich B (1986b) Thermoregulation and flight activity of a satyrine, *Coenonympha inornata* (Lepidoptera: Satyridae). *Ecology* 67: 593–597.

Heinrich B (1993) *The Hot-Blooded Insects: Strategies and Mechanisms of Thermoregulation*. Berlin: Springer-Verlag.

Heinrich B (1996) *The Thermal Warriors: Strategies of Insect Survival*. Cambridge, MA: Harvard University Press.

Heinrich B, Collins SL (1983) Caterpillar leaf damage and the game of hide and seek with birds. *Ecology* 64: 592–602.

Heinrich B, Mommsen TP (1985) Flight of winter moths near 0 °C. *Science* 228: 177–179.

Heinz CA, Feeny P (2005) Effects of contact chemistry and host plant experience in the oviposition behaviour of the eastern black swallowtail butterfly. *Animal Behaviour* 69: 107–115.

Heinz SK, Wissel C, Conradt L, Frank K (2007) Integrating individual movement behaviour into dispersal functions. *Journal of Theoretical Biology* 245: 601–609.

Held C, Spieth H (1999) First experimental evidence of pupal summer diapause in *Pieris brassicae* L.: the evolution of local adaptedness. *Journal of Insect Physiology* 45: 587–598.

Hellmann JJ, Weiss SB, McLaughlin JF, Boggs CL, Ehrlich PR, Launer AE, Murphy DD (2003) Do hypotheses from short-term studies hold in the long-term? An empirical test. *Ecological Entomology* 28: 74–84.

Hengeveld R, Haeck J (1982) The distribution of abundance. I. Measurements. *Journal of Biogeography* 9: 303–316.

Hennig W (1968) *Elementos de una Sistemática Filogenética* (Translation of: *Grundzüge einer Theorie der phylogenetischen Systematik*), 2nd edn. Buenos Aires: Editorial Universitaria de Buenos Aires.

Hennig C, Hausdorf B (2006) A robust distance coefficient between distribution areas incorporating geographic distances. *Systematic Biology* 55: 170–175.

Henriksen HJ, Kreutzer I (1982) *The Butterflies of Scandinavia in Nature*. Odense, Denmark: Skandinavisk Bogforlag.

Hering M (1926) *Die Biologie der Schmetterlinge*. Berlin: Springer-Verlag.

Hermann G (1998) Erfassung von Präimaginalstadien bei Tagfaltern: ein notwendiger Standard für Bestandsaufnahmen zu Planungsvorhaben. *Naturschutz und Landschaftsplanung* 30: 133–142.

Hermann G (1999) Methoden der qualitativen Erfassung von Tagfaltern. In: Settele J, Feldmann R, Reinhardt R (eds.) *Die Tagfalter Deutschlands: Ein Handbuch für Freilandökologen, Umweltplaner und Naturschützer*, Stuttgart, Germany, Eugen Ulmer, pp. 124–143.

Hermann G, Steiner R (1997) Eiablage- und Larvalhabitat des Komma-Dickkopffalters (*Hesperia comma* Linné 1758). *Carolinea* 55: 35–42.

Hermann G, Steiner R (1998) Eiablagehabitat und Verbreitung des Violetten Feuerfalters (*Lycaena alciphron*) in Baden-Württemberg (Lepidoptera, Lycaenidae). *Carolinea* 56: 99–102.

Hermann G, Steiner R (2000) Der Braune Eichen-Zipfelfalter in Baden-Württemberg: ein Beispiel für die extreme Bedrohung von Lichtwaldarten. *Naturschutz und Landschaftsplanung* 32: 271–277.

Hernandez MIM, Benson WW (1998) Small-male advantage in the territorial tropical butterfly *Heliconius sara*

(Nymphalidae): a paradoxical strategy? *Animal Behaviour* **56**: 533–540.

Herrmann O (1926) Ein Hybrid von *Lyc. corydon* × *Lyc. icarus Internationale Entomologische Zeitschrift* **19**: 380–381.

Herting B (1960) Biologie der westpaläarktischen Raupenfliegen (Dipt., Tachinidae). *Monographien zur angewandte Entomologie* **16**: 1–188.

Heslop Harrison JW (1951) A hybrid between *Pieris napi* male and *P. rapae* female (Lep., Pieridae). *Entomologist* **84**: 99–102.

Hesselbarth G, van Oorshot H, Wagener S (1995) *Die Tagfalter der Türkei unter Berücksichtigung der angrenzenden Länder*, 3 vols. Bocholt, Germany: S Wagener.

Hewitt GM (1988) Hybrid zones: natural laboratories for evolutionary studies. *Trends in Ecology and Evolution* **3**: 158–167.

Hewitt GM (1996) Some genetic consequences of ice ages, and their role in divergence and speciation. *Biological Journal of the Linnean Society* **58**: 247–276.

Hewitt GM (1999) Post-glacial re-colonization of European biota. *Biological Journal of the Linnean Society* **68**: 87–112.

Hewitt GM (2000) The genetic legacy of the Quaternary ice ages. *Nature* **405**: 907–913.

Hewitt GM (2001) Speciation, hybrid zones and phylogeography – or seeing genes in space and time. *Molecular Ecology* **10**: 537–549.

Hewitt GM (2004) Genetic consequences of climatic oscillation in the Quaternary. *Philosophical Transactions of the Royal Society of London B* **359**: 183–195.

Hickling R, Roy DB, Hill JK, Fox R, Thomas CD (2006) The distributions of a wide range of taxonomic groups are expanding polewards. *Global Change Biology* **12**: 450–455.

Higgins LG. (1969) Observations sur les *Melanargia* dans le midi de la France. *Alexanor* **6**: 85–90.

Higgins LG, Hargreaves B (1983) *The Butterflies of Britain and Europe*. London: Collins.

Higgins LG, Riley ND (1970) *A Field Guide to the Butterflies of Britain and Europe*. London: Collins.

Higgins LG, Riley ND (1983) *A Field Guide to the Butterflies of Britain and Europe*. London: Collins.

Hilborn R, Mangel M (1997) *The Ecological Detective: Confronting Models with Data: Monographs in Population Biology*. Princeton, NJ: Princeton University Press.

Hilker M, Meiners T (eds.) (2002) *Chemoecology of Insect Eggs and Egg Deposition*. Oxford: Blackwell Science.

Hill MO (1979) *TWINSPAN: A FORTRAN Program for Arranging Multivariate Data in an Ordered Two-Way Table by Classification of the Individuals and Attributes*. Ithaca, NY: Cornell University.

Hill CJ (1989) The effect of adult diet on the biology of butterflies. II. The common crow butterfly, *Euploea core corinna*. *Oecologia* **81**: 258–266.

Hill CJ, Pierce NE (1989) The effect of adult diet on the biology of butterflies: the common imperial blue, *Jalmenus evagoros*. *Oecologia* **81**: 249–257.

Hill RI, Vaca JF (2004) Differential wing strength in *Pierella* butterflies (Nymphalidae, Satyrinae) supports the deflection hypothesis. *Biotropica* **36**: 362–370.

Hill JK, Thomas CD, Lewis OT (1996) Effects of habitat patch size and isolation on dispersal by *Hesperia comma* butterflies: implications for metapopulation structure. *Journal of Animal Ecology* **65**: 725–735.

Hill JK, Thomas CD, Blakeley DS (1999a) Evolution of flight morphology in a butterfly that has recently expanded its geographic range. *Oecologia* **121**: 165–170.

Hill JK, Thomas CD, Lewis OT (1999b) Flight morphology in fragmented populations of a rare British butterfly, *Hesperia comma*. *Biological Conservation* **87**: 277–283.

Hill JK, Thomas CD, Huntley B (1999c) Climate and habitat availability determine 20th century changes in a butterfly's range margin. *Proceedings of the Royal Society of London B* **266**: 1197–1206.

Hill JK, Collingham YC, Thomas CD, Blakeley DS, Fox R, Moss D, Huntley B (2001) Impacts of landscape structure on butterfly range expansion. *Ecology Letters* **4**: 313–321.

Hill JK, Thomas CD, Fox R, Telfer MG, Willis SG, Asher J, Huntley B (2002) Responses of butterflies to twentieth century climate warming: implications for future ranges. *Proceedings of the Royal Society of London B* **269**: 2163–2171.

Hill JK, Dytham C, Hughes CL (2005) Evolutionary changes in expanding butterfly populations. In: Fellowes MDE, Holloway GJ, Rolff J (eds.) *Insect Evolutionary Ecology*, Wallingford, Oxon.: CABI Publishing, pp. 515–533.

Hill JK, Hughes CL, Dytham C, Searle JB (2006) Genetic diversity in butterflies: interactive effects of habitat fragmentation and climate-driven range expansion. *Biology Letters* **2**: 152–154.

Hillis DM, Moritz C, Mable BK (eds.) (1996) *Molecular Systematics*. Sunderland, MA: Sinauer Associates.

Hilpert H (1992) Zur Systematik der Gattung *Ichneumon* Linnaeus, 1758 in der Westpalaearktis (Hymenoptera, Ichneumonidae, Ichneumoninae). *Entomofauna* (Suppl. 6): 1–389.

Hinton HE (1981) *Biology of Insect Eggs*. Oxford: Pergamon Press.

Hinz R (1983) The biology of the European species of the genus *Ichneumon* and related species (Hymenoptera:

Ichneumonidae). *Contributions of the American Entomological Institute* **20**: 151–152.

Hinz R, Horstmann K (2007) Über Wirtsbeziehungen europäischer *Ichneumon*-Arten (Insecta, Hymenoptera, Ichneumonidae, Ichneumoninae). *Spixiana* **30**: 39–63.

Hiroki M, Obara Y (1997) Delayed mating and its cost to female reproduction in the butterfly, *Eurema hecabe*. *Journal of Ethology* **15**: 79–85.

Hiroki M, Obara Y, Kato Y (1998) Changes in age-related reproductive tactics in the female of the butterfly, *Eurema hecabe*. *Naturwissenschaften* **85**: 551–555.

Hirota T, Obara Y (2000) The influence of air temperature and sunlight intensity on mate-locating behavior of *Pieris rapae crucivora*. *Zoological Science* **17**: 1081–1087.

Hirzel A, Arlettaz R (2003) Modelling habitat suitability for complex species distributions by the environmental-distance geometric mean. *Environmental Management* **32**: 614–623.

Hirzel A, Guisan A (2002) Which is the optimal sampling strategy for habitat suitability modelling? *Ecological Modelling* **157**: 331–341.

Hirzel AH, Helfer V, Metral F (2001) Assessing habitat-suitability models with a virtual species. *Ecological Modelling* **145**: 111–122.

Hochberg ME, Thomas JA, Elmes GW (1992) A modelling study of the population dynamics of a large blue butterfly, *M. rebeli*, a parasite of red ant nests. *Journal of Animal Ecology* **61**: 397–409.

Hochberg ME, Clarke RT, Elmes GW, Thomas JA (1994) Population-dynamic consequences of direct and indirect interactions involving a large blue butterfly and its plant and red ant hosts. *Journal of Animal Ecology* **63**: 375–391.

Hochberg ME, Elmes GW, Thomas JA, Clarke RT (1996) Mechanisms of local persistence in coupled host–parasitoid associations: the case model of *Maculinea rebeli* and *Ichneumon eumerus*. *Philosophical Transactions of the Royal Society of London Series B* **351**: 1713–1724.

Hochberg ME, Elmes GW, Thomas JA, Clarke RT (1998) Effects of habitat reduction and persistence of *Ichneumon eumerus* (Hymenoptera: Ichneumonidae), the specialist parasitoid of *Maculinea rebeli* (Lepidoptera: Lycaenidae). *Journal of Insect Conservation* **2**: 59–66.

Hockin DC (1981) The environmental determinants of the insular butterfly faunas of the British Isles. *Biological Journal of the Linnean Society* **16**: 63–70.

Hodgson JG (1993) Commonness and rarity in British butterflies. *Journal of Applied Ecology* **30**: 407–427.

Hodkinson ID (1999) Species response to global environmental change or why ecophysiological models are

important: a reply to Davis *et al*. *Journal of Animal Ecology* **68**: 1259–1262.

Høegh-Guldberg O (1974) Natural pattern variation and the effect of cold treatment in the genus *Aricia* R. L. (Lepidoptera, Lycaenidae) (*Aricia* study no. 14). *Proceedings of the British Entomological and Natural History Society* **7**: 37–44.

Høegh-Guldberg O, Jarvis FVL (1970) Central and North European *Aricia* (Lep.): relationships, heredity, evolution. *Natura Jutlandica* **15**: 1–119.

Hoeting JA, Madigan D, Raftery AE, Volinsky CT (1999) Bayesian model averaging: a tutorial. *Statistical Science* **14**: 382–417.

Hoffmann KH (1985) Metabolic and enzyme adaptation to temperature. In: Hoffmann KH (ed.) *Environmental Physiology and Biochemistry of Insects*, Berlin: Springer-Verlag, pp. 1–32.

Hoffmann AA (1996) Acclimation: increasing survival at a cost. *Trends in Ecology and Evolution* **10**: 1–2.

Hofmeister J, Mihaljevic M, Hosek J (2004) The spread of ash (*Fraxinus excelsior*) in some European oak forests: an effect of nitrogen deposition or successional change? *Forest Ecology and Management* **203**: 35–47.

Holbeck HB, Clause HD, Reddersen J (2000) Selection of nectar sources by butterflies and burnet moths in organic field boundary habitats (Papilionidea, Hesperioidea and Zygaenidae). *Entomologiske Meddelelser* **68**: 47–59.

Holdhaus K (1954) *Die Spuren der Eiszeit in der Tierwelt Europas*. Innsbruck, Austria: Universitätsverlag Wagener.

Hole DG, Perkins AJ, Wilson JD, Alexander IH, Grice PV, Evans AD (2005) Does organic farming benefit biodiversity? *Biological Conservation* **122**: 133–130.

Holl K (1995) Nectar resources and their influence on butterfly communities on reclaimed coal surface mines. *Restoration Ecology* **3**: 76–85.

Holland JD, Bert DG, Fahrig L (2004) Determining the spatial scale of species' response to habitat. *BioScience* **54**: 227–233.

Holloway JD (1980) *Coenonympha darwiniana* Staudinger (Lepidoptera: Satyridae): a hybrid origin? *Entomologist's Gazette* **31**: 195–198.

Holt RD (1997) From metapopulation dynamics to community structure. In: Hanski I, Gilpin ME (eds.) *Metapopulation Biology*, San Diego, CA: Academic Press, pp. 149–165.

Holt RD, Barfield M (2001) On the relationship between the ideal free distribution and the evolution of dispersal. In: Clobert J, Danchin E, Dhondt AA, Nichols JD (eds.) *Dispersal*, Oxford: Oxford University Press, pp. 83–95.

Holt RD, Lawton JH (1993) Apparent competition and enemy-free space in insect host–parasitoid communities. *American Naturalist* **142**: 623–645.

Honek A (1993) Intraspecific variation in body size and fecundity in insects: a general relationship. *Oikos* **66**: 483–492.

Honey MR (1998) [Exhibition report.] *British Journal of Entomology and Natural History* **11**: 113.

Honkavaara J, Koivula M, Korpimaki E, Siitari H, Vittali J (2002) Ultraviolet vision and foraging in terrestrial vertebrates. *Oikos* **98**: 505–511.

Hoole JC, Joyce DA, Pullin AS (1999) Estimates of gene flow between populations of the swallowtail butterfly, *Papilio machaon* in Broadland, UK and implications for conservation. *Biological Conservation* **89**: 293–299.

Hopkins RJ, van Loon JJA (2001) The effect of host acceptability on oviposition and egg accumulation by the small white butterfly, *Pieris rapae*. *Physiological Entomology* **26**: 149–157.

Hopper R (1999) Risk-spreading and bet-hedging in insect population biology. *Annual Review of Entomology* **44**: 535–560.

Horstmann K (1999) Revisionen von Schlupfwespen-Arten III. *Mitteilungen der Münchner Entomologischen Gesellschaft* **90**: 39–50.

Horstmann K (2007) Revision der westpaläarktischen Arten von *Anisobas* Wesmael, 1845 (Hymenoptera, Ichneumonidae, Ichneumoninae). *Entomofauna* **28**: 93–115.

Horstmann K, Fiedler K, Baumgarten HT (1997) Zur Taxonomie und Bionomie einiger Ichneumonidae (Hymenoptera) als Parasitoide westpaläarktischer Lycaenidae (Lepidoptera). *Nachrichtenblatt der bayerischen Entomologen* **46**: 2–7.

Hortal J, García-Pereira P, García-Barros E (2004) Butterfly species richness in mainland Portugal: predictive models of geographic distribution patterns. *Ecography* **27**: 68–82.

Hosmer DW, Lemeshow S (2000) *Applied Logistic Regression*. New York: John Wiley.

Höttinger H (2002) Checklist und Rote Liste der Tagschmetterlinge der Stadt Wien (Lepidoptera: Papilionoidea und Hesperioidea). *Beiträge zur Entomofaunistik* **3**: 103–123.

Houlder D, Hutchinson M, Nix H, McMahon J (2001) *ANUCLIM 5.1 User's Guide*. Canberra: Australian National University, Centre for Resource and Environmental Studies.

Hovestadt T, Poethke HJ (2006) The control of emigration and its consequences for the survival of populations. *Ecological Modelling* **190**: 443–453.

Hovestadt T, Messner S, Poethke HJ (2001) Evolution of reduced dispersal mortality and 'fat-tailed' dispersal kernels in autocorrelated landscapes. *Proceedings of the Royal Society of London B* **268**: 385–391.

Howe PD (2004) The ecological consequences of morphological variation in the common blue butterfly *Polyommatus icarus* (Rott.) in the United Kingdom. Ph.D. thesis, Oxford Brookes University.

Howe PD, Bryant SR, Shreeve TG (2007) Predicting body temperature and activity of adult *Polyommatus icarus* using neural network models under current and projected climate scenarios. *Oecologia* **153**: 857–869.

Huey RB, Kingsolver JG (1989) Evolution of thermal sensitivity of ectotherm performance. *Trends in Ecology and Evolution* **4**: 131–135.

Hughes JB (2000) The scale of resource specialization and the distribution and abundance of lycaenid butterflies. *Oecologia* **123**: 375–383.

Hughes L, Siew-Woon Chang B, Wagner D, Pierce NE (2000) Effects of mating history on ejaculate size, fecundity, longevity, and copulation duration in the ant-tended lycaenid butterfly, *Jalmenus evagoras*. *Behavioral Ecology and Sociobiology* **47**: 119–128.

Hughes CL, Hill JK, Dytham C (2003) Evolutionary trade-offs between reproduction and dispersal in populations at expanding range boundaries. *Proceedings of the Royal Society of London B* (Suppl. 2) **270**: 147–150.

Hughes CL, Dytham C, Hill JK (2007) Modelling and analysing evolution of dispersal in populations at expanding range boundaries. *Ecological Entomology* **32**: 437–445.

Huisman J, Olff H, Fresco LFM (1993) A hierarchical set of models for species response analysis. *Journal of Vegetation Science* **4**: 37–46.

Hula V, Konvička M, Pavlicko A, Fric Z (2004) Marsh Fritillary (*Euphydryas aurinia*) in the Czech Republic: monitoring, metapopulation structure, and conservation of an endangered butterfly. *Entomologica Fennica* **15**: 231–241.

Hunter AF (1995) Ecology, life-history, and phylogeny of outbreak and nonoutbreaking species. In: Cappuccino N, Price PW (eds.) *Population Dynamics: New Approaches and Synthesis*, San Diego, CA: Academic Press, pp. 41–61.

Huntley B (1988) *Europe*. In: Huntley B, Webb T (eds.) *Vegetation History*, Dordrecht, Netherlands: Kluwer, pp. 341–383.

Huntley B, Webb T III (1989) Migration: species' response to climatic variations caused by changes in the Earth's orbit. *Journal of Biogeography* **16**: 5–19.

Huntley B, Berry PM, Cramer W, McDonald A (1995) Modelling present and potential future ranges of some

European higher plants using climate response surfaces. *Journal of Biogeography* **22**: 967–1001.

Huntley B, Green RE, Collingham YC, Hill JK, Willis SG, Bartlein PJ, Cramer W, Hegemeijer WJM, Thomas CD (2004) The performance of models relating species' geographical distributions to climate is independent of trophic level. *Ecology Letters* **7**: 417–426.

Huntzinger M (2003) Effects of fire management practices on butterfly diversity in the forested western United States. *Biological Conservation* **113**: 1–12.

Hurst GDD, Sharpe RG, Broomfield AH, Walker LE, Majerus TMO, Zakharov IA, Majerus MEN (1995) Sexually transmitted disease in a promiscuous insect, *Adalia bipunctata*. *Ecological Entomology* **20**: 230–236.

Hurvich CM, Tsai C (1989) Regression and time series model selection in small sample sizes. *Biometrika* **76**: 297–307.

Hutchinson GE (1957) Concluding remarks. *Cold Spring Harbor Symposia in Quantitative Biology* **22**: 415–427.

Ibrahim KM, Nichols RA, Hewitt GM (1996) Spatial patterns of genetic variation generated by different forms of dispersal during range expansion. *Heredity* **77**: 282–291.

Ide JY (2004) Selection of age classes of *Sasa* leaves by caterpillars of the skipper butterfly – *Thoressa varia* using albo-margination of overwintered leaves. *Journal of Ethology* **22**: 99–103.

Ide J, Kondoh M (2000) Male-female evolutionary game on mate-locating behaviour and evolution of mating systems in insects. *Ecology Letters* **3**: 433–440.

Ilse D (1928) Über den Farbensinn der Tagfalter. *Zeitschrift für vergleichende Physiologie* **8**: 658–692.

Ilse D (1937) New observations on responses to colours in egg-laying butterflies. *Nature* **140**: 544–546.

Ims RA, Yoccoz NG (1997) Studying transfer processes in metapopulations. In: Hanski I, Gilpin ME (eds.) *Metapopulation Biology*, San Diego, CA: Academic Press, pp. 247–265.

Inouye DW (2000) The ecological and evolutionary significance of frost in the context of climate change. *Ecology Letters* **3**: 457–463.

IPCC (Intergovernmental Panel on Climate Change) (2001) *Climate Change 2001: The Scientific Basis*, Contribution of Working Group I to the Third Assessment Report of the Intergovernmental Panel on Climate Change, Houghton JT, Ding Y, Driggs DJ, Noguer M, van der Linden PJ, Dai X, Maskell K, Johnson CA (eds.). Cambridge: Cambridge University Press.

IPCC (2002) *Climate Change and Biodiversity*, Gitay H, Suárez A, Watson RT, Dokken DJ (eds.). Geneva, Switzerland: IPCC.

IPCC (2007) *Climate Change 2007: Synthesis Report*, Contribution of Working Groups I, II and III to the Fourth Assessment Report of the Intergovernmental Panel on Climate Change, Core Writing Team: Pachauri RK, Reisinger A (eds.). Geneva, Switzerland: IPCC.

Isaac NJB, Mallet J, Mace GM (2004) Taxonomic inflation: its influence on macroecology and conservation. *Trends in Ecology and Evolution* **19**: 464–469.

IUCN (1994) *IUCN Red List Categories*. Gland, Switzerland: IUCN Species Survival Commission.

IUCN (2001) *IUCN Red List Categories and Criteria, Version 3.1*. Gland, Switzerland: IUCN Species Survival Commission.

IUCN (2003) *Guidelines for Application of IUCN Red List Criteria at Regional Levels: Version 3.0*. Gland, Switzerland: IUCN Species Survival Commission.

Jackson SM (1980) Changes since 1900 in the distribution of butterflies in Yorkshire and elsewhere in the north of England. *Entomologist's Record* **105**: 139–142.

Jacobs MD, Watt WB (1994) Seasonal adaptation vs. physiological constraint: photoperiod, thermoregulation and flight in *Colias* butterflies. *Functional Ecology* **8**: 366–376.

Jaenike J (1978) On optimal oviposition behaviour in phytophagous insects. *Theoretical Population Biology* **14**: 350–356.

Jaenike J (1986) Feeding behavior and future fecundity in *Drosophila*. *American Naturalist* **127**: 118–123.

Janosi IM, Scheuring I (1997) On the evolution of density dependent dispersal in a spatially structured population model. *Journal of Theoretical Biology* **187**: 397–408.

Jansson R (2003) Global patterns in endemism explained by past climatic change. *Proceedings of the Royal Society of London B* **270**: 583–590.

Janz N (1998) Sex-linked inheritance of host-plant specialization in a polyphagous butterfly. *Proceedings of the Royal Society of London B* **265**: 1–4.

Janz N (2002) Evolutionary ecology of oviposition strategies. In: Hilker M, Meiners T (eds.) *Chemoecology of Insect Eggs and Egg Deposition*, Oxford: Blackwell Science, pp. 349–376.

Janz N (2003a) Sex linkage of host-plant use in butterflies. In: Boggs CL, Watt WB, Ehrlich PR (eds.) *Butterflies: Ecology and Evolution Taking Flight*, Chicago, IL: University of Chicago Press, pp. 229–239.

Janz N (2003b) The cost of polyphagy: oviposition decision time vs. error rate in a butterfly. *Oikos* **100**: 493–496.

Janz N (2005) The relationship between habitat selection and preference for adult and larval food resources in the polyphagous butterfly *Vanessa cardui* (Lepidoptera: Nymphalidae). *Journal of Insect Behavior* **18**: 767–780.

Janz N, Nylin S (1997) The role of female search behaviour in determining host plant range in plant feeding insects: a test of the information processing hypothesis. *Proceedings of the Royal Society of London B* **264**: 701–707.

Janz N, Nylin S (1998) Butterflies and plants: a phylogenetic study. *Evolution* **52**: 486–502.

Janz N, Nylin S, Wedell N (1994) Host plant utilization by the comma butterfly: sources of variation and evolutionary implications. *Oecologia* **99**: 132–140.

Janz N, Bergström A, Sjögren A (2005) The role of nectar sources for oviposition decisions of the common blue butterfly *Polyommatus icarus*. *Oikos* **109**: 535–538.

Janz N, Nylin S, Wahlberg N (2006) Diversity begets diversity: host expansions and the diversification of plant-feeding insects. *BMC Evolutionary Biology* **6**: 4.

Janzen DH (1980) When is it coevolution? *Evolution* **34**: 611–612.

Javois J, Tammaru T (2004) Reproductive decisions are sensitive to cues of life expectancy: the case of a moth. *Animal Behaviour* **68**: 249–255.

Jeanneret P, Schupbach B, Pfiffner L, Herzog F, Walter T (2003) The Swiss agri-environmental programme and its effects on selected biodiversity indicators. *Journal for Nature Conservation* **11**: 213–220.

Jennersten O (1980) Nectar source plant selection and distribution pattern in an autumn population of *Gonepteryx rhamni* (Lep. Pieridae). *Entomologisk Tidskrift* **101**: 109–114.

Jennersten O (1984) Flower visitation and pollination efficiency of some North European butterflies. *Oecologia* **63**: 80–89.

Jennersten O (1988) Pollination of *Dianthus deltoides* (Caryophyllaceae): effects of habitat fragmentation on visitation and seed set. *Conservation Biology* **2**: 359–366.

Jervis MA, Kidd NAC (1986) Host-feeding strategies in hymenopteran parasitoids. *Biological Reviews* **61**: 395–434.

Jervis MA, Boggs CL, Ferns PN (2005) Egg maturation strategy and its associated trade-offs: a synthesis focusing on Lepidoptera. *Ecological Entomology* **30**: 359–375.

Jervis MA, Ferns PN, Boggs CL (2007) A trade-off between female lifespan and larval diet breadth at the interspecific level in Lepidoptera. *Evolutionary Ecology* **21**: 307–323.

Jiggins FM (2002) Widespread 'hilltopping' in *Acraea* butterflies and the origin of sex-role-reversed swarming in *Acraea encedon* and *A. encedana*. *African Journal of Ecology* **40**: 228–231.

Jiggins CD, Mallet J (2000) Bimodal hybrid zones and speciation. *Trends in Ecology and Evolution* **15**: 250–255.

Jiggins FM, Hurst GDD, Majerus MEN (2000) Sex ratio distorting *Wolbachia* causes sex role reversal in its butterfly host. *Proceedings of the Royal Society of London B* **267**: 69–74.

Jiggins CD, Linares M, Mallet J, Naisbit RE, Salazar C, Yang Z (2001) Sex-linked hybrid sterility in a butterfly. *Evolution* **55**: 1631–1638.

Johannesen J, Veith M, Seitz A (1996) A population genetics structure of the butterfly *Melitaea didyma* (Nymphalidae) along a northern distribution range border. *Molecular Ecology* **5**: 259–267.

Johannesen J, Schwing U, Seufert W, Seitz A, Veith M (1997) Analysis of gene flow and habitat patch network for *Chazara briseis* (Lepidoptera: Satyridae) in an agricultural landscape. *Biochemical Systematics and Ecology* **25**: 419–427.

Johansson AS (1951) Studies on the relation between *Apanteles glomeratus* L. (Hym., Braconidae) and *Pieris brassicae* L. (Lepid., Pieridae). *Norsk Entomologisk Tidsskrift* **8**: 145–186.

Johansson J, Bergström A, Janz N (2007) The benefit of additional oviposition targets for a polyphagous butterfly. *Journal of Insect Science* **7**(3). Online: insectsience.org/7.03.

Johnson CG (1969) *Migration and Dispersal of Insects by Flight*. London: Methuen.

Johnson GB (1971) Analysis of enzyme variation in natural populations of the butterfly *Colias eurytheme*. *Proceedings of the National Academy of Sciences, USA* **68**: 997–1001.

Johnson GB (1977) Assessing electrophoretic similarity: the problem of hidden heterogeneity. *Annual Review of Ecology and Systematics* **8**: 309–328.

Johst K, Drechsler M, Thomas J, Settele J (2006) Influence of mowing on the persistence of two endangered large blue butterfly species. *Journal of Applied Ecology* **43**: 333–342.

Jones RE (1977a) Movement patterns and egg distribution in cabbage butterflies. *Journal of Animal Ecology* **46**: 195–212.

Jones RE (1977b) Search behaviour: a study of three caterpillar species. *Behaviour* **60**: 237–259.

Jones RE (1987) Behavioural evolution in the cabbage butterfly (*Pieris rapae*). *Oecologia* **72**: 69–76.

Jones RE (1991) Host selection and oviposition on plants. In: Bailey WJ, Ridsdill-Smith J (eds.) *Reproductive Behaviour of Insects: Individuals and Populations*, London: Chapman and Hall, pp. 108–138.

Jones RE, Hart JR, Bull GD (1982) Temperature, size and egg production in the cabbage butterfly, *Pieris rapae* L. *Australian Journal of Zoology* **30**: 223–231.

Jones KN, Odendaal FJ, Ehrlich PR (1986) Evidence against the spermatophore as paternal investment in checkerspot butterflies (Euphydryas, Nymphalidae). *American Midland Naturalist* **116**: 1–6.

Jones MJ, Lace LA, Hounsome MV, Hamer K (1987) The butterflies and birds of Madeira and La Gomera: taxon cycles and human influence. *Biological Journal of the Linnean Society* **31**: 95–111.

Jordan K (1896) On mechanical selection and other problems. *Novitates Zoologicae* **3**: 426–525.

Jordan K (1905) Der Gegegensatz zwischen geographischer unt nichtgeographischer Variation. *Zeitschrift für wissenschaftliche Zoologie* **83**: 151–210.

Jordano D, Gomáriz G (1994) Variation in phenology and nutritional quality between host plants and its effect on larval performance in a specialist butterfly, *Zerynthia rumina*. *Entomologia Experimentalis et Applicata* **71**: 271–277.

Jordano D, Thomas CD (1992) Specificity of an ant–lycaenid interaction. *Oecologia* **91**: 431–438.

Jordano D, Fernández Haeger J, Rodríguez J (1990) The effect of seed predation by *Tomares ballus* (Lepidoptera: Lycaenidae) on *Astragalus lusitanicus* (Fabaceae): determinants of differences among patches. *Oikos* **57**: 250–256.

Jordano D, Rodríguez J, Thomas CD, Fernández Haeger J (1992) The distribution and density of a lycaenid butterfly in relation to *Lasius* ants. *Oecologia* **91**: 439–446.

Joron M (2005) Polymorphic mimicry, microhabitat use, and sex-specific behaviour. *Journal of Evolutionary Biology* **18**: 547–556.

Joron M, Brakefield PM (2003) Captivity masks inbreeding effects on male mating success in butterflies. *Nature* **424**: 191–194.

Joron M, Mallet JLB (1998) Diversity in mimicry: paradox or paradigm? *Trends in Ecology and Evolution* **13**: 461–466.

Joron M, Jiggins CD, Papanicolaou A, McMillan WO (2006) *Heliconius* wing patterns: an evo-devo model for understanding phenotypic diversity. *Heredity* **97**: 157–167.

Joy MK, Death RG (2004) Predictive modelling and spatial mapping of freshwater fish and decapod assemblages using GIS and neural networks. *Freshwater Biology* **49**: 1036–1052.

Joy J, Pullin AS (1997) The effects of flooding on the survival and behaviour of overwintering large heath butterfly *Coenonympha tullia* larvae. *Biological Conservation* **82**: 61–66.

Joy J, Pullin AS (1999) Field studies on flooding and survival of overwintering large heath butterfly *Coenonympha tullia* larvae on Fenn's and Whixall Mosses in Shropshire and Wrexham, UK. *Ecological Entomology* **24**: 426–443.

Joyce DA, Pullin AS (2003) Conservation implications of the distribution of genetic diversity at different scales: a case study using the marsh fritillary butterfly (*Euphydryas aurinia*). *Biological Conservation* **114**: 453–461.

Juan C, Emerson BC, Oromí P, Hewitt G (2000) Colonization and diversification: towards a phylogeographic synthesis for the Canary Islands. *Trends in Ecology and Evolution* **15**: 104–109.

Julliard C (1948) Les parasites de la chrysalide de *Vanessa urticae* dans la région de Zinal (Valais). *Mitteilungen der schweizerischen Entomologischen Gesellschaft* **21**: 557–565.

Jutzeler D (1989a) Kann das Weibchen von *Plebejus argus* (Linnaeus, 1761) Ameisen riechen? (Lepidoptera, Lycaenidae). *Mitteilungen der Entomologischen Gesellschaft Basel, NF,* **39**: 150–159.

Jutzeler D (1989b) Weibchen der Bläulingsart *Lycaeides idas* L. riechen ihre Wirtsameisen (Lepidotera, Lycaenidae). *Mitteilungen der entomologischen Gesellschaft Basel, NF,* **39**: 95–118.

Jutzeler D (1992) Elevage d'*Erebia epistygne* (Hübner, 1824) et d'*Erebia neoridas* (Boisduval, 1828) (Lepidoptera: Satyridae). *Bulletin de la Société Entomologique de Mulhouse* **1992**: 17–32.

Jutzeler D (1994) Contribution à la connaisance de l'écologie et du développement de *Brenthis hecate* (Schiffermüller, 1775) dans le Sud-Est de la France (Lepidoptera, Nymphalidae). *Bulletin de la Société Entomologique de Mulhouse* **1994**: 5–8.

Jutzeler D, De Bros E (1996) *Satyrus ferula* (Fabricius, 1793) du Valais (Suisse): durée extraordinairement longue de l'eclosion de la chenille au stade L1 (Lepidoptera: Nymphalidae, Satyrinae). *Linneana Belgica* **15**: 315–316.

Jutzeler D, Leestmans R (1994) Les états préimaginaux et l'écologie de *Satyrus actaea* Esper (1780) dans le S.-E. de la France: considérations sur la taxonomie et la géonémie des taxons du genre *Satyrus* (s. str.) (Lepidoptera: Nymphalidae, Satyrinae). *Linneana Belgica* **14**: 275–288.

Jutzeler D, Russo L, De Bros E (1995) Les premiers états de *Melanargia russiae ssp. japygia* (Cyrillo, 1787) de 'Le Murge' (Pouille, I) et recherches sur la variabilité de ce taxon (Lepidoptera: Nymphalidae, Satyrinae). *Linneana Belgica* **15**: 182–188.

Jutzeler D, Grillo N, Russo L, Nardelli U, De Bros E (1996) Position taxonomique et biologie de *Melanargia pherusa* (Boisduval, 1833) de Sicile selon les stades pré-imaginaux (Lepidoptera: Nymphalidae, Satyrinae). *Linneana Belgica* **15**: 203–213.

Jutzeler D, Biermann H, Hesselbarth G, Russo L, Sala G, de Bros E (1997) Etudes sur la biologie, la morphologie, et l'éthologie de *Hipparchia sbordonii* Kudrna, 1984, de l'Isola di Ponza (Latium, Italie) et *Hipparchia neapolitana* du Monte Faito (Campanie, Italie) et compléments sur la biologie d'*Hipparchia leighebi* (Kudrna, 1976) (Lepidoptera: Nymphalidae, Satyrinae). *Linneana Belgica* **16**: 105–132.

Jutzeler D, Casula P, Gascogne-Pees M, Leigheb G, Grill A (2003a) Confirmation du statut spécifique de *Polyommatus gennargenti* (Leigheb, 1987) de Sardaigne comparé à *Polyommatus coridon* de la région de Schaffhouse (CH) par élevage parallèle (Lepidoptera: Lycaenidae) 2ème partie. *Linneana Belgica* **19**: 149–160.

Jutzeler D, Leigheb G, Manil L, Villa R, Volpe G (2003b) Deux espèces de Lycènes negligées de l'espace tyrrhénien: *Lycaeides bellieri* Oberthür (1910) de Sardaigne et de Corse et *Lycaeides villai* sp. nov. de l'Isle d'Elbe (Lepidoptera: Lycaenidae). *Linneana Belgica* **19**: 65–80.

Kadlec T, Beneš J, Jarosik V, Konvička M (2008) Revisiting urban refuges: changes in butterfly fauna of Prague reserves over three decades. *Landscape and Urban Planning* **85**: 1–11.

Kaisila J (1950). Über die vermuteten Bastarde zwischen *Colias hecla sulitelma* Aur. und *C. nastes werdandi* Zett. (Lep., Pieridae). *Annales Entomologica Fennica* **16**: 112–121.

Kaitala A, Wiklund C (1994) Polyandrous female butterflies forage for matings. *Behavioral Ecology and Sociobiology* **35**: 385–388.

Kaitala A, Wiklund C (1995) Female mate choice and mating costs in the polyandrous butterfly *Pieris napi* (Lepidoptera: Pieridae). *Journal of Insect Behavior* **8**: 355–363.

Kammer AE (1970) Thoracic temperature, shivering, and flight in the Monarch butterfly, *Danaus plexippus*. *Zeitschrift für vergleichende Physiologie* **68**: 334–344.

Kammer AE, Bracchi J (1973) Role of the wings in the absorbtion of radiant energy by a butterfly. *Comparative Biochemistry and Physiology A* **45**: 1057–1063.

Kandul NP, Lukhtanov VA, Dantchenko AV, Coleman JWS, Sekercioglu CH, Haig D, Pierce NE (2004) Phylogeny of *Agrodiaetus* Hübner 1822 (Lepidoptera: Lycaenidae) inferred from mtDNA sequences of COI and COII and nuclear sequences of EF1-a: karyotype diversification and species radiation. *Systematic Biology* **53**: 278–298.

Kankare M, Shaw MR (2004) Molecular phylogeny of *Cotesia* Cameron, 1891 (Insecta: Hymenoptera: Braconidae: Microgastrinae) parasitoids associated with Melitaeini butterflies (Insecta: Lepidoptera: Nymphalidae: Melitaeini). *Molecular Phylogenetics and Evolution* **32**: 207–220.

Kankare M, Stefanescu C, van Nouhuys S, Shaw MR (2005a) Host specialization by *Cotesia* wasps (Hymenoptera: Braconidae) parasitizing species-rich Melitaeini (Lepidoptera: Nymphalidae) communities in north-eastern Spain. *Biological Journal of the Linnean Society* **86**: 45–65.

Kankare M, van Nouhuys S, Hanski I (2005b) Genetic divergence among host-specific cryptic species in *Cotesia melitaearum* aggregate (Hymenoptera: Braconidae), parasitoids of checkerspot butterflies. *Annals of the Entomological Society of America* **98**: 382–394.

Karlsson B (1987) Variation in egg weight, oviposition rate and reproductive reserves with female age in a natural population

of the speckled wood butterfly, *Pararge aegeria*. *Ecological Entomology* **12**: 473–476.

Karlsson B (1994) Feeding habits and change of body composition with age in three nymphalid butterflies. *Oikos* **69**: 224–230.

Karlsson B (1995) Resource allocation and mating systems in butterflies. *Evolution* **49**: 955–961.

Karlsson B (1996) Male reproductive reserves in relation to mating system in butterflies: a comparative study. *Proceedings of the Royal Society of London B* **263**: 187–192.

Karlsson B (1998) Nuptial gifts, resource budgets, and reproductive output in a polyandrous butterfly. *Ecology* **79**: 2931–2940.

Karlsson B, Van Dyck H (2005) Does habitat fragmentation affect temperature-related life-history traits? A laboratory test with a woodland butterfly. *Proceedings of the Royal Society of London B* **272**: 1257–1263.

Karlsson B, Wickman P-O (1989) The cost of prolonged life: an experiment on a nymphalid butterfly. *Functional Ecology* **3**: 399–405.

Karlsson B, Wickman P-O (1990) Increase in reproductive effort as explained by body size and resource allocation in the speckled wood butterfly, *Pararge aegeria* (L.). *Functional Ecology* **4**: 609–617.

Karlsson B, Wiklund C (1984) Egg weight variation and lack of correlation between egg weight and offspring fitness in the wall brown butterfly *Lasiommata megera*. *Oikos* **43**: 376–385.

Karlsson B, Wiklund C (1985) Egg weight variation in relation to egg mortality and starvation endurance of newly hatched larvae in some satyrid butterflies. *Ecological Entomology* **10**: 205–211.

Karlsson B, Wiklund C (2005) Butterfly life history and temperature adaptations; dry open habitats select for increased fecundity and longevity. *Journal of Animal Ecology* **74**: 99–104.

Karlsson B, Leimar O, Wiklund C (1997) Unpredictable environments, nuptial gifts and the evolution of sexual size dimorphism in insects: an experiment. *Proceedings of the Royal Society of London B* **264**: 475–479.

Karlsson B, Stjernholm F, Wiklund C (2008) Test of a developmental trade-off in a polyphenic butterfly: direct development favours reproductive output. *Functional Ecology* **22**: 121–126.

Karowe DN (1989) Facultative monophagy as a consequence of prior feeding experience: behavioral and physiological specialization in *Colias philodice* larvae. *Oecologia* **78**: 106–111.

Karsholt O, Razowski J (1996) *The Lepidoptera of Europe: A Distributional Checklist*. Stenstrup, Sweden: Apollo Books.

Kato Y (2005) Geographic variation in photoperiodic response for the induction of pupal diapause in the *Aristolochia*-feeding butterfly *Atrophaneura alcinous*. *Applied Entomology and Zoology* **40**: 347–350.

Kawagoe T, Suzuki N, Matsumoto K (2001) Multiple mating reduces longevity of females of the windmill butterfly *Atrophaneura alcinous*. *Ecological Entomology* **26**: 258–262.

Kay QON (1978) The role of preferential and assortative pollination in the maintenance of flower color polymorphisms. In: Richards AJ (ed.) *The Pollination of Flowers by Insects*, London: Academic Press, pp. 175–190.

Kay QON (1982) Intraspecific discrimination by pollinators and its role in evolution. In: Armstrong JA, Powell JM, Richards AJ (eds.) *Pollination and Evolution*, Sydney, NSW: Royal Botanic Gardens, pp. 9–28.

Kay QON, Lack AJ, Bamber FC, Davies CR (1984) Differences between sexes in floral morphology, nectar production and insect visits in a dioecious species, *Silene dioica*. *New Phytologist* **98**: 515–529.

Kearns CA, Inouye DW, Waser NM (1998) Endangered mutualisms: the conservation of plant–pollinator interactions. *Annual Review of Ecology and Systematics* **29**: 83–112.

Keitt TH, Bjørnstad O, Dixon P, Citron-Pousty S (2002) Accounting for spatial pattern when modeling organism–environment interactions. *Ecography* **25**: 616–625.

Kelber A (1999a) Why 'false' colours are seen by butterflies. *Nature* **402**: 251.

Kelber A (1999b) Ovipositing butterflies use a red receptor to see green. *Journal of Experimental Biology* **202**: 2619–2630.

Kelber A (2001) Receptor based models for spontaneous colour choices in flies and butterflies. *Entomologia Experimentalis et Applicata* **99**: 231–244.

Kelber A, Thunell C, Arikawa K (2001) Polarisation-dependent colour vision in *Papilio* butterflies. *Journal of Experimental Biology* **204**: 2469–2480.

Kemp DJ (2000a) Butterfly contests: neither paradoxical nor contradictory. *Animal Behaviour* **60**: F44–F46.

Kemp DJ (2000b) Contest behavior in territorial male butterflies: does size matter? *Behavioral Ecology* **11**: 591–596.

Kemp DJ (2001) The ecology of female receptivity in the territorial butterfly *Hypolimnas bolina* (L.) (Nymphalidae): implications for mate location by males. *Australian Journal of Zoology* **49**: 203–211.

Kemp DJ (2002a) Butterfly contests and flight physiology: why do older males fight harder? *Behavioral Ecology* **13**: 456–461.

Kemp DJ (2002b) Sexual selection constrained by life history in a butterfly. *Proceedings of the Royal Society of London B* **269**: 1341–1345.

Kemp DJ (2003) Twilight fighting in the evening brown butterfly, *Melanitis leda* (L.) (Nymphalidae): age and residency effects. *Behavioral Ecology and Sociobiology* **54**: 7–13.

Kemp DJ, Alcock J (2003) Lifetime resource utilization, flight physiology, and the evolution of contest competition in territorial insects. *American Naturalist* **162**: 290–301.

Kemp DJ, Krockenberger AK (2002) A novel method of behavioural thermoregulation in butterflies. *Journal of Evolutionary Biology* **15**: 922–929.

Kemp DJ, Krockenberger AK (2004) Behavioural thermoregulation in butterflies: the interacting effects of body size and basking posture in *Hypolimnas bolina* (L.) (Lepidoptera: Nymphalidae). *Australian Journal of Zoology* **52**: 229–236.

Kemp DJ, Rutowski RL (2004) A survival cost to mating in a polyandrous butterfly, *Colias eurytheme*. *Oikos* **105**: 65–70.

Kemp DJ, Wiklund C (2001) Fighting without weaponry: a review of male–male contest competition in butterflies. *Behavioral Ecology and Sociobiology* **49**: 429–442.

Kemp DJ, Wiklund C (2004) Residency effects in animal contests. *Proceedings of the Royal Society of London B* **271**: 1707–1711.

Kemp DJ, Wiklund C, Van Dyck H (2006a) Contest behaviour in the speckled wood butterfly (*Pararge aegeria*): seasonal phenotypic plasticity and the functional significance of flight performance. *Behavioral Ecology and Sociobiology* **59**: 403–411.

Kemp DJ, Wiklund C, Gotthard K (2006b) Life history effects upon contest behaviour: age as a predictor of territorial contest dynamics in two populations of the speckled wood butterfly, *Pararge aegeria* L. *Ethology* **112**: 471–477.

Kennedy JS (1961) A turning point in the study of insect migration. *Nature* **189**: 785–791.

Kephart S, Reynolds RJ, Rutter MT, Fenster CB, Dudash MR (2006) Pollination and seed predation by moths on *Silene* and allied Caryophyllaceae: evaluating a model system to study the evolution of mutualisms. *New Phytologist* **169**: 667–680.

Kerr JT (2001) Butterfly species richness patterns in Canada: energy, heterogeneity, and the potential consequences of climate change. *Conservation Ecology* **5**: 10. Online: http://www.consecol.org/vol5/iss1/art10.

Kerr JT, Southwood TRE, Cihlar J (2001) Remotely sensed habitat diversity predicts butterfly species richness and community similarity in Canada. *Proceedings of the National Academy of Sciences, USA* **98**: 11 365–11 370.

Kéry M, Matthies D, Fischer M (2001) The effect of plant population size on the interactions between the rare plant *Gentiana cruciata* and its specialised herbivore *Maculinea rebeli*. *Journal of Ecology* **89**: 425–427.

Kevan PG, Shorthouse JD (1970) Behavioural thermoregulation by high arctic butterflies. *Arctic* **23**: 268–279.

Keyghobadi N, Roland J, Strobeck C (1999) Influence of landscape on the population genetic structure of the alpine butterfly *Parnassius smintheus* (Papilionidae). *Molecular Ecology* **8**: 1481–1495.

Keyghobadi N, Roland J, Strobeck C (2005) Genetic differentiation and gene flow among populations of the alpine butterfly, *Parnassius smintheus*, vary with landscape connectivity. *Molecular Ecology* **14**: 1897–1909.

Keymer JE, Marquet PA, Velasco-Hernández JX, Levin SA (2000) Extinction thresholds and metapopulation persistence in dynamic landscapes. *American Naturalist* **156**: 478–494.

Kimura M (1968) Evolutionary rate at the molecular level. *Nature* **217**: 624–626.

Kimura J, Masaki S (1977) Brachypterism and seasonal adaptation in *Orgyia thyellina* Butler (Lepidoptera, Lymantriidae). *Kontyu* **45**: 95–106.

Kimura K, Tsubaki Y (1986) Female size and age-specific fecundity in the small white butterfly, *Pieris rapae crucivora* Boisduval (Lepidoptera, Pieridae). *Researches on Population Ecology* **28**: 295–304.

Kindvall O (1995) The impact of extreme weather on habitat preference and survival in a metapopulation of the bush cricket *Metrioptera bicolor* in Sweden. *Biological Conservation* **73**: 51–58.

King M (1993) *Species Evolution: The Role of Chromosome Change*. Cambridge: Cambridge University Press.

King R-S (1998) Dispersal of Karner Blue Butterflies (*Lycaeides melissa samuelis* Nabokov) at Necedah National Wildlife Refuge. *Transactions of the Wisconsin Academy of Sciences Arts and Letters* **86**: 101–110.

Kingsolver JG (1983a) Ecological significance of flight activity in *Colias* butterflies: implications for reproductive strategy and population structure. *Ecology* **64**: 546–551.

Kingsolver JG (1983b) Thermoregulation and flight in *Colias* butterflies: elevational patterns and mechanistic limitations. *Ecology* **64**: 534–545.

Kingsolver JG (1985a) Butterfly thermoregulation: organismic mechanisms and population consequences. *Journal of Research on the Lepidoptera* **24**: 1–20.

Kingsolver JG (1985b) Thermal ecology of *Pieris* butterflies (Lepidoptera: Pieridae): a new mechanism of behavioral thermoregulation. *Oecologia* **66**: 540–545.

Kingsolver JG (1985c) Thermoregulatory significance of wing melanization in *Pieris* butterflies (Lepidoptera: Pieridae): physics, posture, and pattern. *Oecologia* **66**: 546–553.

Kingsolver JG (1987a) Evolution and coadaptation of thermoregulatory behavior and wing pigmentation pattern in pierid butterflies. *Evolution* **41**: 472–490.

Kingsolver JG (1987b) Predation, thermoregulation, and wing color in pierid butterflies. *Oecologia* **73**: 301–306.

Kingsolver JG (1988) Thermoregulation, flight, and the evolution of wing pattern in pierid butterflies: the topography of adaptive landscapes. *American Zoologist* **28**: 899–912.

Kingsolver JG (1995a) Fitness consequences of seasonal polyphenism in western white butterflies. *Evolution* **49**: 942–954.

Kingsolver JG (1995b) Viability selection on seasonally polyphenic traits: wing melanin pattern in western white butterflies. *Evolution* **49**: 932–941.

Kingsolver JG (1996) Experimental manipulation of wing pigment pattern and survival in western white butterflies. *American Naturalist* **147**: 296–306.

Kingsolver JG (1999) Experimental analyses of wing size, flight, and survival in the western white butterfly. *Evolution* **53**: 1479–1490.

Kingsolver JG, Daniel TL (1979) On the mechanics and energetics of nectar feeding butterflies. *Journal of Theoretical Biology* **76**: 167–179.

Kingsolver JG, Koehl MAR (1985) Aerodynamics, thermoregulation, and the evolution of insect wings: differential scaling and evolutionary change. *Evolution* **39**: 488–504.

Kingsolver JG, Moffat RJ (1983) Thermoregulation and the determinants of heat transfer in *Colias* butterflies. *Oecologia* **53**: 27–33.

Kingsolver JG, Schemske D (1991) Path analyses of selection. *Trends in Ecology and Evolution* **6**: 276–280.

Kingsolver JG, Smith SG (1995) Estimating selection on quantitative traits using capture–recapture data. *Evolution* **49**: 384–388.

Kingsolver JG, Srygley RB (2000) Experimental analyses of body size, flight and survival in pierid butterflies. *Evolutionary Ecology Research* **2**: 593–612.

Kingsolver JG, Watt WB (1983) Thermoregulatory strategies in *Colias* butterflies: thermal stress and the limits to adaptation in temporally varying environments. *American Naturalist* **121**: 32–55.

Kingsolver JG, Watt WB (1984) Mechanistic constraints and optimality models: thermoregulatory strategies in *Colias* butterflies. *Ecology* **65**: 1835–1839.

Kingsolver JG, Wiernasz DC (1987) Dissecting correlated characters: adaptive aspects of phenotypic covariation in melanization pattern of *Pieris* butterflies. *Evolution* **41**: 491–503.

Kingsolver JG, Wiernasz DC (1991a) Seasonal polyphenism in wing-melanin pattern and thermoregulatory adaptation in *Pieris* butterflies. *American Naturalist* **137**: 816–829.

Kingsolver JG, Wiernasz DC (1991b) Development, function and the quantitative genetics of wing melanin patterns in *Pieris* butterflies. *Evolution* **45**: 1480–1492.

Kinoshita M, Shimada N, Arikawa K (1999) Colour vision of foraging swallowtail butterfly, *Papilio xuthus*. *Journal of Experimental Biology* **202**: 95–102.

Kirby KJ (2004) A model of a natural wooded landscape in Britain as influenced by large herbivore activity. *Forestry* **77**: 405–420.

Kivinen S, Luoto M, Kussaari M, Helenius J (2006) Multi-species richness of boreal agricultural landscapes: effects of climate, biotope, soil and geographical location. *Journal of Biogeography* **33**: 862–875.

Kleijn D, Baquero RA, Clough Y, Diaz M, De Esteban J, Fernandez F, Gabriel D, Herzog F, Holzschuh A, Johl R, Knop E, Kruess A, Marshall EJP, Steffan-Dewenter I, Tscharntke T, Verhulst J, West TM, Yela JL (2006) Mixed biodiversity benefits of agri-environment schemes in five European countries. *Ecology Letters* **9**: 243–254.

Kleintjes PK, Jacobs BF, Fettig SM (2004) Initial response of butterflies to an overstory reduction and slash mulching treatment of a degraded pinon–juniper woodland. *Restoration Ecology* **12**: 231–238.

Klemann P (1930) Ein neuer Tagfalterhybrid. *Internationale entomologische Zeitschrift* **23**: 438–442.

Kleyer M, Biedermann R, Henle K, Obermaier E, Poethke HJ, Poschlod P, Schröder B, Settele J, Vetterlein D (2007) Mosaic cycles in agricultural landscapes of Northwest Europe. *Basic and Applied Ecology* **8**: 295–309.

Klinger LF (1996) The myth of the classic hydrosere model of bog succession. *Arctic and Alpine Research* **28**: 1–9.

Knapton RW (1985) Lek structure and territoriality in the chryxus arctic butterfly, *Oenis chryxus* (Satyridae). *Behavioral Ecology and Sociobiology* **17**: 389–395.

Knuttel H, Fiedler K (2001) Host-plant-derived variation in ultraviolet wing patterns influences mate selection by male butterflies. *Journal of Experimental Biology* **204**: 2447–2459.

Koch PB (1992) Seasonal polyphenism in butterflies: a hormonally controlled phenomenon of pattern formation. *Zoologische Jahrbucher-Abteilung fur allgemeine Zoologie und Physiologie der Tiere* **96**: 227–240.

Koch PB, Buckmann D (1987) Hormonal control of seasonal morphs by the timing of ecdysteroid release in *Araschnia levana* L. (Nymphalidae, Lepidoptera). *Journal of Insect Physiology* **33**: 823–829.

Koch PB, Behnecke B, Ffrench-Constant RH (2000) The molecular basis of melanism and mimicry in a swallowtail butterfly. *Current Biology* **10**: 591–594.

Koh LP, Sodhi NS (2004) Importance of reserves, fragments and parks for butterfly conservation in a tropical urban landscape. *Ecological Applications* **14**: 1695–1708.

Kohane MJ, Watt WB (1999) Flight-muscle adenylate pool responses to flight demands and thermal constraints in individual *Colias eurytheme* (Lepidoptera, Pieridae). *Journal of Experimental Biology* **202**: 3145–3154.

Kolb G, Scherer C (1982) Experiments on wavelength specific behavior of *Pieris brassicae* L. during drumming and egg laying. *Journal of Comparative Physiology* **149**: 325–332.

Kolev Z, De Prins W (1995) A new species of the 'brown *Agrodiaetus*' complex from the Crimea (Lepidoptera: Lycaenidae). *Phegea* **23**: 119–132.

Kolligs D (2003) *Schmetterlinge Schleswig-Holsteins*. Neumünster, Germany: Wachholtz.

Komonen A (1997) The parasitoid complexes attacking Fininnsh populations of two threatened butterflies, *Euphydryas maturna* and *E. aurinia*. *Baptria* **22**: 105–109.

Komonen A (1998) Host species used in parasitoids of Melitaeini butterflies in southern France. *Baptria* **23**: 194–200.

Komonen A, Penttilä R, Lindgren M, Hanski I (2000) Forest fragmentation truncates a food chain based on an old-growth forest bracket fungus. *Oikos* **90**: 119–126.

Konvička M, Kuras T (1999) Population structure and the selection of oviposition sites of the endangered butterfly *Parnassius mnemosyne* (Lepidoptera: Papilionidae) in the Litovelske Pomoravi, Czech Republic. *Journal of Insect Conservation* **3**: 211–223.

Konvička M, Beneš J, Kuras T (2002a) Behaviour and microdistribution of two sympatric alpine butterflies (*Erebia epiphron* and *E. euryale*): relation to weather, vegetation and time of day. *Biologia* **57**: 223–233.

Konvička M, Nedved O, Fric Z (2002b) Early-spring floods decrease the survival of hibernating larvae of a wetland-inhabiting population of *Neptis rivularis* (Lepidoptera: Nymphalidae). *Acta Zoologica Academiae Scientiarum Hungaricae* **48**: 79–88.

Konvička M, Maradova M, Beneš J, Fric Z, Kepka P (2003) Uphill shifts in distribution of butterflies in the Czech Republic: effects of changing climate detected on a regional scale. *Global Ecology and Biogeography* **12**: 403–410.

Konvička M, Cizek O, Filipova L, Fric Z, Beneš J, Krupka M, Zamecnik J, Dockalova Z (2005) For whom the bells toll: demography of the last population of the butterfly

Euphydryas maturna in the Czech Republic. *Biologia* **60**: 551–557.

Konvička M, Fric Z, Beneš J (2006a) Butterfly extinctions in European states: do socioeconomic conditions matter more than physical geography? *Global Ecology and Biogeography* **15**: 82–92.

Konvička M, Vlasanek P, Hauck D (2006b) Absence of forest mantles creates ecological traps for *Parnassius mnemosyne* (Lepidoptera, Papilionidae). *Nota lepidopterologica* **29**: 145–152.

Kooi RE (1996) Factors influencing food plant acceptance by the tropical butterfly *Bicyclus anynana* (Satyrinae). *Proceedings of the Section Experimental and Applied Entomology, N.E.V. Amsterdam* **7**: 117–122.

Kopper BJ, Charlton RE, Margolies DC (2000) Oviposition site selection by the regal fritillary, *Speyeria idalia*, as affected by proximity of violet host plants. *Journal of Insect Behavior* **13**: 651–665.

Korshunov Y, Gorbunov P (1995) *Dnevnye babochki aziatskoi chasti Rossii. Spravochnik.* [*Butterflies of the Asian part of Russia: A handbook.*] Ekaterinburg, Russia: Ural University Press. [in Russian]

Koschuh A, Fauster R (2005) Der Braune Eichen-Zipferfalter *Satyrium ilicis* (Esper, 1779) (Lepidoptera: Lycaenidae) in der Steiermark (Österreich). *Beiträge der Entomofaunistik* **6**: 65–86.

Kostrowicki AS (1969) *Geography of the Palaearctic Papilionidae (Lep.).* Krakow, Poland: Zaklad Zoologii Systematycznej Polskiej Akademii Nauk.

Králíček M, Gottwald A (1984) *Motýli jihovýchodní Moravy I. [Butterflies of Southeast Moravia I.]* Uherské Hradiště: Museum J. A. Komenského Uherský Brod.

Krauss J, Schmitt T, Seitz A, Steffan-Dewenter I, Tscharntke T (2004a) Effects of habitat fragmentation on the genetic structure of the monophagous butterfly *Polyommatus coridon* along its northern range margin. *Molecular Ecology* **13**: 311–320.

Krauss J, Steffan-Dewenter I, Tscharntke T (2004b) Landscape occupancy and local population size depends on host plant distribution in the butterfly *Cupido minimus*. *Biological Conservation* **120**: 355–361.

Krauss J, Steffan-Dewenter I, Müller CB, Tscharntke T (2005) Relative importance of resource quantity, isolation and habitat quality for landscape distribution of a monophagous butterfly. *Ecography* **28**: 465–474.

Krebs RA, West DA (1988) Female mate preference and the evolution of female-limited batesian mimicry. *Evolution* **42**: 1101–1104.

Kristensen CO (1994) Investigations on the natural mortality of eggs and larvae of the large white *Pieris brassicae* (L.) (Lep., Pieridae). *Journal of Applied Entomology* **117**: 92–98.

Krogh A, Zeuthen E (1941) The mechanism of flight preparation in some insects. *Journal of Experimental Biology* **18**: 1–10.

Kronforst MR, Young LG, Kapan DD, McNeely C, O'Neill RJ, Gilbert LE (2006) Linkage of butterfly mate preference and wing color preference cue at the genomic location of *wingless*. *Proceedings of the National Academy of Sciences, USA* **103**: 6575–6580.

Kroutov V, Mayer MS, Emmel TC (1999) Olfactory conditioning of the butterfly *Agraulis vanillae* (L.) (Lepidoptera, Nymphalidae) to floral but not host-plant odors. *Journal of Insect Behavior* **12**: 833–843.

Kruess A, Tscharntke T (2002a) Grazing intensity and the diversity of grasshoppers, butterflies, and trap-nesting bees and wasps. *Conservation Biology* **16**: 1570–1580.

Kruess A, Tscharntke T (2002b) Contrasting responses of plant and insect diversity to variation in grazing intensity. *Biological Conservation* **106**: 293–302.

Krzywicki M (1967) Fauna Papilionoidea i Hesperioidea (Lepidoptera) Puszczy Bialowieskiej. *Annales Zoologici (Polska Akademia Nauk, Instytut Zoologiczny)* **25**: 1–213.

Kudrna O (1977) *A Revision of the Genus* Hipparchia *Fabricius*. Faringdon, Oxon.: EW Classey.

Kudrna O (1986) Aspects of the conservation of butterflies in Europe. In: Kudrna O (ed.) *Butterflies of Europe*, Vol. 8, Wiesbaden, Germany: Aula, pp. 1–323.

Kudrna O (1996) Mapping European butterflies: handbook for recorders. *Oedippus* **12**: 1–60.

Kudrna O (1998) Die Tagfalterfauna der Rhön. *Oedippus* **15**: 1–158.

Kudrna O (2002) The distribution atlas of European butterflies. *Oedippus* **20**: 1–342.

Kudrna O, Seufert W (1991) Ökologie und Schutz von *Parnassius mnemosyne* (Linnaeus, 1758) in der Rhön. *Oedippus* **2**: 1–44.

Kudrna O, Lukášek J, Slavík B (1994) Zur Erfolgreichen Wiederansiedlung von *Parnassius apollo* (Linnaeus, 1758) in Tschechien. *Oedippus* **9**: 1–37.

Küer A, Fartmann T (2005) Prominent shoots are preferred: microhabitat preferences of *Maculinea alcon* ([Denis, Schiffermüller], 1775) in Northern Germany (Lycaenidae). *Nota lepidopterologica* **27**: 309–319.

Kühn I (2007) Incorporating spatial autocorrelation may invert observed patterns. *Diversity and Distributions* **13**: 66–69.

Kühn I, Brandl R, Klotz S (2004) The flora of German cities is naturally species rich. *Evolutionary Ecology Research* **6**: 749–764.

Kulfan J (1989) Zur Bionomie des Bläulings *Everes decoloratus* (Stgr.) (Lep., Lycaenidae) bei nordwestlicher Grenze ihrer Verbreitung. *Biologia* **44**: 177–184.

Kulfan J (1993) Zum Vorkommen von *Coenonympha hero* (L., 1761) in der Slowakei (Lepidoptera, Satyridae). *Biologia* **48**: 559–560.

Kulfan J, Kulfan M, Zach P, Topp W (1997) Ist der Kleine Fuchs, *Aglais urticae* (Nymphalidae), in Zukunft gefährdet? *Nota lepidopterologica* **20**: 330–334.

Kulfan M, Kalivoda H, Panigaj, L (2004) Distribúcia zástupcov rodu *Erebia* Dalm. (Lepidoptera, Satytidae) vo Vysokých, Západných a Belianskych Tatrách. [Distribution of the genus *Erebia* Dalm. (Lepidoptera, Satytidae) in High, West and Belianske Tatry.] In: *Biosférické rezervácie na Slovensku. -Zvolen: Fakulta ekológie a environmentalistiky TU* (2004) 221–226.

Kunert G, Otto S, Roese USR, Gershenzon J, Weisser WW (2005) Alarm pheromone mediates production of winged dispersal morphs in aphids. *Ecology Letters* **8**: 596–603.

Kuras T, Tuf IH (2005) Vliv borovice kleče na bezobratlé Hrubého Jeseníku. [Influence of Mugo Pine on Invertebrates in the Hrubý Jeseník Mts.] *Živa* **53**: 268–269.

Kuras T, Beneš J, Konvička M (2000a) Differing habitat affinities of four *Erebia* species (Lepidoptera: Nymphalidae, Satyrinae) in the Hruby Jesenik Mts, Czech Republic. *Biologia* **55**: 169–175.

Kuras T, Beneš J, Čelechovský A, Vrabec V, Konvička M (2000b) *Parnassius mnemosyne* (Lepidoptera: Papilionidae) in North Moravia: review of present and past distribution, proposal for conservation. *Klapalekiana* **36**: 93–112.

Kuras T, Beneš J, Konvička M (2001) Behaviour and within-habitat distribution of adult *Erebia sudetica sudetica*, endemic of the Hrubý Jeseník Mts., Czech Republic (Nymphalidae, Satyrinae). *Nota lepidopterologica* **24**: 87–101.

Kuras T, Beneš J, Fric Z, Konvička M (2003) Dispersal patterns of endemic alpine butterflies with contrasting population structures: *Erebia epiphron* and *E. sudetica*. *Population Ecology* **45**: 115–123.

Kuussaari M, Nieminen M, Hanski I (1996) An experimental study of migration in the Glanville fritillary butterfly *Melitaea cinxia*. *Journal of Animal Ecology* **65**: 791–801.

Kuussaari M, Saccheri I, Camara M, Hanski I (1998) Allee effect and population dynamics in the Glanville fritillary butterfly. *Oikos* **82**: 384–392.

Kuussaari M, Singer MC, Hanski I (2000) Local specialization and landscape-level influence on host use in an herbivorous insect. *Ecology* **81**: 2177–2187.

Kuussaari M, van Nouhuys S, Hellmann J, Singer MC (2004) Larval biology of checkerspot butterflies. In: Ehrlich PR, Hanski I (eds.) *On the Wings of Checkerspots: A Model System for Population Biology*, Oxford: Oxford University Press, pp. 138–160.

Labine PA (1968) The population biology of the butterfly *Euphydryas editha*. VIII. Oviposition and its relation to patterns of oviposition in other butterflies. *Evolution* **22**: 799–805.

Lacy RC (2000a) Considering threats to the viability of small populations using individual-based models. *Ecological Bulletins* **48**: 39–51.

Lacy RC (2000b) Structure of the VORTEX simulation model for population viability analysis. *Ecological Bulletins* **48**: 191–203.

Lafferty KD, Gerber LR (2002) Good medicine for conservation biology: the intersection of epidemiology and conservation theory. *Conservation Biology* **16**: 593–604.

Lafranchis, T. (2000) *Les Papillons de jour de France, Belgique et Luxembourg et leurs chenilles.* Mèze, France: Editions Biotope.

Lafranchis T (2004) *Butterflies of Europe: New Field Guide and Key.* Paris: Diatheo.

Lai BCG, Pullin AS (2005) Distribution and conservation of genetic diversity among UK calcareous grassland regions: a case study using insects. *Biodiversity and Conservation* **14**: 3105–3125.

Lake JG, Hughes L (1999) Nectar production and floral characteristics of *Tropaeolum majus* L. grown in ambient and elevated carbon dioxide. *Annals of Botany* **84**: 535–541.

Lalanne-Cassou B, Lalanne-Cassou C (1972) Notes de chasse. *Alexanor* **7**: 240.

Lalanne-Cassou B, Lalanne-Cassou C (1989) Note à propos d'*Erebia serotina* Descimon et de Lesse. *Alexanor* **16**: 52–53.

Lambert DM, Michaux B, White CS (1987) Are species self-defining? *Systematic Zoology* **36**: 196–205.

Lambin X, Aars J, Piertney SB (2001) Dispersal, intraspecific competition, kin competition and kin facilitation: a review of the empirical evidence. In: Clobert J, Danchin E, Dhont AA, Nichols JD (eds.) *Dispersal*, Oxford: Oxford University Press, pp. 110–122.

Lande R (1987) Genetic correlations between the sexes in the evolution of sexual dimorphism and mating preferences. In: Bradbury JW, Andersson MB (eds.) *Sexual Selection: Testing the Alternatives*, New York: John Wiley, pp. 83–94.

Lande R (1988) Genetics and demography in biological conservation. *Science* **241**: 1455–1460.

Lande R (1995) Mutation and conservation. *Conservation Biology* **9**: 782–791.

Lande R (2002) Incorporating stochasticity in Population Viability Analysis. In: Beissinger SR, McCullough DR (eds.) *Population Viability Analysis*, Chicago, IL: University of Chicago Press, pp. 18–40.

Lanta V, Mach J, Holcova V (2006) The effect of dam construction on the restoration succession of spruce mires in the Giant Mountains (Czech Republic). *Annales Botanici Fennici* **43**: 260–268.

Lantero M, Jordana R (1981) Primera cita de *Erebia serotina* Descimon & de Lesse. *Boletín de la Asociación Española de Entomología* **4**: 185–189.

Lanza J, Smith G, Sack S, Cash A (1995) Variation in nectar volume and composition of *Impatiens capensis*: the individual, plant, and population levels. *Oecologia* **102**: 113–119.

Larsen TB (1984) The zoogeographical composition and distribution of the Arabian butterflies (Lepidoptera; Rhopalocera). *Journal of Biogeography* **11**: 119–158.

Larsen TB (1986) Tropical butterflies in the Mediterranean. *Nota lepidopterologica* **9**: 63–77.

Larsson S, Ekbom B (1995) Oviposition mistakes in herbivorous insects: confusion or a step towards a new host plant? *Oikos* **72**: 155–160.

Latimer AM, Wu S, Gelfand AE, Silander JA Jr (2006) Building statistical models to analyze species distributions. *Ecological Applications* **16**: 33–50.

Lattes A, Mensi P, Cassulo L, Balletto E (1994) Genotypic variability in Western European members of the *Erebia tyndarus* group (Lepidoptera, Satyridae). *Nota lepidopterologica* **17** (Suppl. 5): 93–105.

Lauwers K, Van Dyck H (2006) The cost of mating with a non-virgin male in a monandrous butterfly: experimental evidence from the speckled wood, *Pararge aegeria*. *Behavioral Ecology and Sociobiology* **60**: 69–76.

Lavergne S, Molofsky J (2007) Increased genetic variation and evolutionary potential drive the success of an invasive grass. *Proceedings of the National Academy of Sciences, USA* **104**: 3883–3888.

Lavergne S, Thuiller W, Molina J, Debussche M (2005) Environmental and human factors influencing rare plant local occurrence, extinction and persistence: a 115-year study in the Mediterranean region. *Journal of Biogeography* **32**: 799–812.

Lazri B, Barrows EM (1984) Flower visiting and pollen transport by the imported cabbage butterfly (*Pieris rapae*, Lepidoptera: Pieridae) in a highly disturbed urban habitat. *Environmental Ecology* **13**: 574–578.

Le Cessie S, van Houwelingen JC (1992) Ridge estimators in logistic regression. *Applied Statistics* **41**: 191–201.

Le Guyader H (2002) Doit-on abandonner le concept d'espèce? *Courrier de l'environnement de l'INRA* **46**: 51–64.

Le Masurier AD (1994) Costs and benefits of egg clustering in *Pieris brassicae*. *Journal of Animal Ecology* **63**: 677–685.

Leather SR (1984) The effect of adult feeding on the fecundity, weight loss and survival of the pine beauty moth, *Panolis flammea* (D, S). *Oecologia* **65**: 70–74.

Leather SR (1988) Size, reproductive potential and fecundity in insects: things aren't as simple as they seem. *Oikos* **51**: 386–389.

Leather SR (1995) Factors affecting fecundity, fertility, oviposition and larviposition in insects. In: Leather SR, Hardie J (eds.) *Insect Reproduction*, Boca Raton, FL: CRC Press, pp. 143–174.

Leathwick JR, Rowe D, Richardson J, Elith J, Hastie T (2005) Using multivariate adaptive regression splines to predict the distributions of New Zealand's freshwater diadromous fish. *Freshwater Biology* **50**: 2034–2052.

Lebreton JD, Burnham KP, Clobert J, Anderson DR (1992) Modeling survival and testing biological hypotheses using marked animals: a unified approach with case studies. *Ecological Monographs* **62**: 67–118.

Lederer J (1858) Noch einige syrische Schmetterlinge. *Wiener entomologische Monatsschrift* **2**: 138.

Lederer G (1939) Zur Fortpflanzungsbiologie (Paarung) der *Gonepteryx*-Arten (Lep.). In 7th International Congress for Entomology, Berlin, 15–20 August 1938, pp. 808–813.

Lederer G (1960) Verhaltensweisen der Imagines und der Entwicklungsstadien von *Limenitis camilla camilla* L. (Lep. Nymphalidae). *Zeitschrift für Tierpsychologie* **17**: 521–546.

Lederhouse RC (1982) Territorial defense and lek behavior of the black swallowtail butterfly, *Papilio polyxenes*. *Behavioral Ecology and Sociobiology* **10**: 109–118.

Lederhouse RC (1983) Population structure, residency, and weather related mortality in the black swallowtail butterfly, *Papilio polyxenes*. *Oecologia* **59**: 307–311.

Lederhouse RC, Scriber JM (1996) Intrasexual selection constrains the evolution of the dorsal color pattern of male black swallowtail butterflies, *Papilio polyxenes*. *Evolution* **50**: 717–722.

Lederhouse RC, Ayres MP, Scriber JM (1990) Adult nutrition affects male virility in *Papilio glaucus* L. *Functional Ecology* **4**: 743–751.

Lederhouse RC, Codella SG, Grossmueller DW, Maccarone AD (1992) Host plant-based territoriality in the white peacock butterfly, *Anartia jatrophae* (Lepidoptera, Nymphalidae). *Journal of Insect Behavior* **5**: 721–728.

Lee E, Archer DM (1974) Ecology of *Pieris napi* in Britain. *Entomologist's Gazette* **25**: 231–237.

Lee E, Archer DM (1981) Diapause in various populations of *Pieris napi* L. from different parts of the British Isles. *Journal of Research on the Lepidoptera* **19**: 96–100.

Legendre P (1993) Spatial autocorrelation: trouble or new paradigm? *Ecology* **74**: 1659–1673.

Legendre L, Legendre P (1983) *Numerical Ecology*. New York: Elsevier.

Legendre P, Legendre L (1998) *Numerical Ecology*, 2nd edn Amsterdam: Elsevier.

Lehmann A, Overton JM, Austin MP (2002a) Regression models for spatial prediction: their role for biodiversity and conservation. *Biodiversity and Conservation* **11**: 2085–2092.

Lehmann A, Overton JM, Leathwick JR (2002b) GRASP: generalized regression analysis and spatial prediction. *Ecological Modelling* **157**: 189–207.

Lei G, Hanski I (1997) Metapopulation structure of *Cotesia melitaearum*, a specialist parasitoid of the butterfly *Melitaea cinxia*. *Oikos* **78**: 91–100.

Lei G, Hanski I (1998) Spatial dynamics of two competing specialist parasitoids in a host metapopulation. *Journal of Animal Ecology* **67**: 422–433.

Lei GC, Vikberg V, Nieminen M, Kuussaari M (1997) The parasitoid complex attacking the Finnish populations of Glanville fritillary *Melitaea cinxia* (Lep: Nymphalidae), an endangered butterfly. *Journal of Natural History* **31**: 635–648.

Leigh EG (1981) The average lifetime of a population in a varying environment. *Journal of Theoretical Biology* **90**: 213–239.

Leigh TF, Smith RF (1959) Flight activity of *Colias philodice eurytheme* Boisduval in response to its physical environment. *Hilgardia* **28**: 569–624.

Leigheb G (1987) *Lysandra coridon* ssp. *gennargenti* nova (Lepidoptera: Lycaenidae), nuovo Licenide della Sardegna. *Bolletino del Museo regionale de Sciencie Naturale di Torino* **5**: 447–454.

Leimar O (1996) Life history plasticity: influence of photoperiod on growth and development in the common blue butterfly. *Oikos* **76**: 228–234.

Leimar O, Karlsson B, Wiklund C (1994) Unpredictable food and sexual size dimorphism in insects. *Proceedings of the Royal Society of London B* **258**: 121–125.

Leimar O, Hammerstein P, Van Dooren TJM (2006) A new perspective on developmental plasticity and the principles of adaptive morph determination. *American Naturalist* **167**: 367–376.

Lek S, Guégan JF (1999) Artificial neural networks as a tool in ecological modelling: an introduction. *Ecological Modelling* **120**: 65–73.

Lelièvre T (1992) Phylogénie des Polyommatinae et structure génétique de six espèces du genre *Lysandra* Hemming (Lepidoptères Lycaenidae). Ph.D. thesis, Université de Provence.

León-Cortés JL, Cowley MJR, Thomas CD (1999) Detecting decline in a formerly widespread species: how common is the common blue butterfly *Polyommatus icarus*? *Ecography* **22**: 643–650.

León-Cortés JL, Cowley MJR, Thomas CD (2000) The distribution and decline of a widespread butterfly *Lycaena phlaeas* in a pastoral landscape. *Ecological Entomology* **25**: 285–294.

León-Cortés JL, Lennon JJ, Thomas CD (2003a) Ecological dynamics of extinct species in empty habitat networks. I. The role of habitat pattern and quantity, stochasticity and dispersal. *Oikos* **102**: 449–464.

León-Cortés JL, Lennon JJ, Thomas CD (2003b) Ecological dynamics of extinct species in empty habitat networks. II. The role of host plant dynamics. *Oikos* **102**: 465–477.

Leraut P (1990) Contribution à l'étude des Satyrinae de France. *Entomologica Gallica* **2**: 8–19.

Levin D, Berube DE (1972) *Phlox* and *Colias*: the efficiency of a pollination system. *Evolution* **26**: 242–250.

Levin D, Kerster HW (1974) Gene flow in seed plants. *Evolutionary Biology* **7**: 139–220.

Levins R (1968) Some demographic and genetic consequences of environmental heterogeneity for biological control. *Bulletin of the Entomological Society of America* **15**: 237–240.

Lewis AC (1986) Memory constraints and flower choice in *Pieris rapae*. *Science* **232**: 863–865.

Lewis OT, Thomas CD (2001) Adaptation to captivity in the butterfly *Pieris brassicae*, and the implications for ex situ conservation. *Journal of Insect Conservation* **5**: 55–63.

Lewis OT, Thomas CD, Hill JK, Brookes MI, Crane TPR, Graneau YA, Mallet JLB, Rose OC (1997) Three ways of assessing metapopulation structure in the butterfly *Plebejus argus*. *Ecological Entomology* **22**: 283–293.

Lewontin RC (1974) *The Genetic Basis of Evolutionary Change*. New York: Columbia University Press.

Lhonoré J, Lagarde M (1999) Biogéographie, écologie et protection de *Coenonympha oedippus* (Fab., 1787) (Lepidoptera: Nymphalidae: Satyrinae). *Annales de la Société Entomologique de France* **35** (Suppl.): 299–307.

Li JS, Song YL, Zeng ZG (2003) Elevational gradients of small mammal diversity on the northern slopes of Mt. Qilian, China. *Global Ecology and Biogeography* **12**: 449–460.

Lichstein JW, Simons TR, Shriner SA, Franzreb KE (2002) Spatial autocorrelation and autoregressive models in ecology. *Ecological Monographs* **72**: 445–463.

Liegl A, Dolek M (2008) Conservation of coppice with standards for canopy arthropods. In: Floren A, Schmidl J (eds.) *Canopy Arthropod Research in Central Europe: Basic and*

Applied Studies from the High Frontier. Nuremberg, Germany: Bioform, pp. 551–562.

Lienhoop N, Koschel S, Settele J (2005) The other side of the coin: the economic value of butterfly conservation. In: Settele J, Kühn E, Thomas JA (eds.) *Studies on the Ecology and Conservation of Butterflies in Europe*, Vol. 2, *Species Ecology along a European Gradient*: Maculinea *butterflies as a Model*, Sofia/Moscow: Pensoft, pp. 219–220.

Lincoln RJ, Boxshall GA, Clark PF (1983) *A Dictionary of Ecology, Evolution and Systematics*. Cambridge: Cambridge University Press.

Lindbladh M, Niklasson M, Nilsson SG (2003) Long-time record of fire and open canopy in a high biodiversity forest in southeast Sweden. *Biological Conservation* 114: 231–243.

Lindenmayer DB, Cunningham RB, McCarthy MA (1999) The conservation of arboreal marsupials in the montane ash forests of the central highlands of Victoria, South East Australia. VIII. Landscape analysis of the occurrence of arboreal marsupials. *Biological Conservation* 89: 83–92.

Lindfors V (1998) *Butterfly Life History and Mating Systems*. Stockholm: Department of Zoology, Stockholm University.

Little D, Gouhier-Darimont C, Bruessow F, Reymond P (2007) Oviposition by pierid butterflies triggers defense responses in *Arabidopsis*. *Plant Physiology* 143: 784–800.

Littlewood SC (1983) The timing of emergence of parasitic Hymenoptera of *Pieris rapae* (L.) and *Pieris brassicae* (L.). *Entomologist's Record and Journal of Variation* 95: 104–106.

Liu C, Berry PM, Dawson TP, Pearson RG (2005) Selecting thresholds of occurrence in the prediction of species distributions. *Ecography* 28: 385–393.

Loder N, Gaston KJ, Warren PH, Arnold HR (1998) Body size and feeding specificity: macrolepidoptera in Britain. *Biological Journal of the Linnean Society* 63: 121–139.

Loeliger EA, Karrer F (2000) Unusual demonstration of autosomal dominant inheritance of the black coloration of one of America's swallowtails: F_2 broods of the hybrid *Papilio polyxenes asterius* with *Papilio machaon gorganus* (Papilionidae). *Nota lepidopterologica* 23: 40–49.

Loertscher M, Erhardt A, Zettel J (1994) Microdistribution of butterflies in a mosaic-like habitat: the role of nectar sources. *Ecography* 18: 15–26.

Lokki J, Malmström KK, Suomalainen E (1978) Migration of *Vanessa cardui* and *Plutella xylostella* to Spitsbergen in the summer 1978. *Notulae Entomologicae* 58: 121–123.

Lopez JE, Pfister CA (2001) Local population dynamics in metapopulation models: implications for conservation. *Conservation Biology* 15: 1700–1709.

Lorković Z (1941) Die Chromosomenzahlen in der Spermatogenese der Tagfalter. *Chromosoma* 2: 155–191.

Lorković Z (1949) Chromosomen-Vervielfaschung bei Schmetterlingen und ein neuer Fall fünffacher Zahl. *Revue Suisse de Zoologie* 56: 243–249.

Lorković Z (1953) Spezifische, semispezifische und rassische Differenzierung bei *Erebia tyndarus* Esp. *Travaux de l'Institut de Biologie Experimentale de l'Académie Yougoslave* 11–12: 163–224.

Lorković Z (1957) Die Speziationsstufen in der *Erebia tyndarus* Gruppe. *Bioloski Glasnik Periodicum Biologorum* 10: 61–110.

Lorković Z (1958) Some peculiarities of spatially and sexually restricted gene exchange in the *Erebia tyndarus* group. *Cold Spring Harbor Symposia in Quantitative Biology* 23: 319–325.

Lorković Z (1990) The butterfly chromosomes and their application in systematics and phylogeny. In: Kudrna O (ed.) *Butterflies of Europe*, Vol. 2, Wiesbaden, Germany: Aula, pp. 332–396.

Lorković Z (1993) Ecological association of *Leptidea morsei major* (Grund 1905) (Lepidoptera, Pieridae) with the oak forest *Lathyro-quercetum peetraeae* HR-T 1957 in Croatia. *Periodicum Biologorum Zagreb* 95: 455–457.

Lorković Z (1994) *Leptidea reali* Reissinger 1989 (=*lorkovicii* Real 1988), a new European species (Lepid., Pieridae). *Natura Croatica* 2: 1–26.

Lorković Z, de Lesse H (1954a) Expériences de croisements dans le genre *Erebia* (Lépidoptères: Satyridae). *Bulletin de la Société Zoologique de France* 79: 31–39.

Lorković Z, de Lesse H (1954b) Nouvelles découvertes concernant le degré de parenté d'*Erebia tyndarus* Esp. et *E. cassioides* Hohenw. *Lambillionea* 54: 58–68, 78–86.

Lorković Z, de Lesse H (1955) Note supplémentaire sur le groupe d'*Erebia tyndarus* Esp. *Lambillionea* 55: 55–58.

Lorković Z, Kiriakoff SG (1958) Proposed insertion in the 'Règles' of provisions recognising 'semispecies' as a special category for the classification and nomenclature of definite groups of taxa as now proposed to be defined. *Bulletin of Zoological Nomenclature* 15/B (Case 58): 1031–1033.

Losos JB (1996) Ecological and evolutionary determinants of the species–area relation in Caribbean anoline lizards. *Philosophical Transactions of the Royal Society of London B* 351: 847–854.

Louis-Augustin J (1985) Nouvelles captures d'*Erebia serotina* dans les Hautes Pyrénées. *Alexanor* 13: 348.

Louy D, Habel JC, Schmitt T, Assmann T, Meyer M, Müller P (2007) Strongly diverging population genetic patterns of three skipper species: the role of habitat fragmentation and dispersal ability. *Conservation Genetics* **8**: 671–681.

Luck RF, Stouthamer R, Nunney LP (1992) Sex determination and sex ratio patterns in parasitic Hymenoptera. In: Wrensch DL, Ebbert M (eds.) *Evolution and Diversity of Sex Ratio in Haplodiploid Insects and Mites*, New York: Chapman and Hall, pp. 442–476.

Ludwig D (1996) The distribution of populations survival times. *American Naturalist* **147**: 506–526.

Ludwig JA, Reynolds JF (1988) *Statistical Ecology: A Primer on Methods and Computing*. New York: John Wiley.

Ludwig D, Mangel M, Haddad B (2001) Ecology, conservation, and public policy. *Annual Review of Ecology and Systematics* **32**: 481–517.

Lukhtanov VA, Dantchenko AV (2002a) Descriptions of new taxa of the genus *Agrodiaetus* Hübner [1822] based on karyotype investigation (Lepidoptera, Lycaenidae). *Atalanta* **33**: 81–107, 224–225.

Lukhtanov VA, Dantchenko AV (2002b) Principles of the highly ordered arrangement of metaphase I bivalents in spermatocytes of *Agrodiaetus* (Insecta, Lepidoptera). *Chromosome Research* **10**: 5–20.

Lukhtanov VA, Wiemers M, Meusemann K (2003) Description of a new species of the 'brown' *Agrodiaetus* complex from South-East Turkey (Lycaenidae). *Nota lepidopterologica* **26**: 65–71.

Lukhtanov VA, Kandul NP, Plotkin JB, Dantchenko AV, Haig D, Pierce NE (2005) Reinforcement of pre-zygotic isolation and karyotype evolution in *Agrodiaetus* butterflies. *Nature* **436**: 385–389.

Lundgren L (1977) Role of intra and interspecific male–male interactions in *Polyommatus–Icarus* Rott and some other species of blues (Lycaenidae). *Journal of Research on the Lepidoptera* **16**: 249–264.

Lundgren L, Bergström G (1975) Wing scents and scent-released phases in the courtship behavior of *Lycaeides argyrognomon* (Lepidoptera: Lycaenidae). *Journal of Chemical Ecology* **4**: 399–412.

Luoto M, Kuussaari M, Rita H, Salminen J, von Bonsdorff T (2001) Determinants of distribution and abundance in the clouded apollo butterfly: a landscape ecological approach. *Ecography* **24**: 601–617.

Luoto M, Kuussaari M, Toivonen T (2002) Modelling butterfly distribution based on remote sensing data. *Journal of Biogeography* **29**: 1027–1038.

Luoto M, Pöyry J, Heikkinen RK, Saarinen K (2005) Uncertainty of bioclimate envelope models based on the geographical distribution of species. *Global Ecology and Biogeography* **14**: 575–584.

Luoto M, Heikkinen RK, Poyry J, Saarinen K (2006) Determinants of the biogeographical distribution of butterflies in boreal regions. *Jounal of Biogeography* **33**: 1764–1778.

Lushai G, Fjellsted W, Marcovitch O, Aagaard L, Sherratt TN, Allen JA, MacLean N (2000) Application of molecular techniques to non-lethal tissue samples of endangered butterfly populations (*Parnassius apollo* L.) in Norway for conservation management. *Biological Conservation* **94**: 43–50.

Lütolf M, Kienast F, Guisan A (2006) The ghost of past species occurrence: improving species distribution models for presence-only data. *Journal of Applied Ecology* **43**: 802–815.

Lux C. (1987) Répartition en vertical des lépidoptères dans les Alpes Maritimes. Observation de phénomènes naturels. *Bulletin de la Société des Sciences Naturelles* **55**: 3–8.

Lynch M, Walsh B (1998) *Genetics and Analysis of Quantitative Traits*. Sunderland, MA: Sinauer Associates.

Lythgoe KA, Read AF (1998) Catching the Red Queen? The advice of the rose. *Trends in Ecology and Evolution* **13**: 473–474.

Lyytinen A, Alatalo RV, Lindstrom L, Mappes J (2001) Can ultraviolet cues function as aposematic signals? *Behavioral Ecology* **12**: 65–70.

Lyytinen A, Brakefiled PM, Mappes J (2003) Significance of butterfly eyespots as an anti-predator device in ground-based and aerial attacks. *Oikos* **100**: 373–379.

Ma WC (1972) Dynamic of feeding responses in *Pieris brassicae* Linn. as a function of chemosensory input: a behavioural, ultrastructural and electrophysiological study. *Mededelingen Landbouwhogeschool Wageningen* **72**: 1–162.

Mac Nally R (1996) Hierarchical partitioning as an interpretative tool in multivariate inference. *Australian Journal of Ecology* **21**: 224–228.

Mac Nally R (2000) Regression and model-building in conservation biology, biogeography and ecology: the distinction between – and reconciliation of – 'predictive' and 'explanatory' models. *Biodiversity and Conservation* **9**: 655–671.

Mac Nally R (2002) Multiple regression and inference in ecology and conservation biology: further comments on identifying important predictor variables. *Biodiversity and Conservation* **11**: 1397–1401.

Mac Nally R, Fleishman E, Fay JP, Murphy DD (2003) Modelling butterfly species richness using mesoscale environmental variables: model construction and validation for mountain ranges in the Great Basin of western North America. *Biological Conservation* **110**: 21–31.

MacArthur RH, Wilson EO (1967) *The Theory of Island Biogeography*. Princeton, NJ: Princeton University Press.

Mackey BG, Lindenmayer DB (2001) Towards a hierarchical framework for modelling the spatial distribution of animals. *Journal of Biogeography* **28**: 1147–1166.

Maes D, Van Dyck H (1999) *Dagvlinders in Vlaanderen: Ecologie, Verspreiding en Behoud.* Antwerp, Belgium: Stichting Leefmilieu/KBC, Instituut voor Natuurbehoud & Vlaamse Vlinderwerkgroep.

Maes D, Van Dyck H (2001) Butterfly diversity loss in Flanders (north Belgium): Europe's worst case scenario? *Biological Conservation* **99**: 263–276.

Maes D, van Swaay CAM (1997) A new methodology for compiling national Red Lists applied on butterflies (Lepidoptera, Rhopalocera) in Flanders (N. Belgium) and in The Netherlands. *Journal of Insect Conservation* **1**: 113–124.

Maes D, Gilbert M, Titeux N, Goffart P, Dennis RLH (2003) Prediction of butterfly diversity hotspots in Belgium: a comparison of statistically focused and land use-focused models. *Journal of Biogeography* **30**: 1907–1920.

Maes D, Vanreusel W, Talloen W, Van Dyck H (2004) Functional conservation units for the endangered Alcon Blue butterfly *Maculinea alcon* in Belgium (Lepidoptera: Lycaenidae). *Biological Conservation* **120**: 229–241.

Maes D, Bauwens D, De Bruyn L, Anselin A, Vermeersch G, Van Landuyt W, De Knijf G, Gilbert M (2005) Species richness coincidence: conservation strategies based on predictive modelling. *Biodiversity and Conservation* **14**: 1345–1364.

Maes D, Ghesquière A, Logie M, Bonte D (2006) Habitat use and mobility of two threatened coastal dune insects: implications for conservation. *Journal of Insect Conservation* **10**: 105–115.

Magnus D (1950) Beobachtungen zur Balz und Eiablage des Kaisermantels *Argynnis paphia* L. (Lep., Nymphalidae). *Zeitschrift für Tierpsychologie* **7**: 435–449.

Magnus D (1958) Experimentelle Untersuchungen zur Bionomie und ethologiedes Kaisermantels *Argynnis paphia* L. (Lep. Nymph.). I. Über optische Auslöser von Anfliegereaktionen und ihre Bedeutung für das Sichfinden der Geschlecter. *Zeitschrift für Tierpsychologie* **15**: 397–426.

Magnus DBE (1963) Sex limited mimicry. II. Visual selection in the mate choice of butterflies. *Proceedings of the 16th International Congress of Zoology* **4**: 179–183.

Magnuson JJ, Crowder LB, Medvick PA (1979) Temperature as an ecological resource. *American Zoologist* **19**: 331–343.

Maier C (1998) The behaviour and wing morphology of the Meadow Brown butterfly (*Maniola jurtina L.*) in Britain: the influence of weather and location. Ph.D. thesis, Oxford Brookes University.

Maier C, Shreeve TG (1996) Endothermic heat production in three species of Nymphalidae. *Nota lepidopterologica* **18**: 127–137.

Majerus MEN, Brunton CFA, Stalker J (2000) A bird's eye view of the peppered moth. *Journal of Evolutionary Biology* **13**: 155–159.

Makris C (2003) *Butterflies of Cyprus.* Nicosia, Cyprus: Bank of Cyprus Cultural Foundation.

Malcolm SB, Cockrell BJ, Brower LP (1987) Monarch butterfly voltinism: effects of temperature constraints at different latitudes. *Oikos* **49**: 77–82.

Mallet J (1986) Dispersal and gene flow in a butterfly with home range behavior: *Heliconius erato* (Lepidoptera, Nymphalidae). *Oecologia* **68**: 210–217.

Mallet J (1989) The genetics of warning colour in Peruvian hybrid zones of *Heliconius erato* and *H. melpomene*. *Proceedings of the Royal Society of London B* **236**: 163–185.

Mallet J (1995) A species definition for the Modern Synthesis. *Trends in Ecology and Evolution* **10**: 294–299.

Mallet J (2001) Gene flow. In: Woiwood IP, Reynolds DR, Thomas CD (eds.) *Insect Movement: Mechanisms and Consequences*, Proceedings of a Symposium at the Royal Entomological Society, London, Wallingford, Oxon.: CABI Publishing, pp. 337–360.

Mallet J (2005) Hybridization as an invasion of the genome. *Trends in Ecology and Evolution* **20**: 229–237.

Mallet J, Barton NH (1989a) Inferences from clines stabilized by frequency-dependent selection. *Genetics* **122**: 967–976.

Mallet J, Barton NH (1989b) Strong natural selection in a warning-color hybrid zone. *Evolution* **43**: 421–431.

Mallet J, Gilbert LE (1995) Why are there so many mimicry rings? Correlations between habitat, behaviour and mimicry in *Heliconius* butterflies. *Biological Journal of the Linnean Society* **55**: 159–180.

Mallet J, Barton N, Gerardo LM, Jose SC, Manuel MM, Eeley H (1990) Estimates of selection and gene flow from cline width and linkage disequilibrium in *Heliconius* hybrid zones. *Genetics* **124**: 921–936.

Manel S, Dias JM, Ormerod SJ (1999) Comparing discriminant analysis, neural networks and logistic regression for predicting species distributions: a case study with a Himalayan river bird. *Ecological Modelling* **120**: 337–348.

Manel S, Williams HC, Ormerod SJ (2001) Evaluating presence–absence models in ecology: the need to account for prevalence. *Journal of Applied Ecology* **38**: 921–931.

Mangel M (1987) Oviposition site selection and clutch size in insects. *Journal of Mathematical Biology* **25**: 1–22.

Mangel M, Clark CW (1986) Towards a unified foraging theory. *Ecology* **67**: 1127–1138.

Mangel M, Rosenheim JA, Alder FR (1994) Clutch size, offspring performance and integrational fitness. *Behavioral Ecology* **5**: 412–417.

Manil L, Diringer Y (2003) Excursion entomologique en Corse (juin 2003): note sur les conséquences des incendies sur une population de *Papilio hospiton* Géné. *Bulletin des Lépidoptéristes Parisiens* **12**: 42–44.

Manley WBL, Allcard HG (1970) *A Field Guide to the Butterflies and Burnet Moths of Spain.* Hampton: EW Classey.

Manly BFJ (1985) *The Statistics of Natural Selection on Animal Populations.* London: Chapman and Hall.

Manly BFJ, McDonald LL, Thomas DL (2002) *Resource Selection by Animals: Statistical Design and Analysis for Field Studies.* Dordrecht, Netherlands: Kluwer.

Marchi A, Addis G, Hermosa VE, Crnjar R (1996) Genetic divergence and evolution of *Polyommatus coridon gennargenti* (Lepidoptera, Lycaenidae) in Sardinia. *Heredity* **77**: 16–22.

Marcus JM (2005) Jumping genes and AFLP maps: transforming lepidopteran color pattern genetics. *Evolution and Development* **7**: 108–114.

Marrs RH, Williams CT, Frost AJ, Plant RA (1989) Assessment of the effects of herbicide spray drift on a range of plant species of conservation interest. *Environmental Pollution* **59**: 71–86.

Marshall LD (1982) Male nutrient investment in the Lepidoptera: what nutrients should males invest? *American Naturalist* **120**: 273–279.

Marshall EJP, Birnie JE (1985) Herbicide effects on field margin flora. In: *1985 British Crop Protection Conference: Weeds*, Thornton Heath, Surrey: British Crop Protection Council, pp. 1021–1028.

Martín Cano J (1981) Similitudes biológicas y diferencias ecológicas entre *Glaucopsyche alexis* (Poda) y *Glaucopsyche melanops* (Boisduval) (Lep. Lycaenidae). *Boletín de la Estación central de Ecología* **10**: 59–70.

Martín Cano J (1984) Biología comparada de *Lampides boeticus* (L.), *Syntarucus pirithous* (L.) y *Polyommatus icarus* (Rott.) (Lep., Lycaenidae). *Graellsia* **40**: 163–193.

Martín J, Gurrea P (1990) The peninsular effect in Iberian butterflies (Lepidoptera: Papilionoidea and Hesperioidea). *Journal of Biogeography* **17**: 85–96.

Martin LA, Pullin AS (2004a) Host-plant specialisation and habitat restriction in an endangered insect, *Lycaena dispar batavus* (Lepidoptera: Lycaenidae). I. Larval feeding and oviposition preferences. *European Journal of Entomology* **101**: 51–56.

Martin LA, Pullin AS (2004b) Host-plant specialisation and habitat restriction in an endangered insect, *Lycaena dispar batavus* (Lepidoptera: Lycaenidae). II. Larval survival on alternative host plants in the field. *European Journal of Entomology* **101**: 57–62.

Martín FJ, Sanz J, Silva C, Munguira ML (1990) Densidad y distribución espacial de la puesta de *Zegris eupheme* y *Euchloe ausonia* (Lepidoptera, Pieridae) sobre dos especies de crucífera. *Boletín del Grupo Entomológico de Madrid* **5**: 13–23.

Martin JF, Gilles A, Lörtscher M, Descimon H (2002) Phylogenetics and differentiation of the western taxa of the *Erebia tyndarus* group. *Biological Journal of the Linnean Society* **75**: 319–332.

Martin JF, Gilles A, Descimon H (2003) Species concepts and sibling species: the case of *Leptidea sinapis* and *Leptidea reali*. In: Boggs CL, Watt WB, Ehrlich P (eds.) *Butterflies: Ecology and Evolution Taking Flight.* Chicago, IL: University of Chicago Press, pp. 459–476.

Marttila O, Saarinen K, Jantunen J (1997) Habitat restoration and a successful reintroduction of the endangered Baton Blue butterfly (*Pseudophilotes baton schiffermuelleri*) in SE Finland. *Annales Zoologici Fennici* **34**: 177–185.

Marttila O, Saarinen K, Marttila P (2000) Six years from passing bell to recovery: habitat restoration of the threatened Chequered Blue Butterfly (*Scolitantides orion*) in SE Finland. *Entomologica Fennica* **11**: 113–117.

Masaki S (1980) Summer diapause. *Annual Review of Entomology* **251**: 1–25.

Mason WRM (1981) The polyphyletic nature of the genus *Apanteles* Foerster (Hymenoptera: Braconidae): a phylogeny and reclassification of Microgastrinae. *Memoirs of the Entomological Society of Canada* **115**: 1–147.

Matter SF, Roland J, Moilanen A, Hanski I (2004) Migration and survival of *Parnassius smintheus*: detecting effects of habitat for individual butterflies. *Ecological Applications* **14**: 1526–1534.

Matter SF, Roslin T, Roland J (2005) Predicting immigration of two species in contrasting landscapes: effects of scale, patch size and isolation. *Oikos* **111**: 359–367.

Mattson WJ (1980) Herbivory in relation to plant-nitrogen content. *Annual Review of Ecology and Systematics* **11**: 119–161.

Mauricio R, Bowers MD (1990) Do caterpillars disperse their damage? Larval foraging behaviour of two specialist herbivores, *Euphydryas phaeton* (Nymphalidae) and *Pieris rapae* (Pieridae). *Ecological Entomology* **15**: 153–161.

May ML (1979) Insect thermoregulation. *Annual Review of Entomology* **24**: 313–349.

May PG (1985) Nectar uptake rates and optimal nectar concentrations of two butterfly species. *Oecologia* 66: 381–386.

Mayden RL (1997) A hierarchy of species concepts: the denouement in the saga of the species problem. In: Claridge MF, Dawah HA, Wilson MR (eds.) *Species: The Units of Biodiversity*, London: Chapman and Hall, pp. 381–424.

Mayhew PJ (1997) Adaptive patterns of host-plant selection by phytophagous insects. *Oikos* 79: 417–428.

Mayhew PJ (2001) Herbivore host choice and optimal bad motherhood. *Trends in Ecology and Evolution* 96: 161–167.

Mayr E (1942) *Systematics and the Origin of Species*. New York: Columbia University Press.

Mayr E (1946) History of the North American bird fauna. *Wilson Bulletin* 58: 3–41.

Mayr E (1963) *Animal Species and Evolution*. Cambridge, MA: Harvard University Press.

Mayr E (1965) What is a fauna. *Zoologische Jahrbücher Systematik* 92: 473–486.

Mayr E (1982) *The Growth of Biological Thought: Diversity, Evolution, and Inheritance*. Cambridge, MA: Belknap Press of Harvard University Press.

Mayr E (1996) What is a species, and what is not? *Philosophy of Science* 63: 261–276.

Mayr E (2001) *What Evolution Is*. New York: Basic Books.

Mazel R (1982) Exigences trophiques et évolution dans les genres *Euphydryas* et *Melitaea s.l.* (Lep.: Nymphalidae). *Annales de la Société Entomologique de France* 18: 211–227.

Mazel R (1986) Structure et évolution du peuplement d'*Euphydryas aurinia* Rottemburg (Lepidoptera) dans le sud-ouest européen. *Vie Milieu* 36: 205–225.

Mazel R, Leestmans R (1996) Relations biogéographiques, écologiques et taxonomiques entre *Leptidea sinapis* Linné et *L. reali* Reissinger en France, Belgique et régions limitrophes (Lepidoptera: Pieridae). *Linneana Belgica* 8: 317–326.

Mazerolle MJ, Villard MA (1999) Patch characteristics and landscape context as predictors of species presence and abundance: a review. *Ecoscience* 6: 117–124.

McCoy ED (1990) The distribution of insects along elevational gradients. *Oikos* 58: 313–322.

McCullagh P, Nelder JA (1983) *Generalized Linear Models*. London: Chapman and Hall.

McDonald AK, Nijhout HF (2000) The effect of environmental conditions on mating activity of the Buckeye butterfly, *Precis coenia*. *Journal of Research on the Lepidoptera* 35: 22–28.

McIntire EJB, Schultz CB, Crone EE (2007) Designing a network for butterfly habitat restoration: where individuals, populations and landscapes interact. *Journal of Applied Ecology* 44: 725–736.

McLaughlin JF, Hellmann JJ, Boggs CL, Ehrlich PR (2002a) Climate change hastens population extinctions. *Proceedings of the National Academy of Sciences, USA* 99: 6970–6974.

McLaughlin JF, Hellmann JJ, Boggs CL, Ehrlich PR (2002b) The route to extinction: population dynamics of a threatened butterfly. *Oecologia* 132: 538–548.

McNaughton SJ, Tarrants MM (1983) Grass leaf silification: natural selection for an inducible defence against herbivores. *Proceedings of the National Academy of Sciences, USA* 80: 790–791.

McNeely C, Singer MC (2001) Contrasting the roles of learning in butterflies foraging for nectar and oviposition sites. *Animal Behaviour* 61: 847–852.

McPeek MA, Holt RD (1992) The evolution of dispersal in spatially and temporally varying environments. *American Naturalist* 140: 1010–1027.

McWhirter K (1969) Heritability of spot number in Scillonian strains of the Meadow Brown Butterfly (*Maniola jurtina*). *Heredity* 24: 314–318.

Mecke C. (1923) *Colias myrmidone* und Hybrid *myrmidone* m. x *hyale* f. in Ostpreussen. *Internationale Entomologische Zeitschrift* 16: 173–175.

Meek B, Loxton D, Sparks T, Pywell R, Pickett H, Nowakowski M (2002) The effect of arable field margin composition on invertebrate biodiversity. *Biological Conservation* 106: 259–271.

Meglécz E, Solignac M (1998) Microsatellite loci for *Parnassius mnemosyne* (Lepidoptera). *Hereditas* 128: 179–180.

Meglécz E, Pecsenye K, Peregovits L, Varga Z (1997) Effects of population size and habitat fragmentation on the genetic variability of *Parnassius mnemosyne* populations in NE Hungary. *Acta Zoologica Academiae Scientiarum Hungaricae* 43: 183–190.

Meglécz E, Pecsenye K, Varga Z, Solignac M (1998) Comparison of differentiation pattern at allozyme and microsatellite loci in *Parnassius mnemosyne* (Lepidoptera) populations. *Hereditas* 128: 95–103.

Meglécz E, Nève G, Peczenye K, Varga Z (1999) Genetic variations in space and time in *Parnassius mnemosyne* (L.) (Lepidoptera) populations in north-east Hungary: implications for conservation. *Biological Conservation* 89: 251–259.

Meglécz E, Péténian F, Danchin E, Coeur d'Acier AC, Rasplus JY, Faure E (2004) High similarity between flanking regions of different microsatellites detected within each of two species of Lepidoptera: *Parnassius apollo* and *Euphydryas aurinia*. *Molecular Ecology* 13: 1693–1700.

Meglécz E, Anderson A, Bourguet D, Butcher R, Caldas A, Cassel-Lundhagen A, Cœur d'Acier A, Dawson AD, Faure N, Fauvelot C, Franck P, Harper G, Keyghobadi N, Kluetsch C, Muthulakshmi M, Nagaraju J, Patt A, Péténian F, Silvain JF, Wilcock HR (2007) Microsatellite flanking region similarities among different loci within insect species. *Insect Molecular Biology* **16**: 175–185.

Meier K, Kuusemets V, Luig J, Mander Ü (2005) Riparian buffer zones as elements of ecological networks: case study on *Parnassius mnemosyne* distribution in Estonia. *Ecological Engineering* **24**: 531–537.

Meinshausen N (2006) Quantile regression forests. *Journal of Machine Learning Research* **7**: 983–999.

Menéndez R, Gutiérrez D, Thomas CD (2002) Migration and Allee effects in the six-spot burnet moth *Zygaena filipendulae*. *Ecological Entomology* **27**: 317–325.

Menéndez R, González-Megías A, Hill JK, Braschler B, Willis SG, Collingham Y, Fox R, Roy DB, Thomas CD (2006) Species richness changes lag behind climate change. *Proceedings of the Royal Society of London B* **273**: 1465–1470.

Menéndez R, González-Megías A, Collingham Y, Fox R, Roy DB, Ohlemüller R, Thomas CD (2007) Direct and indirect effects of climate and habitat factors on butterfly diversity. *Ecology* **88**: 605–611.

Menéndez R, González-Megías A, Lewis OT, Shaw MR, Thomas CD (2008) Escape from natural enemies during climate-driven range expansion: a case study. *Ecological Entomology* **33**: 413–421.

Menken SBJ (1987) Is the extremely low heterozygosity level in *Yponomeuta rorellus* caused by bottlenecks? *Evolution* **41**: 630–637.

Mennechez G, Schtickzelle N, Baguette M (2003) Metapopulation dynamics of the bog fritillary butterfly: comparison of demographic parameters and dispersal between a continuous and a highly fragmented landscape. *Landscape Ecology* **18**: 279–291.

Mennechez G, Petit S, Schtickzelle N, Baguette M (2004) Modelling mortality and dispersal: consequences of parameter generalisation on metapopulation dynamics. *Oikos* **106**: 243–252.

Mensi P, Lattes A, Salvidio S, Balleto E (1988) Taxonomy, evolutionary biology and biogeography of South West European *Polyommatus coridon* (Lepidoptera: Lycaenidae). *Zoological Journal of the Linnean Society* **93**: 259–271.

Mensi P, Lattes A, Cassulo L, Balleto E (1994) Biochemical taxonomy and evolutionary relationships in *Polyommatus* (subgenus *Agrodiaetus*) (Lepidoptera: Lycaenidae). *Nota lepidopterologica* **17** (Suppl. 5): 105–114.

Merckx T (2005) Habitat fragmentation and evolutionary ecology of movement behaviour in the Speckled Wood Butterfly (*Pararge aegeria* L.). Ph.D. thesis, University of Antwerp.

Merckx T, Van Dyck H (2005) Mate location behaviour of the butterfly *Pararge aegeria* in woodland and fragmented landscapes. *Animal Behaviour* **70**: 411–416.

Merckx T, Van Dyck H (2006) Landscape structure and phenotypic plasticity in flight morphology in the butterfly *Pararge aegeria*. *Oikos* **113**: 226–232.

Merckx T, Van Dyck H (2007) Habitat fragmentation affects habitat-finding ability of the speckled wood butterfly (*Pararge aegeria* L.). *Animal Behaviour* **74**: 1029–1037.

Merckx T, Van Dyck H, Karlsson B, Leimar O (2003) The evolution of movements and behaviour at boundaries in different landscapes: a common arena experiment with butterflies. *Proceedings of the Royal Society of London B* **270**: 1815–1821.

Merckx T, Karlsson B, Van Dyck H (2006) Sex- and landscape-related differences in flight ability under suboptimal temperatures in a woodland butterfly. *Functional Ecology* **20**: 436–441.

Merckx T, Van Dongen S, Matthysen E, Van Dyck H (2008) Thermal flight budget of a woodland butterfly in woodland versus agricultural landscapes: an experimental assessment. *Basic and Applied Ecology* **9**: 433–442.

Mérit X (2000) Diversité et variations chez *Melanargia galathea* (Linné) en France. *Bulletin des Lépidoptéristes Parisiens* **9**: 49–52.

Merrill RM, Gutiérrez D, Lewis OT, Gutiérrez J, Díez SB, Wilson RJ (2008) Combined effects of climate and biotic interactions on the elevational range of a phytophagous insect. *Journal of Animal Ecology* **77**: 145–155.

Metz JAJ, Gyllenberg M (2001) How should we define fitness in structured metapopulation models? Including an application to the calculation of evolutionarily stable dispersal strategies. *Proceedings of the Royal Society of London B* **268**: 499–508.

Mevi-Schütz J, Erhardt A (2003a) Larval nutrition affects female nectar amino acid preference in the map butterfly (*Araschnia levana*). *Ecology* **84**: 2788–2794.

Mevi-Schütz J, Erhardt A (2003b) Effects of nectar amino acids on fecundity of the wall brown butterfly (*Lasiommata megera* L.). *Basic and Applied Ecology* **4**: 413–421.

Mevi-Schütz J, Erhardt A (2004) Mating frequency influences nectar amino acid preference of *Pieris napi*. *Proceedings of the Royal Society of London B* **271**: 153–158.

Mevi-Schütz J, Erhardt A (2005a) Amino acids in nectar enhance butterfly fecundity: a long-awaited link. *American Naturalist* **165**: 411–419.

Mevi-Schütz J, Erhardt A (2005b) Nectar amino acids compensate for poor larval food quality in the map butterfly *Araschnia levana*. *American Naturalist* **165**: 411–419.

Mevi-Schütz J, Goverde M, Erhardt A (2003) Effects of fertilization and elevated CO_2 on larval food and butterfly nectar amino acid preference in *Coenonympha pamphilus* L. *Behavioral Ecology and Sociobiology* **54**: 36–43.

Meyer-Hozak C (2000) Population biology of *Maculinea rebeli* (Lepidoptera: Lycaenidae) on the chalk grasslands of Eastern Westphalia (Germany) and implications for conservation. *Journal of Insect Conservation* **4**: 63–72.

Mialoulis IN, Heilman BD (1998) Butterfly thin films serve as solar collectors. *Annals of the Entomological Society of America* **91**: 122–127.

Mikkola K (1976) On the pairing and number of generations in Finland of *Aglais urticae*. *Nota Entomologica* **56**: 99–102.

Mikkola K, Kononenko VS (1989) Flight year of the alternate-year *Xestia* moths (Lep., Noctuidae) in north-eastern Siberia: a character from the ice ages? *Nota lepidopterologica* **12**: 144–152.

Mikkola K, Spitzer K (1983) Lepidoptera associated with peatlands in central and northern Europe: a synthesis. *Nota lepidopterologica* **6**: 216–229.

Miller WE (1989) Reproductive enhancement by adult feeding: effects of honeydew in imbibed water on the spruce budworm. *Journal of the Lepidopterists' Society* **43**: 167–177.

Miller WE (1997) Intoxicated lepidopterans: how is their fitness affected, and why do they tipple? *Journal of the Lepidopterists' Society* **51**: 277–287.

Miller J, Franklin J, Aspinall R (2007) Incorporating spatial dependence in predictive vegetation models. *Ecological Modelling* **202**: 225–242.

Minnich DE (1955) The responses of caterpillars to sounds. *Journal of Experimental Zoology* **72**: 439–453.

Mitchell FJG (2005) How open were European primeval forests? Hypothesis testing using palaeoecological data. *Journal of Ecology* **93**: 168–177.

Moilanen A (2004) SPOMSIM: software for stochastic patch occupancy models of metapopulation dynamics. *Ecological Modelling* **179**: 533–550.

Moisen GG, Frescino TS (2002) Comparing five modelling techniques for predicting forest characteristics. *Ecological Modelling* **157**: 209–225.

Moisen GG, Freeman EA, Blackard JA, Frescino TS, Zimmermann NE, Edwards TC Jr (2006) Predicting tree species presence and basal area in Utah: a comparison of stochastic gradient boosting, generalized additive models, and tree-based methods. *Ecological Modelling* **199**: 176–187.

Molina JM (1988) Ropalóceros de la Comarca Sierra Norte (Sevilla). I. Faunística. *SHILAP, Revista de Lepidopterología* **16**: 131–139.

Molina JM (2000) Notas sobre el uso del arándano americano (*Vaccinium* × *corymbosum* L.), por *Charaxes jasius* (L., 1767) en el suroeste de Andalucía, España (Lepidoptera: Nymphalidae). *SHILAP, Revista de Lepidopterología* **28**: 91–96.

Molleman F, Grunsven RHA, Liefting M, Zwaan BJ, Brakefield PM (2005) Is male puddling behaviour of tropical butterflies targeted at sodium for nuptial gifts or activity? *Biological Journal of the Linnean Society* **86**: 345–361.

Monserud RA, Leemans R (1992) Comparing global vegetation maps with Kappa statistic. *Ecological Modelling* **62**: 275–293.

Moore SD (1987) Male-biased mortality in the butterfly *Euphydryas editha*: a novel cost of mate acquisition. *American Naturalist* **130**: 306–309.

Moore RA, Singer MC (1987) Effects of maternal age and adult diet on egg weight in the butterfly *Euphydryas editha*. *Ecological Entomology* **12**: 401–408.

Moran PAP (1950) Notes on continuous stochastic phenomena. *Biometrika* **37**: 17–23.

Moretti M, Obrist MK, Duelli P (2004) Arthropod biodiversity after forest fires: winners and losers in the winter fire regime of the southern Alps. *Ecography* **27**: 173–186.

Morris WF, Doak DF (2002) *Quantitative Conservation Biology: Theory and Practice of Population Viability Analysis*. Sunderland, MA: Sinauer Associates.

Moss JE (1933) The natural control of the cabbage caterpillars, *Pieris* spp. *Journal of Animal Ecology* **2**: 210–231.

Moss D, Pollard E (1993) Calculation of collated indices of abundance of butterflies based on monitored sites. *Ecological Entomology* **18**: 77–83.

Moucha J (1951) Poznámka k zeměpisnému rozšíření *Leptidea morsei* Fent. ssp. *major* Lork. *Časopis Československé Společnosti Entomologické* **48**: 181–186.

Mouquet N, Thomas JA, Elmes GW, Clarke RT, Hochberg ME (2005) Population dynamics and conservation of a specialized predator: a case study of *Maculinea arion*. *Ecological Monographs* **75**: 525–542.

Mousseau TA (1997) Ectotherms follow the converse to Bergmann's Rule. *Evolution* **51**: 630–632.

Mousseau TA, Dingle H (1991) Maternal effects in insect life histories. *Annual Review of Entomology* **36**: 511–534.

Mousseau TA, Roff DA (1989) Adaptation to seasonality in a cricket: patterns of phenotypic and genotypic variation in body size and diapause expression along a cline in season length. *Evolution* **43**: 1483–1496.

Mousson L, Nève G, Baguette M (1999) Metapopulation structure and conservation of the cranberry fritillary *Boloria aquilonaris* (Lepidoptera, Nymphalidae) in Belgium. *Biological Conservation* **87**: 285–293.

Müller HJ (1955) Die Saisonformenbildung von *Araschnia levana*: ein photoperiodisch gesteuerte Diapause-Effekt. *Naturwissenschaften* **42**: 134–135.

Munguira ML (1988) Biología y biogeografía de los licénidos ibéricos en peligro de extinción (Lepidoptera, Lycaenidae). Ph.D. thesis, Universidad Autónoma, Madrid.

Munguira ML, Martin J (1993) The conservation of endangered lycaenid butterflies in Spain. *Biological Conservation* **66**: 17–22.

Munguira ML, Martín J (1999) *Action Plan for* Maculinea *Butterflies in Europe*, Nature and Environment No. 97. Strasbourg, France: Council of Europe.

Munguira ML, Thomas JA (1992) Use of road verges by butterfly and burnet populations, and the effect of roads on adult dispersal and mortality. *Journal of Applied Ecology* **29**: 316–329.

Munguira M, Martin J, Perez Valiente M (1994) Karyology and distribution as tools in the taxonomy of Iberian *Agrodiaetus* butterflies (Lepidoptera: Lycaenidae). *Nota lepidopterologica* **17**: 125–140.

Munguira ML, García-Barros E, Martín J (1997a) Plantas nutricias de los licénidos y satirinos españoles (Lepidoptera: Lycaenidae y Nymphalidae). *Boletín Asociación española Entomología* **21**: 29–53.

Munguira ML, Martín J, García-Barros E, Viejo JL (1997b) Use of space and resources in a Mediterranean population of the butterfly *Euphydryas aurinia*. *Acta Oecologica* **18**: 597–612.

Muraoka K, Watanabe M (1994) A preliminary study of nectar production of the field cress, *Rorippa indica*, in relation to the age of its flowers. *Ecological Research* **9**: 33–36.

Murphy D (1984) Butterflies and their nectar plants: the role of the Checkerspot Butterfly *Euphydryas editha* as a pollen vector. *Oikos* **43**: 113–117.

Murphy AH, Winkler RL (1992) Diagnostic verification of probability forecasts. *International Journal of Forecasting* **7**: 435–455.

Murphy DD, Launer AE, Ehrlich PR (1983) The role of adult feeding in the egg production and population dynamics of the checkerspot butterfly *Euphydryas editha*. *Oecologia* **56**: 257–263.

Murphy DD, Menninger MS, Ehrlich P (1984) Nectar source distribution as a determinant of oviposition host species in *Euphydryas chalcedona*. *Oecologia* **62**: 269–271.

Musche M, Anton C, Worgan A, Settele J (2006) No experimental evidence for host ant related oviposition in a parasitic butterfly. *Journal of Insect Behavior* **19**: 631–643.

Myers JH (1985) Effect of physical condition of the host plant on the ovipositional choice of the cabbage white butterfly, *Pieris rapae*. *Journal of Applied Ecology* **54**: 193–204.

Myers AA, Giller PS (1988) *Analytic Biogeography*. London: Chapman and Hall.

Nabokov V (1941) *Lysandra cormion*, a new European butterfly. *Journal of the New York Entomological Society* **49**: 265–267.

Nagaraju J, Jolly MS (1986) Interspecific hybrids of *Antheraea roylei* and *A. pernyi*: a cytogenetic reassessment. *Theoretical and Applied Genetics* **72**: 269–273.

Nagelkerke NJD (1991) A note on general definition of the coefficient of determination. *Biometrika* **78**: 691–692.

Naisbit RE, Jiggins CD, Mallet J (2001) Disruptive sexual selection against hybrids contributes to speciation between *Heliconius cydno* and *H. melpomene*. *Proceedings of the Royal Society of London B* **268**: 1849–1854.

Naisbit RE, Jiggins CD, Linares M, Mallet J (2002) Hybrid sterility, Haldane's rule, and speciation *Heliconius cydno* and *H. melpomene*. *Genetics* **161**: 1517–1526.

Nakanishi Y, Watanabe M, Ito T (2000) Differences in lifetime reproduction output and mating frequency of two morphs of the sulfur butterfly, *Colias erate* (Lepidoptera: Pieridae). *Journal of the Research on the Lepidoptera* **35**: 1–8.

Nakasuji F, Nakano A (1990) Flight activity and oviposition characteristics of the seasonal form of a migrant skipper, *Parnara guttata guttata* (Lepidoptera: Hesperiidae). *Research in Population Ecology* **32**: 227–233.

Nakayama T, Honda K (2004) Chemical basis for differential acceptance of two sympatric rutaceous plants by ovipositing females of a swallowtail butterfly, *Papilio polytes* (Lepidoptera, Papilionidae). *Chemoecology* **14**: 199–205.

Nakayama T, Honda K, Omura H, Hayashi N (2003) Oviposition stimulants for the tropical swallowtail butterfly, *Papilio polytes*, feeding on a rutaceous plant, *Toddalia asiatica*. *Journal of Chemical Ecology* **29**: 1621–1634.

Nakonieczny M, Kędziorski A, Michalczyk K (2007) Apollo Butterfly (*Parnassius apollo* L.) in Europe: its history, decline and perspective of conservation. *Functional Ecosystems and Communities* **2007**: 56–79.

Napolitano M, Descimon H (1994) Genetic structure of French populations of the mountain butterfly *Parnassius mnemosyne* L. (Lepidoptera: Papilionidae). *Biological Journal of the Linnean Society* **53**: 325–341.

Napolitano M, Geiger H, Descimon H (1988) Structure démographique et génétique de quatre populations provençales de *Parnassius mnemosyne* (L.) (Lepidoptera Papilionidae): isolement et polymorphisme dans des populations 'menacées'. *Génétique, Sélection, Evolution* **20**: 51–62.

Naumann CM, Tarmann GM, Tremewan WG (1999) *The Western Palaearctic Zygaenidae*. Stenstrup, Denmark: Apollo Books.

Nazari V, Zakharov EV, Sperling FAH (2007) Phylogeny, historical biogeography and taxonomic ranking of Parnassiinae (Lepidoptera, Papilionidae) based on morphology and seven genes. *Molecular Phylogenetics and Evolution* **42**: 131–156.

Nei M (1972) Genetic distances between populations. *American Naturalist* **106**: 283–291.

Nei M (1978) Estimation of average heterozygosity and genetic distance from a small number of individuals. *Genetics* **89**: 583–590.

Neigel JE (2002) Is F_{ST} obsolete? *Conservation Genetics* **3**: 167–173.

Nekola JC (1998) Butterfly (Lepidoptera: Lycaenidae, Nymphalidae, and Satyridae) faunas of three peatland habitat types in the Lake Superior drainage basin of Wisconsin. *Great Lakes Entomologist* **31**: 27–38.

Nekrutenko YP (1985) *Butterflies of the Crimea: Guide* [in Russian]. Kiev, Ukraine: Naukova Dumka.

Nel J (1984) Note sur les *Pyrgus sidae* Esper: sa plante-hôte et son cycle biologique en Provence (Lep., Hesperiidae). *Alexanor* **13**: 275–281.

Nel J (1985) Note sur la répartition, les plantes-hôtes et le cycle de développement des Pyrginae en Provence (Lep., Hesperiidae). *Alexanor* **14**: 51–63.

Nève G (1996) Dispersion chez une espèce à habitat fragmenté: *Proclossiana eunomia* (Lepidoptera, Nymphalidae). Ph.D. Thesis, Université catholique de Louvain.

Nève G, Meglécz E (2000) Microsatellite frequencies in different taxa. *Trends in Ecology and Evolution* **15**: 376–377.

Nève G, Barascud B, Hughes R, Aubert J, Descimon H, Lebrun P, Baguette M (1996a) Dispersal, colonization power and metapopulation structure in the vulnerable butterfly *Proclossiana eunomia* (Lepidoptera: Nymphalidae). *Journal of Applied Ecology* **33**: 14–22.

Nève G, Mousson L, Baguette M (1996b) Adult dispersal and genetic structure of butterfly populations in a fragmented landscape. *Acta Oecologica* **17**: 621–626.

Nève G, Barascud B, Descimon H, Baguette M (2000) Genetic structure of *Proclossiana eunomia* populations at the regional scale (Lepidoptera, Nymphalidae). *Heredity* **84**: 657–666.

Nève G, Barascud B, Descimon H, Baguette M (2008) Gene flow rise with habitat fragmentation in the bog fritillary butterfly (Lepidoptera, Nymphalidae). *BMC Evolutionary Biology* **8**: 84.

Nève G, Pavličko A, Konvička M (2009) Loss of genetic diversity through spontaneous colonization in the Bog Fritillary butterfly (*Proclossiana eunomia*) in the Czeck Republic (Lepidoptera: Nymphalidae). *European Journal of Entomology* **106**: 11–19.

Nice CC, Shapiro AM (2001) Population genetic evidence of restricted gene flow between host races in the butterfly genus *Mitoura* (Lepidoptera: Lycaenidae). *Annals of the Entomological Society of America* **94**: 257–267.

Nicholls AO (1989) How to make biological surveys go further with generalised linear models. *Biological Conservation* **50**: 51–75.

Nicholls CN, Pullin AS (2003) The effects of flooding on survivorship in overwintering larvae of the large copper butterfly *Lycaena dispar batavus* (Lepidoptera: Lycaenidae), and its possible implications for restoration management. *European Journal of Entomology* **100**: 65–72.

Niculescu EU (1985) Problèmes de morphologie et de taxonomie dans le genre *Erebia* (Lep.: Satyridae). *Bulletin de la Société Entomologique de Mulhouse* **1985**: 1–8.

Nielsen ET (1961) On the habits of the migratory butterfly *Ascia monuste* L. *Biologiske Meddelelser Kongelige Danske Videnskabernes Selskab* **23**: 1–81.

Nielsen MG, Watt WB (1998) Behavioural fitness component effects of the alba polymorphism of *Colias* (Lepidoptera, Pieridae): resource and time budget analysis. *Functional Ecology* **12**: 149–158.

Nielsen MG, Watt WB (2000) Interference competition and sexual selection promote polymorphism in *Colias* (Lepidoptera, Pieridae). *Functional Ecology* **14**: 718–730.

Niemela J, Spence JR (1991) Distribution and abundance of an exotic ground-beetle (Carabidae): a test of community impact. *Oikos* **62**: 351–359.

Nieminen M, Singer MC, Fortelius W, Schöps K, Hanski I (2001) Experimental confirmation that inbreeding depression increases extinction risk in butterfly populations. *American Naturalist* **157**: 237–244.

Nieminen M, Suomi J, van Nouhuys S, Sauri P, Riekkola ML (2003) Effect of iridoid glycoside content on oviposition host plant choice and parasitism in a specialist herbivore. *Journal of Chemical Ecology* **29**: 823–844.

Nijhout HF (1991) *The Development and Evolution of Butterfly Wing Patterns*. Washington, DC: Smithsonian Institution Press.

Nijhout HF (2003) Development and evolution of adaptive polyphenisms. *Evolution and Development* **5**: 9–18.

Nijhout HF, Emlen DJ (1998) Competition among body parts in the development and evolution of insect morphology. *Proceedings of the National Academy of Sciences, USA* **95**: 3685–3689.

Nilsson NH (1930) Synthetische Bastardierungversuche in der Gattung *Salix. Lunds Universitets Årsskrift, N.F.*, **27**: 1–97.

Nilsson LA (1988) The evolution of flowers with deep corolla tubes. *Nature* **334**: 147–149.

Nilsson LA (1998) Deep flowers for long tongues. *Trends in Ecology and Evolution* **13**: 259–260.

Nixon GEJ (1974) A revision of the north-western European species of the *glomeratus*-group of *Apanteles* Förster (Hymenoptera: Braconidae). *Bulletin of Entomological Research* **64**: 453–524.

Nomakuchi S, Masumoto T, Sawada K, Sunahara T, Itakura N, Suzuki N (2001) Possible age-dependent variation in egg-loaded host selectivity of the pierid butterfly, *Anthocharis scolymus* (Lepidoptera: Pieridae): a field observation. *Journal of Insect Behavior* **14**: 451–458.

Noon BR, McKelvey KS (1996) Management of the spotted owl: a case history in conservation biology. *Annual Review of Ecology and Systematics* **27**: 135–162.

Noor MAF (1995) Speciation driven by natural selection in *Drosophila. Nature* **375**: 674–675.

Noor MAF (1999) Reinforcement and other consequences of sympatry. *Heredity* **83**: 503–508.

Norberg U, Leimar O (2002) Spatial and temporal variation in flight morphology in the butterfly *Melitaea cinxia* (Lepidoptera: Nymphalidae). *Biological Journal of the Linnean Society* **77**: 445–453.

Norberg U, Enfjäll K, Leimar O (2002) Habitat exploration in butterflies: an outdoor cage experiment. *Evolutionary Ecology* **16**: 1–14.

Norris MJ (1935) A feeding experiment on the adults of *Pieris rapae. Entomologist* **68**: 125–127.

Norris MJ (1936) The feeding habits of the adult Lepidoptera (Heteroneura). *Transactions of the Royal Entomological Society of London* **85**: 61–90.

Noss RF, Csuti B (1997) Habitat fragmentation. In: Meffe GK, Carroll CR (eds.) *Principles of Conservation Biology*, 2nd Edn, Sunderland, MA: Sinauer Associates, pp. 269–304.

Notaro M, Vavrus S, Liu ZY (2007) Global vegetation and climate change due to future increases in CO_2 as projected by a fully coupled model with dynamic vegetation. *Journal of Climate* **20**: 70–90.

Novak J, Prach K (2003) Vegetation succession in basalt quarries: pattern on a landscape scale. *Applied Vegetation Science* **6**: 111–116.

Novotny V, Tonner M, Spitzer K (1991/1993) Distribution and flight behaviour of the junglequeen butterfly *Stichophtalma louisa* (Lepidoptera Nymphalidae), in an Indochinese montane rainforest. *Journal of Research on the Lepidoptera* **30**: 279–288.

Nowicki P, Settele J, Thomas JA, Woyciechowski M (2005a) A review of population structure of *Maculinea* butterflies. In: Settele J, Kühn E, Thomas JA (eds.) *Studies on the Ecology and Conservation of Butterflies in Europe*, Vol. 2, *Species Ecology along a European Gradient:* Maculinea *Butterflies as a Model*, Leipzig, Germany: Pensoft, pp. 144–149.

Nowicki P, Witek M, Skorka P, Woyciechowski M (2005b) Oviposition patterns in the myrmecophilous butterfly *Maculinea alcon* Denis, Schiffermüller (Lepidoptera: Lycaenidae) in relation to characteristics of foodplants and presence of ant hosts. *Polish Journal of Ecology* **53**: 409–417.

Noyes JS (1982) Collecting and preserving chalcid wasps (Hymenoptera: Chalcidoidea). *Journal of Natural History* **16**: 313–334.

Noyes JS (1990) Chalcid parasitoids. In: Rosen D (ed.) *The Armoured Scale Insects: Their Biology, Natural Enemies and Control*, Amsterdam, Netherlands: Elsevier, pp. 247–262.

Noyes JS (1994) The reliability of published host–parasitoid records: a taxonomist's view. *Norwegian Journal of Agricultural Science* **16** (Suppl.): 59–69.

Nunner A (2006) Zur Verbreitung, Bestandssituation und Habitatbindung des Blauschillernden Feuerfalters (*Lycaena helle*) in Bayern. *Abhandlungen aus dem Westfälischen Museum für Naturkunde* **68**: 153–170.

Nygren GH, Nylin S, Stefanescu C (2006) Genetics of host plant use and life history in the comma butterfly across Europe: varying modes of inheritance as a potential reproductive barrier. *Journal of Evolutionary Biology* **19**: 1882–1893.

Nylin S (1988) Host plant specialization and seasonality in a polyphagous butterfly, *Polygonia c-album* (Nymphalidae). *Oikos* **53**: 381–386.

Nylin S (1989) Effects of changing photoperiods in the life cycle regulation of the comma butterfly, *Polygonia c-album* (Nymphalidae). *Ecological Entomology* **14**: 209–218.

Nylin S (1992) Seasonal plasticity in life history traits: growth and development in *Polygonia c-album* (Lepidoptera: Nymphalidae). *Biological Journal of the Linnean Society* **47**: 301–323.

Nylin S (1994) Seasonal plasticity and life-cycle adaptations in butterflies. In: Danks HV (ed.) *Insect Life-Cycle Polymorphism*, Dordrecht, Netherlands: Kluwer, pp. 41–67.

Nylin S (2001) Life history perspectives on pest insects: what's the use? *Australian Ecology* **26**: 507–517.

Nylin S, Gotthard K (1998) Plasticity in life history traits. *Annual Review of Entomology* **43**: 63–83.

Nylin S, Janz N (1996) Host plant preferences in the comma butterfly (*Polygonia c-album*): do parents and offspring agree? *Ecoscience* **3**: 285–289.

Nylin S, Svärd L (1991) Latitudinal patterns in the size of European butterflies. *Holarctic Ecology* **14**: 192–202.

Nylin S, Wedell N (1994) Sexual size dimorphism and comparative methods. In: Eggleton P, Vane-Wright RI (eds.) *Phylogenetics and Ecology*, London: Academic Press, pp. 253–280.

Nylin S, Wickman P-O, Wiklund C (1989) Seasonal plasticity in growth and development of the speckled wood butterfly, *Pararge aegeria* (Satyrinae). *Biological Journal of the Linnean Society* **38**: 155–171.

Nylin S, Wiklund C, Wickman P-O, García-Barros E (1993) Absence of trade-offs between early male emergence and sexual size dimorphism in a butterfly. *Ecology* **74**: 1414–1472.

Nylin S, Wickman P-O, Wiklund C (1995a) An adaptive explanation for male-biased sex ratios in overwintering monarch butterflies. *Animal Behaviour* **49**: 511–514.

Nylin S, Wickman P-O, Wiklund C (1995b) Life-cycle regulation and life history plasticity in the speckled wood butterfly: are reaction norms predictable? *Biological Journal of the Linnean Society* **55**: 143–157.

Nylin S, Gotthard K, Wiklund C (1996a) Reaction norms for age and size at maturity in *Lasiommata* butterflies: predictions and tests. *Evolution* **50**: 1351–1358.

Nylin S, Janz N, Wedell N (1996b) Oviposition plant preference and offspring performance in the comma butterfly: correlations and conflicts. *Entomologia Experimentalis et Applicata* **80**: 141–144.

Nylin S, Bergström A, Janz N (2000) Butterfly host plant choice in the face of possible confusion. *Journal of Insect Behavior* **13**: 462–482.

Nylin S, Nygren GH, Windig JJ, Janz N, Bergström A (2005) Genetics of host-plant preference in the comma butterfly *Polygonia c-album* (Nymphalidae), and evolutionary implications. *Biological Journal of the Linnean Society* **84**: 755–765.

Obara Y (1964) Mating behaviour of the cabbage white, *Pieris rapae crucivora*. II. The 'mate refusal posture' of the female. *Zoological Magazine* **73**: 175–178.

Oberhauser KS (1988) Male monarch butterfly spermatophore mass and mating strategies. *Animal Behaviour* **36**: 1384–1388.

Oberhauser KS (1989) Effects of spermatophores on male and female monarch butterfly reproductive success. *Behavioral Ecology and Sociobiology* **25**: 237–246.

Oberhauser KS (1992) Rate of ejaculate breakdown and intermating intervals in monarch butterflies. *Behavioral Ecology and Sociobiology* **31**: 367–373.

Oberhauser KS (1997) Fecundity, lifespan and egg mass in butterflies: effects of male-derived nutrients and female size. *Functional Ecology* **11**: 166–175.

Oberthür C (1914) *Etudes de Lépidoptérologie Comparée*, Vol. 10. Rennes, France: C. Oberthür.

O'Brien P, Wolfluss J (1991) Distribution spatio-temporelle de *Coscinocera victoria*, *Coscinocera tigrata carpenteri*, *Coscinocera punctata* Barton, *Coscinocera nigrata* d'Iputupi. In: Perec G (ed.) Cantatrix sopranica *et autres écrits scientifiques*, Paris: Seuil, pp. 35–52.

O'Brien DM, Schrag DP, Martinez del Rio C (2000) Allocation to reproduction in a hawkmoth: a quantitative analysis using stable carbon isotopes. *Ecology* **81**: 2822–2831.

O'Brien DM, Boggs CL, Fogel ML (2004) Making eggs from nectar: the role of life history and dietary carbon turnover in butterfly reproductive resource allocation. *Oikos* **105**: 279–291.

O'Brien DM, Boggs CL, Fogel ML (2005) The amino acids used in reproduction by butterflies: a comparative study of dietary sources using compound-specific stable isotope analysis. *Physiological and Biochemical Zoology* **78**: 819–827.

Öckinger E, Smith HG (2006) Landscape composition and habitat area affects butterfly species richness in semi-natural grasslands. *Oecologia* **149**: 526–534.

Öckinger E, Smith HG (2007) Semi-natural grasslands as population sources for pollinating insects in agricultural landscapes. *Journal of Applied Ecology* **44**: 50–59.

Öckinger E, Hammarstedt O, Nilsson SG, Smith HG (2006) The relationship between local extinctions of grassland butterflies and increased soil nitrogen levels. *Biological Conservation* **128**: 564–573.

Odendaal FJ, Iwasa Y, Ehrlich PR (1985) Duration of female availability and its effect on butterfly mating systems. *American Naturalist* **125**: 673–678.

Odendaal FJ, Turchin P, Stermitz FR (1989) Influence of host-plant density and male harassment on the distribution of female *Euphydryas anicia* (Nymphalidae). *Oecologia* **78**: 283–288.

Odendaal FJ, Eekhout S, Brown AC, Branch GM (1999) Aggregations of the sandy-beach isopod, *Tylos granulatus*: adaptation or incidental-effect? *South African Journal of Zoology* **34**: 180–189.

Ohsaki N (2005) A common mechanism explaining the evolution of female-limited and both-sex Batesian mimicry in butterflies. *Journal of Animal Ecology* **74**: 728–734.

Ohsaki N, Sato Y (1994) Food plant choice of *Pieris* butterflies as a trade-off between parasitoid avoidance and quality of plants. *Ecology* **75**: 59–68.

Ohsaki N, Sato Y (1999) The role of parasitoids in evolution of habitat and larval food plant preference by three *Pieris* butterflies. *Research in Population Ecology* **41**: 107–119.

Oksanen J, Minchin PR (2002) Continuum theory revisited: what shape are species responses along ecological gradients? *Ecological Modelling* **157**: 119–129.

Olden JD, Jackson DA (2002) A comparison of statistical approaches for modelling fish species distributions. *Freshwater Biology* **47**: 1976–1995.

Oliver CG (1972) Genetic differences between English and French populations of the satyrid butterfly *Pararge megaera*. *Heredity* **29**: 307–313.

Oliver CG (1978) Experimental hybridization between the nymphalid butterflies *Phyciodes tharos* and *P. campestris montana*. *Evolution* **32**: 594–601.

Olivier A, van der Poorten D, Puplesiene J, de Prins W (2000) *Polyommatus* (*Agrodiaetus*) *artvinensis*, stat. nov. and *P.* (*A.*) *sigberti*, two vicariant species known so far only from Turkey (Lepidoptera: Lycaenidae). *Phegea* **28**: 57–74.

Olivieri I, Gouyon PH (1997) Evolution of migration rate and other traits. In: Hanski IA, Gilpin ME (eds.) *Metapopulation Biology*, San Diego, CA: Academic Press, pp. 293–323.

Olofsson J, Kitti H, Rautiainen P, Stark S, Oksanen L (2001) Effects of summer grazing by reindeer on composition of vegetation, productivity and nitrogen cycling. *Ecography* **24**: 13–24.

Olofsson J, Hulme PE, Oksanen L, Suominen O (2004) Importance of large and small mammalian herbivores for the plant community structure in the forest tundra ecotone. *Oikos* **106**: 324–334.

Omura H, Honda K (2005) Priority of color over scent during flower visitation by adult *Vanessa indica* butterflies. *Oecologia* **142**: 564–575.

Omura H, Honda K, Hayashi N (1999) Chemical and chromatic bases for preferential visiting by the cabbage butterfly, *Pieris rapae*, to rape flowers. *Journal of Chemical Ecology* **25**: 1895–1906.

Ono H, Yoshikawa H (2004) Identification of amine receptors from a swallowtail butterfly, *Papilio xuthus* L.: cloning and mRNA localization in foreleg chemosensory organ for recognition of host plants. *Insect Biochemistry and Molecular Biology* **34**: 1247–1256.

Oppel S, Schaefer HM, Schmidt V, Schröder B (2004) Habitat selection by the pale-headed brush-finch, *Atlapetes pallidiceps*, in southern Ecuador: implications for conservation. *Biological Conservation* **118**: 33–40.

Orloci L (1972) On objective functions of phytosociological resemblance. *American Midland Naturalist* **88**: 28–55.

Orr HA (1995) The population genetics of speciation: the evolution of hybrid incompatibilities. *Genetics* **139**: 1805–1813.

Orr AG (1999) The big greasy, *Cressida cressida* (Papilionidae). In: Kitching RL, Scheermeyer E, Jones RE, Pierce NE (eds.) *Biology of Australian Butterflies*, Collingwood, Australia: CSIRO Publishing, pp. 115–134.

Orr AG (2002) The sphragis of *Heteronympha penelope* Waterhouse (Lepidoptera: Satyridae): its structure, formation and role in sperm guarding. *Journal of Natural History* **36**: 185–196.

Orsini L, Pajunen M, Hanski I, Savilahti H (2007), SNP discovery by mismatch-targeting of *Mu* transposition. *Nucleic Acids Research* **35**: e44.

Orsini L, Corander J, Alasentie A, Hanski I (2008) Genetic spatial structure in a butterfly metapopulation correlates better with past than present demographic structure. *Molecular Ecology* **17**: 2629–2642.

Osborne PE, Alonso JC, Bryant RG (2001) Modeling landscape-scale habitat use using GIS and remote sensing: a case study with great bustards. *Journal of Applied Ecology* **38**: 458–471.

Ouin A, Aviron S, Dover J, Burel F (2004) Complementation/supplementation of resources for butterflies in agricultural landscapes. *Agriculture, Ecosystems and Environment* **103**: 473–479.

Pagenstecker A (1909) *Die geographische Verbreitung der Schmetterlinge*. Jena: Gustav Fischer.

Päivinen J, Grapputo A, Kaitala V, Komonen A, Kotiaho JS, Saarinen K, Wahlberg N (2005) Negative density-distribution relationship in butterflies. *BMC Biology* **3**: 5.

Pan W, Le CT (2001) Bootstrap model selection in generalized linear models. *Journal of Agricultural, Biological and Environmental Statistics* **6**: 49–61.

Pannekoek J, Van Strien AJ (2005) *TRIM 3 Manual: Trends and Indices for Monitoring Data*. Voorburg, Netherlands: CBS Statistics Netherlands (Freely available via www.ebcc.info.)

Papaj DR, Newsom GM (2005) A within-species warning function for an aposematic signal. *Proceedings of the Royal Society of London B* **272**: 2519–2523.

Parker GA (1974) Assessment strategy and the evolution of fighting behaviour. *Journal of Theoretical Biology* **47**: 223–243.

Parker GA, Courtney SP (1984) Models of clutch size in insect oviposition. *Theoretical Population Biology* **26**: 27–48.

Parker GA, Simmons LW (1996) Parental investment and the control of sexual selection: predicting the direction of sexual competition. *Proceedings of the Royal Society of London B* **263**: 315–321.

Parker PG, Snow AA, Schug MD, Booton GC, Fuerst PA (1998) What molecules can tell us about populations: choosing and using a molecular marker. *Ecology* **79**: 361–382.

Parmesan C (1996) Climate and species' range. *Nature* **382**: 765–766.

Parmesan C (2003) Butterflies as bioindicators for climate change effects. In: Boggs CL, Watt WB, Ehrlich PE (eds.) *Butterflies: Ecology and Evolution Taking Flight*, Chicago, IL: University of Chicago Press, pp. 541–560.

Parmesan C (2006) Ecological and evolutionary responses to recent climate change. *Annual Review of Ecology, Evolution and Systematics* **37**: 637–669.

Parmesan C, Yohe G (2003) A globally coherent fingerprint of climate change impacts across natural systems. *Nature* **421**: 37–42.

Parmesan C, Singer MC, Harris I (1995) Absence of adaptive learning from the oviposition foraging behaviour of a checkerspot butterfly. *Animal Behaviour* **50**: 161–175.

Parmesan C, Ryrholm N, Stefanescu C, Hill JK, Thomas CD, Descimon H, Huntley B, Kaila L, Kullberg J, Tammaru T, Tennent WJ, Thomas JA, Warren MS (1999) Poleward shifts in geographical ranges of butterfly species associated with regional warming. *Nature* **399**: 579–583.

Parmesan C, Gaines S, Gonzalez L, Kaufman DM, Kingsolver J, Peterson AT, Sagarin R (2005) Empirical perspectives on species borders: from traditional biogeography to global change. *Oikos* **108**: 58–75.

Parry DA (1951) Factors determining the temperature of terrestrial arthropods in sunlight. *Journal of Experimental Biology* **28**: 445–462.

Partridge L, Barrie B, Barton NH, Fowler K, French V (1995) Rapid laboratory evolution of adult life-history traits in *Drosophila melanogaster* in response to temperature. *Evolution* **49**: 538–544.

Pashley DP, Ke LD (1992) Sequence evolution in mitochondrial ribosomal and ND1 genes in Lepidoptera: implications for phylogenetic analyses. *Molecular Biology and Evolution* **9**: 1061–1075.

Paterson HEH (1985) The recognition concept of species. In: Vrba ES (ed.) *Species and Speciation*, Pretoria, South Africa: Transvaal Museum, pp. 21–29.

Pauler R, Kaule G, Verhaagh M, Settele J (1995) Untersuchungen zur Autökologie des Schwarzgefleckten Ameisenbläulings, *Maculinea arion* (Linnaeus 1758) (Lepidoptera: Lycaenidae), in Südwestdeutschland. *Nachrichten des entomologischen Vereins Apollo Frankfurt/Main, N.F.* **16**: 147–186.

Paulo OS, Dias C, Bruford MW, Jordan WC, Nichols RA (2001) The persistence of Pliocene populations though the Pleistocene climatic cycles: evidence from the phylogeography of an Iberian lizard. *Proceedings of the Royal Society of London B* **268**: 1625–1630.

Pausas JG, Carreras J, Ferré A, Font X (2003) Coarse-scale plant species richness in relation to environmental heterogeneity. *Journal of Vegetation Science* **14**: 661–668.

Pavlíčko A (1996) Rozšíření perleťovce mokřadního (*Proclossiana eunomia*) na Šumavě a jeho vztah k hospodaření v krajině. *Silva Gabreta* **1**: 197–202.

Pavlíčko A (2001) Vojenský výcvikový prostor Boletice. *Sborník Zlatá Stezka, Sborník Prachatického Muzea* **7**: 283–323.

Pearce JL, Boyce MS (2006) Modelling distribution and abundance with presence-only data. *Journal of Applied Ecology* **43**: 405–412.

Pearce J, Ferrier S (2000) Evaluating the predictive performance of habitat models developed using logistic regression. *Ecological Modelling* **133**: 225–245.

Pearson RG, Dawson TP (2003) Predicting the impacts of climate change on the distribution of species: are bioclimate envelope models useful? *Global Ecology and Biogeography* **12**: 361–371.

Pearson RG, Raxworthy CJ, Nakamura M, Peterson AT (2007) Predicting species distributions from small numbers of occurrence records: a test case using cryptic geckos in Madagascar. *Journal of Biogeography* **34**: 102–117.

Pecsenye K, Bereczki J, Szilágyi M, Varga Z (2007a) High level of genetic variation in *Aricia artaxerxes issekutzi* (Lycaenidae) populations in Northern Hungary. *Nota lepidopterologica* **30**: 225–234.

Pecsenye K, Bereczki J, Tihanyi B, Toth A, Peregovits L, Varga Z (2007b) Genetic differentiation among the *Maculinea* species (Lepidoptera: Lycaenidae) in eastern Central Europe. *Biological Journal of the Linnean Society* **91**: 11–21.

Pe'er G, Saltz D, Thulke HH, Motro U (2004) Response to topography in a hilltopping butterfly and implications for modelling nonrandom dispersal. *Animal Behaviour* **68**: 825–839.

Pellmyr O (1983) Plebeian courtship revisited: studies on the female-produced male behavior-eliciting signals in *Lycaeides idas* courtship (Lycaenidae). *Journal of Research on the Lepidoptera* **21**: 147–157.

Pengelly DH (1961) *Thymelicus lineola* (Ochs.) (Lepidoptera: Hesperiidae) a pest of hay and pasture grasses in southern Ontario. *Proceedings of the Entomological Society of Ontario* **91**: 189–196.

Peñuelas J, Sardans J, Stefanescu C, Parella T, Filella I (2006) *Lonicera implexa* leaves bearing naturally laid eggs of the specialist herbivore *Euphydryas aurinia* have dramatically greater concentrations of iridoid glycosides than other leaves. *Journal of Chemical Ecology* **32**: 1925–1933.

Peppler-Lisbach C, Schröder B (2004) Predicting the species composition of mat-grass communities (Nardetalia) by logistic regression modelling. *Journal of Vegetation Science* **15**: 623–634.

Perceval MJ (1977) *Erebia serotina* Descimon, de Lesse 1953: a possible hybrid. *Entomological Record and Journal of Variation* **89**: 19–20.

Perceval MJ (1995) The African influence on the butterfly populations of southern Spain. *Entomologist's Gazette* **46**: 257–266.

Perez de Gregorio JJ (1979) Nota sobre las especies catalanas del género *Agrodiaetus* Hübner 1823 (Lepidoptera). *Bulletín de la Sociedad Catalana de Lepidopterología* **22**: 7–9.

Perrin N, Goudet J (2001) Inbreeding, kinship, and the evolution of natal dispersal. In: Clobert J, Danchin E, Dhondt AA, Nichols JD (eds.) *Dispersal*, Oxford: Oxford University Press, pp. 123–142.

Péténian F, Nève G (2003) Influence of spatial structure on genetic isolation in *Plebejus argus* populations (Lepidoptera: Lycaenidae). *Hereditas* **138**: 179–186.

Peter W. 1928. Zwitter von *Lycaena argus* L. und *Lycaena icarus* Rott. *Internationale Entomologische Zeitschrift* **22**: 139–140.

Petersen B (1947) *Die Geographische Variation einiger Fennoskandischer Lepidopteren.* Uppsala: Almgvist and Wiksell.

Petersen B (1954) Egg-laying and habitat selection in some *Pieris* species. *Entomologisk Tidskrift* **75**: 194–203.

Peterson MA (1995) Phenological isolation, gene flow and developmental differences among low-elevation and high-elevation populations of *Euphilotes enoptes* (Lepidoptera, Lycaenidae). *Evolution* **49**: 446–455.

Peterson MA, Denno RF (1998) The influence of dispersal and diet breadth on patterns of genetic isolation by distance in phytophagous insects. *American Naturalist* **152**: 428–446.

Petit S, Moilanen A, Hanski I, Baguette M (2001) Metapopulation dynamics of the bog fritillary butterfly: movements between habitat patches. *Oikos* **92**: 491–500.

Pfeifer MA, Andrick UR, Frey W, Settele J (2000) On the ethology and ecology of a small and isolated population of the Dusky Large Blue Butterfly *Glaucopsyche* (*Maculinea*) *nausithous* (Lycaenidae). *Nota lepidopterologica* **23**: 147–172.

Pfeuffer E (2000) Zur Ökologie der Präimaginalstadien des Himmelblauen Bläulings (*Lysandra bellargus* Rottemburg 1775) und des Silbergrünen Bläulings (*Lysandra coridon* Poda 1761), unter besonderer Berücksichtigung der Myrmekophilie. *Berichte des Naturwissenschaftlichen Vereins für Schwaben* **104**: 72–98.

Phillips SJ, Anderson RP, Schapire RE (2005) Maximum entropy modelling of species geographic distributions. *Ecological Modelling* **190**: 231–259.

Pierce N (1995) Predatory and parasitic Lepidoptera: carnivores living on plants. *Journal of the Lepidopterists' Society* **49**: 412–453.

Pierce NE, Easteal S (1986) The selective advantage of attendant ants for the larvae of a lycaenid butterfly, *Glaucopsyche lygdamus*. *Journal of Animal Ecology* **55**: 451–462.

Pierce NE, Kitching RL, Buckley RC, Taylor MFJ, Benbow KF (1987) The costs and benefits of cooperation between the Australian lycaenid butterfly, *Jalmenus evagoras*, and its attendant ants. *Behavioral Ecology and Sociobiology* **21**: 237–248.

Pilson D, Rausher MD (1988) Clutch size adjustment by a swallowtail butterfly. *Nature* **333**: 361–363.

Pimm SL, Jones HL, Diamond J (1988) On the risk of extinction. *American Naturalist* **132**: 757–785.

Pivnick KA, McNeil JN (1985a) Mate location and mating behavior of *Thymelicus lineola* (Lepidoptera:Hesperiidae). *Annals of the Entomological Society of America* **78**: 651–656.

Pivnik KA, McNeil JN (1985b) Effects of nectar concentration on butterfly feeding: measured feeding rates for *Thymelicus lineola* (Lepidoptera: Hesperiidae) and a general feeding model for adult Lepidoptera. *Oecologia* **66**: 226–237.

Pivnick KA, McNeil JN (1986) Sexual differences in the thermoregulation of *Thymelicus lineola* adults (Lepidoptera: Hesperiidae). *Ecology* **67**: 1024–1035.

Pivnick KA, McNeil J (1987a) Diel patterns of activity of *Thymelicus lineola* adults (Lepidoptera: Hesperiidae) in relation to weather. *Ecological Entomology* **12**: 197–207.

Pivnick KA, McNeil JN (1987b) Puddling in butterflies: sodium affects reproductive success in *Thymelicus lineola*. *Physiological Entomology* **12**: 461–472.

Pivnick KA, Lavoiedornik J, Mcneil JN (1992) The role of the androconia in the mating behaviour of the European skipper, *Thymelicus lineola*, and evidence for a male sex pheromone. *Physiological Entomology* **17**: 260–268.

Platt AP, Allen JF (2001) Sperm precedence and competition in doubly-mated *Limenitis arthemis-astyanax* butterflies

(Rhopalocera: Nymphalidae). *Annals of the Entomological Society of America* **94**: 654–663.

Pliske TE (1975) Courtship behavior of the monarch butterfly, *Danaus plexippus* L. *Annals of the Entomological Society of America* **68**: 143–151.

Poethke HJ, Hovestadt T (2002) Evolution of density- and patch-size-dependent dispersal rates. *Proceedings of the Royal Society of London B* **269**: 637–645.

Poethke HJ, Hovestadt T, Mitesser O (2003) Local extinction and the evolution of dispersal rates: causes and correlations. *American Naturalist* **161**: 631–640.

Poethke HJ, Pfenning B, Hovestadt T (2007) The relative contributions of individual- and kin-selection in the evolution of density-dependent dispersal rates. *Evolutionary Ecology Research* **9**: 41–50.

Poff NL (1997) Landscape filters and species traits: towards mechanistic understanding and prediction in stream ecology. *Journal of the North American Benthological Society* **16**: 391–409.

Polcyn DM, Chappell MA (1986) Analysis of heat transfer in *Vanessa* butterflies: effects of wing position and orientation to wind and light. *Physiological Zoology* **59**: 706–716.

Pollard E (1977) A method for assessing changes in the abundance of butterflies. *Biological Conservation* **12**: 115–134.

Pollard E (1979) Population ecology and change in range of the White Admiral butterfly *Ladoga camilla* L. in England. *Ecological Entomology* **4**: 61–74.

Pollard E (1988) Temperature, rainfall and butterfly numbers. *Journal of Applied Ecology* **25**: 819–828.

Pollard E (1991) Synchrony of population fluctuations: the dominant influence of widespread factors on local butterfly populations. *Oikos* **60**: 7–10.

Pollard E, Eversham BC (1995) Butterfly monitoring. II. Interpreting the changes. In: Pullin AS (ed.) *Ecology and Conservation of Butterflies*, London: Chapman and Hall, pp. 23–36.

Pollard E, Moss D (1995) Historical records of the occurrence of butterflies in Britain: examples showing associations between annual number of records and weather. *Global Change Biology* **1**: 107–113.

Pollard E, Yates TJ (1993) *Monitoring Butterflies for Ecology and Conservation: The British Butterfly Monitoring Scheme*. London: Chapman and Hall.

Pollard E, Van Swaay CAM, Yates TJ (1993) Changes in butterfly numbers in Britain and the Netherlands, 1990–91. *Ecological Entomology* **18**: 93–94.

Pollard E, Moss D, Yates TJ (1995) Population trends of common British butterflies at monitored sites. *Journal of Applied Ecology* **32**: 9–16.

Pollard E, Rothery P, Yates TJ (1996) Annual growth rates in newly established populations of the butterfly *Pararge aegeria*. *Ecological Entomology* **21**: 365–369.

Pollard E, Greatorex-Davies JN, Thomas JA (1997) Drought reduces breeding success of the butterfly *Aglais urticae*. *Ecological Entomology* **22**: 315–318.

Pollard E, Woiwod IP, Greatorex-Davies JN, Yates TJ, Welch RC (1998) The spread of coarse grasses and changes in numbers of Lepidoptera in a woodland nature reserve. *Biological Conservation* **84**: 17–24.

Polus E, Vandewoestijne S, Choutt J, Baguette M (2007) Tracking the effects of one century of habitat loss and fragmentation on calcareous grassland butterfly communities. *Biodiversity and Conservation* **16**: 3423–3436.

Ponel P (1997) The response of Coleoptera to late-Quaternary climate changes: evidence from north-east France. In: Huntley B, Cramer W, Morgan AV (eds.) *Past and Future Rapid Environmental Changes*, NATO ASI Series Vol. 147, Berlin: Springer-Verlag, pp. 143–151.

Poole A, Pienkowaki M, McCracken DI, Petretti F, Brédy C, Deffeyes C (eds.) (1998) *Mountain Livestock Farming and EU Policy Development*, Proceedings of the 5 European Forum on Nature Conservation and Pasturalism, Valle d'Aosta, Italy.

Popescu-Gorj A (1974) A rare hybrid within the genus *Erebia* (Lep., Satyridae). *Travaux du Musée d'Histoire Naturelle 'Gregorie Antipa'* **14**: 267–271.

Porter K (1981) The population dynamics of small colonies of the butterfly *Euphydryas aurinia*. Ph.D. Thesis, University of Oxford.

Porter K (1982) Basking behaviour in larvae of the butterfly *Euphydryas aurinia*. *Oikos* **38**: 308–312.

Porter K (1983) Multivoltinism in *Apanteles bignelli* and the influence of weather on synchronisation with its host *Euphydryas aurinia*. *Entomologia Experimentalis et Applicata* **34**: 155–162.

Porter K (1984) Sunshine, sex-ratio and behaviour of *Euphydryas aurinia* larvae. In: Vane-Wright RI, Ackery PR (eds.) *The Biology of Butterflies*, London: Academic Press, pp. 309–311.

Porter K (1992) Eggs and egg-laying. In: Dennis RLH (ed.) *The Ecology of Butterflies in Britain*, Oxford: Oxford University Press, pp. 46–72.

Porter AH (1997) The *Pieris napi/bryoniae* hybrid zone at Pont de Nant, Switzerland: broad overlap in the range of suitable host plants. *Ecological Entomology* **22**: 189–196.

Porter AH (2005) ClineFit: software for the analysis of cline shape. Online: http://www-unix.oit.umass.edu/~aporter/software/

Porter WP, Gates DM (1969) Thermodynamic equilibria of animals with environment. *Ecological Monographs* **39**: 227–244.

Porter AH, Geiger H (1995) Limitations to the inference of gene flow at regional geographic scales: an example from the *Pieris napi* group (Lepidoptera: Pieridae) in Europe. *Biological Journal of the Linnean Society* **54**: 329–348.

Porter AH, Shapiro AM (1990) Lock-and-key hypothesis: lack of mechanical isolation in a butterfly (Lepidoptera: Pieridae) hybrid zone. *Annals of the Entomological Society of America* **83**: 107–114.

Porter K, Steel CA, Thomas JA (1992) Butterflies and communities. In: Dennis RLH (ed.) *The Ecology of Butterflies in Britain*, Oxford: Oxford University Press, pp. 139–177.

Porter AH, Schneider RW, Price BA (1995) Wing pattern and allozyme relationships in the *Coenonympha arcania* group, emphasising the *C. gardetta–darwiniana* contact area at Bellwald, Switzerland (Lepidoptera, Satyridae). *Nota lepidopterologica* **17**: 155–174.

Porter AH, Wenger R, Geiger H, Scholl A, Shapiro AM (1997) The *Pontia daplidice–edusa* hybrid zone in northwestern Italy. *Evolution* **51**: 1561–1573.

Potter CF, Bertin RI (1988) Amino acids in artificial nectar: feeding preferences of the flesh fly *Sarcophaga bullata*. *American Midland Naturalist* **120**: 156–162.

Potts JM, Elith J (2006) Comparing species abundance models. *Ecological Modelling* **199**: 153–163.

Poulsen M (1996) Fluctuating asymmetry of butterflies with decreasing and stable populations. M.Sc. Thesis University of Århus.

Poulton EB (1904a) Notes on the mating behaviour of *Aglais urticae*. *Transactions of the Entomological Society of London* **1903**: xlii–xlv.

Poulton EB (1904b) What is a species? *Proceedings of the Entomological Society of London* **1903**: lxxvii–cxvi.

Powell JA (1980) Evolution of larval food preferences in microlepidoptera. *Annual Review of Entomology* **25**: 133–159.

Powell JA, Mitter C, Farrell B (1998) Evolution of larval food preferences in Lepidoptera. In: Kristensen NP (ed.) *The Lepidoptera*, Vol. 1, *Systematics, Evolution and Biogeography*, Berlin: W. de Gruyter, pp. 403–422.

Pöyry J, Lindgren S, Salminen J, Kuussaari M (2004) Restoration of butterfly and moth communities in semi-natural grasslands by cattle grazing. *Ecological Applications* **14**: 1656–1670.

Pöyry J, Lindgren S, Sahinen J, Kuussaari M (2005) Responses of butterfly and moth species to restored cattle grazing in semi-natural grasslands. *Biological Conservation* **122**: 465–478.

Prasad AM, Iversion LR, Liaw A (2006) Newer classification and regression tree techniques: bagging and random forests for ecological prediction. *Ecosystems* **9**: 181–199.

Pratt C (1986–87) A history and investigation into the fluctuations of *Polygonia c-album* L. *Entomologists' Record and Journal of Variation* **98**: 197–203, 244–250; **99**: 21–27.

Pratt GF, Ballmer GR (1993) Correlations of diapause intensities of *Euphilotes* spp. and *Philotiella speciosa* (Lepidoptera, Lycaenidae) to host bloom period and elevation. *Annals of the Entomological Society of America* **86**: 265–272.

Preiss E, Martin JL, Debussche M (1997) Rural depopulation and recent landscape changes in a Mediterranean region: consequences to the breeding avifauna. *Landscape Ecology* **12**: 51–61.

Prentice IC, Jolly D, BIOME 6000 participants (2000) Mid-Holocene and glacial-maximum vegetation geography of the northern continents and Africa. *Journal of Biogeography* **27**: 507–519.

Presgraves DC (2002) Patterns of postzygotic isolation in Lepidoptera. *Evolution* **56**: 1168–1183.

Preston-Mafham R, Preston-Mafham K (1993) *The Encyclopedia of Land Invertebrate Behaviour*. London: Blandford.

Proctor M, Yeo P, Lack A (1996) *The Natural History of Pollination*. London: HarperCollins.

Prokopy RJ, Owens ED (1983) Visual detection of plants by herbivorous insects. *Annual Review of Entomology* **28**: 40–48.

Prota R (1963) Note morfo-etologiche su *Trogus violaceus* (Mocs.) (Hymenoptera, Ichneumonidae) endoparassita solitario delle larve di *Papilio hospiton* Gené (Lepidoptera, Papilionidae). *Bollettino dell'Istituto di Entomologia della Università di Bologna* **26**: 289–318.

Pullin AS (1986a) Unusual egg-laying strategies of the small tortoiseshell butterfly, *Aglais urticae*. *Entomologists' Record and Journal of Variation* **98**: 9–10.

Pullin AS (1986b) Influence of the food plant, *Urtica dioica*, on larvae development, feeding efficiency, and voltinism of a specialist insect, *Inachis io*. *Holarctic Ecology* **9**: 72–78.

Pullin AS (1986c) Effect of photoperiod and temperature on the life-cycle of different populations of the peacock butterfly *Inachis io*. *Entomologia Experimentalis et Applicata* **41**: 237–242.

Pullin AS (1987a) Adult feeding, lipid accumulation, and overwintering in *Aglais urticae* and *Inachis io* (Lepidoptera, Nymphalidae). *Journal of Zoology* **24**: 631–641.

Pullin AS (1987b) Changes in leaf quality following clipping and regrowth of *Urtica dioica*, and consequences for a specialist insect herbivore, *Aglais urticae*. *Oikos* **49**: 39–45.

Pullin AS (1992) Butterfly overwintering strategies. *Butterfly Conservation News* 52: 43–47.

Pullin AS, Knight TM (2001) Effectiveness in conservation practice: pointers from medicine and public health. *Conservation Biology* 15: 50–54.

Pullin AS, Bálint Z, Balletto E, Buszko J, Coutsis JG, Goffart P, Kulfan M, Lhonoré JE, Settele J, van der Made JG (1998) The status, ecology and conservation of *Lycaena dispar* (Lycaenidae: Lycaenini) in Europe. *Nota lepidopterologica* 21: 94–100.

Pyörnilä M (1976a) Parasitism in *Aglais urticae* (L.) (Lep., Nymphalidae). II. Parasitism of larval stages by tachinids. *Annales Entomologici Fennici* 42: 133–139.

Pyörnilä M (1976b) Parasitism in *Aglais urticae* (L.) (Lep., Nymphalidae). III. Parasitism of larval stages by ichneumonids. *Annales Entomologici Fennici* 42: 156–161.

Pyörnilä M (1977) Parasitism in *Aglais urticae* (L.) (Lep., Nymphalidae). IV. Pupal parasitoids. *Annales Entomologici Fennici* 43: 21–27.

Pywell RF, Warman EA, Sparks TH, Greatorex-Davies JN, Walker KJ, Meek WR, Carvell C, Petit S, Firbank LG (2004) Assessing habitat quality for butterflies on intensively managed arable farmland. *Biological Conservation* 118: 313–325.

Quental TB, Patten MM, Pierce NE (2007) Host plant specialization driven by sexual selection. *American Naturalist* 169: 830–836.

Qui X, Arikawa K (2003) Polymorphism of red receptors: sensitivity spectra of proximal photoreceptors in the small white butterfly *Pieris rapae crucivora*. *Journal of Experimental Biology* 206: 2787–2793.

Quicke DLJ (1997) *Parasitic Wasps*. London: Chapman and Hall.

Quinn GP, Keough MJ (2002) *Experimental Design and Data Analysis for Biologists*. Cambridge: Cambridge University Press.

Quinn RM, Gaston KJ, Roy DB (1997) Coincidence between consumer and host occurrence: macrolepidoptera in Britain. *Ecological Entomology* 22: 197–208.

Quinn RM, Gaston KJ, Roy DB (1998) Coincidence in the distribution of butterflies and their foodplants. *Ecography* 21: 279–288.

Rabasa SG, Gutiérrez D, Escudero A (2005) Egg laying by a butterfly on a fragmented host plant: a multi-level approach. *Ecography* 28: 629–639.

Rabasa SG, Gutiérrez D, Escudero A (2007) Metapopulation structure and habitat quality in modelling dispersal in the butterfly *Iolana iolas*. *Oikos* 116: 793–806.

Rabinowitz D (1981) Seven forms of rarity. In: Synge H (ed.) *The Biological Aspects of Rare Plant Conservation*, New York: John Wiley, pp. 205–217.

Rackham O (1998) Savanna in Europe. In: Kirby KJ, Watkins C (eds.) *The Ecological History of European Forests*, Wallingford, Oxon: CABI Publishing, pp. 1–24.

Rackham O (2003) *Ancient Woodland: Its History, Vegetation and Uses in England*. Dalbeattie, Scotland: Castlepoint Press.

Radlmair S, Plachter H, Pfadenhauer J (1999) Geschichte der landwirtschaftlichen Moornutzung im süddeutschen Alpenvorland: ein Beitrag zur naturschutzfachlichen Leitbilddiskussion. *Natur und Landschaft* 74: 91–98.

Rahbek C (1995) The elevational gradient of species richness: a uniform pattern? *Ecography* 18: 200–205.

Ralls K, Beissinger SR, Cochrane JF (2002) Guidelines for using population viability analysis in endangered-species management. In: Beissinger SR, McCullough DR (eds.) *Population Viability Analysis*, Chicago, IL: University of Chicago Press, pp. 521–550.

Randall MGM (1982) The dynamics of an insect population throughout its altitudinal distribution: *Coleophora alticolella* (Lepidoptera) in northern England. *Journal of Animal Ecology* 51: 993–1016.

Randin CF, Dirnböck T, Dullinger S, Zimmermann NE, Zappa M, Guisan A (2006) Are niche-based species distribution models transferable in space? *Journal of Biogeography* 33: 1689–1703.

Rands MRW, Sotherton NW (1986) Pesticide use on cereal crops and changes in the abundance of butterflies on arable farmland in England. *Biological Conservation* 36: 71–82.

Rankin MA, Burchsted JCA (1992) The cost of migration in insects. *Annual Review of Entomolgy* 37: 533.

Rapoport EH (1982) *Aerography: Geographical Strategies of Species*. Oxford: Pergamon Press.

Ratkiewicz M, Jaroszewicz B (2006) Allopatric origins of sympatric forms: the skippers *Carterocephalus palaemon palaemon*, *C. p. tolli* and *C. silvicolus*. *Annales Zoologici Fennici* 43: 285–294.

Rausher MD (1978) Search image for leaf shape in a butterfly. *Science* 200: 1071–1073.

Rausher MD (1982) Population differentiation in *Euphydryas editha* butterflies: larval adaptation to different hosts. *Evolution* 36: 581–590.

Rausher MD, MacKay DA, Singer MC (1981) Pre- and post-alighting host discrimination by *Euphydryas editha* butterflies: the behavioural mechanism causing clumped distributions of egg clusters. *Animal Behaviour* 29: 1220–1228.

Ravenscroft NOM (1986) *An Investigation into the Distribution and Ecology of the Silver-Studded Blue Butterfly (*Plebejus argus *L.) in Suffolk: An Interim Report*. Ipswich, Suffolk: Suffolk Trust for Nature Conservation.

Ravenscroft NOM (1994a) Environmental influences on mate location in male chequered skipper butterflies, *Carterocephalus palaemon* (Lepidoptera: Hesperiidae). *Animal Behaviour* **47**: 1179–1187.

Ravenscroft NOM (1994b) The ecology of the checkered skipper butterfly *Carterocephalus palaemon* in Scotland. I. Microhabitat. *Journal of Applied Ecology* **31**: 613–622.

Ravenscroft NOM (1994c) The feeding behaviour of *Carterocephalus palaemon* (Lepidoptera: Hesperiidae) caterpillars: does it avoid host defences or maximize nutrient intake? *Ecological Entomology* **19**: 26–30.

Rawlins JE (1980) Thermoregulation by the black swallowtail butterfly, *Papilio polyxenes* (Lepidoptera: Papilionidae). *Ecology* **61**: 345–357.

Ray J (1710) *Historia Insectorum*. London: Churchill.

Ray TA, Andrews CC (1980) Ant butterflies: butterflies that follow army ants to feed on ant-bird droppings. *Science* **210**: 1147–1148.

Read M (1985) The silver-studded blue conservation report. M.Sc. thesis, Imperial College, London.

Réal P (1962) Un phénomène écologique singulier, mais complexe, l'amphiphénotisme, observé chez les piérides. *Annales Scientifiques de l'Université de Besançon, Série II* **17**: 87–95.

Réal P (1988) Lépidoptères nouveaux principalement jurassiens. *Mémoires du Comité de Liaison pour les Recherches Ecofaunistiques dans le Jura* **4**: 1–28.

Reavey D (1992) Egg size, first instar behavior and the ecology of Lepidoptera. *Journal of Zoology* **227**: 277–297.

Reavey D (1993) Why body size matters to caterpillars. In: Stamp NE, Casey TM (eds.) *Caterpillars: Ecological and Evolutionary Constraints on Foraging*, New York: Chapman and Hall, pp. 248–279.

Reavey D, Lawton JH (1991) Larval contribution to fitness in leaf-eating insects. In: Bailey WJ, Ridsdill-Smith J (eds.) *Reproductive Behaviour of Insects: Individuals and Populations*, London: Chapman and Hall, pp. 292–329.

Rebel H (1920) *Lycaena* hybr. *meledamon* m. (*Lycaena meleager* Esp. m. x *Lyc. damon* Schiff. f.). *Verhandlungen der Zoologisch-Botanischen Gesellschaft. Wien* **70**: 75–77.

Rebel H (1930a) *Lycaena* hybr. *bion* Rbl. m. (*L. damon* S. V. m. X *L. icarus* Rott f.). *Verhandlungen der Zoologisch-Botanischen Gesellschaft. Wien* **79**: 38–40.

Rebel H (1930b) *Lycaena* hybr. *corydamon*-m. (*L. corydon* Pod. m. x *L. damon* Schiff. f.). *Verhandlungen der Zoologisch-Botanischen Gesellschaft. Wien* **79**: 33–36.

Reed TM (1985) The number of butterfly species on British islands. *Proceedings of the 3rd Congress of European Lepidoptera*, Cambridge, pp. 146–152.

Reed JM (2004) Recognition behavior based problems in species conservation. *Annales Zoologici Fennici* **41**: 859–877.

Reed DH, Frankham R (2003) Correlation between fitness and genetic diversity. *Conservation Biology* **17**: 230–237.

Reed JM, Murphy DD, Brussard PF (1998) Efficacy of population viability analysis. *Wildlife Society Bulletin* **26**: 244–251.

Reed JM, Mills LS, Dunning JB, Menges ES, McKelvey KS, Frye R, Beissinger SR, Anstett MC, Miller P (2002) Emerging issues in population viability analysis. *Conservation Biology* **16**: 7–19.

Reineking B, Schröder B (2003) Computer-intensive methods in the analysis of species–habitat relationships. In: Breckling B, Reuter H, Mitwollen A (eds.) *Gene, Bits und Ökosysteme: Implikationen neuer Technologien für die ökologische Theorie*, Frankfunt, Germany: Peter Lang, pp. 165–182.

Reineking B, Schröder B (2004) Gütekriterien. *UFZ-Bericht* **9**: 27–37.

Reineking B, Schröder B (2006) Constrain to perform: regularization of habitat models. *Ecological Modelling* **193**: 675–690.

Reinig W (1937) *Die Holarktis*. Jena: Georg Fischer.

Reinig W (1938) *Elimination und Selektion*. Jena: Georg Fischer.

Reissinger E (1989) Checkliste Pieridae Duponchel, 1835, der Westpalaearktis (Europa, Nordwestafrika, Kaukasus, Kleinasien). *Atalanta* **20**: 149–185.

Remington CL (1954) The genetics of *Colias* (Lepidoptera, Pieridae). *Advances in Genetics* **6**: 403–450.

Remington CL (1968) Suture-zones of hybrid interaction between recently joined biotas. *Evolutionary Biology* **1**: 321–428.

Renwick JAA (1989) Chemical ecology of oviposition in phytophagous insects. *Experientia* **45**: 223–228.

Renwick JAA, Chew FS (1994) Oviposition behavior in Lepidoptera. *Annual Review of Entomology* **39**: 377–400.

Renwick JAA, Radke CD, Sachdev-Gupta K, Städler E (1989) Chemical constituents of *Erysimum cheiranthoides* deterring oviposition by the cabbage butterfly, *Pieris rapae*. *Journal of Chemical Ecology* **15**: 2161–2169.

Renwick JAA, Radke CD, Sachdev-Gupta K (1992) Leaf surface chemicals stimulating oviposition by *Pieris rapae* (Lepidoptera: Pieridae) on cabbage. *Chemoecology* **3**: 33–38.

Resetarits WJ Jr (1996) Oviposition site choice and life history evolution. *American Zoology* **36**: 205–215.

Revels R (1994) The rise and fall of the holly blue butterfly. *British Wildlife* **5**: 236–239.

Revels R (2006) More on the rise and fall of the holly blue. *British Wildlife* **17**: 419–424.

Reverdin J (1908) Variétés et aberrations d'*Erebia tyndarus* dans les Alpes de la Suisse. *Bulletin de la Société Lépidoptérologique de Genève* **1**: 192–215.

Reverdin J (1920) Note sur *Melitaea athalia* Rott. et diagnose d'une espèce nouvelle. *Bulletin de la Société Entomologique de France* **25**: 319–321.

Reverdin J (1922) *Melitaea athalia* Rott. et *Melitaea pseudathalia* nova species (?). *Bulletin de la Société Lépidoptérologique de Genève* **5**: 24–46.

Rew LJ, Theaker AJ, Froud-Williams RJ (1992) Nitrogen fertilizer misplacement and field boundaries. *Aspects of Applied Biology: Nitrate and Farming Systems* **30**: 203–206.

Reynolds SE (1990) Feeding in caterpillars: maximising or optimising food acquisition? In: Mellinger J (ed.) *Animal Nutrition and Transport Processes*, Vol. 1, *Nutrition in Wild and Domestic Animals*, Basel, Switzerland: Karger, pp. 106–118.

Rhainds M, Gries G, Chew PS (1997) Adaptive significance of density-dependent ballooning by bagworm larvae, *Metisa plana* (Walker) (Lepidoptera: Psychidae). *Canadian Entomologist* **129**: 927–931.

Rhode K (1992) Latitudinal gradients in species diversity: the search for the primary cause. *Oikos* **65**: 514–527.

Rhode K (1997) The larger area of the tropics does not explain latitudinal gradients in species diversity. *Oikos* **79**: 169–172.

Rhode K (1998) Latitudinal gradients in species diversity: area matters, but how much? *Oikos* **82**: 184–190.

Rhode K (1999) Latitudinal gradients in species diversity and Rapoport's rule revisited: a review of recent work and what can parasites teach us about the causes of the gradients. *Ecography* **22**: 593–613.

Rhode K, Heap M (1996) Latitudinal ranges of teleost fish in the Atlantic and Indo-Pacific oceans. *American Naturalist* **147**: 659–665.

Rhode K, Heap M, Heap D (1993) Rapoport's rule does not apply to marine teleosts and cannot explain latitudinal gradients in species richness. *American Naturalist* **142**: 1–16.

Rice WR, Hostert EE (1994) Laboratory experiments on speciation: what have we learned in forty years? *Evolution* **47**: 1637–1653.

Richards OW (1940) The biology of the small white butterfly (*Pieris rapae*), with special reference to the factors controlling its abundance. *Journal of Animal Ecology* **9**: 243–288.

Richards LJ, Myers JH (1980) Maternal influences on size and emergence time of the cinnabar moth. *Canadian Journal of Zoology* **58**: 1452–1457.

Richardson BJ, Baverstock PR, Adams M (1986) *Allozyme Electrophoresis: A Handbook for Animal Systematics and Population Studies*. London: Academic Press.

Ricketts TH (2001) The matrix matters: effective isolation in fragmented landscapes. *American Naturalist* **158**: 87–99.

Ricklefs RE, Schluter D (eds.) (1993) *Species Diversity in Ecological Communities: Historical and Geographical Perspectives*. Chicago, IL: University of Chicago Press.

Ridley M (1988) Mating frequency and fecundity in insects. *Biological Review* **63**: 509–549.

Ridley M (1989) The incidence of sperm displacement in insects: four conjectures, one corroboration. *Biological Journal of the Linnean Society* **38**: 349–367.

Ries L, Debinski DM (2001) Butterfly responses to habitat edges in the highly fragmented prairies of Central Iowa. *Journal of Animal Ecology* **70**: 840–852.

Ries L, Debinski DM, Wieland ML (2001) Conservation value of roadside prairie restoration to butterfly communities. *Conservation Biology* **15**: 401–411.

Riley ND (1975) *Erebia serotina* Descimon, de Lesse. *Entomological Record and Journal of Variation* **87**: 266.

Ripley BD (1996) *Pattern Recognition and Neural Networks*. Cambridge: Cambridge University Press.

Robertson KA, Monteiro A (2005) Female *Bicyclus anynana* butterflies choose males on the basis of their dorsal UV-reflective eyespot pupils. *Proceedings of the Royal Society of London B* **272**: 1541–1546.

Robinson R (1971) *Lepidoptera Genetics*. Oxford: Pergamon Press.

Robinson R (1990) Genetics of European butterflies. In: Kudrna O (eds.) *Butterflies of Europe*, Vol. 2, *Introduction to Lepidopterology*, Wiesbaden, Germany: Aula, pp. 234–306.

Robinson GS (1999) HOSTS: a database of the hostplants of the world's Lepidoptera. *Nota lepidopterologica* **22**: 35–47.

Rodríguez J, Fernández-Haeger J, Jordano D (1993) Oviposición de *Cyaniris semiargus* (Rottemburg, 1775) (Lycaenidae) sobre *Armeria velutina* (Weñw. ex Boiss., Reuter, 1852) (Plumbaginaceae): distribución espacial de las puestas. *SHILAP, Revista de Lepidopterología* **21**: 19–30.

Rodríguez J, Jordano D, Haeger JF (1994) Spatial heterogeneity in a butterfly–host plant interaction. *Journal of Animal Ecology* **63**: 31–38.

Roemis J, Babendreier D, Wäckers FL, Shanower TG (2005) Habitat and plant specificity of *Trichogramma* egg parasitoids: underlying mechanisms and implications. *Basic and Applied Ecology* **6**: 215–236.

Roer H (1968) Weitere Untersuchungen über die Auswirkungen der Witterung auf Richtung und Distanz der Flüge des Kleinen Fuchses (*Aglais urticae*). *Decheniana* **120**: 313–334.

Roff DA (1980) Optimizing development time in a seasonal environment: the 'ups and downs' of clinal variation. *Oecologia* **45**: 202–208.

Roff D (1983) Phenological adaptation in a seasonal environment: a theoretical perspective. In: Brown VK, Hodek I (eds.) *Diapause and Life Cycle Strategies in Insects*, The Hague: Junk Publishers, pp. 253–270.

Roff DA (1990) The evolution of flightlessness in insects. *Ecological Monographs* **60**: 389–421.

Roff DA (1991) Life-history consequences of bioenergetic and biomechanical constraints on migration. *American Zoologist* **31**: 205–215.

Roff DA (1992) *The Evolution of Life Histories*. New York: Chapman and Hall.

Roff DA (2002) *Life History Evolution*. Sunderland, MA: Sinauer Associates.

Roff DA, Fairbairn DJ (1991) Wing dimorphisms and the evolution of migratory polymorphisms among the Insecta. *American Zoologist* **31**: 243–251.

Roitberg BD, Robertson IC, Tyerman JGA (1999) Vive la variance: a functional oviposition theory for insect herbivores. *Entomologia Experimentalis et Applicata* **91**: 187–194.

Roland J (1982) Melanism and diel activity of Alpine *Colias* (Lepidoptera: Pieridae). *Oecologia* **53**: 214–221.

Roland J (2005) Effect of melanism of alpine *Colias nastes* butterflies (Lepidoptera: Pieridae) on activity and predation. *Canadian Entomologist* **138**: 52–58.

Roland J, Keyghobadi N, Fownes S (2000) Alpine *Parnassius* butterfly dispersal: effects of landscape and population size. *Ecology* **81**: 1642–1653.

Romeis J, Wäckers FL (2002) Nutritional suitability of individual carbohydrates and amino acids for adult *Pieris brassicae*. *Physiological Entomology* **27**: 148–156.

Ronce O, Olivieri I (1997) Evolution of reproductive effort in a metapopulation with local extinctions and ecological succession. *American Naturalist* **150**: 220–249.

Ronce O, Olivieri I (2004) Life history and evolution in metapopulations. In: Hanski I, Gaggiotti OE (eds.) *Ecology, Genetics, and Evolution of Metapopulations*, Amsterdam, Netherlands: Elsevier, pp. 227–257.

Ronce O, Gandon S, Rousset F (2000) Kin selection and natal dispersal in an age-structured population. *Theoretical Population Biology* **58**: 143–159.

Ronce O, Brachet S, Olivieri I, Gouyon PH, Clobert J (2005) Plastic changes in seed dispersal along ecological succession: theoretical predictions from an evolutionary model. *Journal of Ecology* **93**: 431–440.

Root RB, Kareiva PM (1984) The search for resources by cabbage butterflies (*Pieris rapae*): ecological consequences and adaptive significance of Markovian movements. *Ecology* **65**: 147–165.

Root TL, Price JT, Hall KR, Schneider SH, Rosenzweig C, Pounds JA (2003) Fingerprints of global warming on wild animals and plants. *Nature* **421**: 57–60.

Rosenberg RH (1989) Behavior of the territorial species *Limenitis weidemeyerii* (Nymphalidae) within temporary feeding areas. *Journal of the Lepidopterists' Society* **43**: 102–107.

Rosenberg RH, Enquist M (1991) Contest behaviour in Weidemeyer's admiral butterfly *Limenitis weidemeyerii* (Nymphalidae): the effect of size and residency. *Animal Behaviour* **42**: 805–811.

Rosenzweig ML (2003) *Win–win Ecology: How the Earth's Species Can Survive in the Midst of Human Enterprise*. Oxford: Oxford University Press.

Rosenzweig ML (2005) Avoiding mass extinction: basic and applied challenges. *American Midland Naturalist* **153**: 195–208.

Rosenzweig ML, Sandlin EA (1997) Species diversity and latitudes: listening to area's signal. *Oikos* **80**: 172–176.

Roskam JC, Brakefield PM (1999) Seasonal polyphenism in *Bicyclus* (Lepidoptera: Satyridae) butterflies: different climates need different cues. *Biological Journal of the Linnean Society* **66**: 345–356.

Rothschild M (1978) Hell's angels. *Antenna* **2**: 38–39.

Rothschild W, Jordan K (1906) A revision of the American Papilios. *Novitates Zoologicae* **13**: 411–752.

Rothschild M, Schoonhoven L (1977) Assessment of egg load by *Pieris brassicae* (Lepidoptera: Pieridae). *Nature* **266**: 352–355.

Rougeot PC. 1977. Un hybride présumé d'*Anthocaris* grec. *Alexanor* **10**: 50.

Rousset F (2008) GENEPOP'007: a complete re-implementation of the GENEPOP software for Windows and Linux. *Molecular Ecology Resources* **8**: 103–106.

Roy DB, Sparks TH (2000) Phenology of British butterflies and climate change. *Global Change Biology* **6**: 407–416.

Roy DB, Thomas JA (2003) Seasonal variation in the niche, habitat availability and population fluctuations of a bivoltine thermophilous insect near its range margin. *Oecologia* **134**: 439–444.

Roy DB, Rothery P, Moss D, Pollard E, Thomas JA (2001) Butterfly numbers and weather: predicting historical trends in abundance and the future effects of climate change. *Journal of Animal Ecology* **70**: 201–217.

Roze D, Rousset F (2005) Inbreeding depression and the evolution of dispersal rates: a multilocus model. *American Naturalist* **166**: 708–721.

Ruberson JR, Kring TJ (1993) Parasitism of developing eggs by *Trichogramma pretiosum* (Hymenoptera: Trichogrammatidae): host use preference and suitability. *Biological Control* **3**: 39–46.

Ruddiman WF, Raymo ME (1988) Northern Hemisphere climate régime during the past 3 Ma: possible tectonic connections. *Philosophical Transactions of the Royal Society of London B* **318**: 411–430.

Rudner M, Biedermann R, Schröder B, Kleyer M (2007) Integrated grid based ecological and economic (INGRID) landscape model: a tool to support landscape management decisions. *Environmental Modelling and Software* **22**: 177–187.

Ruehle C (1999) Preferences for nectar amino acids and nectar sugars in different Lepidoptera species: *Lysandra bellargus, Polyommatus icarus, Maniola jurtina* and *Autographa gamma*. Ph.D. thesis, University of Basel.

Ruetschi J, Scholl A (1985) Movements of individually marked *Colias palaeno europome* (Lepidoptera, Pieridae) in a habitat consisting of insularlike subsites. *Revue Suisse de Zoologie* **92**: 803–810.

Rundlöf M, Smith HG (2006) The effect of organic farming on butterfly diversity depends on landscape context. *Journal of Applied Ecology* **43**: 1121–1127.

Russwurm ADA (1978) *Aberrations of British Butterflies*. Faringdon, Oxon.: EW Classey.

Rusterholz HP (1998) Interactions of butterflies with their nectar plants under present conditions and under elevated levels of atmospheric CO_2. Ph.D. thesis, University of Basel.

Rusterholz HP, Erhardt A (1997) Preferences for nectar sugars in the peacock butterfly *Inachis io*. *Ecological Entomology* **22**: 220–224.

Rusterholz HP, Erhardt A (1998) Effects of elevated CO_2 on flowering phenology and nectar production of nectar plants important for butterflies of calcareous grasslands. *Oecologia* **113**: 341–349.

Rusterholz HP, Erhardt A (2000) Can nectar properties explain sex-specific flower preferences in the Adonis Blue butterfly *Lysandra bellargus*? *Ecological Entomology* **25**: 81–90.

Rutowski RL (1977) The use of visual cues in sexual and species discrimination by males of the small sulfur butterfly *Eurema lisa*. *Journal of Comparative Physiology* **115**: 61–74.

Rutowski RL (1978a) The courtship behaviour of the small sulphur butterfly, *Eurema lisa* (Lepidoptera: Pieridae). *Animal Behaviour* **26**: 892–903.

Rutowski RL (1978b) The form and function of ascending flights in *Colias* butterflies. *Behavioral Ecology and Sociobiology* **3**: 163–172.

Rutowski RL (1979) The butterfly as an honest salesman. *Animal Behaviour* **27**: 1269–1270.

Rutowski RL (1980a) Courtship solicitation by females of the checkered white butterfly, *Pieris protodice*. *Behavioral Ecology and Sociobiology* **7**: 113–117.

Rutowski RL (1980b) Male scent-producing structures in *Colias* butterflies: function, localization and adaptive features. *Journal of Chemical Ecology* **6**: 13–26.

Rutowski RL (1982) Epigamic selection by males as evidenced by courtship partner preferences in the checkered white butterfly (*Pieris protodice*). *Animal Behaviour* **30**: 108–112.

Rutowski RL (1983) The wing-waving display of *Eurema daira* males (Lepidoptera: Pieridae): its structure and role in successful courtship. *Animal Behaviour* **31**: 985–989.

Rutowski RL (1984) Sexual selection and the evolution of butterfly mating behavior. *Journal of Research on the Lepidoptera* **23**: 125–142.

Rutowski RL (1985) Evidence for mate choice in a sulphur butterfly. *Zeitschrift für Tierpsychologie* **70**: 103–114.

Rutowski RL (1991) The evolution of male mate-locating behavior in butterflies. *American Naturalist* **138**: 1121–1139.

Rutowski RL (1992) Male mate-locating behavior in the common eggfly, *Hypolimnas bolina* (Nymphalidae). *Journal of the Lepidopterists' Society* **46**: 24–38.

Rutowski RL (1997) Sexual dimorphism, mating systems and ecology in butterflies. In: Choe JC, Crespi BJ (eds.) *The Evolution of Mating Systems in Insects and Arachnids*, Cambridge: Cambridge University Press, pp. 257–272.

Rutowski RL (2000a) Postural changes accompany perch location changes in male butterflies (*Asterocampa leilia*) engaged in visual mate searching. *Ethology* **106**: 453–456.

Rutowski RL (2000b) Variation of eye size in butterflies: inter- and intraspecific patterns. *Journal of Zoology* **252**: 187–195.

Rutowski RL (2003) Visual ecology of adult butterflies. In: Boggs CL, Watt WB, Ehrlich PR (eds.) *Butterflies: Ecology and Evolution Taking Flight*, Chicago, IL: University of Chicago Press, pp. 9–25.

Rutowski RL, Gilchrist GW (1987) Courtship, copulation and oviposition in the chalcedon checkerspot, *Euphydryas chalcedona* (Lepidoptera:Nymphalidae). *Journal of Natural History* **21**: 1109–1117.

Rutowski RL, Gilchrist GW (1988) Male mate-locating behavior in the desert hackberry butterfly, *Asterocampa leilia* (Nymphalidae). *Journal of Research on the Lepidoptera* **26**: 1–12.

Rutowski RL, Long CE, Marshall LD, Vetter RS (1981) Courtship solicitation by *Colias* females (Lepidoptera: Pieridae). *American Midland Naturalist* **105**: 334–340.

Rutowski RL, Newton M, Schaefer J (1983) Interspecific variation in the size of the nutrient investment made by male butterflies during copulation. *Evolution* **37**: 708–713.

Rutowski RL, Gilchrist GW, Terkanian B (1987) Female butterflies mated with recently mated males show reduced reproductive output. *Behavioral Ecology and Sociobiology* **20**: 319–322.

Rutowski RL, Gilchrist GW, Terkanian B (1988) Male mate-locating behavior in *Euphydryas chalcedona* (Lepidoptera: Nymphalidae) related to pupation site preferences. *Journal of Insect Behavior* **1**: 277–289.

Rutowski RL, Alcock J, Carey M (1989) Hilltopping in the pipevine Swallowtail (*Battus philenor*). *Ethology* **82**: 244–254.

Rutowski RL, Dickinson JL, Terkanian B (1991) Behavior of male desert hackberry butterflies, *Asterocampa leilia* (Nymphalidae) at perching sites used in mate location. *Journal of Research on the Lepidoptera* **30**: 129–139.

Rutowski RL, Demlong MJ, Leffingwell T (1994) Behavioural thermoregulation at mate encounter site by male butterflies (*Asterocampa leilia*, Nymphalidae). *Animal Behaviour* **48**: 833–841.

Rutowski RL, Terkanian B, Eitan O (1997) Male mate-locating behavior and yearly population cycles in the snout butterfly, *Libytheana bachmanii* (Libytheidae). *Journal of the Lepidopterists' Society* **51**: 197–207.

Rutowski RL, McCoy L, Demlong MJ (2001) Visual mate detection in a territorial male butterfly (*Asterocampa leilia*): effects of distance and perch location. *Behaviour* **138**: 31–43.

Rykiel EJ Jr (1996) Testing ecological models: the meaning of validation. *Ecological Modelling* **90**: 229–244.

Ryrholm VN, Huemer P (1995) Schmetterlingszönosen alpiner Pflanzengesellschaften im Bereich der Sajatmähder (Venedigergruppe, Nationalpark Hohe Tauern) (Lepidoptera). *Carinthia* II: 513–525.

Ryszka H (1949) Ein *Colias* Hybride *ex ovo*. (*Colias hyale* L. m. x *Col. croceus* Fourc. f.). *Wiener Entomologische Rundschau* **1**: 5–7.

Saarinen K, Jantunen J, Valtonen A (2005a) Resumed forest grazing restored a population of *Euphydryas aurinia* (Lepidoptera: Nymphalidae) in SE Finland. *European Journal of Entomology* **102**: 683–690.

Saarinen K, Valtonen A, Jantunen J, Saarnio S (2005b) Butterflies and diurnal moths along road verges: does road type affect diversity and abundance? *Biological Conservation* **123**: 403–412.

Saastamoinen M (2007) Life-history, genotypic and environmental correlates of clutch size in the Glanville fritillary butterfly. *Ecological Entomology* **32**: 235–242.

Saastamoinen M (2008) Heritability of dispersal rate and other life history traits in the Glanville fritillary butterfly. *Heredity* **100**: 39–46.

Saastamoinen M, Hanski I (2008) Genotypic and environmental effect on flight activity and oviposition in the Granville fritillary butterfly. *American Naturalist* **171**: 701–712.

Saccheri IJ, Brakefield PM (2002) Rapid spread of immigrant genomes into inbred populations. *Proceedings of the Royal Society of London B* **269**: 1073–1078.

Saccheri I, Hanski I (2006) Natural selection and population dynamics. *Trends in Ecology and Evolution* **21**: 341–347.

Saccheri IJ, Brakefield PM, Nichols RA (1996) Severe inbreeding depression and rapid fitness rebound in the butterfly *Bicyclus anynana* (Satyridae). *Evolution* **50**: 2000–2013.

Saccheri IJ, Kuussaari M, Kankare M, Vikman P, Fortelius W, Hanski I (1998) Inbreeding and extinction in a butterfly metapopulation. *Nature* **392**: 491–494.

Saccheri IJ, Boggs CL, Hanski I, Ehrlich PR (2004) Genetics of cherspot populations. In: Ehrlich PR, Hanski I (eds.) *On the Wings of Checkerspots: A Model System for Population Biology*, Oxford: Oxford University Press, pp. 199–218.

Saccheri IJ, Lloyd HD, Helyar SJ, Brakefield PM (2005) Inbreeding uncovers fundamental differences in the genetic load affecting male and female fertility in a butterfly. *Proceedings of the Royal Society of London B* **272**: 39–46.

Saether BE, Engen S (2002) Including uncertainties in population viability analysis using population prediction intervals. In: Beissinger SR, McCullough DR (eds.) *Population Viability Analysis*, Chicago, IL: University of Chicago Press, pp. 191–212.

Sala G (1996) *I Lepidopotteri Diurni del Comprensorio Gardesano*. Salò Italy: Societá Editoriale Multimediale.

Salkeld EH (1980) Microtype eggs of some Tachinidae (Diptera). *Canadian Entomologist* **112**: 51–83.

Sánchez-Rodríguez JF, Baz A (1995) The effects of elevation on the butterfly communities of a Mediterranean mountain, Sierra de Javalambre, Central Spain. *Journal of the Lepidopterists' Society* **49**: 192–207.

Sanderson RA, Eyre MD, Rushton SP (2005) Distribution of selected macroinvertebrates in a mosaic of temporary and permanent freshwater ponds as explained by autologistic models. *Ecography* **28**: 355–362.

Sarto V (1998) El taladro de los geranios *Cacyreus marshalli* Butler, 1898 se establece en Francia: nuevos datos sobre su biología. *SHILAP, Revista de Lepidopterología* **26**: 221–227.

Sarto V, Masó A (1991) Confirmación de *Cacyreus marshalli* Butler, 1898 (Lycaenidae, Polyommatinae) como nueva especie para la fauna europea. *Boletín del Servicio Vegetal de Plagas* 17: 173–183.

Sato Y, Yano S, Takabayashi J, Ohsaki N (1999) *Pieris rapae* (Lepidoptera: Pieridae) females avoid oviposition on *Rorippa indica* plants infested by conspecific larvae. *Applied Entomology and Zoology* 34: 333–337.

Sawchik J, Dufrêne M, Lebrun P, Schtickzelle N, Baguette M (2002) Metapopulation dynamics of the bog fritillary butterfly: modelling the effect of habitat fragmentation. *Acta Oecologica* 23: 287–296.

Sawchik J, Dufrêne M, Lebrun P (2005) Distribution patterns and indicator species of butterfly assemblages of wet meadows in southern Belgium. *Belgian Journal of Zoology* 135: 43–52.

SBN (Schweizerischer Bund für Naturschutz, Lepidopterologen-Arbeitsgruppe) (eds.) (1987) *Tagfalter und ihre Lebensräume: Arten, Gefährdung, Schutz*, Vol. 1. Egg, Switzerland: Fotorotar AG.

SBN (Schweizerischer Bund für Naturschutz, Lepidopterologen-Arbeitsgruppe) (eds.) (1997) *Schmetterlinge und ihre Lebensräume: Arten, Gefährdung, Schutz*, Vol. 2. Egg, Switzerland: Fotorotar AG.

Sbordoni V, Foresterio S (1985) *The World of Butterflies*. Poole, Dorset: Blandford Press.

Scali V (1971) Imaginal diapause and gonadal maturation of *Maniola jurtina* (Lepidoptera, Satyrinae) from Tuscany. *Journal of Animal Ecology* 40: 467–472.

Scharff RF (1899) *The History of the European Fauna*. London: Walter Scott.

Scheirs J (2002) Integrating optimal foraging and optimal oviposition theory in plant–insect research. *Oikos* 96: 187–191.

Scheirs J, Zoebisch TG, Schuster DJ, De Bruin L (2004) Optimal foraging shapes host preference of a polyphagous leafminer. *Ecological Entomology* 29: 375–379.

Schluter D (2000) *The Ecology of Adaptive Radiation*. New York: Oxford University Press.

Schluter D (2001) Ecology and the origin of species. *Trends in Ecology and Evolution* 16: 372–380.

Schluter D, Ricklefs RE (1993) Species diversity: an introduction to the problem. In: Ricklefs RE, Schluter D (eds.) *Species Diversity in Ecological Communities: Historical and Geographical Perspectives*, Chicago, IL: University of Chicago Press, pp. 1–10.

Schmidt DJ, Hughes JM (2006) Genetic affinities among subspecies of a widespread Australian lycaenid butterfly, *Ogyris amaryllis* (Hewitson). *Australian Journal of Zoology* 54: 429–446.

Schmitt T (2001) Arealgeschichte von *Erebia medusa* (Lepidoptera, Nymphalidae) im Würm und Postglazial. *Verhandlungen des westdeutschen Entomologentag* 2000: 253–265.

Schmitt T (2007) Molecular biogeography of Europe: Pleistocene cycles and Postglacial trends. *Frontiers in Zoology* 4: 11. DOI: 10.1186/1742–9994–4–11

Schmitt T, Hewitt GM (2004a) Molecular biogeography of the arctic–alpine disjunct burnet moth species *Zygaena exulans* (Zygaenidae, Lepidoptera) in the Pyrenees and Alps. *Journal of Biogeography* 31: 885–893.

Schmitt T, Hewitt GM (2004b) The genetic pattern of population threat and loss: a case study of butterflies. *Molecular Ecology* 13: 21–31.

Schmitt T, Krauss J (2004) Reconstruction of the colonization route from glacial refugium to the northern distribution range of the European butterfly *Polyommatus coridon* (Lepidoptera: Lycaenidae). *Diversity and Distributions* 10: 271–274.

Schmitt T, Rákosy L (2007) Changes of traditional agrarian landscapes and their conservation implications: a case study of butterflies in Romania. *Diversity and Distributions* 13: 855–862.

Schmitt T, Seitz A (2001a) Allozyme variation in *Polyommatus coridon* (Lepidoptera: Lycaenidae): identification of ice-age refugia and reconstruction of post-glacial expansion. *Journal of Biogeography* 28: 1129–1136.

Schmitt T, Seitz A (2001b) Intraspecific structuring of *Polyommatus coridon* (Lycaenidae). *Nota lepidopterologica* 24: 53–63.

Schmitt T, Seitz A (2001c) Influence of the ice-age on the genetics and intraspecific differentiation of butterflies. *Proceedings of the International Colloquium of the European Invertebrate Survey*, Macevol Priory, Arboussols (66-France), 30 August 1999 – 4 September 1999, pp. 16–26.

Schmitt T, Seitz A (2001d) Intraspecific allozymatic differentiation reveals the glacial refugia and the postglacial expansions of European *Erebia medusa* (Lepidoptera: Nymphalidae). *Biological Journal of the Linnean Society* 74: 429–458.

Schmitt T, Seitz A (2002a) Influence of habitat fragmentation on the genetic structure of *Polyommatus coridon* (Lepidoptera: Lycaenidae): implications for conservation. *Biological Conservation* 107: 291–297.

Schmitt T, Seitz A (2002b) Postglacial distribution area expansion of *Polyommatus coridon* (Lepidoptera: Lycaenidae) from its Ponto-Mediterranean glacial refugium. *Heredity* 89: 20–26.

Schmitt T, Seitz A (2004) Low diversity but high differentiation: the population genetics of *Aglaope infausta* (Zygaenidae: Lepidoptera). *Journal of Biogeography* 31: 137–144.

Schmitt T, Varga Z, Seitz A (2000) Forests as dispersal barriers for *Erebia medusa* (Nymphalidae, Lepidoptera). *Basic and Applied Ecology* **1**: 53–59.

Schmitt T, Giessl A, Seitz A (2002) Postglacial colonisation of western Central Europe by *Polyommatus coridon* (Poda 1761) (Lepidoptera: Lycaenidae): evidence from population genetics. *Heredity* **88**: 26–34.

Schmitt T, Giessl A, Seitz A (2003) Did *Polyommatus icarus* (Lepidoptera: Lycaenidae) have distinct glacial refugia in southern Europe? Evidence from population genetics. *Biological Journal of the Linnean Society* **80**: 529–538.

Schmitt T, Varga Z, Seitz A (2005a) Are *Polyommatus hispana* and *Polyommatus slovacus* bivoltine *Polyommatus coridon* (Lepidoptera: Lycaenidae)? The discriminatory value of genetics in the taxonomy. *Organisms, Diversity, Evolution* **5**: 297–307.

Schmitt T, Röber S, Seitz A (2005b) Is the last glaciation the only relevant event for the present genetic population structure of the Meadow Brown butterfly *Maniola jurtina* (Lepidoptera: Nymphalidae)? *Biological Journal of the Linnean Society* **85**: 419–431.

Schmitt T, Cizek O, Konvička M (2005c) Genetics of a butterfly relocation: large, small and introduced populations of the mountain endemic *Erebia epiphron silesiana*. *Biological Conservation* **123**: 11–18.

Schmitt T, Habel JC, Zimmermann M, Muller P (2006a) Genetic differentiation of the marbled white butterfly, *Melanargia galathea*, accounts for glacial distribution patterns and postglacial range expansion in southeastern Europe. *Molecular Ecology* **15**: 1889–1901.

Schmitt T, Hewitt GM, Müller P (2006b) Disjunct distributions during glacial and interglacial periods in mountain butterflies: *Erebia epiphron* as an example. *Journal of Evolutionary Biology* **19**: 108–113.

Schmitt T, Rákosy L, Abadjiev S, Müller P (2007) Multiple differentiation centres of a non-Mediterranean butterfly species in south-eastern Europe. *Journal of Biogeography* **34**: 939–950.

Schmitz H (1994) Thermal characterization of butterfly wings. I. Absorption in relation to different color, surface structure and basking type. *Journal of Thermal Biology* **19**: 403–412.

Schmitz H, Wasserthal LTH (1993) Antennal thermoreceptors and wing thermosensitivity of heliotherm butterflies: their possible role in thermoregulatory behaviour. *Journal of Insect Physiology* **39**: 1007–1019.

Schneider C, Fry GLA (2001) The influence of landscape grain size on butterfly diversity in grasslands. *Journal of Insect Conservation* **5**: 163–171.

Schneider C, Dover J, Fry GLA (2003) Movement of two grassland butterflies in the same habitat network: the role of adult resources and size of the study area. *Ecological Entomology* **28**: 219–227.

Schoonhoven LM, Beerling EAM, Klijnstra JW, van Vugt Y (1990) Two related butterfly species avoid oviposition near each other's eggs. *Experientia* **46**: 526–528.

Schröder B (2000) *Zwischen Naturschutz und theoretischer Ökologie: Modelle zur Habitateignung und räumlichen Populationsdynamik für Heuschrecken im Niedermoor.* Braunschweig, Germany: Institut für Geographie und Geoökologie.

Schröder B (2006) *ROC, AUC Calculation: Evaluating the Predictive Performance of Habitat Models.* (Computer program.) Online: http://brandenburg.geoecology.uni-potsdam.de/users/schroeder/download.html

Schröder B, Reineking B (2004) Modellierung der Art-Habitat-Beziehung: ein Überblick über die Verfahren der Habitatmodellierung. *UFZ-Bericht* **9**/2004: 5–26.

Schröder B, Richter O (1999) Are habitat models transferable in space and time? *Journal for Nature Conservation* **8**: 195–205.

Schröder B, Seppelt R (2006) Analysis of pattern-process-interactions based on landscape models: overview, general concepts, methodological issues. *Ecological Modelling* **99**: 505–516.

Schtickzelle N, Baguette M (2003) Behavioural responses to habitat patch boundaries restrict dispersal and generate emigration–patch area relationships in fragmented landscapes. *Journal of Animal Ecology* **72**: 533–545

Schtickzelle N, Baguette M (2004) Metapopulation viability analysis of the bog fritillary butterfly using RAMAS/GIS. *Oikos* **104**: 277–290.

Schtickzelle N, Le Boulengé E, Baguette M (2002) Metapopulation dynamics of the bog fritillary butterfly: demographic processes in a patchy population. *Oikos* **97**: 349–360.

Schtickzelle N, Choutt J, Goffart P, Fichefet V, Baguette M (2005a) Metapopulation dynamics and conservation of the marsh fritillary butterfly: population viability analysis and management options for a critically endangered species in Western Europe. *Biological Conservation* **126**: 569–581.

Schtickzelle N, WallisDeVries MF, Baguette M (2005b) Using surrogate data in population viability analysis: the case of the critically endangered cranberry fritillary butterfly. *Oikos* **109**: 89–100.

Schtickzelle N, Mennechez G, Baguette M (2006) Dispersal depression with habitat fragmentation in the bog fritillary butterfly. *Ecology* **87**: 1057–1065.

Schtickzelle N, Joiris A, Van Dyck H, Baguette M (2007) Quantitative analysis of changes in movement behaviour within and outside habitat in a specialist butterfly. *BMC Evolutionary Biology* 7: 4.

Schultz CB, Crone EE (2001) Edge-mediated dispersal behavior in a prairie butterfly. *Ecology* 82: 1879–1892.

Schultz CB, Dlugosch KM (1999) Nectar and hostplant scarcity limit populations of an endangered Oregon butterfly. *Oecologia* 119: 231–238.

Schultz CB, Hammond PC (2003) Using population viability analysis to develop recovery criteria for endangered insects: case study of the Fender's Blue butterfly. *Conservation Biology* 17: 1372–1385.

Schurian KG (1977) Eine neue Unterart von *Lysandra coridon* Poda (Lep.: Lycaenidae). *Entomologische Zeitschrift* 87: 13–18.

Schurian KG (1989) Bemerkungen zu "*Lysandra cormion* Nabokov 1941" (Lepidoptera: Lycaenidae). *Nachrichten des entomologischen Vereins Apollo, N.F.* 10: 183–192.

Schurian KG (1991) Nachtrag zu den 'Bemerkungen zu "*Lysandra cormion*"' (Lepidoptera: Lycaenidae). *Nachrichten des entomologischen Vereins Apollo, N.F.* 12: 193–196.

Schurian KG (1997) Freilandexemplare des hybriden *cormion* [=*Polyommatus (Melaegeria) coridon* × *P. (M.) daphnis*] (Lepidoptera: Lycaenidae). *Nachrichten des entomologischen Vereins Apollo, N.F.* 18: 227–230.

Schurian KG, Hoffmann P. (1975). Ein neuer Lycaeniden-Hybrid von *Agrodiaetus damon* Schiff. *A. ripartii* Frr. (Lep., Lycaenidae). *Atalanta* 6: 227–231.

Schurian KG, Hoffmann P (1980) Ein neuer lycaeniden Hybrid: *Agrodiaetus rippartii* Freyer × *Agrodiaetus menalcas* Freyer (Lep. Lycaenidae). *Nachrichten des entomologischen Vereins Apollo, N.F.* 1: 21–23.

Schurian KG, Gascoigne-Pees M, Diringer Y (2006) Contribution to the life-cycle, ecology and taxonomy of *Polyommatus (Lysandra) coridon nufrellensis* Schurian (1977) (Lepidoptera: Lycaenidae). *Linneana Belgica* 20: 180–192.

Schwarz G (1978) Estimating the dimension of a model. *Annals of Statistics* 6: 461–464.

Schwarz M, Shaw MR (1999) Western Palaearctic Cryptinae (Hymenoptera: Ichneumonidae) in the National Museums of Scotland, with nomenclatural changes, taxonomic notes, rearing records and special reference to the British check list. II. Genus *Gelis* Thunberg (Phygadeuontini: Gelina). *Entomologist's Gazette* 50: 117–142.

Schwarz M, Shaw MR (2000) Western Palaearctic Cryptinae (Hymenoptera: Ichneumonidae) in the National Museums of Scotland, with nomenclatural changes, taxonomic notes, rearing records and special reference to the British check list.

III. Tribe Phygadeuontini, subtribes Chiroticina, Acrolytina, Hemitelina and Gelina (excluding *Gelis*), with descriptions of new species. *Entomologist's Gazette* 51: 147–186.

Schwarzwälder B, Lörtscher M, Erhardt A, Zettel J (1997) Habitat utilization by the heath fritillary butterfly, *Mellicta athalia* ssp. *celadussa* (Rott.) (Lepidoptera: Nymphalidae) in montane grasslands of different management. *Biological Conservation* 82: 157–165.

Scoble MJ (1992) *The Lepidoptera: Form, Function and Diversity*. Oxford: Oxford University Press.

Scott JA (1970) Hilltopping as a mating mechanism to aid the survival of low density species. *Journal of Research on the Lepidoptera* 7: 191–204.

Scott JA (1973a) Population biology and adult behavior of the circumpolar butterfly *Parnassius phoebus* F. (Papilionidae). *Entomologica Scandinavica* 4: 161–168.

Scott JA (1973b) Adult behavior and population biology of two skippers (Hesperiidae) mating in contrasting topographic sites. *Journal of Research on the Lepidoptera* 12: 181–196.

Scott JA (1974) Mate-locating behavior of butterflies. *American Midland Naturalist* 91: 103–117.

Scott JA (1975) Flight patterns among eleven species of diurnal Lepidoptera. *Ecology* 56: 1367–1377.

Scott JA (1983) Mate-locating behavior of western North American butterflies. II. New observations and morphological adaptations. *Journal of Research on the Lepidoptera* 21: 177–187.

Scott JA (1986) *The Butterflies of North America: A Natural History and Field Guide*, Stanford, CA: Stanford University Press.

Scriber JM (1988) Tale of the Tiger: Beringian biogeography, binomial classification, and breakfast choices in the *Papilio glaucus* complex of butterflies. In: Spencer KC (ed.) *Chemical Mediation of Coevolution*, San Diego, CA: Academic Press, pp. 291–301.

Scriber JM (1994) Climatic legacies and sex chromosomes: latitudinal patterns of voltinism, diapause, body size, and host-plant selection on two species of swallowtail butterflies at their hybrid zone. In: Danks HV (ed.) *Insect Life-Cycle Polymorphism: Theory, Evolution and Ecological Consequences for Seasonality and Diapause Control*, Dordrecht, Netherlands: Kluwer, pp. 133–172.

Scriber JM, Lederhouse RC (1992) The thermal environment as a resource dictating geographic patterns of feeding specialization of insect herbivores. In: Hunter MR, Ohgushi T, Price PW (eds.) *Effect of Resource Distribution on Animal–Plant Interactions*, New York: Academic Press, pp. 429–466.

Scriber JM, Slanski F (1981) The nutritional ecology of immature insects. *Annual Review of Entomology* 26: 183–211.

Sculley C, Boggs CL (1996) Mating systems and sexual division of foraging effort affect puddling behaviour by butterflies. *Ecological Entomology* **21**: 193–197.

Seehausen O (2003) Hybridization and adaptive radiation. *Trends in Ecology and Evolution* **19**: 198–207.

Seger J, Brockmand HJ (1987) What is bet hedging? *Oxford Surveys in Evolutionary Biology* **4**: 182–211.

Segurado P, Araújo MB (2004) An evaluation of methods for modelling species distributions. *Journal of Biogeography* **31**: 1555–1568.

Segurado P, Araújo MB, Kunin WE (2006) Consequences of spatial autocorrelation for niche-based models. *Journal of Applied Ecology* **43**: 433–444.

Sei M (2004) Larval adaptation of the endangered maritime ringlet *Coenonympha tullia nipisiquit* McDonnough (Lepidoptera: Nymphalidae) to a saline wetland habitat. *Environmental Entomology* **33**: 1535–1540.

Sei M, Porter AH (2003) Microhabitat-specific early-larval survival of the maritime ringlet (*Coenonympha tullia nipisiquit*). *Animal Conservation* **6**: 55–61.

Seitz A (1894) Allgemeine Biologie des Schmetterlinge. III. Die Ernährung. *Zoologische Jahresbericht* **7**: 9–186.

Seko T, Nakasuji F (2004) Effect of egg size variation on survival rate, development and fecundity of offspring in a migrant skipper, *Parnara guttata guttata* (Lepidoptera: Hesperiidae). *Applied Entomology and Zoology* **39**: 171–176.

Seko T, Miyatake T, Fujioka S, Nakasuji F (2006) Genetic and environmental sources of egg size, fecundity and body size in the migrant skipper, *Parnara guttata guttata* (Lepidoptera: Hesperiidae). *Population Ecology* **48**: 225–232.

Selfa J, Bordera S, Anento JL (1994) Data on a Spanish species of Ichneumoninae (Hym., Ichneumonidae). *Norwegian Journal of Agricultural Science* (Suppl. 16): 413 [poster abstract].

Settele J (1998): *Metapopulationsanalyse auf Rasterdatenbasis: Möglichkeiten des Modelleinsatzes und der Ergebnisumsetzung im Landschaftsmaßstab am Beispiel von Tagfaltern.* Stuttgart, Germany: Teubner.

Settele J (2005) How endangered is *Maculinea nausithous*? In: Settele J, Kühn E, Thomas JA (eds.) *Studies on the Ecology and Conservation of Butterflies in Europe*, Vol. 2, *Species Ecology along a European Gradient*: Maculinea *Butterflies as a Model*, Moscow: Pensoft.

Settele J, Henle K (2003) Grazing and cutting regimes for old grassland in temperate zones, in biodiversity conservation and habitat management. In: Gherardi F, Corti C, Gualtieri M (eds.) *Encyclopedia of Life Support Systems*, Oxford: EOLSS, chapter E1-67-03-02. Online: http://www.eolss.net

Settele J, Kühn E (2009) Insect conservation. *Science* **325**: 41–42.

Settele J, Andrick U, Pistorius EM (1992) Zur Bedeutung von Trittsteinbiotopen und Biotopverbund in der Geschichte: das Beispiel des Hochmoorperlmutterfalters (*Boloria aquilonaris* Stichel 1908) und anderer Moorvegetation bewohnender Schmetterlinge in der Pfalz (SW-Deutschland). *Nota lepidopterologica* (Suppl. 4): 18–31.

Settele J, Feldmann R, Reinhardt R (eds.) (2000) *Die Tagfalter Deutschlands.* Stuttgart, Germany: Eugen Ulmer.

Settele J, Hammen V, Hulme P, Karlson U, Klotz S, Kotarac M, Kunin W, Marion G, O'Connor M, Petanidou T, Peterson K, Potts S, Pritchard H, Pysek P, Rounsevell M, Spangenberg J, Steffan-Dewenter I, Sykes M, Vighi M, Zobel M, Kühn I (2005) ALARM: Assessing LArge-scale environmental Risks for biodiversity with tested Methods. *GAIA – Ecological Perspectives for Science and Society* **14**: 69–72.

Seufert P, Fiedler K (1999) Myrmecophily and parasitoid infestation of south-east Asian lycaenid butterfly larvae. *Ecotropica* **5**: 59–64.

Seufert W, Grosser N (1996) A population ecological study of *Chazara briseis* (Lepidoptera, Satyrinae). In: Settele J, Margules C, Poschlod P, Henle K (eds.) *Species Survival in Fragmented Landscapes*, Dordrecht, Netherlands: Kluwer, pp. 268–274.

Seymour AS, Gutiérrez D, Jordano D (2003) Dispersal of the lycaenid *Plebejus argus* in response to patches of its mutualist ant *Lasius niger*. *Oikos* **103**: 162–174.

Shaffer ML (1981) Minimum population sizes for species conservation. *BioScience* **31**: 131–134.

Shaffer ML, Samson FB (1985) Population size and extinction: a note on determining critical population size. *American Naturalist* **125**: 144–152.

Shaffer ML, Watchman LH, Snape III WJ, Latchis IK (2002) Population viability analysis and conservation policy. In: Beissinger SR, McCullough DR (eds.) *Population Viability Analysis*, Chicago, IL: University of Chicago Press, pp. 123–142.

Shapiro AM (1970) Role of sexual behavior in density-related dispersal of pierid butterflies. *American Naturalist* **104**: 367–372.

Shapiro AM (1976) Seasonal polyphenism. *Evolutionary Biology* **9**: 259–333.

Shapiro AM (1981) The pierid red-egg syndrome. *American Naturalist* **117**: 276–294.

Shapiro AM (1984a) Experimental studies on the evolution of seasonal polyphenism. In: Vane-Wright RI, Ackery PR (eds.) *The Biology of Butterflies*, London: Academic Press, pp. 297–307.

Shapiro AM (1984b) The genetics of seasonal polyphenism and the evolution of 'general purpose genotypes' in butterflies. In: Wöhrmann K, Loeschcke V (eds.) *Population Biology and Evolution*, Berlin: Springer-Verlag, pp. 16–30.

Shapiro AM, Carde RT (1970) Habitat selection and competition among sibling species of satyrid butterflies. *Evolution* 24: 48–54.

Sharp MA, Parks DR, Ehrlich PR (1974) Plant resources and butterfly habitat selection. *Ecology* 55: 870–875.

Shaw MR (1975) A rationale for abnormal, male-dominated sex-ratios in adult populations of *Zygaena* (Lep. Zygaenidae). *Entomologist's Record and Journal of Variation* 87: 52–54.

Shaw MR (1978) The status of *Trogus lapidator* (F.) (Hymenoptera: Ichneumonidae) in Britain, a parasite of *Papilio machaon* L. *Entomologist's Gazette* 29: 287–288.

Shaw MR (1981) Parasitism by Hymenoptera of larvae of the white admiral butterfly, *Ladoga camilla* (L.), in England. *Ecological Entomology* 6: 333–335.

Shaw MR (1982) Parasitic control, Section A: General information – Biology of some effective parasites. In: Feltwell J (ed.) *Large White Butterfly: The Biology, Biochemistry and Physiology of* Pieris brassicae *(Linnaeus)*, The Hague: Junk Publishers, pp. 401–407.

Shaw MR (1990) Parasitoids of European butterflies and their study. In: Kudrna O (ed.) *Butterflies of Europe*, Vol. 2, *Introduction to Lepidopterology*, Wiesbaden, Germany: Aula, pp. 449–479.

Shaw MR (1993) An enigmatic rearing of *Dolopsidea indagator* (Haliday) (Hymenoptera: Braconidae). *Entomologist's Record and Journal of Variation* 105: 31–36.

Shaw MR (1994) Parasitoid host ranges. In: Hawkins BA, Sheehan W (eds.) *Parasitoid Community Ecology*, Oxford: Oxford University Press, pp. 111–144.

Shaw MR (1996) Parasitism of *Aricia* species: preliminary results … and a call for help. *Butterfly Conservation News* 62: 14–15.

Shaw MR (1997) *Rearing Parasitic Hymenoptera*. Orpington, Kent: The Amateur Entomologists' Society.

Shaw MR (2002a) Host ranges of *Aleiodes* species (Hymenoptera: Braconidae), and an evolutionary hypothesis. In: Melika G, Thuroczy C (eds.) *Parasitic Wasps: Evolution, Systematics, Biodiversity and Biological Control*, Budapest: Agroinform, pp. 321–327.

Shaw MR (2002b) Experimental confirmation that *Pteromalus apum* (Retzius) (Hym., Pteromalidae) parasitises both leaf-cutter bees (Hym., Megachilidae) and fritillary butterflies (Lep., Nymphalidae). *Entomologist's Monthly Magazine* 138: 37–41.

Shaw MR (2004) *Microgaster alebion* Nixon and its 'var A': description of a new species and biological notes (Hymenoptera: Braconidae, Microgastrinae). *Entomologist's Gazette* 55: 217–224.

Shaw MR (2006) Habitat considerations for parasitic wasps (Hymenoptera). *Journal of Insect Conservation* 10: 117–127.

Shaw MR (in press) *Cotesia* Cameron (Hymenoptera: Braconidae: Microgastrinae) parasitoids of Heliconiinae (Lepidoptera: Nymphalidae) in Europe, with description of three new species. *British Journal of Entomology and Natural History* 22.

Shaw MR, Aeschlimann JP (1994) Host ranges of parasitoids (Hymenoptera: Braconidae and Ichneumonidae) reared from *Epermenia chaerophyllella* (Goeze) (Lepidoptera: Epermeniidae) in Britain, with description of a new species of *Triclistus* (Ichneumonidae). *Journal of Natural History* 28: 619–629.

Shaw MR, Askew RR (1976) Parasites. In: Heath J (ed.) *The Moths and Butterflies of Great Britain and Ireland*, Vol. 1, *Micropterigidae: Heliozelidae*. Colchester, Essex: Harley Books, pp. 24–56.

Shaw MR, Hochberg ME (2001) The neglect of parasitic Hymenoptera in insect conservation strategies: the British fauna as a prime example. *Journal of Insect Conservation* 5: 253–263.

Shaw MR, Horstmann K (1997) An analysis of host range in the *Diadegma nanus* group of parasitoids in Western Europe, with a key to species (Hymenoptera: Ichneumonidae: Campopleginae). *Journal of Hymenoptera Research* 6: 273–296.

Shaw MR, Huddleston T (1991) Classification and biology of braconid wasps (Hymenoptera: Braconidae). *Handbooks for the Identification of British Insects* 7: 1–126.

Shelly TE, Ludwig D (1985) Thermoregulatory behavior of the butterfly *Calisto nubila* (Satyridae) in a Puerto Rican forest. *Oikos* 44: 229–233.

Shields O (1967) Hilltopping: an ecological study of summit congregation behavior of butterflies on a southern California hill. *Journal of Research on the Lepidoptera* 6: 69–178.

Shiojiri K, Takabayashi J, Yano S, Takafuji A (2002) Oviposition preferences of herbivores are affected by tritrophic interaction webs. *Ecology Letters* 5: 186–192.

Shreeve TG (1984) Habitat selection, mate location, and microclimatic constraints on the activity of the speckled wood butterfly *Pararge aegeria*. *Oikos* 42: 371–377.

Shreeve TG (1986a) The effect of weather on the life-cycle of the Speckled Wood Butterfly *Pararge aegeria*. *Ecological Entomology* 11: 325–332.

Shreeve TG (1986b) Egg-laying by the speckled wood butterfly *Pararge aegeria*: the role of female behaviour, hostplant abundance and temperature. *Ecological Entomology* 11: 229–236.

Shreeve TG (1987) Mud-puddling behaviour of the green-veined white butterfly. *Entomologist's Record and Journal of Variation* 99: 27.

Shreeve TG (1990a) The movements of butterflies. In: Kudrna O (ed.) *Butterflies of Europe*, Vol. 2, *Introduction to Lepidopterology*, Wiesbaden, Germany: Aula, pp. 512–532.

Shreeve TG (1990b) Microhabitat use and hindwing phenotype in *Hipparchia semele* (Lepidoptera, Satyrinae): thermoregulation and background matching. *Ecological Entomology* **15**: 201–213.

Shreeve TG (1992a) Monitoring butterfly movement. In: Dennis RHL (ed.) *The Ecology of Butterflies in Great Britain*, Oxford: Oxford University Press, pp. 120–138.

Shreeve TG (1992b) Adult behaviour. In: Dennis R (ed.) *The Ecology of Butterflies in Britain*, Oxford: Oxford University Press, pp. 22–45.

Shreeve TG (1995) Butterfly mobility. In: Pullin AS (ed.) *Ecology and Conservation of Butterflies*, London: Chapman and Hall, pp. 37–45.

Shreeve TG, Dennis RLH (1992) The development of butterfly settling posture: the role of predators, climate, hostplant-habitat and phylogeny. *Biological Journal of the Linnean Society* **45**: 57–69.

Shreeve TG, Mason CF (1980) The number of butterfly species in woodlands. *Oecologia* **45**: 414–418.

Shreeve TG, Smith AG (1992) The role of weather-related habitat use on the impact of the European speckled wood *Pararge aegeria* on the endemic *Pararge xiphia* on the island of Madeira. *Biological Journal of the Linnean Society* **46**: 59–75.

Shreeve TG, Dennis RLH, Pullin AS (1996) Marginality: scale-determined processes and the conservation of the British butterfly fauna. *Biodiversity and Conservation* **5**: 1131–1141.

Shreeve TG, Dennis RLH, Moss D, Roy DB (2001) An ecological classification of British butterflies: ecological attributes and biotope occupancy. *Journal of Insect Conservation* **5**: 145–161.

Sibly RM, Monk K (1987) A theory of grasshopper life cycles. *Oikos* **48**: 186–194.

Sibly RM, Winokur L, Smith RH (1997) Interpopulation variation in phenotypic plasticity in the speckled wood butterfly *Pararge aegeria*. *Oikos* **78**: 323–330.

Silberglied RE (1984) Visual communication and sexual selection among butterflies. In: Vane-Wright R, Ackery PR (eds.) *The Biology of Butterflies*, Princeton, NJ: Princeton University Press, pp. 207–223.

Silberglied RE, Taylor OR (1973) Ultraviolet differences between the sulfur butterflies *Colias eurytheme* and *C. philodice*, and a possible isolation mechanism. *Nature* **241**: 406–408.

Silberglied RE, Taylor OR (1978) Ultraviolet reflection and its behavioural role in the courtship of the sulfur butterflies *Colias eurytheme* and *C. philodice*. *Behavioral Ecology and Sociobiology* **3**: 203–243.

Sillén-Tullberg B (1988) Evolution of gregariousness in aposematic butterflies: a phylogenetic analysis. *Evolution* **42**: 293–305.

Sillén-Tullberg B, Leimar O (1988) The evolution of gregariousness in distasteful insects as a defence against predators. *American Naturalist* **132**: 723–734.

Simmons AD, Thomas CD (2004) Changes in dispersal during species' range expansions. *American Naturalist* **164**: 378–395.

Simpson GG (1943) Turtles and the origin of the fauna of Latin America. *American Journal of Science* **241**: 413–429.

Sims SR (1979) Aspects of mating frequency and reproductive maturity in *Papilio zelicaon*. *American Midland Naturalist* **102**: 36–50.

Singer MC (1972) Complex components of habitat suitability within a butterfly colony. *Science* **176**: 75–77.

Singer MC (1982) Sexual selection for small size in male butterflies. *American Naturalist* **119**: 440–443.

Singer MC (1984) Butterfly–hostplant relationships: host quality, adult choice and larval success. In: Vane-Wright RI, Ackery PR (eds.) *The Biology of Butterflies*, London: Academic Press, pp. 81–88.

Singer MC (2003) Spatial and temporal patterns of checkerspot butterfly–host plant association: the diverse roles of oviposition preference. In: Boggs CL, Watt WB, Ehrlich PR (eds.) *Butterflies: Ecology and Evolution Taking Flight*, Chicago, IL: University of Chicago Press, pp. 207–228.

Singer MC (2004) Measurement, correlates, and importance of oviposition preferences in the life of checkerspots. In: Ehrlich PR, Hanski I (eds.) *On the Wings of Checkerspots: A Model System for Population Biology*, Oxford: Oxford University Press, pp. 112–137.

Singer MC, Parmesan C (1993) Sources of variation in patterns of plant–insect association. *Nature* **361**: 251–253.

Singer MC, Thomas CD (1992) The difficulty of deducing behavior from resource use: an example from hilltopping checkerspot butterflies. *American Naturalist* **140**: 654–664.

Singer MC, Thomas CD (1996) Evolutionary responses of a butterfly metapopulation to human-caused and climate-caused environmental variation. *American Naturalist* **148**: 9–39.

Singer MC, Wee B (2005) Spatial pattern in checkerspot butterfly–host plant association at local, metapopulation and regional scales. *Annales Zoologici Fennici* **42**: 347–361.

Singer MC, Ng D, Vasco D, Thomas CD (1992a) Rapidly evolving associations among oviposition preferences fail to constrain evolution of insect diet. *American Naturalist* **139**: 9–20.

Singer MC, Vasco D, Parmesan C, Thomas CD (1992b) Distinguishing between 'preference' and 'motivation' in food choice: an example from insect oviposition. *Animal Behaviour* **44**: 463–471.

Singer MC, Thomas CD, Parmesan C (1993) Rapid human-induced evolution of insect host associations. *Nature* **366**: 681–683.

Sinha SN, Lakhani KH, Davis BNK (1990) Studies on the toxicity of insecticidal drift to the first instar larvae of the large white butterfly *Pieris brassicae* (Lep: Pieridae). *Annals of Applied Biology* **116**: 27–41.

Sison-Mangus MP, Bernard GD, Lampel J, Briscoe AD (2006) Beauty in the eye of the beholder: the two blue opsins of lycaenid butterflies and the opsin gene-driven evolution of sexually dimorphic eyes. *Journal of Experimental Biology* **209**: 3079–3090.

Sjögren-Gulve P, Hanski I (2000) Metapopulation viability analysis using occupancy models. *Ecological Bulletins* **48**: 53–71.

Skinner SW (1985) Clutch size as an optimal foraging problem for insects. *Behavioral Ecology and Sociobiology* **17**: 231–238.

Slansky FJ (1974) Relationship of larval food-plants and voltinism patterns in temperate butterflies. *Psyche* **81**: 243–253.

Slansky FJ (1976) Phagism relationships among butterflies. *Journal of the New York Entomological Society* **84**: 91–105.

Slansky FJ (1993) Nutritional ecology: the fundamental quest for nutrients. In: Stamp NE, Casey TM (eds.) *Caterpillars: Ecological and Evolutionary Constraints on Foraging*, New York: Chapman and Hall, pp. 29–91.

Slansky FJ, Rodriguez JG (eds.) (1987) *Nutritional Ecology of Insects, Mites, Spiders, and Related Invertebrates*. New York: John Wiley.

Slansky FJ, Scriber JM (1985) Food consumption and utilization. In: Kerkut GA, Gilbert LI (eds.) *Comprehensive Insect Physiology, Biochemistry and Pharmacology*, Vol. 4, Oxford: Pergamon Press, pp. 87–163.

Slatkin M (1993) Isolation by distance in equilibrium and non-equilibrium populations. *Evolution* **47**: 264–279.

Smallegange RC, Everaarts TC, Van Loon JJA (2006) Associative learning of visual and gustatory cues in the large cabbage white butterfly, *Pieris brassicae*. *Animal Biology* **56**: 157–172.

Smallidge PJ, Leopold DJ (1997) Vegetation management for the maintenance and conservation of butterfly habitats in temperate human-dominated landscapes. *Landscape and Urban Planning* **38**: 259–280.

Smelhaus J (1947) *Polyommatus meleager* Esp. × *coridon* Poda (Lep. Lyc.). *Acta Societatis Entomologicae Cecoslovaquiae* **44**: 44–47.

Smelhaus J (1948) *Polyommatus* hybr. *cormion* Nabokov *(meleager* Esp. × *coridon* Poda) (Lep. Lyc.). *Acta Societatis Entomologicae Cecoslovaquiae* **45**: 50–55.

Smid HM, Wang G, Bukovinszky T, Steidle JLM, Bleeker MAK, van Loon JJA, Vet LM (2007) Species-specific acquisition and consolidation of long-term memory in parasitic wasps. *Proceedings of the Royal Society of London B* **274**: 1539–1546.

Smith JM (1966) Sympatric speciation. *American Naturalist* **100**: 637–650.

Smith CC, Fretwell SD (1974) Optimal balance between size and number of offspring. *American Naturalist* **108**: 499–506.

Smith DAS, Owen DF (1997) Colour genes as markers for migratory activity: the butterfly *Danaus chrysippus* in Africa. *Oikos* **78**: 127–135.

Smith RH, Sibly RM, Moller H (1987) Control of size and fecundity in *Pieris rapae*: towards a theory of butterfly life cycles. *Journal of Animal Ecology* **56**: 341–350.

Smouse PE, Peakall R (1999) Spatial autocorrelation analysis of individual multiallele and multilocus genetic structure. *Heredity* **82**: 561–573.

Sneath PHA, Sokal RR (1973) *Numerical Taxonomy: The Principles and Practice of Numerical Classification*. San Francisco, CA: WH Freeman.

Snell-Rood EC, Papaj DR (2006) Learning signals within sensory environments: does host cue learning in butterflies depend on background? *Animal Biology* **56**: 173–192.

Soderstrom B, Svensson B, Vessby K, Glimskar A (2001) Plants, insects and birds in semi-natural pastures in relation to local habitat and landscape factors. *Biodiversity and Conservation* **10**: 1839–1863.

Sokal RR, Crovello TJ (1970) The biological species concept: a critical evaluation. *American Naturalist* **104**: 107–123.

Sokal RR, Rohlf FJ (1981) *Biometry*, 2nd edn. San Francisco, CA: WH Freeman.

Sokal RR, Harding RM, Oden NL (1989) Spatial patterns of human gene frequencies in Europe. *American Journal of Physical Anthropology* **80**: 267–294.

Sonderegger P (2005) *Die Erebien der Schweiz (Lepidoptera: Satyrinae, Genus* Erebia*)*. Brugg, Switzerland: Peter Sonderegger.

Sonntag G (1981) Öko-etologische Untersuchungen zur Sexualbiologie des Schachbrettfalters (*Agapetes galathea* L.) unter besonderer Berücksichtigung thermobiologischer Aspekte. *Zeitschrift für Tierpsychologie* **56**: 169–186.

Soontiens J, Bink FA (1997) Developmental response of *Coenonympha pamphilus* (Lepidoptera: Satyrinae) to differences in nitrogen and water content of grasses.

Proceedings of the Section Experimental and Applied Entomology, N.E.V. **8**: 29–36.

Sotavalta O (1952) The essential factor regulating the wing-stroke frequency of insects in wing mutilation and loading experiments and in experiments at subatmospheric pressure. *Annales Zoologici Societatis Zoologicae-Botanicae Fennicae Vanamo* **15**: 1–67.

Soulé ME (1985) What is conservation biology? *BioScience* **35**: 727–734.

Soulé ME (1987) *Viable Populations for Conservation.* Cambridge: Cambridge University Press.

Southwood TRE (1962) Migration of terrestrial arthropods in relation to habitat. *Biological Reviews of the Cambridge Philosophical Society* **37**: 171–214.

Sparks TH, Yates TJ (1997) The effect of spring temperature on the appearance dates of British butterflies 1883–1993. *Ecography* **20**: 368–374.

Sparks TH, Porter K, Greatorex-Davies JN, Hall ML, Marrs RH (1994) The choice of oviposition sites in woodland by the Duke of Burgundy butterfly *Hamearis lucina* in England. *Biological Conservation* **70**: 257–264.

Speight MR, Hunter MD, Watt AD (1999) *Ecology of Insects: Concepts and Applications.* Oxford: Blackwell Science.

Sperling FAH (1990) Natural hybrids of *Papilio* (Insecta: Lepidoptera): poor taxonomy or interesting evolutionary problem? *Canadian Journal of Zoology* **68**: 1790–1799.

Sperling FAH (1993) Mitochondrial DNA variation and Haldane's Rule in the *Papilio glaucus* and *P. troilus* species groups. *Heredity* **71**: 227–233.

Sperling FAH (1994) Sex-linked genes and species differences in Lepidoptera. *Canadian Entomologist* **126**: 807–818.

Sperling FAH (2003) Butterfly molecular systematics: from species definitions to higher-level phylogenies. In: Boggs CL, Watt WB, Ehrlich P (eds.) *Butterflies: Ecology and Evolution Taking Flight*, Chicago, IL: University of Chicago Press, pp. 431–458.

Spieth HR (2002) Estivation and hibernation of *Pieris brassicae* (L.) in southern Spain: synchronization of two complex behavioral patterns. *Population Ecology* **44**: 273–280.

Spitzer K, Danks HV (2006) Insect biodiversity of boreal peat bogs. *Annual Review of Entomology* **51**: 137–161.

Spitzer K, Jaros J (1993) Lepidoptera associated with the Červené Blato bog (Central Europe): conservation implications. *European Journal of Entomology* **90**: 323–336.

Springer P, Boggs CL (1986) Resource allocation to oocytes: heritable variation with altitude in *Colias philodice eriphyle* (Lepidoptera). *American Naturalist* **127**: 252–256.

Srygley RB (1994) Shivering and its cost during reproductive behaviour in Neotropical owl butterflies, *Caligo* and *Opsiphanes* (Nymphalidae: Brassolinae). *Animal Behaviour* **47**: 23–32.

Srygley RB (2004) The aerodynamic costs of warning signals in palatable mimetic butterflies and their distasteful models. *Proceedings of the Royal Society of London B* **271**: 589–594.

Srygley RB, Chai P (1990a) Flight morphology of Neotropical butterflies: palatability and distribution of mass to the thorax and abdomen. *Oecologia* **84**: 491–499.

Srygley RB, Chai P (1990b) Predation and the elevation of thoracic temperature in brightly colored neotropical butterflies. *American Naturalist* **135**: 766–787.

Srygley RB, Kingsolver JG (1998) Red-wing blackbird reproductive behaviour and the palatability, flight performance, and morphology of temperate pierid butterflies (*Colias*, *Pieris* and *Pontia*). *Biological Journal of the Linnean Society* **64**: 41–55.

Srygley RB, Kingsolver JG (2000) Effects of weight loading on flight performance and survival of palatable Neotropical *Anartia fatima* butterflies. *Biological Journal of the Linnean Society* **70**: 707–725.

Städler E (1984) Contact chemoreception. In: Bell WJ, Gardé RT (eds.) *Chemical Ecology of Insects*, London: Chapman and Hall, pp. 3–35.

Städler E (1986) Oviposition and feeding stimuli in leaf surface waxes. In: Juniper B, Southwood R (eds.) *Insects and the Plant Surface*, London: Edward Arnold, pp. 105–121.

Stallings DB, Stallings VNT, Turner JR, Turner BR (1985) Courtship and oviposition patterns of two *Agathymus* (Megathymidae). *Journal of the Lepidopterists' Society* **39**: 171–176.

Stamp NE (1980) Egg deposition patterns in butterflies: why do some species cluster their eggs rather than deposit them singly? *American Naturalist* **115**: 367–380.

Stamp NE (1981a) Effect of group size on parasitism in a natural population of the Baltimore Checkerspot *Euphydryas phaeton*. *Oecologia* **49**: 201–206.

Stamp NE (1981b) Parasitism of single and multiple egg clusters of *Euphydryas phaeton* (Nymphalidae). *Journal of the New York Entomological Society* **89**: 89–97.

Stamp NE (1993) A temperate region view of the interaction between temperature, food quality, and predators on caterpillar foraging. In: Stamp NE, Casey TM (eds.) *Caterpillars: Ecological and Evolutionary Constraints on Foraging*, New York: Chapman and Hall, pp. 478–508.

Stamp NE, Bowers MD (1990) Variation in food quality and temperature constrain foraging of gregarious caterpillars. *Ecology* **71**: 1031–1039.

Stanton ML (1984) Short-term learning and the searching accuracy of egg-laying butterflies. *Animal Behaviour* **32**: 33–40.

Starnecker G, Burret M, Nissler A (1998) The exposed butterfly pupa of *Zerynthia polyxena* and its attachment to the substrate of the pupation site (Lepidoptera: Papilionidae: Zerynthiinae). *Entomologische Zeitschrift* **108**: 157–164.

Stearns SC (1989) The evolutionary significance of phenotypic plasticity. *BioScience* **39**: 436–445.

Stearns SC (1992) *The Evolution of Life Histories*. Oxford: Oxford University Press.

Stebbins GL (1959) The role of hybridization in evolution. *Proceedings of the American Philosophical Society* **103**: 231–251.

Stefanescu C (2000a) The Catalan Butterfly Monitoring Scheme: the first five years [El Butterfly Monitoring Scheme en Catalunya: los primeros cinco anos]. *Treballs de la Societat Catalana de Lepidopterologia* **15**: 5–48.

Stefanescu C (2000b) Bird predation on cryptic larvae and pupae of a swallowtail butterfly. *Butlletí del Grup Català d'Anellament* **17**: 39–49.

Stefanescu C (2004) Seasonal change in pupation behaviour and pupal mortality in a swallowtail butterfly. *Animal Biodiversity and Conservation* **27**: 25–36.

Stefanescu C, Dantart J (2002) Distribucio i ecologia d'Apatura ilia ((Denis, Schiffermueller), 1775) (Nymphalidae: Apaturinae) a Catalunya. *Butlletí de la Societat Catalana de Lepidopterologia* **88**: 25–56.

Stefanescu C, Peñuelas J, Filella I (2003a) Effects of climatic change on the phenology of butterflies in the northwest Mediterranean Basin. *Global Change Biology* **9**: 1494–1506.

Stefanescu C, Pintureau B, Tschorsnig HP, Pujade-Villar J (2003b) The parasitoid complex of the butterfly *Iphiclides podalirius feisthamelii* (Lepidoptera: Papilionidae) in north-east Spain. *Journal of Natural History* **37**: 379–396.

Stefanescu C, Herrando S, Páramo F (2004) Butterfly species richness in the north-west Mediterranean Basin: the role of natural and human-induced factors. *Journal of Biogeography* **31**: 905–916.

Stefanescu C, Jubany J, Dantart J (2005a) Egg-laying by the butterfly *Iphiclides podalirius* (Lepidoptera, Papilionidae) on alien plants: a broadening of host range or oviposition mistakes? *Animal Biodiversity and Conservation* **29**: 83–90.

Stefanescu C, Peñuelas J, Filella I (2005b) Butterflies highlight the conservation value of hay meadows highly threatened by land-use changes in a Mediterranean area. *Biological Conservation* **126**: 234–246.

Stefanescu C, Peñuelas J, Sardans J, Filella I (2006) Females of the specialist butterfly *Euphydryas aurinia* (Lepidoptera: Nymphalidae: Melitaeini) select the greenest leaves of *Lonicera implexa* (Caprifoliaceae) for oviposition. *European Journal of Entomology* **103**: 569–574.

Stefanescu C, Planas J, Shaw MR (2009) The parasitoid complex attacking coexisting Spanish populations of *Euphydryas aurinia* and *Euphydryas desfontainii* (Lepidoptera: Nymphalidae, Melitaeini). *Journal of Natural History* **43**: 553–568.

Steffan-Dewenter I, Tscharntke T (1997) Early succession of butterfly and plant communities on set-aside fields. *Oecologia* **109**: 294–302.

Steffan-Dewenter I, Tscharntke T (2000) Butterfly community structure in fragmented habitats. *Ecology Letters* **3**: 449–456.

Steffan-Dewenter I, Münzenberg U, Bürger C, Thies C, Tscharntke T (2002) Scale-dependent effects of landscape context on three pollinator guilds. *Ecology* **83**: 1421–1432.

Stehr FW (1987) Order Lepidoptera. In: Stehr FW (ed.) *Immature Insects*, Dubuque, IA: Kendall/Hunt, pp. 288–596.

Steiner R, Hermann G (1999) Freilandbeobachtungen zu Eiablageverhalten und habitat des Wald-Wiesenvögelchens, *Coenonympha hero* (Linnaeus, 1761), an einer Flugstelle in Baden-Württemberg (Lepidoptera: Nymphalidae). *Nachrichten des Entomologischen Vereins Apollo* **20**: 111–118.

Steiner R, Trusch R (2000) Eiablageverhalten und habitat von *Hipparchia statilinus* in Brandenburg (Lepidoptera: Nymphalidae: Satyrinae). *Stuttgarter Beitraege zur Naturkunde, Serie A, Biologie* **606**: 1–10.

Stelter C, Reich M, Grimm V, Wissel C (1997) Modelling persistence in dynamic landscapes: lessons from a metapopulation of the grasshopper *Bryodema tuberculata*. *Journal of Animal Ecology* **66**: 508–518.

Stephens DW, Krebs JR (1986) *Foraging Theory*. Princeton, NJ: Princeton University Press.

Stern VM, Smith RF (1960) Factors affecting egg production and oviposition in populations of *Colias philodicae eurytheme* Boisduval (Lepidoptera: Pieridae). *Hilgardia* **29**: 411–454.

Stevens GC (1989) The latitudinal gradient in geographical range: how so many species coexist in the tropics. *American Naturalist* **133**: 240–256.

Stevens GC (1992) The elevational gradient in altitudinal range: an extension of Rapoport's latitudinal rule to altitude. *American Naturalist* **140**: 893–911.

Stevens M (2005) The role of eyespots as anti-predator mechanisms, principally demonstrated in the Lepidoptera. *Biological Reviews* **80**: 573–588.

Stevens M (2007) Predator perception and the interrelation between different forms of protective coloration. *Proceedings of the Royal Society of London B* **274**: 1457–1464.

Stevens M, Hardman CJ, Stubbins CL (2008) Conspicuousness, not eye mimicry, makes 'eyespots' effective antipredator signals. *Behavioral Ecology* 19: 525–531.

Stevenson RD (1985a) Body size and limits to the daily range of body temperature in terrestrial ectotherms. *American Naturalist* 125: 102–117.

Stevenson RD (1985b) The relative importance of behavioral and physiological adjustments controlling body temperature in terrestrial ectotherms. *American Naturalist* 126: 362–386.

Steyerberg EW, Eijkemans MJC, Harrell FE Jr, Habbema JDF (2000) Prognostic modelling with logistic regression analysis: a comparison of selection and estimation methods in small data sets. *Statistics in Medicine* 19: 1059–1080.

Steyerberg EW, Eijkemans MJC, Harrell FE Jr, Habbema JDF (2001a) Prognostic modelling with logistic regression analysis. *Medical Decision Making* 21: 45–58.

Steyerberg EW, Harrell FE Jr, Borsboom GJJM, Eijkemans MJC, Vergouwe Y, Habbema JDF (2001b) Internal validation of predictive models: efficiency of some procedures for logistic regression analysis. *Journal of Clinical Epidemiology* 54: 774–781.

Stiling P (1988) Density-dependent processes and key factors in insect populations. *Journal of Animal Ecology* 57: 581–594.

Stireman JO, O'Hara JE, Wood DM (2006) Tachinidae: evolution, behaviour, and ecology. *Annual Review of Entomology* 51: 523–555.

Stjernholm F, Karlsson B (2000) Nuptial gifts and use of body resources for reproduction in the green-veined white butterfly *Pieris napi*. *Proceedings of the Royal Society of London B* 267: 807–811.

Stjernholm F, Karlsson B (2006) Reproductive expenditure affects utilization of thoracic and abdominal resources in male *Pieris napi* butterflies. *Functional Ecology* 20: 442–448.

Stjernholm F, Karlsson B, Boggs CL (2005) Age-related changes in thoracic mass: possible reallocation of resources to reproduction in butterflies. *Biological Journal of the Linnean Society* 86: 363–380.

Stockwell DRB (2006) Improving ecological niche models by data mining large environmental datasets for surrogate models. *Ecological Modelling* 192: 188–196.

Stockwell DRB, Peters D (1999) The GARP modeling system: problems and solutions to automated spatial prediction. *International Journal of Geographical Information Science* 13: 143–158.

Stockwell DRB, Peterson AT (2002) Effects of sample size on accuracy of species distribution models. *Ecological Modelling* 148: 1–13.

Stoltze M (1996) *Danske Dagsommerfugle*. Copenhagen: Gyldendal.

Stouthamer R, Breeuwer JAJ, Luck RF, Werren JH (1992) Molecular identification of microorganisms associated with parthenogenesis. *Nature* 361: 66–68.

Stoutjesdijk P, Barkman JJ (1992) *Microclimate, Vegetation and Fauna*. Knivsta, Sweden: Opulus Press.

Strand MR, Vinson SB (1984) Facultative hyperparasitism by the egg parasitoid *Trichogramma pretiosum* (Hymenoptera: Trichogrammatidae). *Annals of the Entomological Society of America* 77: 679–686.

Strauss B, Biedermann R (2005) The use of habitat models in conservation of rare and endangered leafhopper species (Hemiptera, Auchenorrhyncha). *Journal of Insect Conservation* 9: 245–260.

Strauss B, Biedermann R (2007) Evaluating temporal and spatial generality: how valid are species–habitat relationship models? *Ecological Modelling* 204: 104–114.

Stride GO (1956) On the courtship behaviour of *Hypolimnas misippus* L. (Lepidoptera: Nymphalidae), with notes on the mimetic association with *Danaus chrysippus* L. (Lepidoptera). *British Journal of Animal Behaviour* 4: 52–68.

Stride GO (1957) Investigations in the courtship behaviour of the male of *Hypolimnas misippus* L. (Lepidoptera, Nymphalidae), with special reference to the role of visual stimuli. *Animal Behaviour* 5: 153–167.

Stride GO (1958) Further studies on the courtship behaviour of African mimetic butterflies. *Animal Behaviour* 6: 224–230.

Strong DR, Lawton JH, Southwood R (1984) *Insects on Plants: Community Patterns and Mechanisms*. Oxford: Blackwell Scientific Publications.

Stutt AD, Willmer P (1998) Territorial defense in the speckled wood butterflies: do the hottest males always win? *Animal Behaviour* 55: 1341–1347.

Sunnucks P (2000a) Efficient genetic markers for population biology. *Trends in Ecology and Evolution* 15: 199–203.

Sunnucks P (2000b) Reply from P. Sunnucks. *Trends in Ecology and Evolution* 15: 377.

Suomalainen E (1953) The kinetochore and the bivalent structure in the Lepidoptera. *Hereditas* 39: 88–96.

Suomalainen E (1965) On the chromosomes of the geometrid moth genus *Cidaria*. *Chromosoma* 16: 166–184.

Suominen O, Olofsson J (2000) Impacts of semi-domesticated reindeer on structure of tundra and forest communities in Fennoscandia: a review. *Annales Zoologici Fennici* 37: 233–249.

Sutcliffe OL, Thomas CD (1996) Open corridors appear to facilitate dispersal by ringlet butterflies (*Aphantopus hyperantus*) between woodland clearings. *Conservation Biology* **10**: 1359–1365.

Sutcliffe OL, Thomas CD, Moss D (1996) Spatial synchrony and asynchrony in butterfly population dynamics. *Journal of Animal Ecology* **65**: 85–95.

Sutcliffe OL, Thomas CD, Peggie D (1997a) Area-dependent migration by ringlet butterflies generates a mixture of patchy and metapopulation attributes. *Oecologia* **109**: 229–234.

Sutcliffe OL, Thomas CD, Yates TJ, Greatorex-Davies JN (1997b) Correlated extinctions, colonizations and population fluctuations in a highly connected ringlet butterfly metapopulation. *Oecologia* **109**: 235–241.

Sutherland WJ, Gill JA, Norris K (2002) Density-dependent dispersal in animals: concept, evidence, mechanisms and consequences. In: Bullock JM, Kenward RE, Hails RS (eds.) *Dispersal Ecology*, Oxford: Blackwell Science, pp. 134–151.

Suzuki Y (1976) So-called territorial behaviour of the small copper, *Lycaena phlaeas daimio* Seitz (Lepidoptera, Lycaenidae). *Kontyu, Tokyo* **44**: 193–204.

Suzuki Y (1978) Adult longevity and reproductive potential of the small cabbage white, *Pieris rapae erucivora* Boisduval (Lepidoptera -Pieridae). *Applied Entomology and Zoology* **13**: 312–313.

Svärd L (1985) Paternal investment in a monandrous butterfly, *Pararge aegeria*. *Oikos* **45**: 66–70.

Svärd L, Wiklund C (1986) Different ejaculate delivery strategies in first versus subsequent matings in the swallowtail butterfly *Papilio machaon* L. *Behavioral Ecology and Sociobiology* **18**: 325–330.

Svärd L, Wiklund C (1988a) Fecundity, egg weight and longevity in relation to multiple matings in females of the monarch butterfly. *Behavioral Ecology and Sociobiology* **23**: 39–43.

Svärd L, Wiklund C (1988b) Prolonged mating in the monarch butterfly *Danaus plexippus* and nightfall as a cue for sperm transfer. *Oikos* **51**: 351–354.

Svärd L, Wiklund C (1989) Mass and production rate of ejaculates in relation to monandry/polyandry in butterflies. *Behavioral Ecology and Sociobiology* **24**: 395–402.

Svärd L, Wiklund C (1991) The effect of ejaculate mass on female reproductive output in the European swallowtail butterfly, *Papilio machaon*. *Journal of Insect Behavior* **4**: 33–41.

Sweeney A, Jiggins C, Johnsen S (2003) Insect communication: polarized light as a butterfly mating signal. *Nature* **423**: 31–32.

Swets JA (1988) Measuring the accuracy of diagnostic systems. *Science* **240**: 1285–1293.

Swihart SL (1967) Neural adaptations in the visual pathway of certain heliconiinae butterflies, and related forms, to variation in wing coloration. *Zoologica* **51**: 1–14.

Szymura JM, Barton NH (1986) Genetic analysis of a hybrid zone between the fire-bellied toads, *Bombina bombina* and *Bombina variegata*, near Cracow in southern Poland. *Evolution* **40**: 1141–1159.

Szymura JM, Barton NH (1991) The genetic structure of the hybrid zone between the fire-bellied toads *Bombina bombina* and *B. variegata*: comparisons between transects and between loci. *Evolution* **45**: 237–261.

Tabashnik BE (1982) Responses of pest and non-pest *Colias* butterfly larvae to intraspecific variation in leaf nitrogen and water content. *Oecologia* **55**: 389–394.

Taberlet P, Fumagalli L, Wust-Saucy AG, Cosson JF (1998) Comparative phylogeography and postglacial colonization routes in Europe. *Molecular Ecology* **7**: 453–464.

Takanashi T, Hiroki M, Obara Y (2001) Evidence for male and female sex pheromones in the sulfur butterfly, *Eureme hecabe*. *Entomologia Experimentalis et Applicata* **101**: 89–92.

Takeuchi T (2006) Matter of size or matter of residency experience? Territorial contest in a green hairstreak, *Chrysozephyrus smaragdinus* (Lepidoptera: Lycaenidae). *Ethology* **112**: 293–299.

Talloen W, Van Dyck H, Lens L (2004) The cost of melanization: butterfly wing coloration under environmental stress. *Evolution* **58**: 360–366.

Tammaru T, Haukioja E (1996) Capital breeders and income breeders among Lepidoptera: consequences to population dynamics. *Oikos* **77**: 561–564.

Tammaru T, Javois J (2005) When being alive implies being safe: variation in mortality rates can cause oviposition selectivity to increase with age. *Oikos* **111**: 649–653.

Tarasov PE, Volkova VS, Webb T III, Guiot J, Andreev AA, Bezusko LG (2000) Last glacial maximum biomes reconstructed from pollen and plant macrofossil data from northern Eurasia. *Journal of Biogeography* **27**: 609–620.

Tartally A (2005) *Neotypus melanocephalus* (Hymenoptera: Ichneumonidae): the first record of a parasitoid wasp attacking *Maculinea teleius* (Lycaenidae). *Nota lepidopterologica* **28**: 65–67.

Tauber MJ, Tauber CA, Masaki S (1986) *Seasonal Adaptations of Insects*. Oxford: Oxford University Press.

Tavoillot C (1967) Un hybride probable entre *Melanargia russiae* et *M. lachesis*. *Alexanor* **5**: 19–25.

Taylor CM, Hastings A (2005) Allee effects in biological invasions. *Ecology Letters* **8**: 895–908.

Telfer WH (1965) The mechanisms and control of yolk formation. *Annual Review of Entomology* **10**: 161–184.

Telfer WH, Rutberg LD (1960) The effects of blood protein depletion on the growth of the oocytes in the *Cecropia* moth. *Biological Bulletin* **118**: 352–366.

Templado J (1975) La regulación natural de las poblaciones de *Euphydryas aurinia* Rott. (Lep. Nymphalidae). *Boletín de la Estación Central de Ecología* **7**: 77–81.

Templado J (1981) Diapausa y voltinismo en *Euchloe ausonia crameri* Butler. *Eos* **57**: 273–277.

Templeton AR (1989 The meaning of species and speciation: a genetic perspective. In: Otte D, Endler JA (eds.) *Speciation and its Consequences*, Sunderland, MA: Sinauer Associates, pp. 3–27.

Templeton AR (1994) The role of molecular genetics in speciation studies. In: Schierwater B, Streit B, Wagner GP, De Salle R (eds.) *Molecular Ecology and Evolution: Approaches and Applications*, Basel, Switzerland: Birkhäuser, pp. 455–477.

Templeton AR (1998a) Species and speciation: geography, population structure, ecology, and gene trees. In: Howard DJ, Berlocher SH (eds.) *Endless Forms: Species and Speciation*, New York: Oxford University Press, pp. 32–43.

Templeton AR (1998b) Nested clade analyses of phylogeographic data: testing hypotheses about gene flow and population history. *Molecular Ecology* **7**: 381–397.

Ter Braak CJF, Prentice IC (1988) A theory of gradient analysis. *Advances in Ecological Research* **18**: 271–317.

Thirion C (1976) Les Ichneumoninae 'Amblypygi' *sensu* Wesmael, en Belgique. *Bulletin de la Société Entomologique de Belgique* **112**: 29–69.

Thirion C (1981) Les Ichneumoninae (Hymenoptera Ichneumonidae) en Belgique (2éme partie). *Bulletin de la Société Entomologique de Belgique* **117**: 229–254.

Thomas JA (1980) Why did the large blue become extinct in Britain? *Oryx* **15**: 243–247.

Thomas JA (1983a) A quick method for estimating butterfly numbers during surveys. *Biological Conservation* **27**: 195–211.

Thomas JA (1983b) The ecology and conservation of *Lysandra bellargus* (Lepidoptera: Lycaenidae) in Britain. *Journal of Applied Ecology* **20**: 59–83.

Thomas CD (1984a) Oviposition and egg load assessment by *Anthocharis cardamines* (L.) (Lepidoptera: Pieridae). *Entomologist's Gazette* **35**: 145–148.

Thomas JA (1984b) The conservation of butterflies in temperate countries: past efforts and lessons for the future. In: Vane-Wright RI, Ackery PR (eds.) *The Biology of Butterflies*, London: Academic Press, pp. 333–353.

Thomas CD (1985) The status and conservation of the butterfly *Plebejus argus* L. (Lepidoptera, Lycaenidae) in north-west Britain. *Biological Conservation* **33**: 29–51.

Thomas JA (1989) The return of the large blue butterfly. *British Wildlife* **1**: 2–13.

Thomas CD (1991a) Spatial and temporal variability in a butterfly population. *Oecologia* **87**: 577–580.

Thomas JA (1991b) Rare species conservation: case studies of European butterflies. In: Spellerberg IF, Goldsmith FB, Morris MG (eds.) *The Scientific Management of Temperate Communities for Conservation*, Oxford: Blackwell, Scientific Publications, pp. 149–197.

Thomas JA (1993) Holocene climate change and warm man-made refugia may explain why a sixth of British butterflies inhabit unnatural early successional habitats. *Ecography* **16**: 278–284.

Thomas CD (1994) Extinction, colonization, and metapopulations: environmental tracking by rare species. *Conservation Biology* **8**: 373–378.

Thomas JA (1995a) The ecology and conservation of *Maculinea arion* and other European species of large butterfliy. In: Pullin AS (ed.) *Ecology and Conservation of Butterflies*, London: Chapman and Hall, pp. 180–197.

Thomas JA (1995b) The conservation of declining butterfly populations in Britain and Europe: priorities, problems and successes. *Biological Journal of the Linnean Society* **56**: 55–72.

Thomas CD (1995c) Ecology and conservation of butterfly metapopulations in the fragmented British landscape. In: Pullin AS (ed.) *Ecology and Conservation of Butterflies*, London: Chapman and Hall, pp. 46–63.

Thomas JA (1996) *Maculinea arion* (Linnaeus, 1758). In: Helsdingen PJ, van Willemse L, Speight MCD (eds.) *Background Information on Invertebrates of the Habitats Directive and the Bern Convention*, Part I, *Crustacea, Coleoptera and Lepidoptera*. Strasbourg, France: Council of Europe, pp. 157–163.

Thomas CD (2000) Dispersal and extinction in fragmented landscapes. *Proceedings of the Royal Society of London B* **267**: 139–145.

Thomas JA (2005) Monitoring change in the abundance and distribution of insects using butterflies and other indicator groups. *Philosophical Transactions of the Royal Society of London B* **360**: 339–357.

Thomas CD, Abery JCG (1995) Estimating rates of butterfly decline from distribution maps: the effect of scale. *Biological Conservation* **73**: 59–65.

Thomas JA, Bovee KD (1993) Application and testing of a procedure to evaluate transferability of habitat suitability criteria. *Regulated Rivers: Research and Management* **8**: 285–294.

Thomas JA, Elmes GW (1993) Specialized searching and the hostile use of allomones by a parasitoid whose host, the butterfly *Maculinea rebeli*, inhabits ant nests. *Animal Behaviour* **45**: 593–602.

Thomas JA, Elmes GW (1998) Higher productivity at the cost of increased host-specificity when *Maculinea* butterfly larvae exploit ant colonies through trophallaxis rather than by predation. *Ecological Entomology* **23**: 457–464.

Thomas JA, Elmes GW (2001) Food-plant niche selection rather than the presence of ant nests explains oviposition patterns in the myrmecophilous butterfly genus *Maculinea*. *Proceedings of the Royal Society of London B* **268**: 471–477.

Thomas CD, Hanski I (1997) Butterfly metapopulations. In: Hanski I, Gilpin ME (eds.) *Metapopulation Biology: Ecology, Genetics, and Evolution*, San Diego, CA: Academic Press, pp. 359–386.

Thomas CD, Hanski I (2004) Metapopulation dynamics in changing environments: butterfly responses to habitat and climate change. In: Hanski I, Gaggiotti EO (eds.) *Ecology, Genetics and Evolution of Metapopulations*, San Diego, CA: Elsevier, pp. 489–514.

Thomas CD, Jones TM (1993) Partial recovery of a skipper butterfly (*Hesperia comma*) from population refuges: lessons for conservation in a fragmented landscape. *Journal of Animal Ecology* **62**: 472–481.

Thomas CD, Kunin WE (1999) The spatial structure of populations. *Journal of Animal Ecology* **68**: 647–657.

Thomas JA, Lewington R (1991) *The Butterflies of Britain and Ireland*. London: Dorling Kindersley.

Thomas CD, Singer MC (1987) Variation in host preference affects movement patterns within a butterfly population. *Ecology* **68**: 1262–1267.

Thomas CD, Singer MC (1998) Scale-dependent evolution of specialization in a checkerspot butterfly: from individuals to metapopulations and ecotypes. In: Mopper S, Strauss S (eds.) *Genetic Structure and Local Adaptation in Natural Insect Populations*, New York: Chapman and Hall, pp. 343–374.

Thomas JA, Wardlaw JC (1992) The capacity of a *Myrmica* ant nest to support a predacious species of *Maculinea* butterfly. *Oecologia* **91**: 101–109.

Thomas JA, Thomas CD, Simcox DJ, Clarke RT (1986) The ecology and declining status of the silver-spotted skipper butterfly (*Hesperia comma*) in Britain. *Journal of Applied Ecology* **23**: 365–380.

Thomas CD, Hill JK, Lewis OT (1988) Evolutionary consequences of habitat fragmentation in a localized butterfly. *Journal of Animal Ecology* **67**: 485–497.

Thomas JA, Elmes GW, Wardlaw JC, Woyciechowski M (1989) Host specificity among *Maculinea* butterflies in *Myrmica* ant nests. *Oecologia* **79**: 452–457.

Thomas JA, Munguira ML, Martín J, Elmes GW (1991) Basal hatching by *Maculinea* butterfly eggs: a consequence of advanced myrmecophily? *Biological Journal of the Linnean Society* **44**: 175–184.

Thomas CD, Thomas JA, Warren MS (1992) Distributions of occupied and vacant butterfly habitats in fragmented landscapes. *Oecologia* **92**: 563–567.

Thomas JA, Elmes GW, Wardlaw JC (1993) Contest competition among *Maculinea rebeli* butterfly larvae in ant nests. *Ecological Entomology* **18**: 73–76.

Thomas JA, Moss D, Pollard E (1994a) Increased fluctuations of butterfly populations towards the northern edges of species' ranges. *Ecography* **17**: 215–220.

Thomas JA, Morris MG, Hambler C (1994b) Patterns, mechanisms and rates of extinction among invertebrates in the United Kingdom. *Philosophical Transactions of the Royal Society of London B* **344**: 47–54

Thomas CD, Singer MC, Boughton DA (1996) Catastrophic extinction of population sources in a butterfly metapopulation. *American Naturalist* **148**: 957–975.

Thomas JA, Clarke RT, Elmes GW, Hochberg ME (1998a) Population dynamics in the genus *Maculinea* (Lepidoptera: Lycaenidae) In: Dempster JP, McLean IFG (eds.) *Insect Populations in Theory and in Practice*, Dordrecht, Netherlands: Kluwer, pp. 261–290.

Thomas JA, Simcox DJ, Wardlaw JC, Elmes GW, Hochberg ME, Clarke RT (1998b) Effects of latitude, altitude and climate on the habitat and conservation of the endangered butterfly *Maculinea arion* and its *Myrmica* ant hosts. *Journal of Insect Conservation* **2**: 39–46.

Thomas CD, Jordano D, Lewis OT, Hill JK, Sutcliffe OL, Thomas JA (1998c) Butterfly distributional patterns, processes and conservation. In: Mace GM, Balmford A, Ginsberg JR (eds.) *Conservation in a Changing World: Integrating Processes into Priorities for Action*, Cambridge: Cambridge University Press, pp. 107–138.

Thomas JA, Rose RJ, Clarke RT, Thomas CD, Webb NR (1999) Intraspecific variation in habitat availability among ectothermic animals near their climatic limits and their centres of range. *Functional Ecology* **13**: 55–64.

Thomas CD, Bodsworth EJ, Wilson RJ, Simmons AD, Davies ZG, Musche M, Conradt L (2001a) Ecological and evolutionary processes at expanding range margins. *Nature* **411**: 577–581.

Thomas JA, Bourn NAD, Clarke RT, Stewart KE, Simcox DJ, Pearman GS, Curtis R, Goodger B (2001b) The quality and isolation of habitat patches both determine where butterflies persist in fragmented landscapes. *Proceedings of the Royal Society of London B* **268**: 1791–1796.

Thomas JA, Knapp JJ, Akino T, Gerty S, Wakamura S, Simcox DJ, Wardlaw JC, Elmes GW (2002) Parasitoid secretions provoke ant warfare. *Nature* **417**: 505–506.

Thomas JA, Telfer MG, Roy DB, Preston CD, Greenwood JJD, Asher J, Fox R, Clarke RT, Lawton JH (2004) Comparative losses of British butterflies, birds, and plants and the global extinction crisis. *Science* **303**: 1879–1881.

Thompson JN (1988) Evolutionary ecology of the relationship between oviposition preference and performance of offspring in phytophagous insects. *Entomologia Experimentalis et Applicata* **47**: 3–14.

Thompson JN, Pellmyr O (1991) Evolution of oviposition behaviour and host preference in Lepidoptera. *Annual Review of Entomology* **41**: 407–431.

Thomson, G (1980) *The Butterflies of Scotland*. London: Croom Helm.

Thomson G (1987) Enzyme Variation at Morphological Boundaries in *Maniola* and Related Genera (Lepidoptera: Nymphalidae: Satyrinae). Ph.D. thesis, University of Stirling, Scotland.

Thorne JH, O'Brien J, Forister ML, Shapiro AM (2006) Building phenological models from presence/absence data for a butterfly fauna. *Ecological Applications* **16**: 1842–1853.

Thornhill R, Alcock J (1983) *The Evolution of Insect Mating Systems*. Cambridge, MA: Harvard University Press.

Thuiller W (2004) Patterns and uncertainties of species' range shifts under climate change. *Global Change Biology* **10**: 2020–2070.

Thuiller W, Lavorel S, Araújo MB (2005) Niche properties and geographical extent as predictors of species sensitivity to climate change. *Global Ecology and Biogeography* **14**: 347–357.

Thuiller W, Midgley GF, Rougeti M, Cowling RM (2006) Predicting patterns of plant species richness in megadiverse South Africa. *Ecography* **29**: 733–744.

Tibshirani R (1996) Regression shrinkage and selection via the lasso. *Journal of the Royal Statistical Society B* **58**: 267–288.

Tilley RJD (1983) Rearing *Melanargia galathea* (L.) and *M. lachesis* (Huebner) (Lepidoptera: Satyridae). *Entomologist's Gazette* **34**: 9–11.

Tinbergen N, Meeuse BJD, Boerema LK, Varossieau WW (1942) Die Balz des Samtfalters, *Eumenis (=Satyrus) semele* L. *Zeitschrift für Tierpsychologie* **5**: 188–226.

Ting CT, Tsaur SC, Wu CI (2000) The phylogeny of closely related species as revealed by the genealogy of a speciation gene, *Odysseus*. *Proceedings of the National Academy of Sciences, USA* **97**: 5313–5316.

Tolman T, Lewington R (1997) *Butterflies of Britain and Europe*. London: HarperCollins.

Torres-Vila LM, Jennions MD (2005) Male mating history and female fecundity in the Lepidoptera: do male virgins make better partners? *Behavioral Ecology and Sociobiology* **57**: 318–326.

Torres-Vila LM, Rodríguez-Molina MC, Jennions MD (2004) Polyandry and fecundity in the Lepidoptera: can methodological and conceptual approaches bias outcomes? *Behavioral Ecology and Sociobiology* **55**: 315–324.

Toso GG, Balleto E (1977) Una nuova specie del *Agrodiaetus* Hübner (Lepidoptera: Lycaenidae). *Annali del Museo Civico di Storia Naturale 'Giacomo Doria'* **81**: 124–130.

Tracy CR, Christian KA (1986) Ecological relationships among space, time, and thermal niche axes. *Ecology* **67**: 609–615.

Tränkle U, Hehmann M (2002) *Naturschutz und Zementindustrie*, Projektteil 3, *Management-Empfehlungen*. Düsseldorf, Germany: Verlag Bau+Technik.

Tränkle U, Offenwanger H, Röhl M, Hübner F, Poschlod P (2003) *Naturschutz und Zementindustrie*, Projektteil 2, *Literaturstudie*. Düsseldorf, Germany: Verlag Bau+Technik.

Travis JMJ, Dytham C (1998) The evolution of dispersal in a metapopulation: a spatially explicit, individual-based model. *Proceedings of the Royal Society of London B* **265**: 17–23.

Travis JMJ, Dytham C (1999) Habitat persistence, habitat availability and the evolution of dispersal. *Proceedings of the Royal Society of London B* **266**: 723–728.

Travis JMJ, Murrell DJ, Dytham C (1999) The evolution of density-dependent dispersal. *Proceedings of the Royal Society of London B* **266**: 1837–1842.

Traynier RMM (1984) Associative learning in the ovipositional behaviour of the cabbage butterfly, *Pieris rapae*. *Physiological Entomology* **9**: 465–472.

Traynier RMM (1986) Visual learning in assays of sinigrin solution as an oviposition releaser for the cabbage butterfly, *Pieris rapae*. *Entomologia Experimentalis et Applicata* **40**: 25–33.

Tremewan WG (2007) *Ecology, Phenotypes and the Mendelian Genetics of Burnet Moths (Zygaena Fabricius, 1775)*. Wallingford, Oxon.: Gem Publishing Co.

Trexler JC, Travis J (1993) Nontraditional regression analyses. *Ecology* **74**: 1629–1637.

Troiano G, Balleto E, Toso GG (1979) The karyotype of *Agrodiaetus humedasae* Toso, Balleto, 1976. *Bollettino della Società Entomologica Italiana* **111**: 141–143.

Tsubaki Y, Kitching RL (1986) Central place foraging of the Charaxine butterfly, *Polyura pyrrhus* (L.): a case study in a herbivore. *Journal of Ethology* **4**: 59–68.

Tsuji JS, Kingsolver JG, Watt WB (1986) Thermal physiological ecology of *Colias* butterflies in flight. *Oecologia* **69**: 161–170.

Tucker GM, Heath MF (1994) *Birds in Europe: Their Conservation Status*, Birdlife Conservation Series No. 3. Cambridge: Birdlife International.

Tudor O, Dennis RLH, Greatorex-Davies JN, Sparks TH (2004) Flower preferences of woodland butterflies in the UK: nectaring specialists are species of conservation concern. *Conservation Biology* **119**: 397–403.

Tullberg BS, Merilaita S, Wiklund C (2005) Aposematism and crypsis combined as a result of distance dependence: functional versatility of the colour pattern in the swallowtail butterfly larva. *Proceedings of the Royal Society of London B* **272**: 1315–1321.

Tumler G (1885) Lebensdauer der Tagschmetterlinge während des Sommers. *Jahresberichte des westfälischen Provinziale-Vereins für Wissenschaft und Kunst* **31**.

Turchin P (1998) *Quantitative Analysis of Movement: Measuring and Modeling Population Redistribution in Animals and Plants*. Sunderland, MA: Sinauer Associates.

Turner JRG (1963) A quantitative study of a Welsh colony of the Large Heath Butterfly, *Coenonympha tullia* Müller (Lepidoptera). *Proceedings of the Royal Entomological Society of London A* **38**: 101–112.

Turner JRG (1978) Why male butterflies are non-mimetic: natural selection, sexual selection, group selection, modification and seiving. *Biological Journal of the Linnean Society* **10**: 385–432.

Turner JRG (1981) Adaptation and evolution in *Heliconius*: a defense of neodarwinism. *Annual Review of Ecology and Systematics* **12**: 99–121.

Turner JRG (1986) Why are there so few butterflies in Liverpool? Homage to Alfred Russel Wallace. *Antenna* **10**: 18–24.

Turner JRG (1987) The evolutionary dynamics of Batesian and Mullerian mimicry: similarities and differences. *Ecological Entomology* **12**: 81–95.

Turner JRG, Mallet JLB (1996) Did forest islands drive the diversity of warningly coloured butterflies? Biotic drift and the shifting balance. *Philosophical Transactions of the Royal Society of London B* **351**: 835–845.

Turner JRG, Gatehouse CM, Corey CA (1987) Does solar energy control organic diversity? Butterflies, moths and the British climate. *Oikos* **48**: 195–205.

Tutin TG, Heywood VH, Burges NA, Valentine DH, Walters SM, Webb DA (1964) *Flora European*, Vol. 1, Cambridge: Cambridge University Press.

Tutt JW (1910) *Agriades polonus* Zeller, mit Bemerkungen über die bekannten Exemplare dieser Form. *Society of Entomologists* **25**: 3–4.

Tutt JW (1914) *A Natural History of the British Butterflies, their Worldwide Variation and Geographical Distribution*. London: Friedlander.

Tuzov VK, Bogdanov PV, Devyatkin AL, Kaabak LI, Korolev VA, Murzin VS, Samodurov GD, Tarasov EA (1997) *Guide to the Butterflies of Russia and Adjacent Territories*, Vol. 1, *Hesperiidae, Papilionidae, Pieridae, Satyridae*. Moscow: Pensoft.

Tuzov VK, Bogdanov PV, Churkin SV, Devyatkin AL, Danchenko AV, Murzin VS, Samodurov GD, Zhdanko AB (2000) *Guide to the Butterflies of Russia and Adjacent Territories*, Vol. 2, *Libytheidae, Danaidae, Nymphalidae, Riodinidae, and Lycaenidae*. Moscow: Pensoft.

Udvardy MDF (1969) *Dynamic Zoogeography*. New York: Van Nostrand Reinhold.

Ulrich W, Buszko J (2003) Species–area relationships of butterflies in Europe and species richness forecasting. *Ecography* **26**: 365–373.

Ulrich W, Buszko J (2005) Detecting biodiversity hotspots using species–area and endemics–area relationships: the case of butterflies. *Biodiversity and Conservation* **14**: 1977–1988.

Uvarov BP (1931) Insects and climate. *Transactions of the Entomological Society of London* **79**: 1–247.

Vahed K (1998) The function of nuptial feeding in insects: a review of empirical studies. *Biological Reviews of the Cambridge Philosophical Society* **73**: 43–78.

Väisänen R (1992) Distribution and abundance of diurnal Lepidoptera on a raised bog in southern Finland. *Annales Zoologici Fennici* **29**: 75–92.

Väisänen R, Somerma P (1985) The status of *Parnassius mnemosyne* (Lepidoptera, Papilionidae) in Finland. *Notulae Entomologicae* **65**: 109–118.

Väisänen R, Heliövaara K, Somerma P (1994a) Wing variation of *Maculinea arion* (Linnaeus) in Finland (Lepidoptera, Lycaenidae). *Entomologica Fennica* **5**: 139–146.

Vaisänen R, Kuussaari M, Nieminen M, Somerma P (1994b) Biology and conservation of *Pseudophilotes baton* in Finland (Lepidoptera, Lycaenidae). *Annales Zoologici Fennici* **31**: 145–156.

Välimäki P (2005) The effects of reindeer grazing on Lepidopteran communities in two fells in northern Fennoscandia. In: Jokinen M (ed.) *The Effects of Reindeer Herding and Protection in the Malla Nature Reserve*, Metsäntutkimuslaitoksen tiedonantoja 941, pp. 179–227. [In Finnish with English summary.]

Välimäki P, Itämies J (2003) Migration of the Clouded Apollo butterfly *Parnassius mnemosyne* in a network of suitable habitats: effects of patch characteristics. *Ecography* **26**: 679–691.

Vallin A, Jakobsson S, Lind J, Wiklund C (2006) Crypsis versus intimidation: anti-predation defence in three closely related butterflies. *Behavioral Ecology and Sociobiology* **59**: 455–459.

Valtonen A, Saarinen K (2005) A highway intersection as an alternative habitat for a meadow butterfly: effect of mowing, habitat geometry and roads on the ringlet (*Aphantopus hyperantus*). *Annales Zoologici Fennici* **42**: 545–556.

Valtonen AJ, Jantunen J, Saarinen K (2006) Flora and lepidoptera fauna adversely affected by invasive *Lupinus polyphyllus* along road verges. *Biological Conservation* **133**: 389–396.

van Achterberg C (1984) Revision of the genera of Braconinae with first and second metasomal tergites immovably joined (Hymenoptera, Braconidae, Braconinae). *Tijdschrift voor Entomologie* **127**: 137–164.

van Achterberg C (2002) *Apanteles (Choeras) gielisi* spec. nov. (Hymenoptera: Braconidae: Microgastrinae) from The Netherlands and the first report of Trichoptera as hosts of Braconidae. *Zoologische Mededelingen Leiden* **76**: 53–60.

Van der Meijden E, Nisbet RM, Crawley MJ (1998) The dynamics of a herbivore–plant interaction, the cinnabar moth and ragwort. In: Dempster P, McLean IFG (eds.) *Insect Populations in Theory and Practice*. 19th Symposium of the Royal Entomological Society, Dordrecht, Netherlands: Kluwer, pp. 291–309.

Van Dongen S (2006) Fluctuating asymmetry and developmental instability in evolutionary biology: past, present and future. *Journal of Evolutionary Biology* **19**: 1727–1743.

Van Dongen S, Backeljau T, Matthysen E, Dhondt AA (1994) Effects of forest fragmentation on the population structure of the Winter Moth *Operophtera brumata* L. (Lepidoptera, Geometridae). *Acta Oecologica* **15**: 193–206.

Van Dyck H (2003) Mate location: a matter of design? Adaptive morphological variation in the speckled wood butterfly. In: Boggs CL, Watt WB, Ehrlich PR (eds.) *Butterflies: Ecology and Evolution Taking Flight*, Chicago, IL: University of Chicago Press, pp. 353–366.

Van Dyck H, Baguette M (2005) Dispersal behaviour in fragmented landscapes: routine or special movements? *Basic and Applied Ecology* **6**: 535–545.

Van Dyck H, Matthysen E (1998) Thermoregulatory differences between phenotypes in the speckled wood butterfly: hot perchers and cold patrollers? *Oecologia* **114**: 326–334.

Van Dyck H, Matthysen E (1999) Habitat fragmentation and insect flight: a changing 'design' in a changing landscape? *Trends in Ecology and Evolution* **14**: 172–174.

Van Dyck H, Wiklund C (2002) Seasonal butterfly design: morphological plasticity among three developmental pathways relative to sex, flight and thermoregulation. *Journal of Evolutionary Biology* **15**: 216–225.

Van Dyck H, Matthysen E, Dhondt AA (1997a) Mate-locating strategies are related to relative body length and wing colour in the speckled wood butterfly *Pararge aegeria*. *Ecological Entomology* **22**: 116–120.

Van Dyck H, Matthysen E, Dhont AA (1997b) The effect of wing colour on male behavioural strategies in the speckled wood butterfly. *Animal Behaviour* **53**: 39–51.

Van Dyck H, Matthysen E, Wiklund C (1998) Phenotypic variation in adult morphology and pupal colour within and among families of the speckled wood butterfly *Pararge aegeria* (L.). *Ecological Entomology* **22**: 116–120.

Van Dyck H, Oostermeijer JGB, Talloen W, Feenstra V, van der Hidde A, Wynhoff I (2000) Does the presence of ant nests matter for oviposition to a specialized myrmecophilous *Maculinea* butterfly? *Proceedings of the Royal Society of London B* **267**: 861–866.

Van Dyck H, Van Strien AJ, Maes D, Van Swaag CAM (2009) Declines in common, widespread butterflies in a landscape under intense human use. *Conservation Biology* **23**: 957–965.

van Nouhuys S, Ehrnsten J (2004) Wasp behavior that leads to uniform parasitism of a host available only a few hours per year. *Behavioral Ecology* **15**: 661–665.

van Nouhuys S, Hanski I (2000) Apparent competition between parasitoids mediated by a shared hyperparasitoid. *Ecology Letters* **3**: 82–84.

van Nouhuys S, Hanski I (2002) Multitrophic interactions in space: metacommunity dynamics in fragmented landscapes. In: Tscharntke T, Hawkins BA (eds.) *Multitrophic Level Interactions*, Cambridge: Cambridge University Press, pp. 124–147.

van Nouhuys S, Hanski I (2004) Natural enemies of checkerspot butterflies. In: Ehrlich PR, Hanski I (eds.) *On the Wings of Checkerspots: A Model System for Population Biology*, Oxford: Oxford University Press, pp. 161–180.

van Nouhuys S, Hanski I (2005) Metacommunities of butterflies, their host plants and their parasitoids. In: Holyoak M, Leibold MA, Holt RD (eds.) *Metacommunities: Spatial Dynamics and Ecological Communities*, Chicago, IL: University of Chicago Press, pp. 99–121.

van Nouhuys S, Lei G (2004) Parasitoid–host metapopulation dynamics: the causes and consequences of phenological asynchrony. *Journal of Animal Ecology* **73**: 526–535.

van Nouhuys S, Tay WT (2001) Causes and consequences of mortality in small populations of a parasitoid wasp in a fragmented landscape. *Oecologia* **128**: 126–133.

van Nouhuys S, Singer MC, Nieminen M (2003) Spatial and temporal patterns of caterpillar performance and the suitability of two host plant species. *Ecological Entomology* **28**: 193–202.

van Oosterhout C, Zijlstra WG, van Heuven MK, Brakefield PM (2000) Inbreeding depression and genetic load in laboratory metapopulations of the butterfly *Bicyclus anynana*. *Evolution* **54**: 218–225.

van Oosterhout C, van Heuven MK, Brakefield PM (2004) On the neutrality of molecular genetic markers: pedigree analysis of genetic variation in fragmented populations. *Molecular Ecology* **13**: 1025–1034.

Van Swaay CAM (1990) An assessment of the changes in butterfly abundance in the Netherlands during the twentieth century. *Biological Conservation* **52**: 287–302.

Van Swaay CAM (1999) *Beschermingsplan Grote Vuurvlinder 2000–2004.* [Species protection plan Large Copper – in Dutch.] 'S-Gravenhage, Netherlands: Ministerie van Landbouw, Natuurbeheer en Visserij.

Van Swaay CAM (2002) The importance of calcareous grasslands for butterflies in Europe. *Biological Conservation* **104**: 315–318.

Van Swaay CAM (2006) *Basisrapport Rode Lijst Dagvlinders*, Rapport VS2006.002. Wageningen, Netherlands: De Vlinderstichting.

Van Swaay CAM, Van Strien AJ (2005) Using butterfly monitoring data to develop a European grassland butterfly indicator. In: Kühn E, Feldmann R, Thomas JA, Settele J (eds.) *Studies on the Ecology and Conservation of Butterflies in Europe*, Vol. 1, *General Concepts and Case Studies*, Moscow: Pensoft, pp. 106–108.

Van Swaay CAM, Van Strien AJ (2008) *The European Butterfly Indicator for Grassland Species: 1990–2007*, Report VJ2008.22. Wageningen, Netherlands: De Vlinderstichting.

Van Swaay CAM, Warren MS (1999) *Red Data Book of European Butterflies (Rhopalocera)*, Nature and the Environment No. 99. Strasbourg, France: Council of Europe.

Van Swaay CAM, Warren MS (eds.) (2003) *Prime Butterfly Areas in Europe: Priority Sites for Conservation*. Wageningen, Netherlands: National Reference Centre for Agriculture, Nature and Fisheries, Ministry of Agriculture, Nature Management and Fisheries.

Van Swaay CAM, Maes D, Plate C (1997) Monitoring butterflies in The Netherlands and Flanders: the first results. *Journal of Insect Conservation* **1**: 78–81.

Van Swaay CAM, Warren M, Loïs G (2006) Biotope use and trends of European butterflies. *Journal of Insect Conservation* **10**: 189–209.

Van Valen L (1976) Ecological species, multispecies, and oaks. *Taxon* **25**: 233–239.

Van Voorhies WA (1996) Bergmann size clines: a simple explanation for their occurrence in ectotherms. *Evolution* **50**: 1259–1264.

Vandewoestijne S, Baguette M (2002) The genetic structure of endangered populations in the Cranberry Fritillary, *Boloria aquilonaris* (Lepidoptera, Nymphalidae): RAPDs vs. allozymes. *Heredity* **89**: 439–445.

Vandewoestijne S, Baguette M (2004) Demographic versus genetic dispersal measures. *Population Ecology* **46**: 281–285.

Vandewoestijne S, Nève G, Baguette M (1999) Spatial and temporal population structure of the butterfly *Aglais urticae* (Lepidoptera Nymphalidae). *Molecular Ecology* **8**: 1639–1544.

Vandewoestijne S, Martin T, Liégeois S, Baguette M (2004) Dispersal, landscape occupancy and population structure in the butterfly *Melanargia galathea*. *Basic and Applied Ecology* **5**: 581–591.

Vandewoestijne S, Schtickzelle N, Baguette M (2008) Positive correlation between genetic diversity and fitness in a large, well-connected metapopulation. *BMC Biology* **6**: 46.

Vane-Wright RI, Ackery PR (eds.) (1984) *The Biology of Butterflies*. London: Academic Press.

Vanreusel W, Van Dyck H (2007) When functional habitat does not match vegetation types: a resource-based approach to map butterfly habitat. *Biological Conservation* **135**: 202–211.

Vanreusel W, Maes D, Van Dyck H (2007) Transferability of species distribution models: a functional habitat approach for two regionally threatened butterflies. *Conservation Biology* **21**: 201–212.

Varga Z (1977) Das Prinzip der areal-analytischen Methode in der Zoogeographie und die Faunenelement-Einteilung der europäischen Tagschmetterlinge (Lepidoptera: Diurna). *Acta Biologica Debrecina* **14**: 223–285.

Varga Z (1997) Biogeography and evolution of the oreal Lepidoptera in the Palearctic. *Acta Zoologica Academiae Scientiarum Hungaricae* **42**: 289–330.

Vasconcellos-Neto J, Monteiro RF (1993) Inspection and evaluation of host-plant by the butterfly *Mechanitis lysimnia*

(Nymph, Ithomiinae) before laying eggs: a mechanism to reduce intraspecific competition. *Oecologia* **95**: 431–438.

Vaughan IP, Ormerod SJ (2003) Improving the quality of distribution models for conservation by addressing shortcomings in the field collection of training data. *Conservation Biology* **17**: 1601–1611.

Vaughan IP, Ormerod SJ (2005) The continuing challenges of testing species distribution models. *Journal of Applied Ecology* **42**: 720–730.

Veith M, Bahl A, Seitz A (1999) Populationsgenetik im Naturschutz: Einsatzmöglichkeiten und Fallbeispiele. In: Amler K, Bahl A, Henle K, Kaule G, Poschlod P, Settele J (eds.) *Populationsbiologie in der Naturschutzpraxis*, Stuttgart, Germany: Ulmer, pp.112–126.

Vera FWM (2000) *Grazing Ecology and Forest History*. Wallingford, Oxon.: CABI Publishing.

Vera JC, Wheat CW, Fescemyer HW, Fescemyer HW, Frilander MJ, Crawford DL, Hanski I, Marden JH (2008) Rapid transcriptome characterization for a nonmodel organism using 454 pyrosequencing. *Molecular Ecology* **17**: 1636–1647.

Verboom J, Schotman A, Opdam P, Metz JAJ (1991) European nuthatch metapopulations in a fragmented agricultural landscape. *Oikos* **61**: 149–156.

Verbyla DL, Litvaitis JA (1989) Resampling methods for evaluation of classification accuracy of wildlife habitat models. *Environmental Management* **13**: 783–787.

Verity R (1911–1913) *Rhopalocera Palaearctica*. Florence, Italy: R. Verity.

Verity R (1916) Sur deux *Lycaena* confondus sous le nom de *L. (Agriades) coridon* Poda. *Annales de la Société Entomologique de France* **84**: 504–520.

Viejo JL, Martín J (1988) Las mariposas del Macizo Central de Gredos (Lepidoptera: Hesperioidea et Papilionoidea). *Actas Gredos, Boletín Universitario* **7**: 81–93.

Vielmetter W (1958) Physiologie des Verhaltens zur Sonnenstrahlung bei dem Tagfalter *Argynnis paphia* L. I. Untersuchungen im Freiland. *Journal of Insect Physiology* **2**: 13–37.

Viktorov GA (1966) [*Telenomus sokolovi* Mayr (Hymenoptera, Scelionidae) as a secondary parasite of the eggs of *Eurygaster integriceps* Put.]. *Doklady Akademia Nauk SSSR, Biol.* **169**: 741–744. [In Russian.]

Vila M, Vidal-Romani JR, Bjorklund M (2005) The importance of time scale and multiple refugia: incipient speciation and admixture of lineages in the butterfly *Erebia triaria* (Nymphalidae). *Molecular Phylogenetics and Evolution* **36**: 249–260.

Vila M, Lundhagen AC, Thuman KA, Stone JR, Bjorklund M (2006) A new conservation unit in the butterfly *Erebia triaria* (Nymphalidae) as revealed by nuclear and mitochondrial markers. *Annales Zoologici Fennici* **43**: 72–79.

Viloria AL, Pyrcz TW, Wojtusiak J, Ferrer-Paris JR, Beccaloni GW, Sattler K, Lees DC (2003) A brachypterous butterfly? *Proceedings of the Royal Society of London B* (Suppl.) **270**: 21–24.

Vincent JFV (1982) The mechanical design of grasses. *Journal of Material Sciences* **17**: 856–860.

Vincent PJ, Haworth JM (1983) Poisson regression models of species abundance. *Journal of Biogeography* **10**: 153–160.

Vitaz L, Balint Z, Zitnan D (1997) *Polyommatus slovacus* sp. nov. (Lepidoptera, Lycaenidae): the bivoltine relative of *Polyommatus coridon* in Slovakia. *Entomological Problems* **28**: 1–8.

Vogel K (1997) *Sonne, Ziest und Flockenblumen: Was braucht eine überlebensfähige Population des Roten Scheckenfalters (*Melitaea didyma*)?* Göttingen, Germany: Cuvillier Verlag.

Völkl R, Schiefer T, Bräu M, Stettmer C, Binzenhöfer B, Settele J (2008) Auswirkungen von Mahdtermin und -turnus auf Populationen der Ameisen-Bläulinge *Maculinea nausithous* und *Maculinea teleius*: Ergebnisse mehrjähriger Habitatanalysen in Bayern. *Naturschutz und Landschaftsplanung* **40**: 147–155.

von Mentzer E (1960) Über die Spezifizität von *Erebia neleus* Frr. und *Erebia aquitania* Frhst. (Lep.: Satyridae). *Entomologisk Tidskrift* **81**: 77–90.

Vrabec V (1994) *Parnassius mnemosyne*, its population ecology, variability and distribution in Bohemia (Lepidoptera: Papilionidae). M.Sc. thesis, Charles University, Prague.

Vulinec K (1990) Collective security, aggregation by insects as a defence. In: Evans JL, Schmidt JO (eds.) *Insects' Defense: Adaptive Mechanisms of Prey and Predators*. New York: State University of New York Press, pp. 251–288.

Wade PR (2002) Bayesian population viability analysis. In: Beissinger SR, McCullough DR (eds.) *Population Viability Analysis*, Chicago, IL: University of Chicago Press, pp. 213–238.

Waeyenbergh M, Baguette M (1996) First Belgian record of *Cotesia vestalis* (Haliday, 1834) (Hymenoptera, Braconidae), parasite of *Proclossiana eunomia* (Esper, 1799) (Lepidoptera, Nymphalidae), new host species. *Belgian Journal of Zoology* **126**: 81–83.

Wagener S. (1984) *Melanargia lachesis* Hübner 1790, est-elle une espèce différente de *Melanargia galathea* Linnaeus 1758, oui ou non? *Nota lepidopterologica* **7**: 375–386.

Wahl D (1989) A revision of *Benjaminia* (Hymenoptera: Ichneumonidae, Campopleginae). *Systematic Entomology* **14**: 275–298.

Wahl DB, Sime KR (2006) A revision of the genus *Trogus* (Hymenoptera: Ichneumonidae, Ichneumoninae). *Systematic Entomology* **31**: 584–610.

Wahlberg N (1995) One day in the life of a butterfly: a study on the biology of the Glanville Fritillary *Melitaea cinxia*. M.Sc. thesis, University of Helsinki.

Wahlberg N (1997) The life history and ecology of *Melitaea diamina* (Nymphalidae) in Finland. *Nota lepidopterologica* **20**: 70–81.

Wahlberg N (1998) The life history and ecology of *Euphydryas maturna* (Nymphalidae: Melitaeini) in Finland. *Nota lepidopterologica* **21**: 154–169.

Wahlberg N (2001) The phylogenetics and biochemistry of host-plant specialisation in Melitaeine butterflies (Lepidoptera: Nymphalidae). *Evolution* **55**: 522–537.

Wahlberg N, Zimmermann M (2000) Pattern of phylogenetic relationships among members of the tribe Melitaeini (Lepidoptera: Nymphalidae) inferred from mitochondrial DNA sequences. *Cladistics* **16**: 347–363.

Wahlberg N, Moilanen A, Hanski I (1996) Predicting the occurrence of endangered species in fragmented landscapes. *Science* **273**: 1536–1538.

Wahlberg N, Kullberg J, Hanski I (2001) Natural history of some Siberian melitaeine butterfly species (Melitaeini: Nymphalidae) and their parasitoids. *Entomologica Fennica* **12**: 72–77.

Wahlberg N, Klemetti T, Hanski I (2002a) Dynamic populations in a dynamic landscape: the metapopulation structure of the marsh fritillary butterfly. *Ecography* **25**: 224–232.

Wahlberg N, Klemetti T, Selonen V, Hanski I (2002b) Metapopulation structure and movements in five species of checkerspot butterflies. *Oecologia* **130**: 33–43.

Wahlberg N, Ehrlich PR, Boggs CL, Hanski I (2004) Bay checkerspot and Glanville fritillary compared with other species. In: Ehrlich PR, Hanski I (eds.) *On the Wings of Checkerspots: A Model System for Population Biology*, Oxford: Oxford University Press, pp. 219–244.

Wahlberg N, Braby MF, Brower AVZ, de Jong R, Lee MM, Nylin S, Pierce NE, Sperling FAH, Vila R, Warren AD, Zakharov E (2005) Synergistic effects of combining morphological and molecular data in resolving the phylogeny of butterflies and skippers. *Proceedings of the Royal Society of London B* **272**: 1577–1586.

Wakeham-Dawson A, Spurdens P (1994) Anomalous blue butterflies of the genus *Agrodiaetus* Hübner (Lepidoptera: Lycaenidae) in Southern Greece. *Entomologist's Gazette* **45**: 13–20.

Wakeham-Dawson A, Jaksic P, Holloway JD, Dennis RLH (2004) Multivariate analysis of male genital structures in the *Hipparchia semele–muelleri–delattini* complex (Lepidoptera: Nymphalidae, Satyrinae) from the Balkans: how many taxa? *Nota lepidopterologica* **27**: 103–124.

Wallace AR (1865) On the phenomena of variation and geographical distribution as illustrated by the Papilionidae of the Malay region. *Transactions of the Linnean Society of London* **25**: 1–17.

Wallis GP (1994) Population genetics and conservation in New Zealand: a hierarchical synthesis and recommendations for the 1990s. *Journal of the Royal Society of New Zealand* **24**: 143–160.

WallisDeVries MF (2004) A quantitative conservation approach for the endangered butterfly *Maculinea alcon*. *Conservation Biology* **18**: 489–499.

WallisDeVries MF, Raemakers I (2001) Does extensive grazing benefit butterflies in coastal dunes? *Restoration Ecology* **9**: 179–188.

Walther GR, Post E, Convey P, Menzel A, Parmesan C, Beebee TJC, Fromentin JM, Hoegh-Guldberg H, Bairlein F (2002) Ecological responses to recent climate change. *Nature* **416**: 389–395.

Walther GR, Berger S, Sykes MT (2005) An ecological 'footprint' of climate change. *Proceedings of the Royal Society of London B* **272**: 1427–1432.

Waltz AEM, Covington WW (2004) Ecological restoration treatments increase butterfly richness and abundance: mechanisms of response. *Restoration Ecology* **12**: 85–96.

Wang JL, Whitlock MC (2003) Estimating effective population size and migration rates from genetic samples over space and time. *Genetics* **163**: 429–446.

Wang R, Wang Y, Lei G, Xu R, Painter J (2003) Genetic differentiation within metapopulations of *Euphydryas aurinia* and *Melitaea phoebe* in China. *Biochemical Genetics* **41**: 107–118.

Wang R, Wang Y, Chen J, Lei G, Xu R (2004) Contrasting movement patterns in two species of chequerspot butterflies, *Euphydryas aurinia* and *Melitaea phoebe*, in the same patch network. *Ecological Entomology* **29**: 367–374.

Ward KE, Landolt PJ (1995) Influence of multiple matings on fecundity and longevity of female cabbage looper moths (Lepidoptera: Noctuidae). *Annals of the Entomological Society of America* **88**: 768–772.

Warren BCS (1926) Monograph of the tribe Hesperiini. *Transactions of the Entomological Society of London* **74**: 1–170.

Warren BCS (1936) *Monograph of the Genus* Erebia *Dalman*. London: British Museum (Natural History).

Warren BCS (1949) Three hitherto unrecognized European species of *Erebia*. *Entomologist* **82**: 97–105.

Warren BCS (1954) *Erebia tyndarus* Esp. and *Erebia cassioides* Rein. and Hohenw. (Lep.: Satyridae): two distinct species. *Entomologist's Monthly Magazine* **90**: 129–130.

Warren BCS (1981) *Supplement to the Monograph of the Genus* Erebia. Faringdon, Oxon.: EW Classey.

Warren MS (1985) The influence of shade on butterfly numbers in woodland rides, with special reference to the wood white *Leptidea sinapis*. *Biological Conservation* **33**: 147–164.

Warren MS (1987a) The ecology and conservation of the heath fritillary, *Mellicta athalia*. I. Host selection and phenology. *Journal of Applied Ecology* **24**: 467–482.

Warren MS (1987b) The ecology and conservation of the heath fritillary, *Mellicta athalia*. II. Adult population structure and mobility. *Journal of Applied Ecology* **24**: 483–498.

Warren MS (1987c) The ecology and conservation of the heath fritillary, *Mellicta athalia*. III. Population dynamics and the effect of habitat management. *Journal of Applied Ecology* **24**: 499–513.

Warren MS (1991) The successful conservation of an endangered species, the heath fritillary butterfly *Mellicta athalia* in Britain. *Biological Conservation* **55**: 37–56.

Warren MS (1992a) Butterfly populations. In: Dennis RLH (ed.) *The Ecology of Butterflies in Britain*, Oxford: Oxford Scientific Publications, pp. 73–79.

Warren MS (1992b) Conservation research on *Mellicta athalia*, an endangered species in the U.K. In: Pavlicek-van Beek T, Ovaa AH, van der Made JG (eds.) *Future of Butterflies in Europe*, Wageningen, Netherlands: Department of Nature Conservation, Agricultural University of Wageningen, pp. 124–133.

Warren MS (1992c) The conservation of British butterflies. In: Dennis RLH (ed.) *The Ecology of Butterflies in Britain*, Oxford: Oxford University Press, pp. 246–274.

Warren M (1994) The UK status and suspected metapopulation structure of threatened European butterfly, the Marsh Frittilary *Eurodryas aurina*. *Biological Conservation* **67**: 239–249.

Warren MS (1995) Managing local microclimates for the High Brown fritillary, *Argynnis adippe*. In: Pullin AS (ed.) *Ecology and Conservation of Butterflies*, London: Chapman and Hall, pp. 198–210.

Warren MS (2002) Species recovery work on butterflies and moths: an overview. In: Stone D, Tither J, Lacey P (eds.) *English Nature: The Species Recovery Programme*, Peterborough, Cambs: English Nature, pp. 25–31.

Warren MS, Bourn NAD (1997) The impact of grassland management on threatened butterflies in ESAs. In: Sheldrick RD (ed.) *Grassland Management in the Environmentally Sensitive Areas*, Occasional Symposium No. 32, Reading, British Grassland Society, Berks: pp. 138–143.

Warren MS, Key RS (1991) Woodlands: past, present and potential for insects. In: Collins NM, Thomas JA (eds.) *The Conservation of Insects and Their Habitats*, London: Academic Press, pp. 155–211.

Warren MS, Thomas JA (1992) Butterfly responses to coppicing. In: Buckley GP (ed.) *The Ecological Effects of Coppicing*, London: Chapman and Hall, pp. 249–270.

Warren MS, Thomas CD, Thomas JA (1984) The status of the Heath Fritillary butterfly *Mellicta athalia* Rott. in Britain. *Biological Conservation* **29**: 287–305.

Warren MS, Pollard E, Bibby TJ (1986) Annual and long-term changes in a population of the wood white butterfly *Leptidea sinapis*. *Journal of Animal Ecology* **55**: 707–719.

Warren MS, Munguira ML, Ferrin J (1994) Notes on the distribution, habitats and conservation of *Eurodryas aurinia* (Rottenburg) (Lepidoptera: Nymphalidae) in Spain. *Entomologist's Gazette* **45**: 5–12.

Warren MS, Barnett LK, Gibbons DW, Avery MI (1997) Assessing national conservation priorities: an improved *Red List* of British butterflies. *Biological Conservation* **82**: 317–328.

Warren MS, Hill JK, Thomas JA, Asher J, Fox R, Huntley B, Roy DB, Telfer MG, Jeffcoate S, Harding P, Jeffcoate G, Willis SG, Greatorex-Davies JN, Moss D, Thomas CD (2001) Rapid responses of British butterflies to opposing forces of climate and habitat change. *Nature* **414**: 65–69.

Warren M, Brereton T, Wigglesworth T (2005) Do agri-environment schemes help butterflies? Experience from the UK. In: Kühn E, Feldmann R, Thomas JA, Settele J (eds.) *Studies on the Ecology and Conservation of Butterflies in Europe*, Vol. 1, *General Concepts and Case Studies*, Moscow: Pensoft, pp. 121–123.

Wasserthal LT (1975) The role of butterfly wings in regulation of body temperature. *Journal of Insect Physiology* **21**: 1921–1930.

Wasserthal LT (1980) Oscillating hemolymph 'circulation' in the butterfly *Papilio machaon* L. revealed by contact thermography and photocell measurement. *Journal of Comparative Physiology* **139**: 145–163.

Wasserthal LT (1983) Hemolymph flows in the wings of pierid butterflies visualized by vital staining (Insecta, Lepidoptera). *Zoomorphology* **103**: 177–192.

Watanabe M (1988) Multiple matings increase the fecundity of the yellow swallowtail butterfly, *Papilio xuthus* L., in summer generations. *Journal of Insect Behavior* **1**: 17–29.

Watanabe M (1992) Egg maturation in laboratory reared females of the swallowtail butterfly, *Papilio xuthus* L. (Lep. Pap.), feeding on different concentrations solutions of sugar. *Zoological Science* **9**: 133–141.

Watanabe M, Ando S (1993) Influence of mating frequency on life-time fecundity in wild females of the small white *Pieris rapae* (Lepidoptera: Pieridae). *Japanese Journal of Entomology* **61**: 691–696.

Watanabe M, Yamaguchi H (1993) Egg cannibalism and egg distribution of two *Pieris* butterflies, *Pieris rapae* and *P. melete* (Lepidoptera, Pieridae) on a host plant, *Rorippa indica* (Cruciferae). *Japanese Journal of Ecology* **43**: 181–188.

Watanabe M, Nozato K, Kiritani K, Miyai S (1984) Seasonal fluctuations of egg density and survival rate in Japanese black swallowtail butterflies. *Japanese Journal of Ecology* **34**: 271–279.

Watt WB (1968) Adaptive significance of pigment polymorphisms in *Colias* butterflies. I. Variation of melanin pigment in relation to thermoregulation. *Evolution* **22**: 437–458.

Watt WB (1985) Bioenergetics and evolutionary genetics: opportunities for new synthesis. *American Naturalist* **125**: 118–143.

Watt WB (1992) Eggs, enzymes, and evolution: natural genetic variants change insect fecundity. *Proceedings of the National Academy of Sciences, USA* **89**: 10 608–10 612.

Watt WB (2003) Mechanistic studies of butterfly adaptations. In: Boggs CL, Watt WB, Ehrlich PR (eds.) *Butterflies: Ecology and Evolution Taking Flight*, Chicago, IL: University of Chicago Press, pp. 319–352.

Watt WB, Boggs CL (2003) Butterflies as model systems in ecology and evolution: present and future. In: Boggs CL, Watt WB, Ehrlich PR (eds.) *Butterflies: Ecology and Evolution Taking Flight*, Chicago, IL: University of Chicago Press, pp. 603–613.

Watt WB, Hoch PC, Mills SG (1974) Nectar resource use by *Colias* butterflies: chemical and visual aspects. *Oecologia* **14**: 353–374.

Watt WB, Chew FS, Snyder LRG, Watt AG, Rothschild DE (1977) Population structure of pierid butterflies. I. Numbers and movements of some montane *Colias*. *Oecologia* **27**: 1–22.

Watt WB, Cassin RC, Swan MS (1983) Adaptation at specific loci. III. Field behaviour and survivorship differences among *Colias* PGI genotypes are predictable from in vitro biochemistry. *Genetics* **103**: 725–739.

Watt WB, Carter PA, Donohue K (1986) Females' choice of 'good genotypes' as mates is promoted by an insect mating system. *Science* **233**: 1187–1190.

Watt WB, Donohue K, Cater PA (1996) Adaptation at specific loci. IV. Divergence vs. parallelism of polymorphic allozymes in molecular function and fitness: component effects among *Colias* species (Lepidoptera, Pieridae). *Molecular Biology and Evolution* **13**: 699–709.

Watt WB, Wheat CW, Meyer EH, Martin JF (2003) Adaptation at specific loci. VII. Natural selection, dispersal and the diversity of molecular–functional variation patterns among butterfly species complexes (*Colias*: Lepidoptera, Pieridae). *Molecular Ecology* **12**: 1265–1275.

Wätzold F, Lienhoop N, Drechsler M, Settele J (2006) Estimating optimal conservation in agricultural landscapes when costs and benefits of conservation measures are heterogenoeous in space and over time. *UFZ Discussion Papers* **5**/2006: 1–25.

Webb NR (1994) The habitat, the biotope and the landscape. In: Dover JW (ed.) *Fragmentation in Agricultural Landscapes*, Preston, Lancs.: IALE(UK), Myerscough College, pp. 21–29.

Webb NR (1998) The traditional management of European heathlands. *Journal of Applied Ecology* **35**: 987–990.

Webb MR, Pullin AS (1996) Larval survival in populations of the large copper butterfly *Lycaena dispar batavus*. *Ecography* **19**: 279–286.

Webb MR, Pullin AS (1998) Effects of submergence by winter floods on diapausing caterpillars of a wetland butterfly, *Lycaena dispar batavus*. *Ecological Entomology* **23**: 96–99.

Webb MR, Pullin AS (2000) Egg distribution in the large copper butterfly *Lycaena dispar batavus* (Lepidoptera: Lycaenidae): host plant versus habitat mediated effects. *European Journal of Entomology* **97**: 363–367.

Webb NR, Vermaat AH (1990) Changes in vegetational diversity on remnant heathland fragments. *Biological Conservation* **53**: 253–264.

Wedell N (1996) Mate quality affects reproductive effort in a paternally investing species. *American Naturalist* **148**: 1075–1088.

Wedell N (2005) Sperm competition in butterflies and moths. In: Fellowes MDE, Holloway GJ, Rolff J (eds.) *Insect Evolutionary Ecology*, Wallingford, Oxon.: CABI Publishing, pp. 49–81.

Wedell N (2006) Male genotype affects female fitness in a paternally investing species. *Evolution* **60**: 1638–1645.

Wedell N, Karlsson B (2003) Paternal investment directly affects female reproductive effort in an insect. *Proceedings of the Royal Society of London B* **270**: 2065–2071.

Wedell N, Nylin S, Janz N (1997) Effects of larval host plant and sex on the propensity to enter diapause in the comma butterfly. *Oikos* **78**: 569–575.

Wedell N, Wiklund C, Cook PA (2002) Monandry and polyandry as alternative lifestyles in a butterfly. *Behavioral Ecology* **4**: 450–455.

Wee B (2004) Effects of geographic distance, landscape features and host association on genetic differentiation of checkerspot butterflies. Ph.D. thesis, University of Texas, Austin.

Wehling WF, Thompson JN (1997) Evolutionary conservatism of oviposition preference in a widespread polyphagous insect herbivore, *Papilio zelicaon. Oecologia* **111**: 209–215.

Weibull AC, Ostman O (2003) Species composition in agroecosystems: the effect of landscape, habitat and farm management. *Basic and Applied Ecology* **4**: 349–361.

Weibull AC, Bengtsson J, Nohlgren E (2000) Diversity of butterflies in the agricultural landscape: the role of farming system and landscape heterogeneity. *Ecography* **23**: 743–750.

Weibull AC, Ostman O, Granqvist Å (2003) Species richness in agro-ecosystems: the effect of landscape, habitat and farm management. *Biodiversity and Conservation* **12**: 1335–1355.

Weidemann HJ (1995): *Tagfalter: Beobachten, Bestimmen*, 2nd Edn. Augsburg, Germany: Naturbuch-Verlag.

Weiss MR (1997) Innate colour preferences and flexible colour learning in the pipevine swallowtail. *Animal Behaviour* **53**: 1043–1052.

Weiss MR (2001) Vision and learning in some neglected pollinators: beetles, flies moths and butterflies. In: Chittka L, Thomson J (eds.) *Cognitive Ecology of Pollination*, Cambridge: Cambridge University Press, pp. 171–190.

Weiss MR, Papaj DR (2003) Colour learning in two behavioural contexts: how much can a butterfly keep in mind? *Animal Behaviour* **65**: 425–434.

Weiss SB, White RR, Murphy DD, Ehrlich PR (1987) Growth and dispersal of larvae of the checherkspot butterfly *Euphydryas editha. Oikos* **50**: 161–166.

Weiss SB, Murphy DD, White RR (1988) Sun, slope and butterflies: topographic determinants of habitat quality for *Euphydryas editha. Ecology* **69**: 1486–1496.

Weiss SB, Murphy DD, Ehrlich PR, Metzler CF (1993) Adult emergence phenology in checkerspot butterflies: the effects of macroclimate, topoclimate, and population history. *Oecologia* **96**: 261–270.

Weisser WW (2001) The effects of predation on dispersal. In: Clobert J, Danchin E, Dhont AA, Nichols JD (eds.) *Dispersal*, Oxford: Oxford University Press, pp. 180–188.

Weisser WW, Braendle C, Minoretti N (1999) Predator-induced morphological shift in the pea aphid. *Proceedings of the Royal Society of London B* **266**: 1175–1181.

Wellington WG (1965) Some maternal influences on progeny quality in western tent caterpillar *Malacosoma pluviale* (Dyar). *Canadian Entomologist* **97**: 1–14.

Wenger R, Geiger H, Scholl A (1993) Genetic analysis of a hybrid zone between *P. daplidice* (Linnaeus, 1758) and *P. edusa* (Fabricius, 1777) (Lepidoptera: Pieridae) in Northern Italy. *Proceedings of the 4th Congress of the European Society for Evolutionary Biology*, p. 488.

Wenzel M, Schmitt T, Weitzel M, Seitz A (2006) The severe decline of butterflies on western German calcareous grasslands during the last 30 years: a conservation problem. *Biological Conservation* **128**: 542–552.

Wessels KJ, Van Jaarsveld AS, Grimbeek JD, Van Der Linde MJ (1998) An evaluation of the gradsect biological survey method. *Biodiversity and Conservation* **7**: 1093–1121.

West RG (1988) The records of cold stages. *Philosophical Transactions of the Royal Society of London B* **318**: 505–520.

Westerbergh A (2004) An interaction between a specialized seed predator moth and its dioecious host plant shifting from parasitism to mutualism. *Oikos* **105**: 564–574.

Wettstein W, Schmid B (1999) Conservation of arthropod diversity in montane wetlands: effect of altitude, habitat quality and habitat fragmentation on butterflies and grasshoppers. *Journal of Applied Ecology* **36**: 363–373.

Weyh R, Maschwitz U (1982) Individual trail marking by larvae of the scarce swallowtail *Iphiclides podalirius* L. (Lepidoptera; Papilionidae). *Oecologia* **52**: 415–416.

Wheat CW, Watt WB, Pollock DD, Schulte PM (2006) From DNA to fitness differences: sequences and structures of adaptive variants of *Colias* phosphoglucose isomerase (PGI). *Molecular Biology and Evolution* **23**: 499–512.

Wheeler D (1996) The role of nourishment in oogenesis. *Annual Review of Entomology* **41**: 407–431.

White MJD (1973) *Animal Cytology and Evolution*, 3rd edn. Cambridge: Cambridge University Press.

White P, Kerr JT (2006) Contrasting spatial and temporal global change impacts on butterfly species richness during the twentieth century. *Ecography* **29**: 908–918.

Whitlock MC (2001) Dispersal and the genetic properties of metapopulations. In: Clobert J, Danchin E, Dhont AA, Nichols JD (eds.) *Dispersal*, Oxford: Oxford University Press, pp. 273–282.

Whitlock MC (2003) Fixation probability and time in subdivided populations. *Genetics* **164**: 767–779.

Whitlock MC, Barton NH (1997) The effective size of a subdivided population. *Genetics* **146**: 427–441.

Whitlock MC, McCauley DE (1999) Indirect measures of gene flow and migration: $F_{st} \neq 1/(4Nm+1)$. *Heredity* **82**: 117–125.

Whittingham MJ, Stephens, Bradbury BR, Freckleton RP (2006) Why do we still use stepwise modelling in ecology and behaviour? *Journal of Animal Ecology* **75**: 1182–1189.

Wickman P-O (1985a) The influence of temperature on the territorial and mate locating behaviour of the small heath butterfly, *Coenonympha pamphilus* (L.) (Lepidoptera: Satyridae). *Behavioral Ecology and Sociobiology* **16**: 233–238.

Wickman P-O (1985b) Male determined mating duration in butterflies? *Journal of the Lepidopterists' Society* **39**: 341–342.

Wickman P-O (1985c) Territorial defence and mating success in males of the small heath butterfly, *Coenonympha pamphilus* L. (Lepidoptera: Satyridae). *Animal Behaviour* **33**: 1162–1168.

Wickman P-O (1986) Courtship solicitation by females of the small heath butterfly, *Coenonympha pamphilus* (L.) (Lepidoptera: Satyridae) and their behaviour in relation to male territories before and after copulation. *Animal Behaviour* **34**: 153–157.

Wickman P-O (1988) Dynamics of mate-searching behaviour in a hilltopping butterfly, *Lasiommata megera* (L.): the effects of weather and male density. *Zoological Journal of the Linnean Society* **93**: 357–377.

Wickman P-O (1992a) Mating systems of *Coenonympha* butterflies in relation to longevity. *Animal Behaviour* **44**: 141–148.

Wickman P-O (1992b) Sexual selection and butterfly design: a comparative study. *Evolution* **46**: 1525–1536.

Wickman P-O (1997) Dagfjärilslekar. [Butterfly leks.] *Entomologisk Tidskrift* **117**: 73–85.[In Swedish.]

Wickman P-O, Jansson P (1997) An estimate of female mate searching costs in the lekking butterfly *Coenonympha pamphilus*. *Behavioral Ecology and Sociobiology* **40**: 321–328.

Wickman P-O, Karlsson B (1987) Changes in egg colour, egg weight and oviposition rate with the number of eggs laid by wild females of the small heath butterfly, *Coenonympha pamphilus*. *Ecological Entomology* **12**: 109–114.

Wickman P-O, Karlsson B (1989) Abdomen size, body size and the reproductive effort of insects. *Oikos* **56**: 209–214.

Wickman P-O, Rutowski RL (1999) The evolution of mating dispersion in insects. *Oikos* **84**: 463–472.

Wickman P-O, Wiklund C (1983) Territorial defence and its seasonal decline in the speckled wood butterfly (*Pararge aegeria*). *Animal Behaviour* **31**: 1206–1216.

Wickman P-O, Wiklund C, Karlsson B (1990) Comparative phenology of four satyrine butterflies inhabiting dry grasslands in Sweden. *Holarctic Ecology* **13**: 238–246.

Wickman P-O, García-Barros E, Rappe-George C (1995) The location of landmark leks in the small heath butterfly, *Coenonympha pamphilus*: evidence against the hot-spot model. *Behavioral Ecology* **6**: 39–45.

Wiemers M (1995) The butterflies of the Canary Islands: a survey on their distribution, biology and ecology (Lepidoptera: Papilionoidea and Hesperiodea). *Linneana Belgica* **15**: 63–118.

Wiemers M (2003) Chromosome differentiation and the radiation of the butterfly subgenus *Agrodiaetus* (Lepidoptera: Lycaenidae: Polyommatus): a molecular phylogenetic approach. Ph.D. thesis, Bonn University. Online: http://hss. ulb.uni-bonn.de/diss_online/math_nat_fak/2003/ wiemers_martin – URN: http://nbn-resolving.de/urn:nbn: de:hbz:5n-02787.

Wiernasz DC (1989) Female choice and sexual selection of male wing melanin pattern in *Pieris occidentalis* (Lepidoptera). *Evolution* **43**: 1672–1682.

Wiesen B, Krug E, Fiedler K, Wray V, Proksch P (1994) Sequestration of host-plant-derived flavonoids by the lycaenid butterfly *Polyommatus icarus*. *Journal of Chemical Ecology* **20**: 2523–2538.

Wijngaarden PJ, Brakefield PM (2000) The genetic basis of eyespot size in the butterfly *Bicyclus anynana*: an analysis of line crosses. *Heredity* **85**: 471–479.

Wiklund C (1974) Oviposition preferences in *Papilio machaon* in relation to the host plants of the larvae. *Entomologia Experimentalis et Applicata* **17**: 189–198.

Wiklund C (1975) The evolutionary relationship between adult oviposition preferences and larval host plant range in *Papilio machaon*. *Oecologia* **18**: 185–197.

Wiklund C (1977a) Oviposition, feeding and spatial separation of breeding and foraging habitats in a population of *Leptidea sinapis*. *Oikos* **28**: 56–68.

Wiklund C (1977b) Observationer över ägglaggning, födosök och vila hos Donzels blåvinge, *Aricia nicias scandicus*. *Entomologisk Tidskrift* **98**: 1–4.

Wiklund C (1977c) Courtship behaviour in relation to female monogamy in *Leptidea sinapis*. *Oikos* **29**: 275–283.

Wiklund C (1981) Generalist vs. specialist oviposition behaviour in *Papilio machaon* and functional aspects on the hierarchy of oviposition preferences. *Oikos* **36**: 163–170.

Wiklund C (1982a) Behavioural shift from courtship solicitation to mate avoidance in female ringlet butterflies (*Aphanthopus hyperanthus*) after copulation. *Animal Behaviour* **30**: 790–793.

Wiklund C (1982b) Generalist versus specialist utilization of host plants among butterflies. *Proceedings of the 5th*

International Symposium on Insect Plant Relationships, Wageningen, pp. 181–192.

Wiklund C (1984) Egg-laying patterns in butterflies in relation to their phenology and the visual apparency and abundance of their host plants. *Oecologia* **63**: 23–29.

Wiklund C (2003) Sexual selection and the evolution of butterfly mating systems. In: Boggs CL, Watt WB, Ehrlich PR (eds.). *Ecology and Evolution Taking Flight: Butterflies as Model Study Organisms*, Chicago, IL: University of Chicago Press, pp. 67–90.

Wiklund C (2005) Hornet predation on peacock butterflies and ecological aspects on the evolution of complex eyespots on butterfly wings. *Entomologica Fennica* **16**: 266–272.

Wiklund C, Åhrberg C (1978) Host plants, nectar source plants, and habitat selection of males and females of *Anthocharis cardamines* (Lepidoptera). *Oikos* **31**: 169–183.

Wiklund C, Fagerström T (1977) Why do males emerge before females? A hypothesis to explain the incidence of protandry in butterflies. *Oecologia* **31**: 153–158.

Wiklund C, Forsberg J (1985) Courtship and male discrimination between virgin and mated females in the orange tip butterfly *Anthocharis cardamines*. *Animal Behaviour* **34**: 328–332.

Wiklund C, Forsberg J (1991) Sexual size dimorphism in relation to female polygamy and protandry in butterflies: a comparative study of Swedish Pieridae and Satyridae. *Oikos* **60**: 373–381.

Wiklund C, Karlsson B (1984) Egg size variation in satyrid butterflies: adaptive vs. historical, 'Bauplan', and mechanistic explanations. *Oikos* **43**: 391–400.

Wiklund C, Karlsson B (1988) Sexual size dimorphism in relation to fecundity in some Swedish satyrid butterflies. *American Naturalist* **131**: 132–138.

Wiklund C, Persson A (1983) Fecundity, and the relation of egg weight variation to offspring fitness in the speckled wood butterfly *Pararge aegeria* or why don't butterfly females lay more eggs. *Oikos* **40**: 53–63.

Wiklund C, Tullberg B (2004) Seasonal polyphenism and leaf mimicry in the comma butterfly. *Animal Behaviour* **68**: 621–627.

Wiklund C, Eriksson T, Lundberg H (1979) The wood white butterfly *Leptidea sinapis* and its nectar plants: a case of mutualism or parasitism? *Oikos* **33**: 358–362.

Wiklund C, Eriksson T, Lundberg H (1982) On the pollination efficiency of butterflies: a reply to Courtney *et al. Oikos* **38**: 263.

Wiklund C, Persson A, Wickman P-O (1983) Larval aestivation and direct development as alternative strategies in the speckled wood butterfly *Pararge aegeria* in Sweden. *Ecological Entomology* **8**: 233–238.

Wiklund C, Karlsson B, Forsberg B (1987) Adaptive versus constraint explanations for egg-to-body size relationships in two butterfly families. *American Naturalist* **130**: 828–838.

Wiklund C, Nylin S, Forsberg J (1991) Sex-related variation in growth rate as a result of selection for large size and protandry in a bivoltine butterfly (*Pieris napi* L.). *Oikos* **60**: 241–250.

Wiklund C, Wickman P-O, Nylin S (1992) A sex difference in reaction norms for direct/diapause development: a result of selection for protandry. *Evolution* **46**: 519–528.

Wiklund C, Kaitala A, Lindfors V, Abenius J (1993) Polyandry and its effect on female reproduction in the green-veined white butterfly (*Pieris napi* L.). *Behavioral Ecology and Sociobiology* **33**: 25–33.

Wiklund C, Lindfors V, Forsberg J (1996) Early male emergence and reproductive phenology of the adult overwintering butterfly *Gonepteryx rhamni* in Sweden. *Oikos* **75**: 227–240.

Wiklund C, Kaitala A, Wedell N (1998) Decoupling of reproductive rates and parental expenditure in a polyandrous butterfly. *Behavioral Ecology* **9**: 20–25.

Wiklund C, Karlsson B, Leimar O (2001) Sexual conflict and cooperation in butterfly reproduction: a comparative study of polyandry and female fitness. *Proceedings of the Royal Society of London B* **268**: 1661–1667.

Wiklund C, Gotthard K, Nylin S (2003) Mating system and the evolution of sex-specific mortality rates in two nymphalid butterflies. *Proceedings of the Royal Society of London B* **270**: 1823–1828.

Wiklund C, Vallin A, Friberg M, Jakobsson S (2008) Rodent predation on hibernating peacock and small tortoiseshell butterflies. *Behavioural Ecology and Sociobiology* **62**: 379–389.

Wilcockson AJ (2002) The functional significance of wing morphology in Pieris napi. Ph.D. thesis, Oxford Brookes University, Oxford.

Wilcockson AJ, Shreeve TG (2003) The subspecific status of *Pieris napi* (L., 1758) (Pieridae) within the British Isles. *Nota lepidopterologica* **25**: 235–247.

Wiley EO (1981) Phylogenetics: *The Theory and Practice of Phylogenetic Systematics*. New York: John Wiley.

Wilkinson DS (1939) On two species of *Apanteles* (Hym. Brac.) not previously recognised from the western palaearctic region. *Bulletin of Entomological Research* **30**: 77–84.

Williams GC (1966) *Adaptation and Natural Selection: A Critique of Some Current Evolutionary Thought*. Princeton, NJ: Princeton University Press.

Williams EH, Bowers MD (1987) Factors affecting host-plant use by the montane butterfly *Euphydryas gillettii* (Nymphalidae). *American Midland Naturalist* **118**: 153–161.

Williams KS, Gilbert LE (1981) Insects as selective agents on plant vegetative morphology: egg mimicry reduces egg laying in butterflies. *Science* **212**: 467–469.

Willig MR, Kaufman DM, Stevens RD (2003) Latitudinal gradients of biodiversity: pattern, process, scale and synthesis. *Annual Review of Ecology, Evolution and Systematics* **34**: 273–309.

Willmer PG (1982) Microclimate and the environmental physiology of insects. *Advances in Insect Physiology* **16**: 1–57.

Willmer PG (1983) Thermal constraints on activity patterns in nectar-feeding insects. *Ecological Entomology* **8**: 455–469.

Willmer PG (1991) Thermal biology and mate acquisition in ectotherms. *Trends in Ecology and Evolution* **6**: 396–399.

Willmer PG, Unwin DM (1981) Field analyses of insect heat budgets: reflectance, size and heating rates. *Oecologia* **50**: 250–255.

Wilson A (1985) Flavonoid pigments of butterflies in the genus *Melanargia*. *Phytochemistry* **24**: 1685–1691.

Wilson R (ed.) (1999a) *Species: New Interdisciplinary Essays*. Cambridge, MA: MIT Press.

Wilson RJ (1999b) The spatiotemporal dynamics of three lepidopteran herbivores of *Helianthemum chamaecistus*. Ph.D. thesis, University of Leeds, Yorks.

Wilson RJ, Ellis S, Baker JS, Lineham ME, Whitehead RW, Thomas CD (2002) Large-scale patterns of distribution and persistence at the range margins of a butterfly. *Ecology* **83**: 3357–3368.

Wilson RJ, Gutiérrez D, Gutiérrez J, Martínez D, Agudo R, Monserrat VJ (2005) Changes to the elevation limits and extent of species ranges associated with climate change. *Ecology Letters* **8**: 1138–1146.

Wilson RJ, Davies ZG, Thomas CD (2007) Insects and climate change: processes, patterns and implications for conservation. In: Stewart A, Lewis OT, New T (ed.) *Insect Conservation, Biology*, Wallingford, Oxon.: CABI Publishing, pp. 245–279.

Wimmers C (1932) Bastarde unserer europäische Bläulinge. *Entomologische Zeitschrift* **45**: 311–314.

Windig JJ (1991) Quantification of Lepidoptera wing patterns using an image analyser. *Journal of Research on the Lepidoptera* **30**: 82–94.

Windig JJ (1994) Reaction norms and the genetic basis of phenotypic plasticity in the wing pattern of the butterfly *Bicyclus anynana*. *Journal of Evolutionary Biology* **7**: 665–695.

Windig JJ (1998) Evolutionary genetics of fluctuating asymmetry in the peacock butterfly (*Inachis io*). *Heredity* **80**: 382–392.

Windig JJ (1999) Trade-offs between melanization, development time and adult size in *Inachis io* and *Araschnia levana* (Lepidoptera: Nymphalidae)? *Heredity* **82**: 57–68.

Windig J, Lammar P (1999) Evolutionary genetics of seasonal polyphenism in the map butterfly *Araschnia levana* (Nymphalidae: Lepidoptera). *Evolutionary Ecology Research* **1**: 875–894.

Windig JJ, Nylin S (1999) Adaptive asymmetry in the speckled wood butterfly (*Pararge aegeria*)? *Proceedings of the Royal Society of London B* **266**: 1413–1418.

Windig JJ, Nylin S (2002) Genetics of fluctuating asymmetry in pupal traits of the speckled wood butterfly (*Pararge aegeria*). *Heredity* **89**: 225–234.

Windig JJ, Rintamaki PT, Cassel A, Nylin S (2001) How useful is fluctuating asymmetry in conservation biology? Asymmetry in rare and abundant *Coenonympha* butterflies. *Journal of Insect Conservation* **4**: 253–261.

Wintle BA, McCarthy MA, Volinsky CT, Kavanagh RP (2003) The use of Bayesian model averaging to better represent uncertainty in ecological models. *Conservation Biology* **17**: 1579–1590.

Wisnowski JW, Simpson JR, Montgomery DC, Runger GC (2003) Resampling methods for variable selection in robust regression. *Computational Statistics, Data Analysis* **43**: 341–355.

With KA, Crist TO (1995) Critical thresholds in species' responses to landscape structure. *Ecology* **76**: 2446–2459.

Witkowski Z, Adamski P, Kosior A, Plonka P (1997) Extinction and reintroduction of *Parnassius apollo* in the Pieniny National Park (Polish Carpathians). *Biologia* **52**: 199–208.

Woiwod IP, Hanski I (1992) Patterns of density dependence in moths and aphids. *Journal of Animal Ecology* **61**: 619–629.

Woiwod IP, Reynolds DR, Thomas CD (2001) *Insect Movement: Mechanisms and Consequences*. Oxon.: CABI Publishing, Wallingford.

Wojtusiak J (1979) Studies on locomotor activity during the post-embryonic development of Lepidoptera. *Folia Biologica, Kraków* **27**: 305–342.

Wojtusiak J, Raczka P (1983) Silking behaviour of the swallowtail *Papilio machaon* L. and peacock *Vanessa io* L. (Lepidopetra, Papilionidae, Nymphalidae). *Folia Biologica, Kraków* **31**: 39–51.

Wolda H (1987) Altitude, habitat and tropical insect diversity. *Biological Journal of the Linnean Society* **30**: 313–323.

Wolda H (1988) Insect seasonality: why? *Annual Review of Ecology and Systematics* **19**: 1–18.

Wolfinger R, O'Connell M (1993) Generalized linear mixed models: a pseudolikelihood approach. *Journal of Statistical Computation and Simulation* **48**: 233–243.

Wood PA, Samways MJ (1991) Landscape element pattern and continuity of butterfly flight paths in an ecologically landscaped botanical garden, Natal, South Africa. *Biological Conservation* **58**: 149–166.

Wootton RJ (1993) Leading-edge section and asymmetric twisting in the wings of flying butterflies (Insecta, Papilionoidea). *Journal of Experimental Biology* **180**: 105–117.

Wright S (1943) Isolation by distance. *Genetics* **28**: 114–138.

Wright S (1978) *Evolution and the Genetics of Populations*, Vol. 4, *Variability Within and Among Natural Populations*. Chicago, IL: University of Chicago Press.

Wu CI (2001) The genic view of the process of speciation. *Journal of Evolutionary Biology* **14**: 851–865.

Wynhoff I (1998) The recent distribution of *Maculinea* species. *Journal of Insect Conservation* **2**: 15–27.

Wynhoff I (2001) At home on foreign meadows: the reintroduction of two *Maculinea* butterfly species. Ph.D. thesis, University of Wageningen, Netherlands.

Wynhoff I, van Swaay CAM (1995) *Bedreigde en Kwetsbare Dagvlinders in Nederland: Basisrapport met Voorstel voor de Rode Lijst*. Wageningen, Netherlands: De Vlinderstichting.

Wynne IR, Brookes CP (1992) A device for producing multiple deep-frozen samples for allozyme electrophoresis. In: Berry RJ, Crawford TJ, Hewitt GM (eds.) *Genes in Ecology*, Oxford: Blackwell Scientific Publications, pp. 500–502.

Wynne IR, Loxdale HD, Brookes CP (1992) Use of a cellulose acetate system for allozyme electrophoresis. In: Berry RJ, Crawford TJ, Hewitt GM (eds.) *Genes in Ecology*, Oxford: Blackwell Scientific Publications, pp. 494–499.

Yeates DK, Greathead D (1997) The evolutionary pattern of host use in the family Bombyliidae (Diptera): a diverse family of parasitoid flies. *Biological Journal of the Linnean Society* **60**: 149–185.

Yee TW, Mitchell ND (1991) Generalized additive models in plant ecology. *Journal of Vegetation Science* **2**: 587–602.

Yoshino MM (1975) *Climate in a Small Area*. Tokyo: University of Tokyo Press.

Young AM (1983) On the evolution of egg placement and gregariousness of caterpillars in the Lepidoptera. *Acta Biotheoretica* **32**: 43–60.

Young TP (2000) Restoration ecology and conservation biology. *Biological Conservation* **92**: 73–83.

Yu DS, Horstmann K (1997) Catalogue of world Ichneumonidae. *Memoirs of the American Entomological Institute* **58**: 1–1558.

Zakharov EV, Hellmann JJ (2007) Characterization of 17 polymorphic microsatellite loci in the Anise swallowtail, *Papilio zelicaon* (Lepidoptera: Papilionidae), and their amplification in related species. *Molecular Ecology Notes* **7**: 144–146.

Zakharov EV, Hellmann JJ (2008) Genetic differentiation across a latitudinal gradient in two co-occurring butterfly species: revealing population differences in a context of climate change. *Molecular Ecology* **17**: 189–208.

Zakharov EV, Chelomina GN, Zhuravlev YN (2000) Isolation and analysis of DNA from museum specimens of butterflies (Lepidoptera, Papilionidae) with the aid of polymerase chain reaction using arbitrary and universal gene-specific primers. *Russian Journal of Genetics* **36**: 1017–1024.

Zalucki MP (1983) Simulation of movement and egg laying in *Danaus plexippus* (Lepidoptera, Nymphalidae). *Researches on Population Ecology* **25**: 353–365.

Zeller P (1845) *Polyommatus polonus*, eine neue Tagfalterart. *Stettiner Entomologische Zeitung* **3**: 351–354.

Zera AJ, Denno RF (1997) Physiology and ecology of dispersal polymorphism in insects. *Annual Review of Entomology* **42**: 207–230.

Zhang GR, Zimmermann O, Hassan SA (2004) Pollen as a source of food for egg parasitoids of the genus *Trichogramma* (Hymenoptera: Trichogrammatidae). *Biocontrol Science and Technology* **14**: 201–229.

Zimmermann M, Aubert J, Descimon H (1999) Phylogénie moléculaire des Mélitées (Lepidoptera: Nymphalidae). *Comptes Rendus de l'Académie des Sciences de Paris, Série III, Sciences de la Vie* **322**: 429–439.

Zimmermann M, Wahlberg N, Descimon H (2000) A phylogeny of *Euphydryas* checkerspot butterflies (Lepidoptera: Nymphalidae) based on mitochondrial DNA sequence data. *Annals of the Entomological Society of America* **93**: 347–355.

Zimmermen K, Fric Z, Filipová L, Konvička M (2004) Adult demography, dispersal and behaviour of *Brenthis ino* (Lepidoptera: Nymphalidae): how to be a successful wetland butterfly. *European Journal of Entomology* **102**: 699–709.

Zschokke S, Dolt C, Rusterholz HP, Oggier P, Braschler B, Thommen GH, Lüdin E, Erhardt A, Baur B (2000) Short-term responses of plants and invertebrates to experimental small-scale grassland fragmentation. *Oecologia* **125**: 559–572.

Zurell D, Jeltsch F, Dormann CF, Schröder B (2009) Static species distribution models in dynamically changing systems: How good can predictions really be? *Ecography* in press, doi: 10.1111/j.0906-7590.2009.05810.x

Index of scientific names

Keyword index